Frontiers in Electronic Materials

A Collection of Extended Abstracts
of the Nature Conference Frontiers in Electronic Materials,
June 17th to 20th 2012, Aachen, Germany

Edited by
Joerg Heber, Darrell Schlom, Yoshinori Tokura,
Rainer Waser, and Matthias Wuttig

natureconferences

WILEY-VCH Verlag GmbH & Co. KGaA

The Editors

Prof. Dr.-Ing. Rainer Waser
RWTH Aachen
Institut für Elektrotechnik II
Sommerfeldstr. 24
52074 Aachen
Germany

Dr. Jörg Heber
Nature Materials
The Macmillan Building
4, Crinan Street
London N1 9XW
United Kingdom

Prof. Darrell Schlom
Cornell University
Materials Science & Engineering
230, Bard Hall
Ithaca, NY 14853-2201
USA

Prof. Dr. Yoshinori Tokura
University of Tokio
Dept. of Applied Physics
7-3-1 Hongo, Bunkyo-ku
Tokyo 113-8656
Japan

Prof. Dr. Matthias Wuttig
RWTH Aachen
I., Physikalisches Institut
Sommerfeldstr. 14
52056 Aachen
Germany

All books published by **Wiley-VCH** are carefully produced. Nevertheless, authors, editors, and publisher do not warrant the information contained in these books, including this book, to be free of errors. Readers are advised to keep in mind that statements, data, illustrations, procedural details or other items may inadvertently be inaccurate.

Library of Congress Card No.:
applied for

British Library Cataloguing-in-Publication Data
A catalogue record for this book is available from the British Library.

Bibliographic information published by the Deutsche Nationalbibliothek
The Deutsche Nationalbibliothek lists this publication in the Deutsche Nationalbibliografie; detailed bibliographic data are available on the Internet at <http://dnb.d-nb.de>.

© 2012 Wiley-VCH Verlag & Co. KGaA, Boschstr. 12, 69469 Weinheim, Germany

All rights reserved (including those of translation into other languages). No part of this book may be reproduced in any form – by photoprinting, microfilm, or any other means – nor transmitted or translated into a machine language without written permission from the publishers. Registered names, trademarks, etc. used in this book, even when not specifically marked as such, are not to be considered unprotected by law.

Typesetting Authors
Printing and Binding betz-druck GmbH, Darmstadt
Cover Design Th. Pössinger and Grafik-Design Schulz, Fußgönheim

Print ISBN 978-3-527-41191-7

Printed in the Federal Republic of Germany
Printed on acid-free paper

Invited Talks 29

INV 1:	NEW MAGNETIC MATERIALS BASED ON DEFECTS, INTERFACES AND DOPING George A. Sawatzky, Ilya Elfimov, Bayo Lau and Mona Berciu	31
INV 2:	ATOMIC-RESOLUTION ELECTRON SPECTROSCOPY OF INTERFACES AND DEFECTS IN COMPLEX OXIDES D. A. Muller, J. A. Mundy, L. Fitting Kourkoutis, M. P. Warusawithana, J. Ludwig, P. Roy, A. A. Pawlicki, T. Heeg, C. Richter, S. Paetel, M. Zheng, B. Mulcahy, W. Zander, J. N. Eckstein, J. Schubert, J. Mannhart, D. G. Schlom	32
INV 3:	SIGNIFICANCE OF SOLID STATE IONICS FOR TRANSPORT AND STORAGE Joachim Maier	33
INV 4:	ELECTROCHEMICAL DOPING OF OXIDE HETEROSTRUCTURES E. Artacho, N.C Bristowe, P.B. Littlewood, J.M. Pruneda, M. Stengel	35
INV 5:	SWITCHABLE PHOTODIODE EFFECT IN FERROELECTRIC $BiFeO_3$ S-W. Cheong, H. T. Yi, T. Choi, and A. Hogan	36
INV 6:	EXPLORATION OF ELECTRON SYSTEMS AT OXIDE INTERFACES Werner Dietsche, Benjamin Förg, Cameron Hughes, Carsten Woltmann, Thilo Kopp, Florian Loder, Jochen Mannhart, Natalia Pavlenko, Christoph Richter, Ulrike Waizmann, Jürgen Weis	38
INV 7:	THE INFLUENCE OF IMPERFECTIONS ON THE 2DEG TRANSPORT PROPERTIES IN THE $LaAlO_3$-$SrTiO_3$ SYSTEM Guus Rijnders	39
INV 8:	CORRELATED ELECTRONIC MATERIALS: COMPUTATIONAL STUDIES OF MULTIORBITAL MODELS FOR BULK COMPOUNDS AND INTERFACES OF MAGNETIC AND SUPERCONDUCTING MATERIALS Elbio Dagotto	40
INV 9:	ELECTROLYTE GATE INDUCED METALLIZATION OF SEVERAL FACETS (101, 001, 110 and 100) OF RUTILE TiO_2 AND (001) $SrTiO_3$ Stuart S.P. Parkin, Thomas D. Schladt, Tanja Graf, Mingyang Li, Nagaphani Aetukuri, Xin Jiang and Mahesh Samant	42
INV 10:	COMPLEX THERMOELECTRIC MATERIALS G. Jeffrey Snyder	44
INV 11:	PCRAM OPERATION AT DRAM SPEEDS: EXPERIMENTAL DEMONSTRATION AND COMPUTER-SIMULATIONAL UNDERSTANDING D. Loke, T. H. Lee, W. J. Wang, L. P. Shi, R. Zhao, Y. C. Yeo, T. C. Chong, and S. R. Elliott,	45
INV 12:	ELECTRONIC PHASE CHANGE AND ENTROPIC FUNCTIONS IN TRANSITION METAL OXIDES Hidenori Takagi and Seiji Niitaka	46
INV 13:	DISORDER INDUCED METAL-INSULATOR TRANSITION IN PHASE CHANGE MATERIALS T. Siegrist	47
INV 14:	ELECTRONIC PROPERTIES OF THE INTERFACIAL $LaAlO_3$ / $SrTiO_3$ SYSTEM J.-M. Triscone, A. Fête, S. Gariglio, A. Caviglia, D. Li, D. Stornaiuolo, M. Gabay, B. Sacépé, A. Morpurgo, M. Schmitt, C. Cancellieri, P. Willmott	48
INV 15:	EMERGENT PHENOMENA IN TWO-DIMENSIONAL ELECTRON GASES AT OXIDE INTERFACES Susanne Stemmer, Pouya Moetakef, Daniel Ouellette, and S. James Allen	49
INV 16:	GIANT TUNNEL ELECTRORESISTANCE IN FERROELECTRIC TUNNEL JUNCTIONS A. Chanthbouala, V. Garcia, K. Bouzehouane, S. Fusil, X. Moya, S. Xavier, H. Yamada, C. Deranlot, N.D. Mathur, J. Grollier, A. Barthélémy and M. Bibes	50
INV 17:	REVISITING THE HEXAGONAL MANGANITES Nicola Spaldin	51
INV 18:	STUDY OF MAGNETOELECTRIC EFFECTS DUE TO MULTI-SPIN VARIABLES Tsuyoshi Kimura	52

INV 19:	BI-LAYERED RERAM: MULTI-LEVEL SWITCHING, RELIABILITY AND ITS MECHANISM FOR STORAGE CLASS MEMORY AND RECONFIGURATION LOGIC. U-In Chung, Young-Bae Kim, Seung Ryul Lee, Dongsoo Lee, Chang Bum Lee, Man Chang, Kyung min Kim, Ji Hyun Hur, Myoung-Jae Lee, Chang Jung Kim	53
INV 20:	SELF-ORGANIZATION IN ADAPTIVE, RECURRENT, AUTONOMOUS MEMRISTIVE CROSSNETS Konstantin K. Likharev, Dmitri N. Gavrilov, Thomas J. Walls	55
INV 21:	ELECTRIC FIELD CONTROL OF MAGNETIZATION R. Ramesh	57
INV 22:	MAGNETIC SWITCHING OF FERROELECTRIC DOMAINS AT ROOM TEMPERATURE IN A NEW MULTIFERROIC J. F. Scott, D. M. Evans, J. M. Gregg, Ashok Kumar, D. Sanchez, N. Ortega, and R. S. Katiyar	59
INV 23:	CONTROL OF CORRELATED ELECTRONS IN METAL-OXIDE SUPERLATTICES Bernhard Keimer	61
INV 24:	METAL-INSULATOR TRANSITIONS OF CORRELATED ELECTRONS IN OXIDE HETEROSTRUCTURES Masashi Kawasaki	62
INV 25:	THEORETICAL DESIGN OF TOPOLOGICAL PHENOMENA Naoto Nagaosa	64
INV 26:	MAGNETIC RECONSTRUCTIONS IN PEROVSKITE HETEROINTERFACES AND ULTRATHIN FILMS Harold Y. Hwang	65
INV 27:	PROGRESS IN THE ATOMIC SWITCH Masakazu Aono, Tsuyoshi Hasegawa, Kazuya Terabe, Tohru Tsuruoka,	66

Content 5

Nanosessions 69

Nanosession: 2D electron systems - Atomic configurations 71

2DA 1: HIGHLY CONFINED SPIN-POLARIZED TWO-DIMENSIONAL ELECTRON GAS IN $SrTiO_3/SrRuO_3$ SUPERLATTICES 71
Javier Junquera, Pablo García-Fernández, Marcos Verissimo-Alves, Daniel I. Bilc, Philippe Ghosez

2DA 2: FIRST-PRINCIPLES STUDY OF INTERMIXING AND POLARIZATION AT THE $DyScO_3/SrTiO_3$ INTERFACE 73
Kourosh Rahmanizadeh, Gustav Bihlmayer, Martina Luysberg, Stefan Blügel

2DA 3: ATOMIC-SCALE SPECTROSCOPY OF AN OXIDE INTERFACE BETWEEN A MOTT INSULATOR AND A BAND INSULATOR 74
M.-W. CHU, C. P. CHANG, S.-L. Cheng, J. G. Lin, C. H. CHEN

2DA 4: INTERFACE ATOMIC STRUCTURE IN $LaSrAlO_4/LaNiO_3/LaAlO_3$ HETEROSTRUCTURES 75
M. K. Kinyanjui, N. Gauquelin, G. Botton, E. Benckiser, B. Keimer, U. Kaiser

2DA 5: TAILORING THE ELECTRONIC PROPERTIES OF THE LAO/STO INTERFACE BY CONTROLLED CATION-STOICHIOMETRY VARIATION IN STO THIN FILMS 77
Felix Gunkel, Peter Brinks, Sebastian Wicklein, Susanne Hoffmann-Eifert, Regina Dittmann, Mark Huijben, Josée E. Kleibeuker, Gertjan Koster, Guus Rijnders, and Rainer Waser

2DA 6: ELECTROSTATIC DOPING OF A MOTT INSULATOR IN AN OXIDE HETEROSTRUCTURE: THE CASE OF $LaVO_3/SrTiO_3$ 79
F. Pfaff, A. Müller, H. Boschker, G. Berner, G. Koster, M. Gorgoi, W. Drube, G. Rijnders, M. Kamp, D.H.A. Blank, M. Sing, R. Claessen

2DA 7: THEORETICAL STUDY OF ORBITAL-, SPIN- AND CHARGE-RECONSTRUCTION IN LVO/STO HETEROSTRUCTURES 80
Giorgio Sangiovanni, Zhicheng Zhong, Elias Assmann, Peter Blaha, Karsten Held and Satoshi Okamoto

Nanosession: 2D electron systems - Correlation effects and transport 81

2DC 1: TWO-DIMENSIONAL ELECTRON GAS WITH ORBITAL SYMMETRY RECONSTRUCTION AND STRONG EFFECTIVE MASS LOWERING AT THE SURFACE OF $KTaO_3$ 81
A. F. Santander-Syro, C. Bareille, F. Fortuna, O. Copie F. Bertran, A. Taleb-Ibrahimi, P. Le Fèvre, G. Herranz, M. Bibes, A. Barthélémy, P. Lecoeur, J. Guevara, M. Gabay and M. J. Rozenberg

2DC 2: STRAIN MEDIATED LONG-RANGE QUASI-ORDERED DOMAIN STRUCTURES AT THE $SrTiO_3$ (110) SURFACE 83
Zhiming Wang, Fengmiao Li, Sheng Meng, Ulrike Diebold, Jiandong Guo

2DC 3: FIRST-PRINCIPLES STUDY OF THE $LaAlO_3/SrTiO_3$ INTERFACE 84
Fontaine Denis, Philippe Ghosez

2DC 4: FERROMAGNETISM DRIVEN BY $SrTiO_3$ FERROELECTRIC-LIKE LATTICE DEFORMATION IN $LaAlO_3/SrTiO_3$ HETEROSTRUCTURES 85
M. Carmen Muñoz, Jichao C. Li

2DC 5: COHERENT TRANSPORT IN MESOSCOPIC $LaAlO_3/SrTiO_3$ DEVICES 87
Daniela Stornaiuolo, Stefano Gariglio, Nuno J. G. Couto, Alexandre Fête, Andrea D. Caviglia, Gabriel Seyfarth, Didier Jaccard, Alberto F. Morpurgo, and Jean-Marc Triscone

2DC 6: STRONGLY CORRELATED HIGH-MOBILITY ELECTRON GAS AT A MgZnO/ZnO INTERFACE 88
Yusuke Kozuka, Joseph Falson, Denis Maryenko, Atsushi Tsukazaki, Christopher Bell, Minu Kim, Yasuyuki Hikita, Harold. Y. Hwang, Masashi Kawasaki

Nanosession: 2D electron systems - Electronic structure and field effects 89

2DE 1: REVEALING THE FERMI SURFACE OF THE BURIED $LaAlO_3/SrTiO_3$ INTERFACE BY ANGLE RESOLVED SOFT X-RAY PHOTOELECTRON SPECTROSCOPY 89
R. Claessen, G. Berner, H. Fujiwara, M. Sing, C. Richter, J. Mannhart, A. Yasui, Y. Saitoh, A. Yamasaki, Y. Nishitani, A. Sekiyama, S. Suga

2DE 2:	NANOSCALE MODULATION OF THE LOCAL DENSITY OF STATES AT THE INTERFACE BETWEEN $LaAlO_3$ AND $SrTiO_3$ BAND INSULATORS M. Salluzzo, Z. Ristic, I. Maggio Aprile, R. Di Capua, G. M. De Luca, F. Chiarella, M. Radovic	90
2DE 3:	TUNING THE TWO-DIMENSIONAL ELECTRON GAS AT THE $LaAlO_3/SrTiO_3(001)$ INTERFACE BY METALLIC CONTACTS Rossitza Pentcheva, Rémi Arras, Victor G. Ruiz and Warren E. Pickett	92
2DE 4:	FIELD-EFFECT DEVICES UTILIZING OXIDE INTERFACES Christoph Richter, Benjamin Förg, Rainer Jany, Georg Pfanzelt, Carsten Woltmann, Jochen Mannhart	93
2DE 5:	GATE-CONTROLLED SPIN INJECTION AT LAO/STO INTERFACES Henri Jaffrès, N. Reyren, E. Lesne, J.-M. George, C. Deranlot, S. Collin, M. Bibes, and A. Barthélémy	94
2DE 6:	FIELD EFFECT MODULATION OF THE ELECTRON GAS AT THE $LaAlO_3/SrTiO_3$ INTERFACE : A THERMOELECTRIC STUDY Stefano Gariglio, Danfeng Li, Jean-Marc Triscone, Ilaria Pallecchi, Sara Catalano, Alessandro Gadaleta, Daniele Marré, Alessio Filippetti	95
2DE 7:	IS IT POSSIBLE FOR A $La_{0.5}Sr_{0.5}TiO_3$ FERMI LIQUID TO EXIST IN A CONFINED TWO-DIMENSIONAL SYSTEM? X. Wang, Z. Huang, W.M. Lü, D. P. Leusink, A. Annadi, Z.Q. Liu, T. Venkatesan, and Ariando	96

Nanosession: Calorics 99

CAL 1:	MECHANISM OF "PHONON GLASS – ELECTRON CRYSTAL" BEHAVIOUR IN THERMOELECTRIC LAYERED COBALTATE L. Wu, Q. Meng, Ch. Jooss, J. Zheng, H. Inada, D. Su, Q. Li, and Y. Zhu	99
CAL 2:	SIGN REVERSAL OF THE TUNNELING MAGNETO SEEBECK EFFECT Andy Thomas, Markus Münzenberg, Christian Heiliger	101
CAL 3:	AB-INITIO INVESTIGATION OF MAGNETIC *KONBU* PHASES AS NANOSTRUCTURES WITH SPIN-CALORIC-TRANSPORT PROPERTIES Elias Rabel, Phivos Mavropoulos, Alexander Thiess, Rudolf Zeller, Tetsuya Fukushima, Nguyen D. Vu, Kazunori Sato, Hiroshi Katayama-Yoshida, Roman Kovacik, Peter H. Dederichs, Stefan Blügel	102
CAL 4:	CHARGE KONDO EFFECT IN THERMOELECTRIC PROPERTIES OF LEAD TELLURIDE DOPED WITH THALLIUM IMPURITIES Theo Costi, Veljko Zlatic	103
CAL 5:	ELECTRONIC STRUCTURE AND THERMOELECTRIC PROPERTIES OF NANOSTRUCTURED $EuTi_{1-x}Nb_xO_{3-\delta}$ (x = 0.00; 0.02) A.Shkabko, L. Sagarna, S. Populoh, L. Karvonen, A. Weidenkaff	104
CAL 6:	STRONG PHONON SCATTERING AND GLASSLIKE THERMAL CONDUCTIVITY IN CRYSTALLINE PHASE CHANGE MATERIALS F.R.L. Lange, K.S. Siegert and M. Wuttig	105
CAL 7:	TAILORING THERMOPOWER AND CARRIER MOBILITY IN NANOSTRUCTURED HALF-HEUSLERS Pierre F. P. Poudeu*; Julien P. A. Makongo; Pranati Sahoo; Liu Yuanfeng; Xiaoyuan Zhou; Ctirad Uher	107

Nanosession: Topological effects 109

TOP 1:	ANOMALOUS HALL EFFECT IN GRAPHENE DECORATED WITH 5d TRANSITION-METAL ADATOMS Y. Mokrousov, H. Zhang, F. Freimuth, C. Lazo, S. Heinze, S. Blügel	109
TOP 2:	SPIN POLARIZED PHOTOEMISSION FROM Bi_2Te_3 AND Sb_2Te_3 TOPOLOGICAL INSULATOR THIN FILMS L. Plucinski, A. Herdt, G. Bihlmayer, S. Döring, S. Blügel, C.M. Schneider	110
TOP 3:	INSTABILITIES OF INTERACTING ELECTRONS ON THE HONEYCOMB BILAYER Michael Scherer, Stefan Uebelacker, Carsten Honerkamp	111
TOP 4:	FIELD-INDUCED POLARIZATION OF DIRAC VALLEYS IN BISMUTH Zengwei Zhu, Aurélie Callaudin, Benoît Fauqué, Woun Kang, Kamran Behnia	112
TOP 5:	PEIERS DIMERIZATION AT THE EDGE OF 2D TOPOLOGICAL INSULATORS? Gustav Bihlmayer, Hyun-Jung Kim, Jun-Hyung Cho, Stefan Blügel	113

Content

TOP 6:	PREDICTING TOPOLOGICAL SURFACE STATES FROM THE SCATTERING PROPERTIES OF THE BULK Daniel Wortmann, Gustav Bihlmayer, Stefan Blügel	114

Nanosession: Mott insulators and transitions   115

MIT 1:	ELECTRIC-FIELD CONTROL OF THE FIRST ORDER METAL-INSULATOR TRANSITION IN VO_2 Masaki Nakano, Keisuke Shibuya, Daisuke Okuyama, Takafumi Hatano, Shimpei Ono, Masashi Kawasaki, Yoshihiro Iwasa, Yoshinori Tokura	115
MIT 2:	COLOSSAL MAGNETORESISTANCE AND HALF-METAL BEHAVIOR IN THE DOPED MOTT INSULATOR GaV_4S_8 B. Corraze, E. Janod, E. Dorolti, V. Guiot, C. Vaju, H.-J. Koo, E. Kan, M.-H. Whangbo and L. Cario	116
MIT 3:	THE SPIN-STATE AND METAL-INSULATOR TRANTIONS IN LnCoO$_3$ Guoren Zhang, Evgeny Gorelov, Erik Kochand Eva Pavarini	118
MIT 4:	SPIN-SPECTRAL-WEIGHT DISTRIBUTION AND ENERGY RANGE OF THE PARENT COMPOUND La(2)CuO(4) J. M. P. Carmelo, M. A. N. Araújo, S. W. White	119
MIT 5:	SUPERCONDUCTIVITY DRIVEN IMBALANCE OF THE MAGNETIC DOMAIN POPULATION IN $CeCoIn_5$ Simon Gerber, Nikola Egetenmeyer, Jorge Gavilano, Eric Ressouche, Christof Niedermayer, Andrea Bianchi, Roman Movshovich, Eric Bauer, John Sarrao, Joe Thompson, and Michel Kenzelmann	120
MIT 6:	RUBIDIUM SUPEROXIDE: A P-ELECTRON MOTT INSULATOR Roman Kovacik, Claude Ederer, Philipp Werner	121
MIT 7:	HIGH MOBILITY IN A STABLE TRANSPARENT PEROVSKITE Kookrin Char, Kee Hoon Kim, Hyung Joon Kim, Useong Kim, Hoon Min Kim, Tai Hoon Kim, Hyo Sik Mun, Byung-Gu Jeon, Kwang Taek Hong, Woong-Jhae Lee, Chanjong Ju	122

Nanosession: Advanced spectroscopy and scattering   123

SAS 1:	SHEDDING LIGHT ON ARTIFICIAL QUANTUM MATERIALS AND COMPLEX OXIDE INTERFACES WITH ANGLE-RESOLVED PHOTOEMISSION SPECTROSCOPY Kyle M. Shen, Eric J. Monkman, Carolina Adamo, John W. Harter, Daniel E. Shai, Yuefeng Nie, Julia A. Mundy, Alex J. Melville, David. A. Muller, Luigi Maritato, Darrell G. Schlom	123
SAS 2:	HARD AND SOFT X-RAY PHOTOEMISSION STUDIES OF OXIDE MULTILAYER BAND OFFSETS AND OF ELECTRONIC STRUCTURE IN $LaNiO_3/SrTiO_3$ AND $GdTiO_3/SrTiO_3$ G. Conti, A. M. Kaiser, A. X. Gray,, S.Nemsak , A. Bostwick, , A. Janotti, C. G. Van de Walle,J.Son , P.Moetakef , S. Stemmer,S. Ueda , K. Kobayashi, A. Gloskovskii , W. Drube,V.N. Strokov , C.S. Fadley	124
SAS 3:	EVIDENCE FOR Fe^{2+} CONFIGURATION IN Fe:STO THIN FILMS BY X-RAY ABSORPTION SPECTROSCOPY A. Köhl, D. Kajewski, J. Kubacki, K. Szot, Ch. Lenser, P.Meuffels , R. Dittmann, R. Waser, J. Szade	126
SAS 4:	PHOTOELECTRON AND RECOIL DIFFRACTION AT HIGH ENERGIES FOR BULK-SENSITIVE AND ELEMENT-RESOLVED CRYSTALLOGRAPHIC ANALYSIS OF MATERIALS Aimo Winkelmann, Maarten Vos	128
SAS 5:	HARD X-RAY ANGLE-RESOLVED PHOTOEMISSION AS A BULK-SENSITIVE PROBE OF ELECTRONIC STRUCTURE Alexander X. Gray, Christian Papp, Shigenori Ueda, Jan Minár, Lukasz Plucinski, Jürgen Braun, Benjamin Balke, Claus M. Schneider, Warren E. Pickett, Giancarlo Panaccione, Hubert Ebert, Keisuke Kobayashi, and Charles S. Fadley	129
SAS 6:	HAXPEEM – SPECTROSCOPIC IMAGING OF BURIED LAYERS USING HARD X-RAYS C. Wiemann, M. C. Patt, A. Gloskovskii, S. Thiess, W. Drube, M. Merkel, M. Escher, C. M. Schneider	131
SAS 7:	MARIA: THE MODERN NEUTRON REFLECTOMETER OF THE JCNS OPTIMISED FOR SMALL SAMPLE SIZES AND THINS LAYERS Stefan Mattauch[1], Ulrich Rücker[2], Denis Korolkov[1], Thomas Brückel[2]	132

Nanosession: High-resolution transmission electron microscopy — 133

TEM 1: TRANSMISSION ELECTRON MICROSCOPY OF FUNCTIONAL PEROVSKITE OXIDE HETEROSTRUCTURES — 133
Dietrich Hesse

TEM 2: NiO PRECIPITATES IN $LaNiO_3/LaAlO_3$ SUPERLATTICES INDUCED BY A POLAR MISMATCH — 135
Eric Detemple, Quentin M. Ramasse, Wilfried Sigle, Eva Benckiser, Georg Cristiani, Hanns-Ulrich Habermeier, Bernhard Keimer, Peter A. van Aken

TEM 3: MINIMUM ENERGY CONFIGURATION OF SCANDATE/TITANATE INTERFACES: ORDERED INTERFACES — 137
Martina Luysberg, Kourosh Rahmanizadeh, Gustav Biehlmayer, Jürgen Schubert,

TEM 4: RUDDLESDEN–POPPER TYPE FAULTS IN $LaNiO_3/LaAlO_3$ SUPERLATTICES — 138
Eric Detemple, Quentin M. Ramasse, Wilfried Sigle, Georg Cristiani, Hanns-Ulrich Habermeier, Bernhard Keimer, Peter A. van Aken

TEM 5: ATOMIC STRUCTURE OF TRIMERIZATION-POLARIZATION DOMAIN WALLS IN HEXAGONAL $ErMnO_3$ — 140
Myung-Geun Han, Lijun Wu, Toshihiro Aoki, Nara Lee, Seung Chul Chae, Sang-Wook Cheong, and Yimei Zhu

Nanosession: New technologies for scanning probes — 143

NTS 1: CONSTRUCTION AND FIRST RESULTS OF AN STM OPERATING AT MILLI-KELVIN TEMPERATURES — 143
C. R. Ast, M. Assig, M. Etzkorn, M. Eltschka, B. Jäck, and K. Kern

NTS 2: QUANTITATIVE FORCE IMAGING OF THE ATOMS IN EPITAXIALLY GROWN GRAPHENE — 144
M.P.Boneschanscher, Z. Sun, J. van der Lit, P. Liljeroth and

NTS 3: RADIO FREQUENCY OPTIMIZED SCANNING TUNNELING MICROSCOPE FOR THE USE WITH PULSED TUNNELING VOLTAGES — 146
Christian Saunus, Marco Pratzer, Markus Morgenstern

NTS 4: ULTRA COMPACT 4-TIP STM/AFM FOR ELECTRICAL MEASUREMENTS AT THE NANOSCALE — 147
Vasily Cherepanov, Stefan Korte, Marcus Blab, Evgeny Zubkov, Hubertus Junker, Peter Coenen, Bert Voigtländer

NTS 5: NANOSCALE MECHANICAL CHARACTERIZATION OF THIN FILMS with Different TopologiCAL STRUCTURES — 148
Kong-Boon Yeap, Malgorzata Kopycinska-Mueller, Lei Chen, Martin Gall, Ehrenfried Zschech

NTS 6: ELECTRONIC ACTIVATION IN THE $(La_{0.8}Sr_{0.2})CoO_3/(La_{0.5}Sr_{0.5})_2CoO_4$ SUPERLATTICES AT HIGH TEMPERATURE — 150
Yan Chen, Zhuhua Cai, Yener Kuru, Harry L.Tuller and Bilge Yildiz*

NTS 7: THE SEM/FIB WORKBENCH: Automated Nanorobotics system inside of Scanning Electron or Focussed Ion Beam Microscopes — 152
Volker Klocke, Ivo Burkart

Nanosession: Phase change materials — 155

PCA 1: DENSITY FUNCTIONAL THEORY STUDY OF ANDERSON METAL-INSULATOR TRANSITIONS IN CRYSTALLINE PHASE-CHANGE MATERIALS — 155
Wei Zhang, Alexander Thiess, Peter Zalden, Jean-Yves Raty, Rudolf Zeller, Peter H. Dederichs, Matthias Wuttig, Stefan Blügel, Riccardo Mazzarello

PCA 2: LARGE SCALE MOLECULAR DYNAMICS SIMULATIONS OF PHASE CHANGE MATERIALS — 157
Gabriele Cesare Sosso, Giacomo Miceli, Sebastiano Caravati, Davide Donadio, Jörg Behler and Marco Bernasconi

PCA 3: QUANTUM-CHEMICAL ANALYSIS OF ATOMIC MOTION IN Ge-Sb-Te PHASE-CHANGE ALLOYS — 158
Ralf Stoffel, Marck Lumeij, Volker Deringer, Richard Dronskowski

PCA 4: SIMULATION OF RAPID CRYSTALIZATION IN PHASE CHANGE MATERIALS BY MEANS OF PHASE FIELD MODELING — 160
Fatemeh Tabatabaei, Markus Apel, Efim Brener

Content

PCA 5:	EPITAXYAL PHASE CHANGE MATERIALS: GROWTH, STRUCTURE AND PHASE TRANSITION Henning Riechert, Peter Rodenbach, Alessandro Giussani, Karthick Perumal, Michael Hanke, Jonas Laehnemann, Martin Dubslaff, Raffaella Calarco, Manfred Burghammer, Alexander Kolobov and Paul Fons	161
PCA 6:	EFFECT OF CARBON AND NITROGEN DOPING ON THE STRUCTURE AND DYNAMICS OF AMORPHOUS GeTe PHASE CHANGE MATERIAL J. Y. Raty, G. Ghezzi, P. Noé, E. Souchier, S. Maitrejean, C. Bichara, F. Hippert	162

Nanosession: Phase change memories — 163

PCM 1:	EXPLOITING THE MEMRISTIVE-LIKE BEHAVIOUR OF PHASE-CHANGE MATERIALS AND DEVICES FOR ARITHMETIC, LOGIC AND NEUROMORPHIC PROCESSING C D Wright, J A Vázquez Diosdado, L Wang, Y Liu, P Ashwin, K I Kohary, M M Aziz, P Hosseini and R J Hicken	163
PCM 2:	INVERSE TIME-VOLTAGE RELATION OF THRESHOLD SWITCHING IN PHASE CHANGE MATERIALS Marco Cassinerio, Nicola Ciocchini and Daniele Ielmini	165
PCM 3:	INTERPLAY OF DEFECTS AND CHEMICAL BONDING IN THE "GST" FAMILY OF PHASE-CHANGE MATERIALS Volker L. Deringer, Marck Lumeij, Ralf Stoffel, Richard Dronskowski	167
PCM 4:	PHOTONICS-BASED NON-VOLATILE MEMORY DEVICE USING PHASE CHANGE MATERIALS Wolfram H.P. Pernice and Harish Bhaskaran	169
PCM 5:	ROLE OF ACTIVATION ENERGY IN RESISTANCE DRIFT OF AMORPHOUS PHASE CHANGE MATERIALS Martin Salinga, Martin Wimmer, Matthias Käs, Matthias Wuttig	171
PCM 6:	$Ge_2Sb_2Te_5$ LINE TEST-STRUCTURES FOR PHASE-CHANGE NON VOLATILE MEMORIES G. D'Arrigo, A.M. Mio, A. Cattaneo, C. Spinella, A.L. Lacaita and E. Rimini	173
PCM 7:	IN SITU TRANSMISSION ELECTRON MICROSCOPY STUDY OF THE CRYSTALLIZATION OF BITS IN $Ag_4In_3Sb_{67}Te_{26}$ Manuel Bornhöfft, Andreas Kaldenbach, Matthias Wuttig, Joachim Mayer	175

Nanosession: Scanning probe microscopy on oxides — 177

SPO 1:	TEMPLATED ADSORPTION AT THE $Fe_3O_4(001)$ SURFACE: THE EFFECT OF SUBSURFACE CHARGE AND ORBITAL ORDER Gareth S. Parkinson, Zbynek Novotny, Michael Schmid, Ulrike Diebold	177
SPO 2:	EXPLORING ROUTES TO TAILOR THE ELECTRONIC PROPERTIES OF THIN-OXIDE FILMS ON METAL SUPPORTS Xiang Shao, Fernando Stavale, Niklas Nilius	178
SPO 3:	PREPARATION AND CHARACTERIZATION OF THIN MgO FILMS DOPED WITH NITROGEN Martin Grob, Marco Pratzer, M. Ležaić, Markus Morgenstern	179
SPO 4:	SCANNING TUNNELING MICROSCOPY STUDY OF SINGLE-CRYSTALLINE $Sr_3Ru_2O_7$ Bernhard Stöger, Zhiming Wang, Michael Schmid, Ulrike Diebold, David Fobes, Zhiqiang Mao	180
SPO 5:	BIMETALLIC ALLOYS AS MODEL SYSTEMS FOR THE GROWTH OF ULTRATHIN METAL OXIDE FILMS Marco Moors, Séverine Le Moal, Jan Markus Essen, Christian Breinlich, Maria Kesting, Stefan Degen, Aleksander Krupski, Conrad Becker, Klaus Wandelt	181
SPO 6:	ELECTROSTATIC FIELD EFFECT MODULATION OF SHUBNIKOV-DE HAAS OSCILLATIONS IN $LaAlO_3/SrTiO_3$ Nicolas Reyren, Mario Basletić, Manuel Bibes, Cécile Carrétéro, Virginie Trinité, Amir Hamzić and Agnès Barthélémy	183

Nanosession: Logic devices and circuit design — 185

LDC 1:	SELF-RECTIFYING RESISTIVE MEMORY DEVICES Wei Lu, Sung-Hyun Jo, Yuchao Yang	185

LDC 2:	THE DESIRED MEMRISTOR FOR CIRCUIT DESIGNERS Shahar Kvatinsky, Eby G. Friedman, Avinoam Kolodny, and Uri C. Weiser	187
LDC 3:	COMPLEMENTARY RESISTIVE SWITCH-BASED ASSOCIATIVE MEMORY CAPABLE OF FULLY PARALLEL SEARCH FOR MINIMUM HAMMING DISTANCE Omid Kavehei, Stan Skafidas, Kamran Eshraghian	188
LDC 4:	COMPUTATIONAL CONCEPT BASED ON COMPLEMENTARY RESISTIVE SWITCHES Ondrej Šuch, Martin Klimo, Stanislav Foltán, Karol Grondžák	190
LDC 5:	LOGIC OPERATIONS IN PASSIVE COMPLEMENTARY RESISTIVE SWITCH CROSSBAR ARRAYS Eike Linn, Roland Rosezin, Stefan Tappertzhofen, Ulrich Böttger, Rainer Waser	192
LDC 6:	A NON-VOLATILE LOW-POWER ZERO-LEAKAGE NANOMAGNETIC COMPUTING SYSTEM M. Becherer, J. Kiermaier, S. Breitkreutz, I. Eichwald, G. Csaba, D. Schmitt-Landsiedel	194

Nanosession: Neuromorphic concepts — 197

NMC 1:	AN ELECTRONIC VERSION OF PAVLOV`S DOG Hermann Kohlstedt, Martin Ziegler, Rohit Soni, Timo Patelczyk,	197
NMC 2:	NEUROMORPHIC FUNCTIONALITIES OF NANOSCALE MEMRISTORS Ting Chang, Sung-Hyun Jo, Patrick Sheridan, Wei Lu	198
NMC 3:	DEMONSTRATION OF IMPLICITE MEMORY IN ELECTRONIC CIRCUITS BY USING MEMRESISTIVE DEVICES Martin Ziegler, Mirko Hansen, Hermann Kohlstedt	200
NMC 4:	USAGE OF NANOELECTRONIC RESISTIVE SWITCHES WITH NONLINEAR SWITCHING KINETICS IN HYBRID CIRCUITS FOR LINEAR CONDUCTANCE ADAPTATION Arne Heittmann, Tobias G. Noll	201
NMC 5:	Pt/HfO$_2$/TiN/Al ON SiO$_2$ WITH POTENTIAL APPLICATIONS TO MEMORY AND NEUROMORPHIC CIRCUITS Davide Sacchetto, Yusuf Leblebici, Sung-Mo Steve Kang	203
NMC 6:	MEMRISTORS: TWO CENTURIES ON Themistoklis Prodromakis, Christopher Toumazou, Leon Chua	205

Nanosession: Electrochemical metallization memories — 207

ECM 1:	ATOM/ION MOVEMENT CONTROLLED THREE-TERMINAL DEVICE: ATOM TRANSISTOR Tsuyoshi Hasegawa, Yaomi Itoh, Tohru Tsuruoka, Masakazu Aono	207
ECM 2:	QUANTUM CONDUCTANCE OF AGI BASED RESISTIVE SWITCHES: TOWARDS AN ATOMIC SCALE MEMORY Stefan Tappertzhofen, Ilia Valov, Rainer Waser	208
ECM 3:	DYNAMIC GROWTH/DISSOLUTION OF CONDUCTIVE FILAMENT IN OXIDE-ELECTROLYTE-BASED RRAM Ming Liu, Qi Liu, Hangbing Lv, Shibing Long and Yingtao Li	210
ECM 4:	IN-SITU HARD X-RAY PES POLARIZATION MEASUREMENTS OF OXIDE AMORPHOUS FILMS UNDER INTENSE ELECTRICAL FIELD S. YAMAGUCHI,T. TSUCHIYA, S. MIYOSHI,Y. YAMASHITAH. YOSHIKAWA, K. TERABE, and K. KOBAYASHI	211
ECM 5:	SPECTROSCOPIC INVESTIGATION OF Charge Transfer in Electrochemical Metallization Memory CELL Deok-Yong Cho, Ilia Valov, Jan van den Hurk, Stefan Tappertzhofen, Rainer Waser	213
ECM 6:	CHARACTERIZATION OF GERMANIUM SULFIDE THIN FILMS GROWN BY HOT WIRE CHEMICAL VAPOR DEPOSITION Denis Reso, Mindaugas Silinskas, Nancy Frenzel, Marco Lisker, Edmund P. Burte	214
ECM 7:	NANOFILAMENT RELAXATION MODEL FOR SIZE-DEPENDENT RESISTANCE DRIFT IN ELECTROCHEMICAL MEMORIES Seol Choi, Simone Balatti, Federico Nardi and Daniele Ielmini	216

Content

Nanosession: Valence Change Memories - redox mechanism and modelling — 219

VCR 1: ION MIGRATION MODEL FOR RESISTIVE SWITCHING IN TRANSITION METAL OXIDES — 219
D. Ielmini, S. Larentis, S. Balatti, F. Nardi and D. Gilmer

VCR 2: SIMULATION STUDIES OF THE MATERIAL DEPENDENT SWITCHING PERFORMANCE OF VALENCE CHANGE MEMORY CELLS — 221
Stephan Menzel, Astrid Marchewka, Ulrich Böttger, Rainer Waser

VCR 3: ELECTRORESISTANCE VERSUS JOULE HEATING EFFECTS IN MANGANITE THIN FILMS — 223
Ll. Balcells, A. Pomar, R. Galceran, Z. Konstantinovic, L. Peña, B. Bozzo,

VCR 4: ON ELECTROFORMING FOR BIPOLAR SWITCHING — 225
Ilan Riess, Dima Kalaev

VCR 5: ROOM-TEMPERATURE KINETICS OF DEFECT MIGRATION IN NON-FRADAIC Pt/TiO$_2$/Pt CAPACITORS — 227
Hyungkwang Lim, Ho-Won Jang, Cheol Seong Hwang, Doo Seok Jeong

VCR 6: TANTALUM OXIDE ULTRA-THIN FILMS BY METAL OXIDATION FOR APPLICATION IN RESISTIVE RANDOM ACCESS MEMORY (RRAM) — 229
Sebastian Schmelzer, Ulrich Böttger, Rainer Waser

VCR 7: RESISTIVE SWITCHING PHENOMENA IN Li$_x$CoO$_2$ THIN FILMS — 231
Olivier Schneegans, Van Huy Mai, Alec Moradpour, Pascale Auban-Senzier, Claude Pasquier, Kang Wang, Sylvain Franger, Alexandre Revcolevschi, Efthymios Svoukis, John Giapintzakis, Philippe Lecoeur, Pascal Aubert, Guillaume Agnus, Thomas Maroutian, Raphaël Salot, Pascal Chrétien

Nanosession: Valence Change Memories - a look inside — 233

VCI 1: NANASCALE ANALYSIS OF FORMING AND RESISTIVE SWITCHING IN Fe:STO THIN FILM DEVICES — 233
R. Dittmann, R. Muenstermann, I. Krug, D. Park, F. Kronast, A. Besmehn, J. Mayer, C. M. Schneider and Rainer Waser

VCI 2: IN-OPERANDO HAXPES ANALYSIS OF THE RESISTIVE SWITCHING PHENOMENON IN Ti/HfO$_2$-BASED SYSTEMS — 235
Malgorzata Sowinska, Thomas Bertaud, Damian Walczyk, Sebastian Thiess, Christian Walczyk, and Thomas Schroeder

VCI 3: THE OXYGEN VACANCY DISTRIBUTION IN RESISTIVE SWITCHING Fe-SrTiO$_3$ MIM STRUCTURES BY µXAFS — 237
Christian Lenser, Alexei Kuzmin, Aleksandr Kalinko, Juris Purans, Rainer Waser and Regina Dittmann

VCI 4: MULTILEVEL RESISTIVE SWITCHING AND METAL –INSULATOR TRANSITION IN SOLUTION-DERIVED La$_{1-x}$Sr$_x$MnO$_3$ THIN FILMS — 239
C. Moreno, J. Zabaleta, A. Palau, J. Gázquez, N. Mestres, T. Puig, C. Ocal, X. Obradors

VCI 5: PUMP AND RELEASE SCENARIO FOR THE BIPOLAR RESISTIVE SWITCHING OF MEMRISTIVE MANGANITE-METAL INTERFACES — 241
Pablo Levy, N.Ghenzi, M. J. Sanchez, M. J. Rozenberg, P. Stoliar, F. G. Marlasca, and D. Rubi

VCI 6: RESISTIVE SWITCHING IN NiO BASED NANOWIRE ARRAY FOR LOW POWER RERAM — 242
Sabina Spiga, Stefano Brivio, Grazia Tallarida, Daniele Perego, Silvia Franz, Damien Deleruyelle, Christophe Muller

VCI 7: EXPERIMENTAL EVALUATION OF THE TEMPERATURE IN CONDUCTIVE FILAMENTS CREATED IN RESISTIVE SWITCHING MATERIALS — 244
Eilam Yalon, Shimon Cohen, Arkadi Gavrilov, Boris Meyler, Joseph Salzman, and Dan Ritter

Nanosession: Variants of resistive switching — 247

VRS 1: FERROELECTRIC RESISTIVE SWITCHING AT SCHOTTKY-LIKE BiFeO$_3$ INTERFACES — 247
Akihito Sawa, Atsushi Tsurumaki-Fukuchi, Hiroyuki Yamada

VRS 2: HOW CAN WE SWITCH THE RESISTIVITY OF A METALLIC PEROVSKITE OXIDE (SrTiO$_3$:Nb) BY ELECTRICAL STIMULI? — 248
Christian Rodenbücher, Krzysztof Szot, Rainer Waser

VRS 3:	PHYAICAL MECHANISM OF OXYGEN VACANCY MIGRATION IN Pt/Nb:SrTiO$_3$ INTERFACES Shin Buhm Lee, Jong-Bong Park, Myoung-Jae Lee, Tae Won Noh	250
VRS 4:	MECHANISM OF RESISTIVE SWITCHING IN BIPOLAR TRANSITION METAL OXIDES Marcelo J. Rozenberg, María J. Sánchez, Pablo Stoliar, Ruben Weht, Carlos Acha, Fernando Gomez-Marlasca and Pablo Levy	251
VRS 5:	ELECTRIC FIELD INDUCED RESISTIVE SWITCHING IN A FAMILY OF MOTT INSULATORS: TOWARDS A MOTT-MEMRISTOR? L. Cario, B. Corraze, V. Guiot, S. Salmon, J. Tranchant, M.-P. Besland, V. Ta Phuoc, M. Rozenberg, T. Cren, D. Roditchev, E. Janod	253
VRS 6:	INTRINSIC DEFECTS IN TiO$_2$ TO EXPLAIN RESISTIVE SWITCHING DEVICES Dieter Schmeißer, Matthias Richter, Massimo Tallarida	255
VRS 7:	CONDUCTANCE QUANTIZATION IN RESISTIVE SWITCHING Shibing Long, Carlo Gagli, Xavier Cartoixà, Riccardo Rurali, Enrique Miranda, David Jiménez, Julien Buckley, Ming Liu and Jordi Suñé	257

Nanosession: Magnetic interfaces and surfaces — 259

MAG 1:	HIGHLY SPIN-POLARIZED CONDUCTING STATE AT THE INTERFACE BETWEEN NON-MAGNETIC BAND INSULATORS: LaAlO$_3$/FeS$_2$ (001) J. D. Burton and E. Y. Tsymbal	259
MAG 2:	NEW INSIGHTS INTO NANOMAGNETISM BY SPIN-POLARIZED SCANNING TUNNELING MICROSCOPY AND SPECTROSCOPY Dirk Sander, Hirofumi Oka, Safia Ouazi, Sebastian Wedekind, Guillemin Rodary, Pavel Ignatiev, Larissa Niebergall, Valeri Stepanyuk, and Jürgen Kirschner	260
MAG 3:	SELECTIVE ORBITAL OCCUPATION AT MANGANITE INTERFACES INDUCED BY CRYSTAL SYMMETRY BREAKING B. Martínez, S. Valencia, L. Peña, Z. Konstantinovic, Ll. Balcells, R. Galceran, D. Schmitz, F. Sandiumenge, M. Casanove	261
MAG 4:	SCALABLE EXCHANGE BIAS IN LSMO/STO THIN FILMS Daniel Schumacher, Alexandra Steffen, Jörg Voigt, Jürgen Schubert, Hailemariam Ambaye, Valeria Lauter, John Freeland, Thomas Brückel	263
MAG 5:	STRUCTURAL AND MAGNETIC PROPERTIES OF NANOPARTICLE SUPERLATTICES O. Petracic, D. Greving, D. Mishra, M. J. Benitez, P. Szary, G. Badini Confalonieri, A. Ludwig, M. Ewerlin, L. Agudo, G. Eggeler, B.P. Toperverg, and H. Zabel	265
MAG 6:	NANOPARTICLES OF ANTIFERROMAGNETIC AND FERRIMAGNETIC OXIDES AS MAGNETIC HETEROSTRUCTURES Veronica Salgueiriño, Nerio Fontaíña-Troitiño, Ruth Otero-Lorenzo, Sara Liébana-Viñas	266
MAG 7:	SELF ASSEMBLED IRON OXIDE NANOPARTICLES – FROM A 2D POWDER TO A SINGLE MESOCRYSTAL Elisabeth Josten, Erik Wetterskog, Doris Meertens, Ulrich Rücker, German Salazar-Alvarez, Oliver Seeck, Peter Boesecke, Tobias Schulli Manuel Angst, Raphael Hermann, Lennart Bergström, Thomas Brückel	267

Nanosession: Ionics - lattice disorder and grain boundaries — 269

IOL 1:	CHARACTERIZATION OF VACANCY-RELATED DEFECTS IN Fe-DOPED SrTiO$_3$ THIN FILMS USING POSITRON ANNIHILATION LIFETIME SPECTROSCOPY D. J. Keeble, S. Wicklein, G.S. Kanda, W. Egger, and R. Dittmann	269
IOL 2:	*AB INITIO* CALCULATIONS OF DEFECTS IN GALLIUM OXIDE T. Zacherle, P.C. Schmidt, M. Martin	271
IOL 3:	CATION DEFECT ENGINEERING IN STO THIN FILMS BY PLD - VERIFICATION AND IMPLICATIONS ON MEMRISTIVE PROPERTIES S. Wicklein, C. Xu, A. Sambri, S. Amoruso, D.J. Keeble, R.A. Mackie, W. Egger, R. Dittmann	273
IOL 4:	STRUCTURAL RESPONSE OF SINGLE CRYSTAL SrTiO$_3$ ON O-VACANCY MIGRATION IN THERMAL AND ELECTRICAL FIELDS Barbara Abendroth, Juliane Hanzig, Hartmut Stöcker, Florian Hanzig, Ralph Strohmeyer, Solveig Rentrop, Uwe Mühle, Dirk C. Meyer	275
IOL 5:	CRYSTAL- AND DEFECT- CHEMISTRY OF REDUCTION RESISTANT FINE GRAINED THERMISTOR CERAMICS ON BaTiO$_3$-BASIS Christian Pithan, Hayato Katsu, Rainer Waser, Hiroshi Takagi	276

Content

IOL 6: RED-OX DRIVEN POINT DEFECT EQUILIBRIA, ANISOTROPIC CHEMICAL AND THERMAL EXPANSION AND FERROELASTICITY OF ACCEPTOR DOPED LaMO$_3$ PEROVSKITE OXIDE AT THE NANO-SCALE 278
Xinzhi Chen, Julian R. Tolchard, Sverre M. Selbach, Tor Grande

Nanosession: Ionics - redox kinetics, ion transport, and interfaces ... 281

IOR 1: OXYGEN EXCHANGE KINETICS ON PEROVSKITE SURFACES: IMPORTANCE OF ELECTRONIC AND IONIC DEFECTS 281
R. Merkle, L. Wang, Y. A. Mastrikov, E. A. Kotomin, J. Maier

IOR 2: ORDERS OF MAGNITUDE VARIATIONS IN THE ELECTRICAL CONDUCTION PROPERTIES OF ACCEPTOR AND DONOR DOPED STRONTIUM TITANATE ON DOWNSIZING 283
Giuliano Gregori, Piero Lupetin, Joachim Maier

IOR 3: DYNAMIC SIMULATION OF OXYGEN MIGRATION IN TiO$_2$ 285
Jan M. Knaup, Michael Wehlau, Thomas Frauenheim

IOR 4: ATOMISTIC SIMULATION STUDY ON OXYGEN DEFICIENT STRONTIUM TITANATE 286
Marcel Schie, Astrid Marchewka, Roger A. De Souza, Thomas Müller, Rainer Waser

IOR 5: FIRST PRINCIPLE STUDY AND MODELING OF STRAIN-DEPENDENT IONIC MIGRATION IN ZIRCONIA 287
Julian A. Hirschfeld, Hans Lustfeld

IOR 6: INVESTIGATIONS ON THE INTEGRATED CATHODES FOR HIGH ENERGY DENSITY LITHIUM RECHARGEABLE BATTERIES 288
S.B. Majumder, C. Ghanty, R.N. Basu

Nanosession: Spin dynamics ... 291

SDY 1: VORTEX DOMAIN WALL DYNAMICS IN MAGNETIC NANOTUBES 291
Attila Kákay, Ming Yan, Christian Andreas, Felipe García-Sánchez, Riccardo Hertel

SDY 2: SPIN-TORQUE DYNAMICS OF STACKED VORTICES IN MAGNETIC NANOPILLARS 293
Daniel E. Bürgler, Volker Sluka, Alina Deac, Attila Kakay, Riccardo Hertel, Claus M. Schneider

SDY 3: PURE SPIN CURRENTS IN FERROMAGNETIC INSULATOR/NORMAL METAL HYBRID STRUCTURES 295
Matthias Althammer, Mathias Weiler, Franz D. Czeschka, Johannes Lotze, Georg Woltersdorf, Michael Schreier, Stephan Gepraegs, Hans Huebl, Matthias Opel, Rudolf Gross, Sebastian T.B. Goennenwein

SDY 4: FEMTOSECOND SPIN DYNAMICS AND NANOMETER IMAGING WITH LASER-BASED EXTREME ULTRAVIOLET SOURCE 297
Roman Adam, Dennis Rudolf, Alexander Bauer,Christian Weier, Moritz Plötzing, Patrik Grychtol, Chan La-O-Vorakiat, Emrah Turgut,Henry C. Kapteyn, Margaret M. Murnane, Justin M. Shaw, Hans T. Nembach, Thomas J. Silva, Stefan Mathias, Martin AeschlimannandClaus M. Schneider

SDY 5: THEORETICAL STUDY OF ULTRAFAST LASER INDUCED MAGNETIC PRECESSIONS 299
Daria Popova, Andreas Bringer, Stefan Blügel

SDY 6: SPIN RELAXATION INDUCED BY THE ELLIOTT-YAFET MECHANISM IN 5d TRANSITION-METAL THIN FILMS 300
N. H. Long, Ph. Mavropoulos, S. Heers, B. Zimmermann, Y. Mokrousov and S. Blügel

Nanosession: Spin injection and transport ... 301

SIT 1: N-TYPE ELECTRON-INDUCED FERROMAGNETIC SEMICONDUCTOR (In,Fe)As 301
Pham Nam Hai, Le Duc Anh, Daisuke Sakaki, Masaaki Tanaka

SIT 2: ELECTRICAL SPIN INJECTION AND SPIN TRANSPORT IN ZINC OXIDE 303
Matthias Althammer, Eva-Maria Karrer-Müller, Sebastian T.B. Goennenwein, Matthias Opel, Rudolf Gross

SIT 3: SPIN RELAXATION BY IMPURITY SCATTERING: IMPORTANCE OF RESONANT SCATTERING 305
Phivos Mavropoulos, Swantje Heers, Rudolf Zeller, and Stefan Blügel

SIT 4:	EXPERIMENTAL AND THEORETICAL ANALYSIS OF OXYGEN-DEFICIENT EuO THIN FILMS A. Ionescu, M. Barbagallo, P.M.D.S. Monteiro, N.D.M. Hine, J.F.K. Cooper, N.-J. Steinke, J.-Y. Kim, K.R.A. Ziebeck, C.H.W. Barnes, C. J. Kinane, B.R.M. Dalgliesh, T.R. Charlton, S. Langridge, T. Stollenwerk and J. Kroha	306
SIT 5:	DELTA DOPED ANTIFERROMAGNETIC MANGANITES T. S. Santos, B. J. Kirby, S. Kumar, S. J. May, J. A. Borchers, B. B. Maranville, J. Zarestky, S. G. E. te Velthuis, J. van den Brink, B. Nelson-Cheeseman and A. Bhattacharya	307
SIT 6:	SPIN-ORBIT MEDIATED TORQUES IN HETEROSTRUCTURES WITH STRUCTURAL INVERSION ASYMMETRY Frank Freimuth, Yuriy Mokrousov, Stefan Blügel	308

Nanosession: Spin tunneling systems   **311**

STS 1:	MAGNETICALLY ENHANCED MEMRISTOR Mirko Prezioso, Alberto Riminucci, Ilaria Bergenti, Patrizio Graziosi and Valentin A. Dediu	311
STS 2:	NOVEL FUNCTIONALITIES AT INTERFACES IN $La_{0.7}Ca_{0.3}MnO_3$/$PrBa_2Cu_3O_7$/$La_{0.7}Ca_{0.3}MnO_3$ MAGNETIC TUNNEL JUNCTIONS Fabián A. Cuellar, Yaohua Liu, Norbert M. Nemes, Mar Garcia Hernandez, John Freeland, Juan Salafranca, Satoshi Okamoto, Suzanne G. E. te Velthuis, María Varela, Stephen J. Pennycook, Manuel Bibes, Agnes Barthélémy, Zouhair Sefrioui, Carlos Leon, Jacobo Santamaria	313
STS 3:	INTEGRATION OF A MAGNETIC OXIDE DIRECTLY WITH SILICON Martina Müller, C. Caspers, A. X. Gray, A. M. Kaiser, A. Gloskovskii, W. Drube, M. Gorgoi, C. S. Fadley, and C. M. Schneider	315
STS 4:	A SPINTRONIC MEMRISTOR J. Grollier, A. Chanthbouala, J. Sampaio, P. Metaxas, R. Matsumoto, A. Anane, A. V. Khvalkovskiy, V. Cros, A. Fert, K. A. Zvezdin, A. Fukushima, H. Kubota, K. Yakushiji, S. Yuasa	317
STS 5:	MEMRISTIVE MAGNETIC TUNNEL JUNCTIONS AND THEIR APPLICATIONS Andy Thomas, Patryk Krzysteczko, Günter Reiss, Jana Münchenberger, Markus Schäfers	318
STS 6:	ANTIFERROMAGNETIC COUPLING ACROSS SILICON REGULATED BY TUNNELING CURRENTS Rashid Gareev, MaximilianSchmid, Johann Vancea, Christian Back, Reinert Schreiber, Daniel Bürgler, Claus Schneider, Frank Stromberg, Heiko Wende	319
STS 7:	NON-VOLATILE ELECTRICAL CONTROL OF MAGNETISM IN MANGANESE-DOPED ZINC OXIDE Antonio Ruotolo, Xiao Lei Wang, Chi Wah Leung, Rolf Lortz	321

Nanosession: Multiferroic thin films and heterostructures   **323**

MFH 1:	MAGNETOELECTRICALLY INDUCED GIANT TUNNELING ELECTRORESISTANCE EFFECT J. D. Burton, Yuewei Yin, X. G. Li, Young-Min Kim, Albina Y. Borisevich, Qi Li and Evgeny Y. Tsymbal	323
MFH 2:	REVERSIBLE ELECTRICAL SWITCHING OF SPIN POLARIZATION IN MULTIFERROIC Co/$Pb(Zr_{0.2}Ti_{0.8})O_3$/$La_{0.7}Sr_{0.3}MnO_3$ TUNNEL JUNCTIONS Daniel Pantel, Silvana Goetze, Marin Alexe, and Dietrich Hesse	325
MFH 3:	MAGNETOELASTIC AND MAGNETOELECTRIC EFFECTS IN COMPOSITE MULTIFERROIC HYBRID STRUCTURES Stephan Geprägs, Matthias Opel, Sebastian T.B. Goennenwein, Rudolf Gross	327
MFH 4:	ELECTRIC CONTROL OF THE MAGNETIZATION IN $BiFeO_3$/$LaFeO_3$ SUPERLATTICES. Zeila Zanolli, Jacek C. Wojdel, Jorge Iniguez, Philippe Ghosez	329
MFH 5:	THE NEXT STEP ON THE SPIRAL – $TbMnO_3$ THIN FILMS Artur Glavic, Jörg Voigt, Enrico Schierle, Eugen Weschke, Thomas Brückel	330
MFH 6:	MAGNETIC CHIRAL DOMAINS IN MULTIFERROIC THIN FILMS KEEP MEMORY Josep Fontcuberta, I. Fina, L. Fàbrega, X. Martí and F. Sánchez	332
MFH 7:	EXCHANGE BIAS AND MAGNETOELECTRIC COUPLING EFFECTS IN $ZnFe_2O_4$ – $BaTiO_3$ COMPOSITE THIN FILMS Michael Lorenz, Michael Ziese, Gerald Wagner, Pablo Esquinazi, Marius Grundmann	334

Content

Nanosession: Multiferroics - ordering phenomena — 335

MFO 1: CHARGE ORDER IN LUTETIUM IRON OXIDE: AN UNLIKELY ROUTE TO FERROELECTRICITY — 335
J. de Groot, T. Mueller, R.A. Rosenberg, D.J. Keavney, Z. Islam, J.-W. Kim, and M. Angst

MFO 2: CRYSTAL STRUCTURE, PHASE TRANSITION, CHEMICAL EXPANSION AND DEFECT CHEMISTRY OF HEXAGONAL $HoMnO_3$ — 337
Sverre M. Selbach, Kristin Bergum, Amund Nordli Løvik, Julian R. Tolchard, Mari-Ann Einarsrud, Tor Grande

MFO 3: COLLECTIVE MAGNETISM AT FERROELECTRIC DOMAIN WALLS — 339
Weida Wu, Yanan Geng, Y.J. Choi, N. Lee and Sang-Wook Cheong

MFO 4: HARD X-RAY NANOSCALE STRUCTURAL IMAGING OF MULTIFERROIC THIN FILMS — 340
Martin Holt, Stephan Hruszkewycz, Chad Folkman, Robert Winarski, Volker Rose, Paul Fuoss, Ian McNulty

MFO 5: THE STRUCTURE OF THE MULTIFERROIC $BaTiO_3/Fe(001)$ INTERFACE — 342
H.L. Meyerheim, F. Klimenta, A. Ernst, K. Mohseni, S. Ostanin, M. Fechner, S.S. Parihar, I.V. Maznichenko, I. Mertig, and J. Kirschner

MFO 6: INVESTIGATION OF TWO MECHANISMS FOR MULTIFERROICITY IN $PbCrO_3$ BY DFT — 344
Martin Schlipf, Marjana Lezaic

Nanosession: Multiferroics - high transition temperatures — 347

MFT 1: REALIZATION OF FULL MAGNETOELECTRIC CONTROL AT ROOM TEMPERATURE — 347
Kee Hoon Kim, Sae Hwan Chun, Yi Sheng Chai, Byung-Gu Jeon, Kwang Woo Shin, Hyung Joon Kim, Yoon Seok Oh, Ingyu Kim, Ju-Young Park, Suk Ho Lee, Jae-Ho Chung, Jae-Hoon Park

MFT 2: SPIN WAVES AND LATTICE ANOMALY OF $BiFeO_3$ MEASURED BY NEUTRON SCATTERING — 348
Jaehong Jeong, E. A. Goremychkin, T. Guidi, K. Nakajima, Gun Sang Jeon, Shin-Ae Kim, S. Furukawa, Yong Baek Kim, Seongsu Lee, V. Kiryukhin, S-W. Cheong, and Je-Geun Park

MFT 3: TUNING THE MULTIFERROIC PHASE OF CuO WITH IMPURITIES — 350
J. Hellsvik, M. Balistieri, A. Stroppa, A. Bergman, L. Bergqvist, O. Eriksson, S. Picozzi, J. Lorenzana

MFT 4: MULTIFERROICITY AND MAGNETOELECTRICITY IN A DOPED TOPOLOGICAL FERROELECTRIC — 352
Marco Scarrozza, Maria Barbara Maccioni, Alessio Filippetti, and Vincenzo Fiorentini

MFT 5: SEARCH FOR NEW STRAIN-INDUCED MULTIFERROICS WITH HIGH CRITICAL TEMPERATURES — 354
Stanislav Kamba, Veronica Goian, Přemysl Vaněk, Carolina Adamo, Charles M. Brook, Alexander Melville, Nicole A. Benedek, Craig. J. Fennie, June Hee Lee, Karin M. Rabe, Alexei A. Belik, Darrell G. Schlom

Nanosession: Qubit systems — 357

QUB 1: Quantum Electronic Materials — 357
Andrew Briggs

QUB 2: IDENTIFYING CAPACITIVE AND INDUCTIVE LOSS IN LUMPED ELEMENT SUPERCONDUCTING RESONATORS — 359
Martin P. Weides*, Michael R. Vissers, Jeffrey S. Kline, Martin O. Sandberg, and David P. Pappas

QUB 3: SUPERCONDUCTIVITY IN QUASI-1D $LaAlO_3/SrTiO_3$ NANOSTRUCTURES — 361
Joshua Veazey, Guanglei Cheng, Patrick Irvin, Shicheng Lu, Mengchen Huang, Chung Wung Bark, Sangwoo Ryu, Chang-Beom Eom, Jeremy Levy

QUB 4: ON-DEMAND SINGLE ELECTRON TRANSFER BETWEEN DISTANT QUANTUM DOTS — ELECTRON "PING-PONG" IN A SINGLE ELECTRON CIRCUIT — 362
R. P. G. McNeil, M. Kataoka, C. J. B. Ford, C. H. W. Barnes, D. Anderson, G. A. C. Jones, I. Farrer and D. A.Ritchie.

QUB 5: ULTRAFAST ENTANGLING GATES BETWEEN NUCLEAR SPINS USING PHOTO-EXCITED TRIPLET STATES — 364
Vasileia Filidou, Stephanie Simmons, Steven D. Karlen, Feliciano Giustino, Harry L. Anderson, and John J. L. Morton,

QUB 6: NOISE SPECTROSCOPY USING CORRELATIONS OF SINGLE-SHOT QUBIT READOUT — 365
Thomas Fink, Hendrik Bluhm

Nanosession: Superconductivity — 367

- **SUP 1:** INVESTIGATING THE IRON BASED SUPERCONDUCTOR ($FeSe_{0.4}Te_{0.6}$) WITH SPECTROCOPIC-IMAGING SCANNING TUNNELING MICROSCOPE — 367
Stefan Schmaus, Udai Raj Singh, Seth White, Joachim Deisenhofer, Vladimir Tsurkan, Alois Loidl, Peter Wahl

- **SUP 2:** NANOSCALE LAYERING OF ANTIFERROMAGNETIC AND SUPERCONDUCTING PHASES IN $Rb_2Fe_4Se_5$ — 368
Aliaksei Charnukha, Antonija Cvitkovic, Thomas Prokscha, Daniel Proepper, Nenand Ocelic, Andreas Suter, Zaher Salman, Elvezio Morenzoni, Joachim Deisenhofer, Vladimir Tsurkan, Alois Loidl, Bernhard Keimer, and Alexander Boris

- **SUP 3:** RESONANCE MODE IN RARE-EARTH SYSTEMS WITH VALENCE INSTABILITY — 369
Kirill Nemkovski, Pavel Alekseev, Jean-Michel Mignot

- **SUP 4:** MULTI-BAND EFFECTS ON SUPERCONDUCTING INSTABILITIES DRIVEN BY ELECTRON-ELECTRON INTERACTIONS — 371
Stefan Uebelacker, Carsten Honerkamp

- **SUP 5:** FIELD-INDUCED SUPERCONDUCTIVITY IN A LAYERED TRANSITION METAL DICHALCOGENIDE — 372
Jianting Ye, Yijin Zhang, Yoshihiro Iwasa

- **SUP 6:** TOWARDS IDEAL HIGH-T_c JOSEPHSON JUNCTIONS — 373
Yuriy Divin, Irina Gundareva, Matvei Lyatti, Ulrich Poppe

- **SUP 7:** TUNED EPITAXY OF OXIDE HETEROSTRUCTURES — 375
M. I. Faley, U. Poppe, C. L. Jia, O. M. Faley, R. E. Dunin-Borkowski

Nanosession: Interplay between strain and electronic structure in metal oxides — 377

- **ISE 1:** STRAIN EFFECTS ON THE ELECTRONIC SUBBAND STRUCTURE OF $SrTiO_3$ — 377
Vladimir Laukhin, Olivier Copie, Marcelo Rozenberg, Karim Bouzehouane, Éric Jacquet, Manuel Bibes, Agnès Barthélémy, Gervasi Herranz

- **ISE 2:** ELECTRON OCCUPANCY OF 3D-ORBITALS IN MANGANITE THIN FILMS — 379
D. Pesquera, A. Barla, E. Pellegrin, F. Sánchez, F. Bondino, E. Magnano and J. Fontcuberta

- **ISE 3:** SUBSTRATE COHERENCY DRIVEN PHASE SEPERATION AND INTRINSIC ANISOTROPY IN EPITAXIAL $La0_{.67}Ca_{0.33}MnO_3/NdGaO_3(001)$ EPITAXIAL FILMS — 381
Lingfei Wang, Wenbin Wu

- **ISE 4:** THICKNESS DEPENDENCE OF LATTICE DISTORTIONS IN EPITAXIAL FRAMEWORK STRUCTURES OF STRONGLY CORRELATED OXIDES: $La_{2/3}Sr_{1/3}MnO_3/SrTiO_3$ — 382
Felip Sandiumenge, Jose Santiso, Lluís Balcells, Z. Konstantinovic, Jaume Roqueta, Alberto Pomar, Benjamin Martínez

- **ISE 5:** ORBITAL ENGINEERING BY STRAIN IN THIN FILMS OF $La_{1-x}Sr_{1+x}MnO_4$ GROWN BY PULSED LASER DEPOSITION — 384
Mehran Vafaee Khanjani, Philipp Komissinskiy, Mehrdad Baghaie Yazdi, Roberto Krauss, Valentina Bisogni, Jochen Geck, and Lambert Alff

- **ISE 6:** STRUCTURE AND TRANSPORT PROPERTIES OF $SmNiO_3$ THIN FILMS — 385
Flavio Y. Bruno, Konstantin Rushchanskii, Cécile Carretero, Yves Dumont, Marjana Lezaic, Stefan Blügel, Manuel Bibes and Agnès Barthélémy

- **ISE 7:** METAL-INSULATOR TRANSITION AND INTERFACE PHENOMENA IN NICKELATE HETEROSTRUCTURES — 387
Raoul Scherwitzl, Marta Gibert, Pavlo Zubko, Stefano Gariglio, Gustau Catalan, Jorge Iniguez, Marc Gabay, Alberto Morpurgo and Jean-Marc Triscone

Nanosession: Photovoltaics, photocatalysis, and optical effects — 389

- **PPO 1:** PHOTOVOLTAIC ENERGY CONVERSION BASED ON STRONGLY CORRELATED OXIDES — 389
Christian Jooss, Gesine Saucke, Jonas Norpoth, Dong Su and Yimei Zhu

- **PPO 2:** TERAHERTZ AND INFRARED BEHAVIOR OF STRAINED $Sr_{n+1}Ti_nO_{3n+1}$ THIN FILMS WITH RUDDLESDEN-POPPER STRUCTURE — 391
Veronica Goian, Stanislav Kamba, Nathan D. Orloff, Che-Hui Lee, Viktor Bovtun, Martin Kempa, Dmitry Nuzhnyy and Darrell G. Schlom

PPO 3:	FERROELECTRIC ENHANCED CHARGE GENERATION IN SOLAR ENERGY HARVESTING Yeng Ming Lam, Teddy Salim, Theo Schneller	393
PPO 4:	A FACILE PREPARATION AND EXTREMILY FAST PHOTOCATALYTIC PROPERTIES OF OXIDE SEMICONDUCTOR/FERROELECTRIC NANO HETEROSTRUCTURE Huiqing Fan, Pengrong Ren, Xin Wang	395
PPO 5:	LIGHT CONTROLLED AMORPHOUS-Al_2O_3 MEMRISTIVE DEVICES M. Ungureanu, R. Zazpe, F. Golmar, Pablo Stoliar, R. Llopis, F. Casanova, L.E. Hueso	396

Nanosession: Ferroelectric interfaces — 399

FIN 1:	FERROELECTRIC SWITCHING DYNAMICS AT THE NANOSCALE WITH HIGH SPEED SPM Bryan D. Huey	399
FIN 2:	IONIC CHARGE INTERACTIONS WITH FERROELECTRIC SURFACES: POLARIZATION OF ULTRATHIN $PbTiO_3$ WITH CONTROLLED SURFACE COMPENSATION S.K. Streiffer, M.J. Highland, T.T. Fister, D.D. Fong, P.H. Fuoss, Carol Thompson, J.A. Eastman, and G.B. Stephenson	401
FIN 3:	STRAIN TUNNING OF FERROELECTRIC-ANTIFERRODISTORTIVE COUPLING IN $PbTiO_3$/$SrTiO_3$ SUPERLATTICES Pablo García-Fernández, Pablo Aguado-Puente, Javier Junquera	403
FIN 4:	INTERFACE CONTROL OF BULK FERROELECTRIC POLARIZATION D. Yi, P. Yu, W. Luo, J. X. Zhang, M. D. Rossell, C. –H. Yang, G. Singh-Bhalla, S. Y. Yang, Q. He, Q. M. Ramasse, R. Erni, L. W. Martin, Y. H. Chu, S. T. Pantelides, S. J. Pennycook and R. Ramesh	405
FIN 5:	ANALYZING POLARIZATION AND LATTICE STRAINS AT THE INTERFACE OF FERROELECTRIC HETEROSTRUCTURES ON ATOMIC SCAL VIA CS-CORRECTED SCANNING TRANSMISSION ELECTRON M IXROSCOPY (STEM) D. Park, A. Herpers, T. Menke, R. Dittmann and J. Mayer	406
FIN 6:	CONTROL OF CONDUCTION THROUGH DOMAINS AND DOMAIN WALLS IN $BiFeO_3$ THIN FILMS Saeedeh Farokhipoor and Beatriz Noheda	408

Nanosession: Ferroelectrics - new and unusal material systems — 409

FER 1:	CMOS COMPATIBLE FERROELECTRIC MATERIALS BASED ON HAFNIUM OXIDE Thomas Mikolajick, Uwe Schroeder, Johannes Müller, Stefan Müller and Stefan Slesazeck	409
FER 2:	ATOMIC LAYER DEPOSITED Gd-DOPED HfO_2 THIN FILMS: FROM HIGH-K DIELECTRICS TO FERROELECTRICS Christoph Adelmann, Lars-Åke Ragnarsson, Alain Moussa, Joseph A. Woicik, Stefan Müller, Uwe Schroeder, Valeri V. Afanas'ev, and Sven Van Elshocht	411
FER 3:	TEMPERATURE-DEPENDENT ELECTRICAL CHARACTERIZATION OF HAFNIUM OXIDE BASED FERROELECTRIC ULTRA-THIN FILMS U. Böttger, I. Müller, J. Müller, U. Schröder	413
FER 4:	CORRELATION BETWEEN COMPOSITION AND ELASTIC PROPERTIES OF $Ca_xBa_{1-x}Nb_2O_6$ RELAXOR FERROELECTRICS Chandra Shekhar Pandey, Jürgen Schreuer, Manfred Buranekand Manfred Mühlberg	415
FER 5:	LONE PAIR-INDUCED COVALENCY AS THE CAUSE OF TEMPERATURE AND FIELD-INDUCED INSTABILITIES IN BISMUTH SODIUM TITANATE Denis Schütz, Marco Deluca, Werner Krauss, Antonio Feteira, Klaus Reichmann	416
FER 6:	INTERACTION OF POINT DEFECTS AND FERROELECTRIC POLARIZATION IN A LEAD-FREE PIEZOELECTRIC MATERIAL Sabine Körbel, Christian Elsässer	418

Nanosession: Atomic layer deposition — 419

ALD 1:	ATOMIC LAYER DEPOSITION FOR MICROELECTRONIC DEVICES Cheol Seong Hwang*	419

ALD 2:	ATOMIC LAYER DEPOSITION OF SrTiO$_3$ FILMS WITH Cp-BASED PRECURSORS FOR DRAM CAPACITORS Woongkyu Lee, Jeong Hwan Han, Woojin Jeon, YeonWoo Yoo, Changhee Ko, Julien Gatineau and Cheol Seong Hwang	420
ALD 3:	DEPOSITION OF INNOVATIVE MATERIALS BY GAS PHASE TECHNOLOGIES FOR THE SEMICONDUCTOR INDUSTRY B. Gouat, U. Weber, Peter K. Baumann, Michael Heuken, and B. Lu	421
ALD 4:	ATOMIC LAYER DEPOSITION OF TRANSITION METAL OXIDE THIN FILMS FOR RESISTIVE MEMORY APPLICATIONS S. Hoffmann-Eifert, M. Reiners, N. Aslam, I. Kärkkänen, J. H. Kim and R. Waser	423
ALD 5:	ALD PROCESS CONTROL FOR TAILORING THE NANOSTRUCTURE OF TiO$_2$ FILMS FOR RESISTIVE SWITCHING APPLICATIONS M. Reiners, N. Aslam, S. Hoffmann-Eifert, R. Waser	425
ALD 6:	INVESTIGATION OF ATOMIC LAYER DEPOSITION PROPERTIES OF (GeTe$_2$)$_{1-x}$(Sb$_2$Te$_3$)$_x$ PSEUDO-BINARY COMPOUND FOR PHASE CHANGE MEMORY APPLICATION Taeyong Eom, Taehong Gwon, Si Jung Yoo, Moo-Sung Kim, Manchao Xiao, Iain Buchanan, and Cheol Seong Hwang	427

Nanosession: Nanotechnological fabrication strategies 429

NAN 1:	DIRECT PATTERNING OF OXIDE INTERFACE WITH HIGH MOBILITY 2DEG WITHOUT PHYSICAL ETCHING Nirupam Banerjee, Mark Huijben, Gertjan Koster and Guus Rijnders	429
NAN 2:	EUV INTERFERENCE LITHOGRAPHY WITH LABORATORY SOURCES Serhiy Danylyuk, Larissa Juschkin, Sascha Brose, Hyun-Su Kim, Jürgen Moers, Peter Loosen, Detlev Grützmacher	430
NAN 3:	HIGH EFFICIENCY TRANSMISSION MASKS FOR EUV INTERFERENCE LITHOGRAPHY S. Brose, S. Danylyuk, L. Juschkin, K. Bergmann, J. Moers, G. Panaitov, S. Trellenkamp, P. Loosen, D. Grützmacher	432
NAN 4:	SELF-PATTERNED ABO$_3$(001) SUBSTRATES: A PLAYGROUND FOR FUNCTIONAL NANOSTRUCTURES Romain Bachelet, Florencio Sánchez, Carmen Ocal, Josep Fontcuberta	434
NAN 5:	ANOMALOUS GAS SENSING CHARACTERISTICS OF EMBEDDED AND ISOLATED MAGNESIUM ZINC FERRITE NANO-TUBES S.B. Majumder, K. Mukherjee, A. Maity, S. Basu, C. Lang, and M. Topic	435
NAN 6:	ORDERED MESOPOROUS METAL OXIDES BY STRUCUTRE REPLICATION: STRUCTURE-PROPERTY-RELATIONSHIPS AND APPLICATION IN GAS SENSING Thorsten Wagner, Michael Tiemann	437
NAN 7:	WIDTH CONTROL AND OPTICAL NONLINEARITY OF PLATINUM NANOWIRES Yoichi Ogata, and Goro Mizutani	438

Nanosession: Low-dimensional transport and ballistic effects 441

LTB 1:	NON-LINEAR PROPERTIES OF BALLISTIC ELECTRON FOCUSING DEVICES Arkadius Ganczarczyk, Martin Geller, Axel Lorke, Dirk Reuter, Andreas D. Wieck	441
LTB 2:	HALL EFFECT IN AN ASYMMETRIC BALLISTIC CROSS JUNCTION M. Szelong, U. Wieser, M. Knop, U. Kunze, D. Reuter, A. D. Wieck	443
LTB 3:	ELIMINATION OF HOT-ELECTRON THERMOPOWER FROM BALLISTIC RECTIFICATION USING A DUAL-CROSS DEVICE J. F. von Pock, D. Salloch, U. Wieser, U. Kunze, T. Hackbarth	445
LTB 4:	STRUCTURAL INFLUENCES ON ELECTRONIC TRANSPORT IN NANOSTRUCTURES Robert Frielinghaus, K. Sladek, K. Flöhr, L. Houben, St. Trellenkamp, T.E. Weirich, M. Morgenstern, H. Hardtdegen, Th. Schäpers, C.M. Schneider, C. Meyer	447
LTB 5:	TUNNEL-INDUCED SPIN-ANISOTROPY IN QUANTUM DOT SPIN VALVES Maciej Misiorny, Michael Hell, Maarten Wegewijs	449

Content

LTB 6:	CHARGING EFFECTS AND ELECTRON TRANSPORT PHENOMENA ASSOCIATED WITH THE REDOX PROPERTIES OF SELF-ASSEMBLED POLYOXOMETALATE MOLECULES Angeliki Balliou, Antonios Douvas, Dimitrios Velessiotis, Vassilis Ioannou-Sougleridis, Pascal Normand, Panagiotis Argitis, Nikos Glezos	450

Nanosession: Molecular and polymer electronics — 453

MOL 1:	MOLECULAR ELECTRONICS MEETS SPINTRONICS: AN *AB INITIO* EXPLORATION Nicolae Atodiresei, Vasile Caciuc, Predrag Lazic, Stefan Blügel	453
MOL 2:	CHARGE TRAPPING AND ELECTROFORMING IN METAL OXIDE /POLYMER RESISITVE SWITCHING MEMORY DIODES Stefan C. J. Meskers, Benjamin F. Bory, Henrique L. Gomes, René A. J. Janssen, Dago M. de Leeuw	455
MOL 3:	ELECTROFORMING IN LIF/POLYMER RESISITVE SWITCHING MEMORY DIODES AND HOLE INJECTION Benjamin F. Bory, Stefan C. J. Meskers, Henrique L. Gomes, René A. J. Janssen, Dago M. de Leeuw	456
MOL 4:	EFFECT OF HETEROMETALLIC CONTACTS ON CHARGE TRANsPORT N. Babajani, C. Kaulen, M. Homberger, U. Simon, R. Waser, S. Karthäuser	457
MOL 5:	SEMI-EMPIRICAL VS. AB-INITIO CORRELATION EFFECTS: DFT STUDY OF THIOPHENE ADSORBED ON THE Cu(111) SURFACE Martin Callsen, Nicolae Atodiresei, Vasile Caciuc, Stefan Blügel	459
MOL 6:	INTERFACE DIPOLE FORMATION IN DITIOCARBAMATE BASED SURFACE FUNCTIONALIZATIONS Philip Schulz, Tobias Schäfer, Christopher Zangmeister, Dominik Meyer, Christian Effertz, Riccardo Mazzarello, Roger Van Zee and Matthias Wuttig	460

Nanosession: Carbon-based molecular systems — 461

CMS 1:	HYBRIDIZATION OF PARALLEL CARBON NANOTUBE QUANTUM DOTS Carola Meyer, K. Goß, N. Peica, S. Smerat, M. Leijnse, M. R. Wegewijs, C. Thomsen, J. Maultzsch, C. M. Schneider	461
CMS 2:	LOW-TEMPERATURE SCANNING PROBE MICROSCOPY EXPERIMENTS ON ATOMICALLY WELL-DEFINED GRAPHENE NANORIBBONS Peter Liljeroth, Joost van der Lit, Mark P. Boneschanscher, Mari Ijäs, Andreas Uppstu, Ari Harju, Daniël Vanmaekelbergh	463
CMS 3:	CORRELATIONS BETWEEN SWITCHING OF CONDUCTIVITY AND OPTICAL RADIATION OBSERVED IN THIN CARBON FILMS Sergey G.Lebedev	464
CMS 4:	FIRST-PRINCIPLES AND SEMI-EMPIRICAL VAN DER WAALS STUDY OF π-CONJUGATED MOLECULES PHYSISORBED ON GRAPHENE AND A BORON NITRIDE LAYER Caciuc Vasile, Nicolae Atodiresei, Martin Callsen, Predrag Lazic, and Stefan Blügel	465
CMS 5:	FINITE-TEMPERATURE EXACT DIAGONALIZATION STUDY OF THE HUBBARD MOLECULES IN HETEROSTRUCTURES H. Ishida, A. Liebsch	467
CMS 6:	DFT+CI CALCULATIONS OF QUANTUM DOTS IN GRAPHENE NANORIBBONS Tobias Burnus, Gustav Bihlmayer, Daniel Wortmann, Ersoy Şaşıoğlu, Yuriy Mokrousov, Stefan Blügel, Klaus Michael Indlekofer	469

Poster Sessions 471

Poster: Electronic structure, lattice dynamics, and transport 473

ELT 1: FUNDAMENTAL PROPERTIES OF THE SUPERCONDUCTING STATE AT THE $LaAlO_3/SrTiO_3$ INTERFACE 473
Christoph Richter, Hans Boschker, Werner Dietsche, Jochen Mannhart

ELT 2: TORQUE MAGNETOMETRY ON $LaAlO_3$-$SrTiO_3$ HETEROSTRUCTURES 474
M. Brasse, R. Jany, Ch. Heyn, J. Mannhart, M.A. Wilde, D. Grundler

ELT 3: BUILDING UP A STRATEGY TO CONTROL LATTICE THERMAL CONDUCTIVITY IN LAYERED COBALT OXIDES 476
Masahiro TADA, Yohei MIYAUCHI, Masato YOSHIYA, Hideyuki YASUDA

ELT 4: THE EFFECT OF AMMONOLYSIS ON THE STRUCTURE AND THERMOELECTRIC PROPERTIES OF $EuTiO_3$ AND $EuTi_{0.98}Nb_{0.02}O_3$ 477
Leyre Sagarna, Alexandra Maegli, Songhak Yoon, Sascha Populoh, Andrey Shkabko, Anke Weidenkaff

ELT 5: ELECTRICAL TRANSPORT INVESTIGATIONS ON NANOPARTICLE TEST STRUCTURES 478
Silvia Karthäuser, Marcel Manheller, Rainer Waser, Kerstin Blech, Ulrich Simon

ELT 6: FIRST-PRINCIPLES CALCULATION OF MAGNETISM OF SUBSTITUTIONAL TRANSITION IMPURITIES IN BINARY IRON-SELENIUM SYSTEM 480
T. Pengpan, A. Boonthummo

ELT 7: FIRST-PRINCIPLES CALCULATIONS OF PHONON-PHONON INTERACTION IN ROCK-SALT TYPE CRYSTALS 482
Atsushi Togo, Laurent Chaput, Isao Tanaka

ELT 8: MICROSTRUCTURAL CHARACTERIZATION OF VARISTOR CERAMICS AFTER ACCELERATED AGEING WITH DC VOLTAGE. 484
M. A Ramírez, A.Z, Simões, E. Longo, J. A Varela

ELT 9: FIRST-PRINCIPLES ELECTRONIC STRUCTURE OF β-$FeSi_2$ AND FeS_2 SURFACES 485
Pengxiang Xu, Timo Schena, Gustav Bihlmayer, Stefan Blügel

ELT 10: ACCURATE BAND GAPS OF TRANSPARENT CONDUCTING OXIDES WITH A SEMILOCAL EXCHANGE-CORRELATION POTENTIAL 486
A. Thatribud, T. Tungsurat, T. Pengpan

ELT 11: OCTAHEDRAL-TILTING-DEPENDENT STRUCTURE DISTORTION IN EPITAXIAL PEROVSKITE OXIDE FILMS 488
X. L. Tan, P. F. Chen, L. F. Wang, B. W. Zhi, and W. B. Wu

ELT 12: THE $\gamma \rightarrow \alpha$ CHANGE IN CERIUM IS HIDDEN STRUCTURAL PHASE TRANSITION: THEORY AND EXPERIMENT 489
A.V. Nikolaev, K.H. Michel, A.V. Tsvyashchenko, A.I. Velichkov, A.V. Salamatin, L.N. Fomicheva, G.K. Ryasny, A.A. Sorokin, O.I. Kochetov and M. Budzynski

ELT 13: HETEROSTRUCTURES BASED ON EPITAXIAL OXIDE THIN FILMS 491
P. Prieto, M.E. Gómez

ELT 14: INCOHERENT INTERFACES AND LOCAL LATTICE STRAINS IN SOLUTION-DERIVED YBCO NANOCOMPOSITES: A NOVEL VORTEX PINNING MECHANISM 493
T. Puig, X. Obradors, A. Palau, A. Llordés, M. Coll, R. Vlad, J. Gazquez, J. Arbiol, R. Guzmán, A. Pomar, F. Sandiumenge, S. Ricart, V. Rouco, S. Ye, G. Deutscher, D. Chataigner, M. Varela, C. Magen, J. Vanacken, J. Gutierrez, V. V. Moshchalkov

ELT 15: DENSITY INFLUENCE ON AMORPHOUS HfO_2 STRUCTURE: A MOLECULAR DYNAMICS STUDY 495
Giulia Broglia, Monia Montorsi, Luca Larcher, Andrea Padovani.

ELT 16: ELECTRIC FIELD TUNING OF THE QUASIPARTICLE WEIGHT IN THIN FILMS MADE OF STRONGLY CORRELATED MATERIALS 497
D. Nasr Esfahani, L. Covaci and F. M. Peeters

ELT 17: METALLIC STATE INDUCED BY SPIN-CANTING IN LIGHTLY ELECTRON-DOPED $CaMnO_3$ 498
Hiromasa Ohnishi, Taichi Kosugi, Takashi Miyake, Shoji Ishibashi, and Kiyoyuki Terakura

ELT 18:	PRESSURE-INDUCED STRUCTURAL CHANGES AT THE CROSSOVER FROM LOCALIZED TO ITINERANT BEHAVIOUR IN $PrNiO_3$ Marisa Medarde, Thierry Straessle, Vladimir Pomjakushin, María Jesus Martínez-Lope and José Antonio Alonso	500
ELT 19:	MECHANICS MEETS ELECTRONICS IN NANOSCALE: THE MYSTERY OF CURRENT SPIKE AND NANOSCALE-CONFINEMENT R. Nowak, D. Chrobak, W.W. Gerberich, K. Niihara, T. Wyrobek	502
ELT 20:	EFFECT OF THE CAPPING ON THE MANGANESE OXIDATION STATE IN $SrTiO_3/La_{2/3}Ca_{1/3}MnO_3$ INTERFACES AS A FUNCTION OF ORIENTATION S. Estradé, J. M. Rebled, M. G. Walls, F. de la Peña, C. Colliex, R. Córdoba, I. C. Infante, G. Herranz, F. Sánchez, J. Fontcuberta, F. Peiró.	504
ELT 21:	ON THE ELECTRICAL BEHAVIOR OF PLANAR TUNGSTEN POLYOXOMETALATE SELF-ASSEMBLED MONO- AND BI-LAYER JUNCTIONS D. Velessiotis, A. M. Douvas, P. Dimitrakis, P. Argitis, N. Glezos	506
ELT 22:	MECHANICS MEETS ELECTRONICS IN NANOSCALE: NANOSCALE DECONFINEMENT OF SILICON ALTERS ITS PROPERTIES Dariusz Chrobak, William W. Gerberich, Roman Nowak	508
ELT 23:	BREATHING-LIKE MODES IN AN INDIVIDUAL MULTIWALLED CARBON NANOTUBE Carola Meyer, C. Spudat, M. Müller, L. Houben, J. Maultzsch, K. Goss, C. Thomsen, C. M. Schneider	510
ELT 24:	EFFECT OF DIFFERENT ACID TREATMENT OF CARBON NANOTUBES (CNT) ON CNT/TiO_2 NANOCOMPOSITES VIA SOL-GEL METHOD Mohammad Reza Golobostanfard, Hossein Abdizadeh	512
ELT 25:	INTERPLAY OF ELECTRONIC CORRELATIONS AND SPIN-ORBIT INTERACTIONS AT THE ENDS OF CARBON NANOTUBES Manuel J. Schmidt	513
ELT 26:	GRAPHENE CHARGE DETECTOR ON A CARBON NANOTUBE QUANTUM DOT Stephan Engels, Alexander Epping, Carola Meyer, Stefan Trellenkamp, Uwe Wichmann, and Christoph Stampfer	514
ELT 27:	RESISTIVE SWITCHING ON METAL-OXIDE POLYMER MEMORIES Paulo R. F. Rocha, Qian Chen, Asal Kiazadeh, Henrique L. Gomes, Stefan Meskers and Dago de Leeuw	516
ELT 28:	THIOLATED (OLIGO)PHENOTHIAZINES AS PROMISING CANDIDATES FOR FUTURE STORAGE ELEMENTS Michael Paßens, Adam Busiakiewicz, Adam W. Franz, Christa S. Barkschat, Dominik Urselmann, Thomas J. J. Müller, Silvia Karthäuser	517
ELT 29:	SPM-INVESTIGATIONS OF THE SPIROPYRAN – MEROCYANIN PHOTOISOMERIZATION A.Soltow, S. Karthäuser, R. Waser	519
ELT 30:	MEMRISTIVE PHENOMENA OF CONDUCTION POLYMER PEDOT:PSS Fei Zeng, Jing Yang, Zhishun Wang, Yisong Lin and Sizhao Li	521
ELT 31:	A NEW MINIMUM SEARCH METHOD FOR COMPLEX OPTIMIZATION PROBLEMS Julian A. Hirschfeld, Hans Lustfeld	522

Poster: Memristive systems   523

MEM 1:	RESISTIVE SWITCHING CHARACTERISTICS IN HfO_2 THIN FILMS DEPENDING ON THEIR CRYSTALLINE STRUCTURE Jung Ho Yoon, Hyung-Suk Jung, Min Hwan Lee, Gun Hwan Kim, Seul Ji Song, Jun Yeong Seok, Kyung Jean Yoon and Cheol Seong Hwang	523
MEM 2:	DEPOSITION OF CHALCOGENIDE THIN LAYERS BY MAGNETRON SPUTTERING FOR RRAM APPLICATIONS M.-P. Besland, J. Tranchant, E. Souchier, P. Moreau, S. Salmon, B. Corraze, E. Janod, L. Cario	524
MEM 3:	DETERMINISTIC RESISTIVE SWITCHING CONTROL IN HfO_2-BASED MEMORY DEVICES Raúl Zazpe, Mariana Ungureanu, Roger Llopis, Federico Golmar, Pablo Stoliar, Félix Casanova, Luis Eduardo Hueso	526
MEM 4:	INVESTIGATION OF TRANSIENT CURRENTS DURING ULTRA-FAST DATA OPERATION OF TiO_2 BASED RRAM C. Hermes, M. Wimmer, S. Menzel, K. Fleck, V. Rana, M. Salinga, U. Böttger, R. Bruchhaus, M. Wuttig, R. Waser	527

MEM 5:	FAST PULSE FORMING PROCESS FOR TiO$_2$ BASED RRAM NANO-CROSSBAR DEVICES F. Lentz, C. Hermes, B. Rösgen, T. Selle, R. Bruchhaus, V. Rana, R. Waser	529
MEM 6:	CURRENT TRANSPORT MODELING IN OXIDE-BASED RESISTIVELY SWITCHING MEMORY CELLS FOR THE INVESTIGATION OF ELECTROFORMATION Astrid Marchewka, Stephan Menzel, Ulrich Böttger, Rainer Waser	531
MEM 7:	MODELING OF SWITCHING DYNAMICS FOR TiO$_{2-x}$ MEMRISTIVE DEVICES Brian Hoskins, Fabien Alibart, Dmitri Strukov	532
MEM 8:	MULTISTATE MEMORY DEVICES BASED ON FREE-STANDING VO$_2$/TiO$_2$ MICROSTRUCTURES DRIVEN BY JOULE SELF-HEATING Luca Pellegrino , Nicola Manca , Teruo Kanki , Hidekazu Tanaka , Michele Biasotti , Emilio Bellingeri, Antonio Sergio Siri , Daniele Marré	533
MEM 9:	TANTALUM OXIDE BASED MEMRISTIVE DEVICES AB INITIO ELECTRONIC STRUCTURE CALCULATIONS FOR STABILITY, DEFECTS AND DIFFUSION BARRIERS Antonio Claudio M. Padilha, Gustavo Martini Dalpian, Alexandre Reily Rocha	535
MEM 10:	CONCURRENT RESISTIVE AND CAPACITIVE STATE SWITCHING OF NANOSCALE TiO$_2$ MEMRISTORS Themistoklis Prodromakis, Iulia Salaoru, Ali Khiat, Christopher Toumazou	536
MEM 11:	THE MEMORY-CONSERVATION MODEL OF MEMRISTANCE Ella M. Gale	538
MEM 12:	ANALOG AND DIGITAL COMPUTING WITH MEMRISTIVE DEVICES A. Madhavan, G. Adam, F. Alibart, L. Gao, and D.B. Strukov	540
MEM 13:	EFFECT OF VACANCIES ON THE PHASE CHANGE CHARACTERISTICS OF GeSbTe ALLOYS D. Wamwangi, W. Welnic, M. Wuttig	542
MEM 14:	ULTRA LOW POWER CONSUMING, THERMALLY STABLE SULPHIDE MATERIALS FOR RESISTIVE AND PHASE CHANGE MEMRISTIVE APPLICATION Behrad Gholipour, Chung-Che Huang, Alexandros Anastasopoulos, Feras Al-Saab, Brian E. Hayden and Daniel W. Hewak	543
MEM 15:	THERMAL CONDUCTIVITY MEASUREMENTS OF Sb-Te ALLOYS BY HOT STRIP METHOD Rui Lan, Rie Endo, Masashi Kuwahara, Yoshinao Kobayashi, Masahiro Susa	545
MEM 16:	juRS – MASSIVELY PARALLEL REAL-SPACE DFT Paul Baumeister, Daniel Wortmann, Stefan Blügel	547
MEM 17:	FIRST-PRINCIPLES STUDY OF PHASE-CHANGE MATERIALS DOPED WITH MAGNETIC IMPURITIES Riccardo Mazzarello, Yan Li, Wei Zhang, Ider Ronneberger	548
MEM 18:	NUCLEAR RESONANCE SCATTERING IN PHASE-CHANGE MATERIALS Ronnie Simon, Jens Gallus, Dimitrios Bessas, Ilya Sergueev,	550
MEM 19:	DEFECT STATES IN AMORPHOUS PHASE-CHANGE MATERIALS Jennifer Luckas, Pascal Rausch, Daniel Krebs, Peter Zalden, Janika Boltz, Jean-Yves Raty Martin Salinga, Christophe Longeaud and Matthias Wuttig	552
MEM 20:	ELECTRIC FIELD INDUCED LOCAL SURFACE POTENTIAL MODIFICATION AND TRANSPORT BEHAVIORS OF TiO$_2$ SINGLE CRYSTALS Haeri Kim, Dong-Wook Kim, Soo-Hyon Phark, and Seungbum Hong	553
MEM 21:	RESISTIVE SWITCHING IN DIFFERENT FORMING STATES OF Ti/Pr$_{0.48}$Ca$_{0.52}$MnO$_3$ JUNCTIONS C. Park, A. Herpers, R. Bruchhaus, J. Verbeeck, R. Egoavil, F. Borgatti, G. Panaccione, F. Offi, and R. Dittmann	554
MEM 22:	FIRST PRINCIPLES SIMULATIONS OF OXYGEN DIFFUSION IN RRAM MATERIALS Sergiu Clima, Kiroubanand Sankaran, Maarten Mees, Yang Yin Chen, Ludovic Goux, Bogdan Govoreanu, Dirk J.Wouters, Jorge Kittl, Malgorzata Jurczak, Geoffrey Pourtois	556
MEM 23:	OBSERVATION OF A CONDUCTIVE REGION IN THE TiN/HfO$_2$ SYSTEM AFTER RESISTANCE SWITCHING P. Calka, E. Martinez, V. Delaye, D. Lafond, G. Audoit, D. Mariolle, N. Chevalier, H. Grampeix, C. Cagli, V. Jousseaume, C. Guedj	558

Content

MEM 24: TRANSIENT CHARACTERISTICS DURING SET OPERATION OF A Ta_2O_5 SOLID ELECTROLYTE MEMRISTIVE SWITCH — 560
Pragya Shrestha, Adaku Ochia, Kin. P. Cheung, Jason Campbell, Helmut Baumgart and Gary Harris

MEM 25: REMANENT RESISTANCE CHANGES IN METAL-MANGANITE-METAL SANDWICH STRUCTURES — 561
Malte Scherff, Bjoern Meyer, Julius Scholz, Joerg Hoffmann, and Christian Jooss

MEM 26: THEORETICAL STUDY ON THE CONDUCTIVE PATH IN TANTALUM OXIDE ATOMIC SWITCH — 563
Bo Xiao, Tomofumi Tada, Tingkun Gu, Arihiro Tawara, Satoshi Watanabe

MEM 27: CHARACTERISTIC OF LOW TEMPERATURE FABRICATED NONVOLATILE MEMORY DEVICES OF ZN AND SN NANO THIN FILM EMBEDDED MIS — 564
Tai-Fa Young, Ya-Liang Yang, Ting-Chang Chang, Kuang-Ting Hsu, and Chao-Yu Chen.

MEM 28: ELECTROCHEMICAL STUDIES ON Al_2O_3 THIN FILMS FOR RESISTIVE MEMORY APPLICATIONS — 565
A Burkert, I. Valov, G. Staikov, R. Waser

MEM 29: RESISTIVE SWITCHING PHENOMENA IN $Ag-GeS_x$ MEMORY CELLS — 566
Jan van den Hurk, Ilia Valov, Rainer Waser

MEM 30: STATES AND PROCESSES IN NANO-SCALED CATION BASED RRAM CELLS — 568
Ilia Valov, Stefan Tappertzhofen, Jan van der Hurk and Rainer Waser

MEM 31: FIGHTING VARIATIONS IN PT/TIO2-X/PT AND AG/A-SI/PT MEMRISTIVE DEVICES — 570
G. Adam, F. Alibart, L. Gao, B. Hoskins, and D.B. Strukov

MEM 32: A STUDY UPON THE SWITCHING CHARACTERISTICS AT RUPTURED CONDUCTING FILAMENTS REGION IN A $Pt/TiO_2/Pt$ MEMRISTIVE DEVICE — 572
Kyung Jean Yoon, Seul Ji Song, Gun Hwan Kim, Jun Yeong Seok, Jeong Ho Yoon, and Cheol Seong Hwang

MEM 33: STUDY ON RESISTIVE SWITCHING OF BINARY OXIDE THIN FILMS USING SEMICONDUCTING $In_2Ga_2ZnO_7$ ELECTRODE — 573
Jun Yeong Seok, Gun Hwan Kim, Seul Ji Song, Jung Ho Yoon, Kyung Jin Yoon, and Cheol Seong Hwang

MEM 34: BIPOLAR RESISTIVE SWITCHING BEHAVIORS OF PLASMA-ENHANCED ATOMIC LAYER DEPOSITED NiO FILMS ON TUNGSTEN SUBSTRATE — 574
Seul Ji Song, Gun Hwan Kim, Jun Yeong Seok, Kyung Jean Yoon, Jung Ho Yoon and Cheol Seong Hwang*

MEM 35: NONVOLATILE RESISTIVE SWITCHING IN $Au/BiFeO_3$ RECTIFYING JUNCTION — 576
Yao Shuai, Chuangui Wu, Wanli Zhang, Shengqiang Zhou, Danilo Bürger, Stefan Slesazeck, Thomas Mikolajick, Manfred Helm, and Heidemarie Schmidt

MEM 36: DIFFERENT BEHAVIOUR SEEN IN FLEXIBLE TITANIUM DIOXIDE SOL-GEL MEMRISTORS DEPENDENT ON THE CHOICE OF ELECTRODE MATERIAL — 577
Ella Gale, David Pearson, Stephen Kitson, Andrew Adamatzky and Ben de Lacy Costello

MEM 37: MEMRISTIVE COGNITIVE COMPUTING — 579
Eero Lehtonen, Jussi Poikonen, Mika Laiho, Pentti Kanerva

MEM 38: SYNAPTIC PLASTICITY OF ELECTROCHEMICAL CAPACITORS BASED ON TiO_2 — 581
Hyungkwang Lim, Ho-Won Jang, Cheol Seong Hwang, Doo Seok Jeong

MEM 39: SIMULATION OF ASYMMETRIC RESISTIVE SWITCHING IN ELECTROCHEMICAL METALLIZATION MEMORY CELLS — 583
Stephan Menzel, Ulrich Böttger, Rainer Waser

MEM 40: A V_2O_5-BASED RESISTANCE RANDOM ACCESS MEMORY AND IMPROVEMENT OF SWITCHING CHARACTERISTICS BY EMBEDDING A THIN VO_2 INTERFACE LAYER — 584
Xun Cao, Meng Jiang, Feng Zhang, Xinjun Liu, and Ping Jin

MEM 41: METAL–INSULATOR TRANSITION OF ALD VO_2 THIN FILMS FOR PHASE TRANSITION SWITCHING — 585
Kai Zhang, Madhavi Tangirala, Pragya Shrestha, Helmut Baumgart, Salinporn Kittiwatanakul, Jiwei Lu, Stuart Wolf, Venkateswara Pallem and Christian Dussarrat

MEM 42: SWITCHING AND LEARNING IN Ni-DOPED GRAPHENE OXIDE THIN FILMS — 586
S. Pinto, R. Krishna, C. Dias, G. Pimentel, G. N. P. Oliveira, J. M. Teixeira, P. Aguiar[3], E. Titus, J. Gracio, J. Ventura, and J. P. Araujo

Poster: Spin-related phenomena 589

SRP 1: CHARGE CARRIER-MEDIATED FERROMAGNETISM IN $FeSb_{2-x}Sn_xSe_4$ 589
Honore Djieutedjeu; Kulugammana G.S. Ranmohotti; Nathan J. Takas; Julien P. A. Makongo; Xiaoyuan Zhou; Ctirad Uher; N. Haldolaarachchige; D.P. Young; Pierre F. P. Poudeu*

SRP 2: CHARACTERIZATION OF THE ATOMIC INTERFACE IN HIGH-QUALITY Fe_3O_4/ZnO HETEROSTRUCTURES 590
O. Kirilmaz, M. Paul, A. Müller, D. Kufer, M. Sing, R. Claessen, S. Brück, C. Praetorius, K. Fauth, M. Kamp, P. Audehm, E. Goering, J. Verbeeck, H. Tian, G. Van Tendeloo, N.J.C. Ingle, M. Przybylski, M. Gorgoi

SRP 3: ULTRATHIN Fe OXIDES FILMS: STRUCTURAL, ELECTRONIC AND MAGNETIC PROPERTIES UNDER REDUCED DIMENSIONS 592
Bernal, Iván, Gallego, Silvia

SRP 4: STRUCTURAL DEPENDENCE OF MAGNETIC PROPERTIES IN TWO DIMENSIONAL NICKEL NANOSTRIPS: MAGNETIC ANOMALY AND MAGNETIC TRANSITION IN NANOSTRIPS 593
Vikas Kashid, Vaishali Shah, H. G. Salunke, Y. Mokrousov and S. Blügel

SRP 5: SPIN BLOCKADE AND MAGNETIC EXCHANGE IN THE LAYERED COBALTATES $La_{1.5}A_{0.5}CoO_4$ (A = Ca, Sr, or Ba) 595
Dirk Fuchs, Michael Merz, Levin Dieterle, Stefan Uebe, Peter Nagel, Stefan Schuppler, Dagmar Gerthsen, and Hilbert von Löhneysen

SRP 6: EXPERIMENTAL VERIFICATION OF CHIRAL MAGNETIC ORDERS: CHIRAL MAGNETIC SOLITON LATTICE IN CHIRAL HELIMAGNET 596
Yoshihiko Togawa, Tsukasa Koyama, Shigeo Mori, Yusuke Kousaka,

SRP 7: FREQUENCY DEPENDENT MAGANOTRANSPORT IN $Sm_{0.6}Sr_{0.4}MnO_3$: USNUAL POSITIVE AND NEGATIVE MAGNETORESISTANCE 598
Ramanathan MAHENDIRAN

SRP 8: FIELD INDUCED SPIN-REORIENTATION AND STRONG SPIN-CHARGE-LATTICE COUPLING IN $EuFe_2As_2$ 600
Y. Xiao, Y. Su, S. Nandi, S. Price, Th. Brückel

SRP 9: ANISOTROPY OF SPIN RELAXATION IN hcp OSMIUM AND bcc TUNGSTEN 601
B. Zimmermann, P. Mavropoulos, S. Heers, N. H. Long, S. Blügel, Y. Mokrousov

SRP 10: SPIN-WAVE DYNAMICS IN TETRAGONAL FeCo ALLOYS 602
Ersoy Sasioglu, Christoph Friedrich, Stefan Blügel

SRP 11: ROLE OF INTERFACES IN MANGANITE PHYSICS 603
V. Moshnyaga

SRP 12: ISOTHERMAL ELECTRIC CONTROL OF EXCHANGE BIAS NEAR ROOM TEMPERATURE 604
Christian Binek Xi He, Yi Wang, N.Wu, Aleksander L. Wysocki, Takashi Komesu, Uday Lanke, Anthony N. Caruso, Elio Vescovo, Kirill D. Belashchenko, Peter A. Dowben

SRP 13: STRAIN ENGINEERING OF MULTIFERROIC PHASE TRANSITIONS AND ORDER PARAMETERS IN $BiFeO_3$ 606
Daniel Sando, Arsène Agbelele, Christophe Daumont, Jean Juraszek, Maximilien Cazayous, Ingrid Infante, Wei Ren, Sergey Lisenkov, Cécile Carretero, Laurent Bellaiche and Brahim Dkhil, Agnès Barthélémy and Manuel Bibes

SRP 14: OBSERVATION AND EFFECT OF MAGNETIC DOMAINS IN LATERAL SPIN VALVES 608
Xianzhong Zhou, Julius Mennig, Frank Matthes, Daniel E. Bürgler, Claus M. Schneider

SRP 15: THICKNESS EVOLUTION OF THE STRAIN IN PCMO THIN FILMS 610
Anja Herpers, Chanwoo Park, Ricardo Egoavil, Jo Verbeeck and Regina Dittmann

SRP 16: EFFECT OF MIXED ORTHORHOMBIC/HEXAGONAL STRUCTURE ON MAGNETIC ORDERING IN STRONTIUM DOPED YTTERBIUM MANGANITES 612
A.I. Kurbakov, I.A. Abdel-Latif, V.A. Trunov, H. U. Habermeier4, A. Al-Hajry, A.L. Malyshev, V.A. Ulyanov

SRP 17: QUANTUM TRANSPORT THROUGH TOPOLOGICAL SPIN TEXTURE IN CHIRAL HELIMAGNET 614
J. Kishine, A.S. Ovchinnikov, I.V. Proskurin

Content

SRP 18:	COMPARATIVE INVESTIGATION OF Sb_2Te_3 NANOSTRUCTURE PROPERTIES WITH RESPECT TO THEIR PRODUCTION METHOD T. Saltzmann, S.Rieß, J. Kampmeier, G. Mussler, T. Stoica, B. Kardynal, U. Simon, H. Hardtdegen	615
SRP 19:	TOPOLOGICAL INSULATORS FROM THE VIEW POINT OF CHEMISTRY C. Felser, L. Müchler, S. Chadov, B. Yan, J. Kübler, HJ Zhang, and SC Zhang	617
SRP 20:	PROBING TWO TOPOLOGICAL SURFACE BANDS OF ANTIMONY-TELLURIDE BY SPIN-POLARIZED PHOTOEMISSION SPECTROSCOPY Christian Pauly, Gustav Bihlmayer, Marcus Liebmann, Martin Grob, Alexander Georgi, Dinesh Subramaniam, Markus Scholz, Jaime Sánchez-Barriga, Andrei Varykhalov, Stefan Blügel, Oliver Rader, Markus Morgenstern	618
SRP 21:	QUASIPARTICLE CORRECTIONS AND SURFACE STATES OF TOPOLOGICAL INSULATORS Bi_2Se_3, Bi_2Te_3 AND Sb_2Te_3. Irene Aguilera, Christoph Friedrich, Gustav Bihlmayer, Stefan Blügel	620
SRP 22:	ELECTRIC FIELD INDUCED SWITCHING OF MAGNETIZATION IN $CoPt_3/BaTiO_3$ HETEROSTRUCTURES Konstantin Z. Rushchanskii, Felipe Garcia-Sanchez, Riccardo Hertel, Stefan Blügel, Marjana Ležaić	622
SRP 23:	EVIDENCE OF PARAELECTROMAGNON-LIKE EXCITATIONS IN THz SPECTRA OF HEXAGONAL $YMnO_3$ SINGLE CRYSTAL Veronica Goian, Stanislav Kamba,Christelle Kadlec, Petr Kužel, Konstantin Z. Rushchanskii,Marjana Ležaić, Roman V. Pisarev	623
SRP 24:	ELECTRICAL BEHAVIOR OF $Bi_{0.95}Nd_{0.05}FeO_3$ THIN FILMS GROWN BY THE SOFT CHEMICAL METHOD. C. R. Foschini, F. Moura, M. A. Ramirez, J. A. Varela, E. Longo, A.Z. Simões*	625
SRP 25:	THEORETICAL INVESTIGATION OF THE MAGNETOELECTRIC PROPERTIES OF $MnPS_3$ Diana Iușan, Kunihiko Yamauchi, Kris Delaney, Silvia Picozzi	626
SRP 26:	HIGH TEMPERATURE MULTIFERROIC COMPOUNDS: FROM $BiFeO_3$ TO Bi_2FeCrO_6 Jian Yu, Linlin Zhang, and Xianbo Hou	628
SRP 27:	HAADF STEM TOMOGRAPHY OF FERRIMAGNETIC $Fe_xCo_{(3-x)}O_4$ NANOSTRUCTURES EMBEDDED IN HIGHLY ORDERED ANTIFERROMAGNETIC Co_3O_4 MESOPOROUS TEMPLATES. Lluís. Yedra, Sònia Estradé, Eva. Pellicer, Moisés Cabo, Alberto López-Ortega, Marta Estrader, Josep Nogués, Dolors Baró, Zineb Saghi, Paul A. Midgley and Francesca Peiró	629
SRP 28:	NON-COLLINEAR MAGNETISM IN 3d-5d ZIGZAG CHAINS: TIGHT BINDING MODEL AND AB-INITIO CALCULATIONS Vikas Kashid, Timo Schena, Bernd Zimmermann, Vaishali Shah, H. G. Salunke, Y. Mokrousov and S. Blügel	631

Poster: Polar dielectrics, optics, and ionics — 633

POL 1:	DIRECT OBSERVATION OF TRANSIENT NEGATIVE CAPACITANCE IN DOMAIN WALL OF FERROELECTRIC THIN FILMS Yu Jin Kim, Min Hyuk Park, Han Joon Kim, Doo Seok Jeong, Anquan Jiang, and Cheol Seong Hwang	633
POL 2:	STRONTIUM TITANATE ULTRA-THIN FILM CAPACITORS ON SILICON SUBSTRATES FOR APPLICATION IN DYNAMIC RANDOM ACCESS MEMORY (DRAM) Sebastian Schmelzer, Ulrich Böttger, Rainer Waser	634
POL 3:	ENHANCEMENT OF FERROELECTRIC POLARIZATION BY INTERFACE ENGINEERING X. Liu, H. Lu, J. D. Burton, Y. Wang, C. W. Bark, Y. Zhang, D. J. Kim, A. Stamm, P. Lukashev, D. A. Felker, C. M. Folkman, P. Gao, M. S. Rzchowski, X. Q. Pan, C. B. Eom, A. Gruverman, and E. Y. Tsymbal	636
POL 4:	INFLUENCE OF ADDITIVES WITH LOW MELTING TEMPERATURES ON STRUCTURE, MICROSTRUCTURE, PHASE TRANSITIONS, DIELECTRIC AND PIEZOELECTRIC PROPERTIES OF $BiScO_3 - PbTiO_3$ CERAMICS E.D. Politova, G.M. Kaleva, A.V. Mosunov, N.V. Sadovskaya, A.G. Segalla	638
POL 5:	CONTROL OF β- AND γ-PHASE FORMATION IN ELECTROACTIVE P(VDF-HFP) FILMS BY SILVER NANO-PARTICLE DOPING Dipankar Mandal, Karsten Henkel, Suken Das, Dieter Schmeißer	640

POL 6:	SYNTHESIS AND CHARACTERIZATION OF NANOSTRUCTURED MATERIALS FOR REMOVAL OF EXHAUST GASES K.S. Abdel Halim, M.H.Khedr, A.A.Farghali, M.I. Nasr, N.K.Soliman	642
POL 7:	GROWTH AND CHARACTERIZATION OF MAGNETO-OPTICAL CERIUM-DOPED YTTRIUM IRON GARNET FILMS ON SILICON NITRIDE FOR NONRECIPROCAL PHOTONIC DEVICE APPLICATIONS Mehmet Onbasli, T. Goto, D. H. Kim, L. Bi, G. F. Dionne, C. A. Ross	644
POL 8:	PHOTELECTROCHEMICAL WATER SPLITTING BY CHEMICAL SOLUTION DEPOSITED NANOSTRUCTURE AND COMPOSITION ENGINEERED HEMATITE FILMS Theodor Schneller, Simon Goodwin, Rainer Waser	646
POL 9:	A CHEMICAL APPROACH TO THE ESTIMATE OF THE OPTICAL BAND GAP AND BOWING PARAMETER IN MIXED d,d-METAL OXIDES Francesco Di Quarto, Francesco Di Franco, Monica Santamaria	648
POL 10:	BAND ALIGNMENT ENGINEERING WITH LIQUID DIELECTRICS Hongtao Yuan, Y. Ishida, K. Koizumi, H. Shimotani, K. Kanai, K. Akaike, Y. Kubozono, A. Tsukazaki, M. Kawasaki, S. Shin, Y. Iwasa	650
POL 11:	THE BEHAVIOUR OF OXYGEN VACANCIES IN THE PEROVSKITE OXIDE STRONTIUM TITANATE AND AT ITS EXTENDED DEFECTS Roger A. De Souza, Veronika Metlenko, Henning Schraknepper, Amr Ramadan	652
POL 12:	DOPED CERIA: A DFT AND MONTE CARLO STUDY B. O. H. Grope, S. Grieshammer, J. Koettgen, M. Martin	653
POL 13:	OXYGEN VACANCY FORMATION ON (110) TiO_2 SURFACE - A FIRST PRINCIPLES STUDY Taizo Shibuya, Kenji Yasuoka, Susanne Mirbt and Biplab Sanyal	655
POL 14:	FINITE SIZE EFFECT OF PROTON CONDUCTIVITY OF AMORPHOUS OXIDE THIN FILM AND ITS APPLICATION YO HYDROGEN-PERMEABLE MEMBRANE FUEL CELL Yoshitaka Aoki, Manfred Martin	656
POL 15:	GRAIN BOUNDARIES IN PROTON-CONDUCTING $BaZrO_3$ PEROVSKITES R. Merkle, M. Shirpour, B. Rahmati, W. Sigle, P.A. van Aken, J. Maier	657
POL 16:	FAST ELECTROMIGRATION OF OXYGEN VACANCIES IN IMPLANTED RUTILE (TiO_2) N.A. Sobolev, A.M. Azevedo, V.V. Bazarov, E.R. Zhiteytsev, R.I. Khaibullin	658
POL 17:	TAILOR-MADE COMPLEX OXIDE THIN FILMS AS PROTON CONDUCTING ELECTROLYTES FOR LOW TEMPERATURE OPERATING SOLID OXIDE FUEL CELLS David Griesche, Theodor Schneller, Rainer Waser	659
POL 18:	EFFECT OF THE SUBSTITUTION OF OXYGEN BY NITROGEN ON THE CRYSTAL CHEMISTRY OF La-DOPED $BaTiO_3$ Yusuke Otsuka, Christian Pithan, Jürgen Dornseiffer, Rainer Waser	661
POL 19:	METALLIC ELECTROLYTE COMPOSITES IN THE FRAMEWORK OF THE BRICK-LAYER MODEL H. Lustfeld, C. Pithan, M. Reißel	663

Poster: Advances in technology and characterization   665

ATC 1:	DETECTION OF FILAMENT FORMATION IN FORMING-FREE RESISTIVELY SWITCHING $SrTiO_3$ DEVICES WITH Ti TOP ELECTRODES S. Stille, Ch. Lenser, R. Dittmann, A. Köhl, I. Krug, R. Muenstermann, J. Perlich, C. M. Schneider, U. Klemradt, and R. Waser	665
ATC 2:	FREE-ELECTRON FINAL-STATE CALCULATIONS FOR THE INTERPRETATION OF HARD X-RAY ANGLE-RESOLVED PHOTOEMISSION L. Plucinski, J. Minar, J. Braun, A. X. Gray, S. Ueda, Y. Yamashita, K. Kobayashi, H. Ebert, C.S. Fadley, and C.M. Schneider	667
ATC 3:	GROWTH PRESSURE CONTROL OF ELECTRONIC INTERFACE PROPERTIES IN LAO/STO HETEROSTRUCTURES: NEW INSIGHT FROM HIGH-ENERGY SPECTROSCOPY M. Sing, A. Müller, H. Boschker, F. Pfaff, G. Berner, S. Thiess, W. Drube, G. Koster, G. Rijnders, D.H.A Blank, M. Sing, R. Claessen	668
ATC 4:	ANISOTROPIC MAGNETOELASTIC COUPLING IN FERROPNICTIDES Haifeng Li	669

Content

ATC 5:	DEPTH-RESOLVED ARPES OF BURIED LAYERS AND INTERFACES VIA SOFT X-RAY STANDING-WAVE EXCITATION Alexander X. Gray, Jan Minár, Lukasz Plucinski, Mark Huijben, Alexander M. Kaiser, Slavomír Nemšák, Giuseppina Conti, Aaron Bostwick, Eli Rotenberg, See-Hun Yang, Jürgen Braun, Guus Rijnders, Dave H. A. Blank, Susanne Stemmer, Claus M. Schneider, Juergen Braun Hubert Ebert, and Charles S. Fadley	670
ATC 6:	STRUCTURING COMPLEX OXIDE HETEROSYSTEMS VIA E-BEAM LITHOGRAPHY Carsten Woltmann, Hans Boschker, Rainer Jany, Christoph Richter, Jochen Mannhart	672
ATC 7:	LASER HEATING OF OXIDE SUBSTRATES: CHALLENGES AND SOLUTIONS T. Heeg, W.Stein	673
ATC 8:	WET CHEMICAL ETCHING OF STRONTIUM RUTHENIUM OXIDE THIN FILMS BY OXIDATION Dieter Weber, Róza Vőfély, Yuehua Chen, Ulrich Poppe	674
ATC 9:	EFFECT OF ALD PROCESSING AND TOP ELECTRODES ON ZrO_2 THIN FILMS STRUCTURAL AND RESISTIVE SWITCHING CHARACTERISTICS Irina Kärkkänen, Mikko Heikkilä, Jaakko Niinistö, Mikko Ritala, Markku Leskelä, Susanne Hoffmann-Eifert, Rainer Waser	675
ATC 10:	RESISTIVE SWITCHING STUDY ON Nb_2O_5 THIN FILMS OF DIFFERENT THICKNESS AND MORPHOLOGY GROWN BY PVD AND ALD N. Aslam, M. Reiners, T. Blanquart, H. Mähne, J. Niinistö, M. Leskelä, Mikko Ritala, T. Mikolajick, S. Hoffmann-Eifert and R. Waser	676
ATC 11:	CALCULATION OF LORENZ TRANSITION ELECTRON MICROSCOPY DIFFRACTION MAP FOR A CHIRAL MAGNETIC SOLITON LATTICE Yoshihiko Togawa, Jun-ichiro Kishine, Alexander Ovchinnikov, Igor Proskurin	678
ATC 12:	CHARGE DENSITY WAVE STUDY IN $Dy_5Ir_4Si_{10}$ M. H. Lee, C. H. Chen, M.-W. Chu, H. D. Yang,	680
ATC 13:	PREPARING INAS NANOWIRES FOR FUNCTIONALIZED STM TIPS Kilian Flöhr, H. Yusuf Günel, Kamil Sladek, Robert Frielinghaus, Hilde Hardtdegen, Marcus Liebmann, Thomas Schäpers, and Markus Morgenstern	681
ATC 14:	FROM THE MICROSCALE TO THE NANOSCALE: EDX ANALYSIS WITH HIGH SPATIAL RESOLUTION USING SILICON DRIFT DETECTORS (SDD) Tobias Salge, Igor Nemeth, Meiken Falke	682
ATC 15:	NANOPULSED FIELD INDUCED CONTROL OF HYBRID MATERIALS AND ITS FUNCTIONARYTY Tadachika Nakayama, Roman Nowak, Koichi Niihara	684
ATC 16:	EUV actinic mask blank defect inspection: results and status of concept realization Aleksey Maryasov, Stefan Herbert, Larissa Juschkin, Anke Aretz, Rainer Lebert	685
ATC 17:	INVESTIGATIONS OF METALL-FILMS ABSORPTION ON COBALT WITH NVIDIA CUDA TECHNOLOGY Sergey Seriy	687
ATC 18:	AN OXIDE MBE SYSTEM FOR QUASI IN-SITU NEUTRON REFLECTOMETRY STUDIES Sabine Pütter, Alexandra Steffen, Markus Waschk, Alexander Weber, Stefan Mattauch, Thomas Brückel	688
ATC 19:	EXPLORING THE RESOLUTION LIMIT OF THE TALBOT LITHOGRAPHY WITH EUV LIGHT H. Kim, S. Danylyuk, L. Juschkin, K. Bergmann, P. Loosen	689
ATC 20:	NON-STOICHIOMETRIC HfO_{2-x} THIN FILMS Milias Crumbach, Manfred Martin	691

Invited Talks

NEW MAGNETIC MATERIALS BASED ON DEFECTS, INTERFACES AND DOPING

George A. Sawatzky, Ilya Elfimov, Bayo Lau and Mona Berciu
Physics dept and Max Planck-UBC Centre for Quantum Materials
University of British Columbia Vancouver Canada

Ideas based on theory and some experiments will be presented regarding possible new magnetic materials based on extended and point defects (1), interface engineering (2), anion substitution in oxides and hole and electron doping of oxides. The concentration will be on rather ionic oxides mostly not involving conventional magnetic elements. Special attention will also be placed on surface and interface effects involving polar surfaces as well as on the role of doped holes in O 2p in charge transfer gap oxides. O 2p holes play an extremely important role in the magnetism and superconductivity of oxides and new results will be presented regarding the ferromagnetic exchange coupling they introduce in transition metal oxides(3) and the interplay between transport properties, magnetic order and the general phase diagrams of materials involving O 2p holes either in the so called self doped case of stochiometric oxides or in chemically substituted systems and cation or anion vacancies. We also present exact results on the spin polaron formation(4) and charge propagation of doped Fermions in Ferromagnetic lattices and the pairing interaction due to the magnetic background.

1. I. S. Elfimov, S. Yunoki, and G. A. Sawatzky PRL 89, 216403, (2002)
2. N. Pavlenko, T. Kopp, E.Y. Tsymbal, G.A. Sawatzky, and J. Mannhart PRB 85, 020407, (2011)
3. Bayo Lau, Mona Berciu and George A. Sawatzky, PRL 106, 036401 (2011)
4. Berciu and G. A. Sawatzky, PRB 79, 195116 (2009)

ATOMIC-RESOLUTION ELECTRON SPECTROSCOPY OF INTERFACES AND DEFECTS IN COMPLEX OXIDES

D. A. Muller[1], J. A. Mundy[1], L. Fitting Kourkoutis[1], M. P. Warusawithana[2], J. Ludwig[2], P. Roy[2], A. A. Pawlicki[2], T. Heeg[3], C. Richter[4], S. Paetel[4], M. Zheng[5], B. Mulcahy[5], W. Zander[6], J. N. Eckstein[5], J. Schubert[6], J. Mannhart[4], D. G. Schlom[3]

[1] School of Applied and Engineering Physics, Cornell University,
[2] Department of Physics and NHMFL, Florida State University,
[3] Department of Materials Science and Engineering, Cornell University,
[4] Experimentalphysik VI, University of Augsburg,
[5] Department of Physics, University of Illinois at Urbana – Champaign,
[6] Inst. of Bio and Nanosystems IBN1-IT and JARA-FIT, Research Centre Jülich

Electron energy loss spectroscopy (EELS) in a new generation of aberration-corrected electron microscopes provides direct images of the local physical and electronic structure of a material at the atomic scale [1]. The sensitivity and resolution can extend to imaging single dopant atoms or vacancies in their native environments. The detection and control of interface defects using EELS, closely-coupled with atomically-precise growth methods, has enabled the realization of interface-stabilized emergent ground states, including a 2D metal at the $LaTiO_3/SrTiO_3$ interface; a 2D superconductor between a $LaAlO_3$ and $SrTiO_3$; and, by eliminating extended 2D defects, ferromagnetic manganites a few unit cells thick - well below the widely-assumed critical thickness for ferromagnetism and conductivity in manganite systems. In each case, the detection and control of defects has proven crucial to distinguishing between intrinsic and extrinsic interface effects. This is well illustrated at the $LaAlO_3/SrTiO_3$ interface. After controlled experiments effectively eliminate the extrinsic effects that have been suggested as possible mechanisms of conductivity, electron microscopy reveals that defect compensation at the interface is different for A-site vs B-site rich systems, and the stoichiometry is key to the existence of the interface 2-dimentional electron gas.

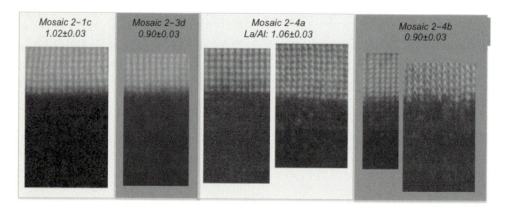

Figure 1: Spectroscopic maps of La at $LaAlO_3/SrTiO_3$ interfaces. The maps are arranged in order of apparent La interdiffusion and labeled by the La/Al ratio. Samples with La/Al > 1 (light grey) showed insulating interfaces while La/Al<1 (dark grey) were conducting. As shown, there is no correlation between the La interdiffusion and the transport properties.

[1] D. A. Muller, L. F. Kourkoutis, M. Murfitt, J. H. Song, H. Y. Hwang, J. Silcox, N. Dellby, O. L. Krivanek, *Science* **319**, 1073 (2008).

SIGNIFICANCE OF SOLID STATE IONICS FOR TRANSPORT AND STORAGE

Joachim Maier

Max Planck Institute for Solid State Research, Stuttgart, Germany

The role of ionic point defects and their impact on electronic charge carriers are considered form a thermodynamic point of view. Particular emphasis is laid on the influence of interfaces. The thermodynamic correspondence between ionic and electronic excitations is highlighted by Fig. 1.

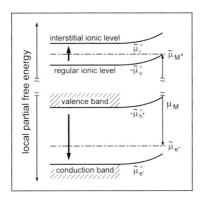

Figure 1: Ionic and electronic disorder in bulk and at boundaries. According to Ref. [1].

The relations are set out using four examples. Example 1 discusses ionic and electronic transport in the bulk for typical transition metal oxide under full equilibrium at high temperatures and links the situation to a typical low temperature situation where the ionic point defects are frozen. This example highlights the significance of point defects and preparation conditions for electronic transport at room temperature [2,3].

Example 2 refers to the effect of grain boundaries on the ionic and electronic charge carrier distribution. The effects can be well understood by ionically dominated space charge phenomena. The mixed conductor $SrTiO_3$ is considered here. Such space charge effects culminate in giant variations of n-, p- and oxygen ion conductivity in nanocrystalline $SrTiO_3$ (see Fig. 2) solely due to the small grain size. The effect can be well explained as fully mesoscopic phenomena as a function of temperature and oxygen partial pressure [4].

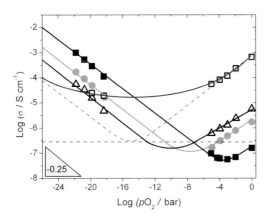

Figure 2: Conductivity versus oxygen partial pressure P measured at 544 °C for microcrystalline (open squares) and nanocrystalline $SrTiO_3$ (filled squares: 30nm) as a function of oxygen partial pressure. The p-type (n-type) conductivity is decreased (increased) by 3 orders of magnitude and the ion conductivity decreased by 6 orders of magnitude. According to Ref. [4].

Example 3 refers to an ionic conductor system and treats heterophase contacts [6]. MBE-grown multilayers of CaF$_2$ and BaF$_2$ (as well as composites) are investigated as a function of layer thickness ranging from ~1 μm (semi-infinite conditions) to ~1 nm (mesoscopic conditions) (see Fig. 3).

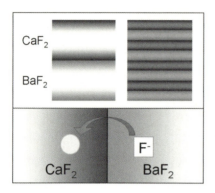

Figure 3: In CaF$_2$-BaF$_2$-heterostructures thermodynamic contact equilibrium demands fluorine ion redistribution (bottom).This leads to charge carrier accumulation in semi-infinite space charge zones (top, l. h. s.) or for very small spacings to a fully mesoscopic situation (top, r. h. s.). According to Ref. [5].

The greatly increased F$^-$ conductivity on size reduction can be well understood in terms of ionic charge transfer at the boundaries and shows the significance of "nanoionics" as a field of fundamental importance for energy research. The consideration of the full interfacial thermodynamics paved the way for "composite electrolytes" as a new class of solid electrolyte [6].

Example 4: Space charge effects do not only lead to drastically altered transport properties, they can also lead to drastically altered storage properties. As an example we consider the nano-composite Li$_2$O/Ru in which quite an amount of lithium can be stored even though none of the individual phases can do this (Fig. 4).

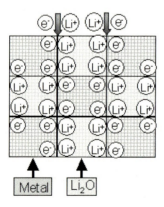

Figure 4: The Li$_2$O/M stores Li via job-sharing. This mechanism forms the bridge between an electrostatic composite and a battery electrode if the space charge profiles overlap. According to Ref. [7]

The reason lies in this novel "job-sharing" storage at the interface: Li$_2$O stores Li$^+$, Ru stores e^-. This mechanism forms the bridge between an electrostatic and chemical capacitor [6,7].

[1] J. Maier, *Physical Chemistry of Ionic Materials. Ions and Electrons in Solids*, Wiley, 2004.
[2] R. Waser, J. Am. Ceram. Soc.**74**, 1934 (1991).
[3] J. Maier, Phys. Chem. Chem. Phys. **5**, 2164 (2003).
[4] P.Lupetin, G. Gregori, and J. Maier, Angew. Chem. Int. Ed. **49**, 10123 (2010).
[5] N. Sata, K. Eberman, K. Eberl, and J. Maier, Nature **408**, 946 (2000).
[6] J. Maier, Nature Materials **4**, 805 (2005); J. Maier, Prog. Solid St. Chem.**23**, 171 (1995).
[7] J. Maier, Faraday Discussions **134**, 51 (2007).

ELECTROCHEMICAL DOPING OF OXIDE HETEROSTRUCTURES

E. Artacho[1,2,6,7], N.C Bristowe[1,2], P.B. Littlewood[5], J.M. Pruneda[8], M. Stengel[3,4]

[1]Theory of Condensed Matter, Cavendish Laboratory,University of Cambridge, Cambridge CB3 0HE, UK;
[2]Department of Earth Sciences, University of Cambridge, Downing Street, Cambridge CB2 3EQ, UK;
[3]ICREA - Institució Catalana de Recerca i Estudis Avan,cats, 08010 Barcelona, Spain;
[4]Institut de Ciéncia de Materials de Barcelona (ICMAB-CSIC), Campus UAB, 08193 Bellaterra, Spain;
[5]Physical Sciences and Engineering, Argonne National Laboratory, Argonne, Illinois 60439, USA;
[6]CIC Nanogune, and Donostia International Physics Center DIPC, Tolosa Hiribidea 76, 20018 San Sebastian, Spain;
[7]Basque Foundation for Science Ikerbasque, 48011 Bilbao, Spain;
[8]Centre d'Investigación en Nanociéncia i Nanotecnologia (CSIC-ICN), Campus UAB, 08193 Bellaterra, Spain

Heterostructure oxides have emerged as a promising avenue to control electronic functionality by precise control of chemistry on the atomic scale. In particular, the possibility of utilising built-in electric fields in polar structures presents the possibility of modulation doping of two- dimensional electron gases (2DEG) in oxides. A case in point is the generation of carriers at the polar interface between LaAlO3 (LAO) and SrTiO3 (STO).

To progress, the origin of the 2DEG must be understood. While in pristine materials, electron transfer to counter the polar catastrophe would be expected[1], often defects, non-stoichiometry and in particular surface O vacancies are the source. We find that surface redox reactions, in particular surface O vacancies, are thermodynamically stable under most growth conditions, using a phenomenological model supported by first principles calculations which is in agreement with spectroscopic data[2]. Pristine systems will likely require changed growth conditions or modifed materials with a lower vacancy free energy.

In a related problem, but against expectations, robust switchable ferroelectricity has been recently observed under open-circuit electrical boundary conditions in nm thick ferroelectric films. First-principles calculations show that the pristine system does not polarize and instead we propose electrochemical ferroelectric switching as the phenomenon being observed[3]. If not exceeding its bulk value, the ferroelectric polarization of the film adapts to the bound charge generated on its surface by redox processes when poling the film.

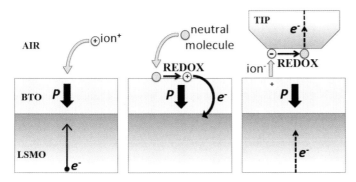

FIG. 1: Schematic illustration of the conventional (left) and redox (center) mechanisms for ferroelectric screening in the absence of a top electrode. The presence of a biased tip can promote an alternative redox mechanism that provides an external circuit for the screening electrons (right). (From Ref 4)

The interplay of electrochemistry and modulation doping offers opportunities to speculate about novel nanoscale electrochemical and storage applications.

1 N.C. Bristowe, E. Artacho, and P.B. Littlewood, Phys. Rev. B 80 045425 (2009)
2 N. C. Bristowe, P. B. Littlewood, Emilio Artacho, Phys. Rev. B 83, 205405 (2011)
3 V. Garcia, S. Fusil, K. Bouzehouane, S. Enouz-Vedrenne, N. Mathur, A. Barth_el_emy, and M. Bibes, Nature 460, 81 (2009).
4 N. C. Bristowe, Massimiliano Stengel, P. B. Littlewood, J. M. Pruneda, Emilio Artacho, Phys. Rev. B 85, 024106 (2012)

SWITCHABLE PHOTODIODE EFFECT IN FERROELECTRIC BiFeO$_3$

S-W. Cheong, H. T. Yi, T. Choi, and A. Hogan

Rutgers Center for Emergent materials and Department of Physics and Astronomy, Rutgers University, Piscataway, NJ 08854, USA

Using photodiodes and solar cells, solar energy can be converted directly into electric current flow and light illumination can modulate the current. The rectification at a p-n junction originates from carrier diffusion across the p-n barrier, associated with an internal electric field. Recently, to enhance the optical device and solar energy harvest functionalities, other types of photovoltaic cells have been widely researched. For example, the bulk photocurrent flow created in ferroelectrics such as LiNbO$_3$ by UV light illumination depends upon the direction of ferroelectric polarization, leading to a ferroelectric photovoltaic (FPV) effect [1,2]. The main drawback of this fascinating FPV effect is the tiny magnitude of photocurrent in typically-insulating ferroelectrics. A recent breakthrough has been made in ferroelectric BiFeO$_3$ (BFO) in which large FPV current is induced by visible light illumination [3]. On the other hand, the mechanism for this remarkable FPV effect in BFO is largely unknown, and technical exploitation has to be pursued.

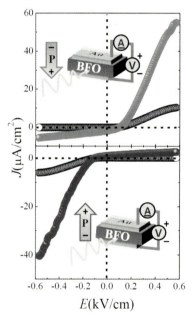

Figure 1: I-V curves with downward polarization (top) and upward polarization (bottom). Black open circles are for the I-V curves in dark, and red and blue filled circles indicate current under green laser illumination. Cartoon shows the schematic of our experiment setup.

From comprehensive investigation of the photodiode effect in BFO, we found that significant rectification and ferroelectric photovoltaic effects exist in BFO, and the direction of the rectification and photovoltaic current is reversely switchable by large external voltages (Fig. 1). The forward bias direction is along the ferroelectric polarization direction while the photovoltaic current direction is against the polarization direction (Fig. 2). The polarization clearly plays an essential role on the rectification and photovoltaic effects. On the other hand, we unveiled that polarization flipping at low temperatures is not sufficient to switch the rectification direction and near-room-temperature poling is necessary for the switching, indicating that electromigration of defects such as oxygen vacancies is important for the switching [4]. The rectification effect is consistent with the presence of Schottky-to-Ohmic contacts, and the Schottky-to-Ohmic contacts appear to be switchable with external voltages. The switching of Schottky-to-Ohmic contacts results from a combined effect of polarization flipping and electromigration of defects (oxygen vacancies) from one contact to the other contact. We also found that the bulk absorption across the bulk band gap determines the photovoltaic effect, and external quantum efficiency of our simple device can be as large as ~1.5 % at ~2.9 eV. By engineering the band gap, carrier concentration, and device configurations, this fascinating switchable photovoltaic effect needs to be further exploited for novel technologies such as fast readout of ferroelectric state, ferroelectric solar cells, or ferroelectric sensors.

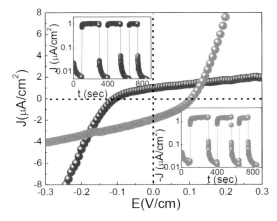

Fig. 2 Expanded view of the *I-V* curves for two different polarization orientations. Insets show zero-bias photocurrent with repetition of light on and off. Blue and red colored data are for upward and downward polarizations, respectively.

[1] A. M. Glass, D. von der Linde, and T. J. Negran, Appl. Phys. Lett. **25**, 233 (1974).
[2] G. Dalba *et al.*, Phys. Rev. Lett. **74**, 988 (1995).
[3] T. Choi *at al.*, Science **324**, 63 (2009).
[4] H. T. Yi *et al.*, Adv. Mater. **23**, 3403 (2011)

EXPLORATION OF ELECTRON SYSTEMS AT OXIDE INTERFACES

Werner Dietsche[1], Benjamin Förg[2], Cameron Hughes[1], Carsten Woltmann[1], Thilo Kopp[2], Florian Loder[2], <u>Jochen Mannhart</u>[1], Natalia Pavlenko[2], Christoph Richter[1], Ulrike Waizmann[1], Jürgen Weis[1]

[1]Max Planck Institute for Solid State Research, 70569 Stuttgart, Germany
[2]Experimental Physics VI, Augsburg University, 86135 Augsburg, Germany

Induced by quantum phenomena, oxide interfaces offer a fascinating potential to create novel electron systems. In this presentation, we will report on recent progress we have made in exploring the fundamental properties of the 2D electron liquid at $LaAlO_3$-$SrTiO_3$ interfaces.

Furthermore, we will analyse the potential of such 2D systems for use in future electronic devices. We have fabricated, for example, field-effect devices that utilize the two-dimensional electron liquid generated at the bilayers' n-type $LaAlO_3$-$SrTiO_3$ interface as drain-source channels and the $LaAlO_3$ layers as gate dielectrics (Fig. 1). With gate voltages well below 1V, the devices are characterized by voltage gain and current gain [1].

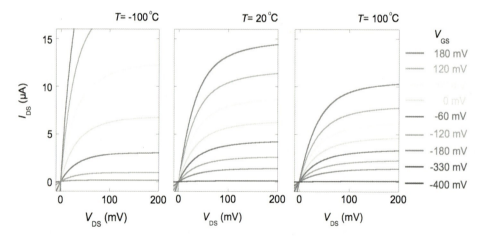

Figure 1: I_{DS} (V_{DS})-characteristics of a device measured in four-point configuration at -100, 20, and 100 °C. The measurement was done on a device with a channel length of 40 μm and a channel width of 1600 μm (from Ref. [1]).

[1] B. Förg, C. Richter, and J. Mannhart, Appl. Phys. Lett. **100**, 053506 (2012).

THE INFLUENCE OF IMPERFECTIONS ON THE 2DEG TRANSPORT PROPERTIES IN THE LaAlO$_3$-SrTiO$_3$ SYSTEM

Guus Rijnders[1]

[1] University of Twente, MESA+ Institute for Nanotechnology, Enschede, the Netherlands

Within this contribution, I will focus on the creation and annihilation of oxygen vacancies in LaAlO$_3$/SrTiO$_3$ heterostructures, as well as their influence on the transport properties of the 2DEG created at the interface between LaAlO$_3$ and SrTiO$_3$. The manifestation of quantum behavior in two dimensional electron gases in semiconducting heterostructures and their progressive complexity towards fractional quantum Hall effect went hand-in-hand with the efforts to remove the effect of impurity scattering. For oxide materials, history is repeating itself and to date sample quality is reaching levels where quantum behavior starts to become accessible. To really understand the ground state of two dimensional electron gases in oxide LaAlO$_3$-SrTiO$_3$ systems, the influence of imperfections on the 2DEG should be investigated and minimized.

I will show that, due to redox reactions, oxygen vacancies can be created within SrTiO$_3$ heterostructures. As an example, metallic conducting interfaces are observed in such heterostructures with various overlayers of amorphous LaAlO$_3$, SrTiO$_3$ and yttria-stabilized zirconia (YSZ) films. Whereas, an insulating heterointerface is found when the overlayer is an amorphous La$_{7/8}$Sr$_{1/8}$MnO$_3$ film. I will present evidence that the interfacial conductivity results from the oxygen vacancies on the SrTiO$_3$ substrate side due to the exposure of the substrate surface to reactive species of film growth, see figure 1. Although the energy of the arriving species has been suggested to be responsible for the creation of defects, the chemical reactivity of these species at the substrate surface has not been considered yet. Our results [1] show that the latter mechanism is an important source for the creation of mobile charge carriers in SrTiO$_3$-based oxide heterostructures.

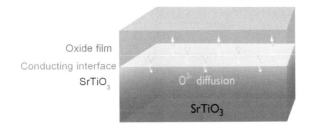

Figure 1: Schematic representation for the oxygen ions outward diffusion induced interfacial conduction in SrTiO$_3$-based heterostructures during growth of the oxide films at room temperature.

On the other hand, impurity scattering can heavily reduce the carrier mobility in epitaxial heterostructures. I will show that the impurity level can be significantly suppressed by defect engineering, resulting in an increase of the carrier mobility, allowing, for instance, the observation of quantum transport. We used SrTiO$_3$-LaAlO$_3$ heterostructures with epitaxial capping layers, in which the latter plays an important role in the suppression of defect scattering.

Within the contribution, growth and properties of the complex oxide heterostructures will be presented, with a focus on the underlying mechanism of defect engineering.

[1] *Nanoletters* **11** (9), (2011) 3774–3778

CORRELATED ELECTRONIC MATERIALS: COMPUTATIONAL STUDIES OF MULTIORBITAL MODELS FOR BULK COMPOUNDS AND INTERFACES OF MAGNETIC AND SUPERCONDUCTING MATERIALS

Elbio Dagotto[1]

[1]Department of Physics and Astronomy, University of Tennessee, Knoxville, Tennessee 37996, USA, and Materials Science and Technology Division, Oak Ridge National Laboratory, Oak Ridge, Tennessee 37831, USA

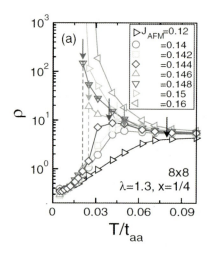

The current status of computer simulations of model Hamiltonians for correlated electronic materials will be reviewed. The focus will be on the Mn-oxides with the colossal magneto-resistance (CMR) and the novel Fe-based high critical temperature superconductors. It will be shown that state-of-the-art computer simulations involving Monte Carlo studies of double-exchange models with Jahn-Teller phonons do display the CMR effect. This is exemplified in the figure on the right, reproduced from [1], where the calculated resistivity vs. temperature (in units of the hopping amplitude t_{aa}) is shown (with $T/t_{aa} \sim 0.09$ being room temperature). Varying the super-exchange coupling J_{AFM}, the results interpolate between those of a ferromagnetic metallic state to those corresponding to a competing insulator [1]. The interpolation, with an increasing resistivity that abruptly drops to a metallic state with decreasing temperature, is in excellent agreement with experiments carried out for a variety of rare-earth elements in the perovskite chemical formula, as discussed in [1].

In addition, it will be argued that multiple-orbital Hubbard models for pnictides and chalcogenides have magnetic order with the correct wavevector in the undoped-limit ground state, and pairing tendencies in several competing channels upon doping. For example the figure on the left shows the phase diagram of a three-orbital Hubbard model obtained via a mean-field approximation, reproduced from [2]. Shown are three regions: a non-magnetic metallic state at small Hubbard repulsion U, an insulator (with orbital order) at large U, and an intermediate phase with magnetic

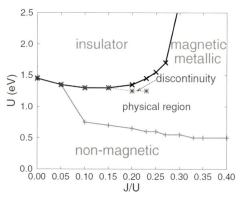

and metallic characteristics that seem to fit nicely the properties of the pnictides. In fact, the region shown in yellow, centered at a Hund coupling J which in units of U is approximately 0.25, corresponds to a regime where the theoretical results agree quantitatively with those of neutron and photo-emission scattering, as discussed in [3]. A combination of localized and itinerant features appears to be needed for properly describing these exotic high-Tc superconductors that are in the "intermediate" coupling range between the weak and strong coupling limits, as exemplified in the sketched shown on the right reproduced from [3]. In this sketch, the yellow region is a magnetically ordered state that starts at small U (bandwidth W units) in a regime where Fermi surface nesting works, followed by an intermediate region where the pnictides may be located, and then ending in a large U insulating region where Heisenberg models with localized spins provide an accurate representation of the physics.

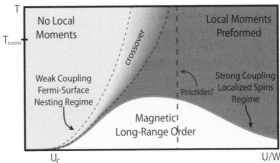

The presentation will continue by addressing next the case of superlattices made of strongly correlated materials, as in the figure shown on the left, with emphasis on results for magnetic compounds, such as large- and low-bandwidth manganites (i.e. LMO/SMO, LMO/CMO), and also manganites/ferroelectrics. The focus will be on the states that are stabilized at the interfaces, some of which do not have an analog in the materials that form the superlattice when in bulk form, as reported in [4].

Time allowing, interesting results obtained using computational technique in quasi one dimensional systems will also be discussed. They correspond to the propagation of excitons, as those created in photovoltaic devices, in the framework of Hubbard insulators, as exemplified in the figure on the right that is reproduced from [5]. A real-time analysis can follow the "holon" and "doublon" as they propagate and eventually cross the boundary from the insulator to the metallic leads.

It is concluded that the use of computational techniques applied to model Hamiltonian systems is crucial for the analysis of a variety of interesting materials, with magnetic, superconducting, and multiferroic properties.

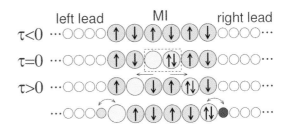

[1] C. Sen et al., Phys. Rev. Letters **105**, 097203 (2010).
[2] Q. Luo et al., PRB **82**, 104508 (2010).
[3] P. Dai, J. Hu, and E. Dagotto, Nature Physics (submitted).
[4] R. Yu et al., Phys Rev B **80**, 125115 (2009).
[5] L. Dias et al., Phys. Rev. B **81**, 125113 (2010).

Note: Different parts of the research reported in this presentation have been supported by the Department of Energy and the National Science Foundation of the USA.

ELECTROLYTE GATE INDUCED METALLIZATION OF SEVERAL FACETS (101, 001, 110 and 100) OF RUTILE TiO$_2$ AND (001) SrTiO$_3$

<u>Stuart S.P. Parkin</u>[1], Thomas D. Schladt[1], Tanja Graf[1], Mingyang Li[1,2], Nagaphani Aetukuri[1,3], Xin Jiang[1] and Mahesh Samant[1]

[1]IBM Almaden Research Center, San Jose, California, USA;
[2]Department of Physics, Stanford University, Stanford, California, USA; [3]Department of Materials Science and Engineering, Stanford University, Stanford, California, USA

The electric field induced metallization of insulating oxides is a powerful means of exploring and creating novel electronic states. Recently large internal electric fields from polar surfaces have been used to create emergent metallic, superconducting and magnetic states at interfaces between two insulating oxides[1]. However, the origin of the metallicity is a subject of considerable debate. Moreover, relying on the interface polar discontinuity to create the electric field restricts the interface orientation to that in which the surface of the polar material has an uncompensated charge. Electrolyte gating, on the other hand, can be applied to any crystal facet, and allows for varying electric fields and associated induced carrier densities. We have used electrolyte gating to study four different facets of rutile TiO$_2$. Two of these, namely (101) and (001), show clear evidence of metallization, with a disorder-induced metal-to-insulator transition at low temperatures, whereas two other facets, (110) and (100), show no substantial effects (see Figure 1). This facet-dependent metallization can be correlated with the energy of formation of oxygen vacancies on the respective facet[2], thereby clearly showing that gate induced charge transfer effects are not the sole origin of the metallicity. The orientation dependence of electrolyte gating is a novel way of distinguishing purely electrostatic field effects from electric field induced modifications of the surface structure and stoichiometry.

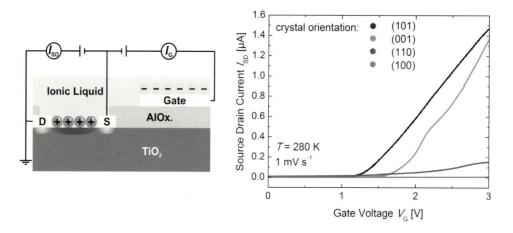

Figure 1: Left: Schematic diagram of the EDLT device configuration, showing the lateral gate electrode and the source and drain contacts to the surface of the TiO$_2$. Right: Source-drain current I_{SD} as a function of gate voltage V_G for four crystal orientations of TiO$_2$.

Invited Talks 43

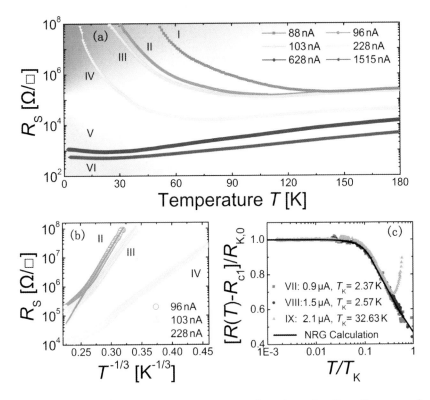

Figure 2: Top: Sheet resistance versus temperature curves for various gate voltages V_G corresponding to the source-drain currents I_{SD} shown in the inset to the figure. Bottom: Left: Sheet resistance plotted versus $1/T^{1/3}$ in the non-percolative regime, and Right: Normalized sheet resistance versus temperature normalized to the Kondo temperature T_K showing a resistance upturn and a temperature independent resistance as T→ 0.

The role of disorder in the structure of the gated electrolyte is explored in studies of the electrolyte gate-induced conductance at the surface of single crystalline $SrTiO_3$ (001). By varying the gate voltage and, thereby, the induced carrier density, we find two distinct transport regimes. At high carrier densities, a percolative, metallic state is induced in which, at low temperatures, signatures of a Kondo effect are clearly observed. Specifically, an upturn in resistance saturates to a constant value from below ~10% of the Kondo temperature down to the lowest temperatures measured (20 mK). At lower carrier densities, the resistance decreases from below the freezing temperature of the ionic liquid dielectric to a broad minimum below which the resistance diverges to very high values at low temperatures. In this regime the resistance increase can be well described by a variable range hopping model in two dimensions. We speculate that this results from non-percolative transport that likely results from inhomogeneous electric fields due to imperfect ordering of the ions at the frozen liquid/ oxide interface. Further evidence of the non-percolative and percolative regimes is provided by their non-linear and linear current versus voltage behaviours, respectively, as well the much larger magnetoconductance in the non-percolative regime.

[1] H. Y. Hwang et al. Nat. Mater. **11**, 103 (2012).
[2] B. J. Morgan and G. W. Watson, J. Phys. Chem. C **113**, 7322 (2009).

COMPLEX THERMOELECTRIC MATERIALS

G. Jeffrey Snyder

California Institute of Technology, Pasadena, California 91125 USA

The widespread use of thermoelectric generators has been limited by the low material efficiency of the thermoelectric material. A number of strategies for *Complex Thermoelectric Materials* [1] with higher Thermoelectric figure of merit, zT, are being actively studied at Caltech.

Complex electronic band structures provide mechanisms to achieve high zT in thermoelectric materials through *band structure engineering*. High zT is obtained p-type PbTe and PbSe which contains both light and heavy valence bands that can be engineered to achieve high valley degeneracy which leads to an extraordinary peak zT of nearly 1.8 at 750K [2].

Figure 1: A thermoelectric generator directly converts heat into electricity with no moving parts. The long term reliability of these systems has encouraged NASA to use thermoelectric generators in many space probes since the 1960s (up to 35 years unattended). Today, thermoelectrics are being considered for terrestrial applications such as automotive and industrial waste heat recovery as well as solar-electricity generation.

Complex crystal structures that enable relatively low thermal conductivity have lead to several new classes of thermoelectric materials. Ca_3AlSb_3, Ca_3AlSb_3 and $Yb_{14}AlSb_{11}$ are complex Zintl compounds containing differently connected $AlSb_4$ tetrahedra that obtain zT near 1 at high temperatures. Fast diffusing or 'liquid-like' elements in the complex materials Zn_4Sb_3 [3] and Cu_2Se [4] provide additional mechanisms to scatter and otherwise inhibit phonon heat conductivity. The principles of Zintl chemistry facilitates the search for new complex materials and the tuning of known thermoelectric materials with earth abundant, non-toxic elements [5]

Finally, the incorporation of nanometer sized particles reduces thermal conductivity from long mean-free-path phonons. This principle has been successfully demonstrated in PbTe with large nanoscale precipitates (>100nm) that can be independently doped with La (n-type) or Na (p-type). The synthesis of nanoscale composites can be controlled with the aid of equilibrium phase diagrams (experimental or theoretically determined) to produce microstructure of varying composition and length scale [6].

[1] G. J. Snyder, E. S. Toberer. "Complex thermoelectric materials" *Nature Materials* **7**, p 105 - 114 (2008)
[2] Y. Z. Pei, G. J. Snyder, et al. "Convergence of Electronic Bands for High Performance Bulk Thermoelectrics"*Nature* **473**, p 66 (2011)
[3] H. Liu, X. Shi, G. J. Snyder, et al. "Liquid-like Copper Ion Thermoelectric Materials" *Nature Materials*, doi:10.1038/nmat3273 (2012)
[4] G. J. Snyder, et al., "Disordered Zinc in Zn_4Sb_3 with Phonon Glass, Electron Crystal Thermoelectric Properties" *Nature Materials*, **Vol 3**, p. 458 (2004)
[5] E. S. Toberer. A. F. May, G. J. Snyder, "Zintl Chemistry for Designing High Efficiency Thermoelectric Materials" *Chemistry of Materials* **22**, p 624 (2010)
[6] D.L. Medlin and G.J. Snyder "Interfaces in Bulk Thermoelectric Materials" *Current Opinion in Colloid & Interface Science* **14**, 226 (2009)

PCRAM OPERATION AT DRAM SPEEDS: EXPERIMENTAL DEMONSTRATION AND COMPUTER-SIMULATIONAL UNDERSTANDING

D. Loke,[1,2,3,*] T. H. Lee,[1,*] W. J. Wang,[2] L. P. Shi,[2,†] R. Zhao,[2] Y. C. Yeo,[4] T. C. Chong,[5] and S. R. Elliott,[1,†]

[1]Department of Chemistry, University of Cambridge, Lensfield Road, Cambridge CB2 1EW, UK.
[2]Data Storage Institute, A*STAR, DSI Building, 5 Engineering Drive 1, Singapore 117608.
[3]NUS Graduate School for Integrative Sciences and Engineering, 28 Medical Drive, Centre for Life Sciences #05-01, Singapore 117456.
[4]Department of Electrical and Computer Engineering, National University of Singapore, 1 Engineering Drive 3, Singapore 117576.
[5]Singapore University of Technology & Design, 20 Dover Drive, Singapore 138682.

Phase-change random access memory (PCRAM) is one of the leading candidates for next-generation, non-volatile electronic data-storage devices, in which data bits are stored in terms of different structural states of a memory material (e.g. $Ge_2Sb_2Te_5$ – GST), i.e. either crystalline or amorphous, each having a different electrical resistivity. Switching between these two metastable memory states is achieved by the application of suitable voltage pulses. This new non-volatile memory technology is scalable beyond the current size limitations of silicon MOSFET-based 'flash' memory, and is now starting to appear in consumer products, e.g. Samsung smart-phones. However, the present writing (crystallization) speed of GST (ca 10ns) has been insufficiently fast to enable PCRAM to replace volatile DRAM with a *non-volatile* equivalent, for which switching speeds of less than 1ns are required.

We have controlled the crystallization kinetics of a phase-change material (GST) by the application of a constant low voltage, via pre-structural ordering (incubation) effects. An ultrafast crystallization speed of 500 ps was achieved, the first time that the 1ns barrier has been broken for PCRAM devices[1]. High-speed reversible switching using 500 ps pulses has also been demonstrated. *Ab initio* molecular-dynamics simulations have been performed to reveal the phase-change kinetics in PCRAM devices, and the structural origin of the incubation-assisted increase in crystallization speed has been identified. This paves the way to achieve a "universal electronic memory", capable of non-volatile operations at GHz data-transfer rates.

[1] D. Loke *et al.*, Science (to be published).

ELECTRONIC PHASE CHANGE AND ENTROPIC FUNCTIONS IN TRANSITION METAL OXIDES

Hidenori Takagi[1,2] and Seiji Niitaka[2]

[1]Department of Physics, University of Tokyo, Bunkyo-ku, Tokyo 113-0022, Japan;
[2]RIKEN Advanced Science Institute, Wako, Japan

The electric and magnetic properties of transition metal oxides (TMO) are often dominated by electrons in d-orbitals. Large Coulomb repulsion between electrons accommodated in the spatially constrained d-orbitals tends to block the motion of electrons from one atom to another, and the electrons are highly entangled. Just like interacting atoms and molecules, the entangled electrons, called correlated electrons, form solid (insulator), liquid (metal), and superfluid (superconductor) states inside the solid. The presence of the three degrees of freedom attached to electrons – charge, spin and orbital, enrich these electronic phases further. These rich electronic phases compete with each other in a delicate balance. Even a minute perturbation can induce a phase change, giving rise to a dramatic response to external fields. This is the hallmark of phase change functions in transition metal oxides, useful as sensors, memories and for signal conversion [1].

In this talk, we would like to discuss the application of phase change concept to entropic functions rather than the long discussed functions mentioned above. Partly because of the multiple degrees of freedom, the complicated electronic phases in transition metal oxides are often highly entropic. The high entropy can manifest itself in a large entropy (enthalpy) change associated with the electronic phase change. VO_2 is known to show a paramagnetic metal (liquid) to a nonmagnetic insulator (solid) transition around room temperature, where we indeed observed a large entropy change per volume, comparable to that of water-ice transition. This can be utilized as a "solid" electronic ice pack of which melting temperature is "tunable" upon doping. We can construct electronic icepack working at 10 C, which for example may be used to preserve human tissues during surgery. The large electric entropy coupled with electric current can be utilized for thermoelectric conversion. The oxide thermoelectrics, Na_xCoO_2, can be viewed as a realization of such scenario. Other possible application of the large electronic entropy will be discussed.

[1] H.Takagi and H.Y.Hwang, Science 327 (2010) 1601.

DISORDER INDUCED METAL-INSULATOR TRANSITION IN PHASE CHANGE MATERIALS

T. Siegrist[1,2]

[1]Department of Chemical and Biomedical Engineering, FAMU-FSU College of Engineering, Tallahassee, FL, USA; [2]1. Physikalisches Institut (IA), RWTH Aachen, Aachen, Germany

Phase change materials that reversibly switch between amorphous and crystalline states show large optical and electronic property contrasts. Rewritable DVDs and Blu-ray discs are based on such materials, while the large change in electronic transport properties with resistivity contrast of up to six orders in magnitude is used in a new class of non-volatile data storage devices. Such devices hold great potential for future miniaturization, and with fast read/write operations, they may find applications as universal data storage devices in mobile applications.

While the amorphous state is characterized by saturated covalent bonds, the crystalline phase forms resonant bonds. This bonding mechanism can account for the high electronic polarizabilities which characterize crystalline phase change materials. Interestingly, the relevant electronic states also govern the charge transport in the crystalline phase, leading to unique transport properties including a high degree of electronic localization, in those phase change materials, which are characterized by a high degree of disorder. An initially amorphous thin film of $Ge_1Sb_2Te_4$ that is annealed in steps up to 340°C shows a transition to a crystalline phase with a high degree of disorder at 145°C, where the sheet resistance drops by two orders of magnitude. Even though the crystalline material is a degenerate semiconductor, its temperature dependence of the resistivity shows a non-metallic behavior. Stepwise annealing to higher temperatures changes the temperature coefficient of the resistivity from negative to positive, indicative of a metal-insulator transition (MIT) (Figure 1)

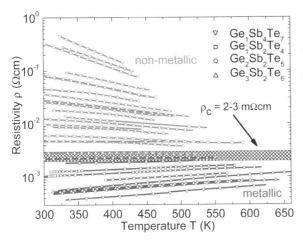

Figure 1: Systematic change of resistivity for four different Ge-Sb-Te phase change alloys annealed in steps of 20°C: $Ge_1Sb_4Te_7$, $Ge_1Sb_2Te_4$, $Ge_2Sb_2Te_5$, and $Ge_3Sb_2Te_6$. Only cooling data are shown, as obtained from the step annealed samples. All Ge-Sb-Te phases display the same behavior, with a critical resistivity of about 2-3 mΩcm where the temperature coefficient changes sign for all four alloys. Ref. [1].

Other phase change materials along the pseudo-binary $(GeTe)$-(Sb_2Te_3) line exhibit the same behavior, with a critical resistivity of 2-3mΩcm where the temperature coefficients of the resistivity change sign. The phase change materials therefore represent a unique system that is governed by localization effects induced by strong disorder and weak electron correlations. This universal behavior seems to be responsible for the high level of reproducibility of the resistance switching crucial to the application in non-volatile memory devices.

[1] T. Siegrist et al., Nature Mater. 10(3), 202-208 (2011).

ELECTRONIC PROPERTIES OF THE INTERFACIAL $LaAlO_3/SrTiO_3$ SYSTEM

J.-M. Triscone[1], A. Fête[1], S. Gariglio[1], A. Caviglia[1], D. Li[1], D. Stornaiuolo[1], M. Gabay[2], B. Sacépé[1], A. Morpurgo[1], M. Schmitt[3], C. Cancellieri[3], P. Willmott[3]

[1]DPMC, University of Geneva, 24 quai E.-Ansermet, CH-1211 Geneva 4, Switzerland;
[2]Laboratoire de Physique des Solides, Université de Paris Sud, 91405 Orsay, Cedex, France;
[3]Paul Scherrer Institut, CH-5232 Villigen, Switzerland

Oxide materials display within the same family of compounds a variety of exciting electronic properties ranging from ferroelectricity to ferromagnetism and superconductivity. These systems are often characterized by strong electronic correlations, complex phase diagrams and competing ground states. This competition makes these materials very sensitive to external parameters such as pressure or magnetic field. An interface, which naturally breaks inversion symmetry, is a major perturbation and one may thus expect that electronic systems with unusual properties can be generated at oxide interfaces [for reviews, see 1-3]. A striking example is the interface between $LaAlO_3$ and $SrTiO_3$, two good band *insulators*, which was found in 2004 to be conducting [4], and, in some doping range, superconducting with a maximum critical temperature of about 200 mK [5]. The characteristics observed in the normal and superconducting states are consistent with a two-dimensional electronic system. The thickness of the electron gas is found to be a few nanometers at low temperatures. This electron gas with low electronic density, typically $5\ 10^{13}$ electrons/cm^2, and naturally sandwiched between two insulators is ideal for performing electric field effect experiments allowing the carrier density to be tuned. Such an approach revealed the sensitivity of the normal and superconducting states to the carrier density. In particular, the electric field allows the tuning of the critical temperature between 200 mK and 0 K and thus the on-off switching of superconductivity. The system phase diagram reveals a superconducting pocket with an underdoped and an overdoped regime [6]. A large, interfacially generated, tunable spin-orbit coupling and a remarkable correlation between the spin-orbit coupling strength and the system phase diagram are other hallmarks of this fascinating system [7].

Here I will describe recent experiments aiming at determining the origin of the electron gas. I will then discuss superconductivity and the phase diagram of the system, magnetotransport in "standard" and in recently obtained high mobility samples that display Shubnikov de Haas (SdH) oscillations [8].

[1] J. Heber, Nature **459**, 28 (2009).
[2] J. Mannhart and D. Schlom, Science **327**, 1607 (2010).
[3] P. Zubko, S. Gariglio, M. Gabay, P. Ghosez, and J.-M. Triscone, Annual Review : Condensed Matter Physics **2**, 141 (2011).
[4] A. Ohtomo, H. Y. Hwang, Nature **427**, 423 (2004).
[5] N. Reyren, S. Thiel, A. D. Caviglia, L. Fitting Kourkoutis, G. Hammerl, C. Richter, C. W. Schneider, T. Kopp, A.-S. Ruetschi, D. Jaccard, M. Gabay, D. A. Muller, J.-M. Triscone and J. Mannhart, Science **317**, 1196 (2007).
[6] A. Caviglia, S. Gariglio, N. Reyren, D. Jaccard, T. Schneider, M. Gabay, S. Thiel, G. Hammerl, J. Mannhart, and J.-M. Triscone, Nature **456**, 624 (2008).
[7] A.D. Caviglia, M. Gabay, S. Gariglio, N. Reyren, C. Cancellieri, and J.-M. Triscone, Physical Review **104**, 126803 (2010).
[8] A.D. Caviglia, S. Gariglio, C. Cancellieri, B. Sacépé, A.Fête, N. Reyren, M. Gabay, A.F. Morpurgo, J.-M. Triscone, Physical Review Letters **105**, 236802 (2010).

EMERGENT PHENOMENA IN TWO-DIMENSIONAL ELECTRON GASES AT OXIDE INTERFACES

Susanne Stemmer[1], Pouya Moetakef[1], Daniel Ouellette[2], and S. James Allen[2]

[1]Materials Department, University of California, Santa Barbara, California, 93106-5050, USA
[2]Department of Physics, University of California, Santa Barbara, California, 93106-9530, USA

Two-dimensional electron gases at oxide interfaces have attracted significant attention because they can exhibit unique properties, such as strong electron correlations, superconductivity and magnetism. In the first half of the presentation we will highlight the importance of materials quality and deposition methods in achieving the desired control over these phenomena, as needed for novel electronic devices: similar to what has long been accepted in the semiconductor device community, only low-energetic deposition techniques, such as molecular beam epitaxy (MBE), can produce electronic device-quality materials. For example, we demonstrate record electron mobilities in $SrTiO_3$ thin films grown by MBE, which exceed even those of single crystals. We show that these high-quality MBE films allow for the study of quantum oscillations in two-dimensional electron gases in $SrTiO_3$ films and that these oscillations are much more pronounced than those currently observed in structures with $SrTiO_3$ substrates.

In the second half of the presentation, we will discuss emergent phenomena at interfaces between band insulators, such as $SrTiO_3$, and strongly correlated (Mott) materials, such as the rare earth titanates ($RTiO_3$, where R is a trivalent rare earth ion), or the rare earth nickelates ($RNiO_3$). $SrTiO_3/RTiO_3$ interfaces are particularly interesting, because both the oxygen and Ti sublattices are continuous across the interface. An interfacial fixed polar charge arises because of a polar discontinuity at the interface. This interfacial charge can be compensated by a two-dimensional electron gas, residing in the bands of the Mott and/or band insulator and bound to the interface by the fixed interface charge. In this presentation, we report on intrinsic electronic reconstructions, of approximately 1/2 electron per surface unit cell at a prototype Mott/band insulator interface between $GdTiO_3$ and $SrTiO_3$, grown by molecular beam epitaxy. The sheet carrier densities of all $GdTiO_3/SrTiO_3$ heterostructures containing more than one unit cell of $SrTiO_3$ are approximately ½ electron per surface unit cell (or 3×10^{14} cm^{-2}), independent of layer thicknesses and growth sequences. Unlike the more commonly studies $LaAlO_3/SrTiO_3$ interface, these carrier densities closely meet the electrostatic requirements for compensating the fixed charge at these polar interfaces. We will report on electron correlation effects, such as magnetism, in the extremely high carrier density $SrTiO_3$ quantum wells that can be obtained using these interfaces. We will discuss the coexistence of emergent phenomena, in particular ferromagnetism and superconductivity, in electron gases in $SrTiO_3$. Models of the charge distribution and measurements of transport coefficients, such as the Seebeck effect, and of the optical conductivity provide insights into the nature of the two-dimensional electron gas, the importance of band alignments, background doping and the occupancy of subbands that are derived from the Ti d-states. We will also discuss new experimental approaches to probe the Mott insulating state, using modulation doping with heterointerfaces for electrostatic control of large carrier densities in Mott materials. Finally, we will discuss the potential for new device applications of complex oxide heterostructures.

GIANT TUNNEL ELECTRORESISTANCE IN FERROELECTRIC TUNNEL JUNCTIONS

A. Chanthbouala[1], V. Garcia[1], K. Bouzehouane[1], S. Fusil[1], X. Moya[2], S. Xavier[3], H. Yamada[4], C. Deranlot[1], N.D. Mathur[2], J. Grollier[1], A. Barthélémy[1] and <u>M. Bibes</u>[1]*

[1] Unité Mixte de Physique CNRS/Thales, 1 Av. A. Fresnel, Campus de l'Ecole Polytechnique, 91767 Palaiseau (France) and Université Paris-Sud, 91405 Orsay (France)
[2] Department of Materials Science, University of Cambridge, Cambridge, CB2 3QZ, UK
[3] Thales Research and Technology, 1 Av. A. Fresnel, Campus de l'Ecole Polytechnique, 91767 Palaiseau (France)
[4] National Institute of Advanced Industrial Science and Technology (AIST), Tsukuba, Ibaraki 305-8562 (Japan)

Because it is spontaneous, stable and electrically switchable the polarization of ferroelectrics is an excellent state variable for non-volatile data storage. In addition, polarization reversal can be as fast as tens of ps [1] and only dissipates the modest power associated with polarization charge switching (with current densities typically lower than 10^4 A/cm²). When ferroelectrics are made as thin as a few nm, they can be used as tunnel barriers and the tunneling current is influenced by the polarization direction [2] enabling a simple non-destructive readout of the polarization state.

In this talk, we will show how the tunnel resistance can vary by more than two orders of magnitude upon polarization switching in highly-strained ultrathin $BaTiO_3$ tunnel barriers. This strong electroresistance effect can be probed using a conductive AFM tip as the top electrode [3], or using solid-state submicron pads. Such ferroelectric tunnel junctions show large, stable, reproducible and reliable tunnel electroresistance, with resistance switching related to ferroelectric polarisation reversal [4]. They thus emerge as an alternative to other resistive memories, with the additional advantage of not being based on voltage-induced migration of matter at the nanoscale, but on a purely electronic mechanism. Importantly, switching can be as fast as a few ns, and I will present data on the dynamical response of ferroelectric junctions, and their analysis with standard models of polarization reversal [5].

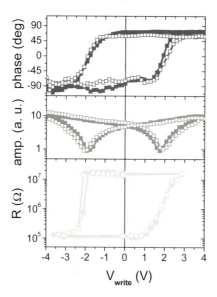

Figure 1: Ferroelectric switching versus resistive switching. Out-of-plane PFM phase (top) and amplitude (center) measurements on a typical gold/cobalt/BTO/LSMO ferroelectric tunnel junction. Bottom: Dependence of the junction resistance on the amplitude of the write voltage pulse measured in remanence (V_{read} ~100 mV) after applying successive write voltage pulses of 100 µs. The open and filled circles represent two different scans to show reproducibility. Adapted from Ref. [4].

[1] D.S. Rana et al. *Adv. Mater.* **21**, 2881 (2009)
[2] E.Y. Tsymbal and H. Kohlstedt, *Science* **313** (2006)
[3] V. Garcia et al, *Nature* **460**, 81 (2009)
[4] A. Chanthbouala et al, *Nature Nanotech.* **7**, 101 (2012)
[5] Y. Ishibashi & Y.N. Takagi, *J. Phys. Soc. Jpn.* **31**, 506 (1970) ; H. Orihara et al, *J. Phys. Soc. Jpn.* **63**, 1601 (1994)

REVISITING THE HEXAGONAL MANGANITES

Nicola Spaldin

Materials Theory, ETH Zurich, Wolfgang-Pauli-Strasse 27, 8093 Zurich, Switzerland
nicola.spaldin@mat.ethz.ch

The hexagonal manganite multiferroics, of which $YMnO_3$ is the prototype, have recently been shown to exhibit a fascinating ferroelectric domain structure [1] which is a consequence of their improper geometric ferroelectricity [2,3]. Here we discuss how this results in topologically protected ferroelectric vortices providing a model system to test theories of cosmic string formation in the early universe. We show how first-principles electronic structure calculations are contributing to the design of test experiments, as well as understanding the coupling between ferroelectricity and magnetism, and the behavior of the ferroelectric domain walls.

[1] Fennie, C. J. & Rabe, K. M., *Ferroelectric transition in $YMnO_3$ from first principles*. Phys. Rev. B 72, 100103 (2005).
[2] Van Aken, B. B., Palstra, T. T. M., Filippetti, A. & Spaldin, N. A., *The origin of ferroelectricity in magnetoelectric $YMnO_3$*. Nature Mater. 3, 164-170 (2004).
[3] Choi, T. et al. Insulating interlocked ferroelectric and structural antiphase domain walls in multiferroic $YMnO_3$. Nature Mater. 9, 253-258 (2010)

STUDY OF MAGNETOELECTRIC EFFECTS DUE TO MULTI-SPIN VARIABLES

Tsuyoshi Kimura

Division of Materials Physics, Graduate School of Engineering Science, Osaka University, Toyonaka, Osaka 506-8531, Japan

Interest in the study of magnetoelectric (ME) effects, magnetic control of electric polarization P or electric control of magnetization M has been reinvigorated since the discovery of spin driven ferroelectricity and giant ME effects in some spin-spiral magnets. Usually, the ME effect can exist in crystals with ordered spin structures having peculiar magnetic symmetries. In ME materials, additional multi-spin variables often play an important role in their ME properties. A well-known multi-spin variable which couples spins with P is "vector spin chirality", defined as $\kappa = (S_i \times S_j)$ where S_i and S_j denote spins at the sites i and j. The most successful microscopic mechanisms for the contribution of the vector spin chirality to the spiral-spin driven ferroelectricity are the so-called "spin current" and "inverse Dzyaloshinskii-Moriya" mechanisms. Another known ME active multi-spin variable is toroidal moment t which is described as the outer product of the displacement of magnetic ions from the center position r_i and their spins S_i; i.e., $t \propto \Sigma_i r_i \times S_i$. The sign of t changes under time reversal and space inversion operation, and ME effects in several compounds such as $LiCoPO_4$ have been discussed in terms of the toroidal ordering.

In this presentation, I show recent progress on the study of ME effects in several magnetoelectrics in which the above-mentioned multi-spin variables play important roles. (For example, low-field ME effect at room temperature in hexaferrite compounds such as $Sr_3Co_2Fe_{24}O_{41}$ [1], ferromagnetic and ferroelectric nature in an olivine compound Mn_2GeO_4 [2], and antisymmetric off-diagonal ME effects in a spin-glass system, ilmenite $(Ni,Mn)TiO_3$ [3].)

This work has been done in collaboration with Y. Hiraoka, Y. Yamaguchi, T. Honda, T. Ishikura, K. Okumura, Y. Kitagawa, H. Nakamura, Y. Wakabayashi, M. Soda, T. Asaka, T. Nakano, Y. Nozue, Y. Tanaka, S. Shin, J. S. White, and M. Kenzelmann.

[1] T. Kimura, Annu. Rev. Condens. Matter Phys. 3, 93 (2012).
[2] J. S. White, T. Honda, K. Kimura, T. Kimura, Ch. Niedermayer, O. Zaharko, A. Poole, B. Roessli, and M. Kenzelmann, Phys. Rev. Lett. 108, 077204 (2012).
[3] Y. Yamaguchi, T. Nakano, Y. Nozue, and T. Kimura, Phys. Rev. Lett. 108, 057203 (2012).

BI-LAYERED RERAM: MULTI-LEVEL SWITCHING, RELIABILITY AND ITS MECHANISM FOR STORAGE CLASS MEMORY AND RECONFIGURATION LOGIC.

U-In Chung[1], Young-Bae Kim[1], Seung Ryul Lee[1], Dongsoo Lee[1], Chang Bum Lee[1], Man Chang[1], Kyung min Kim[1], Ji Hyun Hur[1], Myoung-Jae Lee[1], Chang Jung Kim[1]

[1] Samsung Advanced Institute of Technology, San 14-1, Nongseo-dong, Giheung-gu, Yongin-si Gyeonggi-do, 446-712, South Korea, E-mail) uin.chung@samsung.com

The age of fast-moving information and computer technology is driven by silicon CMOS technology. Even though there seems to be no fundamental limit to scaling current devices to below 10 nanometers, there has been a shift toward functional diversification. Among several devices that could add more function, oxide based devices have been kept in focus because of its possibility of emulating organic brain functions and reconfigurable functions on Si circuits in addition to basic storage operations. Although research on ReRAM has continued to be reported [1-2], a reliable memory performance has not yet been presented, because of lacking material architecture and an insufficient understanding of the switching mechanism. In this presentation, a new model and architecture will be proposed based on the movement of oxygen vacancies in a bi-layered metal oxide.

Figure 1 shows the basic device structure with the equivalent bi-layered switching element ReRAM circuit architecture. The oxygen concentration of the oxygen exchange layer (OEL), was controlled by plasma oxidation of the TaOx surface. We propose a model [3-5] in which resistance varies through the movement of oxygen vacancies into or out of the conductive paths formed in the OEL. Figure 2 demonstrates the experimental and the calculated DC voltage sweep properties of typical ReRAM materials (SZO, PCMO and TaO_x). Our model was able to successfully predict the I-V characteristics of these resistive switching materials.

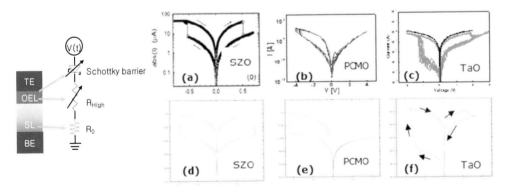

Figure 1: A bi-layered TaOx/Ta2O5 ReRAM structure with an equivalent circuit. The basic device structure is composed of bottom electrode (BE)/ self-compliance layer (SL)/oxygen exchange layer (OEL)/top electrode (TE).

Fig. 2. Experimental [(a), (b) and (c)] and calculated [(d), (e) and (f)] results for SZO, PCMO and TaO system, respectively.

Figure 3 indicates that the resistive switching is originated from the formation and rupture of conducting path which is consisted of nano-sized Ta-rich oxide clusters in Ta_2O_5 layer.

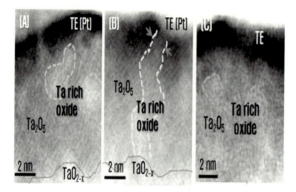

Figure 3 : (A), (B) and (C) display the in-situ high resolution scanning transmission electron microscopy (HR-STEM) images at pristine state, low resistance state (LRS) and high resistance state (HRS), respectively, in which, the LRS and HRS are formed in TEM apparatus, sequentially.

Figure 4 shows that the cell stably operates over 1E7 cycles for all the resistance levels.

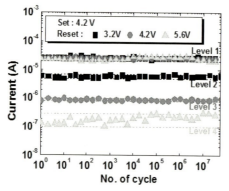

Figure 4: Endurance performance in 2 bits/cell operation using Constant Signal Pulse Programming (CSPP).

To realize bi-layered ReRAM as a storage, another bit selection switch is necessary. Among many selection methods, the antiserial connection is one of the many solutions for bit selection [6]. Figure 5 shows antiserial connection and its switching result.

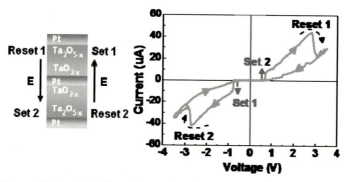

Figure 5: The structure of an antiserial architecture and its operation. It shows the required switching region in between -0.7 and +0.7 volt which inhibits unwanted switching.

In conclusion, the bi-layered switching element can provide the key element for the future of next generation non-volatile storage class memory, reconfigurable logic and neuromorphic circuit toward very high speed and extremely low energy information processing.

[1] C. H. Ho, et al, Symp. VLSI Tech., p. 228-229 (2007).
[2] M. J. Lee, et al, Nano Lett., 9, 1476-1481 (2009).
[3] J. H. Hur, et al, Phys. Rev. B, 82, 155321 (2010).
[4] Y. B. Kim, et al, Symp. VLSI Tech., p. 52-53 (2011).
[5] M. J. Lee et al., Nature Mater. 10, 625 (2011)
[6] Eike Linn, et al., Nature Mater. 9, 403(2010)

SELF-ORGANIZATION IN ADAPTIVE, RECURRENT, AUTONOMOUS MEMRISTIVE CROSSNETS

Konstantin K. Likharev[1], Dmitri N. Gavrilov[1], Thomas J. Walls[1]

[1]Stony Brook University, Stony Brook, NY 11794-3800, U.S.A.

CrossNets [1-3] are analog neuromorphic networks based on hybrid CMOS/nano-crossbar circuits with two-terminal memristive devices at each crosspoint. Such networks are believed to be the first plausible hardware basis for overcoming the mammalian cortical circuitry in density per unit area (at comparable connectivity $M \sim 10^4$), while far exceeding them in speed, at manageable power consumption [2, 3]. However, reaching a comparable cognitive functionality of CrossNets is a grand challenge, toward which only the first humble steps have been made so far [3].

Two crucial aspects of neuromorphic networks are their adaptation (plasticity) and recurrency (internal feedback). In this work we have addressed a simple problem which is a natural background for network interaction with incoming information: what happens to a recurrent, adaptive network if it is left alone (is autonomous)? The problem was considered within two most prominent neural network models:

(i) a firing-rate model with quasi-Hebbian weight adaptation which may be implemented in CrossNets using the stochastic multiplication technique [3, 4], and

(ii) a spiking model with spike-time-dependent plasticity (STDP) which may be also simply implemented in CrossNets, with very small area overhead [3, 5].

Previous work [1] had shown that in recurrent firing-rate CrossNets with fixed, random synaptic weights, an increase of somatic gain g beyond certain threshold value g_t leads to self-excitation of random localized spots – some of them static, some oscillatory (at $g \gg g_t$, chaotic). In this work we demonstrate that quasi-Hebbian plasticity leads to the transformation of the excitation spots into self-contained domains with static, periodic internal structure. (The structure is the same within one domain, but may be different between the domains.) Domain boundaries gradually expand into the passive CrossNet regions (Fig. 1a). Running into each other, the domains compete for space, the largest of them (and those with the simplest internal structure) suppressing others (Fig. 1b).

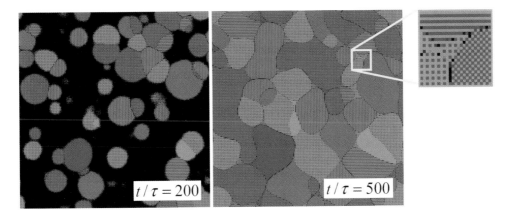

Figure 1: Domain growth and competition in firing-rate CrossNets of size $N = 256\times256$ with $M = 24$ at (a) an early stage, and (b) a late stage of time evolution.

Analysis of the equations describing firing-rate CrossNets has confirmed that at $g \gg g_t$, the network can sustain stable domains with *any* periodic internal structure. This feature hardly bodes well for possible information processing by such networks, so that the future development of cognitive functionality should probably focus on systems with $g < g_t$.

For spiking CrossNets the picture is rather different. Here the regulating role of somatic gain is played by the spiking threshold x_t for the action potential $x(t)$ of each soma. Our simulations have shown that spiking CrossNets with random, fixed weights w (either positive or negative, enabling both excitation and inhibition) may sustain self-excitation if x_t is below certain critical value x_c, with the average spiking rate $\langle r \rangle$ (per cell per relaxation time τ) close to 0.3 at $x_t \to x_c$, and gradually increasing with the further decrease of x_t. Both the minimum rate and the critical value x_c have been found to be in reasonable (~15%) agreement with an approximate analytical theory (asymptotically exact at $\langle r \rangle \ll 1$) based on the Smoluchowski equation. This fact, as well as essentially flat auto- and cross-correlation functions, confirm the hypothesis of essentially random spiking in the absence of adaptation (Fig. 2a).

Turning on STDP adaptation, with a global stabilization of the average firing rate, leads to a noticeable change of the synaptic weight distribution (Fig. 2b) and the auto- and cross-correlation functions. However, more specific correlation function(s) still need to be defined and evaluated to understand the exact nature of the observed self-organization. Our hope is to report more definite results at the meeting.

The work was supported in part by NSF and AFOSR. Useful discussions with D. Hammerstrom, R. O'Reilly, C. Segal, and D. Strukov are gratefully acknowledged.

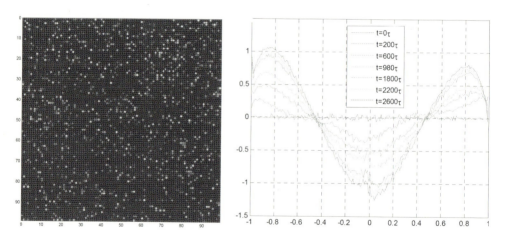

Figure 2: (a) A snapshot of spiking in a CrossNet of size $N = 100 \times 100$ with connectivity $M = 960$, and (b) change of the synaptic weight histogram with time at STDP adaptation (1 = 10% of max).

[1] Ö. Türel and K. Likharev, Int. J. Cir. Theory & Appl. **31**, 37 (2003).
[2] J. H. Lee et al., in: *Adv. in Neural Inf. Proc. Syst.* **18**, MIT Press, 2006, pp. 755-762.
[3] K. K. Likharev, Sci. Adv. Mater. **3**, 322 (2011).
[4] J. H. Lee and K. K. Likharev, in: *Proc. WCCI/IJCNN*, IEEE Press, 2006, pp. 5026-5034.
[5] K. K. Likharev, J. Nanoel. & Optoel. **3**, 203 (2008).

ELECTRIC FIELD CONTROL OF MAGNETIZATION

R. Ramesh

University of California, Berkeley

Complex oxides have fascinated the scientific community for years due to the rich physics and the often unique phenomena displayed by these materials. More specifically, complex perovskite oxides have been studied intensively as of the past few decades due to the vast collection of functional electronic phases observed in this class of materials (magnetism, ferroelectricity, superconductivity, highly correlated electron behavior). There is a set of complex oxide perovskites, known as multiferroics, which possess both a ferroelectric polarization and a (anti)ferromagnetic order. Furthermore, a multiferroic is deemed magnetoelectric when these two orders are coupled together. Due to the ability to use an electric field to switch the remnant states of the ferroelectric and magnetic orders, magnetoelectric multiferroics have been proposed as a potential solution to the energy losses faced by field of spintronics.

The energy dissipation in spintronics applications is the result of resistive losses that come during the writing of the magnetic state (i.e. a reversal of the magnetization direction). Writing of the magnetic state is done with either an externally applied magnetic field or injecting a large current density into a magnetic tunnel junction. Recent developments have shown that an applied voltage to a magnetic tunnel junction with an MgO barrier can reduce the energy needed to reverse a magnetization by modulating its anisotropy. However a biasing magnetic field is still needed to make this assisted switching possible and would be required to change sign in order to make this assisted switching reversible. In contrast, a reversal of magnetization requiring only the application of an electric field can lead to the low-power writing of magnetic devices. Using multiferroics, this idea becomes possible. To support both a magnetic order and a ferroelectric polarization, magnetoelectric multiferroics are typically antiferromagnetic insulators and thus cannot serve as a standalone spintronic material. We circumvent this issue by exchange coupling a ferromagnetic metal to a multiferroic in a heterostructure which then enables the low-power control of a magnetic layer using only an applied electric field, (a). Previous approaches based on this concept have seen limited success by only achieving rotations ($\leq 90°$) of the magnetization upon applying an electric field. To pave the way to new low-power devices, the more desirable electric-field driven magnetization reversal must be achieved and read out with a small current.

In this presentation we will discuss our work that investigates the magnetic coupling heterostructures of the room temperature multiferroic – $BiFeO_3$ and a ferromagnetic metal - $Co_{.90}Fe_{.10}$ in the light of $BiFeO_3$ domain structure, magnetometry and X-ray PEEM studies (b,c). How this coupling is used to demonstrate a reversal of the magnetization of the $Co_{.90}Fe_{.10}$ layer using only an in-plane applied electric field at room temperature in a magnetotransport-based device is also discussed. Using magnetotransport measurements (AMR), this electrically-driven 180° reorientation of the magnetization is non-volatile and reversible, (d). In an effort to further demonstrate the potential for low-power consumption spintronics, a vertically applied electric field is used to reverse the magnetization; requiring only 7 volts.

With these results, we push towards the realization an all electrical magnetoelectric memory device. By pinning the bottom layer of a GMR device with $BiFeO_3$ we work towards establishing the electric field control of resistance state of the GMR device. Contrary to the contemporary method of controlling the resistance state of a GMR device (magnetic field switching of the free layer), our approach envisions the purely electrical control of a spin valve by manipulating the magnetization that is coupled to a magnetoelectric multiferroic.

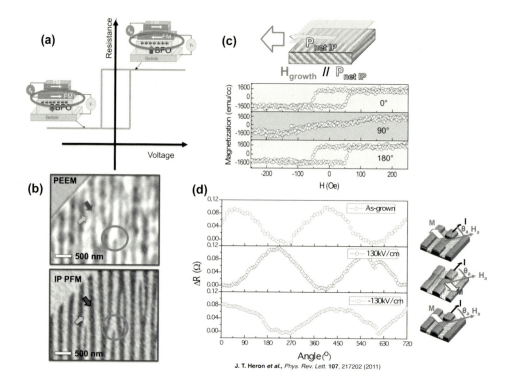

J. T. Heron *et al.*, *Phys. Rev. Lett.* **107**, 217202 (2011)

MAGNETIC SWITCHING OF FERROELECTRIC DOMAINS AT ROOM TEMPERATURE IN A NEW MULTIFERROIC

J. F. Scott[1], D. M. Evans[2], J. M. Gregg[2], Ashok Kumar[3,4], D. Sanchez[3], N. Ortega[3], and R. S. Katiyar[3]

[1]Dept. Physics, Cavendish Lab., Cambridge University, Cambridgem UK
[2]Dept. of Physics, Queen's University, Belfast, Northern Ireland, UK
3 Speclab, Dept. Physics, University of Puerto Rico, San Juan, P.R., USA

We have prepared sintered ceramic specimens of ball-milled ceramics of formula $Pb(Fe,Ta,Zr,Ti)O_3$ and measured their electrical and magnetic properties.[1] This perovskite oxide is prepared by mixing 30-40% $PbFe_{1/2}Ta_{1/2}O_3$ ["PFT"] with 70-60% $PbZr_{1/2}Ti_{1/2}O_3$ ["PZT"] and gives a single-phase crystal with very high-temperature ferroelectricity. Although pure PFT exhibits long-range magnetic ordering onlyup to 150K, it is known to have weak ferromagnetism due to Fe clustering up to ca. 400K. As a result, single-phase mixtures of PFT/PZT are multiferroic at room temperature. There is only one other known room-temperature multiferroic – $BiFeO_3$ – and our new material exhibits far lower electrical conductivity and dielectric loss (ca. 1%) for device applications. Several other materials such as CuO are multiferroic slightly below room temperature, sometimes requiring a small dc field.

We have carefully analyzed our specimens via EDX (Fig.1), TEM (Fig.2), Raman spectroscopy, and other techniques and confirm than any second phase must be in amounts << 1%. This is too small to explain the measured magnetization at 295K and cannot explain the switching results below. In our initial work we were unable to see either a linear magnetoelectric effect or magnetoelectric switching, due to the measurement area extending over many domains. However, in the present work (Fig.3) we demonstrate good magnetoelectric switching at room temperature: In particular the ferroelectric domains measured via PFM are switched using a very small bar magnet (rare earth, ca. 0.1 Tesla). The direction of H was normal to the plane of the domains.

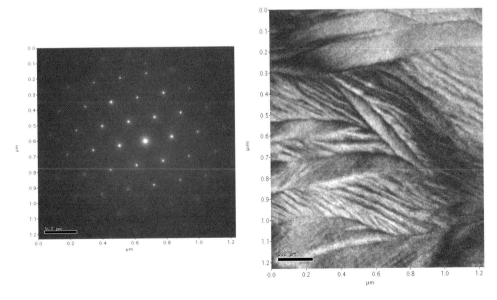

Fig. 1. X-ray diffraction pattern.   Fig.2. TEM pattern.

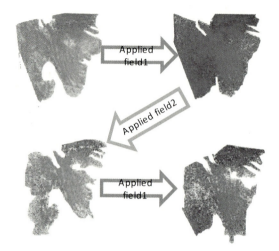

Fig.3: Magnetic switching:

Dark areas of PFM images. Apply field 1 and cause dark area to grow

Apply opposite field to cause dark area to shrink

Can cause 'switching' of dark area by using different magnetic field.

[1] D. Sanchez, N. Ortega, A. Kumar, R. S. Katiyar, J. F. Scott, AIP Adv. 1, 042169 (2011).

CONTROL OF CORRELATED ELECTRONS IN METAL-OXIDE SUPERLATTICES

Bernhard Keimer

Max Planck Institute for Solid State Research, Stuttgart, Germany

We will outline recent results of an experimental program aimed at controlling the phase behavior of correlated electrons through the synthesis and characterization of metal-oxide superlattices, with particular emphasis on copper and nickel oxides. Control parameters include the occupation of transition-metal d-orbitals [1] and the dimensionality of the electron system [2]. In particular, we will demonstrate control of the electron-phonon interaction in cuprate superlattices [3], and of the spin density wave polarization in nickelate superlattices [4]. These results also highlight the power of resonant x-ray scattering [1,4], spectral ellipsometry [2], and Raman scattering [3] as microscopic probes of the electron system in metal-oxide heterostructures and superlattices.

[1] E. Benckiser, M. W. Haverkort, S. Brück, E. Goering, S. Macke, A. Frañó, X. Yang, O. K. Andersen, G. Cristiani, H. U. Habermeier, A. V. Boris, I. Zegkinoglou, P. Wochner, H. J. Kim, V. Hinkov, and B. Keimer, Nature Mater. **10**, 189 (2011).
[2] A. V. Boris, Y. Matiks, E. Benckiser, A. Frano, P. Popovich, V. Hinkov, P. Wochner, M. Castro-Colin, E. Detemple, V. K. Malik, C. Bernhard, T. Prokscha, A. Suter, Z. Salman, E. Morenzoni, G. Cristiani, H.-U. Habermeier, and B. Keimer, Science **332**, 937 (2011).
[3] N. Driza, S. Blanco-Canosa, M. Bakr, S. Soltan, M. Khalid, L. Mustafa, K. Kawashima, G. Christiani, H.-U. Habermeier, G. Khaliullin, C. Ulrich, M. Le Tacon, and B. Keimer, submitted.
[4] A. Frano, E. Schierle, M. W. Haverkort, Y. Lu, M. Wu, S. Blanco-Canosa, Y. Matiks, A. V. Boris, P. Wochner, G. Cristiani, G. Logvenov, H.U. Habermeier, V. Hinkov, E. Benckiser, E. Weschke, and B. Keimer, in preparation.

METAL-INSULATOR TRANSITIONS OF CORRELATED ELECTRONS IN OXIDE HETEROSTRUCTURES

Masashi Kawasaki[1,2]

[1]Quantum Phase Electronics Center (QPEC) and Department of Applied Physics, University of Tokyo, Tokyo 113-8656, Japan
[2]Correlated Electron Research Group (CERG) and Cross-Correlated Materials Research Group (CMRG), RIKEN Advanced Science Institute, Wako 351-0198, Japan

There have been numbers of examples where metal-insulator transition is induced at the interfaces of correlated oxides by tiny stimuli such as electric field, photo-irradiation, and magnetic field [1]. These phenomena provide potential functionalities for the application of non-volatile memories, switches, sensors, and photovoltaics. In addition, these interfaces are excellent playground to explore new physics. We have studied numbers of heterostructures comprised of transition metal oxides and found some similarities to but also limitations for the rigid band model to describe the connection of two different electronic structures [2]. Examples of carrier accumulation or depletion at the interfaces by electric field, charge transfer, and photo irradiation are discussed to elucidate characteristics of each oxide. The system includes half-doped manganites [3], robust Mott insulators of manganites and cuprates [4], spin-peierls vanadium oxide [5] and correlated electrons in band insulators [6,7].

Figure 1 shows an example of the rigid band view of hetero-junctions. Photocurrent, optical absorption, and capacitance-voltage measurements enabled us to map out the band lineup [2]. From the bias dependence of photocurrent, one can readily extract an acceptor concentration of 4×10^{19} cm^{-3}, depletion layer width of 3 nm, and minority carrier diffusion length of 3 nm in LaMnO$_3$ [4].

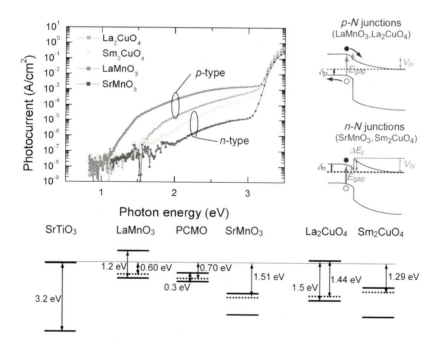

Figure 1: Photocurrent spectra (upper left) of hetero-junctions of Mot insulators and a doped band insulator (SrTiO$_3$). The band diagrams can be classified as p-N and n-N junctions (upper right). From the analysis, band lineup (bottom) is drawn.

If one can design a hetero-junction that can generate photocurrent from a narrow charge-gap (Δ) insulator such as half-doped manganites with charge-orbital ordered structure, one may be able to construct a multiple-carrier generation solar cell. The concept of the solar cell is given in Figure 2. Photo-induced insulator-metal transition could be regarded as a sort of multiplication process. Theoretically, this process is verified to occur [8].

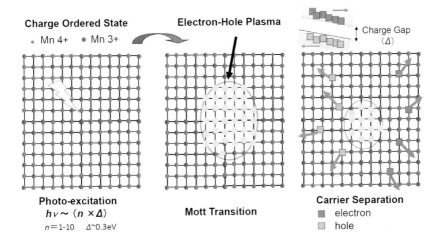

Figure 1: Schematic representation of the concept of correlated electron solar cell. There are many cases, in which multiplication of photo-generated carriers occurs in Mott insulators. If the carriers can be extracted, it may be possible to devise a high efficiency solar cell.

Vanadium oxide is another material that is promising for possible devices. The insulator metal transition can be easily tuned by doping [9], photo-irradiation [5] and electric field [10]. On the other hand it is elucidated to be very robust against quantum confinement [11]. The band lineup of VO_2 and other materials are also discussed in view of solar cell and transistor application.

[1] As a review see for example, H. Y. Hwang et al., Nature. Mater. **11**, 103 (2012).
[2] M. Nakamura et al., Phys. Rev. B **82** 201101 (2011)
[3] M. Nakamura et al., Adv. Mater **22**, 500 (2011)
[4] J. Fujioka et al., J. Appl. Phys. **111**, 016107 (2012)
[5] K. Shibuya et al., Phys. Rev. B **84**, 165108 (2011)
[6] Y. Kozuka et al., Phys. Rev. B **84**, 033304 (2011)
[7] D. Maryenko et al., Phys. Rev. Lett. in press. arXive 1203.3349
[8] W, Koshibae et al., Phys. Rev. Lett. **103**, 266402 (2009); Euro. Phys. Lett. **94**, 27003 (2011)
[9] K. Shibuya et al., Appl. Phys. Lett., **96**, 022102 (2010)
[10] M. Nakano et al., submitted
[11] K. Shibuya et al., Phys. Rev. B **82**, 205118 (2010)

THEORETICAL DESIGN OF TOPOLOGICAL PHENOMENA

Naoto Nagaosa[1,2]

[1]Department of Applied Physics, The University of Tokyo, Japan
[2]Cross-Correlated Materials Research Group (CMRG), and Correlated Electron Research Group (CERG), RIKEN-ASI, Japan

The topological classification of the electronic states in solids has recently revealed the rich possibilities of materials and phenomena, and attracts much attention both from the scientific and technological viewpoints. Especially the topological insulator (TI) and topological superconductor (TSC) are the new states of matter characterized by the topological invariants and reveal their novel properties at the edge/surface. Topological periodic table is now established to classify all the possible topologically nontrivial states including at the textures such as dislocations. Based on these recent developments, the next step is to design theoretically the materials and their functions using for example the artificial structures.

In this talk, I will present the principles for this design, and show some of the examples which include (i) the correlated TI and TSC at oxide super-lattice [1,2,3], (ii) noncentrosymmetric TI and their quantum critical phenomena in Bi-based compounds [4,5], and (iii) correlated Majorana fermion in TSC [6,7]. Especially, the role of dimensional reduction is stressed, which enhances the topological singularity and also the electron correlation. These works have been done in collaboration with Y. Tanaka, A. Yamakage, S. Okamoto, X. Di, Y. Ran, W. Zhu, S. Bahramy, S. Nakosai, D. Isobe, D. Asahi and R. Arita.

[1] Di. Xiao et al., Nature Communications 2, 596 (2011).
[2] S. Nakosai et al., Phys. Rev. Lett. In press (2012)
[3] H. Hwang et al, Nature Materials. 11, 103 (2012).
[4] K. Ishizaka et al., Nature Materials, 10, 521 (2011).
[5] M. S. Bahramy et al., Nature Communications 3, 679 (2012)
[6] Y. Asano et al., Phys. Rev. Lett. 105, 056402 (2010).
[7] A. Yamakage et al., Phys. Rev. Lett. 106, 246601 (2011)

MAGNETIC RECONSTRUCTIONS IN PEROVSKITE HETEROINTERFACES AND ULTRATHIN FILMS

Harold Y. Hwang[1]

[1]Departments of Applied Physics and Photon Science, Stanford University and SLAC National Accelerator Laboratory

Complex oxides are fascinating systems which host a vast array of unique phenomena, such as high temperature (and unconventional) superconductivity, 'colossal' magnetoresistance, all forms of magnetism and ferroelectricity, as well as (quantum) phase transitions and couplings between these states. In recent years, there has been a mini-revolution in the ability to grow thin film heterostructures of these materials with atomic precision. With this level of control, the boundary conditions at oxide surfaces and interfaces can be used to form new electronic phases [1]. Between two insulators, for example, metallic, superconducting, and magnetic states can be induced. Here we focus on two examples of magnetic reconstructions we observe in perovskite heterostructures (Figure 1); the inhomogeneous ferromagnetic patches which occur in LaAlO$_3$/SrTiO$_3$ [2-4], and the thickness evolution of the magnetic structure in ultrathin manganite films [5].

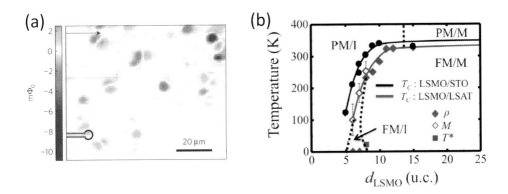

Figure 1: (a) Scanning SQUID magnetometry image of a 10 u.c. LaAlO$_3$/SrTiO$_3$ heterostructure [3]. (b) Thickness phase diagram of ultrathin La$_{0.7}$Sr$_{0.3}$MnO$_3$ thin films [5].

This work done in collaboration with C. Bell, J. A. Bert, M. H. Burkhardt, H. Durr, Y. Hikita, M Hosoda, M.A. Hossain, B. Kalisky, C.-C. Kao, B. Kim, B. G. Kim, M. Kim, B. B. Klopfer, D Kwon, J.-S. Lee, K. A. Moler, H. K. Sato, A. Scholl, J. Stohr, and T. Yajima.

[1] H. Y. Hwang, Y. Iwasa, M. Kawasaki, B. Keimer, N. Nagaosa, and Y. Tokura, Nature Mater., 11 (2012) 103.
[2] A. Brinkman, M. Huijben, M. van Zalk, J. Huijben, U. Zeitler, J. C. Maan, W. G. van der Wiel, G. Rijnders, D. H. A. Blank, and H. Hilgenkamp, Nature Mater., 6 (2007) 493.
[3] J. A. Bert, B. Kalisky, C. Bell, M. Kim, Y. Hikita, H. Y. Hwang, and K. A. Moler, Nature Phys., 7 (2011) 767.
[4] B. Kalisky, J. A. Bert, B. B. Klopfer, C. Bell, H. K. Sato, M. Hosoda, Y. Hikita, H. Y. Hwang, and K. A. Moler, arXiv:1201.1063.
[5] B. Kim, D. Kwon, T. Yajima, C. Bell, Y. Hikita, B. G. Kim, and H. Y. Hwang, Appl. Phys. Lett., 99 (2011) 092513.

PROGRESS IN THE ATOMIC SWITCH

<u>Masakazu Aono</u>[1], Tsuyoshi Hasegawa[1], Kazuya Terabe[1], Tohru Tsuruoka[1], Takeo Ohno[1], and Toshitsugu Sakamoto[2]

[1]WPI Center for Materials Nanoarchitectonics (MANA), National Institute for Materials Science (NIMS), Tsukuba, Japan; [2]Low-power Electronics Association and Project (LEAP), Tsukuba, Japan

The atomic switch is generally known as such nanoscale switching devices that make ON/OFF switching by the growth and shrinkage of a conduction path composed of metal atoms (in contrast with other nanoscale switching devices collectively called the resistive switch in which a conduction path is formed by anion [e.g. oxygen ion] vacancies, etc.). Actually, the atomic switch has many more interesting functionalities depending on its structure and constituent materials (see Fig. 1).

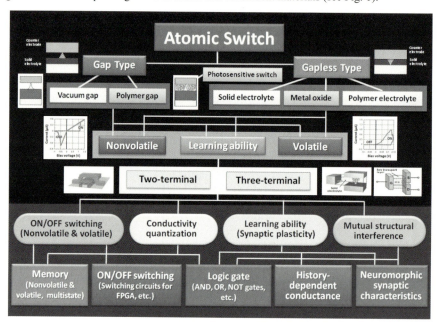

Figure 1: Various types of the atomic switch, which have different structures and constituent materials.

The atomic switch was first developed as a **nanoscale, two-terminal, nonvolatile** switch with a nanoscale vacuum gap between a **solid-electrolyte** (Ag_2S) electrode and a simple-metal counter electrode, i.e. a **gap-type** atomic switch [1, 2]; if necessary, a **volatile** atomic switch can be made [3]. It has been found later that the vacuum gap can be filled with soft organic molecules [4] and if the molecules are photoconductive, a photosensitive atomic switch can be made, where ON/OFF switching is controlled by photons [4]. The switching mechanism of the gap-type atomic switch has been studied in detail [5-7].

Soon after the development of the gap-type atomic switch, we developed a **gapless- type** atomic switch without a gap between a solid-electrolyte electrode (Cu_2S was used) and a simple-metal counter electrode [8-11]; this gapless-type atomic switch is advantageous for practical application. We have also found that the solid electrolyte in the gapless-type atomic switch can be a **polymer-based electrolyte** (e.g. poly-ethylene + $AgClO_4$) [12], suggestinh that a flexible two-dimensional atomic switch array can be fabricated. Moreover, it has been found that the electrolyte in the gapless-type atomic switch can be replaced by a **metal oxide** (e.g. Ta_2O_5) [13-17]; the metal oxide is not a solid electrolyte but works as an **ion transport layer**. The switching mechanism of this ion-transport-layer atomic switch has been studied in detail [18-21].

We have succeeded to develop **three-terminal atomic switches (transistors)** using a solid electrolyte (Cu_2S) [22, 23] or an ion-transport layer (Ta_2O_5) [24, 25]. Interest-ingly, an atomic transistor using Ta_2O_5 can be operated in **either volatile or non-volatile modes** by simply controlling applied voltage [24].

Interestingly, we have revealed that the two-terminal gap-type atomic switch exhibits **learning ability** [26, 27]; namely, the conductivity of the switch can have inter-mediate values between the OFF and ON conductivities, depending on the history of input signals. More interestingly, the atomic switch show interesting **characteristics similar to a synapse** in neural network [28-30]; such characteristics are also observed in a certain gapless-type atomic switch [31]. On the basis of these results, we have been developing **neuromorphic circuits** made of atomic switches [28, 31, 32].

The studies described above have been partially reviewed in Refs. [33-37].

[1] K. Terabe et al., *Riken Review* No. 37 (July, 2001) 7.
[2] K. Terabe et al., *Nature* 433 (2005) 47.
[3] T. Hasegawa et al., to be published.
[4] T. Hino et al., *Small* 6 (2010) 1745.
[5] A. Nayak et al., *J. Phys. Chem. Lett.* 1 (2010) 604.
[6] A. Nayak et al., *Appl. Phys. Lett.* 98 (2011) 233501.
[7] I. Valov et al., *Nature Mater.*, in press.
[8] T. Sakamoto et al., *Appl. Phys. Lett.* 82 (2003) 3032.
[9] S. Kaeriyama et al., *IEEE J. Solid-State Circuits* 40 (2005) 168.
[10] N. Banno et al., *IEICE Trans. Electron.* E89-C (2006) 1492.
[11] N. Banno et al., *IEEE Trans. Electron Devices* 55 (2008) 3283.
[12] S.-W. Wu et al., *Adv. Func. Mater.* 21 (2011) 93.
[13] T. Sakamoto et al., *Appl. Phys. Lett.* 91 (2007) 092110.
[14] N. Banno et al., *Appl. Phys. Lett.* 97 (2010) 113507.
[15] M. Tada et al., *IEEE Trans. Electron Devices* 57 (2010) 1987.
[16] Y. Tsuji et al., *Appl. Phys. Lett.* 96 (2010) 023504.
[17] N. Banno et al., *Jpn. J. Appl. Phys.* 50 (2011) 074201.
[18] T. Tsuruoka et al., *Nanotechnology* 21 (2010) 425205.
[19] T. Tsuruoka et al., *Nanotechnology* 22 (2011) 379502.
[20] T. Tsuruoka et al., *Adv. Func. Mater.* 22 (2012) 70.
[21] A. Nayak et al., *Nanotechnology* 22 (2011) 235201.
[22] T. Sakamoto et al., *IEDM Technical Digest* (2005) 475.
[23] T. Sakamoto et al., *Appl. Phys. Lett.* 96 (2010) 252104.
[24] T. Hasegawa et al., *Appl. Phys. Express* 4 (2011) 015204.
[25] H. Kawaura et al., *Electronics and Communications in Japan* 94 (2011) 55.
[26] T. Hasegawa et al., *Adv. Mater.* 22 (2010) 1831.
[27] T. Hasegawa et al., *Appl. Phys. A* 102 (2011) 811.
[28] T. Ohno et al., *Nature Mater.* 10 (2011) 591.
[29] T. Ohno et al., *Appl. Phys. Lett.* 99 (2011) 203108.
[30] A. Nayak et al., submitted. [31] R. Yang et al., submitted.
[32] A. Stieg et al., *Adv. Mater.* 24 (2012) 286.
[33] R. Waser, M. Aono, *Nature Mater.*, 6 (2007) 833.
[34] T. Hasegawa et al., *MRS Bulletin* 34 (2009) 929.
[35] M. Aono, T. Hasegawa, *Proc. IEEE* 12 (2010) 2228.
[36] T. Hino et al., *Sci. Technol. Adv. Mater.* 12 (2011) 013003.
[37] T. Hasegawa et al., *Adv. Mater.* 24 (2012) 252.

Nanosessions

2D electron systems - Atomic configurations (2DA)	71
2D electron systems - Correlation effects and transport (2DC)	81
2D electron systems - Electronic structure and field effects (2DE)	89
Calorics (CAL)	99
Topological effects (TOP)	109
Mott insulators and transitions (MIT)	115
Advanced spectroscopy and scattering (SAS)	123
High-resolution transmission electron microscopy (TEM)	133
New technologies for scanning probes (NTS)	143
Phase change materials (PCA)	155
Phase change memories (PCM)	163
Scanning probe microscopy on oxides (SPO)	177
Logic devices and circuit design (LDC)	185
Neuromorphic concepts (NMC)	197
Electrochemical metallization memories (ECM)	207
Valence Change Memories - redox mechanism and modelling (VCR)	219
Valence Change Memories - a look inside (VCI)	233
Variants of resistive switching (VRS)	247
Magnetic interfaces and surfaces (MAG)	259
Ionics - lattice disorder and grain boundaries (IOL)	269
Ionics - redox kinetics, ion transport, and interfaces (IOR)	281
Spin dynamics (SDY)	291
Spin injection and transport (SIT)	301
Spin tunneling systems (STS)	311
Multiferroic thin films and heterostructures (MFH)	323
Multiferroics - ordering phenomena (MFO)	335
Multiferroics - high transition temperatures (MFT)	347
Qubit systems (QUB)	357
Superconductivity (SUP)	367
Interplay between strain and electronic structure in metal oxides (ISE)	377
Photovoltaics, photocatalysis, and optical effects (PPO)	389
Ferroelectric interfaces (FIN)	399
Ferroelectrics - new and unusual material systems (FER)	409
Atomic layer deposition (ALD)	419
Nanotechnological fabrication strategies (NAN)	429
Low-dimensional transport and ballistic effects (LTB)	441
Molecular and polymer electronics (MOL)	453
Carbon-based molecular systems (CMS)	461

HIGHLY CONFINED SPIN-POLARIZED TWO-DIMENSIONAL ELECTRON GAS IN SrTiO$_3$/SrRuO$_3$ SUPERLATTICES

Javier Junquera[1], **Pablo García-Fernández**[1], **Marcos Verissimo-Alves**[1], **Daniel I. Bilc**[2], **Philippe Ghosez**[2]

[1]Departamento de Ciencias de la Tierra y Física de la Materia Condensada, Universidad de Cantabria, Santander, Spain;
[2]Physique Théorique de Matériaux, Université de Liège, Sart Tilman, Liège, Belgium

We report first-principles characterization of the structural and electronic properties of (SrTiO$_3$)$_5$/(SrRuO$_3$)$_1$ superlattices [1]. The most important results that can be drawn from our simulations are: (i) we show that the system exhibits a two-dimensional electron gas (2DEG). The mechanism to create the 2DEG is totally new and based on the difference in electronegativity between Ru and Ti. Therefore, it is not related with the standard ones (polar catastrophe, atomic interdiffusion, or presence of oxygen vacancies) invoked to explain its genesis in the paradigmatic SrTiO$_3$/LaAlO$_3$ interface. (ii) The 2DEG is localized on the $4d$ orbitals of Ru in the SrRuO$_3$ layer and not on the Ti atoms of SrTiO$_3$ as in most of the previously reported examples of 2DEG at oxide interfaces. (iii) The 2DEG is extremely confined, and has the ultimate thickness of one atomic layer. (iv) The 2DEG is intrinsically magnetic. The magnetic properties do not depend on extrinsic vacancies or defects. Every interface in the superlattice behaves as a minority-spin half-metal ferromagnet, with a magnetic moment of $\mu = 2.0\ \mu_B$/SrRuO$_3$. (v) The shape of the electronic density of states, half-metallicity, and magnetism are explained in terms of a simplified tight-binding model, considering only the t_{2g} orbitals plus the bidimensionality of the system and the strong electron correlations. (vi) At the pure methodological level, our results present the novelty of the use, for the first time, of the B1-WC hybrid functional that mixes the generalized gradient approximation of Wu and Cohen with 16% of exact exchange within the B1 scheme.

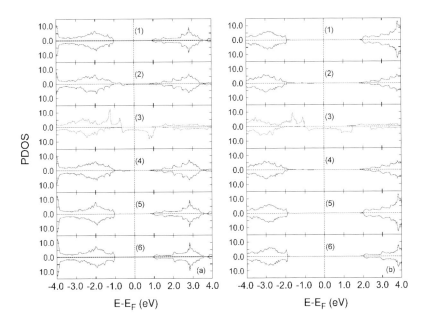

Figure 1: Layer by layer projected density of states (PDOS) on the atoms at the SrBO$_3$ layers [B = Ti (in blue) or Ru (in red)] within the LSDA+U (left panel), and the B1-WC hybrid functional (right panel) for the corresponding relaxed SrTiO$_3$/SrRuO$_3$ superlattice.

As already observed in other confined 2DEG [2] these interfaces exhibit an extremely large Seebeck coefficient, of the order of 1700 μV/K. However, this enhancement of the Seebeck coefficient does not reflect on the power factor, defined as the numerator of the thermoelectric figure of merit. The reason is that, wherever the Seebeck coefficient is large, the electrical conductivity is very small or even zero.

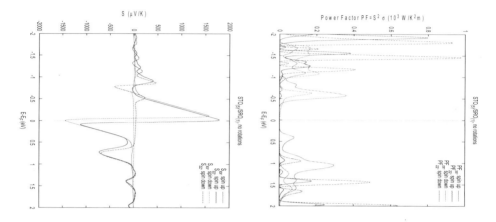

Figure 2: Seebeck coefficient (left panel) and power factor (right panel) for the majority (up) and minority (down) spin channels in $(SrTiO_3)_5/(SrRuO_3)_1$ superlattices.

[1] M. Verissimo-Alves, P. García-Fernández, D. I. Bilc, Ph. Ghosez, and J. Junquera, Phys. Rev. Lett. **108**, 107003 (2012).
[2] H. Ohta et al., Nat. Mater. **6**, 129 (2007).

FIRST-PRINCIPLES STUDY OF INTERMIXING AND POLARIZATION AT THE DyScO$_3$/SrTiO$_3$ INTERFACE

Kourosh Rahmanizadeh[1], Gustav Bihlmayer[1], Martina Luysberg[2], Stefan Blügel[1]

[1]Peter *Grünberg* Institut (PGI-1) & Institute for Advanced Simulation (IAS-1), Forschungszentrum Jülich and JARA, 52425 Jülich, Germany;
[2]Peter Grünberg Institut (PGI-5) & Ernst Ruska Centre, Forschungszentrum Jülich, 52425 Jülich, Germany

Exploring oxide interfaces is an attractive challenge, due to the emerging novel properties that do not exist in the corresponding parent bulk compounds. E.g. joining two simple band insulators, the polar LaAlO$_3$ and the non-polar SrTiO$_3$, can induce conductivity at the interface. We carried out density functional theory (DFT) calculations based on the full potential linearized augmented planewave (FLAPW) method as implemented in the FLEUR code [1] for studying the polar to non-polar interface of DyScO$_3$ and SrTiO$_3$. Due to the polar discontinuity, arising from nominally charged DyO or ScO$_2$ layers, sharp interfaces induce a strong ferroelectric-like polarization in the SrTiO$_3$, while in off-stoichiometric interfaces this discontinuity can be avoided and no such polarization can be found. In both scenarios the interface remains insulating with only a small reduction of the bandgap. Our calculations show that chemically mixed interfaces are energetically more favorable than sharp ones [2]. Experimental evidence for intermixing and an insulating behaviour was reported for the DyScO$_3$ / SrTiO$_3$ interface [3].

The work was conducted under the auspices of the IFOX consortia under grant agreement NMP3-LA-2010-246102.

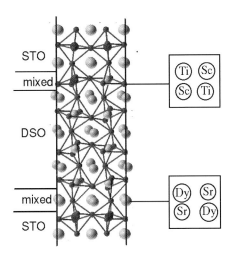

Figure 1: DyScO$_3$ (110) / SrTiO$_3$ (001) mixed interface. There are two mixed layers at the interfaces: Ti/Sc on one side and on the Dy/Sr in other side. The figure shows the structure with oxygen octahedra, with Sc and Ti atoms in their centre and spheres outside the oxygen octahedra representing the Dy and Sr ions.
Our DFT calculations explore also different configurations of the Dy and Sr atoms within the mixed interface plane. The calculated ground state configuration is confirmed by new experimental observations.

[1] www.flapw.de
[2] K.Rahmanizadeh et al., Phys. Rev. B **85**, 075314 (2012)
[3] M. Luysberg et al., Acta Materialia **57**, 3192 (2009)

ATOMIC-SCALE SPECTROSCOPY OF AN OXIDE INTERFACE BETWEEN A MOTT INSULATOR AND A BAND INSULATOR

M.-W. CHU[1], C. P. CHANG[1,2], S.-L. Cheng[1,2], J. G. Lin[1], C. H. CHEN[1]

[1]Center for Condensed Matter Sciences, National Taiwan University, Taipei 106, Taiwan;
[2]Department of Materials Science and Engineering, National Taiwan University, Taipei 106, Taiwan

Using pulsed laser deposition, we have grown the oxide heterostructure of Mott-insulating $(Nd_{0.35}Sr_{0.65})MnO_3$ on band-insulating $SrTiO_3$ with the primitive interface configuration of $(MnO_2)^{0.35-}$-$(Nd_{0.35}Sr_{0.65}O)^{0.35+}$-$(TiO_2)^0$-$(SrO)^0$, as shown in Figure 1. The primitive polar discontinuity across the interface leads one to expect a conductive interface in the classical electronic-reconstruction context [1], while an electrically insulating character was characterized. Using atomically-resolved electron energy-loss spectroscopy (EELS) in an aberration-corrected scanning transmission electron microscope (STEM), we further investigated this insulating interface. Through atomic-column-by-atomic-column STEM-EELS spectroscopy and atomic-plane-by-atomic-plane chemical quantifications, extensive interdiffusion across the primitive interface was unambiguously observed and found to arise from an electronic origin to diminish the polar discontinuity across the interface. The charge transfer expected in electronic reconstruction to render a metallic interface [1] is, therefore, quenched. With the STEM-EELS quantifications, the lattice-by-lattice stoichiometry was also evaluated and the corresponding local-lattice off-stoichiometry induced by the interdiffusion is, in effect, microscopically associated with the strongly-correlated phase diagram of parent $(Nd_{1-x}Sr_x)MnO_3$. The profound correlations physics addressing the insulating interface were elaborately discussed in addition to the quenched electronic-reconstruction aspect.

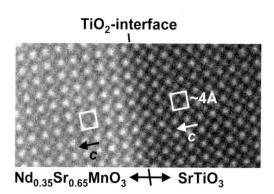

Figure 1: STEM high-angle annular dark-field image of the interface of the [001]-oriented $(Nd_{0.35}Sr_{0.65})MnO_3$ film (~20-nm-thick) grown on the (001)-TiO_2 plane of $SrTiO_3$. Left and right boxes, the lattices of the respective materials with a cell dimension of ~4 Å along the projections.

[1] H. Y. Hwang et al., Nature Mater. **11**, 103 (2012).

INTERFACE ATOMIC STRUCTURE IN LaSrAlO$_4$/LaNiO$_3$/LaAlO$_3$ HETEROSTRUCTURES

M. K. Kinyanjui [1], N. Gauquelin [2], G. Botton [2], E. Benckiser [3], B. Keimer [3], U. Kaiser [1]

[1] Central Facility of Electron Microscopy, University of Ulm, Ulm, Germany; [2] Brockhouse Institute for Materials Research and Canadian Centre for Electron Microscopy, McMaster University, Hamilton, Ontario Canada; [3] Max Planck Institute for Solid State Research, Stuttgart, Germany

A number of novel phenomena including 2-D electron gases, charge ordering, electronic and atomic reconstructions and metal insulator transitions [1, 2] have been observed at complex oxide interfaces. These interfacial phenomena have been mostly attributed to electronic interactions arising at the interface and are directly related to the atomic and electronic structure of the interface. From an experimental point of view it is crucial to understand the relationship between the observed phenomena with the chemical, electronic, and atomic structure at the interface. We have used transmission electron microscopy (TEM) techniques including atomically resolved high-angle annular dark-field scanning TEM (HAADF-STEM), electron energy loss spectroscopy (EELS), and high-resolution electron microscopy (HRTEM) to study the interface atomic structure in LaNiO$_3$ (LNO) – LaAlO$_3$ (LAO) superlattice grown on LaSrAlO$_4$ (LSAO) substrate [3]. Since, the lattice constant of the (LSAO) substrate is less than the pseudo-cubic lattice constants of LNO and LAO, the superlattice is expected to be under compressive epitaxial strain. Model calculations on LNO layers have shown epitaxial strain and confinement of the layers in a superlattice geometry can be used to manipulate orbital occupancy and that the electronic structure matches that of the copper oxide high-temperature superconductors. This system is therefore considered to be a potential candidate for 'orbitally engineered' superconductivity [3,4]

Figure1 (a) shows a STEM-HAADF image of the LSAO-LNO-LAO layers in the [1 0 0] zone axis orientation. The image intensity in the HAADF-STEM images is proportional to $Z^{1.7}$ (Z = atomic number) which allows one to obtain information about the chemical composition at the interface. The La/Sr atoms (Z = 57 & 38) show the brightest contrast followed by Ni atoms (Z =28) and Al atoms (Z= 13). The black and white rectangles in show the LSAO-LNO and LNO-LAO interfaces respectively.

Figure 1(b) and (c) show 2-D Gaussian fits of the atomic column positions at the LSAO-LNO and LNO-LAO interface respectively.

The distribution of the lattice parameters at the LSAO-LNO interface is shown in fig. 1(d). The lattice mismatch between the substrate and the layers results in a tetragonal strain of the lattice parameters in LNO layers (lattice parameter c > b). Based on atomically resolved STEM-HAADF imaging and chemical mapping of the LSAO-LNO and LNO-LAO interfaces, we observe the presence of a LAO perovskite layer between the LSAO substrate and the LNO layers, and atomically rough interfaces at both interfaces.

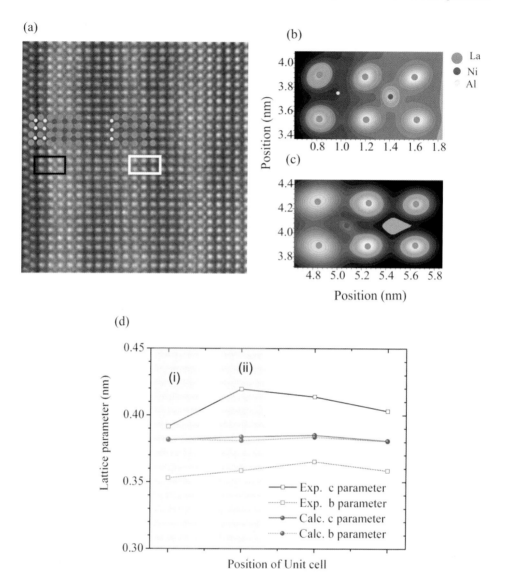

Figure 1: (a) atomically resolved STEM-HAADF image of the LaSrAlO$_4$-LaNiO$_3$-LaAlO$_3$ layers. The black rectangle shows the LaSrAlO4-LaNiO$_3$ interface. The white rectangle shows the LaNiO$_3$-LaAlO$_3$ interface (b) 2-D Gaussian fitting of the atomic column positions at the LaSrAlO$_4$ -LaNiO$_3$ interface (black rectangle in fig. 1(a)). (c) 2-D Gaussian fitting of the atomic column positions at the LaNiO$_3$- LaAlO$_3$ interface (white rectangle in fig. 1(a)). (d) Lattice parameter distribution at the LaSrAlO$_4$-LaNiO$_3$ interface. The regions marked (i) and (ii) show the lattice parameters of the perovksite block and LNO layer at the interface region respectively.

[1] J. Mannhart et. al., Science 327, 1607 (2010).
[2] P. Zubko et. al., Annu.Rev. Cond-Mat-Phys. 2, 141 (2011).
[3] J. Chaloupka et. al., Phys. Rev. Lett. 100, 016404 (2008).
[4] E. Benckiser et. al., Nature Materials 10, 189 (2011).

TAILORING THE ELECTRONIC PROPERTIES OF THE LAO/STO INTERFACE BY CONTROLLED CATION-STOICHIOMETRY VARIATION IN STO THIN FILMS

Felix Gunkel[1], Peter Brinks[2], Sebastian Wicklein[1], Susanne Hoffmann-Eifert[1], Regina Dittmann[1], Mark Huijben[2], Josée E. Kleibeuker[2], Gertjan Koster[2], Guus Rijnders[2], and Rainer Waser[1]

[1]Peter Gruenberg Institut and JARA-FIT, Forschungszentrum Juelich, 52425 Juelich, Germany; [2]MESA+ Institute for Nanotechnology, University of Twente, 7500 AE Enschede, The Netherlands

The role of defects is a focus of the ongoing discussion about the electronic properties of the conducting interface between the two insulators LaAlO$_3$ (LAO) and SrTiO$_3$ (STO). Besides the model of charge transfer due to the polar nature of LAO, it is generally accepted that defects in vicinity of the LAO/STO interface can have a large impact on the electronic properties. At the extreme, the interdiffusion of La-ions on Sr-sites (La$_{Sr}^{\cdot}$), and the creation of oxygen vacancies (V$_O^{\cdot\cdot}$) within the STO lattice, have even been considered as sole origin of the interface conduction.

In this study, the LAO/STO-interface will be discussed from a defect chemical point of view. It will be shown that cation vacancies in the STO lattice, i.e. Sr-vacancies (V$_{Sr}^{''}$) and Ti-vacancies (V$_{Ti}^{''''}$), have to be considered in order to draw a complete picture of the LAO/STO-interface (Fig. 1). In addition, we will present how a controlled variation of the cation-stoichiometry of STO thin films can be utilized to tailor the electronic properties of the LAO/STO-interface.

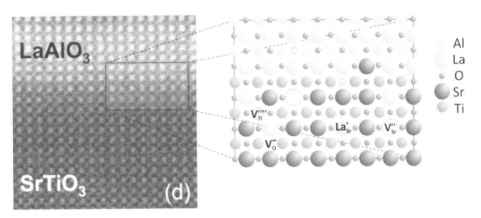

Figure 1: HAADF-image of the LAO/STO-interface (from Ref. [1]) and a schematic illustration of possible defects in the STO lattice close to the interface: Oxygen vacancies (V$_O^{\cdot\cdot}$); interdiffused La-dopants (La$_{Sr}^{\cdot}$); Sr-vacancies (V$_{Sr}^{''}$); Ti-vacancies (V$_{Ti}^{''''}$)

The conducting LAO/STO-interfaces are fabricated by pulsed laser deposition (PLD) of LAO on TiO$_2$-terminated STO single crystal substrates, and by the growth of LAO and STO on LSAT single crystal substrates.

The defect structure of the LAO/STO-interface is investigated by means of high temperature conductance measurements in equilibrium with the surrounding atmosphere (HTEC). In the investigated temperature range (950K-1100K), the decisive defect equilibria at the LAO/STO-interface are activated, and hence, strive for a well-defined thermodynamic equilibrium state which is related to the ambient oxygen pressure (pO$_2$). The resulting conductance characteristics contain information about the chemical reactions, which take place at the LAO/STO-interface, and the corresponding defect concentrations. Using this method, it can be excluded that mobile oxygen vacancies are the sole origin of the conducting interface [1]. In fact, the temperature- and oxygen partial pressure independent plateau region, which is found for reducing atmospheres, indicates the presence of immobile interfacial donor states (D˙). In addition, HTEC measurements on LAO/STO systems grown on LSAT substrates reveal a decrease of the interfacial electron density proportional to pO$_2^{-1/4}$ for oxidizing conditions (Fig. 2). This indicates the formation of Sr-vacancies, which act as acceptor-type electron traps in the vicinity of the LAO/STO-interface [2].

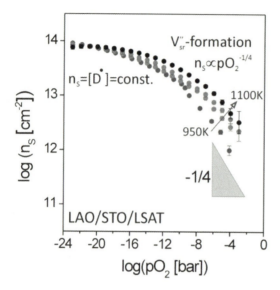

Figure 2: pO$_2$-dependence of the sheet carrier density, n$_S$, of the LAO/STO interface on LSAT obtained from HTEC measurements (from Ref. [2]); A pO$_2$- and temperature independent region is found for reducing conditions indicating the presence of interfacial donor states ([D˙]). For oxidizing atmosphere, n$_S$ decreases proportional to pO$_2^{-1/4}$ which can be attributed to the formation of acceptor-like Sr-vacancies.

For a closer investigation, the controlled insertion of acceptor-type cationic defects in STO is exploited as a potential tool to manipulate the electronic properties of the LAO/STO-interface. For PLD-grown STO thin films, the cation stoichiometry can be artificially modulated by the variation of the laser energy during the growth process [3]. The non-stoichiometry of STO thin films manifests itself as a finite expansion of the unit cell volume, which can be probed by X-ray diffraction (XRD) measurements on homoepitaxial films. This expansion of the STO c-axis lattice parameter, Δc, can directly be related to the incorporation of cation vacancies. As a result, Δc vanishes for the stoichiometric film and increases with increasing non-stoichiometry for Sr-rich and Ti-rich STO films.

Under the variation of the laser energy during the growth of STO, the sheet resistance, R$_S$, of the LAO/STO/LSAT-heterostructures shows a comparable behavior. For stoichiometric STO, R$_S$ exhibits a minimum, while the sheet resistance is increased for Ti-rich as well as for Sr-rich STO layers. This result indicates that the interface conduction can be affected by V$_{Sr}''$ (for Ti-rich films) as well as by V$_{Ti}''''$ (for Sr-rich films), both of which act as acceptor-type defects in STO. The variation of the STO cation-stoichiometry can hence be used to adjust the conductance of the LAO/STO-interface.

[1] F.Gunkel et al., Appl. Phys. Lett. 97, 012103 (2010).
[2] F.Gunkel et al., Appl. Phys. Lett. 100, 052103 (2012).
[3] D. J. Keeble et al., Phys. Rev. Lett. 105, 226102 (2010).

ELECTROSTATIC DOPING OF A MOTT INSULATOR IN AN OXIDE HETEROSTRUCTURE: THE CASE OF LaVO$_3$/SrTiO$_3$

F. Pfaff[1], A. Müller[1], H. Boschker[2], G. Berner[1], G. Koster[2], M. Gorgoi[3], W. Drube[4], G. Rijnders[2], M. Kamp[5], D.H.A. Blank[2], M. Sing[1], R. Claessen[1]

[1]Experimentelle Physik 4, Universität Würzburg, Germany; [2]Inorganic Materials Science Group, University of Twente, 7500 AE Enschede, The Netherlands; [3]Helmholtz Zentrum Berlin (BESSY II), Berlin, Germany; [4]HASYLAB, DESY, Hamburg, Germany; [5]Technische Physik, Universität Würzburg, Germany

The discovery of a two-dimensional electron gas at the interface of the two band insulators LaAlO$_3$ (LAO) and SrTiO$_3$ (STO) above a critical overlayer thickness of three monolayers (ML) of LAO [1] has triggered intense investigations of oxide heterostructures. The idea of an electronic reconstruction leading to the conductivity in LAO/STO suggests that in general an interface electron gas can be generated in a hybrid system by growing a polar overlayer on a non-polar substrate. Indeed, a similar thickness induced phase transition from insulating to conducting can be observed for the Mott insulator LaVO$_3$ (LVO) grown on STO [2]. In contrast to LAO/STO, in LVO/STO both B-cations of the perovskite structure (Ti and V) are multivalent and therefore can host additional electrons taking part in the electronic reconstruction.

We have grown LVO/STO heterostructures by pulsed laser deposition and established their high structural quality by x-ray diffraction and reflectivity as well as transmission electron and atomic force microscopy. Transport measurements confirm the thickness-dependent insulator-metal transition of the interface. For a more detailed investigation of the chemical and electronic structure across the interface we performed non-destructive depth profiling by hard x-ray photoelectron spectroscopy [3]. In contrast to LAO/STO the interface charge is found to reside at the V – not at the Ti – sites, evidencing that the Mott insulator LVO is electrostatically doped at the interface. This opens up a clean way to study band-filling controlled Mott transitions [4] in the absence of disorder effects often induced by chemical dopants.

[1] A. Ohtomo and H.Y. Hwang, Nature **427**, 423 (2004)
[2] Y. Hotta et al., Phys. Rev. Lett. **99**, 236805 (2007)
[3] M. Sing et al., Phys. Rev. Lett. **102**, 176805 (2009)
[4] M. Imada et al., Rev. Mod. Phys. **70**, 1039 (1998)

THEORETICAL STUDY OF ORBITAL-, SPIN- AND CHARGE-RECONSTRUCTION IN LVO/STO HETEROSTRUCTURES

Giorgio Sangiovanni[1], Zhicheng Zhong[2], Elias Assmann[2], Peter Blaha[2], Karsten Held[2] and Satoshi Okamoto[3]

[1]University of Würzburg, Würzburg, Germany;
[2]Vienna University of Technology, Vienna, Austria;
[3]ORNL, Oak Ridge (TN), USA

We study n- and p-type interfaces between $SrTiO_3$ (STO) and the Mott insulator $LaVO_3$ (LVO) by means of Density Functional Theory (DFT) calculations. One of the reasons why we look at this heterostructure is to verify the fascinating possibility of using it as absorbing material for solar cells. This may open new horizons in the fast-growing research on renewable energy.

LVO/STO is characterized by two d electrons on vanadium and therefore correlation-driven physical phenomena like antiferromagnetism or sizeable quasiparticle renormalization are expected. To understand this physics we extract maximally localized Wannier functions for the d electrons on vanadium [1]. An example of the calculated t_{2g} orbitals is shown in Figure 1. With the information coming from the Wannier projection we get crucial information on the hopping processes and on the level positions at the interface.

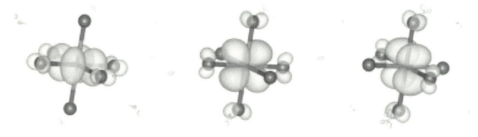

Figure 1: t_{2g} Wannier functions around a vanadium atom calculated for bulk $LaVO_3$. The red spheres represent oxygen atoms.

At the University of Würzburg LVO/STO heterostructures are now grown by means of Pulsed Laser Deposition. Experiments and theory are therefore challenged to study LVO/STO in detail in order to disclose the nature of the electron gas at the interface. In particular we will address the interesting question whether the mobile carriers are located in STO, like in the case of $LaAlO_3/SrTiO_3$ or instead in the LVO part.

[1] J. Kuneš, et al., Comp. Phys. Comm. **181**, 1888 (2010).

TWO-DIMENSIONAL ELECTRON GAS WITH ORBITAL SYMMETRY RECONSTRUCTION AND STRONG EFFECTIVE MASS LOWERING AT THE SURFACE OF KTaO$_3$

A. F. Santander-Syro[1,2], C. Bareille[1], F. Fortuna[1], O. Copie[3] F. Bertran[4], A. Taleb-Ibrahimi[4], P. Le Fèvre[4], G. Herranz[5], M. Bibes[6], A. Barthélémy[6], P. Lecoeur[7], J. Guevara[8,9], M. Gabay[10] and **M. J. Rozenberg**[10]

[1]CSNSM, Université Paris-Sud and CNRS/IN2P3 Bâtiments 104 et 108, 91405 Orsay cedex, France;
[2]Laboratoire Physique et Etude des Matériaux, UMR-8213 ESPCI/UPMC/CNRS 10 rue Vauquelin, 75231 Paris cedex 5, France;
[3]Universität Würzburg, Experimentelle Physik VII, 97074 Würzburg, Germany;
[4]Synchrotron SOLEIL, L'Orme des Merisiers, Saint-Aubin-BP48, 91192 Gif-sur-Yvette, France;
[5]Institut de Ciència de Materials de Barcelona, CSIC, Campus de la UAB, 08193 Bellaterra, Catalunya, Spain;
[6]Unité Mixte de Physique CNRS/Thales, Campus de l'Ecole Polytechnique, 1 Av. A. Fresnel, 91767 Palaiseau, France and Université Paris-Sud, 91405 Orsay, France;
[7]Institut d'Electronique Fondamentale, Université Paris-Sud, Bâtiment 220, 91405 Orsay, France;
[8]Gerencia de Investigación y Aplicaciones, CNEA, 1650 San Martín, Argentina;
[9]Universidad Nacional de San Martín, CNEA, 1650 San Martín, Argentina;
[10]Laboratoire de Physique des Solides, Université Paris-Sud, Bâtiment 510, 91405 Orsay, France.

The discovery of a high mobility two-dimensional electron gas (2DEG) at the interface of two non-magnetic wide-bandgap oxide insulators, SrTiO3 and LaAlO3 [1], triggered a burst of activity that led to more stunning findings, such as superconductivity [2] and large, spin-orbit mediated, magnetoresistance [3–5]. Despite the rapid progress in this area of research, a main issue that remains unanswered is the physical origin of these interfacial 2DEGs [6–9]. This was underscored by the surprising discovery of a 2DEG that is formed at the surface of vacuum-cleaved single crystals of SrTiO3 (STO) [10, 11], which opened a new avenue for the understanding and fabrication of 2DEGs in transition-metal oxides. It was argued that this novel 2DEG arises from surface oxygen vacancies formed when fracturing the crystal in vacuum, and it was put in evidence that different forms of electron confinement at the surface of STO may lead to essentially the same 2DEG [15]. The perspective of creating 2DEGs at the surface of multifunctional oxides which may inherit the properties of their parent compounds is exciting. In the present work, we realize a 2DEG at the vacuum-cleaved surface of KTaO3 (KTO).

Our main results demonstrate that: *(i)* the emergence of a 2DEG at an oxide surface, without any external confining field, is not specific to STO, thus likely quite general to undoped insulating perovskites [10] and *(ii)* that the spin-orbit coupling remains active in the 2DEG, completely *reconstructing* the orbital symmetries of the 2DEG's subbands with respect to the bulk conduction bands.

The theroretical results of Fig.1 compare the orbital content of the schematic bandstructure of KTO in the case of bulk with spin-orbit coupling with respect of the situation when the system becomes confined to 2D. The results illustrate how the effect of confinement leads to orbital mixing reconstruction of the bands, and how the band repulsion leads to a dramatic lowering of the 2DEG effective masses.

The experimental results shown in Fig.2 demonstrate the orbital mixing reconstruction expected from the calculations of Fig. 1: The lower band is extremely sensitive to the light polarization, showing is strong d_xy character, while the upper one is rather immune, showing its strong mixing of d_xz and d_yz characters. The experimental effective masses, lower than the bulk masses, are in excellent agreement with the theoretical masses of the confined system (Figs. 1b-c).

A key insight from our analysis is that the spin-orbit coupling energy of KTO is of the same order than the small Fermi energy of the 2DEG realized at the surface. This, and the subbands emerging from confinement leads to a truly new physical system with respect to that of the ground state of the bulk.

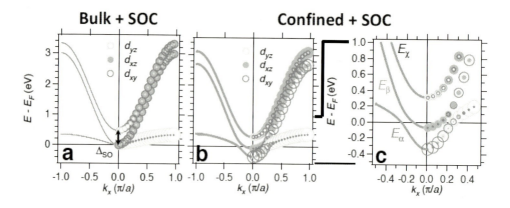

Fig.1 (a) Tigh-binding model calculations for bulk KTaO3 (red lines, left). The nearest neighbor hopping parameters were $t = 1.5$ eV and $t' = 0.15$ eV for the light and heavy bands, respectively. The Fermi level is arbitrarily set to the conduction band minimum. For each band, the weights of the d_{xy}, d_{xz} and d_{yz} orbital characters are proportional to the sizes of the red, blue and green circles, respectively (right). (b-c) Effect of confinement on subbands $n = 1$ (grey lines) and on their orbital characters (circles). The Fermi level on the 2DEG is set to fit the experimental band filling. The effective masses at Γ along ΓX are $m^*_{\Gamma X}(E\alpha) = 0.25 m_e$ and $m^*_{\Gamma X}(E\beta) = 0.5 m_e$ (m_e is the free electron mass).

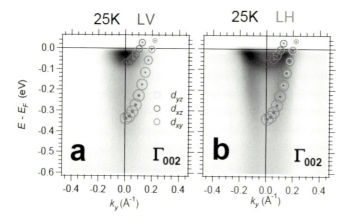

Figure 2: ARPES maps around Γ_{002} along the (010) direction using respectively LH (a) and LV (b) polarized photons at 32 eV. For the $n = 1$ subbands, the projections on the d_{xy}, d_{xz} and d_{yz} characters, extracted from the tight-binding fit of figure 1(b), are also shown (open circles).

[1] A. Ohtomo & H. Y. Hwang, *Nature* **427**, 423 (2004).
[2] N. Reyren *et al. Science* **317**, 1196-1199 (2006).
[3] A. Brinkman *et al. Nature Mater.* **6**, 493-496 (2007).
[4] A. D. Caviglia *et al. Phys. Rev. Lett.* **104**, 126803 (2010).
[5] M. Ben Shalom *et al.*, *Phys. Rev. Lett.* **104**, 126802 (2010).
[6] S. Okamoto & A. J. Millis, *Nature* **428**, 630 (2004).
[7] A. Kalabukhov *et al.*, *Phys. Rev. B* **75**, 121404 (2007).
[8] G. Herranz *et al.*, *Phys. Rev. Lett.* **98**, 216803 (2007).
[9] W. Siemons *et al.*, *Phys. Rev. Lett.* **98**, 196802 (2007).
[10] A. F. Santander-Syro *et al.*, *Nature* **469**, 189 (2011).
[11] W. Meevasana *et al.*, *Nature Mater.* **10**, 114 (2011).

STRAIN MEDIATED LONG-RANGE QUASI-ORDERED DOMAIN STRUCTURES AT THE SrTiO$_3$ (110) SURFACE

Zhiming Wang[1,2], Fengmiao Li[1], Sheng Meng[1], Ulrike Diebold[2], Jiandong Guo[1]

[1]Beijing National Laboratory for Condensed Matter Physics & Institute of Physics, Chinese Academy of Sciences, Beijing, P. R. China;
[2]Institute of Applied Physics, Vienna University of Technology, Vienna, Austria.

The surfaces and interfaces of SrTiO$_3$ have attracted intense interest in many fields ranging from solid-state physics to surface chemistry and oxide electronics. Here we report a scanning tunneling microscopy (STM) and density functional theory (DFT) calculations study of the SrTiO$_3$(110) surface. Recent results indicate a two-dimensional electron gas (2DEG) at the LaAlO$_3$/SrTiO$_3$(110) interface [1], similar to the more commonly studied (001) oriented interface of this system. The SrTiO$_3$(110) surface is polar, composed of alternatively-stacked (SrTiO)$^{4+}$ and (O$_2$)$^{4+}$ layers. It tends to form a series of reconstructions [2], which are well explained by models consisting of added TiO$_4$ tetrahedra [3,4], see Fig. 1.

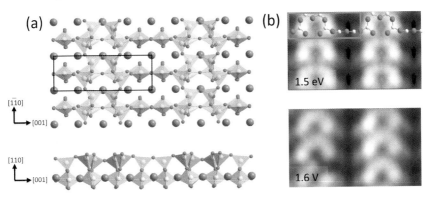

Figure 1, (a) Structural model of the SrTiO$_3$(110) - (4×1) reconstruction in top- and side view. A layer of corner-sharing TiO$_4$ tedrahedra is added on top of the SrTiO-termination. (b) Simulated (top) and experimental (bottom) STM images of SrTiO$_3$(110) - (4×1).

Within this reconstruction two different types of anti-phase domain boundaries are formed when two degenerate antiparallel (4×1) domains are linked. A vacancy-type boundary is formed that relieves the large surface tensile stress caused by the lattice mismatch between the TiO$_4$ added layer and the SrTiO$_3$. At the other type of domain boundary Sr adatoms are adsorbed that compensate the surface polarity. The vacancy and Sr adatom domains exhibit a long-range quasi-ordering and wave-like stripes along the [1-10] and [001] directions, respectively. The domain patterns along [1-10] arise from a competition between the short-range attractive interaction between atoms, related to the boundary energy, and a long-range repulsive interaction between domain boundaries mediated by elastic deformations of the surface and substrate. Along the [001] direction there is a short-range weak attractive interaction between vacancies, giving rise to the formation of the wave-like stripes.

This work was supported by the CMOST and Austrian Science Fund (FWF project F45).

[1] A. Annadi, X. Wang, K. Gopinadhan, W. M. Lv, Z. Q. Liu, A. Roy Barman, A. Srivastava, S. Saha, S. Dhar, H. Hilgenkamp, T. Venkatesan, and Ariando, Bulletin of the American Physical Society **Volume 57, Number 1**, (2012).
[2] Z. Wang, F. Yang, Z. Zhang, Y. Tang, J. Feng, K. Wu, Q. Guo and J. Guo, Phys. Rev. B **83**, 155453 (2011).
[3] J. A. Enterkin, A. K. Subramanian, B. C. Russell, M. R. Castell, K. R. Poeppelmeier, and L. D. Marks, Nat Mater **9**, 245 (2010).
[4] F. Li, Z. Wang, S. Meng, Y. Sun, J. Yang, Q. Guo and J. Guo, Phys.Rev.Lett. **107**, 036103 (2011).

FIRST-PRINCIPLES STUDY OF THE LaAlO$_3$/SrTiO$_3$ INTERFACE

Fontaine Denis[1], Philippe Ghosez[1]

[1]Physique Théorique des Matériaux, Université de Liège, Belgium

In spite of many years of intensive researches, there are still many open questions concerning the properties of the LaAlO$_3$/SrTiO$_3$ (LAO/STO) n-type interface which, beyond a critical LAO thickness, exhibits a confined distribution of charge carriers at the interface (2-dimensional electron gas, 2DEG).[1] These questions concern both the intrinsic and/or extrinsic origin of the 2DEG and its confinement properties. Recent work have questioned the presence of an electric field within LAO assigning the conductivity to extrinsic effects in place of the initial intrinsic polar catastrophe scenario. Our first-principles calculations, using the B1-WC hybrid functional, in support to experimental measurements both on SrTiO$_3$/LaAlO$_3$/vacuum stacks [2] and on SrTiO$_3$/(LaAlO$_3$)$_x$ (SrTiO$_3$)$_{1-x}$ stacks [3] for various solid-solution compositions x strongly support the intrinsic origin of the 2DEG. Independently, the confinement properties of the 2DEG at the interface have also been studied at the first-principles level.[4] This work has been supported by the European project OxIDes.

[1] P. Zubko, S. Gariglio, M. Gabay, Ph. Ghosez and J.-M. Triscone, Ann. Rev. Cond. Matt. Phys. **2**, 141 (2011).
[2] C. Cancellieri, D. Fontaine, S. Gariglio, N. Reyren, A. D. Caviglia, A. Fête, S. J. Leake, S. A. Pauli, P. R. Willmott, M. Stengel, Ph. Ghosez, J.-M. Triscone. Phys. Rev. Lett. **107**, 056102 (2011)
[3] M.L. Reinle-Schmitt, C. Cancellieri, D.F. Li, D. Fontaine, S. Gariglio, M. Medarde, E. Pomjakushina, C.W. Schneider, Ph. Ghosez, J.-M. Triscone, and P.R. Willmott, arXiv:1112.3532v1
[4] P. Delugas, A. Filippetti, V. Fiorentini, D. I. Bilc, D. Fontaine and Ph. Ghosez. Phys. Rev. Lett. **106**, 166807 (2011).

FERROMAGNETISM DRIVEN BY SrTiO$_3$ FERROELECTRIC-LIKE LATTICE DEFORMATION IN LaAlO$_3$/SrTiO$_3$ HETEROSTRUCTURES

M. Carmen Muñoz[1], Jichao C. Li[2]

[1]Instituto de Ciencia de Materiales de Madrid, CSIC, Madrid, Spain;
[2]School of Physics, Shandong University, Jinan, Shandong, China

The conducting two dimensional electron gas (2DEG) formed at the LaO/TiO2 interface between LaAlO3 and SrTiO3 insulating oxides exhibits collective electronic properties such as superconductivity and magnetism[1-5]. Based on first-principles density functional calculations, we address the origin of the ferromagnetic phase and investigate the nature of the electronic states at the interface of ideal defect-free (001) LaAlO3/SrTiO3 heterostructures. We show that the charge transfer to the Ti t_{2g} conduction band, induced by the polar mismatch at the interface, combined with the incipient ferroelectric character of the SrTiO3, lead to a substantial lattice deformation. The ferroelectric-like distortion of the SrTiO3 slab yields differential occupation of the t_{2g} states and generates the spontaneous spin-polarization of the lowest d_{xy} orbital parallel to the interface and the emergence of a ferromagnetic ground state.

Our results indicate that all the heterostructures presents a ferroelectric-like distortion of the Ti-O octahedron with negatively charged O and positively charged Ti ions moving inwards and outward from the interface layer, respectively. The lattice deformation is confined to the SrTiO3 slab, it takes the largest amplitude at the interface plane and decays slowly in deeper SrTiO3 layers. Figure. 1(a) illustrates the dispersion of the occupied conduction band (CB) for a representative relaxed (LaAlO3)$_{3.5}$/(SrTiO3)$_{8.5}$ superlattice. The CB is characterized by a set of parabolic subbands corresponding to the t_{2g} Ti states, which due to the crystal field split in lower d_{xy} non-degenerated and upper d_{xz-yz} two-fold degenerated subbands. In addition, the lowest d_{xy} state shows a large exchange splitting, however higher occupied d_{xy} or degenerated d_{xz-yz} subbands have an almost negligible spin splitting, Different occupancy of the majority and minority spin states leads to a net spin polarization and a ferromagnetic phase is induced. The calculated magnetic moment, ≈0.25 µB per two-dimensional unit cell, mostly localized in the interface atomic plane is in good agreement with the 0.3-0.4 µB inferred from torque magnetometry [4].

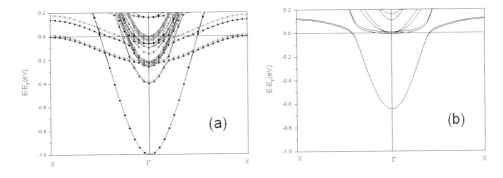

Figure 1: Conduction band for the (LaAlO3)$_{3.5}$/(SrTiO3)$_{8.5}$ superlattice. Black and grey lines represent the majority and minority spin states, respectively. The energy zero corresponds to the Fermi level. The (a) panel shows the conduction band of the relaxed structure and the (b) panel that of the SL without relaxation.

Figure 1(b) also represents the CB dispersion for the same superlattice without lattice deformation. Surprisingly, there is not exchange splitting and only the lowest spin-degenerated *dxy* subbands, completely confined at the interface layer, is occupied. This demonstrates the crucial role of lattice deformation on the appearance of the magnetic ground state at the LaO/TiO2 interface. In fact, this strongly localized charge must be unstable and in order to stabilize the system a ferroelectric-like structural distortion of the SrTiO3 is induced, which in turns alters the relative partial occupation of *t2g* orbitals and gives rise to the emergence of a spin ordered ground state.

Summarizing, the onset of the magnetic phase is induced by the lattice distortion. We predict the existence of three different types of electrons, magnetic two-dimensional (2D) *dxy* electrons confined to the TiO2 interface plane, and non-magnetic higher-lying *dxy* and *dxz-yz* quasi-2D electrons, which spread over several SrTiO3 layers. All *dxy* electrons, both magnetic and non-magnetic, have a 2D-light effective mass (m*) while the Bloch *dxz-yz* states possess a highly anisotropic heavy m*. Hence, only light-mass electrons confined to the interface plane give rise to the ferromagnetic state, while the light electrons with a wave-function more extended in the SrTiO3 slab and therefore susceptible of contribute to the superconducting phase, do not present a significant spin-polarization. Their different characteristics explain the coexistence in the same sample of superconductivity and magnetism [2-5]. Furthermore, since ferroelectric distortions and orbital magnetism are coupled, the magnetic state can be controlled by an electric field.

[1] S. A. Brinkman et al., Nature Materials **6**, 493 (2007).
[2] Ariando, X. Wang et al., Nature Communications **2**, 188 (2011).
[3] D. A. Dikin, et al., Phys. Rev. Lett. **107**, 056802 (2011).
[4] L. Li, C. Richter, J. Mannhart, and R. C. Ashoori, Nature Physics **7**, 762 (2011).
[5] J. A. Bert et al., Nature Physics **7**, 767 (2011).

COHERENT TRANSPORT IN MESOSCOPIC LaAlO$_3$/SrTiO$_3$ DEVICES

Daniela Stornaiuolo, Stefano Gariglio, Nuno J. G. Couto, Alexandre Fête, Andrea D. Caviglia, **Gabriel Seyfarth, Didier Jaccard, Alberto F. Morpurgo, and Jean-Marc Triscone**
DPMC, University of Geneva, Switzerland

The 2-dimensional electron gas (2DEG) present at the interface between LaAlO$_3$ (LAO) and SrTiO$_3$ (STO) [1] is emerging as a unique system in the panorama of oxide electronics for the variety of phenomena therein observed. The possibility to tune its transport properties from superconductivity to weak localization using field effect [2] makes this heterostructure particularly interesting for the investigation of quantum transport in nano-devices.

We set up a fabrication technique, using electron beam lithography, for the realization of mesoscopic channels based on LAO/STO interfaces. Bridges with lateral dimensions down to 500 nm display a metallic behavior similar to the one observed in large scale structures.

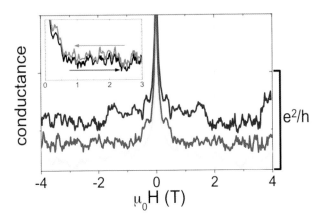

Figure 1: Magneto conductance curves (shifted for clarity) measured at 50 mK for different gate voltages. The gate voltage was tuned in order to modulate the zero field sheet resistance from 7.5 kΩ (top curve) to 19 kΩ (bottom curve). The inset shows the reproducibility of the fluctuations when reversing the direction of the magnetic field sweep.

At low temperatures, magnetotransport measurements display features due to the lateral confinement of the 2DEG. Figure 1 shows magnetoconductance curves of a 2 μm long and 1 μm wide bridge measured at 50 mK for three different gate voltages. These curves show a weak-anti-localization peak at low magnetic field and, at larger field, fluctuations that are reproducible upon reversal of the field sweep direction, as can be seen in the inset. Their amplitude is on the scale of e^2/h and is modulated by the applied gate voltage, in accordance with the sheet resistance. Increasing the temperature progressively suppresses the fluctuations.

We relate these observations to phase coherent transport in a channel with dimensions comparable with the phase coherence length. Analysis of the data within the theory of universal conductance fluctuations yields an estimate of the phase coherence length of 150 nm, in good agreement with the value obtained from the fit of the magnetoconductance peak using the weak localization theory in the presence of spin-orbit interaction.

These results provide clear evidences of a phase coherent transport, and therefore of the effective lateral confinement of the electron gas in these structures.

[1] A. Ohtomo and H.Y. Hwang, Nature (London) 427, 423 (2004)
[2] N. Reyren, *et al.*, Science 317, 1196 (2007); A. Caviglia, *et al.*, Nature 456, 624 (2008); C. Bell, *et al.*, Phys. Rev. Lett. 103, 226802 (2009)

STRONGLY CORRELATED HIGH-MOBILITY ELECTRON GAS AT A MgZnO/ZnO INTERFACE

Yusuke Kozuka[1], Joseph Falson[1], Denis Maryenko[2], Atsushi Tsukazaki[1,3], Christopher Bell[4], Minu Kim[4], Yasuyuki Hikita[4], Harold. Y. Hwang[2,4], Masashi Kawasaki[1,2]

[1]Department of Applied Physics and Quantum-Phase Electronics Center, University of Tokyo, Bunkyo-ku, Tokyo, Japan; [2]Correlated Electron Research Group, RIKEN Advanced Science Institute, Wako, Japan; [3]PRESTO, Japan Science and Technology Agency, Tokyo, Japan; [4]Department of Applied Physics and Stanford Institute for Materials and Energy Sciences, Stanford University, Stanford, California, USA

Oxide interfaces have shown a variety of intriguing phenomena such as metallic conduction between band insulators, interface magnetism, and interface superconductivity. Among them, high-mobility electrons at MgZnO/ZnO provide an interconnecting field between correlated transition metal oxides and conventional semiconductors. Although ZnO is one of the II-VI semiconductors, similar to other oxides, its high electronic polarizability gives a large effective mass, leading to a strongly correlated state in thus high-mobility electron system.

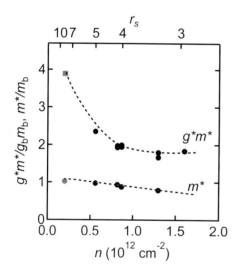

Figure 1: Spin susceptibility (g^*m^*) and effective mass (m^*) as a function of carrier density, normalized by the bulk values. Top axis denotes r_s parameter, which is a ratio of Coulomb energy and the Fermi energy, and measure of correlation. g^*m^* is measured by the coincidence technique, where the period modulation of the Shubnikov-de Haas oscillations are monitored at several fixed angles between the conduction plane and the direction of the magnetic field. The rotation of the sample adjusts the ratio of Landau level splitting and Zeeman splitting. m^* is estimated by measuring the temperature dependence of the Shubnikov-de Haas amplitude.

In this study, we measured the spin susceptibility (g^*m^*) of the two-dimensional electron gas at the MgZnO/ZnO interface, using a coincidence technique, where g^* is the effective g-factor and the m^* is the effective mass relative to the bare electron mass. Figure 1 shows the spin susceptibility as a function of carrier density. At low carrier density, g^*m^* is enhanced by a factor of four compared with the bulk value, indicating strong correlation effects [1]. m^* is separately estimated from the temperature dependence of the Shubnikov-de Haas oscillations. We found a dominant increase in the g-factor rather than m^*, which is suggestive of a proximity to a ferromagnetic state. Given recent observation of fractional quantum Hall effect at this interface [2], the current results shed new light on the problem of spin polarization in the fractional states.

[1] A. Tsukazaki et al., Phys. Rev. B **78**, 233308 (2008); Y. Kozuka et al., Phys. Rev. B **85**, 075302 (2012).
[2] A. Tsukazaki et al., Nature Mater. **9**, 889 (2010).

REVEALING THE FERMI SURFACE OF THE BURIED LaAlO$_3$/SrTiO$_3$ INTERFACE BY ANGLE RESOLVED SOFT X-RAY PHOTOELECTRON SPECTROSCOPY

R. Claessen, G. Berner[1], H. Fujiwara[2], M. Sing[1], C. Richter[3,4], J. Mannhart[4], A. Yasui[5], Y. Saitoh[5], A. Yamasaki[6], Y. Nishitani[6], A. Sekiyama[2], S. Suga[2]

[1]Physikalisches Institut, Universität Würzburg, Germany; [2]Graduate School of Engineering Science, Osaka University, Toyonaka, Japan; [3]Institute of Physics, Experimental Physics VI, Universität Augsburg, Germany; [4]MPI-FKF Stuttgart, Germany; [5]Condensed Matter Science Division, Japan Atomic Energy Agency, Sayo, Hyogo, Japan; [6]Faculty of Science and Engineering, Konan University, Kobe, Japan

The origin of a high-mobility two-dimensional electron system (2DES) at the interface of LaAlO$_3$/SrTiO$_3$ (LAO/STO) heterostructures is still subject of intense research activities in both theory and experiment [1 - 4]. Its sudden appearance above a critical LAO thickness suggests an electronic reconstruction as the driving mechanism, in which electronic charge is transferred from the surface to the interface to avoid excessive build-up of an electric potential gradient across the polar LAO overlayer. Additionally, extrinsic interface doping by cation intermixing or oxygen vacancies may also play a role. The 2DES is expected to be formed by Ti 3d electrons, as Ti is the only multivalent ion in LAO/STO to accommodate the extra interface charge, as indeed evidenced by recent hard x-ray photoelectron spectroscopy (HAXPES) of the Ti 2p core levels [4]. Unfortunately, a direct observation of the conducting Ti 3d states by HAXPES is hindered by unfavorable photoexcitation cross-sections.

Much better spectroscopic access to the interface states can be achieved by resonant photoemission using soft x-rays at the Ti 2p-3d absorption edge, which strongly enhances Ti 3d sensitivity at still moderately high probing depth. With additional angular resolution it should thus become possible to probe the momentum-dependent band structure of the interface. We have applied the method to a conducting LAO/STO heterostructure and report here the direct observation of the metallic Ti 3d-derived 2DES. In particular, we have been able to map out the conduction band dispersion and the Fermi surface topology, representing the first such measurement for a buried interface. The obtained band structure and Fermi surface is consistent with density functional theory calculations [3].

[1] A. Ohtomo and H.Y. Hwang, Nature (London) 27, 423 (2004).
[2] S. Thiel et al., Science 313, 1942 (2006).
[3] Z. Popovich et al., Phys. Rev. Lett. 101, 256801 (2008)
[4] M. Sing et al., Phys. Rev. Lett. 102, 176805 (2009)

NANOSCALE MODULATION OF THE LOCAL DENSITY OF STATES AT THE INTERFACE BETWEEN LaAlO$_3$ AND SrTiO$_3$ BAND INSULATORS

M. Salluzzo[1], Z. Ristic[2], I. Maggio Aprile[2], R. Di Capua[1], G. M. De Luca[1], F. Chiarella[1], M. Radovic[3]

[1] CNR-SPIN, Complesso MonteSantangelo via Cinthia, I-80126 Napoli, Italy
[2] Département de Physique de la Matière Condensée, University of Geneva, 24 Quai Ernest-Ansermet, CH-1211 Geneva 4, Switzerland
[3] LSNS - EPFL, PH A2 354 (Batiment PH) Station 3 CH-1015 Lausanne, Switzerland

The discovery of a high-mobility two-dimensional electron liquid (2DEL) at the interfaces between polar LaAlO$_3$ (LAO) and non-polar SrTiO$_3$ (STO) insulators [1] has motivated a wide research activity due to their intriguing physical properties [2-4]. Moreover, this breakthrough generated new paradigms and challenges in condensed matter physics, requiring non-conventional approaches for their understanding. In their seminal work [1], Ohtomo and Hwang suggested that the appearance of conductivity at the LAO/STO is due to an electronic stabilization to avoid the divergence of the electrostatic potential by a transfer of electrons to the STO conduction band at the interface. However, this picture is debated in particular in view of the possible role of oxygen and cation defects in the phenomenon.

From a theoretical point of view, a pure electronic reconstruction of the LAO/STO interface is sufficient to stabilize the system [5]. In particular, to eliminate the divergence of the electrostatic potential, a fraction of titanium ions have to change their valence from Ti^{4+} to Ti^{3+}. Unfortunately, the appearance of Ti^{3+} is not a sufficient proof of the electronic reconstruction mechanism, since dopants in STO are known to induce partially occupied 3d titanium states. As a consequence, in view of the similarities with the properties of doped STO, finding direct experimental proofs of an electronic reconstruction in the LAO/STO system is quite challenging.

Here we present a scanning tunneling microscopy/spectroscopy (STM/STS) study of the local density of states of the buried LAO/STO interface with nanometer resolution. We performed experiments on samples characterized by insulating or metallic character, i.e. we studied bilayers composed by 2uc and 4uc thick LAO films deposited on TiO$_2$ terminated STO single crystals. STM/STS shows that a distinctive characteristic of metallic LAO/STO bilayers is the presence a short-range quasi-periodic pattern (6x8 nm^2), not commensurate with lattice [6], which is observed in both topographic as-well-as purely electronic maps [Fig.1]. Spectroscopy maps, with nm resolution, suggests that the superstructure is due to the combination of an electronic and structural reconstruction effects. By comparing normalized (dI/dV)/(I/V) data with LDA+U calculations [5], we conclude that the superstructure is associated to a quasi-ordered arrangement of the 3d orbitals. These results are in agreement with X-ray absorption spectroscopy (XAS) studies [6] that showed the LAO/STO interfaces are characterized by an orbital reconstruction removing the orbital degeneracy of titanium 3d states and pushing down the 3d$_{xy}$ orbitals.

It is worth mentioning that the presence of a short range ordered superstructure has no straightforward explanation in the framework of present theoretical models. Calculations performed on LAO/STO superlattices including strong on site repulsion potential U [5], showed that a stable configuration of the system consists in a charge-orbital 2x2 checkerboard pattern of filled 3d$_{xy}$ states. Our data are not fully in agreement with these predictions, considering the much larger period of the superstructure observed. Moreover, our experimental results suggest that the topmost interface layer present a considerable amount of localized charges. We propose that the conducting behaviour of the LAO/STO system is related to mobile electrons that are transferred in deeper layers, which are not accessible to the STM or give a reduced contribution in the tunnelling process.

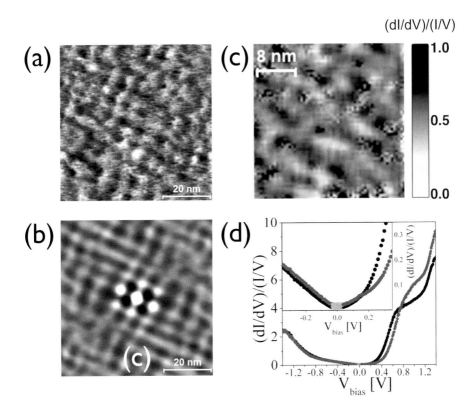

Figure 1: STM/STS showing the short range order superstructure at the metallic interface of 4uc LAO/STO bilayer.
(a) Topography image;
(b) autocorrelation image of (a) enhancing the periodic structure;
(c) Spectroscopy map acquired at $V_{bias}=+0.4$ V (unoccupied states);
(d) the overall differences between normalized dI/dV spectra measured on dark (black circles) and light blue (blue circles) regions of the spectroscopy map shown in (c). See Ref.[6] for details.

[1] A. Othomo, H. Y. Hwang, Nature 427, 423426 (2004).
[2] A. D. Caviglia, S. Gariglio, C. Cancellieri, B. Sacepe, A. Fete, N. Reyren, M. Gabay, A. F. Morpurgo, and J.-M. Triscone, Phys. Rev. Lett. 105, 378380 (2010).
[3] S. Thiel, G. Hammerl, A. Schmehl, C. W. Schneider, and J. Mannhart, Science 313, 1942 (2006).
[4] A. Brinkman, M. Huijben, M. Van Zalk, J. Huijben, U. Zeitler, J. C. Maan, W. G. Van Der Wiel, G. Rijnders, D. H. A. Blank, and H. Hilgenkamp, Nature Mater. 6, 493 (2007).
[5] R. Pentcheva, W. E. Pickett, Phys. Rev B 74, 035112 (2006).
[6] Z. Ristic, R. Di Capua, G. M. De Luca, F. Chiarella, G. Ghiringhelli, J. C. Cezar, N. B. Brookes, C. Richter, J. Mannhart and M. Salluzzo, EPL 93, 17004 (2011).
[7] M. Salluzzo, J. C. Cezar, N. B. Brookes, V. Bisogni, G. M. De Luca, C. Richter, S. Thiel, J. Mannhart, M. Huijben, A. Brinkman, G. Rijnders, and G. Ghiringhelli, Phys. Rev. Lett. 102, 166804 (2009).

TUNING THE TWO-DIMENSIONAL ELECTRON GAS AT THE LaAlO$_3$/SrTiO$_3$(001) INTERFACE BY METALLIC CONTACTS

Rossitza Pentcheva[1], Rémi Arras[2], Victor G. Ruiz[1] and Warren E. Pickett[1]

[1]Ludwig-Maximilians University, Munich, Germany; [2]University of California at Davis, CA, U.S.A.

Density functional theory (DFT) calculations for a representative series of metallic overlayers M on LaAlO$_3$/SrTiO$_3$(001) (M=Na, Al, Ti, Fe, Co, Pt, Cu, Ag, Au) reveal broad variation in the electric field within the polar LaAlO$_3$ film and in the carrier concentration at the interface: For Al and Ti metal contacts the electric field is eliminated, leading to a suppression of the thickness-dependent insulator-to-metal transition observed in uncovered films. Independent of the LaAlO$_3$ thickness, both the surface and the interface are metallic, with an enhanced carrier density at the interface relative to LaAlO$_3$/SrTiO$_3$(001) after the metallization transition. For transition (Fe, Co, Pt) and noble metal contacts (Cu, Ag, Au) a finite and even enhanced (Au) internal electric field develops within LaAlO$_3$. The relationship between the band alignment, Schottky barrier height and the size of work function of the metal on LaAlO$_3$ provides guidelines how to control the carrier concentration at the LaAlO$_3$/SrTiO$_3$ interface by the choice of the metal contact [1].

Funding by the DFG SFB/TR80 and a grant for computational time at the Leibniz Rechenzentrum Garching are gratefully acknowledged.

[1] R. Arras, V. G. Ruiz, W.E. Pickett and R. Pentcheva, Phys Rev. B **85**, 125404 (2012).

FIELD-EFFECT DEVICES UTILIZING OXIDE INTERFACES

<u>Christoph Richter</u>[1,2], Benjamin Förg[2], Rainer Jany[2], Georg Pfanzelt[1], Carsten Woltmann[1], Jochen Mannhart[1]

[1]Max Planck Institute for Solid State Research, 70569 Stuttgart, Germany; [2]Experimental Physics VI, Augsburg University, 86135 Augsburg, Germany

Using $LaAlO_3$-$SrTiO_3$ bilayers, we have fabricated all-oxide field-effect devices[1] that utilize the two-dimensional electron liquid generated at the bilayers' n-type interfaces as drain-source channels and the $LaAlO_3$ layers as gate dielectrics. Figures 1 and 2 present a typical device layout and current-voltage characteristic. I will present such devices and discuss their properties as well as their perspective applications in science and technology.

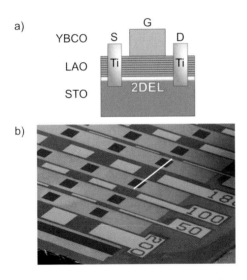

Figure 1: (Color online) Sketch of a cross section of a device (a) and electron microscope image of a typical sample (b). The colors were added. The horizontal lines within the $LaAlO_3$-layer (LAO) symbolize the 9 monolayers of $LaAlO_3$, the standard thickness of gate dielectric in this study, grown on a $SrTiO_3$ substrate (STO). The narrow, straight line (white) denotes the location of the cross section shown in (a); the numbers indicate the gate widths in microns. The two-dimensional electron liquid shown in (b) in pale gray is also present under the $YBa_2Cu_3O_7$ (YBCO) gates (dark gray).

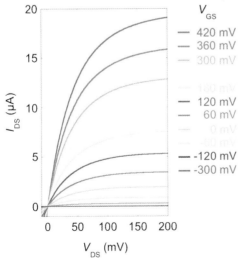

Figure 2: (Color online) Gate-voltage (V_{GS})-dependent I_{DS} (V_{DS}) characteristics of a device (channel length 60 µm, channel width 1600 µm) measured in four-point configuration at room temperature.

[1] B. Förg, C. Richter and J. Mannhart, Appl. Phys. Lett. **100**, 053506 (2012)

GATE-CONTROLLED SPIN INJECTION AT LAO/STO INTERFACES

<u>Henri Jaffrès</u>, N. Reyren, E. Lesne, J.-M. George, C. Deranlot, S. Collin, M. Bibes, and A. Barthélémy

Unité Mixte de Physique CNRS/Thales, 1 Av. A. Fresnel, 91767 Palaiseau, France and Université Paris-Sud, 91405 Orsay, France

Future spintronics devices will be built from elemental blocks allowing the electrical injection, propagation, manipulation and detection of spin-based information. Owing to their remarkable multifunctional and strongly correlated character, oxide materials already provide such building blocks for charge-based planar devices such as ferroelectric field-effect transistors, as well as for spin-based vertical devices like magnetic tunnel junctions, with giant responses in both cases. In an attempt to bridge these two areas, we report results of electrical spin injection at the high-mobility quasi-two-dimensional electron system (2-DES) that forms at the $LaAlO_3/SrTiO_3$ interface [1]. Based on recent techniques adapted from semiconductors, we analyze the voltage variation associated with the precession of the injected spin accumulation at LAO/STO interfaces from a Co source driven by perpendicular or transverse magnetic fields in a three-point geometry (Hanle and inverted Hanle effect –Fig.1-) [2-3]. The influence of bias and back-gate voltages reveals that the spin accumulation signal is amplified by resonant tunneling through localized states in the $LaAlO_3$ strongly coupled to the 2-DES by direct tunneling transfer [4]. We will give in the end to general rules for a massive spin injection in the conductive channel vs. the hierarchy of the different relevant resistances coming into play.

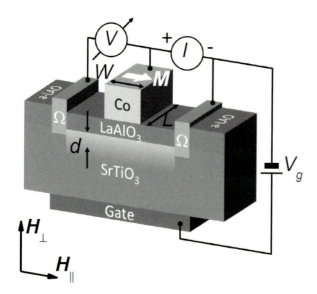

Figure 1: Sketch of the devices for 3-T Hanle measurements in both Voigt (Hperp) and Faraday (Hpara) geometries showing up both electrical Hanle (spin depolarization=loss of spin accumulation=voltage drop) and inverted Hanle effects (spin repolarization=increase of spin accumulation= voltage rise). A gate voltage Vg can be applied between the back of the STO crystal and the LAO/STO 2-DES. `a-LAO' stands for amorphous LaAlO3, `' for ohmic contact, and the 2-DES is shown in green. M is the in-plane Co electrode magnetization.

[1] A. Ohtomo, and H. Y. Hwang, *Nature* **427**, 423 (2004).
[2] M. Tran et al., Phys Rev. Lett. 102, 036601 (2009).
[3] S.P. Dash *et al.*, *Phys. Rev. B* **84**, 054410 (2011).
[4] N. Reyren et al., to appear in Phys. Rev. Lett.

FIELD EFFECT MODULATION OF THE ELECTRON GAS AT THE LaAlO$_3$/SrTiO$_3$ INTERFACE : A THERMOELECTRIC STUDY

Stefano Gariglio[1], Danfeng Li[1], Jean-Marc Triscone[1], Ilaria Pallecchi[2], Sara Catalano[2], Alessandro Gadaleta[2], Daniele Marré[2], Alessio Filippetti[3]

[1]DPMC, University of Geneva, Geneva, Switzerland
[2]CNR-SPIN and Dipartimento di Fisica, Università di Genova, Genova, Italy
[3]CNR-IOM, Cagliari, Italy

From the recent intensive research on the 2D electron gas that forms at the interface between the insulating oxides LaAlO$_3$ and SrTiO$_3$ [1], a complex picture of the interfacial electronic band structure is emerging, where many sub-bands with different energy dispersions are present [2]. In this work we explore the thermoelectric properties of this system, which may benefit from the good thermoelectric performances of SrTiO$_3$ and a possible further enhancement due to quantum confinement [3]. Using field effect, we demonstrate that the Seebeck coefficient can be tuned by more than two orders of magnitude, up to few tens of mV/K at 4 K, in the carrier depletion regime. Moreover, oscillations of the thermoelectric coefficient are observed, a possible signature of the filling of successive energy levels inside the interface quantum well.

[1] Zubko et al., Ann. Rev. Cond. Matter Phys. **2**, 141 (2011).
[2] Delugas et al., Phys. Rev. Lett. **106**, 166807 (2011).
[3] Pallecchi et al., Phys. Rev. B 81, 085414 (2010).

IS IT POSSIBLE FOR A La$_{0.5}$Sr$_{0.5}$TiO$_3$ FERMI LIQUID TO EXIST IN A CONFINED TWO-DIMENSIONAL SYSTEM?

X. Wang[1,2], Z. Huang[1,3], W.M. Lü[1,3], D. P. Leusink[4], A. Annadi[1,2], Z.Q. Liu[1,2], T. Venkatesan[1,2,3], and Ariando[1,2]

[1]NUSNNI-Nanocore, National University of Singapore, 117411 Singapore; [2]Department of Physics, National University of Singapore, 117542 Singapore; [3]Department of Electrical and Computer Engineering, National University of Singapore, 117576 Singapore; [4]Faculty of Science and Technology and MESA+ Institute for Nanotechnology, University of Twente, Twente 7500 AE, The Netherlands

Behind the recent fascinating observations of the two dimensional electron gas (2DEG) at the LaAlO$_3$/SrTiO$_3$ (LAO/STO) interface [1-8], polarization catastrophe is the commonly believed mechanism resulting in an ultrathin Ti^{3+} layer at the surface of the STO substrate. A comparable system can also be realized by growing 50% La doped STO films on the STO substrate [9].

In this study, high quality ultrathin 50% La doped SrTiO$_3$ (LSTO) films were grown layer-by-layer by pulsed laser deposition on STO substrates. The transport properties of LSTO films of various thicknesses were investigated. In contrast to the expected metallic nature of LSTO, an abrupt transition from a metallic to a highly insulating state was observed when the thicknesses of the LSTO thin films were reduced from 6 to 5 uc as shown in Fig. 1. The transition is characterized by a large conductance jump of more than six orders of magnitude. Samples with an LSTO thickness larger than 6 uc show a T^2 dependence of the resistance. This indicates a Fermi liquid behavior dominated by electron-electron scattering. On the other hand, samples with an LSTO thickness less than 6 uc show a highly insulating behavior. The abrupt metal-insulator transition with the absence of an intermediate semiconducting state can possibly be attributed to different mechanisms such as band bending, strain or dimensionality evolution.

To investigate this, various experiments were conducted. A Scanning Tunneling Microscopy measurement was performed on the 7 uc thick LSTO film. Figure 2 shows a uniform profile with visible substrate terraces, suggesting that the insulating state is not due to the existence of the inhomogeneity of the conducting and insulating phases. In Fig. 2b, the Hall resistance shows a clear nonlinear behavior for LSTO films with a thickness above 10 uc suggesting two types of carriers. The Hall resistance becomes linear when the film thickness is further reduced, suggesting only one type of carrier is playing a role in the conduction. Electric field effect experiments with back gating through the substrate were also performed (Fig. 2c). For LSTO films with a thickness above 10 uc, a negative gate voltage tunes the number of carrier types from two to one. For LSTO films with a thickness of 6 uc, the conductance was enhanced by a positive gate voltage, while it was significantly suppressed by a negative gate voltage. Interestingly, a Rashba-like effect was also observed in magnetoresistance measurements. This can be understood by the coexistence of both 2D interface and 3D bulk conducting channels which have very different magnetic and electric responses.

In summary, above the STO surface we observed three different uniform regions: insulating, interface and bulk region. We believe that the conducting Fermi liquid LSTO cannot exist in a strictly 2 dimensional system, resulting in the emergence of an insulating phase below 6 uc thickness. Furthermore, these results can shed light on the interface and surface reconstruction of complex oxides.

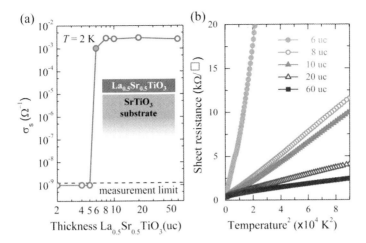

Figure 1: Abrupt metal-insulator transition and metallic Fermi liquid behaviors. (a), 5 to 6 uc critical thickness for LSTO on STO with conductance changes of more than 6 orders of magnitude at 2K. The orange dot at 6 uc indicates the instability of the conductance. Samples with an LSTO thickness of 6 uc show either metallic or highly insulating, never an intermediate semiconducting phase. (b), The quadratic relationship between resistance and temperature indicates a Fermi liquid behavior.

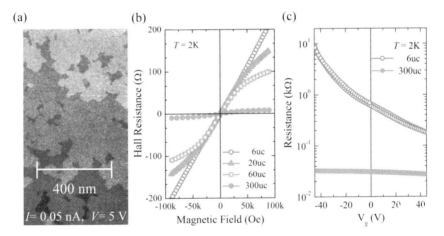

Figure 2: Selected experiments on exploring the mechanism of conductivity in ultrathin LSTO. (a), Scanning Tunneling Microscopy image on 7 uc LSTO with $I = 0.05$ nA and $V = 5$ V at room temperature. (b), Nonlinear Hall resistance for LSTO with different thicknesses. (c), Electric field effect for two different thicknesses of LSTO at 2K using a back gate.

[1] A. Ohtomo & H. Y. Hwang, Nature **427**, 423-426 (2004).
[2] A. Brinkman et al., Nature Mater. **6**, 493-496 (2007).
[3] C. Cen et al., Nature Mater. **7**, 298-302 (2008).
[4] N. Reyren et al., Science **317**, 1196-1199 (2007).
[5] A. D. Caviglia et al., Nature **456**, 624-627 (2008).
[6] Ariando et al., Nature Comm. 2, 188 (2011).
[7] L. Li, C. Richter, J. Mannhart, & R. C. Ashoori, Nature Phys. **7**, 762-766 (2011).
[8] J. A. Bert et al., Nature Phys. **7**, 767-771 (2011).
[9] Y. Tokura et al., Phys. Rev. Lett. **70**, 2126-2129 (1993).

MECHANISM OF "PHONON GLASS – ELECTRON CRYSTAL" BEHAVIOUR IN THERMOELECTRIC LAYERED COBALTATE

L. Wu[1], Q. Meng[1], Ch. Jooss[2], J. Zheng[3], H. Inada[4], D. Su[1], Q. Li[1], and Y. Zhu[1]

[1]Brookhaven National Laboratory, Upton, NY 11973, USA; [2]University of Goettingen, Germany; [3]Xiamen University, China; [4]Hitachi High Technology, Japan

Measurement of local disorder and lattice vibration is of great importance for understanding mechanisms whereby thermoelectric materials efficiently convert heat to electricity. High thermoelectric power requires minimizing thermal conductivity while keeping electric conductivity high. "Phonon glass - electron crystal" is an approach for enhancing phonon scattering through specific structural disorder that also retains sufficient electron mobility and a high Seebeck coefficient. Among the cobaltates $(Ca_2CoO_3)_{0.62}CoO_2$ (often approximated as $Ca_3Co_4O_9$) is of particular interest with its *ZT* above one at high temperature, thus it represents a model system for understanding the enhanced thermoelectric power in misfit-layered structures. The $Ca_3Co_4O_9$ consists of two interpenetrating subsystems of a CdI_2-type CoO_2 layer which shows strong electronic correlations and a distorted tri-layered rock-salt-type Ca_2CoO_3 block, being incommensurate along the b-axis. While the decrease in thermal conductivity in disordered layered structures is established experimentally, the underlying physics for efficient reduction of phonon thermal conductivity remains elusive.

In this presentation, we report a novel method we developed based on scanning transmission electron microscopy (STEM) to reliably measure local disorder and lattice vibration in $Ca_3Co_4O_9$ to shed light on the underlying electronic and vibrational mechanisms on thermal conductivity. Unlike diffraction analysis that derives overall displacement in the two sublattices from intensity of Bragg reflections, we directly measure the atomic displacement in real space that enables us to refine *independently* the static displacement and atomic vibration of adjacent ordered CoO_2 and disordered Ca_2CoO_3 layers, crucial to revealing their different nature in phonon scattering.

Figure 1 shows the simultaneously acquired high-angle annular-dark-field (HAADF) and medium-angle annular-dark-field (MAADF) image pairs of the $Ca_3Co_4O_9$ crystal. The cations in CoO_2, CoO and CaO layers are clearly visible in both imaging modes (Fig. 1b). The layer intensity falls in the order of CoO_2, CoO and CaO in the HAADF image because of the decrease of atomic mass in the atomic columns. The MAADF image, however, shows atomic column intensity of the CoO layer higher than that of the CoO_2 layer, i.e., a contrast reversal, due to the significant effect of local static and thermal displacement on the image intensity. We quantified the measurements by compare experiment with calculations based on the multislice method with frozen phonon approximation and tested various models [1]. We found that the incommensurate displacive modulation and enhanced local atomic vibration are largely confined within its Ca_2CoO_3 blocks. With our accurate displacement measurement and structure refinement as well as systematic thermal conductivity calculations, we concluded that the static displacement-induced scattering and resonance scattering are the main origin of the reduced thermal conductivity in the system. Our density function theory (DFT) calculations (Fig. 2) further revealed that the total projected density of states (pDOS) of Co d-orbital in the ordered CoO_2 layer is by a factor of three higher than that of Co d-orbital in the disordered CoO layer, offering insight into the account of the "phonon glass" (CoO) and "electron crystal" (CoO_2), and of the variation in local magnetic moment being the possible magnetic origin of the high Seebeck effect in the material [2].

[1] Y. Miyazaki et al., *J. Phys. Soc. Jpn* **71**, 491 (2002).
[2] Work was supported by the U.S. DOE, under Contract No. DE-AC02-98CH10886.

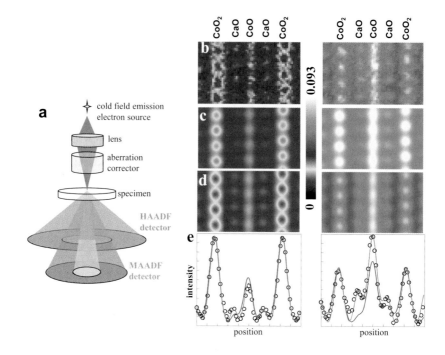

Figure 1: (a) Schematic of simultaneous acquisition of HAADF (collection angle: 114-608 mrad) and MAADF (46-104 mrad) image pairs to determine the displacive modulation in $(Ca_2CoO_3)_{0.62}CoO_2$. (b -d) STEM image pairs (left: HAADF, right: MAADF) viewed along the [010] direction. (b) Experimental image; (c,d) calculated images with Miyazaki's relaxed model [1]; (c) and our refined structure model (d). The intensity values as color legend are shown in the middle. (e) Profiles by averaging the intensities along the [100] direction (vertical direction in the figure). The open circles, and blue and red lines are projected intensity profiles along the vertical direction normalized with the CoO_2 layers from (b), (c) and (d), respectively. The goodness-of-the-fit (χ^2) for blue and red lines are 2.01 and 0.87 for HAADF, and 3.59 and 0.95 for MAADF, respectively.

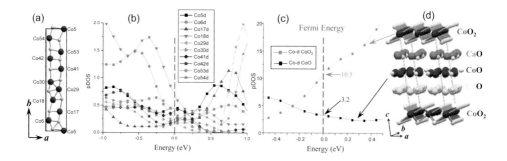

Figure 2: (a) Displacive modulation of CoO layer in Ca_2CoO_3 subsystem, (b) pDOS of d-states of each Co (labelled in (a) in CoO layer), and (c) total pDOS of Co d-states of CoO layer and CoO_2 layer. The Fermi energy is marked by the dashed line. (d) Structure model of our refined $(Ca_2CoO_3)_{0.62}CoO_2$.

CAL 2

SIGN REVERSAL OF THE TUNNELING MAGNETO SEEBECK EFFECT

Andy Thomas,[1] Markus Münzenberg,[2] Christian Heiliger[3]

[1]Bielefeld University, Thin films and physics of nanostructures, Bielefeld, Germany
[2]Göttingen University, I. Physikalisches Institut, Göttingen, Germany
[3]Gießen University, I. Physikalisches Institut, Gießen, Germany

Recently, different Seebeck voltages were also observed for the parallel and the antiparallel configurations in magnetic tunnel junctions, called tunneling magneto Seebeck effect [1,2]. We report on the sign reversal of the tunneling magneto Seebeck effect depending on the base temperature of the device as predicted by Czerner et al. [3].

Figure 1 depicts the tunneling magneto Seebeck effect for two different temperatures. The voltages are different for the different, relative magnetization alignments. At 340 K, the voltage decreases from parallel to antiparallel alignment. If we increase the temperature to 475 K, we observed the opposite behavior: The voltage increases from parallel to antiparallel alignment.

Now, we can repeat the measurements for various laser powers and, therefore, base temperatures and calculate the tunneling magneto Seebeck ratio. If we compare this data with the predicted values of Czerner et al. [12], we find a good agreement of the experimental values with the ab-initio calculations.

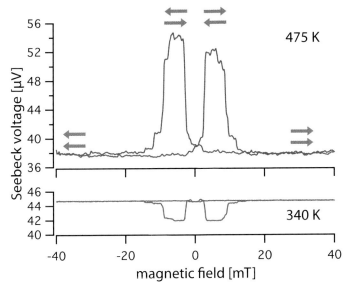

Figure 1: Tunneling magneto Seebeck effect in an MgO based magnetic tunnel junction for two different base temperatures of 340 K and 475 K. The temperature gradient of the two electrodes is in the order of 0.1 K. The red arrows indicate the magnetic orientation of the electrodes.

[1] N. Liebing, S. Serrano-Guisan, K. Rott, G. Reiss, J. Langer, B. Ocker, and H. W. Schumacher: Tunneling Magnetothermopower in Magnetic Tunnel Junction Nanopillars, Phys. Rev. Lett. **107**, 177201 (2011).
[2] M. Walter, J. Walowski, V. Zbarsky, M. Münzenberg, M. Schäfers, D. Ebke, G. Reiss, A. Thomas, P. Peretzki, M. Seibt, J. S. Moodera, M. Czerner, M. Bachmann, and C. Heiliger: Seebeck effect in magnetic tunnel junctions, Nature Mat. **10**, 742 (2011).
[3] M. Czerner, M. Bachmann, and C. Heiliger: Spin caloritronics in magnetic tunnel junctions: Ab initio studies, Phys. Rev. B **83**, 132405 (2011).

AB-INITIO INVESTIGATION OF MAGNETIC *KONBU* PHASES AS NANOSTRUCTURES WITH SPIN-CALORIC-TRANSPORT PROPERTIES

Elias Rabel[1], Phivos Mavropoulos[1], Alexander Thiess[1], Rudolf Zeller[1], Tetsuya Fukushima[2], Nguyen D. Vu[2], Kazunori Sato[2], Hiroshi Katayama-Yoshida[2], Roman Kovacik[1], Peter H. Dederichs[1], Stefan Blügel[1]

[1] Peter Grünberg Institut and Institute for Advanced Simulation, Forschungszentrum Jülich and JARA, D-52425 Jülich;
[2] Department of Materials Engineering Science, Graduate School of Engineering Science, Osaka University, 1-3 Machikaneyama, Toyonaka, Osaka 560-8531, Japan

The term "Konbu-phase" (Konbu is a Japanese term for a type of seaweed) was coined to name certain columnar structures in alloys where spinodal or binodal decomposition is favoured. The idea is that the decomposition of the alloy can be controlled during epitaxial growth by seeding the initial layers with the dopant, so that dopants of the next layers will be preferentially situated above their previous likes. In this way columnar structures of sub-nanometer diameter are created. Recent experiments have shown interesting effects such as giant Peltier cooling in a submicron-sized CuNi/Au junction, which are possibly related to the presence of these phases [1].

From a theoretical perspective an *ab initio* treatment of a few thousand atoms is the minimal requirement, which we are able to meet with the recently developed Korringa-Kohn-Rostoker Green-function program KKRnano. Motivated by the experimentally found spin-thermoelectric properties, we investigate the electronic and magnetic structure of Konbu phases in (Zn,Cr)Te and CuNi. Kinetic Monte-Carlo simulation results provide valuable insight into the spatial structure of these phases and serve as input for our *ab initio* calculations.

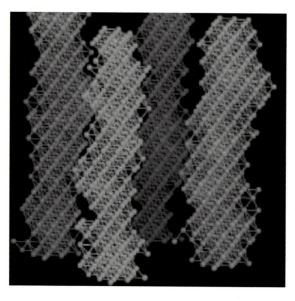

Figure 1: Simulated structure of a Konbu-phase in (Zn,Cr)Te. Cr atoms tend to cluster when placed in a ZnTe matrix due to the low solubility of Cr in ZnTe. The column-like structures are composed of CrTe in the zinc-blende structure (same as the surrounding matrix) and occur during deposition due to an initial seeding of the first ZnTe layers by Cr impurities.

Funding by the DFG Priority Programme 1538 "Spin Caloric Transport" is gratefully acknowledged.

[1] A. Sugihara et al., Appl. Phys. Express **3**, 065204 (2010).

CHARGE KONDO EFFECT IN THERMOELECTRIC PROPERTIES OF LEAD TELLURIDE DOPED WITH THALLIUM IMPURITIES

Theo Costi[1], Veljko Zlatic[2]

[1] Peter Gruenberg Institut and Institute for Advanced Simulation, Research Centre Juelich, 52425 Juelich, Germany; [2] Institute of Physics, POB 304 Zagreb, Croatia

PbTe is one of the most interesting thermoelectric materials. When doped with a small concentration x of Tl impurities, dynamic valence fluctuations set in with increasing x. For x > 0.3%, the Tl ions enter a charge Kondo state. This is believed to account for the unusual superconducting and normal state properties of this system. We studied the thermoelectric properties of PbTe doped with a small concentration x of Tl impurities by combining ab-initio information on PbTe with a negative-U Anderson model of the Tl impurities [1], and solved the latter using the NRG method [2]. The resulting charge Kondo effect accounts for the unusual doping and temperature dependence of the carrier density (see Figure) as well as the Kondo anomalies in the low temperature resistivity and the doping dependence of the mobility.

The charge Kondo effect also provides a mechanism for enhanced thermoelectric power due to a correlation-induced asymmetry in the spectral function about the Fermi level [5]. Depending on whether the charge Kondo thermoelectric power is p-type or n-type (relative to the large p-type band contribution of PbTe) one finds temperature ranges where the thermoelectric power of PbTe can be enhanced or reduced by the charge Kondo effect. In particular, this competition leads to sign changes at low temperature which have been measured in experiments [6].

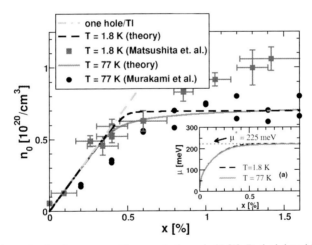

Figure 1: Hole carrier density n_0 versus Tl concentration x in % [1]. Dashed-dotted line: expected behaviour if Tl acts as an acceptor for all x. Symbols: Hall number measurements [3,4]. Solid and dashed lines: theoretical calculations at T=77 and T=1.8K within the negative-U Anderson model [1]

For x<0.3% Tl ions act as acceptors. For x>0.3%, the Tl ions dynamically fluctuate between monovalent acceptor and trivalent donor states, so on average they neither donate nor accept electrons, leading to the saturation of the carrier density ("self-compensation" effect), and the pinning of the Fermi level to 225meV [shown in inset (a)]

[1] T. A. Costi and V. Zlatic, Phys. Rev. Letters, **108**, 036402 (2012).
[2] R. Bulla, T. A. Costi and T. Pruschke, Rev. Mod. Phys., **80**, 395 (2008).
[3] Y. Matsushita et al., Phys. Rev. Letters, **94**, 157002 (2005).
[4] H. Murakami et al., Physica (Amsterdam), **273C**, 41 (1996).
[5] S. Andergassen, T. A. Costi, and V. Zlatic, Phys. Rev. B **84**, 241107 (2011).
[6] M. Matusiak et al., Phys. Rev. B **80**, 220403 (2009).

ELECTRONIC STRUCTURE AND THERMOELECTRIC PROPERTIES OF NANOSTRUCTURED EuTi$_{1-x}$Nb$_x$O$_{3-\delta}$ (x = 0.00; 0.02)

A. Shkabko[1], L. Sagarna[1], S. Populoh[1], L. Karvonen[1], A. Weidenkaff[1]

[1] Empa, Solid State Chemistry and Catalysis, Ueberlandstrasse 129, 8600 Duebendorf, Switzerland

Thermoelectric properties of polycrystalline nanostructured samples (grain size ~ 200 nm) of EuTi$_{1-x}$Nb$_x$O$_{3-\delta}$ (x = 0.00; 0.02) were measured in the temperature range from 8 K to 1040 K. Photoelectron spectroscopy measurements (XPS) at room temperature were performed and related to the electronic transport behavior of the samples. The valence-band structure is formed by O 2p (~ 5.5 eV) and Eu 4f (~ 2.53 eV) states, with the latter sharply peaked near the Fermi level, thus fulfilling the condition for a high Seebeck coefficient ($S \propto (\partial \ln f(E) / \partial E)_{E=Ef}$). In gap states are present in both samples, being more abundant in the sample with x = 0.02, which leads to an enhanced electronic conductivity of EuTi$_{0.98}$Nb$_{0.02}$O$_{3-\delta}$ compared to EuTiO$_{3-\delta}$ in the whole temperature range. The high porosity of the sintered bodies (relative densities between 42% and 54%) reduce the lattice thermal conductivity, enhancing the thermoelectric power. Maximum ZT (Figure of Merit) of 0.4 is reached for EuTi$_{0.98}$Nb$_{0.02}$O$_{3-\delta}$ at T = 1040 K and ZT ~ 0.3 for EuTiO$_{3-\delta}$ at the same temperature.

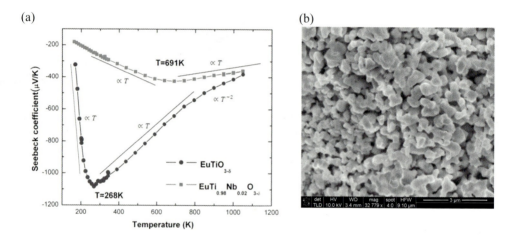

Figure 1. (a) Temperature dependence of the Seebeck coefficient (S) of the samples. Maximum negative values of the Seebeck coefficient $S(T = 268 K) = -1084$ µV K^{-1} and $S(T = 691 K) = -421$ µV K^{-1} are reached for samples with x = 0.00 and x = 0.02, respectively. The origin of the peak in the curve of EuTiO$_{3-\delta}$ is associated with a phonon drag effect, taking place in the low-temperature regime where phonon-electron interactions are dominating, (b) Scanning electron micrograph of the ball milled and pessed pellet of EuTi$_{0.98}$Nb$_{0.02}$O$_{3-\delta}$ (relative density of 49%)

STRONG PHONON SCATTERING AND GLASSLIKE THERMAL CONDUCTIVITY IN CRYSTALLINE PHASE CHANGE MATERIALS

F.R.L. Lange, K.S. Siegert and M. Wuttig

1.Physikalisches Institut (IA) and JARA-FIT, RWTH Aachen University

Phase-change materials (PCMs) are a unique class of materials that can be reversibly switched from an amorphous to a crystalline phase on extremely short time scales. The structural transition is stimulated thermally either optically by a laser pulse or electronically by a short voltage pulse. As this transition is accompanied by a large change in optical reflectivity and electrical resistance PCMs are technologically exploited in data storage applications [1,2]. The electrical switching on a nanosecond timescale [3] makes PCMs attractive for future non-volatile random access memory applications (PCRAM).

In their meta-stable rocksalt phase PCMs possess extraordinary physical properties such as resonant bonding [4] and pronounced structural disorder. The latter is due to a random distribution of Ge, Sb and vacancies on one lattice site which is in the following referred to as configurational disorder. The other lattice site is fully occupied by Te atoms. In addition, there are also significant atomic displacements away from the positions of a perfect rocksalt lattice. This combination of displacements and configurational disorder has already been shown to have a profound impact on the transport of charge carriers. In particular, as shown by Siegrist *et al.* [5], the disorder was sufficient to lead to charge carrier localization, a rare case of disorder-induced localization in a three-dimensional solid. This disorder should also have a pronounced impact on the transport of heat. Therefore a systematical investigation of the effects of structural disorder on thermal transport properties has been undertaken with the goal to identify the degree and origin of phonon localization for different phase change materials.

We present data on the thermal conductivity κ of several sputtered phase-change films which have a stoichiometry along the pseudo-binary line between GeTe and Sb_2Te_3. For these different compounds, the thermal conductivity was determined as a function of temperature as well as annealing conditions [6].

Measurements at room-temperature reveal an irreversible enhancement in κ with annealing temperature for most ternary Ge-Sb-Te alloys. Furthermore the absolute value of κ shows a clear dependence on stoichiometry.

In addition, low temperature measurements down to 50 K for rocksalt structure $Ge_1Sb_2Te_4$ reveal a glasslike thermal conductivity close to the theoretical minimum for disordered crystals. Further data analysis using the Debye model gives insight into microscopic origin of the relevant scattering mechanisms.

In conclusion, it is shown that the strength of phonon scattering is affected by two independent parameters: stoichiometry and annealing temperature. This is illustrated in figure 1. Firstly, by stoichiometric changes, i.e. moving towards higher GeTe content (vertical axis), the amount of statistically distributed scattering centers on the cation sub-lattice is reduced. Secondly, successive heating of the material leads to a structural ordering of the atoms and vacancies into layers and a transition to a stable hexagonal phase (horizontal axis). Both approaches lead to a significant enhancement of κ.

This tailoring of the thermal transport properties of PCMs is valuable to realize advanced data storage devices. The material's thermal properties are crucial to reduce the power consumption as well as the crosstalk between neighboring memory cells.

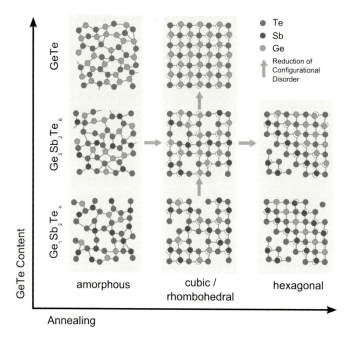

Figure 1: Two dimensional cut through the crystal structure of several phase change materials. Decreasing vacancy content (decreasing Sb concentration) and increasing vacancy ordering, both depicted by orange arrows, lead to a reduction of configurational disorder and a concomitant increase of thermal conductivity.

Furthermore, since many GeSbTe-based PCMs display a metal to insulator transition at higher annealing temperatures [3], the question arises about a possible decoupling of electron and phonon transport properties. Tailoring of both transport mechanisms independently from each other could lead to the realization of a phonon glass-electron crystal like material (PGEC), which bears a high potential for thermoelectric applications.

[1] Ovshinsky, S.R. Reversible electrical switching phenomena in disordered structures. *Physical Review Letters* **21**, (1968).
[2] Wuttig, M. & Yamada, N. Phase-change materials for rewriteable data storage. *Nature materials* **6**, 824–832 (2007).
[3] Bruns, G., Merkelbach, P., Schlockermann, C., Salinga, M. & Wuttig, M. Nanosecond switching in GeTe phase change memory cells. *Applied Physics Letters* **95**, (2009).
[4] Shportko, K. *et al.* Resonant bonding in crystalline phase-change materials. *Nature Materials* **7**, 653–658 (2008).
[5] Siegrist, T. *et al.* Disorder-induced localization in crystalline phase-change materials. *Nature Materials* **10**, 202–208 (2011).
[6] Siegert, K.S., Lange F.R.L. & Wuttig, M. to be published

TAILORING THERMOPOWER AND CARRIER MOBILITY IN NANOSTRUCTURED HALF-HEUSLERS

Pierre F. P. Poudeu[1,*]; Julien P. A. Makongo[1]; Pranati Sahoo[1]; Liu Yuanfeng[1]; Xiaoyuan Zhou[2]; Ctirad Uher[2]

1) Laboratory for Emerging Energy and Electronic Materials, Materials Science and Engineering Department, University of Michigan, Ann Arbor, 48109, USA
2) Department of Physics, University of Michigan, Ann Arbor, MI, 48109, USA
* Corresponding author. Tel: +1-734-763-8436; Fax: +1-734-763-4788   Email: ppoudeup@umich.edu (PFPP)

One of the major roadblocks to large improvements in the thermoelectric figures of merit (ZT) of leading candidate thermoelectric materials such as the Bi_2Te_3, PbTe, $CoSb_3$ and half-Heusler (HH) based systems remains the difficulty in making meaningful simultaneous improvements in both the electrical conductivity (σ) and thermopower (S) of these materials through doping and/or substitutional chemistry. In conventional semiconductors, both materials parameters (S and σ) are fundamentally coupled adversely through the concentration, n, of charge carriers. Therefore, the maximization of one parameter by tuning the carrier concentration (n) via doping and/or substitutional chemistry inevitably results in the minimization of the other. Here, we show that by coherently embedding sub-ten nanometer scale inclusions within a semiconducting half-Heusler matrix, large enhancements of the thermopower (S) and the mobility (μ) can be achieved simultaneously in both n-type and p-type nanocomposites[1-3]. The enhancement in thermopower originates from large reductions in the effective carrier density (n) coupled presumably with an increase in the carrier effective mass (m*). The surprising enhancement in the mobility is attributed to an increase in the mean-free time (τ) between scattering events (phonon-electron scattering, ionized-impurity scattering and electron – electron scattering). Using X-ray powder diffraction, electron microscopy, and electronic transports data, we will discussed the mechanism of phase formation and transformation, at the sub-ten nanometer scale, in bulk half-Heusler (HH) matrix and the mechanism by which the embedded nanostructures regulate electronic charge transport within the semiconducting HH matrix to achieve unprecedented combinations of physical properties such as, large enhancements in the carrier mobility (μ), thermopower (S) and electrical conductivity (σ) simultaneously with drastic decrease in thermal conductivity (κ) at high temperatures. Emphasis will be placed on the n-type $Zr_{0.25}Hf_{0.75}Ni_{1+x}Sn_{1-y}Bi_y$ and $Ti_{0.1}Zr_{0.9}Ni_{1+x}Sn$, and the p-type $Ti_{0.5}Zr_{0.5}Co_{1+x}Sb$ nanocomposites.

(1) Makongo, J. P. A.; Misra, D. K.; Zhou, X.; Pant, A.; Shabetai, M. R.; Su, X.; Uher, C.; Stokes, K. L.; Poudeu, P. F. P. *J. Am. Chem. Soc.* **2011**, *133*, 18843.
(2) Liu, Y.; Makongo, J. P. A.; Zhou, X.; Uher, C.; Poudeu, P. P. F. *in preparation* **2012**.
(3) Sahoo, P.; Makongo, J. P. A.; Zhou, X.; Uher, C.; Poudeu, P. P. F. *in preparation* **2012**.

TOP 1

ANOMALOUS HALL EFFECT IN GRAPHENE DECORATED WITH 5d TRANSITION-METAL ADATOMS

Y. Mokrousov, H. Zhang, F. Freimuth, C. Lazo, S. Heinze, S. Blügel

Peter Grünberg Institut and Institute for Advanced Simulation, Forschungszentrum Jülich and JARA, 52425 Jülich, Germany
Institute for Theoretical Physics and Astrophysics, University of Kiel, D-24098 Kiel, Germany

Modern spintronics puts significant hope into applications based on graphene nanostructures. Graphene, which can be successfully created, deposited on various substrates and doped with different atoms, bares various fascinating properties, especially with respect to transport applications. Namely, the characteristic Dirac shape of graphene's bands is at the heart of various studies, theoretical as well as experimental, which deal with the topological transport in different materials, in particular, topological insulators.

Here we would like to outline our work on magnetic and transport properties of deposited on graphene heavy transition-metal adatoms, Fig. 1 (right). We demonstrate that most of the 5d transition-metal adatoms display strong magnetism and gigantic values of the magnetocrystalline anisotropy energy. Further, we show that in such hybrid systems a strong magneto-electric response can be utilized for the purposes of topological spin-polarized transport, specifically, the anomalous Hall effect. We analyze in detail the intrinsic contribution to the anomalous Hall effect stemming from the non-trivial Berry curvature of valence electrons in the system, see Fig. 1(left), and relate the properties of the anomalous Hall conductivity to the evolution of the electronic bands as a function of the band filling and electric field. Remarkably, we predict that the graphene with electronic structure modified by heavy adatoms can exhibit the so-called quantum anomalous Hall effect, intensively sought for experimentally.

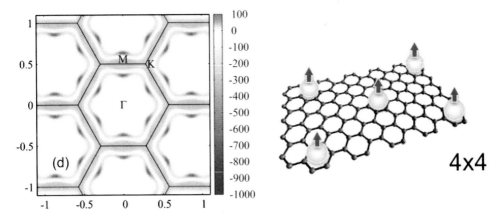

Figure 1: Left: distribution of the Berry curvature for W adatoms on graphene (4x4 geometry).
Right: The pictorial view of the studied adatoms on graphene in 4x4 geometry.

[1] A. H. Castro Neto *et al.*, Rev. Mod. Phys. **81**, 109 (2009)
[2] H. Zhang, C. Lazo, S. Blügel, S. Heinze and Y. Mokrousov, Phys. Rev. Lett. **108**, 056802 (2012)

SPIN POLARIZED PHOTOEMISSION FROM Bi_2Te_3 AND Sb_2Te_3 TOPOLOGICAL INSULATOR THIN FILMS

L. Plucinski[1], A. Herdt[1,2], G. Bihlmayer[3], S. Döring[2], S. Blügel[3], C.M. Schneider[1,2]

[1]Peter Grünberg Institute PGI-6, Forschungszentrum Jülich, Jülich, Germany;
[2]Fakultät für Physik, Universität Duisburg-Essen, Duisburg, Germany;
[3]Peter Grünberg Institute PGI-1, Forschungszentrum Jülich, Jülich, Germany.

The surface electronic properties of the important topological insulators Bi_2Te_3 and Sb_2Te_3 are shown to be robust under an extended surface preparation procedure which includes exposure to atmosphere and subsequent cleaning and recrystallization by an optimized *in-situ* procedure under ultra-high vacuum conditions. Clear Dirac-cone features are displayed in high-resolution angle-resolved photoemission spectra from the resulting samples, indicating insensitivity of the topological surface state to cleaning-induced surface roughness and stoichiometry [1].

Spin polarized photoemission spectra from these surfaces show up to 30% in plane spin polarization in the Dirac cone near the Fermi level, which is consistent with our *ab initio* theoretical results and with previous predictions [2] which find spin polarization in the order of 40-50% averaged over the surface quintuple layer. We also find a non-zero out of plane spin polarization component in Bi_2Te_3 hexagram Fermi surface qualitatively consistent with the Dirac cone warping model [3].

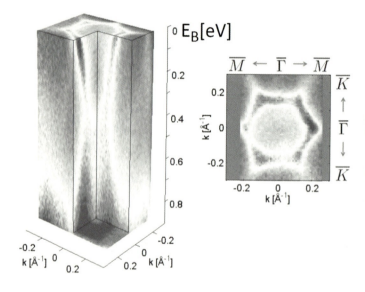

Figure 1: Angle-resolved photoemission results from the Bi_2Te_3 thin film performed using 21.2 eV photons (HeI discharge) on the sample kept at 15K [1]. Left panel shows the 3D band structure with the warped Dirac cone and right panel shows the hexagram structure of Fermi surface.

Our results confirm the robustness of the mechanism of topological protection due to the spin-orbit coupling induced band character inversion. Furthermore they suggest that, due to spin-orbit entanglement, in Bi_2Te_3 and Sb_2Te_3 systems this mechanism produces spin polarization of only 50% or lower at the Fermi level, which has important consequences for possible applications of these materials in spintronics.

[1] L. Plucinski, G. Mussler, J. Krumrain, A. Herdt, S. Suga, D. Grützmacher, and C. M. Schneider, Appl. Phys. Lett. 98, 222503 (2011),
[2] O. V. Yazyev, J. E. Moore, and S. G. Louie, Phys. Rev. Lett. 105, 266806 (2010),
[3] L. Fu, Phys. Rev. Lett. 103, 266801 (2009).

INSTABILITIES OF INTERACTING ELECTRONS ON THE HONEYCOMB BILAYER

Michael Scherer[1], Stefan Uebelacker[1], Carsten Honerkamp[1]

[1]Institute for Theoretical Solid State Physics, RWTH Aachen University, D-52056 Aachen, Germany

We investigate the instabilities of interacting electrons on the honeycomb bilayer by means of the functional renormalization group for a range of interactions up to the third-nearest neighbor. Besides a novel instability toward a gapless charge-density wave we find that using interaction parameters as determined by ab-initio calculations for graphene and graphite puts the system close to the boundary between antiferromagnetic and quantum spin Hall instabilities. Importantly, the energy scales for these instabilities are large such that imperfections and deviations from the basic model are expected to play a major role in real bilayer graphene, where interaction effects seem to be seen only at smaller scales. We therefore analyze how reducing the critical scale and small doping of the layers affect the instabilities.

FIELD-INDUCED POLARIZATION OF DIRAC VALLEYS IN BISMUTH

Zengwei Zhu[1], Aurélie Callaudin[1], Benoît Fauqué[1], Woun Kang[2], Kamran Behnia[1]

[1]LPEM(CNRS-UPMC) -Ecole Supérieure de Physique et de Chimie Industrielles, Paris, France ; [2] Department of Physics, Ewha Womans University, Seoul 120-750, Korea

Electrons are offered a valley degree of freedom in presence of particular lattice structures. The principal challenge in the field of ``valleytronics" is to lift the valley degeneracy of electrons in a controlled way. In the case of graphene, a number of methods to generate a valley-polarized flow of electrons have been proposed, which are yet to be experimentally realized.

In bulk semi-metallic bismuth, the Fermi surface includes three cigar-shaped electron valleys lying almost perpendicular to the high-symmetry axis known as the trigonal axis. The in-plane mass anisotropy of each valley exceeds 200 as a consequence of Dirac dispersion, which drastically reduces the effective mass along two out of the three orientations.

According to our very recent study of angle-dependent magnetoresistance in bismuth, a flow of Dirac electrons along the trigonal axis is extremely sensitive to the orientation of in-plane magnetic field. The effect is visible even at room temperature. Thus, a rotatable magnetic field can be used as a valley valve to tune the contribution of each valley to the total conductivity. At high temperature and low magnetic field, the three valleys are interchangeable and the three-fold symmetry of the underlying lattice is respected. As the temperature is decreased or the magnetic field increased, this symmetry is spontaneously lost.

PEIERS DIMERIZATION AT THE EDGE OF 2D TOPOLOGICAL INSULATORS?

Gustav Bihlmayer[1], Hyun-Jung Kim[2], Jun-Hyung Cho[2], Stefan Blügel[1]

[1]Peter Grünberg Institut and Insitute for Advanced Simulation, Forschungszentrum Jülich and JARA, Jülich, Germany; [2]Department of Physics and Research Institute for Natural Sciences, Hanyang University, Seoul, Republic of Korea

In the last years, the edge states of two-dimensional topological insulators (2D-TIs) attracted considerable interest as they support dissipationless spin-currents. A Bi(111) bilayer was one of the first systems identified as 2D-TI [1,2]. Recently, it was proposed that the zigzag-edge of a Bi(111) bilayer is unstable with respect to a Peiers dimerization [3], a phenomenon that occurs quite general in one-dimensional structures. This proposal was based on an ab-initio investigation without taking spin-orbit coupling into account.

In this work, we investigate the effect of spin-orbit coupling on the atomic structure of zigzag Bi(111) and Sb(111) nanoribbons. Employing the full-potential linearized augmented planewave (FLAPW) method [4], we study possible dimerization effects for these edges.

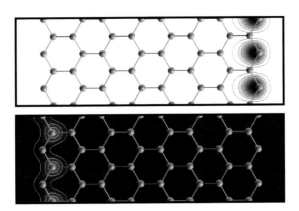

figure 1: Structure of a zigzag Bi(111) bilayer nanoribbon without relaxation effects taken into account. At the edges, the spin density one branch of the topologically protected states is shown in a plane slightly above the nanoribbon. In the upper panel, the majority spin-direction is shown in black on white, while the minority spin direction is shown in the lower panel white on black with the same spacing of the isolines. The spin density is shown for a **k** point near the zone boundary. At the corresponding –**k** point, the state is localized at the opposite edge.

From our investigation we conclude that the topological protection of the states in the Bi ribbon actually prevents the Peierls mechanism to get effective, since the opening of a Peierls gap at the zone boundary is forbidden by time-reversal symmetry. We compare the situation to the Sb structure, where spin-orbit effects are less pronounced. But also in the topologically trivial case of the Sb(111) bilayer ribbon we find a suppression of the dimerization due to spin-orbit effects.

[1] S. Murakami, Phys. Rev. Lett. **97**, 236805 (2006)
[2] M. Wada, F. Freimuth, G.Bihlmayer, and S. Murakami, Phys. Rev. B **83**, 121310(R) (2011)
[3] L. Zhu, T. Zhang, and J. Wang, J. Phys. Chem. C **114**, 19289 (2010)
[4] For the implementation in the FLEUR package see http://www.flapw.de

PREDICTING TOPOLOGICAL SURFACE STATES FROM THE SCATTERING PROPERTIES OF THE BULK

Daniel Wortmann, Gustav Bihlmayer, Stefan Blügel

Peter Grünberg Institut and Institute for Advanced Simulation, Forschungszentrum Jülich and JARA, 52425 Jülich, Germany

The protected states localized at surfaces and interfaces of topological insulators are a consequence of the electronic structure of the bulk. Their peculiar features like the typical spin-structure makes them an interesting field of basic research with possible applications in spintronics. While simple models can already capture many aspects of the surface bands, only ab-initio calculations are able to predict the details of the band dispersion and the response of the states to the potential details at the surface. However, such calculations are computationally very demanding and it can be difficult to distinguish the relative influence of the properties of the bulk and the surface.

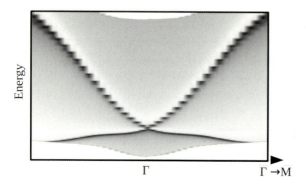

Figure 1: Contour plot of the embedding self-energy for Bi_2Se_3. The white areas at the bottom and the top of the plot indicate the valence and the conduction band, hence only energies within the bandgap are considered and the plot is restricted to a single line of k-values up to approx. 15% of the line ΓM.
Two poles are visible which are degenerate at the Gamma point and which are closely connected to the well-known surface states in Bi_2Se_3.

We demonstrate how the surface states can be efficiently simulated by means of the embedding self-energy as obtained in the Green function embedding technique[1] implemented in the FLEUR code[2]. The embedding self-energy can be understood as a generalized logarithmic derivative and contains all information required to analyze the consequences of the bulk electronic structure on the surface band-structure. The band inversion due to the spin-orbit interaction in a topological insulator introduces poles in the embedding self-energy which connect the valence and the conduction band. We will show how these poles lead to the formation of surface states and connect the spin-texture of these states to the embedding self-energy.

[1] D. Wortmann, H. Ishida, S. Blügel. Phys. Rev. B **66**, 075113 (2002)
[2] http://www.flapw.de

ELECTRIC-FIELD CONTROL OF THE FIRST ORDER METAL-INSULATOR TRANSITION IN VO_2

<u>Masaki Nakano</u>[1], Keisuke Shibuya[1], Daisuke Okuyama[1], Takafumi Hatano[1], Shimpei Ono[1,2], Masashi Kawasaki[1,3], Yoshihiro Iwasa[1,3], Yoshinori Tokura[1,3,4]

[1]Correlated Electron Research Group (CERG) and Cross-correlated Materials Research Group (CMRG), RIKEN Advanced Science Institute, Wako 351-0198, Japan; [2]Central Research Institute of Electric Power Industry, Komae 201-8511, Japan; [3]Quantum-Phase Electronics Center and Department of Applied Physics, University of Tokyo, Tokyo 113-8656, Japan; [4]ERATO, Japan Science and Technology Agency, Tokyo 102-0075, Japan

Conventional electrostatic field-effect induces two-dimensional mobile carriers at a surface of solids, where the channel thickness is limited to a nanometer scale or even less due to the presence of the fundamental electrostatic screening effect. Here we show that this conventional picture is no longer valid for some class of materials having inherent collective interactions between electrons and lattices. We prepared field-effect transistors based on VO_2 thin films with use of a recently developed electric-double-layer transistor gating technique, and found that the electrostatically induced carriers at a channel surface drive all preexisting localized carriers of 10^{22} cm^{-3} even inside a bulk to motion, leading to the emergence of the three-dimensional metallic ground state. This non-local switching of the electronic state is achieved by applying just around 1 V. Furthermore, a novel non-volatile memory like character emerges in a voltage-sweep measurement originating from the first-order nature, which is basically operable at room temperature. Our results demonstrate a conceptually new field-effect device, broadening the concept of electric-field control further to macroscopic phase control beyond the fundamental screening limit.

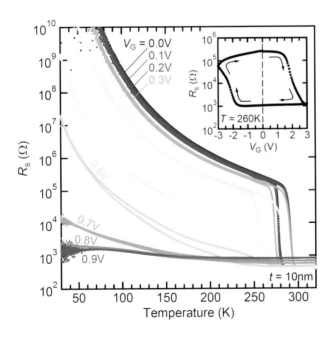

Figure 1:
The temperature dependence of the four-terminal sheet resistance (R_s) for a 10 nm-thick strained VO_2 film with different gate voltages (V_G) ranging from 0.0 V to 0.9 V in 0.1 V steps, showing an electric-field induced metal-insulator transition. The inset shows the V_G dependence of the R_s taken at T = 260 K, showing an electric hysteresis loop.

COLOSSAL MAGNETORESISTANCE AND HALF-METAL BEHAVIOR IN THE DOPED MOTT INSULATOR GaV$_4$S$_8$

B. Corraze[1], E. Janod[1], E. Dorolti[2], V. Guiot[1], C. Vaju[1], H.-J. Koo[3], E. Kan[4], M.-H. Whangbo[4] and L. Cario[1]

[1] Institut de Matériaux Jean Rouxel (IMN), Université de Nantes, CNRS, 2 rue de la Houssinière, BP3229, 44322 Nantes, France
[2] Faculty of Physics, Babes-Bolyai University, Mihail Kogalniceanu, Nr. 1, 400084 Cluj-Napoca, Romania
[3] Department of Chemistry and Research Institute of Basic Science, Kyung Hee University, Seoul 130-701, Korea

During the last quarter of century, magnetoresistance has been at the heart of major scientific breakthroughs, such as the discoveries of giant magnetoresistance (GMR) [1],[2] in magnetic multilayer or of Colossal magnetoresistance (CMR)[3],[4] in different materials. The GMR and the closely related Tunneling Magnetoresistance (TMR) appear only in ferromagnetic-insulator-ferromagnetic multilayers or in granular systems, where grain boundaries play the role of the thin insulator layer. An important prerequisite is that the ferromagnetic materials used in GMR/TMR devices are spin polarized, *i.e.* that only one spin direction has non-zero Density of State (DOS) at the Fermi energy. However, the number of materials with a full polarization at E_F, called half-metal ferromagnets, is rather scarce. On the other hand, in comparison with the well established physics behind the GMR effect, the mechanism inducing the Colossal Magnetoresistance (CMR) are comparatively more debated. In the archetypical manganites perovskites, CMR is related to the coexistence of several ingredients such as double exchange and electronic phase separation [5]. This complex combination of parameters has hindered so far the definition of a simple strategy aiming to obtain new materials with optimized CMR properties.

We report here a study of the narrow gap Mott insulator GaV$_4$S$_8$, a member of the AM$_4$X$_8$ (A = Ga, Ge; M = V, Nb, Ta; X = S, Se) [6] chalcogenide lacunar spinel family. These compounds show exceptional electronic properties such superconductivity under pressure [7] and resistive switching under electric field [8]. However, despite their attractive electronic properties, the consequences of electronic doping have been only poorly studied in these Mott insulators. In a first attempt to tune their electronic filling, we have engaged a study of the solid solution GaTi$_{4-x}$V$_x$S$_8$. We show that a full solid solution GaTi$_{4-x}$V$_x$S$_8$ ($0 \leq x \leq 4$) exists between the Mott insulator GaV$_4$S$_8$ [9] and GaTi$_4$S$_8$ the first non-Mott insulator member of the AM$_4$X$_8$ family [10]. Our magnetic and transport studies show that a metal-insulator transition, driven by electronic filling, disorder and electronic correlation, occurs within the series around $x = 1$, concomitantly with a ferromagnetic instability. GaTi$_3$VS$_8$ is therefore a ferromagnet on the verge of a metal-insulator transition. Our band structure calculations suggest that this compound is a half-metal ferromagnet, *i.e.* with a finite DOS in the spin up bands and a gap in the spin down bands (see Fig.1-a). This prediction is confirmed by our magneto-transport data which shows that granular GaTi$_3$VS$_8$, prepared as sintered pellet, displays a strong negative magnetoresistance reaching 22 % at 2K (see Fig.1-b).

Besides substitution on the transition metal M site, electronic doping of the AM$_4$X$_8$ compounds can be also achieved by heterovalent substitutions on the A site. We therefore explore the consequence of both *n*-type (substitution $Ga^{3+}_{1-x}Ge^{4+}_x V_4 S_8$) and *p*-type (substitution $Ga^{3+}_{1-x}Zn^{2+}_x V_4 S_8$) doping on single crystals of the ferromagnetic Mott insulator GaV$_4$S$_8$. Fig.1-(c-d) show that a bulk, colossal and negative magnetoresistance (MR) in *n*-doped Ga$_{1-x}$Ge$_x$V$_4$S$_8$. Conversely the hole-doped system Ga$_{1-x}$Zn$_x$V$_4$S$_8$ does not display any negative MR. This discovery corresponds to the first experimental validation of a theoretical prediction published 25 years ago [11] and valid for all multiband ferromagnetic Mott insulators. This work therefore points out that the carrier type is a crucial parameter for apparition of CMR in doped FM Mott insulator.

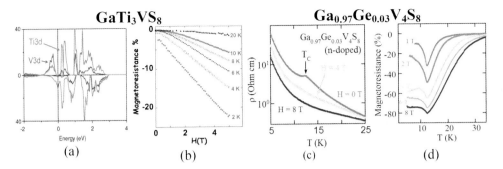

Figure 1: V and Ti partial DOS (a) and magnetoresistance MR *vs.* H (b) of GaTi$_3$VS$_8$. Resistivity *vs.* T (c) and MR *vs.* T (d) of Ga$_{0.97}$Ge$_{0.03}$V$_4$S$_8$.

[1] M.N. Baibich *et al.*, Phys. Rev. Lett. **61**, 2472 (1988).
[2] G. Binasch, P. Grunberg, F. Saurenbach, W. Zinn, Phys. Rev. B **39**, 4828 (1989).
[3] S. Jin *et al.*, Science **264**, 5157 (1994).
[4] A.P. Ramirez, J. Phys.:Condens. Matter **9**, 8171 (1997).
[5] A.J. Millis, P.B. Littlewood, B.I. Shraiman, Phys. Rev. Lett. **74**, 5144 (1995). A. Moreo, S. Yunoki, E. Dagotto, Science **283**, 2034 (1999). E. Dagotto, T. Hotta, A. Moreo, Phys. Reports **344**, 1 (2001). E. Dagotto, Science **309**, 257 (2005)
[6] H. Ben Yaich *et al.*, J. Less-Common Met. **102**, 9 (1984).
[7] M M Abd-Elmeguid *et al.*, Phys Rev Lett. **93** 026401 (2004). R. Pocha *et al.*, JACS **127**, 8732 (2005).
[8] C. Vaju *et al.*, Adv. Mater. **20**, 2760 (2008). V. Dubost *et al.*, Adv. Func. Mat. **19**, 2800 (2009). L. Cario et al., Adv. Mater. **22**, 5193 (2010).
[9] E. Dorolti *et al.*, JACS **132**, 5704 (2010).
[10] C. Vaju *et al.*, Chem. Mater. **20**, 2382-2387 (2008).
[11] P. Muntyanu *et al.*, Russian Physics Journal **29**, 959 (1986).

THE SPIN-STATE AND METAL-INSULATOR TRANTIONS IN $LnCoO_3$

<u>Guoren Zhang</u>[1], Evgeny Gorelov[1], Erik Koch[2] and Eva Pavarini[1]

[1]Institute for Advanced Simulation and JARA, Forschungszentrum Jülich, 52425 Jülich, Germany; [2]German Research School for Simulation Sciences, 52425 Jülich, Germany

In this work, we investigate the origin of the spin-state and metal-insulator transitions in the series $LnCoO_3$ (Ln = Y and Rare Earth), cobaltates with a perovskite structure (Figure 1). The origins of these phenomena have been debated for decades [1,2,3]. By means of Wannier orbitals constructed *ab-initio* in the local-density approximation (LDA), we analyze several energy scales as functions of pressure, temperature and Ln^{3+} ionic radius. To investigate the effects of correlations, we additionally perform many-body calculations in the LDA+dynamical mean-field approximation. With these combined approaches, we identify[4] the crucial material-specific parameters.

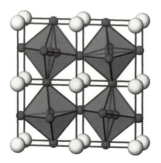

Figure1: The structure of perovskites $LnCoO_3$ [4].

[1] J. B. Goodenough, J. Phys. Chem. Solids **6**, 287 (1958).
[2] M. A. Korotin, S. Yu. Ezhov, I. V. Solovyev, V. I. Anisimov, D. I. Khomskii, G. A. Sawatzky, Phys. Rev. B **54**, 5309 (1996).
[3] A. Podlesnyak, S. Streule, J. Mesot, M. Medarde, E. Pomjakushina, K. Conder, A. Tanaka, M. W. Haverkort, and D. I. Khomskii, Phys. Rev. Lett. **97**, 247208 (2006).
[4] G. Zhang, E. Gorelov, E. Koch and E. Pavarini, to be published.

SPIN-SPECTRAL-WEIGHT DISTRIBUTION AND ENERGY RANGE OF THE PARENT COMPOUND La(2)CuO(4)

J. M. P. Carmelo[1,2], **M. A. N. Araújo**[3], **S. W. White**[4]

[1]Institut für Theoretische Physik III, Universität Stuttgart, D-70550 Stuttgart, Germany;
[2]CFUM, Campus Gualtar, P-4710-057 Braga, Portugal;
[3]CFIF, IST, Technical University of Lisbon, P-1049-001 Lisbon, Portugal;
[4]Department of Physics and Astronomy, University of California, Irvine, CA 92617, USA

The spectral-weight distribution in recent neutron scattering experiments on the parent compound La(2)CuO(4) [1], which are limited in energy range to about 450 meV, is studied in the framework of the Hubbard model on the square lattice [2].

Surprisingly, these experimental studies revealed that the high-energy spin waves are strongly damped near momentum [π,0] and merge into a momentum-dependent continuum.

Here we address the issues raised by the new inelastic neutron scattering results using a combination of a number of theoretical and numerical approaches, including, in addition to standard many-electron treatments [3], a new spinon approach for the spin excitations [4,5] and density matrix renormalization group calculations for Hubbard cylinders [6].

We show that the Hubbard model does describe the new neutron scattering results [2]. In particular, at momentum [π,0] the continuum weight energy-integrated intensity is found to vanish or be extremely small, consistently with the corresponding spin-wave intensity behaviour.

Furthermore, we find that beyond 450 meV, the spectral weight is mostly located around momentum [π,π] and extends to about 566 meV, suggesting directions for future experiments.

[1] N. S. Headings, S. M. Hayden, R. Coldea, and T. G. Perring, Phys. Rev. Lett. **105**, 247001 (2010).
[2] M. A. N. Araújo, J. M. P. Carmelo, M. J. Sampaio, and S. R. White, arXiv:1201.5940.
[3] N. M. Peres and M. A. N. Araújo, Phys. Rev. B **65**, 132404 (2002).
[4] J. M. P. Carmelo, Nucl. Phys. B **824**, 452 (2010); Nucl. Phys. B **840**, 553 (2010), Erratum.
[5] J. M. P. Carmelo, Ann. Phys. **327**, 553 (2012).
[6] S. Yan, D. A. Huse, and S. R. White, Science **332**, 1173 (2011).

SUPERCONDUCTIVITY DRIVEN IMBALANCE OF THE MAGNETIC DOMAIN POPULATION IN $CeCoIn_5$

<u>Simon Gerber</u>[1], Nikola Egetenmeyer[1], Jorge Gavilano[1], Eric Ressouche[2], Christof Niedermayer[1], Andrea Bianchi[3], Roman Movshovich[4], Eric Bauer[4], John Sarrao[4], Joe Thompson[4], and Michel Kenzelmann[5]

[1] Laboratory for Neutron Scattering, Paul Scherrer Institut, Villigen, Switzerland;
[2] INAC/SPSMS-MDN, CEA/Grenoble, Grenoble, France;
[3] Département de Physique, Université de Montréal, Montréal, Québec, Canada;
[4] Condensed Matter and Thermal Physics, Los Alamos National Laboratory, Los Alamos, New Mexico, USA;
[5] Laboratory for Developments and Methods, Paul Scherrer Institut, Villigen, Switzerland

Using high-field neutron diffraction, we have discovered magnetic order inside the superconducting Q-phase of $CeCoIn_5$ for magnetic fields H parallel to the [1 0 0] and [1 1 0] crystallographic direction [1,2]. A novel multicomponent ground state was found, where magnetism and superconductivity co-operate: both orders collapse simultaneously at the upper critical field (see Fig. 1). An experimentally indistinguishable, incommensurate spin-density wave with a propagation vector $Q_1 = (q, -q, 1/2)$ emerges for both field directions inside the superconducting phase (see inset of Fig. 1). Magnetic order is thus modulated along the line nodes of the d_{x2-y2}-wave superconducting order parameter, suggesting that it is driven by electron nesting.

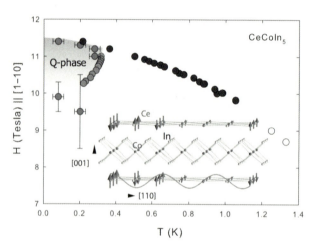

Figure 1: Phase diagram of $CeCoIn_5$ with $H \parallel$ [1 1 0]: Magnetic order is only observed in the superconducting Q-phase. Both orders collapse simultaneously at the phase boundary, separating the superconducting from the normal phase (full and open dots). The insect depicts the structure of the spin-density wave (taken from [1]).

For magnetic fields $H \parallel$ [1 1 0] we unexpectedly find an unequal population of the magnetic domains modulated with $Q_1 = (q, -q, 1/2)$ and $Q_2 = (q, q, 1/2)$, where only the former is populated [3]. The population of a single magneto-superconducting domain can be directly controlled and switched by changing the magnetic field direction from $H \parallel$ [1 1 0] to [1 -1 0].

Although the two domains are different in symmetry for H $\parallel$ [1 1 0] & [1 -1 0], the unequal population cannot be explained by magnetic spin anisotropy on a microscopic level. It appears that the magnetic field direction relative to the superconducting nodes plays a decisive role in the population of the magnetic domains, suggesting that the domain selection is purely electronically driven and mediated by superconductivity.

[1] M. Kenzelmann *et. al.*, Science **321**, 1652 (2008).
[2] M. Kenzelmann *et. al.*, Phys. Rev. Lett. **104**, 127001 (2010).
[3] S. Gerber *et. al.*, unpublished.

RUBIDIUM SUPEROXIDE: A P-ELECTRON MOTT INSULATOR

Roman Kovacik[1], Claude Ederer[2], Philipp Werner[3]

[1]Institute for Advanced Simulation, Forschungszentrum Jülich, 52425 Jülich, Germany; [2]Materials theory, ETH Zuerich, Zuerich, Switzerland; [3]Department of Physics, Fribourg University, Fribourg, Switzerland

Recently, p-electron magnetism has received great attention as alternative option for spintronic applications. The "p-magnetism" is often defect-induced and systematic studies are hampered by poor reproducibility and wide spread in experimental data. It is therefore desirable to study intrinsic p-magnetism in pure bulk materials.

We present results of a combined density functional theory + dynamical mean field theory (DFT+DMFT) study for rubidium superoxide (RbO_2), an insulating antiferromagnet where magnetic properties arise from partially filled oxygen p orbitals. For the high-symmetry tetragonal structure, we calculate the Hamiltonian in the basis of maximally localized Wannier functions [1] with antibonding π^* character, which is then solved within DMFT using a continuous-time quantum Monte Carlo solver [2].

We construct a metal-insulator phase diagram as function of temperature and Hubbard interaction parameters U and J. For realistic values of U and J [3], we find that RbO_2 is a paramagnetic insulator at room temperature (Figure 1). We also find indications for orbital order (Figure 2) at low temperatures ($T \approx 30$ K) in agreement with our previous DFT study [4]. Furthermore, we discuss differences between the realistic Hamiltonian and the one based on the semicircle density of states.

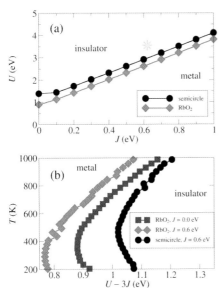

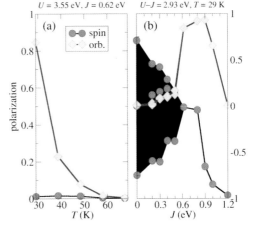

Figure 1: Diagrams of the metal-insulator transition. (a) Critical U as function of J at room temperature ($T \approx 290$). The point corresponding to the realistic U and J parameters [3] is shown as green star. (b) Critical temperature as function of $U-3J$.

Figure 2: (a) Temperature dependence of spin and orbital polarization for fixed U and J. (b) Spin and orbital polarization as function of J for $U-J=2.93$ eV and $T \approx 29$ K. Results shown here are for the realistic Hamiltonian.

[1] I. Souza, N. Marzari, and D. Vanderbilt, Phys. Rev. B **65**, 035109 (2001).
[2] E. Gull et al., Rev. Mod. Phys. **83**, 349 (2011).
[3] I. V. Solovyev, New J. Phys. **10**, 013035 (2008).
[4] R. Kovacik and C. Ederer, Phys. Rev. B **80**, 140411(R) (2009).

HIGH MOBILITY IN A STABLE TRANSPARENT PEROVSKITE

Kookrin Char[1], Kee Hoon Kim[2], Hyung Joon Kim[2], Useong Kim[1], Hoon Min Kim[1], Tai Hoon Kim[2], Hyo Sik Mun[1], Byung-Gu Jeon[2], Kwang Taek Hong[2], Woong-Jhae Lee[2], Chanjong Ju[1]

[1]MDPL, Department of Physics and Astronomy, Seoul National University, Seoul 151-747, Korea;
[2]CeNSCMR, Department of Physics and Astronomy, Seoul National University, Seoul 151-747, Korea

Transparent electronic materials with high electrical mobility are increasingly of demand for a variety of optoelectronic applications, ranging from passive conductive leads to active thin-film transistors. $BaSnO_3$ is a semiconducting oxide with a large band gap of 3.1 eV. We will show that a few percent La-doped $BaSnO_3$ with a simple cubic perovskite structure retains high-level of optical transparency with a high electrical mobility of n-type carriers at room temperature.

In single crystals, the mobility at room temperature was found to be as large as 300 cm^2/V sec in the doping rate of $8\times10^{19}/cm^3$, leading to a resistivity of 0.24 mΩ cm. In epitaxial films, the mobility, the doping rate, and the resulting resistivity were 70 cm^2/V sec, $4.4\times10^{20}/cm^3$, and 0.21 mΩ cm, respectively.

Doping dependence of the mobility implies that the single crystals have dominant ionized dopant scattering while the epitaxial thin films have their mobility limited by dislocations. We attribute the high mobility of $(Ba,La)SnO_3$ to the excellent stability of SnO_2 layers and its doping possibility away from the SnO_2 layers.

We also show that the oxygen atoms in $(Ba,La)SnO_3$ are extremely stable, showing very little change of resistance after a thermal cycle to 530 °C in air. The $(Ba,La)SnO_3$ system thus offers great potential for realizing transparent high-frequency high-power functional devices.

SHEDDING LIGHT ON ARTIFICIAL QUANTUM MATERIALS AND COMPLEX OXIDE INTERFACES WITH ANGLE-RESOLVED PHOTOEMISSION SPECTROSCOPY

<u>Kyle M. Shen</u>[1], Eric J. Monkman[1], Carolina Adamo[2], John W. Harter[1], Daniel E. Shai[1], Yuefeng Nie[1], Julia A. Mundy[3], Alex J. Melville[2], David. A. Muller[3], Luigi Maritato[2,4], Darrell G. Schlom[2]

[1]Department of Physics, Cornell University, Ithaca NY, USA; [2]Department of Materials Science and Engineering, Cornell University, Ithaca NY, USA; [3]Department of Applied and Engineering Physics Science and Engineering, Cornell University, Ithaca NY, USA; [3]Universita di Salerno and CNR-SPIN, 84084 Fisciano (SA), Italy

Engineering interfaces between complex oxides has proven to be a powerful technique for tuning their electronic and magnetic properties. To fully understand how these interfaces can control these electronic properties, one requires advanced spectroscopic tools to uncover their electronic structure. Angle-resolved photoemission spectroscopy (ARPES) is the leading tool for probing energy and momentum resolved electronic structure. To understand the physics of these complex oxide interfaces, we have developed an approach which combines state-of-the-art oxide molecular beam epitaxy with high-resolution ARPES. As one example, I will describe our work on digital manganite superlattices ($[LaMnO_3]_{2n}$ / $[SrMnO_3]_n$), comprised of alternating $LaMnO_3$ and $SrMnO_3$ blocks. Our ARPES measurements reveal that by controlling the separation between the $LaMnO_3$-$SrMnO_3$ interfaces, we can drive the interfacial quasiparticle states from 3D ferromagnetic metal, to a 2D polaron liquid, and finally to a pseudogapped ferromagnetic insulator. I will also describe some of our work on thin films of the "infinite layer" cuprate $Sr_{1-x}La_xCuO_2$, which we can stabilize epitaxially, allowing us to address fundamental issues regarding the asymmetry between doping electrons and holes in the high-Tc cuprates, as well as our work on $SrTiO_3$ and Sr_2TiO_4-based thin films and interfaces.

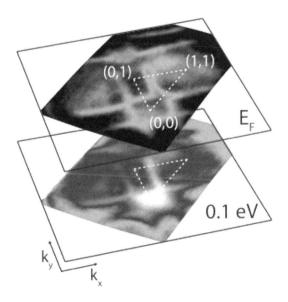

Figure 1: Spectral intensity maps of a $[LaMnO_3]_2$ / $[SrMnO_3]_1$ superlattice at the Fermi level (E_F) and 100 meV below the Fermi level, showing large hole-like d_{x2-y2} pockets around the (π,π) point, and a primarily electron-like d_{z2-r2} pocket around $\boldsymbol{k} = (0,0)$.

HARD AND SOFT X-RAY PHOTOEMISSION STUDIES OF OXIDE MULTILAYER BAND OFFSETS AND OF ELECTRONIC STRUCTURE IN LaNiO$_3$/SrTiO$_3$ AND GdTiO$_3$/SrTiO$_3$

G. Conti[1,2], A. M. Kaiser,[1,2,3] A. X. Gray,[1,2,3], S.Nemsak [1,2], A. Bostwick,[5], A. Janotti[4],
C. G. Van de Walle[4],J.Son [4], P.Moetakef [4], S. Stemmer,[4] S. Ueda [6], K. Kobayashi[6],
A. Gloskovskii [7], W. Drube[7],V.N. Strokov [8], C.S. Fadley[1,2]

[1]Department of Physics, University of California, Davis, California 95616, USA
[2]Materials Sciences Division, Lawrence Berkeley National Laboratory, Berkeley, California 94720, USA
[3]Peter-Grünberg-Institut PGI-6, Forschungszentrum Ju¨lich, 52425 Ju¨lich, Germany
[4]Materials Department, University of California, Santa Barbara, California 93106, USA
5Advanced Light Source, Lawrence Berkeley National Laboratory, Berkeley, California 94720, USA
[6] NIMS Beamline Station at SPring-8, National Institute for Materials Science, Hyogo 679-5148, Japan
[7]DESY Photon Science, Deutsches Elektronen-Synchrotron, DE-22603 Hamburg, Germany
[8] Paul Scherrer Institute, 5232 Villigen – Switzerland

Valence band offsets (VBOs) and the detailed electronic structure of complex oxide heterostructures containing strongly-correlated materials are critical quantities controlling charge transport and quantum confinement, and are of particular relevance in the design of adaptive electronic devices based upon interfaces between two or more materials(1,2).

VBO determinations using soft-x-ray photoemission (photon energy hv ≤ 1500eV) in such complex oxide heterostructures present a challenge because of the charge accumulation at the surface caused by a photoemission, and perhaps also an insufficient probing depth to clearly resolve the interface. In this work, we demonstrate that the use of hard x-ray photoemission spectroscopy (HXPS or HAXPES) with photon energies hv ≥2000eV allows one to avoid, or at least minimize such charging effects, as well as to probe more deeply buried interfaces. We report VBO result for two multilayers (SrTiO$_3$/LaNiO$_3$)x10 and (SrTiO$_3$/GdTiO$_3$)x20 determined by SXPS (hv=833eV) and HXPS (hv=5950 eV). By using HXPS, the charging effects were minimal and the values of the VBOs so obtained, as summarized in Figure 1, are very close to those estimated by first-principles calculations based on density functional theory.

In addition, we have also investigated other electronic properties of the SrTiO$_3$/LaNiO$_3$ superlattice [3]. It has been demonstrated that, although the superlattice structure is based on two insulators, the interface shows a metallic conductivity. This effect is thought to be related to interfacial properties. A suitable method allowing depth-resolved electronic structure studies of buried interfaces is standing-wave-excited soft x-ray photoemission spectroscopy (SW-XPS). Rocking curves of core-level and valence band spectra have been used to derive layer-resolved spectral functions, revealing a suppression of electronic states near the Fermi level in the multilayer as compared to bulk LaNiO$_3$. Further analysis shows that the suppression of these states is not homogeneously distributed over the LaNiO$_3$ layers but is more pronounced near the interfaces. Origins of this effect and its relationship to a previously observed metal-insulator-transition in ultrathin LaNiO$_3$ films are discussed.

In a preliminary study, we have applied the SW-HXPS method to study the electrical properties of the heterostructure with interfaces between the band insulators SrTiO$_3$ and Mott–Hubbard insulator GdTiO$_3$, which recently have attracted attention for tailoring a wide range of emergent phenomena associated with strong electron correlation. We were able to obtain standing-wave rocking curves at the x-ray energy close to the Gd 3d resonance at hv=1182eV. Figure 2 shows the results of the SW-XPS experiment. For each constituent element of the superlattice the strongest core level was chosen. O 1s, Ti 2p$_{3/2}$, Sr 3d$_{5/2}$,and C 1s spectra were collected for each angular step. A first attempt at modelling these rocking curves with a set of x-ray optical calculations using the nominal sample parameters of 20 bilayers of SrTiO$_3$ (8 uc: 3nm) and GdTiO$_3$ (2 uc: 0.7nm) yielded satisfactory agreement. Subsequent measurements will take advantage of this to do standing-wave excited ARPES (see abstract by A.X. Gray).

Nanosession: Advanced spectroscopy and scattering

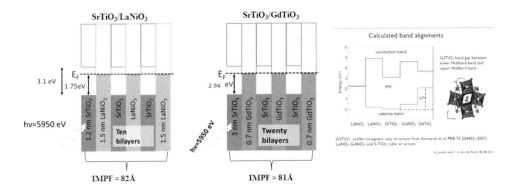

Figure1: A) The HXPS-determined valence band offset for a (SrTiO$_3$/LaNiO$_3$)x10 multilayer is 1.75eV; and that for (SrTiO$_3$/GdTiO$_3$)x20 is 2.94eV. B) Calculated band offset using a first-principles method based on density functional theory.

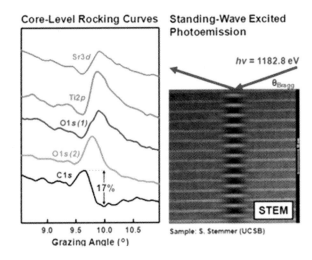

Figure 2: Rocking curve data from a standing wave XPS experiment at the Gd3d resonance hv = 1182.8 eV. The rocking curves of the core levels Sr3d, Ti2p, O1s and C1s are reported as a function of incidence angle.

Supported by the U.S. Dept. of Energy under Contract No. DE-AC02-05CH11231 and ARO MURI Grant W911-NF-09-1-0398.

1   Low-dimensional Mott material: transport in ultrathin epitaxial LaNiO3 films, J. Son, et al. Appl. Phys. Lett. 99, 232116 (2011).
2   Suppression of near-Fermi level electronic states at the interface in a LaNiO3/SrTiO3 superlattice, A. M. Kaiser, Phys. Rev. Lett. 107, 116402 (2011).

EVIDENCE FOR Fe^{2+} CONFIGURATION IN Fe:STO THIN FILMS BY X-RAY ABSORPTION SPECTROSCOPY

A. Köhl[1], D. Kajewski[2], J. Kubacki[2], K. Szot[1,2], Ch. Lenser[1], P. Meuffels[1], R. Dittmann[1], R. Waser[1], J. Szade[2]

[1] Peter Grünberg Institute 7, Forschungszentrum Jülich, Germany;
[2] A Chełkowski Institute of Physics, University of Silesia, Katowice, Poland,

Strontiumtitanate ($SrTiO_3$) is a promising material for future use in oxide electronics due to the intrinsic electronic properties. Furthermore doping gives the opportunity to widely tune these characteristics. For implementation in electronic devices the use of this material in thin film form is inevitable. For this study we performed laser pulsed deposition (PLD) to grow epitaxial thin films of Iron-doped $SrTiO_3$. Due to the strongly kinetically limited growth conditions, the formation of defects and the incorporation of foreign atoms into the crystal lattice differs considerable from single crystals and bulk ceramics.

In order to investigate the Fe oxidation state in the thin films we performed X-ray absorption spectroscopy (XAS) with the use of two different detection methods. The Auger Electron Yield (AEY) is extremely surface sensitive while the Total Electron Yield (TEY) has an information depth of several nm.

The general shape of the Fe-L edge spectra of $Fe:SrTiO_3$ can be described by two main peaks at 708.3eV and 710eV plus a low energy shoulder within the L_3 edge and at least three components within the L_2 edge (see Fig. 1). We concentrate our discussion on the main peaks of the L_3 edge which show a prominent change in intensity ratio comparing AEY and TEY data of the Fe:STO samples.

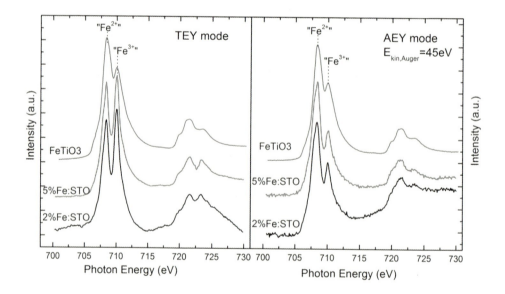

Figure 1: Fe L-edge XAS spectra for Fe doped STO films and bulk $FeTiO_3$ measured in TEY mode (left) and AEY mode (right) after annealing to T=300°C. The Peaks at 708.3eV and 710eV are assigned to mainly Fe^{2+} and Fe^{3+} contributions respectively.

In order to analyse the experimental data we performed atomic multiplet calculations with the CTM4XAS program by E. Stavitski and F.M.F de Groot [1]. From the comparison of our data with the calculation we can conclude, that Iron in the thin films is in a mixture of Fe^{2+} and Fe^{3+} states. In a simplified picture the first peak at 708.3eV is mainly connected to Fe^{2+} states, while the second peak at 710eV mainly steams from Fe^{3+} contributions. Therefore the observed change of the peak ratio between AEY and TEY data can be attributed to an increased Fe^{2+} contribution in the vicinity of the surface.

For further validation of this interpretation we included a bulk mineral $FeTiO_3$ sample in our study. In literature many evidence is found that natural ilmenite shows a combination of Fe^{2+} and Fe^{3+} states due to Fe_2O_3 precipitations [2]. The close similarity of the $FeTiO_3$ spectra to the thin film samples in terms of peak position and shape supports the assumption of mixed valence states in the Fe:STO films.

Regarding the Fe-doping we want to point out that no differences due to the Fe concentration is visible for the Fe:STO thin films. We therefore conclude that the solubility limit in our thin films is not in the studied concentration range.

Additionally we performed XAS measurements at the thin film samples after annealing under UHV conditions. A clear relative increase of the Fe^{2+} component can be observed during annealing from 150°C to 600°C (see Fig. 2). In contrast, the titanium and oxygen absorption spectra did not show any changes upon annealing of the samples or comparison of different detection methods.

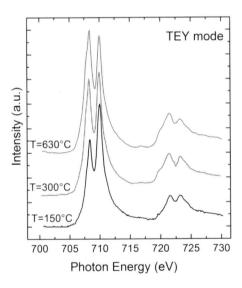

Figure 2: Fe XAS spectra for the 5% Fe-doped STO film after thermal treatment in UHV obtained with the TEY method.

Our findings of Fe^{2+} contributions in the thin films are in contrast to literature data on bulk Fe-doped $SrTiO_3$ which report a mixture of 3+ and 4+ oxidation states [3]. This could be explained by a large amount of oxygen vacancies induced in the film during growth. This model is consistent with the increase of Fe^{2+} states during annealing under reducing conditions which would further increase the oxygen vacancy concentration. It would give an interpretation for the enhanced Fe^{2+} contributions at the surface as the oxygen removal takes place at the UHV-sample interface. The large variations in the Fe spectra compared to the invariance of Ti and Oxygen spectra illustrates nicely that Fe-doping can be used as a sensitive probe for the amount of oxygen vacancies.

[1] E. Stavitski, F. M. F. de Groot, Micron, **41**, 687–694 (2010).
[2] T. Droubay et al., Journal of Electron Spectroscopy and Related Phenomena, **84**, 159–169, (1997).
[3] R. Merkle, J. Maier, Angewandte Chemie International Edition **47**, 3874–3894 (2008).

PHOTOELECTRON AND RECOIL DIFFRACTION AT HIGH ENERGIES FOR BULK-SENSITIVE AND ELEMENT-RESOLVED CRYSTALLOGRAPHIC ANALYSIS OF MATERIALS

Aimo Winkelmann,[1] Maarten Vos[2]

[1] Max-Planck-Institut für Mikrostrukturphysik, Halle(Saale), Germany
[2] Research School of Physics and Engineering, Australian National University, Canberra, Australia

Hard X-ray photoelectron spectroscopy (HAXPES) is a useful tool for the analysis of complex materials and their interfaces due to the increased bulk-sensitivity of this method. For element-resolved crystallographic analysis, diffraction effects in HAXPES can be used. In addition to HAXPES, many of the phenomena observed in high-energy photoemission can also be studied in backscattering experiments of external electrons from surfaces, even at higher energies than is currently possible in HAXPES. The theoretical modelling of both x-ray photoelectron diffraction (XPD) under hard x-ray excitation and backscattered electron diffraction is possible by using the Bloch wave approach of the dynamical theory of electron diffraction.

Photoelectron and backscattered electron diffraction patterns for energies above about 1 keV are dominated by Kikuchi bands which are created by the dynamical scattering of electrons from lattice planes [1]. The origin of the fine structure in such bands can be analysed from the point of view of atomic positions in the unit cell with respect to the properties of the Bloch waves of the diffracted electrons. The profiles and positions of the element-specific photoelectron Kikuchi bands are found to be sensitive to lattice distortions and the position of impurities or dopants with respect to lattice sites [1].

An intriguing phenomenon in high-energy electron scattering is the recoil effect [2]. Here we present a study of sapphire (Al_2O_3), combining both phenomena, diffraction and recoil, in a scattering experiment using 35 keV electrons [3]. For sapphire, the difference in recoil energy allows us to determine if an incident electron was backscattered from an aluminium or from an oxygen atom. The angular electron distributions obtained in such measurements are expected to be a strong function of the recoiling lattice site, as shown below for the calculated element-resolved (11-20) Kikuchi bands. These element-specific recoil diffraction features, based on dynamical theory of high energy electron diffraction, are confirmed by the experiment.

Our observations open up new possibilities for local, element-resolved crystallographic analysis using high-energy electrons. Moreover, our results shed light on the processes by which, with increasing recoil energy, quasi-elastically backscattered electron waves are effectively decoupled from an incident beam with respect to their phase. While the incident and the outgoing pathways are still largely governed by coherent forward scattering, we show that an experimentally identifiable source of dephasing is introduced by nuclear recoil. This explains the transition from discrete spot-like patterns in coherent low-energy electron diffraction (LEED) patterns to the Kikuchi patterns dominating at higher energies.

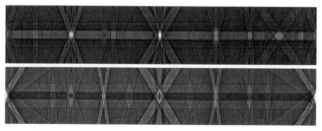

Fig. 1: Sapphire : Aluminium (11-20) Kikuchi Band 35kV, based on Dynamical Simulation

Fig. 2: Sapphire: Oxygen (11-20) Kikuchi Band 35kV, based on Dynamical Simulation

[1] A. Winkelmann, C.S. Fadley, F. J. Garcia de Abajo, *New Journal of Physics* **10**, 113002 (2008)
[2] M.Vos, M.R. Went, Y. Kayanuma, S. Tanaka, Y. Takata, & J. Mayers, Phys. Rev B **78** 043201 (2008)
[3] A. Winkelmann, M.Vos, *Physical Review Letters* **106**, 085503 (2011)

SAS 5

HARD X-RAY ANGLE-RESOLVED PHOTOEMISSION AS A BULK-SENSITIVE PROBE OF ELECTRONIC STRUCTURE

Alexander X. Gray[1,2,3], Christian Papp[1,2,4], Shigenori Ueda[5], Jan Minár[6], Lukasz Plucinski[7], Jürgen Braun[6], Benjamin Balke[8], Claus M. Schneider[7], Warren E. Pickett[1], Giancarlo Panaccione[9], Hubert Ebert[6], Keisuke Kobayashi[5], and <u>Charles S. Fadley</u>[1,2]

[1]Department of Physics, University of California, Davis, California, USA;
[2]Materials Sciences Division, Lawrence Berkeley National Laboratory, Berkeley, California, USA;
[3]Stanford Institute for Materials and Energy Science, SLAC National Accelerator Laboratory, Menlo Park, California, USA;
[4]Lehrstuhl für Physikalische Chemie II, Universität Erlangen-Nürnberg, Erlangen, Germany;
[5]NIMS Beamline Station at SPring-8, National Institute for Materials Science, Hyogo, Japan;
[6]Department of Chemistry and Biochemistry, Ludwig Maximillian University, Munich, Germany;
[7]Peter-Grünberg-Institut PGI-6, Forschungszentrum Jülich GmbH, Jülich, Germany;
[8]Institut für Anorganische und Analytische Chemie, Johannes Gutenberg-Universität, 55099 Mainz, Germany;
[9]Istituto Officina dei Materiali IOM-CNR, Lab. TASC, Trieste, Italy.

Traditional angle-resolved photoemission spectroscopy (ARPES) may, for some systems, be too strongly influenced by surface effects to be a useful probe of bulk electronic structure. This shortcoming suggests that going to higher photon energies with hard x-ray excitation and the resulting larger electron inelastic mean-free paths will provide a more accurate picture of bulk electronic structure. We will present the first experimental data for hard x-ray ARPES (HARPES) at energies of 3.2 and 6.0 keV [1,2]. The systems discussed are W, as a model transition-metal system to illustrate basic principles, GaAs as a typical semiconductor, and (Ga,Mn)As, as a technologically-relevant ferromagnetic semiconductor material, to illustrate the potential broad applicability of this new technique. We have investigated the effects of photon wavevector on wavevector conservation and assessed methods for the removal of phonon-associated smearing of features and photoelectron diffraction effects. The experimental results are compared to free-electron final-state model calculations and more precise one-step photoemission theory including matrix element effects [3]. For (Ga,Mn)As, the results permit settling a controversy surrounding the nature of the electronic states produced by the Mn dopant [2]. Some likely future applications areas will be discussed.

Supported by the U.S. Dept. of Energy under Contract No. DE-AC02-05CH11231.

[1] A. X. Gray *et al.*, Nature Materials **10**, 759 (2011).
[2] A. X. Gray *et al.*, submitted to Nature Materials.
[3] J. Braun *et al.*, Phys. Rev. B **82**, 024411 (2010).

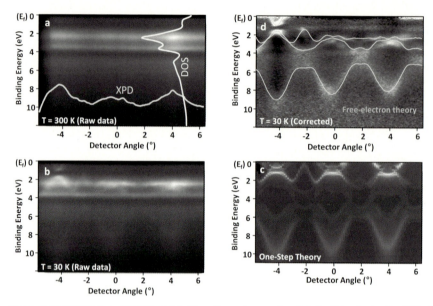

Figure 1. HARPES from W(110) at 6 keV. a. Room temperature data. **b.** Data obtained at 30 K. Phonon effects are suppressed at low temperature and dispersive features are now apparent. **c**, Data from **b**, normalized via a 2-step process so as to remove the DOS and XPD effects (yellow and green curves in **a**, thus enhancing dispersive valence-band features. The solid curves superimposed on the experimental data are the results of band-structure-to-free-electron final-state model calculations. **d**, One-step photoemission calculations of the HARPES spectra taking into account matrix-element effects.

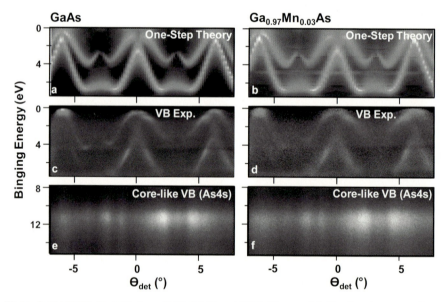

Figure 2. HARPES from GaAs and (Ga,Mn)As at 3.2 keV. a and **b** Results of one-step theory HARPES calculations for GaAs and $Ga_{0.97}Mn_{0.03}As$ respectively. **c** and **d** Experimental HARPES images obtained for GaAs and $Ga_{0.97}Mn_{0.03}As$ respectively. **e** and **f** Core-like and mostly As 4s-in-character band at -11 eV for the two materials, exhibiting non-dispersive characteristic XPD modulations.

HAXPEEM – SPECTROSCOPIC IMAGING OF BURIED LAYERS USING HARD X-RAYS

C. Wiemann[1], M. C. Patt[1], A. Gloskovskii[3], S. Thiess[4], W. Drube[4], M. Merkel[5], M. Escher[5], C. M. Schneider[1,2]

[1]Peter-Grünberg-Institut (PGI-6), Forschungszentrum Jülich GmbH, D-52425 Jülich, Germany;
[2]Fakultät für Physik and Center for Nanointegration Duisburg-Essen (CeNIDE), Universität Duisburg-Essen, D-47048 Duisburg, Germany;
[3]Institut für Anorganische Chemie und Analytische Chemie, Johannes Gutenberg-Universität Mainz, D-55128 Mainz;
[4]DESY Photon Science, Deutsches Elektronen-Synchrotron, D-22603 Hamburg, Germany; [5]Focus GmbH, D-65510 Hünstetten, Germany

Novel electronic devices like resistively switching memory elements consist of several stacked layers of different materials. The topmost layer is usually a metal electrode that is used to electrically contact the active material. For the investigation of the switching process it is necessary to make use of an analytical method with a sufficiently high probing depth to be able to access the active material, which is buried underneath the electrode layer. Additionally, a decent lateral resolution is needed because the formation of conductive channels can be expected to happen very locally.

Photoelectron spectroscopy with hard x-rays (HAXPES) has been proven to reach probing depths of more than 15 nm [1]. On the other hand, photoemission electron microscopy (PEEM) is a well-established technique for high-resolution spectroscopic imaging using soft x-rays already [2]. Combining both methods results in a spectroscopy tool that offers both high spatial and energy resolution together with the higher photoelectron escape depth at higher kinetic energies, which is of great use for the study of layered structures.

In this contribution we present imaging spectroscopy at high kinetic energies using hard x-rays. The measurements reported were performed recently at beamline P09 (DESY, Hamburg) using an energy-filtering PEEM of type NanoEsca [3]. To reach the higher retardation needed to operate at kinetic energies of several keV, the sample holder as well as the microscope's objective lens had to be modified in close cooperation with the manufacturer to allow for higher voltages. We will demonstrate the feasability of imaging at high kinetic energies at a reasonable spatial resolution as well as spatially resolved spectra from buried layers.

[1] K. Kobayashi, Nucl. Instr. Methods A 601, 32 (2009)
[2] A. Locatelli, E. Bauer, J. Phys. Condens. Matter 20, 093002 (2008)
[3] M. Escher, N. Weber, M. Merkel, et al., J. Electron Spectrosc. 144, 1179 (2005)

MARIA: THE MODERN NEUTRON REFLECTOMETER OF THE JCNS OPTIMISED FOR SMALL SAMPLE SIZES AND THINS LAYERS

Stefan Mattauch[1], Ulrich Rücker[2], Denis Korolkov[1], Thomas Brückel[2]

[1]JCNS, Forschungszentrum Jülich GmbH, Garching, Germany
[2]JCNS, Forschungszentrum Jülich GmbH, Jülich, Germany

Nowadays one is able to nanostructure samples in all three dimensions. Particularly interesting is here that also the spin diffusion length of electrons in metals and oxides is on the same length scale (nm) as these nanostructured samples, and thus new effects and even maybe new physics can be expected. New pattern formations in the nanoworld like non-collinear magnetism at the surface, the influence of the substrate on the sample, different orderings at the surface and in the depth of nanostructures and proximity of the superparamagnetic limit are some examples to be addressed. Another intriguing field of research are the properties at interfaces between oxide materials. Here we can expect superconductivity between insulating layers, magnetism between non magnetic layers or ferro-magnetism between anti-ferromagnetic layers.

The JCNS has installed the new, high-intensity reflectometer MARIA in the neutron guide hall of the FRM II reactor in Garching. This instrument uses a velocity selector for the monochromatization of the neutron beam, an elliptically focussing guide to increase the flux at the sample position and a double-reflecting supermirror polarizer to polarize the entire cross-section of the beam delivered by the neutron guide.

Unique features of MARIA include (i) vertical focussing with an elliptic guide from 170 mm down to 10 mm at the sample position, (ii) reflectometer and GISANS mode, (iii) polarization analysis over a large 2d position sensitive detector as standard, (iv) adjustable wavelength spread from 10 to 1 % by a combination of velocity selector and chopper, (v) flexible sample table using a Hexapod for magnetic field and low temperature sample environment and (vi) in-situ sample preparation facilities. Together with a 400 x 400 mm² position sensitive detector and a time-stable ³He polarization analyzer based on in-situ Spin-Exchange Optical Pumping (SEOP), the instrument is dedicated to investigate specular reflectivity and off-specular scattering from magnetic layered structures down to the monolayer regime. In addition the GISANS option can be used to investigate lateral correlations in the nm range. This option is integrated into the reflectometer's collimation, so it can be chosen during the measurement without any realignment.

MARIA is a state of the art reflectometer at a constant flux reactor. It gives you the opportunity to investigate easily reflectivity curves in a dynamic range of up to 7-8 orders of magnitude, off-specular scattering, GISANS and even simple SANS measurement. We will discuss how MARIA can help you to investigate the depth resolved vector information of your magnetic samples.

TEM 1

TRANSMISSION ELECTRON MICROSCOPY OF FUNCTIONAL PEROVSKITE OXIDE HETEROSTRUCTURES

Dietrich Hesse

Max Planck Institute of Microstructure Physics, D-06120 Halle, Germany

Functional oxide heterostructures consisting of layers made from perovskite-type ferroelectric (e.g., $BaTiO_3$, $PbZr_{0.2}Ti_{0.8}O_3$) and ferromagnetic (e.g., $SrRuO_3$, $LaSr_{0.7}Mn_{0.3}O_3$) materials are most interesting in view of their respective ferroic or even multiferroic properties. The latter critically depend on the microstructure of the layers and the involved interfaces. Here, the indispensible role of transmission electron microscopy (TEM) for in-depth investigations of functional heterostructures will be highlighted, reporting recent work on the role of lattice defects and closure domains for the properties of ferroelectric heterostructures, as well as on the relevance of atomic interface structures for the properties of multiferroic tunnel junctions.

(1) TEM results proved essential to explain the observed switching kinetics of ferroelectric $PbZr_{0.2}Ti_{0.8}O_3$ (PZT) nanocapacitors of as low as 70 nm lateral size (diameter), studied in view of their potential suitability for non-volatile solid-state memories. These nanocapacitors were prepared by stencil-based pulsed laser deposition, *cf.* [1]. Their switching kinetics were investigated using switching pulse-length dependent piezoresponse force microscopy (PFM), observing the formation and movement of ferroelectric 180° domains by PFM [2]. Applying a switching pulse to a nanocapacitor, it was found that after nucleation of a single 180° domain the vertical movement of the head-head coupled bottom section of the domain boundary was much more rapid than the lateral spreading of the head-tail coupled sections. This finding could be explained on the base of previous TEM observations of the atomic structure of most similar 180° boundaries in $SrTiO_3$(STO)/PZT/STO model heterostructures [3]. Here, the 180° boundaries were observed by negative-spherical abberation coefficient imaging (NCSI) in high-resolution TEM (HRTEM) using an objective abberation-corrected transmission electron microscope at ER-C Jülich: Head-head coupled 180° boundaries were ten times thicker (ca. 4 nm) than head-tail boundaries (ca. 0.4 nm). Assuming corresponding boundary mobility differences (from the analogy with magnetic domain walls, the mobility of which is the higher, the thinner the walls are) it can be concluded that head-head coupled sections of ferroelectric 180° boundaries move much more rapidly than head-tail coupled sections, thus explaining the above mentioned PFM observations.

(2) NCSI imaging allows not only to establish the positions of the cations within a perovskite lattice, but also those of the oxygen ions [4]. This allows the quantitative evaluation of HRTEM micrographs of ferroelectric materials in terms of the direction and modulus of the ferroelectric dipoles of each unit cell. In this way, e.g., the influence of a single lattice dislocation in the STO substrate on the local value of the ferroelectric polarization of a deposited PZT thin film was investigated, proving that directly above the dislocation the polarization value is reduced by almost 50 % [5]. Another application of this technique is shown in Figure 1a from [6]: A closure domain implying a *continuous rotation* of ferroelectric dipoles in PZT is shown within a STO/PZT/$SrRuO_3$(SRO)/STO heterostructure close to the bottom PZT interface. Here, the upper scale bar (0.5 nm) refers to the lateral scale, whereas the lower scale bar (40 pm) is related to the cation-anion shifts causing the dipoles. The continuous rotation of ferroelectric dipoles in perovskites contradicts standard textbook knowledge, but had recently been predicted by *ab initio* theory to occur in PZT heterostructures close to their interfaces, *cf.* [7]. The present observation, as well as a similar observation in $BiFeO_3$ heterostructures [8], clearly confirm those predictions.

(3) The switching of the ferroelectric polarization in a multiferroic tunneling junction (MFTJ) of type Co/PZT/La$_{0.7}$Sr$_{0.3}$MnO$_3$(LSMO)/STO with a thickness of the PZT tunneling barrier of 3.2 nm was recently shown to modify the spin polarization of the tunneling electrons [9]. As a consequence, polarization switching by an applied elelctrical voltage pulse not only changes the *magnitude* of the tunnel magnetoresistance (TMR), but most importantly also its *sign*. This effect is reversible, making it possible to repeatedly switch the magnetic characteristics of the MFTJ by an electrical signal. It can be proposed that the quality and detailed atomic structure of the Co/PZT interface is responsible for this new effect. Figure 1b (inset of c) shows an overview image of a MFTJ dot, and Figure 1c is a corresponding HRTEM image of a similar MFTJ. Detailed studies of the Co/PZT interface structure are under way in order to understand the mechanism of switching.

Overall, the presentation will demonstrate the preeminent role of TEM for the understanding of the functional properties of perovskite-type functional oxide heterostructures.

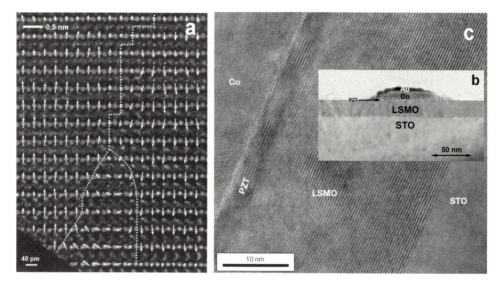

Figure 1: (a) Map of dipoles - based on an NCSI HRTEM image - in a PbZr$_{0.2}$Ti$_{0.8}$O heterostructure showing the continuous rotation of dipoles (from [6]); (b) overview image of a Co/PZT/LSMO multiferroic tunnel junction (MFTJ) on SrTiO$_3$ (STO) covered with a gold (Au) cap, and (c) HRTEM image of a similar MFTJ (from [9]).

[1] H. Han, Y. Kim, M. Alexe, D. Hesse, and W. Lee, Advanced Materials **23**, 4599 (2011).
[2] Y. Kim, H. Han, W. Lee, S. Baik, D. Hesse, and M. Alexe, Nano Letters **10**, 1266 (2010).
[3] C.-L. Jia, S.-B. Mi, K. Urban, I. Vrejoiu, M. Alexe, and D. Hesse, Nature Mat. **7**, 57 (2008).
[4] C.L. Jia, M. Lentzen, and K. Urban, Science **299**, 870 (2003).
[5] C.L. Jia, S.B. Mi, K. Urban, I. Vrejoiu, M. Alexe, and D. Hesse, Phys. Rev. Lett. **102**, 117601 (2009).
[6] C.-L. Jia, K.W. Urban, M. Alexe, D. Hesse, and I. Vrejoiu, Science **331**, 1420 (2011).
[7] S. Prosandeev and L. Bellaiche, Phys. Rev. B **75**, 172109 (2007).
[8] C.T. Nelson, B. Winchester, Y. Zhang, S.-J. Kim et al., Nano Letters **11**, 828 (2011).
[9] D. Pantel, S. Goetze, D. Hesse, and M. Alexe, Nature Mat. (26 February, 2012), doi: 10.1038/nmat3254.

NiO PRECIPITATES IN LaNiO$_3$/LaAlO$_3$ SUPERLATTICES INDUCED BY A POLAR MISMATCH

Eric Detemple[1], **Quentin M. Ramasse**[2], **Wilfried Sigle**[1], **Eva Benckiser**[3], **Georg Cristiani**[3], **Hanns-Ulrich Habermeier**[3], **Bernhard Keimer**[3], **Peter A. van Aken**[1]

[1]Max Planck Institute for Intelligent Systems, Stuttgart, Germany; [2]SuperSTEM Laboratory, STFC Daresbury, Warrington, United Kingdom; [3]Max Planck Institute for Solid State Research, Stuttgart, Germany

Heterostructures of transition metal oxides show intriguing phenomena such as the ordering of charge, spin and orbitals which can give rise to multiferroicity or superconductivity and makes them interesting candidates for future electronic devices [1]. Among them, LaNiO$_3$ (LNO) is a promising representative due to its strongly correlated conduction electrons. It was reported that in ultrathin LNO films or LNO-based superlattices the phase behavior and the transport properties depend on the thickness of the films or the single layers of the superlattice [2].

Since the properties of these systems are very sensitive to minor changes, it is evident that all heterostructures should be very accurate. However, if a polar and a non-polar material are combined, a "perfect" system is not stable because of energetic reasons (polar catastrophe) so it has to alter, e.g. by interdiffusion or electronic reconstruction [3]. In the case of the interface between LNO and SrTiO$_3$ (STO), a Ni valence state of 2+ was reported besides the trivalent Ni of bulk LNO concluding the formation of an interfacial LNO layer with a Ni valence state of 2+ which hinders the polar catastrophe [4].

We have studied the microstructure of LaNiO$_3$/LaAlO$_3$ superlattices in dependence of the polarity of the used substrate (polar (La,Sr)AlO$_4$ (LSAO) or non-polar STO) by transmission electron microscopy in combination with electron energy-loss spectroscopy (EELS). A high-angle annular dark field image of a film on a STO substrate and elemental EELS maps of La, Ni, and Al (Figure 1) demonstrate the high precision of the superlattices.

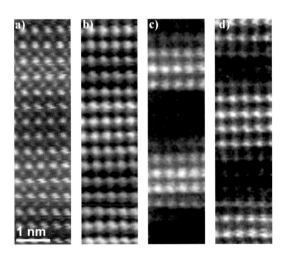

Figure 1: LaNiO$_3$/LaAlO$_3$ superlattice grown on SrTiO$_3$: a) a high-angle annular dark field image and atomically resolved elemental maps recorded by electron energy-loss spectroscopy: b) lanthanum, c) nickel and d) aluminum.

In the case of a non-polar substrate like STO, which results in polar mismatch with LNO, the films contain nanometer-sized precipitates which form directly at the interface between the substrate and the initial LNO layer (Figure 2a). The elemental EELS maps in Figure 2 show a strong enhancement of the Ni signal with a simultaneous lack of La and Al within the precipitate. The fine structure of the O K absorption edge and the measurement of the lattice parameter with atomically resolved images show that these precipitates are NiO. However, no NiO precipitates were found if a polar substrate like LSAO was used so that the formation of these precipitates seems to be triggered by the polarity of the substrate [5].

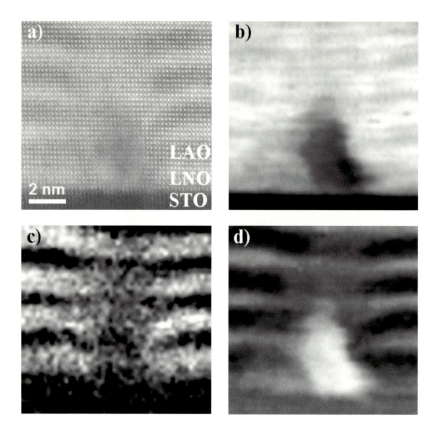

Figure 2: a) A high-angle annular dark field image showing a NiO precipitate which is embedded in the superlattice. It forms directly at the interface to the $SrTiO_3$ substrate. b) lanthanum, c) aluminum and d) nickel maps recorded by electron energy-loss spectroscopy.

It should always be considered that a polar discontinuity in heterostructures of transition metal oxides can cause the formation of small amounts of secondary phases whose electric properties can significantly differ from the properties of the components of the superlattice so that they can strongly affect measurements on such heterostructures and superlattices. Transmission electron microscopy and EELS are excellent tools to rule out the presence of secondary phases at buried interfaces which is in general very challenging with standard experiments because of their small volume fraction.

[1] H.Y. Hwang et al., Nature Mater. **11**, 103 (2012).
[2] A.V. Boris et al., Science **332**, 937 (2011).
[3] J. Mannhart and D. G. Schlom, Science **327**, 1607 (2010).
[4] J. Liu et al., Appl. Phys. Lett. **96**, 133111 (2010).
[5] E. Detemple et al., Appl. Phys. Lett. **99**, 211903 (2011).

MINIMUM ENERGY CONFIGURATION OF SCANDATE/TITANATE INTERFACES: ORDERED INTERFACES

Martina Luysberg[1], Kourosh Rahmanizadeh[2], Gustav Biehlmayer[2], Jürgen Schubert[3],

[1]Peter Grünberg Institut (PGI-5) & Ernst Ruska Centre, Forschungszentrum Jülich, 52425 Jülich, Germany
[2]Peter Grünberg Institut (PGI-1) & Institute for Advanced Simulation (IAS-1), Forschungszentrum Jülich and JARA, 52425 Jülich, Germany;
[3]Peter Grünberg Institut (PGI-9) and JARA, Forschungszentrum Jülich, 52425 Jülich, Germany

Unlike the polar interface between $LaAlO_3$ and $SrTiO_3$, where a conducting interface has been discovered [1], the polar $DyScO_3/SrTiO_3$ system remains electrically insulating. This can be attributed to an off-stoichiometry of ions of different valence at the interface between e.g. $DyScO_3$ and $SrTiO_3$, which counteracts the interface dipoles arising from the polar discontinuity [1]. Here, we investigate the structural and chemical composition of this interface by high-resolution scanning transmission electron microscopy, atomic-resolution spectroscopic imaging and ab-initio calculations.

Figure 1 displays the result of ab initio calculations for different interface models of the $DyScO_3/SrTiO_3$ interface [2], i.e. Dy atoms (silver) and Sr atoms (golden) are arranged in various configurations within the (x,y) interface plane. The lowest energy configuration, which is shown along the [010] direction on the right of Fig.1, matches the experiment: The high-resolution high angle annular dark field image (Fig 2), recorded with an probe corrected Titan 80-300 electron microscope, reveals one interfacial layer, where every second atom column of the Dy/Sr sublattice is brighter, i.e. these atom columns consist mainly of Dy.

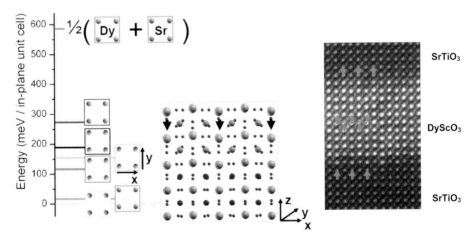

Figure 1: Calculated energy per in-plane unit cell for various interface models. The minimum energy configuration is displayed along the [010] direction on the right, which is also the viewing direction of the HAADF image in Figure 2

Figure 2 HAADF image of $DyScO_3/SrTiO_3$. interfaces reveals an ordering, i.e. atom columns of brighter contrast are detected (arrows)

[1] M. Luysberg, M. Heidelmann, L. Houben, M. Boese, T. Heeg, J. Schubert, and M. Roeckerath Acta Materialia **57**, 3192 (2009)
[2] K. Rahmanizadeh, G. Bihlmayer, M. Luysberg, and S. Blügel Phys. Rev. B 85, 075314 (2012)

RUDDLESDEN–POPPER TYPE FAULTS IN LaNiO$_3$/LaAlO$_3$ SUPERLATTICES

Eric Detemple[1], Quentin M. Ramasse[2], Wilfried Sigle[1], Georg Cristiani[3], Hanns-Ulrich Habermeier[3], Bernhard Keimer[3], Peter A. van Aken[1]

[1]Max Planck Institute for Intelligent Systems, Stuttgart, Germany
[2]SuperSTEM Laboratory, STFC Daresbury, Warrington, United Kingdom
[3]Max Planck Institute for Solid State Research, Stuttgart, Germany

Heterostructures and superlattices of transition metal oxides are potential candidates for a new generation of electronic devices because exciting phenomena can occur at their interfaces which are absent in the bulk components. In addition, epitaxial strain, the layer thickness and external fields allow tuning their properties [1]. However, besides these intrinsic parameters, the microstructure of the entire films can also affect the properties, e.g. in form of dislocations or planar faults [2,3].

Here we report about the characterization of LaNiO$_3$/LaAlO$_3$ superlattices (LNO/LAO SLs) with two or four unit cell thick single-layers, which were grown on (La,Sr)AlO$_4$ (LSAO) substrate by pulsed laser deposition. Transmission electron microscopy in combination with electron energy-loss spectroscopy (EELS) is used to study the defect structure on the atomic scale for different layer thicknesses. The films contain extended planar faults whose origin are surface steps of the LSAO substrate resulting in a relative vertical lattice translation on the two sides of the surface step as envisaged in Figure 1a by the overlaid LNO unit cells. In addition, the elemental EELS map (Figure 1b) shows that one NiO plane is missing along the planar fault (see black arrow). Such a structure is known from Ruddlesden–Popper phases (RP phases) which is why these defects are called RP faults and can be classified as a two dimensional (2D) defect [4].

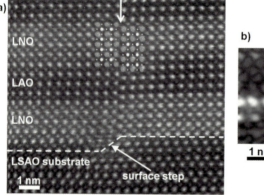
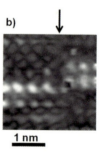

Figure 1: a) A HAADF image of a planar RP fault (marked by the upper arrow) starting at a surface step of the LSAO substrate. The LNO lattice is overlaid (La is blue, Ni yellow and O grey) to illustrate the shift of the two sides against each other. b) Elemental EELS map showing a RP fault (La is blue, Ni yellow, and Al red).

Beside these extended planar RP faults, rectangular blocks exist in the films as can be seen in the high-angle annular dark field (HAADF) image in Figure 2a. The intensity of the atom columns is very similar within the blocks indicating that the average atomic number is similar because the atom columns are mixed, as shown in Figure 2b. For example, the atom column marked by an arrow consists of Al and La atoms. The source of these cuboid-shaped blocks is a strongly localized stacking fault (see blue arrow in Figure 2b) which propagates in growth direction through the film. The zigzag arrangement of La atoms along the borders of the block is characteristic of RP faults showing that the blocks are terminated by RP faults on all faces. Therefore we call this defect type in the following 3D RP fault. In contrast to the planar RP faults which are induced by the surface steps of the substrate, no correlation with the substrate could be found for the 3D RP faults which therefore rather seem to be the result of small irregularities, i.e. the formation of stacking faults, during the growth process. The phase behavior of LNO/LAO SLs differs in dependence of the thickness of the individual layers because of the confinement of the conduction electrons in the thinner single layers [5]. However, both defect structures are found irrespective of the single-layer thickness of the SL. Hence, the microstructure cannot be the decisive factor for the property change of SLs with different single-layer thickness.

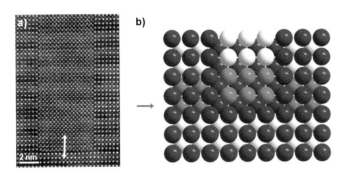

Figure 2: a) A HAADF image showing a single block which consists of columns with similar brightness. b) Cross-section of the TEM sample (the electron propagation direction is from left to right) in the direction of the yellow arrow in a). La atoms are blue, Ni yellow, Al red, and O is neglected. The arrow points to the stacking fault which induces the growth of the 3D RP fault.

In conclusion, we have found two different types of RP faults in LNO/LAO SLs: extended 2D defects which are induced by surface steps of the LSAO substrate and nanometer-sized 3D defects whose origin are local stacking faults. However, it turned out that thickness-dependent differences in the phase behaviour of LNO/LAO SLs are not caused by the defect structure but they are an inherent phenomenon of the SLs.

[1] J. Mannhart and D.G. Schlom, Science **327**, 1607 (2010).
[2] S. Thiel et al., Phys. Rev. Lett. **102**, 046809 (2009).
[3] H. Li et al., Appl. Phys. Lett. **81**, 4398 (2002).
[4] R.J.D. Tilley, J. Solid State Chem. **21**, 293 (1977).
[5] A.V. Boris et al., Science **332**, 937 (2011).

ATOMIC STRUCTURE OF TRIMERIZATION-POLARIZATION DOMAIN WALLS IN HEXAGONAL ErMnO$_3$

Myung-Geun Han[1], Lijun Wu[1], Toshihiro Aoki[2], Nara Lee[3], Seung Chul Chae[3], Sang-Wook Cheong[3], and Yimei Zhu[1]

[1]Brookhaven National Laboratory, Upton, NY, USA; [2]JEOL USA, Inc., Peabody, MA, USA; [3]Rutgers University, Piscataway, NJ, USA

Hexagonal ReMnO$_3$ (Re: rare-earth elements such as Y, Er, Ho, etc.) has been received intense research interest thanks to their coexistence of electric and magnetic orderings and possible applications for nonvolatile data storage devices. These crystals are geometric ferroelectrics since spontaneous polarization is induced by structural trimerization of Mn ions and buckling of Re ions due to ionic size mismatch between Re ions and Mn ions. Considering two possible directions of polarization along the c-axis ("+" being parallel to the c-axis and "-" being antiparallel to the c-axis) and three antiphases (α, β, and γ) as a result of Mn trimerization, total 6 distinctive domains (α+, α-, β+, β-, γ+, γ-) are predicted in a single crystalline hexagonal ReMnO$_3$. Recently, Choi *et al.* reported that the trimerization domain walls were interlocked with ferroelectric domain walls and the two types of domain walls (APBI + FEB and APBII + FEB) were emerging from topological defects such as vortex and antivortex with winding orders of α+, β-, γ+, α-, β+, γ- and α+, γ-, β+, α-, γ+, β-, respectively[1]. These interlocked trimerization-polarization domain walls are believed to be topologically-protected because even with application of large electric fields, no complete switching, i.e., no paired-domain-wall annihilation, is observed.

In this study, using aberration-corrected scanning transmission electron microscopy (STEM) the direct observation of local ionic distortions associated with two types of trimerization-polarization domain walls in hexagonal ErMnO$_3$ is performed. STEM images in Fig. 1 show two distinct types of ferroelectric domain walls interlocked with trimerization domain walls, denoted as type A (Fig. 1(a) and 1(c)) and type B (Fig. 1(b) and 1(d)), respectively. An apparent difference between the two types of the domain walls shown in Fig. 1 is the wall width: 1/3[210] for type-A and 2/3[210] for type-B, respectively, in the [100] projection. Here, domain wall width is defined as the separation between two distinctive unit cells for each neighboring domain. Domain walls do not perfectly run in parallel to the c-axis, indicating these walls are either positively charged in a head-to-head configuration (Fig. 1(a) and 1(c)), or negatively charged in a tail-to-tail configuration (Fig. 1(b) and 1(d)).

Fig. 2 summarizes structural and electrostatic charge configurations of trimerization-polarization domain walls around vortex and antivortex in hexagonal ErMnO$_3$, based on experimental observations shown in Fig. 1. Characteristic lattice translations are determined for each trimerization-polarization domain wall, which characterize the phase shift between unit cells of neighboring trimerization domains across domain walls. For the domain walls emerging from a vortex, the unit cells are shifted by a vector, 1/3[-110], in the *ab* plane. For the domain walls associated with an antivortex, the phase-shift vector is reversed to -1/3[-110] due to the reversed winding order, as shown in Fig. 2. This work implies that the interactions between the neighboring domain walls in terms of local phase discontinuity and electrostatic charges may play a significant role in domain switching behavior of hexagonal ErMnO$_3$.

This work is supported by the U.S. Department of Energy, Office of Basic Energy Science, Division of Materials Science and Engineering, under Contract number DE-AC02-98CH10886. The work at Rutgers was supported by National Science Foundation DMR-11004484.

Nanosession: High-resolution transmission electron microscopy

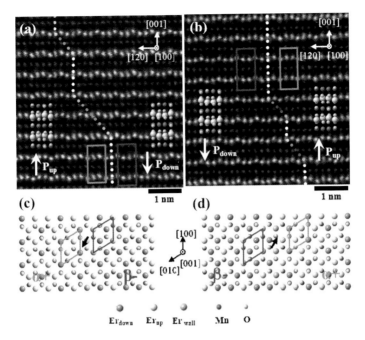

Figure 1: Annular dark-field STEM images of the two types of domain walls in the [100] projection: (a) type A and (b) type B. Domain walls shown in dotted lines separates two neighboring domains with opposite polarization. Charged segments of domain walls are indicated with red (positive charges in head-to-head configuration) and blue (negative charges in tail-to-tail configuration) colors. Atomic models and unit cells (blue and red rectangles) are also shown. Atomic models of two types of domain walls seen along the [001] axis are shown in (c) and (d). Er ions are either displaced upward (brown, Er_{up}) or downward (yellow, Er_{down}) along the [001] direction. Er ions located at domain wall are shown as light brown. Unit cells for each domain are shown in red and blue rhombuses. Black arrows at domain wall represent the phase shift vector for each domain walls.

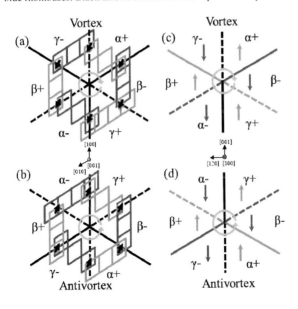

Figure 2: Schematics showing phase shift vectors (black arrows) for domain walls associated vortex (a) and antivortex (b) in the ab plane. Red and blue rhombuses highlight the unit cell for each trimerization domain. Solid lines represent APBI + FEBs and dotted-lines APBII + FEBs (see Ref. 1). Cross-sectional view (in the [100] projection) of vortex (c) and antivortex (d). Red and blue arrows indicate the polarization directions. Solid and broken lines represent APBI + FEB and APBII + FEBs, respectively. Blue, red, and black lines indicate negatively-charged (tail-to-tail), positively-charged (head-to-head), and neutral domain walls, respectively. Grey arrows indicate the winding order for vortice and antivortice.

[1] T. Choi et al., Nature Materials, **9**, 253 (2010).

CONSTRUCTION AND FIRST RESULTS OF AN STM OPERATING AT MILLI-KELVIN TEMPERATURES

C. R. Ast[1], M. Assig[1], M. Etzkorn[1], M. Eltschka[1], B. Jäck[1], and K. Kern[1]

[1]Max-Planck-Institute for Solid State Research, Heisenbergstrasse 1, Stuttgart, Germany

Progress in research is intimately connected with progress in developing novel and improved experimental methods. Since its birth in 1980s, scanning tunneling microscopy has continuously developed into one of the most important research tools in surface science. In order to push its limits further, we have constructed a scanning tunneling microscope operating at 15mK (mK-STM). At these temperatures, we were able to measure an effective spectroscopic energy resolution of about 19µeV (=3.5kT) using a superconducting Al tip on a Cu(111) surface (see Fig. 1). In addition, the mK-STM operates in high magnetic fields up to 14T and is connected to an ultra-high vacuum compatible sample preparation chamber. This allows us to study extremely low lying energy transitions and interactions in, for example, elemental superconductors which generally feature a very low transition temperature or spin-flip transitions. We will discuss the specifications of the mK-STM as well as the challenges during design and construction. We will present first results on the Zeeman splitting of the superconducting gap and give an outlook on new scientific frontiers.

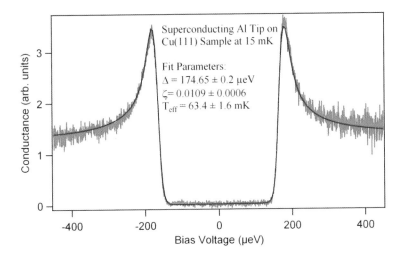

Figure 1: Superconducting gap measured with the mK-STM using an Al tip on a Cu(111) sample. The data was fitted using a model function for a superconducting gap convoluted with the derivative of the Fermi function to accommodate an effective temperature T_{eff} broadening. The parameters ☐ and ☐ are the gap and the orbital depairing, respectively.

QUANTITATIVE FORCE IMAGING OF THE ATOMS IN EPITAXIALLY GROWN GRAPHENE

M.P.Boneschanscher[1], Z. Sun, J. van der Lit[1], P. Liljeroth[2] and D. Vanmaekelbergh[1]

[1]Condensed Matter and Interfaces, Debye Institute for Nanomaterials Science, Utrecht University, PO Box 80000, 3508 TA Utrecht, the Netherlands
[2]Low Temperature Laboratory, Aalto University School of Science, 00076 Aalto, Finland

Scanning probe microscopy is an indispensable tool in order to study both the atomic structure as well as the electronic properties of graphene in real space. A crucial question regarding the atomic structure is how to relate the periodic contrast in the acquired images to the true position of the C atoms and the hollow sites [1].

Using non-contact atomic force microscopy with very small modulation amplitudes (~150pm) we measured the frequency shift df over epitaxially grown graphene on Ir(111) at various tip-sample distances d_{ts}. This was done both using a reactive, metallic tip as well as a non-reactive, carbon monoxide (CO) terminated tip. By simultaneous measurement of the current we could link all features in df to their position within the moiré unit cell. Furthermore we measured spectra of df vs d_{ts} at various positions on the surface.

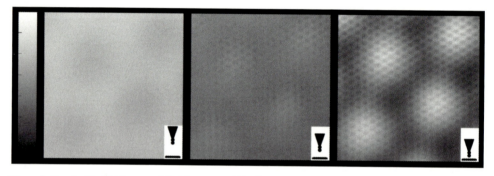

Figure 1: Constant height images of the frequency shift df recorded at different tip-sample distances d_{ts} with a CO terminated tip. All panels are 5x5nm in size, color legend ranges from -50 to -120Hz (dark) with divisions at every 10Hz. The tip-sample distance d_{ts} was decreased by ~50pm between the different panels.

In figure 1, the frequency shift df recorded with a CO terminated tip is plotted for different tip sample distances. At large d_{ts} the only features recorded resemble the attraction (more negative frequency shift, dark) that the CO terminated tip experiences on the on top positions of the moiré unit cell (left panel), where d_{ts} is reduced due to the outward buckling of the graphene. When d_{ts} is lowered, an additional feature is observed in the form of a chicken-wire pattern of repulsive interaction (less negative frequency shift, bright) on the on-top positions of the moire pattern (middle panel). We attribute this to the C atoms forming the backbone of the graphene. At very small d_{ts} this repulsive interaction is observed over the entire graphene substrate (right panel). Furthermore the interaction on the on top positions of the moiré pattern has changed from attractive to repulsive due to the small d_{ts}.

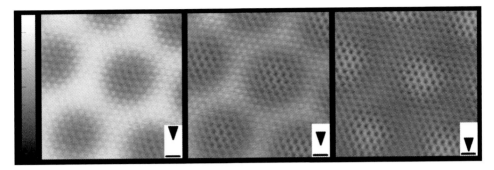

Figure 2: Constant height images of the frequency shift df recorded at different tip-sample distances d_{ts} with a metallic tip. All panels are 5x5nm in size, color legend ranges from -40 to -120Hz (dark) with divisions at every 10Hz. The tip sample distance d_{ts} was decreased by ~35pm between the different panels.

In figure 2 the frequency shift df recorded with a metallic tip is plotted for different tip sample distances. A remarkable difference with respect to the CO terminated case is the atomic scale contrast observed at large d_{ts}. The metallic tip first experiences a significant attractive interaction on top of the carbon backbone (left panel) before it changes to repulsive at very small d_{ts} (right panel). In the middle panel the contrast changes from repulsive on the C atoms making up the on top positions of the moiré, to attractive on the C atoms in the low-lying parts of the moiré unit cell.

We quantified the different contributions by measuring spectra of df vs d_{ts} on different C atoms and hollow sites within the moiré unit cell and integrating this to the force. Strikingly, we found that although the absolute force that the metallic and CO terminated tip experience from the graphene are comparable, the force contrast between the hollow and on-top C positions is significantly larger with the metallic tip. This is further demonstrated in the onset of the atomic scale contrast which occurs at larger d_{ts} with the metallic tip.

In summary, we have quantified the force-distance curves above the different sites of epitaxial graphene both with a reactive tip and a CO-terminated tip. This led to an unambiguous determination of the atomic and hollow positions on graphene, it will allows to study the chemistry of the edges, and understand the nature of the forces on a quantum mechanical basis.

[1] M. Ondracek et al., Phys. Rev. Lett. **106**, 176101 (2011).

RADIO FREQUENCY OPTIMIZED SCANNING TUNNELING MICROSCOPE FOR THE USE WITH PULSED TUNNELING VOLTAGES

Christian Saunus[1], Marco Pratzer[1], Markus Morgenstern[1]

[1]II. Physikalisches Institut, RWTH Aachen University, Aachen, Germany

Short voltage pulses can be used in a pump-probe approach to increase the temporal resolution of a scanning tunneling microscope (STM) [1]. We combine a home built radio frequency (RF) optimized STM with a pulse generator to apply the currently shortest voltage pulses to a tunneling junction. The microscope is designed to work in a 4 K UHV system with a 7 T rotatable magnetic field. Since the pulses are applied to the tunneling tip we do not need RF-optimized samples for using this technique [2]. The pulse width is determined by a pulse superposition technique which utilizes the non-linear I(V)-characteristic of HOPG. We compare the measurement with a simulation of the data to determine a minimal voltage pulse width of 120 ps which seems to be only limited by our pulse source.

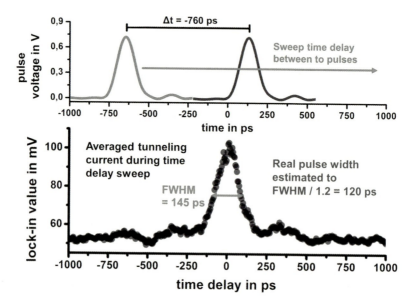

Figure 1: Pulse superposition measurement using two 120 ps pulses. Upper part: Oscilloscope image of the two pulses directly at the output of the pulse generator. The summed voltage of the single pulses is lead to the tunneling tip and generates a tunneling current with respect to the sample's I(V)-characteristic. Lower part: A pulse superposition measurement using the two pulses shown in the upper part. The time delay between the two pulses is swept between -1 and 1 ns and the average tunneling current is measured by a lock-in amplifier. The non-linearity of the HOPG I(V) characteristic leads to an increased average current during pulse overlap. A simulation of the measured data proves the actual pulse width to be 120 ps. Hence, the pulse was not distorted on its way to the tunneling junction. This gives rise to the assumption that the pulse width and therefore the possible time resolution is only limited by the pulse generator's bandwidth.

[1] S. Loth et al., Science **329**, 1628 (2010).
[2] Ian Moult et al., Appl. Phys. Lett. **98**, 233103 (2011).

ULTRA COMPACT 4-TIP STM/AFM FOR ELECTRICAL MEASUREMENTS AT THE NANOSCALE

Vasily Cherepanov[1], Stefan Korte[1], Marcus Blab[1], Evgeny Zubkov[1], Hubertus Junker[1], Peter Coenen[1], Bert Voigtländer[1]

[1]Peter Grünberg Institut (PGI-3), Forschungszentrum Jülich, 52425 Jülich, Germany, and JARA-Fundamentals of Future Information Technology

We constructed a 4-tip STM/AFM where four independent STM units are integrated on a diameter of 50 mm. The coarse positioning of the tips is done under the control of an optical microscope or an SEM in vacuum. Fine positioning is done using STM scanning. The modular and compact design allows building a four tip STM as small as a single tip STM.

It was possible to construct the multi-tip STM this small due to a new kind of nanopositioner, the Koala Drive, which has diameter less than 2.5 mm and length smaller than 10 mm and serves as an STM coarse positioning device in our multi-tip STMs. Alternating movements of springs move a tube which holds the STM tip or AFM sensor. This new operating principle provides a smooth travel and avoids shaking which is intrinsically present for nanopositioners based on inertial motion with saw tooth driving signals. The Koala Drive makes the scanning probe microscopy design ultra compact and leads accordingly to a high mechanical stability.

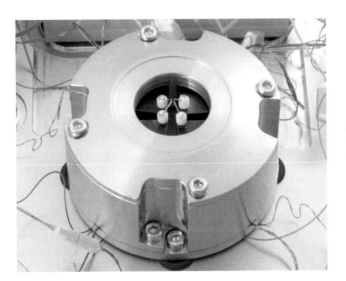

Figure 1: Ultra compact multi tip STM with 50 mm outer diameter.

Here we present the design of the multi tip microscope and show several measurement examples to demonstrate the device performance, such as four point measurements on silicide nanowires and graphene samples.

NANOSCALE MECHANICAL CHARACTERIZATION OF THIN FILMS with Different TopologiCAL STRUCTURES

Kong-Boon Yeap[1], Malgorzata Kopycinska-Mueller[1,2], Lei Chen[3], Martin Gall[1], Ehrenfried Zschech[1]

[1] Fraunhofer Institute for Nondestructive Testing, Dresden, Germany
[2] Technische Universität Dresden, Dresden, Germany
[3] Institute of High Performance Computing, Singapore

Porous organosilicate (OSG) glasses with low dielectric constant (k) are used as insulating material between metal on-chip interconnects to ensure high performance of microelectronic products. However, the introduction of porosity that reduces the dielectric permittivity reduces the Young's modulus to a level that is critical for process integration and reliability. Thus, understanding of the effect of the pore topology on small-scale deformation mechanisms is essential. To study the elastic properties of a new class of low-k porous films [1], two techniques were used, nanoindentation (NI) at small indentation depths (below 20 nm) [2] and atomic force acoustic microscopy (AFAM) [3]. The measurements performed at nanoscale provide information about the pore deformation mechanisms in addition to the modulus values. Based on the experimental results applying the NI and the AFAM techniques, we have shown for the first time that the pore topology has a significant effect on the Young's modulus of porous materials. Particularly, OSG glasses with translation-symmetric pores and with constant pore size show a significantly higher elastic modulus compared to films with randomly distributed pores [4].

Indentation studies were performed for two types of porous materials, OSG glasses with broad pore size distribution (CVD-OSG) and with narrow pore size distribution and ordered pore arrangement (SA-OSG). The results of the indentation measurements performed on CVD-OSG films follow the Hertzian elastic contact model. The thin film material with broad pore size distribution is elastically deformed without densification. The experiments performed on SA-OSG films show deformation mechanisms that include elastic stiffening of the material and elastic hysteresis, corresponding to pores densification and pore-wall buckling (Fig. 1A-1C). These observations can be explained by the deformation mechanism with the lowest energy consumption, i. e. the bending of siloxane bonds [5]. The ordered pore arrangement of SA-OSG films causes a deformation which is connected with pore densification, which is different compared to the CVD-OSG material. The values of the indentation modulus determined for the SA-OSG films are two times larger than those determined for the CVD-OSG films with the same chemical composition and the same level of porosity (25 %) (Fig. 2A). The results of the FE modeling show that the porous structure in the CVD-OSG films can be described as random-overlapping solid spheres. In case of the SA-OSG samples with a porosity $p < 0.3$, the pore arrangement is that of ordered non-overlapping spherical pores. For samples with $p > 0.3$, the arrangement of the pores changes continuously to that of the random-overlapping spherical pores. TEM studies of the SA-OSG thin film with $p > 0.3$ confirm the prediction of the FE model (Fig. 2B) [4].

The combination of the indentation measurements at scales comparable to the length scales of pores arrangements and the predictions of FE modeling provide information invaluable for the design of the porous films and their integration into the on-chip interconnect stack of future microelectronic devices.

Nanosession: New technologies for scanning probes 149

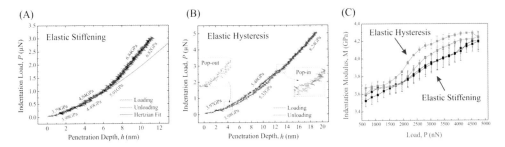

Fig.1: (A-B) NI load-penetration (P-h) curves for SA-OSG ($p=0.4$) showing (A) elastic stiffening of the material and (B) elastic hysteresis. (C) Dependence of the indentation modulus on the applied static load recorded using the AFAM method confirming the NI results.

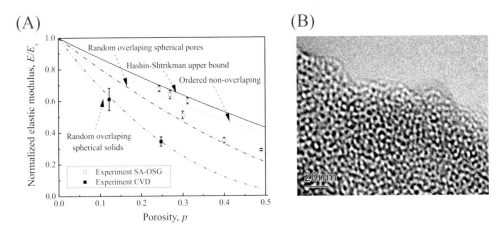

Fig.2: (A) Plot of normalized elastic modulus against porosity (E/E_s vs. p) for different models. (B) TEM image of a thermally cured self-assembled OSG glass (SA-OSG) film with $p>0.3$ [4].

[1] K. Landskron, B.D. Hatton, D.D. Perovic, G.A. Ozin, Science 302, 226 (2003).
[2] E. Zschech, K. B. Yeap, R. Huebner, Int. Conf. AMC, San Diego (2011).
[3] U. Rabe, Applied scanning probe methods II, Springer, Berlin, p. 37 (2006).
[4] K.B. Yeap, M. Kopycinska-Mueller, L. Chen, S. Mahajan, M. Phillips, W. R. Bottoms, Y. Chen, J.J. Vlassak, M. Gall, E. Zschech, in preparation.
[5] J.M. Knaup, H. Li, J.J. Vlassak, E. Kaxiras, Phys. Rev. B 83, 054204 (2011).

ELECTRONIC ACTIVATION IN THE $(La_{0.8}Sr_{0.2})CoO_3/(La_{0.5}Sr_{0.5})_2CoO_4$ SUPERLATTICES AT HIGH TEMPERATURE

Yan Chen,[1] Zhuhua Cai,[1] Yener Kuru,[1,2] Harry L. Tuller[2] and Bilge Yildiz[1]*

[1]Laboratory for Electrochemical Interfaces, Department of Nuclear Science and Engineering, Massachusetts Institute of Technology, 77 Massachusetts Avenue, Cambridge, MA 02139, USA, * byildiz@mit.edu.
[2]Department of Materials Science and Engineering, Massachusetts Institute of Technology, 77 Massachusetts Avenue, Cambridge, MA 02139, USA.

The interfaces between the perovskite $La_{0.8}Sr_{0.2}CoO_3$ (LSC_{113}) and Ruddlesden-Popper $(La_{0.5}Sr_{0.5})_2CoO_4$ (LSC_{214}) phases have been reported to accelerate oxygen reduction kinetics [1-4]. To obtain a microscopic level understanding of such enhancement, a combination of *in situ* Scanning Tunneling Microscopy / Spectroscopy (STM/STS) and Focused Ion Beam (FIB) milling was used to probe the local electronic structure - information that can be directly tied to the local oxygen reduction activity - near the interfaces and within the layers.

High quality multilayer (ML) superlattices made of LSC_{113} and LSC_{214} with a 20 nm modulation thickness were pulsed-laser-deposited on $SrTiO_3$ substrates. By using grazing incidence FIB milling, the buried interfaces and layers of the ML structure were exposed to ambient as illustrated in Fig. 1 (a-b) to enable their characterization with STM/STS [5]. Fig 1 (c) shows the Scanning Electron Microscopy image of FIB cutting region. A modified variable-temperature scanning tunneling microscope (VT-STM) was used to probe the surface morphology and to obtain surface electronic structure information with high spatial resolution at elevated temperatures and in oxygen environment.

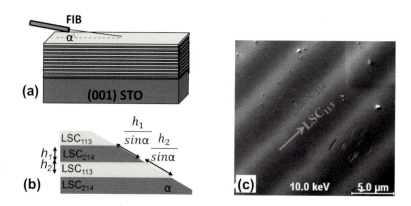

Fig. 1: (a) Illustration of the FIB milling process on $LSC_{214/113}$ multilayer structure with a shallow incidence angle α. (b) The cross-sectional view of the multilayer after the grazing incidence FIB milling, which permits the magnification and visualization of the inner layers and the interfaces. (c) Secondary electron image of the $LSC_{113/214}$ multilayer cross-section after the FIB milling process.

The topography images and tunneling current maps taken on the ML cross section from room temperature (RT) to 300 °C at 10^{-3} mbar of oxygen pressure are shown in Fig. 2. At RT, electronic structure on the LSC_{113} and on the LSC_{214} layers clearly differs from each other (Fig. 2 (a)). This electronic structure contrast was found to decrease with temperature (Fig. 2 (b) and (c)). The LSC_{113} electronic structure in the ML behaves similar to its single-phase counterpart from RT to high temperatures. On the other hand, the electronic structure of LSC_{214} in the ML at the high temperatures differs significantly from its single phase counterpart by exhibiting a large density of states near the Fermi level similar to that on LSC$_{113}$ (Fig. 2 (b) and (c)). The possible mechanisms for how the presence of LSC$_{113}$ influences the LSC$_{214}$ electronic structure through a coupling at their interface include: the migration of oxygen defects from one phase to the other [6]; electron injection from the LSC$_{113}$ oxygen vacancy states [7, 8] into the LSC$_{214}$ layer; and structural changes near the interface. These mechanisms are being investigated in our ongoing research.

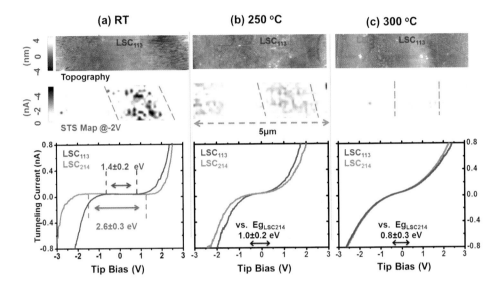

Fig. 2 Topography and tunneling current maps (at V$_{tip}$= -2 V) from room temperature (RT) to 300 °C in 10^{-3} mbar oxygen pressure. (a) at RT: a large energy gap exists in the tunneling spectra on both the LSC$_{113}$ and LSC$_{214}$ layers in the ML structure; the contrast shown in the tunneling current map represents the distinct electronic structures between the LSC$_{113}$ and LSC$_{214}$ phases in this system. (b) at 250 °C: no apparent energy gap is present in the tunneling spectra of LSC$_{113}$ and LSC$_{214}$ phase, the spatial contrast in the tunneling conductance decreases compared to that at RT. (c) at 300°C: no apparent energy gap is present in the tunneling spectra of LSC$_{113}$ and LSC$_{214}$ phase, and no difference between the electronic structure of LSC$_{113}$ and LSC$_{214}$ layers is detected within the bias range shown in the I-V plots and the tunneling current maps. The energy gap values stated for LSC$_{214}$ surface (Eg$_{LSC214}$) on the I-V plots in (b) and (c) are measured on the LSC$_{214}$ single phase film surfaces under the same conditions, and contrasted here to the results obtained from LSC$_{214}$ in the ML structure.

Our results put forward that such electronic activation of the LSC$_{214}$ is one possible reason for the vastly accelerated oxygen reduction kinetics by lowering the electronegativity and facilitating the charge transfer to oxygen near the LSC$_{113}$/LSC$_{214}$ interface. This new knowledge is important for advancing our understanding of the role of dissimilar interfaces in determining the catalytic, as well as electronic, ionic and magnetic properties of oxide hetero-structures, and for the design of new functional material architectures.

[1] M. Sase, et al., Solid State Ion., 178 1843.(2008)
[2] M. Sase, et al., J. Electrochem. Soc., 155 B793.(2008)
[3] K. Yashiro, et al., Electrochem. Solid State Lett., 12 B135.(2009)
[4] E.J. Crumlin, et al., J. Phys. Chem. Lett., 1 3149.(2010)
[5] Y. Kuru, et al., Adv. Mater., 23 4543.(2011)
[6] Y. Chen, et al., Nano Lett., 11 3774 (2011)
[7] N.A. Deskins, et al., J. Phys. Chem. C, 114 5891 (2010)
[8] Z. Cai, et al., J. Am. Chem. Soc., 133 17696.(2011)

THE SEM/FIB WORKBENCH: Automated Nanorobotics system inside of Scanning Electron or Focussed Ion Beam Microscopes

Volker Klocke[1], Ivo Burkart[1]

[1] Klocke Nanotechnik GmbH, Pascalstr. 17, 52076 Aachen, Germany

A good part of the understanding about material functions and process technologies was developed through preparation, handling and assembly of materials under light microscopes. But material properties and functionalities also depend on structure dimensions that are smaller than the wavelength of light. With light microscopes it is natural for everybody to use toolsets like tweezers, knives, hooks, probes and several different measurement tools. Without such handling, manipulation and manufacturing tools many present-day products and methods would not exist. No wristwatch, no in vitro fertilization, no mini-gearbox, just to mention a few.

The operators of SEM, FIB or Dual Beam systems generally work without toolsets and call it natural, although the wavelength limit of light is no physical boundary for using such tools. Reason therefore is the disconnected closed loop operation between human eyes and hands that enable complex operations under light microscopes without thinking.

The SEM/FIB Workbench is the first system substituting this hand-eye coordination effectively with nano-precision in a SEM/FIB. It can be imagined how technology in general can be pushed, when in-SEM/FIB tools can be used as easy as in air under a light microscope. The two main aspects of this new system, the development of its Nanorobotics technology and the applications enabled by it, are described in this paper.

Aspect 1, development of the technology: In general the success of in-SEM/FIB Nanorobotics depends on the co-operation of several important modules in one global system. The main developments include:

• Nanomanipulators in automation, for movement of tools, sample handling, etc.,

• Several different end-effectors for nano- probing, cutting, cleaning, force distance or wear measurements, gripping, sorting or material preparation and processing,

• Automatic in-situ tip cleaning process, e.g. for continuous nano-cutting,

• Automatic 3D position detection of all tools and SEM/FIB-samples up to 3D sample topography measurements,

• SEM picture assisted haptic interface by "Live Image Positioning",

• Modular configuration & teaching of nano-analytical or nano-handling processes,

• One common automation control for Nanorobotics and SEM/FIB.

After realizing these technical demands the SEM/FIB Workbench enabled also for non-professionals the secure and easy usage of toolsets within a SEM/FIB: from manual up to fully automated usage, as job-shop or in high throughput for industry[1].

Aspect 2, development of a series of new applications in one system: Expanding the SEM/FIB to a material processing system and a nano-analytical workbench opens the door to new applications, from material research over biology & bionics, pharmacy, tribology, environmental geology, forensic research and semiconductor technology up to nanofabrication and production [6].

Several examples of these new interdisciplinary research and development fields will be described during the presentation, together with the invitation to participate at an actual research network forming further new applications by proof of concept studies.

Nanosession: New technologies for scanning probes 153

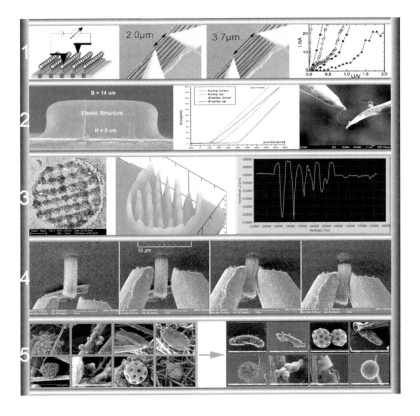

Figure 1. includes the rows:
1) Nano-Probing of Gold55-Clusters arranged in chains [2]: The electrical conductivity along these gold chains is measured over different distances showing quantum effects.
2) Nano-tribology and similar measurements can be performed in high resolution [3].
3) SEM image and 3D-Topography measured quantitatively by the "Dimensional SEM"[4].
4) Gripping of a rigid CNT bundle with high force and separation from the ground [5],
5) Particle sorting from source area (left image set) to clean target area (right image set)

A few examples of SEM/FIB Workbench applications are highlighted in FIG 1. Although these examples may raise the impression of a review about different machines and their usage this is not the case: Described is the development of one SEM/FIB Workbench that is about as universal as a stereo light microscope operated by a qualified user with some tools in his hand.

[1] D. Morrant, EIEx Magazine of European Innovation Exchange, 1 (2009), p. 6.
[2] G. Schmid, M. Noyong, Colloid Polym Sci., (2008) DOI 10.1007/s00396-008-1866-2
[3] Seong Chu Lim, Keun Soo Kim, Dep. of Physics, Sungkyunkwan University, Korea (2002)
[4] Described in detail at www.3D-Nanofinger.com
[5] C.-H. Ke1, H.D. Espinosa, Journal of the Mechanics and Physics of solids, 53 (2005) p. 1314
[6] Supported by European Commission, IST and Ziel2.NRW, see: www.nanohand.eu, www.hydromel-project.eu, www.nanowerkbank.com.

DENSITY FUNCTIONAL THEORY STUDY OF ANDERSON METAL-INSULATOR TRANSITIONS IN CRYSTALLINE PHASE-CHANGE MATERIALS

Wei Zhang[1], Alexander Thiess[2], Peter Zalden[3], Jean-Yves Raty[4], Rudolf Zeller[2], Peter H. Dederichs[2], Matthias Wuttig[3,5], Stefan Blügel[2,5], Riccardo Mazzarello[1,5]

[1] Institute for Theoretical Solid State Physics, RWTH Aachen, Germany;
[2] Institute for Solid State Physics, Forschungszentrum Jülich, Germany;
[3] I. Institute of Physics, RWTH Aachen, Germany;
[4] Condensed Matter Physics, B5, University of Liege, Belgium;
[5] JARA – Jülich-Aachen Research Alliance, Germany

The study of metal-insulator transitions (MITs) in crystalline solids is a subject of utmost importance, both from the fundamental point of view and for its obvious relevance to the understanding of the transport properties of a vast class of technologically useful materials, including doped semiconductors.

Very recently, a MIT was observed [1] in crystalline $Ge_1Sb_2Te_4$, upon increasing the annealing temperature: remarkably, this transition was shown to be exclusively due to disorder. $Ge_1Sb_2Te_4$ is a Chalcogenide phase-change material: this class of materials is considered to be a promising candidate for non-volatile memories of next generation [2], due to their ability to switch rapidly and reversibly between the crystalline and the amorphous phase upon heating and to the pronounced resistivity contrast between the two phases.

Experimentally, $Ge_1Sb_2Te_4$ undergoes a cubic to hexagonal structural transition by increasing the annealing temperature, before the MIT occurs: the cubic phase is always insulating at low temperatures, whereas the hexagonal phase displays the MIT [1].

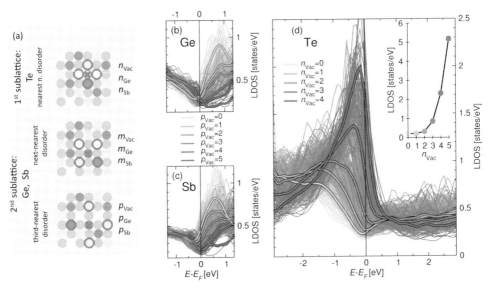

Figure 1: Local density of p states of Ge, Sb and Te atoms in a $Ge_{125}Sb_{250}Te_{500}$ unrelaxed supercell. Te atoms are grouped with respect to the number of nearest-neighbor vacancies (n_{Vac}), whereas Ge and Sb atoms are grouped according to the number of third nearest-neighbor vacancies (p_{Vac}). One can clearly see that the presence of neighboring vacancies shifts the Te p states upwards in energy, close to E_F. Structural relaxation does not change this picture qualitatively, but for the opening of a pseudo gap above E_F.

We have carried out a comprehensive computational study of the structural and electronic properties of crystalline $Ge_1Sb_2Te_4$, which has shed light on the microscopic mechanisms driving this transition [3]. For this purpose, a large number of big models (containing up to 3584 atoms) of cubic GST and hexagonal GST containing different amount of disorder in the form of a) randomly arranged vacancies and b) substitutional Ge/Sb disorder were generated and optimized using Density Functional Theory. The KKRNano [4] and CP2K [5] packages were employed for these simulations.

We show that, in the cubic phase, the electronic states responsible for the transport are exponentially localized inside regions having large vacancy concentrations: this finding is corroborated by local density of states calculations (Fig. 1) and inspection of relevant wave functions near the Fermi energy (Fig. 2). The transition to the metallic state is driven by the formation of ordered vacancy layers. Substitutional disorder and structural distortion do not play a crucial role in this MIT.

The ordering of vacancies is shown to bring about a significant gain in energy and to be responsible for the structural transition as well.

These results provide a step towards the goal of controlling the degree of disorder in these systems so as to tune their resistance in a reproducible fashion, which might lead to the development of devices of new concept based on multiple resistance states.

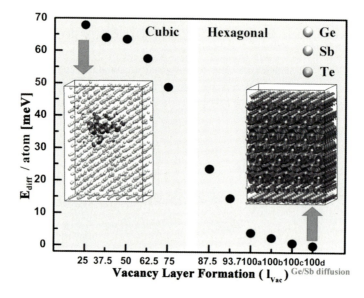

Figure 2: Total energy per atom of 1008 atoms models of the disordered cubic phase of GST (first point on the left), perfect hexagonal GST (first point on the right) and several intermediate structures with varying degree of disorder. The formation of vacancy layers yields a large reduction in energy, of the order of 50 meV per atom. The two insets show a snapshot of the HOMO state of the random cubic GST (left) and hexagonal GST (right). The first state is exponentially localized and has an Inverse Participation Ratio (IPR) value of $4.4 \cdot 10^{-2}$, i.e. it is localized on about 20-30 atomic sites. The second state is completely delocalized: its IPR value is $1.4 \cdot 10^{-3}$.

[1] T. Siegrist, P. Jost, H. Volker, M. Woda, P. Merkelbach, C. Schlockermann, and M. Wuttig, Nature Mater. **10**, 202 (2011).
[2] M. Wuttig and N. Yamada, Nature Mater. **6**, 824 (2007).
[3] W. Zhang, A. Thiess, P. Zalden, J.-Y. Raty, R. Zeller, P. H. Dederichs, M. Wuttig, S. Blügel, R. Mazzarello, to be submitted.
[4] A. Thiess, R. Zeller, M. Bolten, P. H. Dederichs, and S. Blügel, submitted.
[5] J. VandeVondele et al., Comput. Phys. Commun. **167**, 103 (2005).

LARGE SCALE MOLECULAR DYNAMICS SIMULATIONS OF PHASE CHANGE MATERIALS

Gabriele Cesare Sosso[1], Giacomo Miceli[2], Sebastiano Caravati[3], Davide Donadio[4], Jörg Behler[5] and Marco Bernasconi[1]

[1]Department of Materials Science, University of Milano-Bicocca, Milano, Italy;
[2]Institute of Theoretical Physics, Ecole Polytechnique Federale de Lausanne (EPFL),CH-1015 Lausanne, Switzerland;
[3]Department of Chemistry and Applied Biosciences, ETH Zurich, USI Campus, Lugano, Switzerland;
[4]Max Planck Institute for Polymer Research, Ackermannweg 10, 55128 Mainz, Germany;
[5]Lehrstuhl für Theoretische Chemie, Ruhr-Universität Bochum, Bochum, Germany.

In the last few years atomistic simulations based on density functional theory have provided useful insights on the properties of phase change materials (see ref. [1] for a review). However, several key issues are presently beyond the reach of fully ab-initio simulations. A route to overcome the limitations in system size and time scale of ab-initio molecular dynamics is the development of classical interatomic potentials. Behler and Parrinello [2] recently developed empirical interatomic potentials with close to ab-initio accuracy for elemental carbon, silicon and sodium by fitting large ab-initio databases within a neural network (NN) scheme. In general, a NN method is a non-linear technique that allows fitting any function to arbitrary accuracy and does not require any knowledge about the functional form of the underlying problem. By means of this technique, we have developed a classical interatomic potential for GeTe which is one of the compounds under scrutiny for applications in phase change memories.

Simulations with the NN potential are from four to five order of magnitude faster than ab-initio ones for 4000-atom models. We will present results of NN simulations on the properties of liquid and amorphous GeTe including thermal conductivity and the homogeneous and heterogeneous crystallization of the amorphous.

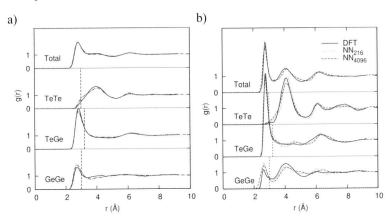

Figure 1: Total and partial pair correlation functions of a) liquid and b) amorphous GeTe from a NN molecular dynamics simulation at 1000K (liquid) and 300 K (amorphous) with 4096-atom and 216-atom cells, compared with results from ab-initio simulations at the same temperatures using a 216-atom cell1 [3]. The vertical lines are the interatomic distance thresholds used to define the coordination numbers, 3.0, 3.22 and 3.0 Å for Ge-Ge, Ge-Te and Te-Te bonds, respectively.

[1] D. Lencer, M. Salinga, and M. Wuttig, Advan. Mat. 23, 2030 (2011).
[2] J. Behler and M.Parrinello, Phys. Rev. Lett. 14, 146401 (2007).
[3] R. Mazzarello, S. Caravati, S. Angioletti-Uberti, M. Bernasconi, and M. Parrinello, Phys. Rev. Lett. 104, 085503 (2010); erratum 107, 039902 (2011).

QUANTUM-CHEMICAL ANALYSIS OF ATOMIC MOTION IN Ge-Sb-Te PHASE-CHANGE ALLOYS

Ralf Stoffel[1], Marck Lumeij[1], Volker Deringer[1], Richard Dronskowski[1]

[1]RWTH Aachen University, Institute of Inorganic Chemistry, Landoltweg 1, 52056 Aachen, Germany

Phase-change materials are nowadays used for writing and storing information on common data carriers due to their ability to switch between a crystalline and an amorphous state [1–4]. This switching behavior between the two phases is certainly accompanied by atomic motions since not only the long-range order changes due to the reversible transition between an amorphous and crystalline phase; in addition, there are clear indications that the atoms also experience a change with respect to their local order. The crystalline state, which can be described as a more or less distorted rock-salt structure, is characterized by a sixfold octahedral coordination of all the atoms. It has been shown that within the amorphous phase the coordination number lowers substantially [5,6]. Therefore a massive atomic rearrangement will occur upon amorphization/crystallization, which of course implies atomic movement. On the one hand we investigate the atomic displacement within GST-based phase-change materials by looking at the actual pathways of the atoms, and we calculate the activation barriers and search for transition states.

Exemplarily, Figure 1 shows how one Ge atom in rocksalt-structured GeTe moves from one octahedron into a neighboring empty octahedron via a tetrahedral surrounding. The corresponding energy profile is also included, and it exhibits two transition states; in both one Ge atom passes a triangle of three Te atoms. This energy profile surely changes whenever germanium atoms are erased (that is, vacancies are created) or when Ge is replaced by Sb. The influence of the substitution of Ge by either Sb or vacancies on the energy profile and the minimum energy pathways is investigated here.

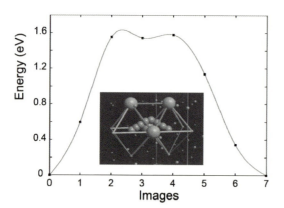

Figure 1: Energy profile and activation barrier for a movement of one Ge atom into a neighboring octahedral void. Ge atoms are in brown and Te in blue. The calculation is performed using a 64-atoms unit cell (32 Ge and 32 Te), but only the relevant atoms are shown. For reasons of clarity, the remaining ones have been graphically minimized.

On the other hand we have calculated the quasiharmonic lattice vibrations [7] via the *ab initio* force-constant method [8] by taking small displacements of the atoms into account. This gives access to temperature-dependent thermodynamic properties and also allows to have a look at the dynamical stability or instability of the used structures via those eigenvectors which belong to special vibrational states.

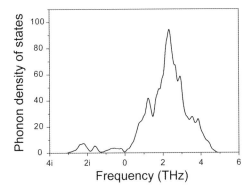

Figure 2: DFT-based phonon density of states for a hypothetical GST-124 structure.

As an example, Figure 2 shows the phonon density of states for an ordered, cubic phase of the composition GeSb$_2$Te$_4$ (GST-124). It is obvious that there are imaginary vibrational modes. Recent work deals with the structural changes due to the vibrational eigenvectors with the largest imaginary frequencies. Distorting the structure along those, we expect to find reasonable structures for modelling the thermodynamic and vibrational properties of the GST phase-change materials from first principles.

[1] D. Lencer, M. Salinga, M. Wuttig, *Adv. Mater.* **2011**, *23*, 2030.
[2] M. Wuttig, W. Bensch, *Chem. unserer Zeit* **2010**, *44*, 92.
[3] S. Raoux, M. Wuttig (Eds.): Phase Change Materials: Science and Applications, Springer, New York **2009**.
[4] D. Lencer, M. Salinga, B. Grabowski, T. Hickel, J. Neugebauer, M. Wuttig, *Nature Mater.* **2007**, *7*, 972.
[5] J. Akola, R. O. Jones, *J. Phys: Condens. Matter* **2008**, *20*, 465103.
[6] R. Mazzarello, S. Caravati, S. Angioletti-Uberti, M. Bernasconi, M. Parrinello, *Phys. Rev. Lett.* **2010**, *104*, 088503.
[7] R. Stoffel, C. Wessel, M.-W. Lumey, R. Dronskowski, *Angew. Chem. Int. Ed.* **2010**, *49*, 5242.
[8] A. Togo, F. Oba, I. Tanaka, *Phys. Rev. B* **2008**, *78*, 134106-1-9.

SIMULATION OF RAPID CRYSTALIZATION IN PHASE CHANGE MATERIALS BY MEANS OF PHASE FIELD MODELING

Fatemeh Tabatabaei[1], Markus Apel[2], Efim Brener[1]

[1]Peter Grünberg Institut (PGI-2), Forschungszentrum Jülich, 52428, Jülich
[2]Access e.V., RWTH Aachen, 52072 Aachen

Phase change materials (PCM), recently used extensively in non-volatile rewritable memory devices, undergo a stable, rapid and reversible transition between a crystalline and an amorphous atomic structure. The ordered crystalline and the disordered amorphous state can be utilized for the data recording in memory applications. To obtain a quantitative understanding of the kinetics of writing and erasing data, gaining insights into the energy transport and phase boundary movement during the phase transformation is aimed. One of the governing parameters for the transformation kinetics is the mobility of the liquid-solid interface. However, only qualitative knowledge and models for it are available and a direct measurement of the interface mobility coefficient is in general not possible.

We apply phase field modeling as a continuum simulation technique in order to study rapid crystallization processes in AgInSbTe and to derive the temperature dependent mobility by matching the simulated crystallization time with experimental values. In particular the spatio-temporal evolution of the crystallization of a molten area in a PCM layer stack was simulated. The simulation model was adapted to experimental conditions recently used [1] for the measurement of the crystallization kinetics by a laser pulse technique. Simulations are performed for different substrate temperatures, i. e. for temperatures close to the melting point of AgInSbTe down to the glass temperature when the amorphous state is involved. The effect of the interface mobility on the solidification kinetics at large undercooling where the interface crystallization is controlled by the mobility is investigated. An Arrhenius type temperature dependent kinetic coefficient instead of a constant value, which is usually used in standard phase field models, is considered. The coefficients for the Arrhenius function were derived by fitting the simulation results to the experimental data for the crystallization velocity. The Arrhenius function for the mobility can be used to calculate the crystallization time at higher temperatures where thermal transport starts to effect the crystallization time and isothermal conditions are not valid any more for the interpretation of the experiments.

[1] Andreas Kaldenbach, *PhD thesis (2012)*, RWTH Aachen

EPITAXYAL PHASE CHANGE MATERIALS: GROWTH, STRUCTURE AND PHASE TRANSITION

Henning Riechert[1], Peter Rodenbach[1], Alessandro Giussani[1], Karthick Perumal[1], Michael Hanke[1], Jonas Laehnemann[1], Martin Dubslaff[1], Raffaella Calarco[1], Manfred Burghammer[2], Alexander Kolobov[3] and Paul Fons[3]

[1]Paul-Drude Institut für Festkörperelektronik, Hausvogteiplatz 5-7, 10117 Berlin, Germany,
[2]European Synchrotron Radiation Facility, 6, rue Jules Horowitz, 38043 Grenoble Cedex 09, France and
[3]National Institute of Advanced Industrial Science and Technology (AIST), Tsukuba Central 4, Higashi 1-1-1 Tsukuba Ibaraki, Japan 305-8562

The rapid expansion of the Internet has lead to the consumption of vast amounts of electricity in huge server farms. Significant savings in energy and the corresponding environmental benefits could be achieved by faster and denser non-volatile memories. Phase change materials, and especially Ge–Sb–Te (GST) alloys are leading candidates for such an innovative generation of non-volatile memory. Only recently our group, uniquely worldwide, demonstrated the successful growth of high-quality epitaxial single-crystalline GST layers.

The epitaxial growth of GST was carried out on different substrates to obtain films with a unique crystalline orientation. Both nearly lattice-matched and strongly mismatched substrates were employed. Samples are transformed from the single crystalline into the amorphous phase and back to the single crystalline state by means of 180 femtosecond duration laser pulses. We used nanometer x-ray diffraction to investigate the detailed crystalline state in the as grown and switched film.

We also present the growth by molecular beam epitaxy and the structural characterization of GeTe / Sb_2Te_3 superlattices. Those layers are expected to display impressive switching characteristics substantially surpassing those of the composite material with the same average composition.

The successful growth of epitaxial, single-crystalline phase-change layers as opposed to polycrystalline films opens up the possibility to investigate fundamental materials properties as the interpretation of experimental results is not complicated anymore by the presence of grain boundaries. Moreover, designing materials with new functional properties for which spatial perfection is of crucial importance has now become a realistic opportunity.

EFFECT OF CARBON AND NITROGEN DOPING ON THE STRUCTURE AND DYNAMICS OF AMORPHOUS GeTe PHASE CHANGE MATERIAL

J. Y. Raty[1], G. Ghezzi[2], P. Noé[2], E. Souchier[2], S. Maitrejean[2], C. Bichara[3], F. Hippert[4]

[1]Physics Department, University of Liège, Sart-Tilman, Belgium;
[2]CEA-Leti, Minatec, Grenoble, France;
[3]CINaM-CNRS, Marseille, France;
[4]LMPG (CNRS, Grenoble INP), Grenoble, France and LNCMI-CNRS, Grenoble, France

Chalcogenide phase change materials (PCMs), such as GeTe and $Ge_2Sb_2Te_5$, are excellent candidates for use in non volatile embedded resistive memories thanks to rapid phase transformation (10th of ns) between amorphous and crystalline states and to large variation of resistivity between the two states. In memory cells, switching from the crystalline (low resistivity) to the amorphous (high resistivity) phase is obtained by applying a short and high current pulse that locally melts the material. A longer and lower pulse is used for crystallization. Memories based on PCMs offer high scalability, fast programming and good cyclability. However, to allow operation at relatively high temperatures for embedded applications, it is crucial to improve the stability of the amorphous phase. Carbon doping, and to a lesser extent nitrogen doping, have been shown to increase significantly the crystallization temperature of GeTe, with a beneficial increase of the retention time.

By combining ab initio molecular dynamics and X-ray scattering experiments, we show that carbon deeply modifies the structure of the amorphous phase through long carbon chains, tetrahedral and triangular units centered on carbon (figure 1). A clear signature of these units is the appearance of an additional interatomic distance around 3.3 Å in the pair correlation function [1]. Besides, the first Ge-Ge and Ge-Te distances are almost not affected by carbon doping. Inclusion of nitrogen also modifies the amorphous GeTe structure. Differences between the effects of carbon and nitrogen will be discussed.

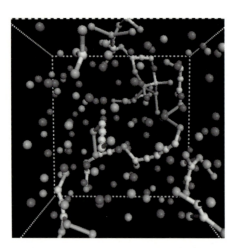

Figure 1: Instantaneous snapshot of a GeTeC configurations at 300 K. Only the bonds between C atoms (light grey, green online) and their first Ge neighbors (dark grey, orange online) are plotted. Te atoms are plotted in dark grey.

Besides, by combining Fourier Transform Infrared (FTIR) spectroscopy and ab initio results, we show that the inclusion of carbon or nitrogen strongly affects the vibrational modes of amorphous GeTe both at low and high frequency. We will relate these effects to the enhanced stability of carbon- and nitrogen-doped amorphous GeTe.

[1] G. Ghezzi et al , Applied Physics Lett.**99**, 151906 (2011).

EXPLOITING THE MEMRISTIVE-LIKE BEHAVIOUR OF PHASE-CHANGE MATERIALS AND DEVICES FOR ARITHMETIC, LOGIC AND NEUROMORPHIC PROCESSING

C D Wright[1], J A Vázquez Diosdado[1], L Wang[2], Y Liu[1], P Ashwin[1], K I Kohary[1], M M Aziz[1], P Hosseini[1] and R J Hicken[1]

[1] College of Engineering, Mathematics and Physical Sciences, University of Exeter, Exeter, England;
[2] Nanchang Hangkong University, Nanchang 330063, P. R. China

Resistors with memory, or memristors, the so-called missing circuit element linking charge and flux and first expounded by Chua in the early 1970s[1] and 're-discovered' in 2008[2] have generated much recent excitement due to their potential to provide remarkable and wide-ranging functionality including the ability to perform arithmetic[3], logic [4] and even implement artificial synapses[5,6]. Memristive-like behaviour has been reported in a number of materials and device structures, most commonly perhaps in metal-insulator-metal (MIM) resistance switching memory cells, such as the well-known transition metal oxide (e.g. TiO_2, VO_2 etc) or Perovskite-oxide (e.g. $PrCaMnO_3$) based devices, and in ionic-conduction devices based on the formation of Ag filaments or the movement of Ag 'fronts' in a sulphide or silicon matrix (for excellent reviews see Refs. [7,8]). Phase-change memories based on chalcogenide alloys also provide a 'resistor with a memory its of previous excitation history', and so might also be used to implement memristive-like devices and systems; however, their potential in this respect has received very little attention to date. Here we therefore describe the operation of various forms of phase-change memristive devices and systems, evaluating their fundamental behaviour and outlining their possible advantages and disadvantages compared to more usual implementations. One particular advantage that deserves special mention is that phase-change devices are both electrically and optically active and that signals can be transferred relatively simply between these two domains. This leads us to the design of new types of memory-reflector (or 'memflector') devices (see Fig. 1) and we elucidate the properties of such devices and outline their potential applications. We also introduce and demonstrate experimentally a new and simple approach to providing non-volatile logic operations, specifically AND and OR Boolean logic, using phase-change devices (see Fig. 2). We show that quite extensive logic operations (e.g. a 512 input AND) can be performed by a single phase-change cell of nanometric size. Our phase-change logic operates in the electrical domain, but the same concept could be used in the optical domain to implement optical logic (or indeed to implement mixed-domain logic with excitation in the electrical domain and detection in the optical domain, or vice versa). We also demonstrate, experimentally and theoretically, that phase-change devices can provide arithmetic processors capable of fast and efficient computations of addition, multiplication, subtraction and division, again working in both the electrical and optical domains, or indeed in a mixed-mode configuration. Finally, we use the natural accumulation property of phase-change materials and devices, the same property that allows us to perform arithmetic calculations with phase-change cells[3], to implement a simple integrate-and-fire phase-change neuron circuit.

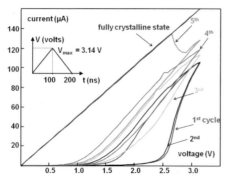 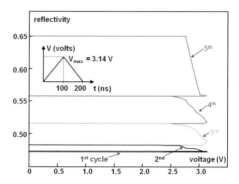

Figure 1: Characteristic electrical I-V response (left) to successive voltage excitations (shown inset is the excitation voltage for a single cycle) for a growth-dominated phase-change memristive-like device (upon completion of the 5th cycle the cell is completely crystallised). Note that in contrast to the behaviour of 'classical' memristors (such as TiO_2 stuctures[2]) the 'up' curve of one cycle of the phase-change I-V response does not follow the 'down' curve of the preceding cycle; this is due to the field-dependent conductivity of the amorphous phase leading to significantly different temperatures being experienced by the phase-change layer during the 'down' and 'up' curves of successive cycles. On the *right* we show the optical reflectivity of the phase-change layer during the 5 successive voltage excitation cycles (a 400 nm wavelength illumination was assumed and the optical reflectivity calculated using effective medium theory); here it can be seen that the 'up' curve for one cycle does follow the 'down' curve of the previous cycle, this is because we are monitoring reflectivity rather than resistance, so the electrical field dependence of conductivity is no longer playing a dominant role.

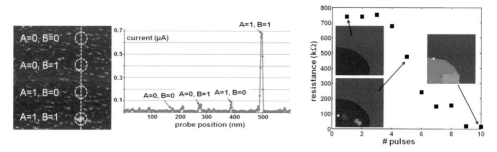

Figure 2: Phase-change logic with an a-C/$Ge_2Sb_2Te_5$/a-C/TiN 'device' with the top electrode being the tip of a CAFM (conducting atomic force microscope). *Left* shows CAFM current image (with 1V tip-sample voltage) after the implementation of a 2-input AND operation; the image is 375 nm x 375 nm and the positions where logic operations were performed are indicated by the dashed circles each 50 nm in diameter (high conductivity regions appear light). *Middle* shows CAFM line scan through the centre of the dotted circles and clearly reveals the successful implementation of AND logic. On the *right* we show the (simulated) resistance and crystal structure of a conventional mushroom-type phase-change cell initially in the RESET state and subject to a succession of identical excitation pulses (each ~1.1V in amplitude and of 60 ns duration). The cell 'accumulates' the input excitation energy and switches to a low resistance state after the receipt of a certain number of pulses. This behaviour is used to provide arithmetic functionality (addition, subtraction, multiplication and division - for further details see Ref. [3]) and to implement a simple integrate-and-fire phase-change neuron circuit.

[1] L.O. Chua, IEEE Trans. Circuit Theory **18**, 507 (1971)
[2] D.B. Strukov et al, Nature **453**, 80 (2008)
[3] C.D.Wright et al, Adv. Mater. **23**, 3408 (2011)
[4] J. Borghetti et al, Nature **464**, 873 (2010)
[5] D Kuzum et al, Nano Lett., doi:10.1021/nl201040y (2011)
[6] M Suri et al, IEDM 2011 Proceedings, doi/10.1109/IEDM.2011.6131488 (2011)
[7] Y.V. Pershin, M. Di Ventra, Advances in Physics **69**, 145 (2011)
[8] R. Waser et al, Adv. Mater. **21**, 2632 (2009)

INVERSE TIME-VOLTAGE RELATION OF THRESHOLD SWITCHING IN PHASE CHANGE MATERIALS

Marco Cassinerio, Nicola Ciocchini and Daniele Ielmini

Dipartimento di Elettronica e Informazione, Politecnico di Milano, Milano, Italy

Chalcogenide materials enables several novel device concepts, such as phase change memories (PCMs), optical disks and electronic *threshold* switches for high-current select devices in crossbar arrays [1]. A key requirement in all these applications is the stability of the threshold voltage V_T, marking the onset of threshold switching in the chalcogenide glass [2]. To understand the physical mechanism of threshold switching, the time dependence of threshold voltage must be analyzed in detailed.

This work studies the time dependence of V_T in PCM devices based on $Ge_2Sb_2Te_5$ (GST), demonstrating an inverse time-dependence of V_T. Data were collected from PCM devices in 90 and 180 nm provided by Micron [3]. The PCM resistance R can be electrically tuned into different values by changing the structural phase of the GST from amorphous to crystalline. Conduction in the amorphous-phase device shows an abrupt transition, called threshold switching, from high to low resistivity at the threshold voltage V_T. Threshold switching is driven by the high electric field (around 0.4 MVcm^{-1}), exciting electrons to high energies and enhancing conductivity [2].

We characterized V_T in response to the application of a triangular pulse with variable rising slope β = dV/dt. The PCM was connected to a load resistance R_L and V_T was revealed probing the voltage across the PCM (inset of Fig. 1). Fig. 1 shows the measured V_T as a function of β for two PCM device sizes. V_T first decreases for increasing β, than increases above β = 10^6 Vs^{-1}. Such behavior was verified on several samples by repeating the experiments at increasing, decreasing or random β, to rule out possible artifacts related to PCM degradation and/or electromigration [4].

The β-dependence of V_T in Fig. 1 is anomalous, in that previous experiments above 10^4 Vs^{-1} and theoretical studies have shown that V_T increases monotonically with β [5]. The V_T increase with β is due to the probabilistic nature of threshold switching: As β increases, the shorter time available for switching must be compensated by a larger switching probability P(t), hence a larger voltage. Based on a simple model, the threshold switching conditions reads:

$$\int_0^{t_T} P(t)dt = \int_0^{t_T} P_0 e^{V/V_0} dt = \eta, \quad (1)$$

where we assumed an exponentially increasing switching probability $P(t) = P_0 e^{V/V_0}$, where P_0 and V_0 are constants. Eq. (1) states that switching takes place at a time t_T, corresponding to the cumulative probability P(t) to reach a value η (e.g. η = ½). Substituting V = βt in Eq. (1), we obtain $V_T = V_0 \log(\beta\eta/P_0V_0)$, which accounts for the logarithmic increase of V_T for high β in Fig. 1c.

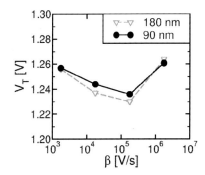

Figure 1: Measured V_T as a function of the pulse rising slope β = dV/dt. Unexpectedly, V_T decreases for increasing β below 10^6 Vs^{-1}.

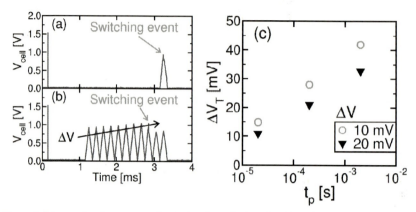

Figure 2: Measured voltage across the device for a single pulse (a) and a sequence of multiple pulses of increasing amplitude (b), and the difference ΔV_T between V_T values measured under multiple and single pulse (c). ΔV_T is positive due to the larger acceleration induced by multiple pulses. ΔV_T also increases with t_P and ΔV.

The unexpected V_T decrease for small β can be explained by an acceleration of structural relaxation (SR) of the amorphous chalcogenide phase induced by the applied electrical pulse at low β. Since SR results in enhanced resistivity and higher threshold voltage [6], the pulse-accelerated SR causes a V_T increase. As a result, V_T is found higher for long pulses, which caused SR acceleration for a longer time. To verify this interpretation, we compared the V_T values measured by applying either (i) a single triangular pulse or (ii) a sequence of pulses with the same β as the single pulse but with increasing amplitude, as shown in Fig. 2a and b. The multiple-pulse operation is intended to enhance the SR-acceleration close to V_T with respect to the single pulse. The delay between the initial reset pulse and the switching was kept approximately the same, to rule out differences in V_T due to different drift time [6].

Fig. 2c shows the difference ΔV_T between the average V_T probed by multiple pulses and the one obtained by a single pulse, as a function of the triangular pulse width t_P. Data indicate $\Delta V_T > 0$, which supports the larger SR achieved through the multiple pulse. Also note that ΔV_T increases with the pulse width, with ΔV_T extrapolating to zero at $t_P^* = 1$ μs. This explains the V_T behavior below $\beta \approx 1$ V/$t_P^* = 10^6$ Vs^{-1} in Fig. 1c. Data are shown for two values of the increment ΔV of the pulse amplitude in the pulse sequence: ΔV_T is higher for the smaller $\Delta V = 10$ mV, since in this case the number of pulses before switching was twice that for $\Delta V = 20$ mV, thus causing a larger pulse-induced acceleration of SR. These data confirm our interpretation of the inverse V_T dependence on pulse time in terms of pulse-accelerated SR.

The authors acknowledge the Micron-Agrate technical team for discussions and for providing PCM experimental samples. This work was supported in part by Fondazione Cariplo (Grant 2010-0500).

[1] S. Raoux, et al., MRS Bull. 37, 118 (2012).
[2] D. Ielmini, Phys. Rev. 78, 035308 (2008).
[3] F. Pellizzer, et al., Symp. VLSI Tech. Dig. 122 (2006).
[4] B. Rajendrah, Symp. VLSI Tech. Dig. 96 (2008).
[5] S. Lavizzari, et al., IEEE Trans. Electron Devices 57, 1047 (2010).
[6] D. Ielmini, et al., Appl. Phys. Lett. 92, 193511 (2008).

INTERPLAY OF DEFECTS AND CHEMICAL BONDING IN THE "GST" FAMILY OF PHASE-CHANGE MATERIALS

Volker L. Deringer[1], Marck Lumeij[1], Ralf Stoffel[1], <u>Richard Dronskowski</u>[1]

[1]RWTH Aachen University, Institute of Inorganic Chemistry, Landoltweg 1, 52056 Aachen, Germany

While the semiconductor industry traditionally strives to produce chemically pure and defect-free materials, the presence of such *structural* defects, in particular, is vital for the very functionality of a different class of materials, namely, phase-change alloys. We focus here on the ternary Ge–Sb–Te alloys (or "GST", in short), which excel in reversible but permanent information storage and have allowed for inventions such as the DVD-RAM or Blu-Ray™ disk. The influence of defects on the electronic and crystalline structure of GST alloys has been studied theoretically by some groups, but only for explicit compositions [1-7]. In this work, we intend to study the full compositional range of GST by *ab initio* calculations, which involves a systematic scan according to the scheme in Figure 1.

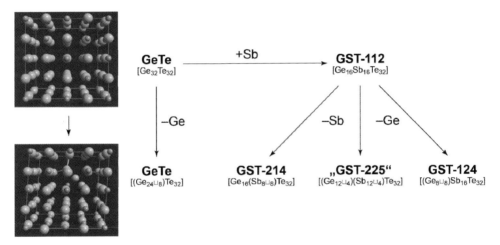

Figure 1: Schematic drawing of a computational model, derived from a 64-atom supercell, which alows facile treatment of a wide range of GST alloys (some examples are given). Supercell compositions are given below in square brackets. The unit cells of $Ge_{32}Te_{32}$ and $Ge_{24}Te_{32}$ are shown on the left side. Ge atoms are in brown and Te in blue.

Here we present results for the line $Ge_{32}Te_{32} \rightarrow (Ge_{24}\square_8)Te_{32}$, which corresponds to the experimentally verified vacancy formation in GeTe; the observed vacancy concentration was 8–10% [2,8,9], which will be put to test by computational means. Furthermore, we perform an in-depth bonding analysis using the well-established COHP technique [10]. This is exemplarily shown in Figure 2: the formation of vacancies depletes antibonding (i.e., destabilizing) states in the crystal's electronic structure, and the remaining bonds become more stable. The delicate interplay between the number and the strength of bonds is shown to be the driving force behind the vacancy-formation process.

Beyond GeTe, we present some results for the cornerstone phases GST-112, GST-214, GST-225, and GST-124, a logical extension of our previous work [11]. It is shown that the physico-chemical rules derived for GeTe may seamlessly be applied to ternary GST alloys as well. A more quantitative assessment of bond strengths (as measured by the integrated COHP) is provided for the GST-124 and GST-214 materials to investigate preferences in expelling Ge versus Sb from a GST structure.

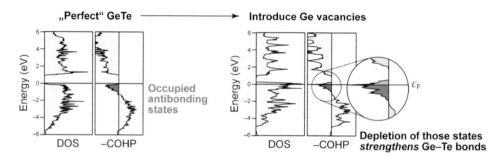

Figure 2: Changes in electronic structure and chemical bonding upon going from a $Ge_{32}Te_{32}$ to a Ge-deficient $Ge_{24}Te_{32}$ supercell, as elucidated by TB-LMTO-ASA computations. COHP curves are drawn in the conventional way, namely, such that bonding contributions fall to the right of the energy axis, and antibonding contributions to the left. A horizontal line marks the Fermi level ε_F, which has been chosen as the energy zero. (V.L.D., unpublished results.)

Finally, we give a detailed outlook on the structural information of the investigated compounds, which includes bond lengths, vacancy distribution, electronic structure and defect levels.

[1] J. L. F. Da Silva, A. Walsh, H. Lee, *Phys. Rev. B* **2008**, *78*, 224111.
[2] A. H. Edwards, A. C. Pineda, P. A. Schultz, M. G. Martin, A. P. Thompson, H. P. Hjalmarson, C. J. Umrigar, *Phys. Rev. B* **2006**, *73*, 045210.
[3] J. H. Eom, Y. G. Yoon, C. Park, H. Lee, D.-S. Suh, J.-S. Noh, Y. Khang, J. Ihm, *Phys. Rev. B* **2006**, *73*, 214202.
[4] G. Lee, S.-H. Jhi, *Phys. Rev. B* **2008**, *77*, 153201.
[5] S. Caravati, M. Bernasconi, T. D. Kühne, M. Krack, M. Parrinello, *J. Phys.: Condens. Matter* **2009**, *21*, 255501.
[6] X. Q. Liu, X. B. Li, L. Zhang, Y. Q. Cheng, Z. G. Yan, M. Xu, X. D. Han, S. B. Zhang, Z. Zhang, E. Ma, *Phys. Rev. Lett.* **2011**, *106*, 025501.
[7] Z. Sun, J. Zhou, R. Ahuja, *Phys. Rev. Lett.* **2006**, *96*, 055507.
[8] F. Tong, X. S. Miao, Y. Wu, Z. P. Chen, Z. Tong, X. M. Cheng, *Appl. Phys. Lett.* **2010**, *97*, 261904.
[9] A. V. Kolobov, J. Tominaga, P. Fons, T. Uruga, *Appl. Phys. Lett.* **2003**, *82*, 382.
[10] R. Dronskowski, P. E. Blöchl, *J. Phys. Chem.* **1993**, *97*, 8617.
[11] M. Wuttig, D. Lusebrink, D. Wamwangi, W. Wełnic, M. Gilleßen, R. Dronskowski, *Nature Mater.* **2007**, *6*, 122.

PHOTONICS-BASED NON-VOLATILE MEMORY DEVICE USING PHASE CHANGE MATERIALS

Wolfram H.P. Pernice[1] and **Harish Bhaskaran**[2]

[1]Karlsruhe Institute of Technology, 76344 Eggenstein-Leopoldshafen, Germany
[2]School of Engineering, University of Exeter, Exeter EX4 4QF UK;

We propose an integrated, all-photonic memory employing phase-change materials in combination with silicon nitride waveguides. We show that it is possible to have sub-nanosecond, high speed, multi-level memories in nanophotonic circuits employing the commonly used chalcogenide, $Ge_2Sb_2Te_5$ (GST) as the memory element. Our theoretical predictions show high sensitivity to the degree of crystallization in an integrated photonic device.

Integrated photonic memories with long-term storage capability, sub-nanosecond read and write times and easily integrated with CMOS processes have been of immense interest for future optical computing. Photonic memories realized so far have made use of transient phenomena in optical resonators [1] and mechanical resonators [2]. Herein, we propose a chalcogenide-based, integrated, non-volatile photonic memory element with the ability for sub-nanosecond reading and writing, while still retaining data for several years. Multilevel recording can be achieved in such integrated circuits, thereby paving the way not only for ultra-dense photonic memories, but also for all-optical computing using non von Newman architectures [3].

Phase change memories made their commercial debut in optical storage discs, such as DVDs and Blue-Ray discs owing to a large difference in the reflectivity between the amorphous and crystalline states of the material. The transformation between these two states is readily achieved by an optical pulse of high power, while reading is achieved at lower optical power. Because their optical properties also change significantly upon switching, phase-change materials provide a new route towards optically tunable photonic circuits. Herein, we analyse the architecture depicted schematically in Fig.1a.

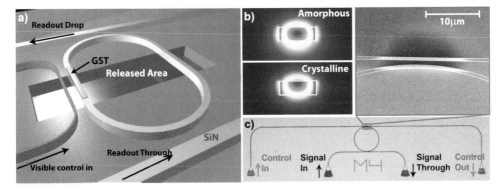

Fig.1a) Schematic of the proposed photonic memristor architecture using optical microring resonators. b) Modal profiles of the free-standing waveguide with GST in the amorphous and crystalline state. c) Sample fabricated device with control and signal waveguides coupled to a microring. Inset: enlarged view of the released area.

It consists of a partially suspended microring resonator coupled to nanophotonic waveguides similar to devices used for nanomechanical sensing [4]. The suspended portion of the waveguide (made from stoichiometric silicon nitride) is covered with a thin layer of GST, which undergoes reversible transformations between crystalline and amorphous states upon application of thermal energy. The thermal energy occurs as a consequence of the power dissipated in the waveguide due to coupling to an additional control port, thus providing a tunable resonator. The suspension in the waveguide ensures that this region reaches a temperature sufficient to ensure crystallization/ amorphization of the phase change materials (see modal profiles in Fig.1b). In order to achieve efficient heat generation, the wavelength control light is in the visible wavelength range at 700 nm, where GST shows strong optical material absorption. The waveguides are connected to optical input/output ports as shown in the device image in Fig.1c. The released part of the waveguide is illustrated in the SEM image in the inset. Assuming telecom wavelengths around 1550 nm, we use an optimized waveguide cross section of 800 × 350 nm^2, which allows for efficient coupling between the propagating mode and the GST top layer. The GST top layer is assumed to cover a suspended waveguide section of 2μm length. Because of the refractive index change induced by the phase transformation inside the GST, the resonance condition of the microring is modified. The refractive index at 1550nm input wavelength changes from $4.52 + i*0.15$ to $6.95 + i*1.82$ when the layer is switched from the amorphous to the crystalline state.

The transmission profile of the device as a function of the degree of crystallization is shown in Figure 2a, whereby we find that the device has a sensitivity of well over a hundred levels using currently available telecom wavelength photodiode technology. We assume that the resonator is critically coupled when GST is in the amorphous state, thus close to no light is transmitted past the resonator. When the layer is in the crystalline state the strong absorption within the film significantly changes the condition for critical coupling into the resonator, leading to high optical intensity in the through port.

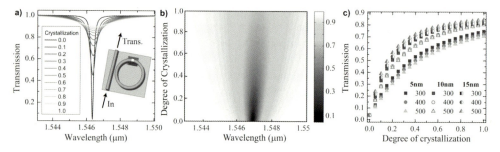

Fig 2. a) Transmission dependence on crystallization at resonance. b) Transmission (contours) dependence on wavelength (x-axis) and degree of crystallization (y-axis). c) Device transmittance for various combinations of film thickness (5nm-15nm) and waveguide height (300nm-500nm).

This is illustrated in Fig.2b, where the dependence of the transmittance in the through-port is plotted as a function of wavelength and degree of crystallization. Depending on the GST thickness the resonance spectrum can be broadened for wideband optical operation. As shown in Fig.2c, by varying the crystal structure of the GST layer, the transmission past the device can be continuously tuned from 0 to 80%.

The proposed resonator structure can be efficiently scaled by connecting a series of devices to a nanophotonic bus. Because GST maintains its crystalline state after the control light has been turned off, the device is inherently non-volatile. Thus our architecture provides the ingredients for scalable, all-optical processing on a chip.

[1] F. Xia et al., Nat. Photon. **1**, (2007).
[2] M. Bagheri et al., Nat. Nanotech. **6**, 726 (2011).
[3] C.D. Wright et al., Adv. Mater. **23**, 3408 (2011).
[4] M. Li et al., Nature **456**, 480 (2008).

ROLE OF ACTIVATION ENERGY IN RESISTANCE DRIFT OF AMORPHOUS PHASE CHANGE MATERIALS

Martin Salinga[1], Martin Wimmer[1], Matthias Käs[1], Matthias Wuttig[1]

[1] I. Physikalisches Institut (IA) and JARA-FIT, RWTH Aachen University, Aachen, Germany

In recent years electronic memories based on phase change materials have matured into a technology that is realized in commercial devices. This unique class of materials [1] combines long-term stability of the involved states, which is necessary for non-volatile memories, with switching speeds that reach into the nanosecond range [2]. The latter enables not only a universal memory [3, 4] that is as fast as DRAM, but it is also favorable when aiming for green IT by minimizing the energy consumption per switching event [5, 6]. Furthermore the scalability of phase change memories also looks advantageous compared to alternative technologies [7]. Finally the pronounced change of resistance upon crystallization is very attractive since it facilitates the realization of multilevel storage multiplying data densities on phase change chips [8]. However, this concept faces an obstacle as the resistance of the amorphous phase drifts over time [9]. This could lead to a loss of data if an intermediate state would unintentionally leave its pre-defined resistance window after a certain time.

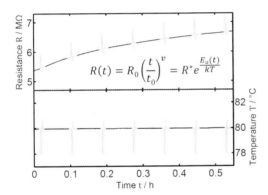

Figure 1: Temporal drift of the resistance in the amorphous phase change material $Ge_2Sb_2Te_5$ can be described by an empirical power law (black lines). Brief interruptions of the annealing by accurate temperature ramps (spikes) are used to directly monitor the evolution of the activation energy for conduction Ea.

$$R(t) = R_0 \left(\frac{t}{t_0}\right)^v = R^* e^{\frac{E_a(t)}{kT}}$$

Studies in various labs have provided empirical evidence that the temporal evolution of the resistance in amorphous phase change materials follows a power law (see formula and black curve in fig. 1). The physical mechanism underlying this process, however, is still discussed controversially. Some researchers see the fundamental reason for resistance drift in mechanical stresses inside the amorphous material caused by the pronounced density contrast between the different phases [10]. Others discuss it as a generic structural relaxation common to amorphous semiconductors in general [11]. When speculating about the physics behind resistance drift it is typically assumed that it must be accompanied by a rise of the activation energy for conduction (Ea as defined in the formula inside of fig.1) [12]. To confirm or refute this assumption we have studied experimentally how the activation energy for conduction actually changes while the resistance drifts. Therefore we have designed an experimental setup that allows the precise control of the sample temperature in order to briefly interrupt annealing/drift experiments by accurate temperature ramps (spikes in fig.1). From an analysis of the latter the activation energy of conduction is determined. The results are compared with the change of activation energy, which would be expected if resistance drift could solely be ascribed to an increase of activation energy. Preliminary results for $Ge_2Sb_2Te_5$ (fig. 2) show indeed a close correlation between resistance drift and the drift of the activation energy for conduction over a long period of time.

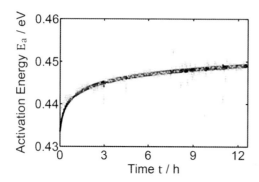

Figure 2: To test the assumption of resistance drift being ruled by a rise of activation energy, Ea is measured over twelve hours (bright dots with error bars) by varying the temperature as depicted in fig.1. Linking the power-law for temporal drift with the Arrhenius law (formula in fig.1) it is calculated from the measured resistances at a constant temperature (black lines in fig.1) how the activation energy for conduction needs to change over time if it was solely responsible for the resistance drift (black curve).

To obtain a systematic understanding of resistance drift, different families of materials have been investigated. The comparison of these different compounds helps to unravel the microscopic origin of resistance drift.

1. Lencer, D., et al., *A map for phase-change materials.* Nature Materials, 2008. **7**(12): p. 972-977.
2. Bruns, G., et al., *Nanosecond switching in GeTe phase change memory cells.* Applied Physics Letters, 2009. **95**(4).
3. Lankhorst, M.H.R., B.W.S.M.M. Ketelaars, and R.A.M. Wolters, *Low-cost and nanoscale non-volatile memory concept for future silicon chips.* Nature Materials, 2005. **4**(4): p. 347-352.
4. Wuttig, M., *Phase-change materials - Towards a universal memory?* Nature Materials, 2005. **4**(4): p. 265-266.
5. Salinga, M. and M. Wuttig, *Phase-Change Memories on a Diet.* Science, 2011. **332**(6029): p. 543-544.
6. Xiong, F., et al., *Low-Power Switching of Phase-Change Materials with Carbon Nanotube Electrodes.* Science, 2011. **332**(6029): p. 568-570.
7. Raoux, S., et al., *Phase-change random access memory: A scalable technology.* Ibm Journal of Research and Development, 2008. **52**(4-5): p. 465-479.
8. Nirschl, T., et al., *Write strategies for 2 and 4-bit multi-level phase-change memory.* 2007 Ieee International Electron Devices Meeting, Vols 1 and 2, 2007: p. 461-464.
9. Pirovano, A., et al., *Low-field amorphous state resistance and threshold voltage drift in chalcogenide materials.* Ieee Transactions on Electron Devices, 2004. **51**(5): p. 714-719.
10. Braga, S., A. Cabrini, and G. Torelli, *Dependence of resistance drift on the amorphous cap size in phase change memory arrays.* Applied Physics Letters, 2009. **94**(9).
11. Boniardi, M. and D. Ielmini, *Physical origin of the resistance drift exponent in amorphous phase change materials.* Applied Physics Letters, 2011. **98**(24).
12. Boniardi, M., et al., *A physics-based model of electrical conduction decrease with time in amorphous Ge(2)Sb(2)Te(5).* Journal of Applied Physics, 2009. **105**(8).
13. Krebs, D., et al., *Impact of DoS changes on resistance drift and threshold switching in amorphous phase change materials.* Journal of Non-Crystalline Solids, 2012. **in press**.

$Ge_2Sb_2Te_5$ LINE TEST-STRUCTURES FOR PHASE-CHANGE NON VOLATILE MEMORIES

G. D'Arrigo[1], A.M. Mio[1], A. Cattaneo[2], C. Spinella[1], A.L. Lacaita[2] and E. Rimini[1]

[1]IMM-CNR, VIII Strada 5, I-95121 Catania, Italy
[2]Dipartimento di Elettronica e Informazione, Politecnico di Milano, piazza L. da Vinci 32, I-20133 Milano, Italy

Chalcogenide alloys, in particular $Ge_2Sb_2Te_5$ (GST), have been used to realize phase-change memories (PCMs) [1,2]. They are the best candidates to evolve NOR-type floating-gate (FG) memories beyond the 45-nm technology node [3-5]. In this work, we have fabricated and characterized GST line test-structure memories, i.e. GST thin films structured into lines that at both sides end in pads of increased area [6]. Test-structures were fabricated by Electron Beam Lithography (EBL) and Inductively Coupled Plasma. Patterned lines were ≈400nm long, ≈20nm thick and ≈30nm wide. An overlayer of hydrogen silsesquioxane (HSQ) has been also patterned for passivation. The samples have been annealed at 220°C to reduce access resistance to 10 kOhm. The structures have been successful switched between SET and RESET states with more than two decade resistivity contrast, with SET pulses as short as 500ns and RESET pulses 300ns long and a melt current of ≈120 μA.

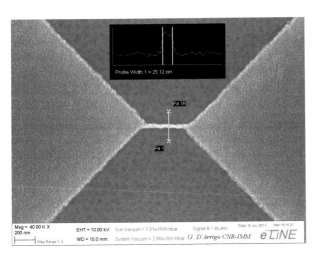

Figure 1: SEM micrograph of a single GST line after the plasma etching of GST. The e-beam lithography was done using the HSQ resist with 55 nm in thickness and several test was done to reduce the proximity effect in the exposition. The structure was switched between SET and RESET states using pulses of 500ns and 300 ns long.

A recovery mechanism, permitting the restore of a set stuck failure with a slow I-V ramp, has been also investigated. For a cell in the initial set min state, a current hysteresis of about 50 μA has been measured during I-V sweep up and sweep down (voltage ramp 0÷2.7V, pulse duration ≈300ns), probably due to stoichiometry variation during phase switching. The direct access to the active region of the cell allows to correlate programming characteristics and their electrical-induced variations with structural, morphological and chemical measurements. The structure is a good "work bench" able to indicate how the system changes in function of the cycling.

The absence of metal plug in direct contact with the GST line allows a better evaluation of the electrical current induced mass transport. We have measured by EDX the changes in chemical composition along pulsed lines (50-200 cycles) and we have found an accumulation of Te in the correspondence of the anode [7]. We present also an alternative approach to study the physics behavior of single cell, in a inverted mushroom configuration, with a plug contacts below the 20 nm in diameter and we show the preliminary results.

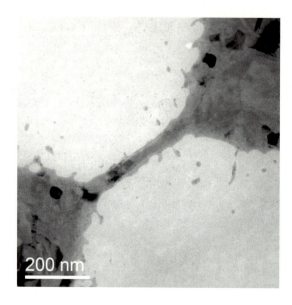

Figura 2 TEM micrograph (bright field) of a single GST line after cycling. The darker region inside the structure indicates a GST line width of about 30 nm.

[1] I. Friedrich, V. Weidenhof, W. Njoroge, P. Franz, and M. Wuttig, J. Appl. Phys. 87, 4130(2000).
[2] M. Wuttig and N. Yamada, Nature Mater., 6, 824 (2007).
[3] S. Lai, "Current status of the phase change memory and its future," in IEDM Tech. Dig., 2003, pp. 255–258.
[4] A. Pirovano, A. L. Lacaita, D. Merlani, A. Benvenuti, F. Pellizzer, and R. Bez, "Electronic switching effect in phase-change memory cell," in IEDM Tech. Dig., 2002, pp. 923–926.
[4] G.Servalli, A 45nm Generation Phase Change Memory Technology, IEDM09-113 (2009).
[5] I. S. Kim, S. L. Cho, D. H. Im, E. H. Cho, D. H. Kim, G. H. Oh, D. H. Ahn, S.O. Park, S. W. Nam, J. T. Moon, C. H. Chung, "High performance PRAM cell scalable to sub-20nm technology with below 4F2 cell size, extendable to DRAM applications," VLSI Technology (VLSIT), 2010 Symposium, pp.203-204, (15-17 June 2010).
[6] M. H. R. Lankhorst, B. W. S. M. M. Ketelaars, R. A. M. Wolters, Nature Mater. 4, 347 (2005).
[7] C. Kim, D. Kang, Tae-Yon Lee, K. H. P. Kim, Y.-S. Kang, J. Lee, S.-W. Nam, K.-B. Kim, Y. Khang, Appl. Phys. Lett. 94, 193504 (2009).

IN SITU TRANSMISSION ELECTRON MICROSCOPY STUDY OF THE CRYSTALLIZATION OF BITS IN $Ag_4In_3Sb_{67}Te_{26}$

Manuel Bornhöfft[1,2], Andreas Kaldenbach[3], Matthias Wuttig[3], Joachim Mayer[1,2]

[1]Central Facility for Electron Microscopy, Aachen, Nordrhein-Westfalen, Germany;
[2]Ernst Ruska-Centre, Nordrhein-Westfalen, Germany; [3]I. Physikalisches Institut (IA), Nordrhein-Westfalen, Germany

Phase-change materials are promising candidates for non-volatile data-storage applications. Already used in rewritable optical data-storage [1], they are also candidates for non-volatile electronic memory applications [2]. The understanding of crystallization kinetics of the phase-change materials is mandatory to develop reliable and fast phase-change data-storage devices, which can surpass actual data-storage technology. A recent topic of interest is the role of nucleation and growth in phase-change materials at different conditions.

In this work, in situ-methods in a transmission electron microscope are used to observe the crystallization of round amorphous marks (bits) in a crystalline matrix of the phase-change material $Ag_4In_3Sb_{67}Te_{26}$. The in situ-methods employed are crystallization by in situ-heating and in situ-irradiation by a focused electron beam of the microscope. The bits are produced by laser radiation. The laser beam heats locally the phase-change material above the melting point. After terminating the laser pulse the heated area is melt quenched. The phase-change layer is embedded in a supporting multilayer stack on a silicon substrate (Figure 1(a)). The silicon substrate in the observed area is removed mechanically by dimple grinding and through etching with potassium hydroxide. The supporting layers are amorphous and the phase-change layer is crystallized through heating. In one set of experiments, the samples were heated in situ with a Phillips PW 6592 heating holder. In the in situ-heating experiment a sample is heated from room temperature to 70 °C and heated to 110 °C after thermal drift has faded. The crystallization of the bit is then observed in a "FEI Tecnai F20" in brightfield imaging mode. To crystallize locally selected parts of a bit, a bit is exposed to the focused electron beam of the "FEI Tecnai F20", until the local area has crystallized. The bit is then examined in brightfield imaging mode and with electron diffraction. In both experiments an accelerating voltage of 200 kV was used.

The in situ heated bits crystallize through inward growth from the crystalline rim towards the middle of the bits. The crystalline rim grows homogeneously into the bit. The crystallization of a complete bit takes around 30 minutes after heating starts. No nucleation is observed inside the bits. Bending contours created during crystallization follow approximately the direction of crystal growth. Even with these bending contours the crystallized area shows little difference to its surrounding crystalline matrix (Figure 1(b)-(f)).

The focused electron beam produces a round crystalline area inside the bit. The crystallized area also shows bending contours, but they lack a directional preference (Figure 2(a)). The size of the crystallized area and the time to crystallize depends on the spot size of the electron beam. To crystallize a round area with 0.16 μm in diameter, it has to be exposed to the focused electron beam for approximately 35 seconds. The spot size was manually controlled and slightly bigger then the crystallized area. Electron diffraction proves that the area is crystalline and consists of only a few grains. This confirms that it is crystallised primarily by growth dominated processes after the first nuclei have formed (Figure 2(b)).

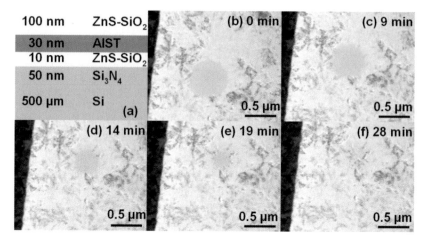

Figure 1: (a) Schematic image of the multilayer stack on the silicon substrate before etching. (b)-(f) Transmission electron microscope brightfield images of an amorphous mark heated in situ to 110 °C in crystalline $Ag_4In_3Sb_{67}Te_{26}$. Time is related to the start of heating. The size of the amorphous mark decreases over time.

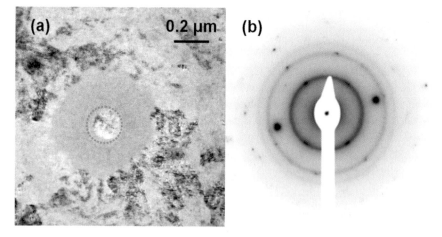

Figure 2: (a) Transmission electron microscope brightfield image of an amorphous mark exposed to a focused electron beam. A round area in the middle of the amorphous mark has crystallized. The dashed circle marks the position and size of **the** diffraction aperture. (b) Inverted diffraction pattern of the crystallized area.

With the in situ transmission electron microscopy experiments, it could be shown that the crystallization occurs primarily by growth for $Ag_4In_3Sb_{67}Te_{26}$. A homogeneous decrease in diameter of the bit until complete crystallization is directly observable in the transmission electron microscope by in situ-heating at 110°C. A focused electron beam also leads to crystallization dominated by growth [3].

[1] E. Meinders in "Phase-change Media and Recording", ed. A. Mijiritskii et al, (Springer, Berlin) 2 (2006).
[2] M. Wuttig and N. Yamada, Nature Materials **6**, 824 (2007).
[3] The authors gratefully acknowledge funding from the "SFB 917 Nanoswitches".

TEMPLATED ADSORPTION AT THE Fe_3O_4(001) SURFACE: THE EFFECT OF SUBSURFACE CHARGE AND ORBITAL ORDER

Gareth S. Parkinson, Zbynek Novotny, Michael Schmid, Ulrike Diebold

Institute for Applied Physics, TU Wien, Vienna, Austria

The (001) surface of the half-metallic ferrimagnet Fe_3O_4 is terminated by a $(\sqrt{2}\times\sqrt{2})R45°$ reconstruction over a wide range of chemical potentials [1]. DFT+U calculations predict this reconstruction to result from the coupling of subsurface charge and orbital order to a lattice distortion [1, 2]. This leads to a surface resembling the insulating bulk phase found after cooling through the Verwey metal-insulator transition at 125 K; a surface band gap has been measured experimentally [3]. In this presentation we report STM and XPS results that show how a strong preference for adsorption above Fe^{2+}-like subsurface cations creates an adsorption template at the Fe_3O_4(001) surface. We will demonstrate that this preference determines the structure of several reduced surface terminations [4], and can be used to create dense arrays of Au adatoms.

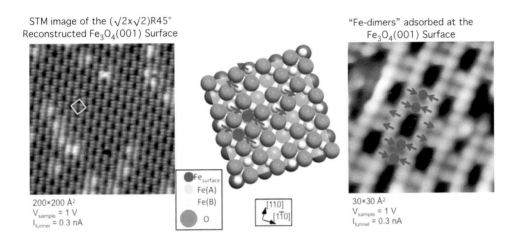

Figure 1: (left) STM images of the $(\sqrt{2}\times\sqrt{2})R45°$ reconstructed Fe_3O_4(001) surface exhibit an undulating appearance related to a lattice distortion in the surface layer. (centre) Ball model of the "Fe dimer" structure formed after 1 ML Fe deposition. (Right) STM image of the Fe-dimer surface.

[1] R. Pentcheva, F. Wendler, H. L. Meyerheim, W. Moritz, N. Jedrecy, and M. Scheffler, Phys. Rev. Lett **94**, 126101 (2005).
[2] Z. Lodziana, Phys. Rev. Lett **99**, 206402 (2007).
[3] K. Jordan, A. Cazacu, G. Manai, S. F. Ceballos, S. Murphy, and I. V. Shvets, Phys. Rev. B **74**, 085416 (2006).
[4] G. S. Parkinson, Z. Novotny, P. Jacobson, M. Schmid, and U. Diebold, Surf. Sci. Lett. **605**, L42 (2011).

EXPLORING ROUTES TO TAILOR THE ELECTRONIC PROPERTIES OF THIN-OXIDE FILMS ON METAL SUPPORTS

Xiang Shao, Fernando Stavale, Niklas Nilius

Fritz-Haber-Institut der Max-Planck-Gesellschaft, Faradayweg 4-6, D-14195 Berlin, Germany

Thin oxide films grown on metal supports offer additional degrees of freedom to tailor their physical and chemical properties that are not available for the respective bulk materials. The interface characteristic may be altered by choosing a suitable combination of metal support and oxide overlayer. By this means, parameters like the metal-oxide binding distance and the amount of interfacial charge transfer can be varied, which in turn affects the band alignment and the work-function of the thin-film system. Both effects can be reinforced by adding suitable promoters to the metal-oxide interface, for instance charge donors such as lithium.

Whereas interface-effects fade away if the film thickness exceeds 5 ML, long ranged modifications may be achieved by introducing defects and dopants into the oxide matrix. This allows us to create charge centres at adjustable positions and with variable concentrations in the oxide lattice. Electron traps might be generated by incorporating defects with formal positive charges (e.g. F^+ colour centres) or under-valent dopants (e.g. alkali metals ions), while electron donors are formed by inserting high-valent impurities. Electron transfer processes into or out of the charge-centres changes the electronic structure of the oxide films and may be exploited to tailor the adsorption and reaction behaviour of the oxide surface.

We have demonstrated the effects discussed above by means of scanning tunnelling microscopy and spectroscopy experiments performed on a variety of thin oxide films, e.g. MgO, CaO and SiO_2 in the presence of different alkali and transition metal dopants.

PREPARATION AND CHARACTERIZATION OF THIN MgO FILMS DOPED WITH NITROGEN

Martin Grob[1], **Marco Pratzer**[1], **M. Ležaić**[2], **Markus Morgenstern**[1]

[1]II. Physikalisches Institut B, Otto-Blumenthal-Straße, RWTH Aachen University and JARA-FIT, 52074 Aachen, Germany; [2]Peter-Grünberg Institut, Forschungszentrum Jülich and JARA, 52425 Jülich, Germany

Theoretical calculations [1] predict nitrogen-doped MgO being a ferromagnetic half metal. SQUID measurements exhibit a coercivity of 26 mT at 2.2% N-concentration and T = 10 K [2]. Therefore we investigated MgO and $MgO_{1-x}N_x$ films on Mo(001) by scanning tunneling microscopy (STM) and spectroscopy (STS) at room temperature. The films were prepared by evaporation of magnesium in oxygen and oxygen/nitrogen atmosphere. A nitrogen concentration of up to x = 6% has been achieved for films with a thickness of up to 11 monolayers. Similarities and differences of pristine MgO and $MgO_{1-x}N_x$ were observed by STM. The electronic structure measured by STS differs in a significant way, showing additional electronic states appearing in the band gap of MgO. These states are compared to DFT calculations taking different N-impurities into account.

The influence of the Mo substrate taking affect as a catalyzer for N-dissociation is discussed by a systematic observation of the nitrogen concentration with respect to the film thickness.

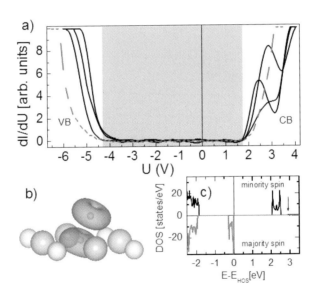

Figure 1: a) dI/dU spectra measured by STS on a 11 ML high $MgO_{0.96}N_{0.04}$ island at several positions (straight lines) and on a 11 ML thick pristine MgO island (dashed line); average film thickness in both cases: 7 ML
(U_{stab} = 3 V, I_{stab} = 0.5 nA, U_{mod} = 40 mV).
b) Calculated charge density of the unoccupied N-induced states of a N-N dimer at the surface. Large and small spheres show Mg and O respectively.
c) Calculated density of states (DOS) for N-N dimer at MgO surface, HOS: highest occupied state; all states between -1.2 eV and 3 eV are N-induced; conduction band minimum is marked by arrow.

[1] P. Mavropoulos, M. Ležaić, and S. Blügel, Phys. Rev. B 80, 184403 (2009).
[2] C. H. Yang, Ph. D. thesis, Stanford University, Stanford 2010.

SCANNING TUNNELING MICROSCOPY STUDY OF SINGLE-CRYSTALLINE $Sr_3Ru_2O_7$

Bernhard Stöger[1], Zhiming Wang[1], Michael Schmid[1], Ulrike Diebold[1], David Fobes[2], Zhiqiang Mao[2]

[1]Institute of Applied Physics, Vienna University of Technology, Vienna, Austria;
[2]Department of Physics and Engineering Physics, Tulane University, New Orleans, LA, USA

Perovskite oxides play an important role as cathodes in solid oxide fuel cells (SOFC) and in catalysis. Investigating surface defects such as oxygen vacancies and the adsorption of relevant molecules helps gaining more insight into the physics behind SOFCs and catalytic processes.

High quality $Sr_3Ru_2O_7$ (SRO) single crystals were grown using the floating zone technique. $Sr_3Ru_2O_7$ is part of the ruthenate Ruddlesden-Popper series $Sr_{n+1}Ru_nO_{3n+1}$, which have a layered structure (Fig. 1). We investigated the surface of SRO by means of Scanning Tunneling Microscopy (STM) at 78 K and at 6 K. The single crystals were cleaved in ultra-high vacuum at 150 K, which results in large terraces. To identify which atoms can be observed in STM, the surface of $Sr_3Ru_2O_7$ was compared with the one of $(Sr_{0.95}Ca_{0.05})_3Ru_2O_7$. Furthermore, we have characterized the defects that are present at the as-cleaved surfaces (Fig. 2), and how reactive they are if exposed to CO and O_2. CO binds to defects at the surface, and, possibly also to apical oxygen atoms at the perfect surface. This work was supported by the Austrian Science Fund (FWF project F45).

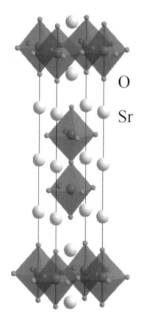

Figure 1: Atomic structure of $Sr_3Ru_2O_7$. The Ru atoms are located in the center of each octahedron.

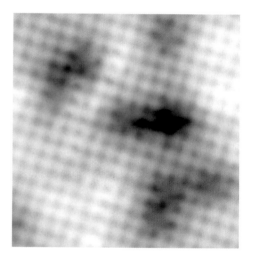

Figure 2: STM image of the freshly cleaved $Sr_3Ru_2O_7$, Sr atoms are seen as elevations in the STM image. Some defects can be also observed.

4.5×4.5 nm², -0.3V/ 0.134 nA, 6 Kelvin

BIMETALLIC ALLOYS AS MODEL SYSTEMS FOR THE GROWTH OF ULTRATHIN METAL OXIDE FILMS

Marco Moors[1,2], Séverine Le Moal[1,3], Jan Markus Essen[1], Christian Breinlich[1], Maria Kesting[4], Stefan Degen[1], Aleksander Krupski[5], Conrad Becker[6], Klaus Wandelt[1]

[1]Institute of Physical and Theoretical Chemistry, University of Bonn, Wegelerstraße 12, D-53115 Bonn, Germany;
[2]Peter Grünberg Institute, Forschungszentrum Jülich, Wilhelm-Johnen-Straße, 52425 Jülich, Germany;
[3]Institute of Technology, University of Orsay, rue Georges Clemenceau, 91405 Orsay Cedex, France;
[4]Institute of Technology, University of Karlsruhe, Hermann-von-Helmholtz-Platz 1, 76344 Eggenstein-Leopoldshafen, Germany;
[5]Institute of Experimental Physics, University of Wrocław, pl. Maksa Borna 9, 50-204 Wrocław, Poland;
[6]Centre Interdisciplinaire de Nanoscience de Marseille, CNRS - UPR 3118, associated to Université de la Méditerranée and Université Paul Cézanne, Campus de Luminy, Case 913, F-13288 Marseille Cedex 09, France

Direct oxidation of reactive metals often results in the growth of rather thick and atomically unordered oxide films. A more promising approach is the use of intermetallic phases consisting of a less and a highly reactive element, which enable the controlled formation of well defined ultrathin oxide layers of the element with the more negative heat of formation. In this work the growth of atomically ordered titanium oxide and aluminum oxide films with a thickness of only one or two monolayers has been studied on $Pt_3Ti(111)$ and $Ni_3Al(111)$, respectively. The multitude of highly reproducible oxide phases and their easy access together with their distinctive adsorptive properties are great advantages of these alloy surfaces.

The morphology as well as the surface composition of ultrathin titanium oxide films grown on a $Pt_3Ti(111)$ surface have been investigated as a function of oxidation temperature and oxygen exposure. Altogether four different ordered oxide phases with both rectangular and hexagonal symmetry have been observed caused by the competitive influence of the hexagonal symmetry of the substrate and the favored rectangular oxide symmetry [1]. Depending on the used oxygen dose the exposure of oxygen at 1000 K results in the formation of two commensurate oxide phases, the zigzag-like z'-TiO_x phase and the wagonwheel-like w'-TiO_x phase. These stable phases offer interesting catalytic and conducting properties as it has already been shown for the reactive adsorption of carbon monoxide and the deposition of palladium clusters [2, 3].

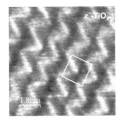 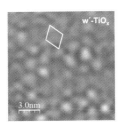  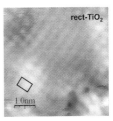

Fig. 1: Atomically resolved STM images of four different titanium oxide phases grown on $Pt_3Ti(111)$.

Additionally, at very high oxygen doses of more than 4000 L and more moderate temperatures (750 - 900 K) two incommensurate oxide phases have been found, the also zigzag-like z-TiO_x phase and the rect-TiO_2 phase. These thicker films are only metastable in a rather small temperature range between 800 and 900 K. Annealing to higher temperatures results in an immediate surface reduction. A complete oxidation to TiO_2 is not possible by using oxygen partial pressures up to 10^{-5} mbar and oxidation temperatures up to 1000 K. All structures are in good agreement with former studies of TiO_x films deposited on Pt(111) [4] and show a high degree of surface order.

The Ni$_3$Al(111) single crystal surface can be used to grow a well defined aluminum oxide film with a thickness of only two atomic layers and a distinctive long range order [5]. At an oxidation temperature of 1000 K a hexagonal oxide phase with astonishing behavior in STM measurements is formed. Depending on the used bias voltage three different structures can be observed indicating a complex electronic structure of the system [6].

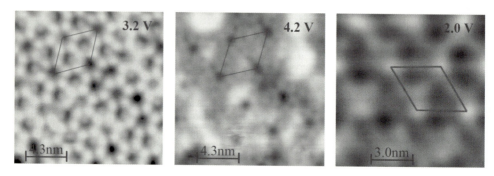

Fig. 2: STM images of a hexagonal aluminum oxide film grown on Ni$_3$Al(111) at different bias voltages.

At U_{Bias} = 3.2 V a hexagonal arrangement of hollows is observed, where every hollow is surrounded by a hexagonal ring of bright dots. This 'network structure' has a lattice vector of 2.4 nm. However, a closer inspection of this structure indicates that not all hollows have the same depth. Some hollows are by 40 pm deeper than the others. At U_{Bias} = 4.2 V the deeper hollows form a regular arrangement, which is the basis of a hexagonal unit cell with a lattice vector of 4.16 nm, the so called 'dot structure'. At U_{Bias} = 2.0 V this structure undergoes a contrast reversal. The 'network structure' represents the real topography of the film, whereas the 'dot structure' is a purely electronic effect. Nevertheless, both have significant influence on the adsorptive properties as for example the deposition of palladium and manganese clusters has shown. While the latter ones orientate along all hollows of the network structure [7], palladium is only affected by the hollows of the 'dot structure' [8].

[1] C. Breinlich, M. Kesting, M. Moors, S. Le Moal, C. Becker, K. Wandelt, in preparation.
[2] S. Le Moal, M. Moors, J.-M. Essen, C. Becker, K. Wandelt, Surf. Sci. **604**, 1637 (2010).
[3] C. Breinlich, M. Kesting, M. Moors, S. Le Moal, C. Becker, K. Wandelt, in preparation.
[4] F. Sedona, G. A. Rizzi, S. Agnoli, F. X. Labrès i Xamena, A. Papageorgiou, D. Ostermann, M. Sambi, P. Finetti, K. Schierbaum, G. Granozzi, J. Phys. Chem. B **109**, 24411 (2005).
[5] A. Rosenhahn, J. Schneider, C. Becker, K. Wandelt, J. Vac. Sci. Technol. A **18**, 1923 (2000).
[6] S. Degen, A. Krupski, M. Kralj, A. Langner, C. Becker, M. Sokolowski, K. Wandelt, Surf. Sci. Lett. **576**, L57 (2005).
[7] C. Becker, K. von Bergmann, A. Rosenhahn, J. Schneider, K. Wandelt, Surf. Sci. **486**, 443 (2000).
[8] S. Degen, C. Becker, K. Wandelt, Faraday Discuss. **125**, 343 (2003).

ELECTROSTATIC FIELD EFFECT MODULATION OF SHUBNIKOV-DE HAAS OSCILLATIONS IN LaAlO$_3$/SrTiO$_3$

Nicolas Reyren[1], Mario Basletić[2], Manuel Bibes[1], Cécile Carrétéro[1], Virginie Trinité[1], Amir Hamzić[2] and Agnès Barthélémy[1]

[1]Unité Mixte de Physique CNRS/Thales, Palaiseau, France; [2]Department of Physics, University of Zagreb, Zagreb, Croatia.

The dimensionality of an electronic system influences its physical properties and the possible applications that it can find. The case of the conducting interface between LaAlO$_3$ (LAO) and SrTiO$_3$ (STO) is hence of great interest [1]. We investigate this question by measuring the magneto-resistance (MR) of the LAO/STO interface in different magnetic field configurations and for different gate voltages. In particular, we study the presence of oscillations in the MR attributed to Shubnikov-de Haas effect, and how they evolve as the system is doped with the gate. Previous report of Shubnikov-de Haas oscillations in confined LAO/STO interfaces [2-4] and in reduced STO [5] did not exhibit the change in behavior that we report here [6].

Figure 1 shows Shubnikov-de Haas oscillations for different gate voltage and for the magnetic field applied out-of-plane or in-plane at 300 mK. At low carrier concentration (-40V), the oscillations have one unique frequency, observable only for out-of-plane magnetic field. This indicates the presence of one (or several degenerated) two-dimensional (2D) band with a high mobility (>1000 cm^2/Vs).

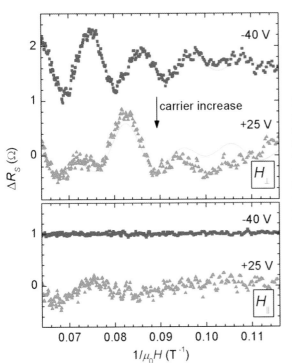

Figure 1: After subtraction of a polynomial background, the resistance at 300 mK is plotted as a function of the reciprocal of the magnetic field, revealing the Shubnikov-de Haas oscillations. In the upper panel, the field is applied along the normal of the conducting LAO/STO interface (out-of-plane). In the lower panel the field in applied parallel to the current, along the interface (in-plane). The different curves are for different gate voltages, corresponding to a carrier doping, from –40V to +25V.

As the carrier concentration increases with the gate voltage, the oscillations become more complicated and, for the highest doping level shown here, +25V, some oscillations seem to appear for parallel magnetic field. This could indicate that new bands are filled with a more three-dimensional (3D) character.

Determining the precise shape of the Fermi surface would require much more measurements, but we can already infer that the system dimensionality can be tuned from 2D to 3D by applying a gate voltage and that this change in dimensionality is probably related to the filling of an extra band.

[1] A. Ohtomo and H. Y. Hwang, Nature **427**, 423 (2004).
[2] M. Ben Shalom et al, Physical Review Letters **105**, 206401 (2010).
[3] A. D. Caviglia et al, Physical Review Letters **105**, 236802 (2010).
[4] M. Huijben et al, arXiv:1008.1896 (2010).
[5] G. Herranz et al, Physical Review Letters **98**, 216803 (2007).
[6] N. Reyren et al, *in preparation*.

SELF-RECTIFYING RESISTIVE MEMORY DEVICES

Wei Lu[1], Sung-Hyun Jo[1], Yuchao Yang[1]
[1]University of Michigan, Ann Arbor, MI 48109, United States

Resistive memory (RRAM) is based on two-terminal resistive switches whose states can be reconfigured with external pulses. RRAM has attracted significant interest recently as a promising candidate for future high-density, high-performance memory applications [1-7]. However a significant challenge for RRAM is "sneak paths" formed among cells in the interconnected passive network [2], and 'selector devices' are needed to break the parasitic paths [3-6]. As a result, a single RRAM cell is essentially composed of two components: the memory switching element that provides data storage and the 'selector device' (transistor or diode) that regulates current flow. Unfortunately, finding a selector device such as a diode that can be fabricated at low-temperatures and can still provide sufficient programming and read currents at the < 20nm scale is extremely challenging [5]. As a result, the scaling and development of RRAM cells are to a large extend limited by the development of the selector device instead of just the memory element performance.

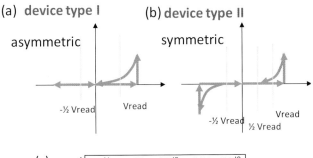

Figure 1. (a)-(b) Two possible device structures to suppress the sneak current paths. The vertical lines represent the read voltage applied to the target cell and ½ read voltage applied to unselected cells. (c) Response of a type I device to different negative voltage pulses. The device remains in the on-state when experienced to small negative biases, as verified from the subsequent read pulse and can be erased with larger negative pulses. From [6].

Instead of relying on external diodes as the selector device, a more ideal approach may be using a memory element with an intrinsic current-rectifying behavior. This approach offers current regulation but does not suffer from the voltage divider effect and the fabrication issues related to the use of external diodes. Indeed, two types of self-rectifying devices may be used and have been demonstrated experimentally, as schematically illustrated in Fig. 1a-b. Although the I-V characteristics look different, both types of devices are in fact essentially governed by the same underlying principle – that the information is stored by the internal configuration of the device which is in turn affected by the read process. As a result, a measurable current can only be obtained at the right read conditions even if the device is in the "1" state. Since the cells in the sneak paths are either experiencing a lower voltage or reverse-biased during read, the self-rectifying behaviors shown in Fig. 1a-b can thus effectively block the sneak currents without the use of additional elements such as transistors or diodes.

In the case of approach I, we note that even though the current is suppressed at small negative biases, the device still remains in the 1 state as evident from the fact that reading the device status at small positive read voltage still results in a high on-conductance (Fig. 1c). The true off-state can be achieved with a much larger negative erase voltage. This effect has been observed in cells having "unstable" filaments [6]. The current-rectifying effect is not due to physical diodes such as Schottky-diode formation, but instead explained by the real-time reconfiguration of the filament at different bias conditions without completely erasing the filament.

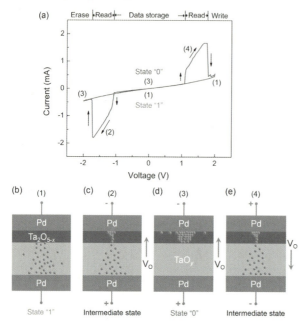

Figure 2. (a) I–V curve of a type II device based on Pd/Ta_2O_{5-x}/TaO_y/Pd structure. (b–e) Schematic illustration of complementary resistive switching processes in a single cell. (b-e) correspond to the 4 states (1-4) in (a), respectively. The different states are determined by the internal oxygen vacancy distributions in the Ta_2O_{5-x} and TaO_y layers. From [9].

Approach II is essentially based on the concept of complementary resistive switching (CRS) cells [8]. However, instead of physically having two cells connected back to back, the "0" and "1" states are represented by different internal configurations within the same device, as schematically illustrated in Fig. 2b-e. This type of device has been demonstrated recently by us in a Ta_2O_{5-x}/TaO_y stack [9] having inert electrodes. Both states "1" or "0" have a very resistive, oxygen-vacancy deficient layer so they both provide high resistance. However, the different configurations can be read out with the right combination of read voltage polarity and amplitude which brings the "1" state into a conducting configuration (e.g. Fig. 2c) but not for the "0" state.

Significant challenges still remain for both approaches. For approach I, the main challenge is the retention since the operation is based on unstable filaments. For approach II, since read involves the movement of large amount of ions the read endurance is a significant concern. For both approaches, precisely controlling the internal configurations is needed. This may cause problems in the shrinkage of the operation window for large memory arrays. Continued fundamental understanding of the device operations, coupled with careful design, engineering and material optimizations are required to bring these approaches closer to applications.

[1] R. Waser, M. Aono, Nature Mater. 6, 833-840 (2007).
[2] ITRS ERD/ERM 2010 Future Memory Devices Workshop Summary.
[3] S. H. Jo, K.-H. Kim, W. Lu, Nano Lett. 9, 870-874 (2009).
[4] M.-J. Lee, et al., Adv. Funct. Mater. 19, 1587 (2008).
[5] A. Chen, et al. IEDM Tech. Dig., pp. 765-768 (2005).
[6] K.-H. Kim et al. Appl. Phys. Lett. 96, 053106 (2010).
[7] D. Strukov, et al. Proc. Natl. Acad. Sci. 106, 20155-20158 (2009).
[8] E. Linn, R. Rosezin, C. Kügeler, R. Waser, Nat. Mater. 9, 403 (2010).
[9] Y. Yang, P. Sheridan, W. Lu, submitted.

THE DESIRED MEMRISTOR FOR CIRCUIT DESIGNERS

Shahar Kvatinsky[1], Eby G. Friedman[2], Avinoam Kolodny[1], and Uri C. Weiser[1]

[1]Technion – Israel Institute of Technology, Haifa 3200, Israel; [2]University of Rochester, Rochester, NY 14627, USA.

Memristors hold promise for use in diverse applications such as memory, logic, analog circuits, and neuromorphic systems. Different applications require different characteristics from the memristor. Understanding the desired characteristics for different applications can therefore assist device physicists in targeting the required behavior when fabricating memristive devices, potentially optimizing these devices for different applications. In this presentation, the desired characteristics for different applications are discussed from the viewpoint of the electronic circuit design process.

Logic and memory are digital applications, in which the resistance of the memristor typically represents a binary value. A low resistance is typically considered as a 'logical one' and a high resistance is treated as a 'logical zero'. In these applications, memristors can be employed for computation and control, as well as data storage. In memory applications, it is also possible to write more than one bit into a single memristor if the resistance of the device is quantized to multiple levels. In these applications, the difference among different data must be carefully determined. It is desirable to provide a non-destructive read mechanism, but read operations may induce drift in the stored state. The device design process therefore should consider the tradeoff between speed and robustness due to this state drift phenomenon. A preferred memristor would therefore be highly nonlinear, with a well defined and abrupt threshold between the two distinct states.

In applications using analog circuits and neuromorphic systems, however, the resistance typically requires a continuous value. Memristors can be used as configurable devices where the resistance of the device is initialized by a specific procedure, separate from typical circuit operation; during regular circuit operation, the memristor behaves as a simple resistor. In these applications, it is desirable for the memristor to behave as a nonlinear nondestructive device, similar to the read mechanism in digital applications. In neural networks, memristors mimic the role of synapses, such that each device may interact with other devices throughout the operation of the circuit.

Memristors can also be used as computational elements in analog circuits, such as analog counters. In these circuits, it is desirable for the memristor to maintain a linear behavior, where the local current changes the resistance of the memristor.

Several additional characteristics are also important and are discussed in this presentation: low power consumption, good scalability, long data retention, high endurance, and manufacturing and voltage compatibility with conventional CMOS.

COMPLEMENTARY RESISTIVE SWITCH-BASED ASSOCIATIVE MEMORY CAPABLE OF FULLY PARALLEL SEARCH FOR MINIMUM HAMMING DISTANCE

Omid Kavehei[1], Stan Skafidas[1], Kamran Eshraghian[2,3]

[1]Centre for Neural Engineering, Department of Electrical and Electronic Engineering, The University of Melbourne, Victoria 3010, Australia.
[2]Department of Electrical and Electronic Engineering, Chungbuk National University, South Korea,
[3]iDataMap Corporation, Australia

Emergence of new materials continue to influence the way nanoelectronics-based components are engineered and hence become the facilitator to revise old concepts towards realization of new and novel architectures that otherwise would have been inconceivable. Architectures for image recognition and speech, classification, intelligent database search engines, and flexible decision making processes heavily depend upon associative memories. A challenge in implementing large-scale physical associative memories is the limitations imposed by integration density. Reduction in transistor feature size has resulted in significant increase in power consumption attributed to leakage current. Furthermore, high-performance matching operation demands for faster devices and in order to achieve this, devices have to be scaled into nano and sub-nano regimes.

In this paper we address the key issues in system architecture for the above and suggest an alternative approach based on Complementary Resistive Switch (CRS) that has been reported to mitigate sneak current-path problems of large and fully passive nanocrossbar arrays [1]. Based on the nondestructive readout mechanism using capacitance of non-uniform anti-serially connected metal-insulator-metal capacitors in a CRS structure [2] we have formulated a novel approach in the implementation of a nonvolatile CRS associative memory. The CRS device and the nondestructive readout technique are utilized to implement associative memories having fully parallel search capability for minimum Hamming distance detection problems. It is the first time that application of CRS devices has been applied to matching systems.

Fig. 1 demonstrates the structure of the CRS device that we have implemented for an associative memory. The CRS structure for nondestructive readout is illustrated in Fig. 1(a), where capacitance of the resistive device A, C_A, is much less than B, C_B. In addition, resistance of A and B are always complementary, which mandates if R_A=LRS (low resistance state), R_B=HRS (high resistance state) and vice versa. Therefore, impedance of the CRS device's state are complementary, either HRS/LRS or LRS/HRS, different capacitances will be seen at output. Parameter values were extracted from [2], C_A=0.3 pF, C_B=0.9 pF, C_{in}=7 pF, and C_{out}=24 pF. Fig. 1(b) shows SPICE simulation results of the readout using 3 V input pulses for a single CRS device in two different states, LRS/HRS and HRS/LRS, which is in agreement with the experimental data in Figure 5(a) in [2]. For device illustrated in Fig. 1(c) an input vector will result in an output that corresponds to the Hamming distance between the input vector and stored pattern. In this design a vector of no input produces an ambiguous output. To avoid this we augment the input vector with its complementary to ensure that the input is never always zero. To illustrate the viability of this device, we have implemented an array of 64 x 8 CRS devices. We have intentionally made the input code match with the first row. The result, shown in Fig. 1(d), is evident that this output is higher than the other (grouped) outputs. The winner-take-all (WTA) stage then produces the output that is clearly distinguishable from other outputs (Fig. 1(e)).

Nanosession: Logic devices and circuit design

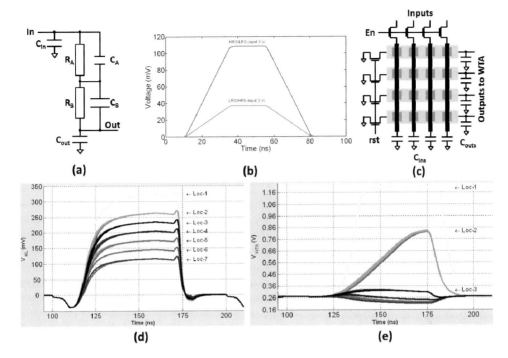

Figure 1. (a) CRS device. (b) Simulation result for a single device capacitance readout. (c) A 4 × 4 schematic of main items involve in the design of associative memory. (d) Outputs of the array before WTA. Loc-x shows the grouping for the outputs. Loc-1 only contains one (first row) output. (e) Outputs after WTA.

[1] Linn, E., Rosezin, R., Kügeler, C., and Waser, R., "Complementary resistive switches for passive nanocrossbar memories," *Nature Materials*, **9**(5), pp. 403-406, 2010.
[2] Tappertzhofen, S., Linn, E., Nielen, L., Rosezin, R., Lentz, F., Bruchhaus, R., Valov, I., Böttger, U., and Waser, R., "Capacity based nondestructive readout for complementary resistive switches," *Nanotechnology*, **22**, art. no. 395203, 2011.

COMPUTATIONAL CONCEPT BASED ON COMPLEMENTARY RESISTIVE SWITCHES

Ondrej Šuch[1], Martin Klimo[2], Stanislav Foltán[2], Karol Grondžák[2]

[1] Slovak Academy of Sciences, Banská Bystrica, Slovakia ; [2] University of Žilina, Žilina, Slovakia

The breakthrough paper [1] by HP research team has brought intense focus on applications of memristive effects in computational circuitry. With the benefit of hindsight a great variety of materials have been found to exhibit memristive effect [2].

In most cases, the material exhibits two distinct states with significantly different resistances, $R_{ON} \ll R_{OFF}$. Switching between these two states is achieved by applying bias in one or the other direction. Let us call a memristor device exhibiting such behavior a bipolar bilevel memristor.

There is a second important effect in play and that is (an almost exponential) inverse dependence of time needed to switch based on the magnitude of applied bias [3], [4]. This effect plays a fundamental role in memristive memories, where a low level bias is used to detect the state of a memristive device without disturbing its state.

Applications of memristive effects extend beyond memories. An implementation of binary implication using two memristors have been demonstrated by Borghetti et al [5], as well as Rosezin et al. [6] using CRS (complementary resistive switches). Klimo and Šuch [7] showed that CRS consisting of *ideal* bipolar bilevel memristors implement min, and max operations, fuzzy logic analogs of classical Boolean logic.

It is a fundamental challenge to find the proper theoretical framework for designing circuits to take advantage of memristive effects. For instance, L. Valiant argues that many basic functions (algorithms) of the brain can be implemented in the neuroidal architecture [8] based solely on Boolean logic. Our working hypothesis is that although Boolean logic is fundamental to higher level reasoning and provides a basis for today's computers, *information processing by brain is better modeled by fuzzy logic, and therefore it should be a key part for brain inspired computational concept.* Fuzzy logic circuits based on CRS will create an inference engine in a specialized or a general purpose fuzzy logic system [9].

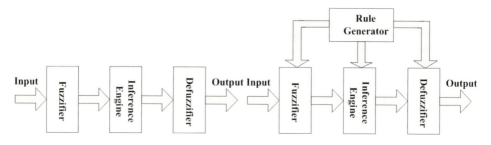

Figure 1: Specialized and general purpose fuzzy logic system

This concept is demonstrated by a wide success of fuzzy controllers, recognition systems, time series forecasting etc. [10]. We would like show, that CRS based circuits are suitable for its implementation. Moreover, joining of computational power and *memory properties* of memristor based fuzzy logic circuits gives an added value to fuzzy logic functions and opens new challenges to brain inspired computational concept.

To this end we have considered two distinct algebraic models. Firstly, we are considering computational power of circuits consisting solely of min, max and average operators. All of these can be implemented using (ideal) bilevel bipolar memristive switches. An apparent omission of implication and negation is not really limiting, if one uses M. Blum's trick [11] and allows as additional inputs also negations of processed data.

Secondly, we are considering circuits whose fundamental operators are min, max, and comparison. The need for a comparison operator arises naturally in the context of speech processing.

Both these models have to be taken as first approximations only. The first caveat is that the dependence of switching time on bias may interfere with real-time computations using such circuits. The second issue is that the workable depth of min-max circuits is greatly affected by R_{OFF}/R_{ON} ratio [7], [12]. Even with these simplifications, design of circuits in both models is a nontrivial discrete optimization problem requiring at present thousands of hours of computing time, if no heuristics are employed.

We are testing the computational power of circuits of both kinds on classical discrimination problems in speech recognition [13]. For instance, a fuzzy logic circuit topology was searched using evolutionary programming and Figure 2 illustrates the improvement in recognition percentage over training duration. Let us note that the final fuzzy logic circuit contains over 4000 memristors.

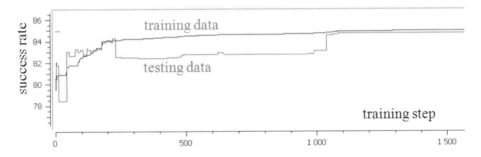

Figure 2: Success percentage during training of CRS based fuzzy logic circuit for vowel "a"/"o" recognition

[1] D. Strukov et al, Nature, **453**, 80-83 (2008)
[2] L. Chua, Appl. Phys. A, **102**, 765-783 (2011)
[3] M. Pickett et al, J. Appl. Phys. , **106**. (2009)
[4] G. Medeiros-Ribeiro et al., Nanotechnology, **22**, (2011)
[5] J. Borghetti, Nature, **464**, 873-876, (2010)
[6] R. Rosezin et al., IEEE Electron Device Letters, **32**, 710-712 (2011)
[7] M. Klimo et al, http://arxiv.org/abs/1110.2074
[8] L. Valiant, Circuits of the mind, Oxford Univ. Press (2000)
[9] A. Mendel, Proceedings of IEEE, **83**, 345-377 (1995)
[10] H. J. Zimmermann, Practical applications of fuzzy technologies, Kluwer (1999)
[11] M. Blum et al, CMU Technical Report, CMU-CS-02-11
[12] K. Likharev, J. of Nanoelectronics and Optoelectronics, **3**, 203-230 (2008)
[13] T. Hastie et al, Elements of Statistical Learning, Springer, (2003)

LOGIC OPERATIONS IN PASSIVE COMPLEMENTARY RESISTIVE SWITCH CROSSBAR ARRAYS

Eike Linn[1], Roland Rosezin[2], Stefan Tappertzhofen[1], Ulrich Böttger[1], Rainer Waser[1,2]

[1]Institut für Werkstoffe der Elektrotechnik II, RWTH Aachen University, Germany;
[2]Peter Grünberg Institut 7, Forschungszentrum Jülich GmbH, Jülich, Germany

Most logic concepts use resistive switches as programmable interconnects, e.g., CMOL [1]. A major drawback of any logic approach requiring programmed (low-resistive) cells is the occurrence of parasitic currents, reducing feasible array sizes [2]. By use of complementary resistive switches (CRS) the sneak path problem can be solved [2] since CRS cells are always high-resistive, but novel approaches to realize logic functions are required. Since CRS cells are favorable devices for high-density passive crossbar memory arrays [3], every logic concept that requires high-density memories is a potential area of application for CRS-based crossbar arrays. As such, it is straight forward to suggest CRS-based arrays for LUTs in FPGA logic blocks [4]. An alternative approach for logic implementations with resistive switches was suggested in [5], focusing on the implication (IMP) property. In general, the implication p IMP q, which is also called material implication or material conditional, is true for any value of p and q except for $p =$ '1' and $q =$ '0'. In [5] three elements, two bipolar resistive switches (BRS) and a load resistor, are used to form an IMP operation, while in [6] only one BRS or CRS is applied. Here we introduce a special operation scheme to realize a CRS 'stateful' logic, implementing 14 out of 16 possible logic functions with a single CRS cell. In Fig. 1, the corresponding CRS state machine representation as well as several read schemes are depicted. For arbitrary logic functionality, a stacked or folded CRS concept is suggested as well, realizing all 16 logic functions in $4F^2$ with two CRS devices. By considering CRS cells as state machines, logic operations can be conducted in crossbar memory, storing the calculation result directly to the memory. Since memory and logic operations can be performed in the same crossbar array, new reconfigurable computer architectures are feasible.

[1] K. K. Likharev and D. B. Strukov, "CMOL: Devices, Circuits, and Architectures," *Introducing Molecular Electronics*, vol. 680, pp. 447-477, 2006.
[2] E. Linn, R. Rosezin, C. Kügeler, and R. Waser, "Complementary Resistive Switches for Passive Nanocrossbar Memories," *Nat. Mater.*, vol. 9, pp. 403-406, 2010.
[3] R. Rosezin, E. Linn, L. Nielen, C. Kügeler, R. Bruchhaus, and R. Waser, "Integrated Complementary Resistive Switches for Passive High-Density Nanocrossbar Arrays," *IEEE Electron Device Lett.*, vol. 32, pp. 191-193, 2011.
[4] A. Dehon, "Nanowire-Based Programmable Architectures," *ACM Journal on Emerging Technologies in Computing Systems*, vol. 1, pp. 109-162, 2005.
[5] J. Borghetti, G. S. Snider, P. J. Kuekes, J. J. Yang, D. R. Stewart, and R. S. Williams, "'Memristive' switches enable 'stateful' logic operations via material implication," *Nature*, vol. 464, pp. 873-876, 2010.
[6] R. Rosezin, E. Linn, C. Kügeler, R. Bruchhaus, and R. Waser, "Crossbar Logic Using Bipolar and Complementary Resistive Switches," *IEEE Electron Device Lett.*, vol. 32, pp. 710-712, 2011.

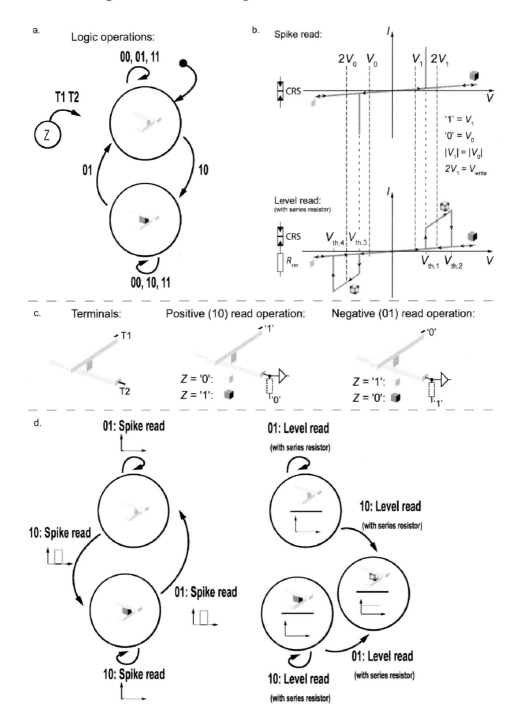

Figure 1: (a) State machine for logic operation (b) Two possible read schemes (c) Terminal configuration for read 10 and read 01. (d) State machines for spike read and level read, respectively, showing read 10 as well as read 01.

A NON-VOLATILE LOW-POWER ZERO-LEAKAGE NANOMAGNETIC COMPUTING SYSTEM

M. Becherer[1], J. Kiermaier[1], S. Breitkreutz[1], I. Eichwald[1], G. Csaba[2], D. Schmitt-Landsiedel[1]

[1]Technische Universität München, D-80333 München, Germany.
[2]University of Notre-Dame, Center for Nanoscience and Technology, Indiana, USA

Nowadays, Boolean algebra processing is exclusively performed by charge-based logic. In general purpose processors, the volatile computing states have to be backed by non-volatile memories like solid-state flash or hard-disk-drives. By contrast, nanomagnetic logic (NML) devices based on magnetic field-coupling promise a paradigm shift in nanoscale computing: Information is both propagated and processed by magnetostatic interaction of single-domain ferromagnetic dots and stripes [1] [2]. NML devices are inherently non-volatile, have zero leakage, and provide a disruptive approach for radiation-hard, dense and highly parallel digital information processing.

Our NML circuits are formed from planar, bistable ferromagnetic islands fabricated from Co/Pt and Co/Ni multilayer films. The logic states 0 and 1 are represented by the perpendicular magnetization state of the single-domain nanomagnets. Metallic interconnects to the individual computing elements are superfluous. Hence, an ultimately scaled programmable NAND/NOR gate fits in a 100 nm x 100 nm area. Figure 1 shows the major NML building blocks.

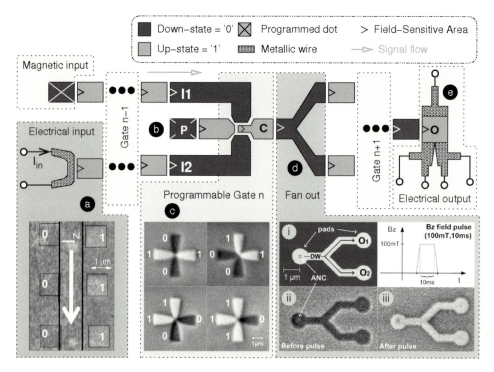

Figure 2 The components of our integrated NML system, including (a) electrical input, (b) programmable input, (c) programmable logic gate, (d) fan-out structure and (e) electrical output.

a) Inputs to a gate can be electric or magnetic [3] and hence facilitate a flexible architecture, e.g. for pattern recognition in a streaming data flow.

b) A universal 3-input majority gate can be programmed to function as NAND or NOR gate depending on the magnetization of the programming dot P. Programming can be easily performed during run-time.

c) The magnetic force microscopy (MFM) measurements of a Majority gate show the correct function for different input patterns [4]. We achieve this by creation of artificial nucleation centers (ANCs) with FIB irradiation: The signal flow direction from the inputs to the output dot is dictated by locally changing the magnetic anisotropy of the dots. Reverse influence from the output to the inputs is prevented yielding non-reciprocal signal flow in spatially homogeneous field excitation [5].

d) A branching structure to realize signal fan-outs. The MFM measurements (i-iii) show, that the magnetization is switched in a deterministic way for both outputs, cloning the magnetic state of the input [5].

e) The electrical output is shown schematically. The Extraordinary Hall-effect is applied for sensing, where the sign of the differential current of the sensor is dictated by the magnetic polarization of the output dot O [6]. The output state of the magnet is directly sensed without blocking coupling fields from its neighbor. GMR and MTJ structures are also prospective candidates for electrical read-out.

A magnetic field power-clock is used to control the synchronous switching behavior of the nanomagnets. We found that it is most efficiently implemented by a globally applied magnetic field generated by on-chip or off-chip inductors. As the clocking is provided by a global field e.g. from an RC oscillator, a synchronous clocking without skew and jitter and with adiabatic energy recovery can be generated easily.

In summary, NML devices with out-of-plane media may turn field-coupled logic into a mainstream family of non-volatile, zero-leakage, interconnect-free and low-power computing circuits.

[1] G. Csaba, W. Porod, and A. I. Csurgay. A computing architecture composed of field-coupled single-domain nanomagnets clocked by magnetic fields. *International Journal of Circuit Theory and Applications*, 31:67–82, 2003.
[2] M. Becherer, G. Csaba, R. Emling, W. Porod, P. Lugli, and D. Schmitt-Landsiedel. Field-coupled nanomagnets for interconnect-free nonvolatile computing. In *Digest Technical Papers IEEE International Solid-State Circuits Conference, ISSCC*, pages 474–475, February 2009.
[3] J. Kiermaier, S. Breitkreutz, G. Csaba, D. Schmitt-Landsiedel, and M. Becherer. Electrical Input Structures for Nanomagnetic Logic Devices. *Journal of Applied Physics*, 111(7):E341, 2012.
[4] S. Breitkreutz, J. Kiermaier, I. Eichwald, X. Ju, G. Csaba, D. Schmitt-Landsiedel, and M. Becherer. Majority gate for nanomagnetic logic with perpendicular magnetic anisotropy. In *Accepted for publication at INTERMAG, Vancouver*, 2012.
[5] S. Breitkreutz, J. Kiermaier, S. V. Karthik, G. Csaba, D. Schmitt-Landsiedel, and M. Becherer. Controlled reversal of Co/Pt dots for nanomagnetic logic applications. *Journal of Applied Physics*, 111(7):A715, 2012.
[6] M. Becherer, J. Kiermaier, S. Breitkreutz, G. Csaba, X. Ju, J. Rezgani, T. Kießling, C. Yilmaz, P. Osswald, P. Lugli, and D. Schmitt-Landsiedel. On-chip Extraordinary Hall-effect sensors for characterization of nanomagnetic logic devices. *Solid-State Electronics, Selected papers of ESSDERC 2009*, 54:1027–1032, 2010.

AN ELECTRONIC VERSION OF PAVLOV'S DOG

Hermann Kohlstedt[1], Martin Ziegler[1], Rohit Soni[1], Timo Patelczyk[1],
Marina Ignatov[1], Thorsten Bartsch[2], Paul Meuffels[3]

[1]Nanoelektronik, Technische Fakultät, Christian-Albrechts-Universität zu Kiel, Kiel 24143, Germany
[2]Klinik für Neurobiologie, Universitätsklinikum, Schleswig-Holstein, Christian-Albrechts-Universität zu Kiel, Kiel 24105, Germany
[3]Institut Festkörperforschung, Forschungszentrum Jülich GmbH, Jülich, 52425 Germany

Neuromorphic plasticity is the basic platform for learning in biological systems and is considered as the unique concept in vertebrates brains to outperforming today's most powerful digital computers when it comes to cognitive and recognitions tasks. An emerging task in the field of neuromorphic engineering is to mimic neural pathways via elegant technological approaches to close the gap between biological and digital computing. In this respect, functional, memresistive devices are considered as promising candidates with yet unknown benefit for neuromorphic circuits.

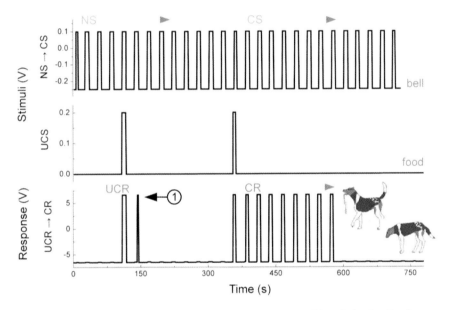

NS: Neutral Stimulus (Bell), CS: Conditioned Stimulus, UCS: unconditioned stimulus (Food),
CR: Conditioned Response, UCR: Unconditioned Response

Figure 1: Experimental demonstration of associative memory using a memresistive device. If the neutral stimulus V_{NS} (upper curve) merges the unconditional stimulus V_{UCS} (lower curve) the resistance of the system is enhanced. After two sequences the circuitry learned to associate the neutral stimuli (Bell) with that of the unconditioned stimulus (Food), affecting the output of the comparator or in other words, the alertness of the dog.

Here, we demonstrate that a single $Pt/Ge_{0.3}Se_{0.7}/SiO_2/Cu$ memresistive device implemented in an analogue circuitry mimic non-associative and associative types of learning. For Pavlovian conditioning, different threshold voltages for the memresistive device and a comparator were essential. Similarities to neurobiological correlates as well as software based neural networks will be discussed.

[1] M. Ziegler et al., will be published in Adv. Funct. Mater.

NEUROMORPHIC FUNCTIONALITIES OF NANOSCALE MEMRISTORS

Ting Chang[1], Sung-Hyun Jo[1], Patrick Sheridan[1], Wei Lu[1]

[1]Department of Electrical Engineering and Computer Science, University of Michigan, Ann Arbor, Michigan 48109, United States

Neurons and synapses together make up neural networks, empowering humans to learn, think, and remember. Human brain contains ~10^{10} neurons, each connecting to ~10^{4} other ones. At each connection is a junction called synapse, comprising ~10^{14} synapses in a human brain. The synaptic weight associated with each synapse can be modulated according to the present information received from the connected neurons and its past history. This weight in turn determines the transmission between these two neurons. In other words, the brain architecture ensures that processing (learning) and storage (memory) occur locally and simultaneously, in contrast to the "bottleneck" that the von Neumann architecture faces due to limited traffic bandwidth between the processing and storage units.

Recently, it has been demonstrated that "memristors" are ideal candidates for implementing artificial synapses [1, 2]. The memristors in these works are realized as two-terminal solid-state devices with adjustable resistance (conductance) that depends on external inputs and internal states [3]. They are simple in structure, yet provide conductance modulation with a memory effect. These demonstrations successfully satisfied many merits of synapses that cannot be fulfilled with von Neumann computers, such as connectivity, network density, power consumption, and adjustable weights with memory. The aim of this report is to present the most essential functionalities of biological synapses and their realization in nanoscale metal-oxide memristive devices.

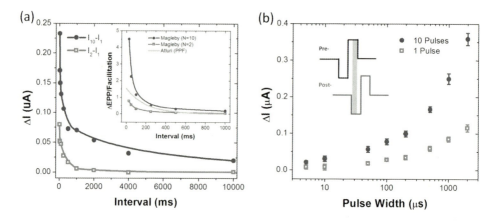

Figure 1: Resistive switching in WOx memristors (a) Rate-dependency. Inset – comparable effects observed in biological experiments [6]. (b) Pulse width-dependency. Each box chart contains the statistics from 10 measurements. Inset – an example of spiking waveform for implementing STDP.

The memristive device studied here consists of two metal electrodes sandwiching a tungsten-oxide (WOx) resistive switching layer formed in a crossbar structure. Resistive switching effects are attributed to the redistribution of oxygen vacancies in the WOx film. Oxygen vacancies act as positively-charged donors that drift under electric fields so the device conductance can be tuned in a bipolar fashion; that is, a positive (negative) bias on the top electrode with respect to the bottom electrode causes the conductance to increase (decrease). In addition, an inherent diffusion term is present which results in the decay of conductance with time. Details of device fabrication and modeling of switching mechanisms can be found in Ref.4.

Spike rate, timing, and associativity are important factors in the determination of synaptic plasticity [5]. Figure 1(a) shows the effect of different stimulation frequency (plotted in terms of stimulation interval) [6]. Due to the natural decay mechanism, the smaller the stimulation interval, the stronger the conductance change, and vice versa.

Spike timing considers the order and time delay between the pre- and post-synaptic neuron spikes on both ends of a synapse. Figure 1b shows the switching uniformity using voltage pulses with various pulse widths. With appropriate spiking waveform and timing design (e.g. inset), spike-timing-dependent plasticity (STDP) [7] can be easily achieved without complicated schemes.

Since every neuron communicates with thousands of other neurons, the networks and firing patterns can become exceedingly sophisticated. Associativity is the ability of a synapse to correlate cooperative firing of related neurons, and has been proven feasible using a memristor emulator [8]. With reliable devices at hand, demonstration of associativity with simple hardware setup becomes straightforward.

Moreover, it has been shown that there are distinct memory regimes in such devices [6]. As the number of stimulation increases, memory effect strengthens, supported by not only the enhanced conductance but also the prolonged retention time (Figure 2).

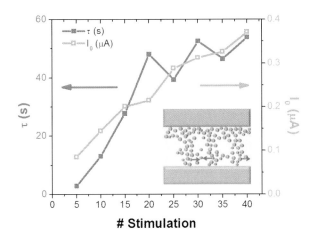

Figure 2: Short-term to long-term memory. Both the conductance (represented by I_0) and retention time (τ) increase with the number of stimulation pulses. Below and above 20 stimulations, the retention time changed remarkably. Inset is an illustration of the oxygen vacancy distribution inside the WOx thin film. Adapted from [6].

In summary, important learning rules and memory behaviors have been reported experimentally using memristor devices. However, how these effects interact with one another to construct functional and reliable networks remains unclear. Future topics also include process integration, circuit design, and fault-tolerant algorithms.

[1] D. B. Strukov et al., Nature **453**, 80 (2008).
[2] S. H. Jo et al., Nano Lett. **10**, 1297 (2010).
[3] L. O. Chua, IEEE Trans. Circuit Theory **18**, 507 (1971).
[4] T. Chang et al., Appl. Phys. A: Mater. Sci. Process **102**, 857 (2011).
[5] P. J. Sjostrom et al., Neuron **32**, 1149 (2001).
[6] T. Chang et al., ACS Nano **5**, 7669 (2011).
[7] G. Q. Bi et al., J. Neurosci. **18**, 10464 (1998).
[8] Y. V. Pershin et al., Neural Netw. **23**, 881 (2010).

DEMONSTRATION OF IMPLICITE MEMORY IN ELECTRONIC CIRCUITS BY USING MEMRESISTIVE DEVICES

Martin Ziegler[1], Mirko Hansen[1], Hermann Kohlstedt[1]

[1] Institut für Elektrotechnik und Informationstechnik, Technische Fakultät der Christian-Albrechts-Universität zu Kiel, D-24143 Kiel, Germany

For biological systems Eric Kandel was able to get access to the fundamentals of learning and memory by applying a radical reductionistic strategy.[1] At this, the principles of implicit memory in mammal brains can be understood on the cellular level, where learning alters the function and structure of neurons and their interconnection strength.

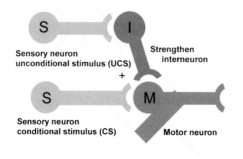

Figure 1: Neural circuity layout for implicit memory: A neural mediating circuit for associative learning requires two sensory neurons (S), a strengthening inter-neuron (I), and a motor-neuron (M). A similar strategy is used to demonstrate associative learning with a memresistive device. More information can be found in Ref. [2].

A similar reductionistic strategy is applied to demonstrate all forms of implicit memory by using memresistive devices. It will be shown that a memresistive device must exhibit a threshold voltage to be considered as a realistic substitute for basic building blocks in nerve cells. As a consequence, different threshold voltages for a memresistive device and an analogue electronic circuitry are essential to mimic non-associative and associative types of learning.

At this respect, analogue circuitries with memresistive devices, which enable to mimic non-associative and associative types of learning, are presented and similarities to neurobiological correlates of learning will be discussed in the framework of hebbian learning rule, plasticity, and long-term potentiation.

[1] E. R. Kandel, in Cellular Basis of Behavior-An Introduction to behaviour Neurobiology, W. H. Freeman and Company 1976.
[2] M. Ziegler et al., will be published in: Adv. Funct. Mater.

USAGE OF NANOELECTRONIC RESISTIVE SWITCHES WITH NONLINEAR SWITCHING KINETICS IN HYBRID CIRCUITS FOR LINEAR CONDUCTANCE ADAPTATION

Arne Heittmann[1], Tobias G. Noll[1]

[1]Chair of Electrical Engineering and Computer Systems, RWTH Aachen University, Schinkelstrasse 2, D-52066 Aachen, Germany

It has been shown experimentally as well as by examination of theoretical models that the on-resistance of nanoelectronic resistive switches is determined by the applied voltage-current profile during switching from the off-state to the on-state. The multi-level capability combined with the high scalability makes these devices attractive for realizing dynamic and non-volatile memory elements in applications such as neuromorphic circuits, adaptive filters, programmable references, memories and logic.

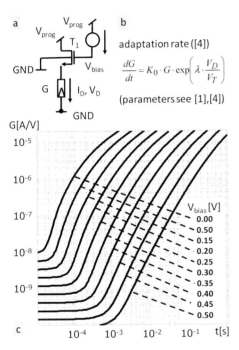

Figure 1: **a.** Source follower consisting of an N-channel transistor T_1 and a resistive switch modeled by G. The bias voltage determines the operating range of T_1. For large bias voltages the transistor operates close to the subthreshold domain providing an exponential I-V-characteristics while for smaller biases the relationship between current I_D and voltage V_D shifts towards a power law [5]. Combined with the linear conductance G a unique operating point emerges for any G which constitutes the particular characteristics of V_D driving the conductance adaptation. **b.** Dynamics of conductance adaptation for resistive switches based on electrochemical metallization. **c.** Development of conductance G over time t for different bias voltages and V_{prog}=1.2V, W/L=240nm/240nm. For $G>10^{-8}$ A/V each course of G has a characteristics which can be approximated by a power law $G \sim (t-t_0)^\alpha$. The exponent α is determined by λ and the operating point of the transistor and is strictly limited by $\alpha \leq 1/\lambda \cdot n$ with n as non-ideality factor of the subthreshold current of T_1 (here: n=1.3). For a particular bias voltage α=1 can be obtained.

In adaptive systems (such as neuromorphic circuits or adaptive filters) conductances of resistive switches acting as dynamic parameters have to be adapted, i.e. the increase (or decrease) of a particular conductance has to be realized with limited magnitude. In highly integrated systems it is still challenging to control the conductance adaptation of a resistive switch very precisely as the switching process typically depends on a rapidly acting kinetics [1]. Usual means to control the switching kinetics are active feedback circuits [2,3] or current mirrors [4]. Both concepts have limitations with respect to circuit size or dynamic operating range.

Here, a simple circuit is proposed for conductance adaptation which combines small circuit size with a large dynamic operating range. The circuit is based on an N-channel MOS transistor which is connected to a resistive switch in a source-follower configuration, cf. Fig. 1a. A bias-voltage controls the operating point of the transistor and determines a particular non-linear I-V characteristics which controls the adaptation dynamics of the resistive switch by feedback of the device voltage V_D.

The circuit performance was evaluated by circuit simulation using *Spectre* as circuit simulator and transistor device models valid for a 40nm CMOS technology. For the resistive switch a physical model [1] was used which describes the switching kinetics of cells based on electrochemical metallization. The I-V characteristics of the resistive switch is linear as the resistance is based on tunnelling.

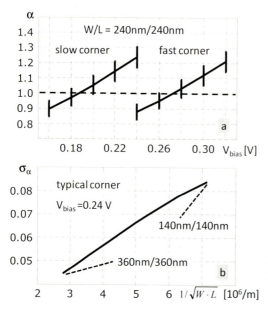

Figure 2: **a.** Variation of α (cf. Fig.1) under the consideration of parameter variations of transistor T_1 using 2000 monte carlo simulations for each examined bias voltage. In each run a different parameter set for T_1 was drawn from a random distribution modeling the measured parameter distributions of transistors implemented in a 40nm CMOS technology. For the typical case the nominal bias voltage was set to 240 mV resulting in a nominal exponent of α=1.0. The variation of α is shown for worst case conditions (slow process corner and fast process corner). The optimal bias depends on the corner case and varies by 80mV. Small variations of the bias voltage result in an almost linear variation of α. **b.** for fixed W/L-ratio of the transistor and different gate areas the variation of the standard deviation of α is shown. The standard deviation depends linearly on the reciprocal square root of the gate area.

The combined circuit provides an approximate power law for the adaptation of G: $G \sim (t-t_0)^\alpha$ (t: time) if the transistor continuously drives G, cf. Fig.1c. The exponent α is limited by the non-ideality factor n of the subthreshold current of T_1 and the factor γ, cf. Fig.1b. Smaller values of α are possible, and particularly α=1 can be obtained for a specific V_{bias} which makes the adaptation rate of G *independent* on G. A linear adaptation rate for G is particularly interesting since it allows the distribution of conductances on a linear scale.

The robustness of the circuit with respect to variability of the transistor parameters was finally examined using monte carlo simulation, cf. Fig.2. The results show that for reasonable transistor sizes the exponent α varies by less than 15% for a $3\sigma_\alpha$- interval.

[1] S.Menzel, U.Böttger, and R.Waser, "Simulation of Multilevel Switching in Electrochemical Metallization Memory Cells", Journal of Applied Physics Vol. 111, pp.014501-014501-5, Jan. 2012
[2] N.Papandreou, H.Pozidis, A.Pantazi, A.Sebastian, M.Breitwisch, C.Lam, E.Eleftheriou, "Programming Algorithms for Multilevel Phase-Change Memory", Proc. IEEE International Symposium on Circuits and Systems, pp. 329-332, May 2011
[3] W.Yi, F.Perner, M.S.Qureshi, H.Abdalla, M.D.Pickett, J.J.Yang, M.-X. M.Zhangm G.Medeiros-Ribeiro, S.Williams: "Feedback write scheme for memristive switching devices", Applied Physics A 102, pp. 973-982, 2011
[4] A.Heittmann, T.G.Noll, "Limits of Writing Multivalued Resistances in Passive Nanoelectronic Crossbars Used in Neuromorphic Circuits", Accepted for GLSVLSI 2012, May3-4, Salt Lake City, Utah, USA, 2012
[5] S.C.Liu, J.Kramer, G.Indiveri, T.Delbrück, R.Douglas, "Analog VLSI: Circuits and Principles", Cambridge, MA: MIT Press, 2002

Pt/HfO$_2$/TiN/Al ON SiO$_2$ WITH POTENTIAL APPLICATIONS TO MEMORY AND NEUROMORPHIC CIRCUITS

Davide Sacchetto[1], Yusuf Leblebici[1], Sung-Mo Steve Kang[1,2]

[1]Ecole Polytechnique Fédérale de Lausanne, Institute of Electrical Engineering, Lausanne, Vaud, Switzerland;
[2]University of California– Santa Cruz, Department of Electrical Engineering, Santa Cruz, CA, USA

We present the resistive switching characteristics of the Pt/HfO$_2$/TiN/Al memory cells with TiN/Al top electrode and their potential for non-volatile memory and neuromorphic circuits, in particular forming artificial synapses following the Hebbian rule of learning based on spike-rate dependent plasticity (SRDP).

The devices are fabricated on an oxidized Si substrate having 500 nm thick SiO$_2$. The Pt bottom electrode is deposited by sputtering, followed by Atomic Layer Deposition of 10 nm HfO$_2$ using tetrakis(ethylmethylamino)hafnium – TEMAH and H$_2$O precursors at 200°C. Finally a 40 nm/200 nm thick TiN/Al bi-layer is patterned by lift-off to form 100 μm x 100 μm square electrodes (see inset a) in Figure 1). Similarly, dense crossbar arrays of 64 bits were built using a double lift-off mask patterned with electron beam lithography, demonstrating the potential of this technology for dense memory arrays (see inset b) in Figure 1). The devices are then electrically formed using 500 μs voltage pulses of linearly increasing amplitudes. Typical forming-voltage pulses of -4.5 V have been applied. Pulse voltage sweeps demonstrating bipolar resistive switching (BRS) mechanism are shown in Figure 1 with excellent reading cycle retention of more than 10^5 cycles when V_{read}<200 mV at room temperature (see Figure 2). The use of a thick TiN/Al bi-layer prevents that bipolar resistive switching can be affected by the release of O$_2$ in the atmosphere through the top electrode but rather be modulated by the movement of oxygen vacancies within the HfO$_2$. This is consistent with the relatively low LRS/HRS resistance ratio and with the excellent stability of resistance switching upon cycling [1]. As previously reported by *Goux et al.* [2], the Pt/HfO$_2$/TiN cell can be programmed into intermediate resistance (IR) states, owing to a gradual RESET switching. Interestingly, using a large V_{read} of 300 mV leads to a monotonic increase of the low resistive state (LRS) resistance, eventually saturating at about 30,000 reading cycles into an IR state close to the HRS state which is then stable for the next 60,000 cycles (see Figure 2). This characteristic can be utilized to emulate the SRDP of neural synapses and can be applied to realize memristor-based neuromorphic circuits [3].

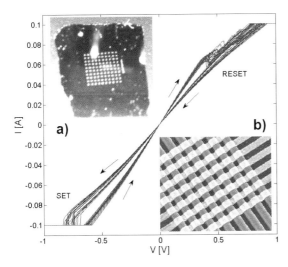

Figure 1: Stable bipolar resistive switching of 50 cycles is formed applying 500 μs large pulses of -4.5 V. During forming the current compliance was set to 100 μA. Stable BRS is achieved with voltage pulse sweeps of 1 ms and with current compliance set to 100 mA. The SET and RESET conditions are obtained at -0.75 V and 0.4 V, respectively.

Inset a) shows the Si die of the measured devices. Inset b) shows a 64bit crossbar array of memristors built with double lift-off mask patterning and electron beam lithography.

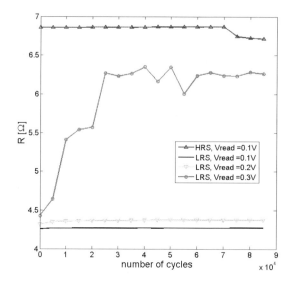

Figure 2: Read retention of low resistive state (LRS) for different reading voltages. The LRS is compared with read retention of the high resistive state (HRS) at $V_{read} = 1V$. Measurements are carried out with Agilent B1500 using 2 ms voltage pulses of 0.1V to 0.3V after programming the resistive state to LRS. Retention time decreases with increasing amplitude of the V_{read} pulse. In particular, $V_{read} = 0.3V$ results in a gradual change of the resistance state, with a transition from LRS to IR after about 30,000 reading cycles. The IR state is then stable for the next 60,000 cycles. This feature may be harnessed to emulate Hebbian learning.

[1] Goux, L.; Czarnecki, P.; Chen, Y.-Y. ; Pantisano, L.; Wang, X. P.; Degraeve, R.; Govoreanu, B.M; Jurczak, M.; Wouters, D. J. ; Altimime, L.;"Evidences of oxygen-mediated resistive-switching mechanism in TiN\HfO$_2$\Pt cells" Applied Physics Letters, vol.97, pp.243509, 2010.
[2] Goux, L.; Chen,Y.-Y. ; Pantisano, L.; Wang, X.-P. ; Groeseneker,G. ; Jurczak, M. ; Wouters, D.J. ; "On the Gradual Unipolar and Bipolar Resistive Switching of TiN\HfO$_2$\Pt Memory Systems," Electrochemical and Solid-State Letters, vol.13, no.6, pp.G54-G56, 2010.
[3] Rachmuth, G.; Shouval, H.Z.; Bear, M.F.; and Poon, C.S.; "A biophysically-inspired neuromorphic model of spike rate- and timing- dependent plasticity," Proceedings of the National Academy of Scinces (PNAS), vol.108, no.49, pp.E1266-E1274, 2011.

MEMRISTORS: TWO CENTURIES ON

Themistoklis Prodromakis[1], Christopher Toumazou[1], Leon Chua[2,3]

[1]Centre for Bio-Inspired Technology, Department of Electrical and Electronics Engineering, Imperial College London, London SW7 2AZ, UK;
[2]Department of EECS, University of California, Berkeley, CA 94720, USA;
[3]Technical University of Munich, Germany;

Memristors are recently attracting significant interest as they are showcased as prospective solutions for memory and emerging ICT applications as well as effectual model systems for studying unprecedented physical mechanisms supported by distinct materials and architectures at the nanoscale. Nonetheless, memristive phenomena have been witnessed well before HP's work on TiO_2-based resistive switching elements[1] and it is demonstrated here that the oldest known memristor is more than two centuries old.

The functional properties of memristors were first documented by Chua[2] and later on Chua and Kang[3], with their main fingerprint being a pinched-hysteresis loop when subjected to a bipolar periodic signal. Hysteresis is typically noticed in systems/devices that possess certain inertia, causing the value of a physical property to lag behind changes in the mechanism causing it; manifesting memory[4]. Such causes are typically associated with irreversible rate-dependent electro- or thermo-dynamic changes that are contingent on both the present as well as the past environment[5]. Particularly in the case of nanoscale memristors, this inertia can be associated with the displacement of mobile ions or oxygen vacancies[6,7], the formation and rupture of conductive filaments[8] or even the phase-transitions of an active core[9]. And despite the fact that these mechanisms have a more substantial effect in lamella devices, similar attributes can also be supported by considerably larger ionic systems, contingent upon the extend of the stimulating cause, the nature of the pertinent ions and the barrier medium that governs their kinetics.

Discharge lamps are recognized as being rather dynamic since the time required for ionization and deionization to take place, depends not only upon the instantaneous current flow, but also upon the current, which has previously flowed through it as well as the rate of change of current. When a potential difference (PD) is applied across a gas-discharge tube that is sufficiently larger than the static ignition potential of the tube, the current flow increases rapidly. However, the deionization time is observed to be substantially larger than the time that it takes to decrease this current flow abruptly, rendering an increase in the tube's resistance as deionization occurs. This dynamic response has been observed for tungsten filaments, high-pressure mercury-vapour lamps, low-pressure mercury tubes, discharge tubes and sodium tubes as illustrated in Fig. 1 respectively.

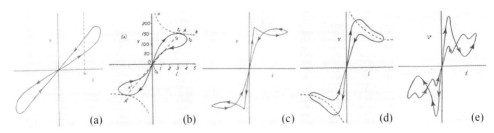

Figure 1: Dynamic characteristics of (a) tungsten filament, (b) high-pressure mercury-vapour lamp, (c) low-pressure mercury tube, (d) discharge tube and (e) sodium tube. Reproduced from [10].

The stratified discharge in a vacuum tube is essentially a magnified version of the thermionic emission as exhibited by the electric arc, which is considered as the predecessor of discharge tubes. The discovery of the electric arc has been ascribed to Sir Humphry Davy[11], the same year Volta discovered the battery[12]. In fact the discovery of the voltaic pile facilitated the establishment of substantial potentials, which in turn enabled Davy to observe continuous luminous filaments without intermittent gaps[13]. Clearly, when these filaments were supported, a substantial amount of current could percolate through the barrier separating the carbon electrodes of the electric arc. This current was shown later that obeys a non-linear trend[14] and most importantly exhibits a rate-limiting response[15] that could give rise to a pinched-hysteresis loop of similar nature to nowadays nanoscale memristors.

Quite often a solution to a problem already exists within a distinct framework. This is indeed the case with the memristor that has not been missing, as reported by HP[1]. Instead, the memristor is merely a description of a basic phenomenon of nature that manifests itself in various dissipative devices, organic or inorganic. In conclusion, Davy's contributions are far more reaching, as he set the foundations for lighting but also analogue computing, both in the early days when vacuum tubes were used and hopefully in the future as memristor-based computing emerges.

1. Strukov, D. B. *et al.* The missing memristor found. *Nature* **453**, 80–83 (2008).
2. Chua, L. Memristor - Missing Circuit Element. *IEEE Trans. Circuit Theory* **CT18**, 507–& (1971).
3. Chua, L. O. & Kang, S. M. Memristive devices and systems. *Proceedings of the IEEE* **64**, 209–223 (1976).
4. Pershin, Y. V. & Di Ventra, M. Memory effects in complex materials and nanoscale systems. *Advances in Physics* **60**, 145–227 (2011).
5. Chua, L. Resistance switching memories are memristors. *Appl. Phys. A* **102**, 765–783 (2011).
6. Yang, J. J. *et al.* Memristive switching mechanism for metal/oxide/metal nanodevices. *Nature Nanotech* **3**, 429–433 (2008).
7. Lee, M.-J. *et al.* A fast, high-endurance and scalable non-volatile memory device made from asymmetric Ta2O5−x/TaO2−x bilayer structures. *Nature Materials* **10**, 625–630 (2011).
8. Kwon, D.-H. *et al.* Atomic structure of conducting nanofilaments in TiO2 resistive switching memory. *Nature Nanotech* **5**, 148–153 (2010).
9. Wuttig, M. & Yamada, N. Phase-change materials for rewriteable data storage. *Nature Materials* **6**, 824–832 (2007).
10. Francis, V. *Fundamentals of discharge tube circuits*. (Methuen: London, 1948).
11. Davy, H. Additional experiments on Galvanic electricity. in a letter to Mr. Nicholson, *Journal of Natural Philosopy, Chemistry, and the Arts* **4**, 326–328 (1800).
12. Volta, A. On the Electricity Excited by the Mere Contact of Conducting Substances of Different Kinds. In a Letter from Mr. Alexander Volta, FRS Professor of Natural Philosophy in the University of Pavia, to the Rt. Hon. Sir Joseph Banks, Bart. KBPRS. *Philosophical Transactions of the Royal Society of London* **90**, 403–431 (1800).
13. Davy, H. Philosophical transactions of the Royal Society of London. **XIV**, 232 (1810).
14. Blondlot, M. R. ÉLECTRICITÉ: Recherches sur la transmission de l"électricité àfaible tension par Vintermédiaire de l"air chaud. *COMPTES RENDUS* **CIV**, 283 (1887).
15. Duddell, W. On rapid variations in the current through the direct-current arc. *Electrical Engineers, Journal of the Institution of* **30**, 232–267 (1901).

ATOM/ION MOVEMENT CONTROLLED THREE-TERMINAL DEVICE: ATOM TRANSISTOR

Tsuyoshi Hasegawa[1,2], Yaomi Itoh[1,2], Tohru Tsuruoka[1,2], Masakazu Aono[1]

[1]WPI-MANA, National Institute for Materials Science (NIMS), Tsukuba, Japan;
[2]Japan Science and Technology Agency (JST), CREST, Tokyo, Japan

Atomic switch is a nanoionic-device that controls the diffusion of metal ions/atoms and their reduction/oxidation processes in the switching operation to form/annihilate a conductive path [1-3]. Since metal atoms can provide a highly conductive channel even if their cluster size is in the nanometer scale, atomic switches may enable downscaling to smaller than the 11 nm technology node, which is a great challenge for semiconductor devices. Atomic switches also possess novel characteristics, such as high on/off ratios, very low power consumption and non-volatility. Although two-terminal devices work as logic devices such as the crossbar circuit, three-terminal devices, in which signal line and control line are separated, are advantageous for the logic applications.

We recently developed an atom movement controlled three-terminal device: 'Atom Transistor' [4]. It operates by bringing metal cations from the gate electrode, which form a conductive channel between the source and drain electrodes. Schematic illustration of the operation is shown in Figure 1. It possess novel characteristics, such as the dual functionality of selective volatile and nonvolatile operations, very small power consumption (pW), and a high on/off ratio (10^6 (volatile operation) to 10^8 (nonvolatile operation)), in addition to being compatible with CMOS processes, which enables their use in the development of computing systems that fully utilize highly-integrated CMOS technology. It is also expected to achieve the nonvolatile logic operations.

In the presentation, present status of the development of Atom Transistor as well as its future will be discussed.

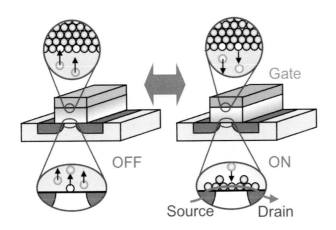

Figure 1: Schematic illustration of the operating mechanism of the Atom Transistor.

[1] K. Terabe et al., Nature, **433** (2005) 47.
[2] M. Aono and T. Hasegawa, Proc. IEEE, **98** (2010) 2228.
[3] T. Hasegawa et al., Adv. Mater., DOI: 10.1002/adma.201102597.
[4] T. Hasegawa et al., Appl. Phys. Express, **4** (2011) 15204.

QUANTUM CONDUCTANCE OF AGI BASED RESISTIVE SWITCHES: TOWARDS AN ATOMIC SCALE MEMORY

Stefan Tappertzhofen[1], Ilia Valov[2], Rainer Waser[1,2]

[1]Institut für Werkstoffe der Elektrotechnik II, RWTH Aachen University, Germany; [2]Peter Grünberg Institut 7, Forschungszentrum Jülich GmbH, Jülich, Germany

Reaching the downscaling limit within the next years, conventional Flash memory technology needs to be replaced by new emerging memory devices [1]. Recently, non-volatile Resistive Switches based on nanoionic redox phenomena attracted high attention [2]. Their benefits are low power consumption, fast switching and high scalability. In case of Electrochemical Metallization (ECM) cells the resistive switching effect is attributed to an electrochemical formation and dissolution of a nanoscale metallic filament within a solid electrolyte sandwiched between two different electrodes.

Typically, amorphous oxides or higher chalcogenides are used as electrolytes which do not initially contain mobile metallic ions being required for ionic transport and the filament formation [3]. Hence, to enable resistive switching ions need to be first injected into the insulator (eg. by thermal diffusion) [4]. In contrast, silver iodide (AgI) initially contains of mobile metallic ions which makes AgI a promising candidate for fast resistive switching applications. However, pattern transfer of AgI is challenging since it is rather sensitive towards conventional process steps such as UV lithography. Furthermore, the very nature of the contact resistance of the filament and the electrodes is yet unknown. Tunneling and metallic contacts are both discussed [5].

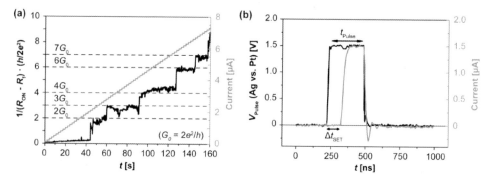

Figure 1. Resisitve Switching of AgI based microcrossbars [6] (a) Quantized conductances have been observed by current sweeping. The conductivity has been obtained from measurement resistance R_{ON} after subtraction of a nearly constant filament resistance R_f in series. We assume $R_f \approx 800\ \Omega$ which fits well to the resistance of a nanoscale Ag wire within the AgI film. (b) Short pulse measurements with a pulse lengths of $t_{pulse} = 250$ ns.

In this study microcrossbar structured Electrochemical Metallization cells based on silver iodide were fabricated and analyzed in terms of the resistive switching effect. The switching behavior implies the existence of a controllable quantized conductance higher than 78 µS which can be identified as a multiple of the single atomic point contact conductivity giving the prospect of an ultimate atomic scale memory. This has been further approved by current sweeps (Fig. 1a) and comprehensive cumulative statistics. Short pulse measurements shown in Fig. 1b were performed to study the switching speed [6]. Resistive switching within at least in 10 ns was observed. The switching kinetics of AgI based ECM cells show a high nonlinearity feasible for memory applications.

[1] Y. Fujisaki, Jpn. J. Appl. Phys. **49**, 100001/1 (2010).
[2] R. Waser and M. Aono, Nat. Mater. **6**, 833 (2007).
[3] I. Valov, R. Waser, J. R. Jameson, and M. N. Kozicki, Nanotechnology **22**, 254003/1 (2011).
[4] S. Tappertzhofen, S. Menzel, I. Valov, and R. Waser, Appl. Phys. Lett. **99**, 203103 (2011).
[5] S. Menzel, U. Böttger, and R. Waser, J. Appl. Phys. **111**, 014501 (2012).
[6] S. Tappertzhofen, I. Valov, and R. Waser, Nanotechnology, in press (2012).

DYNAMIC GROWTH/DISSOLUTION OF CONDUCTIVE FILAMENT IN OXIDE-ELECTROLYTE-BASED RRAM

Ming Liu, Qi Liu, Hangbing Lv, Shibing Long and Yingtao Li

Laboratory of Nano-Fabrication and Novel Devices Integrated Technology, Institute of Microelectronics, Chinese Academy of Sciences, Beijing 100029, China

Electric field or current induced resistance switching (RS) effect is an intriguing phenomenon that forms the basis for potential applications in fast speed and ultra-high density nonvolatile memories. An essential issue for continued device research in this field is to uncover the physical mechanism of RS process, which still remains elusive due to the lack of direct experimental evidence. Here, we present an effective approach to capture the microstructure changes of Cu (or Ag)/ZrO_2/Pt systems at atomic resolutions when adding electrical signals on the device by using *in-situ* transmission electron microscope (TEM) technology. On the basis of this approach, we directly verify the existence of conductive filament (CF) and address several unresolved fundamental issues related to the RS effect, including the starting point of CF growth/dissolution, the direction of CF growth/dissolution, the number of filaments formed under the SET process, and the degree of CF dissolution under the RESET process. We find that the CFs start growth from anode (Ag or Cu) toward the cathode (Pt), which is contrary to the general belief of electrochemical metallization theory. We suggest that the differences of cations solubility and diffusion coefficient between traditional solid-electrolyte and oxide-electrolyte materials should be taken into consideration. Based on the results, a modified microscopic mechanism based on the local redox reaction inside the oxide-electrolyte system is suggested to account for observed RS effect. It is noteworthy that the methodology reported in here can be easily extended to other RRAM systems, which will guide us to understand the origin nature of RS behaviors more clearly and optimize the performances of RRAM device with effective methods.

IN-SITU HARD X-RAY PES POLARIZATION MEASUREMENTS OF OXIDE AMORPHOUS FILMS UNDER INTENSE ELECTRICAL FIELD

S. YAMAGUCHI[1], T. TSUCHIYA[1], S. MIYOSHI[1], Y. YAMASHITA[2]
H. YOSHIKAWA[2], K. TERABE[3], and K. KOBAYASHI[2]

[1] Dept. of Materials Engg., The University of Tokyo, Tokyo, Japan;
[2] Synchrotron X-ray Station at SPring-8, National Institute of Materials Science, Sayo, Japan;
[3] International Center for Materials Nanoarchtectonics, National Institute of Materials Science, Tsukuba, Japan

Novel nonvolatile switching devices, termed *atomic switch*[1] or *atom transistor*[2], using local ion migration are receiving attention as a emerging breakthrough for the miniaturization limitation of nanoelectronics device known as Moor's law. Those devices utilize both ion migration and reduction and oxidation (RedOx) reactions to form/dissolve metallic nano-filaments under applied electrochemical potential for turning on and off. The phenomena have been discussed on the basis of the solid-state electrochemical reaction model, and a carrier modulation by local polarization, termed non-stoichiometry-induced carrier modification (NICM)[2] has been proposed for sulfide system and verified by electrochemical measurements. Because of material compatibility and device processing issues, the devices using oxide electrolytes, such as amorphous tantalum oxide (a-TaO_x) for a heterocation migration media is receiving much attention for commercialization. To understand much complex situations with low electric conductance of oxide system, an in-situ polarization experiments by Hard X-ray Photoemission Spectroscopy (HX-PES) using synchrotron radiation as a light source has been made on the Metal-Insulator-Metal (MIM) capacitor type cell. Discussion on the similarity and difference will be made through a comparison with Cu_2S system, a typical high electronic conductance system with intrinsic ionic species in electrolyte.

30-100 nm thick a-TaO_x films with various x values have been fabricated using a reaction sputtering method by controlling vol% of O_2 gas in Ar-O_2 gas mixture supplied during sputtering. A small amount of D_2O was added to the reaction gas mixture for the preparation of hydrated form of a-TaO_x. The Rutherford Back Scattering (RBS) analysis on the samples fabricated shows a significant x variation ranging from 1.9 to 3.0. Such enormous variation is realized by the variations of TaO_5 and TaO_6 fractions and the multiplicity factors of oxide ions by the extent of polymerization, estimated from the data by Raman Spectroscopy and EXAFS measurements. ESCA measurements have revealed the Fermi energy (E_F) shift (or, chemical shift) about 1.5 eV in the composition range of $1.9<x<3.0$ with maintaining the rigid band feature. The E_F sift from oxidation to reduction limit can be defined as RedOx window, which normally corresponds approximately to the energy band gap, but the present value for a-TaO_x is less than half of the one measured experimentally. XPS (AXIS-His, Shimadzu) and O1s XAS at BL-19B in Photon Factory (KEK, Tsukuba) shows that the formation of donor state in reduced form with $x<2.0$, but no hole or acceptor state in the band gap region at higher x composition[4].

As shown in **Fig. 1**, the in-situ measurements at BL-15XU in SPring-8 facility on C/a-TaO_x electrode revealed the continuous occurrence of the reduction reaction detected from the E_F shift, but asymmetric behavior to the oxidation side; the E_F shift stays at constant value upon polarization to oxidation side followed by a sudden jump at higher applied voltage (termed here as breakdown). All the samples regardless of the metal electrode and x value approach to the reduction limit similarly, while the E_F shift to the oxidation limit is strongly dependent on the x value. The High temperature Diffuse Reflectance FT-IR, TDS, and RBS indicate the presence of protons, suggesting a possible role of protons as mediator in the electrochemical polarization, in which protons contribute asymmetric way to RedOx reaction with a similar reduction behavior regardless of x value by accumulation of protons to the cathode area from all other region and x-dependent oxidation limit due to complete depletion of protons.

A sample with C/a-TaO$_x$ dissolved with Cu by a thermal oxidation of Cu deposited film shows apparent migration of Cu$^+$ and less mobility of Cu^{2+}, both contributing to the enlargement of a-TaO$_x$ RedOx window. The in-situ measurements on Cu/a-TaOx electrode confirm the oxidation of Cu electrode and the E_F shift expansion that is strongly dependent on the x value. The E_F shift shows a similar reduction behavior but very strong x dependent oxidation limit. The present results on a-TaO$_x$ suggest the same conclusion to the case of the cell with Cu$_2$S electrolyte, that the matrix RedOx reaction governs the oxidative dissolution of Cu at anode and the reduction precipitation at cathode. In other word, RedOx pair reaction by an ion migration to supplement the Cu oxidation/reduction reaction is always necessary for every electrochemical reaction. The comparison between oxide and sulfide systems will be further made based on the in-situ measurements using HX-PES.

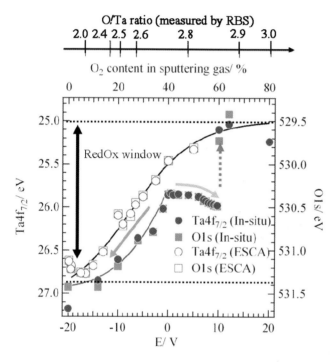

Fig. 1 E dependence of the Ta4f$_{7/2}$ and O1s binding energies at C/a-TaO$_x$ interface. The parallel variation of Ta4f$_{7/2}$ and O1s binding energies in nonstoichiometric a-TaO$_x$[4] fabricated by controlling vol%O$_2$ in Ar+O$_2$ gas mixture during the reaction sputtering is superimposed in **Fig. 1** for reference with additional upper horizontal axis for x in a-TaO$_x$ (O/Ta ratio) estimated from the RBS results.

[1] K. Terabe, et al., Nature, **433** (2005), 47-50.
[2] T. Hasegawa et al., APeX **4** (2011) 015204.
[3] T. Tsuchiya et al., APeX, **2**, (2009) 055002.
[4] T. Tsuchiya et al., PCCP **13**, (2011) 17013

SPECTROSCOPIC INVESTIGATION OF Charge Transfer in Electrochemical Metallization Memory CELL

Deok-Yong Cho[1], Ilia Valov[1,2], Jan van den Hurk[1], Stefan Tappertzhofen[1], Rainer Waser[1,2]

[1]IWE2 & JARA-FIT, RWTH Aachen University, Aachen, Germany;
[2]PGI 7 & JARA-FIT, Forschungszentrum Jülich GmbH, Jülich, Germany

The electrochemical metallization memory (ECM) cell is a class of resistive random access memory (RRAM) cell consisting of an active electrode e.g. Ag or Cu, a solid electrolyte thin film e.g. SiO_2 or GeS_x to transport the cations and an inert counter electrode e.g. Pt [1]. A chemical dissolution of the active electrode atoms into the electrolyte matrix markedly increases the conductivity modifying the transport properties. Therefore, the chemistry of the active electrode atoms as well as the electrolyte atoms should be examined.

In this presentation, we show the results of spectroscopic analyses on the chemistry in the most exploited solid electrolyte, Ag-"doped" amorphous GeS_x. The orbital states of Ge, S, and Ag were investigated using X-ray absorption spectroscopy (XAS). Ag was deposited by an equivalent thickness of 5 nm on 60 nm-thick GeS_x (x=1.6 and 2.2).

With examining the edge shift upon doping in the XAS spectra, we found that Ag is indeed ionized ($Ag^{+\delta}$; $0<\delta<1$) suggesting full dissolution of Ag into the electrolyte.

Figure 1 (reproduced from Ref. [2]) shows the Ag L_3-edge XAS spectra of the two GeS_x films with x =1.6 and 2.2 after Ag deposition, a thin Ag foil and a SiO_2/Ag film. To enhance the visibility, the derivatives of the original spectra are shown. From the energy of the maximal slope, the chemistry of Ag ions in each sample can be compared with each other. In the case of GeS_x/Ag, the valence value of the Ag ions was found to be nonzero but less than +1, in contrast to the null valency in the cases of SiO_2/Ag and Ag foil.

It was further shown in the analyses of Ge L-edge and S K-edge XAS that the electronic charges from Ag preferably transfer to the Ge ions not to the S ions.

These experimental results demonstrate conclusively that the ionization of Ag and its movement under electric field is indeed essential for the resistive switching behaviors of the ECM cell. This nondestructive methodology can be easily extended to other ECM systems such as the SiO_2/Cu or $GeSe_2$/Ag, as to serve as a guide for profound understanding of the chemical reactions and resistive switching behaviors in this new class of RRAM material.

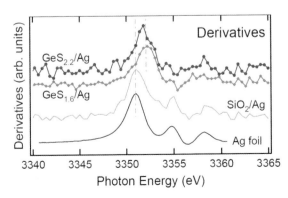

Figure 1: (Reproduced from Ref. [2]) Derivatives of Ag L_3-edge XAS spectra of two GeS_x films with x =1.6 and 2.2 after Ag deposition, a thin Ag foil and a SiO_2/Ag film. The energy of maximal intensity tells the valence value or each sample. It is clearly shown the Ag ions in GeS_x should have larger valences than Ag foil and SiO_2/Ag (zero valence). This evidences the ionization and dissolution of Ag in the solid electrolyte GeS_x.

[1] I. Valov, R. Waser, J. R. Jameson, M. N. Kozicki, *Nanotechnology* **2011**, *22*, 254003/1.
[2] D.-Y. Cho, I. Valov, J. van den Hurk, S. Tappertzhofen, R. Waser, unpublished.

CHARACTERIZATION OF GERMANIUM SULFIDE THIN FILMS GROWN BY HOT WIRE CHEMICAL VAPOR DEPOSITION

Denis Reso[1], Mindaugas Silinskas[1], Nancy Frenzel[1], Marco Lisker[2], Edmund P. Burte

[1]Otto-von-Guericke-University Magdeburg, Saxony-Anhalt, Germany; [2]IHP, Frankfurt (Oder), Brandenburg, Germany

The non-volatile electrochemical metallization (ECM) memory is particularly promising as it requires only a very low operating power [1]. Silver-germanium-sulfide compounds are known to be suitable solid electrolytes (SE) in ECM cells [2]. Since the confinement of the SE is necessary to build high density arrays of the memory cells, a deposition method is needed that allows the conformal filling of small scale structures. In this work, the conformal deposition of the germanium sulfide SE backbone is demonstrated using hot wire metalorganic chemical vapor deposition (HWCVD). The influence of the main deposition parameters (deposition temperature, chamber pressure, hydrogen content) was investigated with respect to growth rate, film composition and filling capabilities. Three dimensional ECM cells were fabricated by electron beam evaporation of a silver top electrode and the subsequent diffusion of silver into the CVD germanium sulfide in order to test the electrical functionality of the SE.

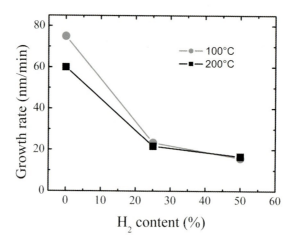

Figure 1: Growth rate of Ge-S films in dependence on the hydrogen content of the hydrogen-nitrogen atmosphere in the HWCVD reactor at a chamber pressure of 600 Pa.

The precursors were injected in short pulses into continuously flowing nitrogen streams. Hydrogen was added as a reactant gas with different percentages with respect to the total flow. A clear influence of the hydrogen content on the growth rate can be observed.

Amorphous films were deposited using a HWCVD reactor equipped with a pulse mode precursor injection system. The precursors for germanium and sulfur were tetraallylgermanium $((C_3H_5)_4Ge)$ and propylene sulfide (C_3H_6S), respectively. Film thicknesses were measured using spectroscopic ellipsometry, surface profiling and cross-sectional scanning electron micrographs. The elemental film composition was analyzed by energy dispersive X-ray analysis (EDX). The bond structure of the films was investigated by Raman spectroscopy. Surface characterization was performed using atomic force microscopy (AFM). The Ge-S film structure and its variation due to thermal annealing were examined using X-ray diffraction (XRD) in order to detect possible crystalline structures.

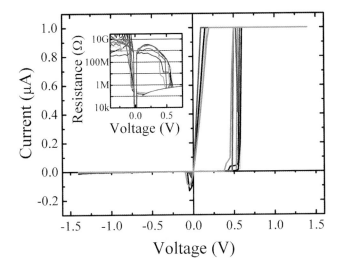

Figure 2: Switching characteristic of a sample 3D-ECM cell based on $Ge_{16}S_{84}$ grown by HWCVD on a structured substrate.

The main diagram shows the current dependent on several voltage double sweeps. Each double sweep started at -1.4 V with the cell in the highly resistive state. In the first sweep, the voltage was increased up to 1.4 V in steps of 25 mV every 240 ms. At about 0.5 V the cell switched to the low resistive state. In the back sweep, the voltage was reduced in the same manner ending at -1.4 V in the highly resistive state, again. The back switch appeared at about -0.1 V.

The inset shows the corresponding resistance graph calculated from the data in the main graph.

The CVD growth rate varied depending on the deposition parameters: it increased with increasing chamber pressures but decreased with increasing temperatures. In addition, higher hydrogen content resulted in a reduced growth rate, too. Certain amount of hydrogen in the chamber atmosphere is beneficial, because it can help to reduce the carbon incorporation into the deposited films. However, it can affect the growth rate by etching the films during the deposition (see Figure 1). Conformal hole filling with smooth and dense films can be obtained particularly at lower growth rates, whereas at high growth rates the films tend to cover the holes and exhibit a soft, porous and rough structure.

Three dimensional ECM cells were electrically characterized and showed a typical resistive switching behavior (see Figure 2). Clearly, two different resistive states can be observed with a separation of three to four orders of magnitude. The resistivity of the films was found to be comparable to similar films deposited by conventional physical deposition methods [3]. Due to the large separation of the resistances, a reduction of the current compliance limit down to 0.2 µA was possible. The ability of repeated switching was confirmed by the application of 10^5 program/erase cycles.

[1] D. Kamalanathan et al., Nanotechnology 22, 254017 (2011).
[2] I. Valov et al., Nanotechnology 22, 254003 (2011).
[3] F. Wang et al., Solid State Electron. 61, 33 (2011).

NANOFILAMENT RELAXATION MODEL FOR SIZE-DEPENDENT RESISTANCE DRIFT IN ELECTROCHEMICAL MEMORIES

Seol Choi, Simone Balatti, Federico Nardi and Daniele Ielmini

Dipartimento di Elettronica e Informazione, Politecnico di Milano, Milano, Italy

Electrochemical memory (ECM) devices have drawn a great deal of attention as a promising nonvolatile memory technology due to their high speed, large resistance window and low switching energy [1-3]. Resistive switching of ECM is based on the electrochemical deposition and dissolution of a nanoscale conductive filament (CF) [4]. The CF size, dictating the resistance R of the ECM in the low-resistance (set) state, can be controlled by the maximum current during set, or compliance current I_C. Thanks to the large R window, ECM is suitable for multilevel cell (MLC) operation. However, R stability against noise [5] or drift must be clarified and understood.

We have studied resistance stability in ECM devices in the set state. Experimental data were collected on GeS-based ECM devices with Ag top electrode and W bottom electrode, provided by Adesto Technologies Corp. [6]. Fig. 1a shows the I-V curves obtained by DC sweep at room temperature. The device was set by positive voltage at $V_{set} \approx 0.3$ V with $I_C = 100$ µA and reset by negative voltage. To evaluate the R stability in the short term (about 1 s) after set, we measured R = dV/dI along the positive sweep from V_{set} to 0 and along the negative sweep from 0 to V_{reset}. The negative-sweep resistance R_- was larger than the positive-sweep resistance R_+, due to the resistance drift during the delay time of 1 s between the two sweeps. These findings were confirmed by positive-only sweeps and long-term drift analysis.

Fig. 2a and b show the measured ΔR (a) and the corresponding normalized $\Delta R/R$ (b) as a function of R. The figure shows both short-term data as in Fig. 1b and data from long-term data (drift time = 10^3 s). Two different regimes are observed below/above a transition resistance $R^* \approx 10^4$ Ω: For large CFs with small R, ΔR is proportional to R^2, thus $\Delta R/R$ increases linearly with R. For R above around 10^4 Ω, corresponding to small CFs at low I_C, ΔR increases linearly with R, thus $\Delta R/R$ remains constant. Both short- and long-term data show these two regimes, although with different amplitudes.

The size-dependent drift in Fig. 2a and b can be explained by the surface relaxation model for a CF with effective length L in Fig. 2c and d. Defects at the CF surface experience a relaxation accompanied by a switch in their charge state, e.g. neutral to negative [7]. The point-charge can affect the CF resistivity ρ within a Coulombic interaction range ΔA, where the carrier density is depleted (Fig. 2c). In the remaining part of the cross section A-ΔA, ρ is unaffected due to the point-charge screening [7]. The resistance after defect relaxation can thus be approximated as $R_{A-\Delta A}$, that is the resistance of a CF with cross section A-ΔA. Thus ΔR is approximately given by:

$$\Delta R \approx \rho \frac{L}{A-\Delta A} - \rho \frac{L}{A} \approx \rho \frac{L}{A}\left(1+\frac{\Delta A}{A}\right) - \rho \frac{L}{A} = \rho \frac{L\Delta A}{A^2} \propto R^2$$

which is consistent with the low-R regime in Fig. 2a. For full depletion, ρ is affected in the whole cross section, thus leading to $\Delta R \approx \Delta \rho L/A \propto R$, as in Fig. 2a for R > R*.

An analytical model for the resistance drift due to surface-defect relaxation was developed and used to calculated ΔR and ΔR/R in Fig. 2a and b. In the calculations for t = 1 s, we assumed a cylindrical CF with resistivity increase by a factor 1.3, a Coulombic range of 1 nm, ρ = 100 µΩcm and L = 10 nm, representing the effective length of the CF constriction. For t = 10^3 s, we used larger ρ enhancement (2x) and Coulombic range (Δϕ = 1.5 nm) to account for the larger number of relaxing defects. The model accounts for ΔR and its dependence on the R above/below the transition $R^* = 10^4 \, \Omega$. This corresponds to a critical CF size of about 1 nm, dictated by the Debye length for charge screening. The model might be extended to random telegraph noise (RTN), which also affects the R stability in nanoscale conductive paths [5].

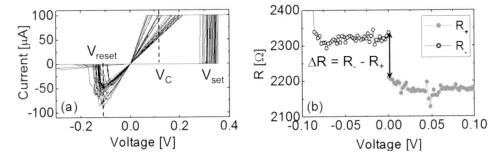

Figure 1: Measured I-V curves of an ECM device (a) and measured $R = dV/dI$ in the set state as a function of voltage during a set/reset cycle (b).

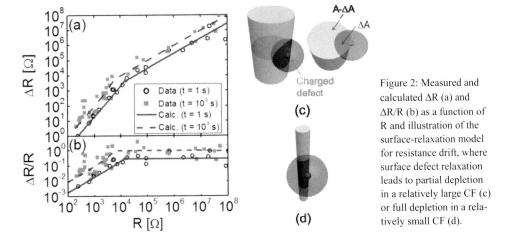

Figure 2: Measured and calculated ΔR (a) and $\Delta R/R$ (b) as a function of R and illustration of the surface-relaxation model for resistance drift, where surface defect relaxation leads to partial depletion in a relatively large CF (c) or full depletion in a relatively small CF (d).

The authors would like to acknowledge the Adesto Technologies Corp. technical team for providing experimental samples and for useful discussions. This work was supported in part by Adesto Technologies Corp. and Fondazione Cariplo (grant 2010-0500).

[1] M. N. Kozicki, et al., IEEE Trans. Nanotechnology 5, 535 (2006).
[2] C. Schindler, et al., Appl. Phys. Lett. 92, 122910 (2008).
[3] N. Derhacobian, et al., Proc. IEEE 98, 283 (2010).
[4] R. Waser and M. Aono, Nature Mater. 6, 833 (2007).
[5] R. Soni, et al., J. Appl. Phys. 107, 024517 (2010).
[6] C. Gopalan, et al., Solid-State Electronics 58, 54 (2011).
[7] D. Ielmini, et al., Appl. Phys. Lett. 96, 053503 (2010).

ION MIGRATION MODEL FOR RESISTIVE SWITCHING IN TRANSITION METAL OXIDES

D. Ielmini[1], S. Larentis[1], S. Balatti[1], F. Nardi[1] and D. Gilmer[2]

[1]Dipartimento di Elettronica e Informazione, Politecnico di Milano, Milano, Italy
[2]SEMATECH, Front End Process and Emerging Technologies, Austin, TX, USA

Oxide-based resistive switching memories (RRAMs) [1] or memristors [2] based on metal oxides have been recently shown to have excellent properties in terms of switching speed [3], endurance [4] and low programming energy [5]. Bipolar resistive switching is due to ion migration at the nanoscale, where the voltage induces an increase in the local electric field and temperature through Joule heating [5, 6]. Although there is general agreement regarding the nanoionic nature of resistive switching, numerical models capable of describing ionic motion under the effects of electric field and temperature have not yet been shown.

This work presents a numerical model for ion migration in bipolar resistive switching devices. Experimental data were collected on metal-insulator-metal (MIM) devices based on HfO_x, provided by SEMATECH. Fig. 1 shows the measured I-V characteristics of a HfO_x MIM, indicating initial forming, set and reset processes under bipolar switching. Forming and set processes reveal the creation of a nanoscale conductive filament (CF), characterized by a low resistance R, while the reset transition represents the dissolution of the CF and the recovery of a high R.

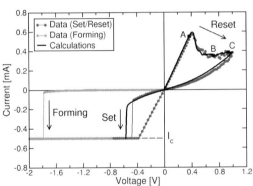

Figure 1: Measured and calculated I-V curves for HfO_x showing forming, set and reset transitions.

To model set and reset mechanisms, we described the electronic, thermal and ionic conduction through HfO_x, through the solution of three differential equations, namely (i) Poisson equation for electrons (equivalent to Ohm's law), (ii) Fourier equation for heat conduction and (iii) drift/diffusion equation for ionic transport. In the latter, we assumed Einstein relationship between ion mobility μ and diffusivity $D = \mu kT/q$. Both D and μ were assumed temperature-activated through the Arrhenius law with an activation energy $E_A = 1$ eV [5]. The differential equations were solved in Comsol for a 3D geometry, reduced to 2D through the use of cylindrical symmetry. Simulation results obtained by the numerical model are shown in Fig. 1 in comparison to experimental data, for the set and reset transitions. Calculations account for both the abrupt set transition, responsible for the formation of the CF, and the reset transition, where resistance increases gradually above $V_{reset} = 0.4$ V.

Fig. 2 shows the calculated contour plot of the concentration n_D of metallic dopants (oxygen vacancies and/or excess Hf) during the reset transition in correspondence of bias points A, B and C along the reset transition in Fig. 1. Fig. 3 shows the 2D maps of calculated n_D (a) and of the temperature T (b) as a function of cylindrical coordinates r (radial) and z (vertical) for states A to C (left to right). The thickness t_{ox} of the HfO_x is 20 nm in the vertical direction, while a radius of 7 nm was assumed for the CF to best account for the resistance value and the set/reset parameters.

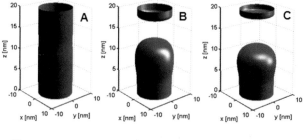

Figure 2: Contour plot of calculated doping density $n_D = 0.6 \times 10^{21}$ cm^{-3} for states A, B and C.

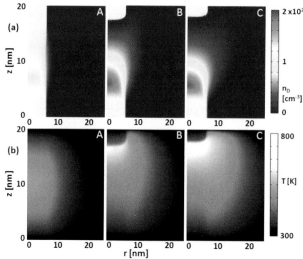

Figure 3: Maps of calculated dopant density n_D (a) and T (b) during reset at bias points A, B and C (left to right) in Fig. 1. Axis represent the r (radial coordinate) and z (vertical coordinate) in the simulated MIM structure. Note that heating is increasingly localized within the depleted gap, leading to self-limiting reset in Fig. 1.

Calculations show that, at V_{reset} (bias point A), the temperature reaches about 600 K at $z = t_{ox}/2$ along the CF, thus triggering the thermally-activated ion migration toward the cathode (bottom electrode). This results in a depletion of dopants for $z > t_{ox}/2$ and in the creation of an undoped gap which increases in length as the voltage is increased from A to C. Simulation results explains the analog nature of the reset transition: As the CF depletion starts, the electric field and temperature becomes increasingly localized within the gap, thus reducing the potential drop and the temperature along the remaining conductive regions of the CF, as shown by the temperature (b) maps in Fig. 3. The reduction of field and temperature suppresses further ion migration, which can thus be sustained only by an increase of the applied voltage.

The model can predict the reset transition as a function of pulse width and set transition as a function of the programmed reset state. We also applied the model to describe the complementary resistive switching in a HfO$_x$, which could be explained by ion migration from anode to cathode and the opening of a gap at the cathode side [7, 8]. The model is thus suitable for scaling and reliability prediction in RRAMs.

The authors would like to acknowledge SEMATECH for providing experimental samples. Discussions with G. Bersuker, P. Kirsch, K. S. Min and G. Spadini are also acknowledged. This work was supported in part by Intel and Fondazione Cariplo (grant 2010-0500).

[1] Y. Watanabe, et al., Appl. Phys. Lett. 78, 3738 (2001).
[2] D. B. Strukov, et al., Nature 45, 80 (2008).
[3] A. C. Torrezan, et al., Nanotechnology 22, 485203 (2011).
[4] M. J. Lee, et al., Nature Materials 10, 625 (2011).
[5] D. Ielmini, IEEE Trans. Electron Devices 58, 4309 (2011).
[6] K. Szot, et al., Nature Materials 5, 312 (2006).
[7] E. Linn, et al., Nature Materials 9, 403 (2010).
[8] F. Nardi, et al., IEDM Tech. Dig. 709 (2011).

SIMULATION STUDIES OF THE MATERIAL DEPENDENT SWITCHING PERFORMANCE OF VALENCE CHANGE MEMORY CELLS

Stephan Menzel[1], **Astrid Marchewka**[1], **Ulrich Böttger**[1], **Rainer Waser**[1,2]

[1]Institut für Werkstoffe der Elektrotechnik 2, RWTH Aachen University, 52074 Aachen, Germany
[2]Peter Grünberg Institut (PGI-7), Forschungszentrum Jülich, 52425 Jülich, Germany

Memristive device concepts based on the valence change memory (VCM) effect have attracted great attention for future nonvolatile memories. VCM type switching has been observed in various transition metal oxides [1],[2]. Typically, VCM cells consist of an active interface, where the resistive switching takes place, a mixed ionic-electronic conducting oxide and an ohmic counter electrode. The switching mechanism in this class of ReRAMs is attributed to the voltage driven movement of oxygen vacancies and a subsequent redox reaction at the nanoscale, which leads to a change of the electronic conductivity in front of the active electrode [1, 3]. The switching kinetics of such a cell is highly nonlinear. This means that a low voltage is sufficient to switch the cell from the low resistive state (LRS) to the high resistive state (HRS) and vice versa within a few seconds, while a high voltage is necessary to achieve switching in the nanosecond regime. Recently, we could identify temperature accelerated drift of oxygen vacancies as the origin of this nonlinearity by using an electro-thermal simulation model [4, 5]. Due to the thermal nature of the switching kinetics, the device performance can be improved by good thermal confinement.

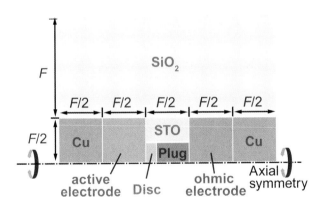

Figure 1: Investigated cell geometry. The feature size is F = 20 nm and the plug/ disc radius is 10 nm. The disc thickness is 5 nm.

In this study, we investigate how the thermal confinement is affected by the choice of materials and device dimensions. Particularly, the impact on switching currents and energy consumption is discussed. For this we apply the derived electro-thermal simulation model [4] to an integrated nanosized VCM cell. Here, the 2D axisymmetric device geometry shown in Figure 1 is considered. It comprises an *n*-conducting cylindrical region ("plug"), which has grown during -electroforming, and a disc shaped region ("disc") in front of the active electrode, where the actual switching takes place. The thermal properties as well as the dimension of specific layers are varied and the switching performance is simulated according to the model in [4]. As an example, the impact of the thermal conductivity on the SET current, switching energy, SET time and disc temperature is shown in Figure 2. Using this methodology important design rules for material selection and device geometry are deduced and discussed.

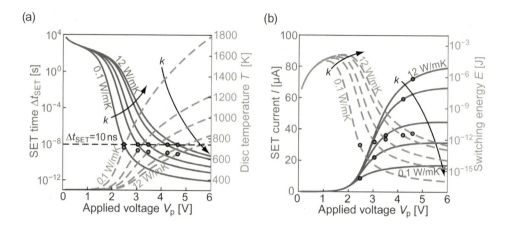

Figure 2: (a) Set times (solid lines) and mean disc temperature (dashed lines) and (b) set current (solid lines) and switching energy (dashed lines) vs. applied voltage for different thermal conductivities of the insulating layer including disc and plug. The circles represent data points referring to a 10 ns set time.

[1] R. Waser et al., Adv. Mater., 21 (2009) 2632.
[2] H. Akinaga et al., Proc. IEEE, 98 (2010) 2237.
[3] J. J. Yang et al., Nat. Nanotechnol., 3 (2008) 429.
[4] S. Menzel et al., Adv. Funct. Mater., 21 (2011) 4487.
[5] C. Hermes et al., IEEE Electron Device Lett., 32 (2011) 1116.

ELECTRORESISTANCE VERSUS JOULE HEATING EFFECTS IN MANGANITE THIN FILMS

Ll. Balcells, A. Pomar, R. Galceran, Z. Konstantinovic, L. Peña, B. Bozzo, F. Sandiumenge and B. Martinez

Instituto de Ciencia de Materiales de Barcelona – CSIC. Campus UAB. E- 08193 Bellaterra, Spain

Non-volatile memories in production today are based mainly in using charge storage as the memory mechanism, however this technology is approaching their technical and physical limits and a new generation of memory devices is currently under development. One of the possibilities for replacing this charge storage based memory devices is using the resistive switching (RS) phenomena, i.e., the change of resistance by using a pulse of current or voltage. RS have been observed in a variety of materials[1,2] however, the observed switching behavior seems to differ depending on the material. In the case of manganites RS was initially associated to the existence of phase separation phenomena[3], but later it was also observed in manganites without phase separation such as the case of $La_{2/3}Sr_{1/3}MnO_3$ (LSMO). Nevertheless, in spite of the work already done the driving mechanism behind the RS phenomena in manganites is still to be unveiled.

In this work we report on the measurement of I(V) characteristic curves in LSMO epitaxial thin films prepared by RF sputtering. I(V) curves have been measured in LSMO microbridges as that shown in Fig. 1. With this topology we avoid the metal-insulator-metal geometry typically used for measuring RS phenomena and we also minimize interface-related effects. We have analyzed I(V) curves as a function of the base temperature and magnetic field. I(V) are non linear and they exhibit a jump from a low resistance state (LRS) to a high resistance state (HRS) which, in principle, precludes electroforming effects. However I(V) curves exhibit origin symmetry and they are almost reversible except for a small irreversibility around the jump from the LRS to the HRS.

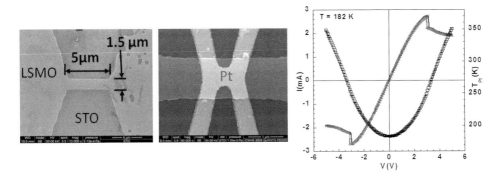

Figure 1: *Left*: Image of one of the microbridges used for I(V) measurements. *Middle*: image of the Pt thermometer evaporated on top of the LSMO microbridge. A 30 nm thick LAO layer have been deposited to isolate Pt from LSMO. *Right*: I(V) characteristic curve and variation of the actual temperature of the sample while doing the I(V) measurement.

All these features of I(V) curves strongly suggest that Joule self-heating effects are of relevance in our samples. To check this possibility we have microfabricated a Pt thermometer on top of the LSMO bridge (see Fig. 1) to have access to the actual temperature of the samples while measuring the I(V) curve. Our results make evident that a huge increase of the temperature of the samples takes place when current increases, thus Joule self-heating effects are responsible for most of the increase in resistance according to the evolution of the R(T) curve of the LSMO. However, the jump from the LRS to the HRS cannot be properly described by Joule heating effects alone.

[1] R. Waser et al., Adv. Mater. **21**, 2632 (2009).
[2] A. Sawa, Materials Today. **11**, 28 (2008)
[3] A. Asamitsu et al. Nature. **388**, 50 (1997)

ON ELECTROFORMING FOR BIPOLAR SWITCHING

Ilan Riess, **Dima Kalaev**

Physics Department, Technion-IIT, Haifa, 32000, Israel

Systems of the form Metal1|MIEC|Metal2 (MIEC, mixed ionic electronic conductor) are considered. Bipolar switching may occur when an a-symmetry exist. The a-symmetry may be built into the device by a difference in the two electrodes but can also be induced in quasi symmetric devices during application of a current. The outcome of simulated electroforming processes is presented in the form of the defects distributions of the mobile point defects and the corresponding I-V relations. The structural a-symmetries considered are: a difference in the work function of the two electrodes and a difference in the chemical potential of the mobile component between the two sides. In quasi symmetric cells the two metal electrodes have the same composition but a difference in the rate of material exchange with the surrounding is considered. This results in symmetry breaking and polarization when a current is applied. Long electroforming processes which allow generation and redistribution of ionic defects are considered and approximated by a dc state.

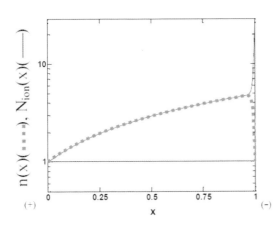

Figure 1: Electrons concentration n, and singly charged mobile donors concentration, N_{ion} vs. position x in an MIEC under a relative low voltage of $9k_BT/q$ with a polarity as indicated. The device Metal1|MIEC|Metal2 is symmetric except for the rate for material exchange at the two electrodes, with the one on the right being lower. x is normalized with respect to the MIEC length. n and N_{ion} are normalized with respect to their common value under equilibrium. The Metal|MIEC contact potentials are assumed to vanish. n and N_{ion} are enhanced under voltage. Local neutrality is observed throughout most of the sample.

The ionic defect distribution under a relative low applied voltage is shown in Fig. 1 assuming a difference of a factor 1000 in the rate of material exchange at the two electrodes. The device is otherwise symmetric and the mobile ions concentration under equilibrium is taken to be uniform. Under current an increase in the overall ions concentration is observed. A similar increase is seen also for the electrons concentration throughout most of the MIEC. The increase is a-symmetric. While at the blocking electrode (on the right) the ions concentration is high, a region of relative low concentration exist at the other end. This region could serve as the active layer in a later switching step.

Without loss of generality we shall refer to oxygen vacancies in an oxide MIEC and electrons in the conduction band. The reduction of the oxide generates oxygen vacancies which are native donors. The level of reduction needs to be quite high as in reality it occurs in filaments only. For the filaments to significantly lower the MIEC resistance the conductivity of the filament has to be high, metallic. This is obtained under high defects concentration leading to a Mott transition. We thus shall discuss enforced high deviations from stoichiometry

Memory appears if the relaxation from the induced state is slow, i.e. the ions (and perhaps also the electrons/holes) have high potential barriers to cross. This means that either the temperature is raised during electroforming to allow going over the barrier (and then lowered to quench the electroformed state) or the barrier is lowered by a high applied electrical driving force. We considered lowering the barrier under a high applied voltage gradient, assuming that the temperature stays constant.

In a recent publication we considered similar effects but only for relative low driving forces.[1] They present qualitatively the defects distributions and I-V relations relevant to the case under consideration. However, we shall discuss extended simulations to high driving forces which are more applicable to electroforming.

[1] D. Kalaev and I. Riess, Solid State Ionics, **212**, 26 (2012).

ROOM-TEMPERATURE KINETICS OF DEFECT MIGRATION IN NON-FRADAIC Pt/TiO₂/Pt CAPACITORS

Hyungkwang Lim[1,2], Ho-Won Jang[1], Cheol Seong Hwang[2], <u>Doo Seok Jeong</u>[1]

[1]Electronic Materials Research Centre, Korea Institute of Science and Technology, Seoul, Republic of Korea;
[2]WCU Hybrid Materials Program, Department of Materials Science and Engineering, and Inter-university Semiconductor Research Centre, Seoul National University, Seoul, Republic of Korea

In transition metal oxides (TMOs), the migration of point defects and redox reactions associated with them are conjectured to play a key role in hysteretic current versus voltage (I-V) behavior. I-V hysteresis is utilized in non-volatile memories, often described as resistive random access memories (RRAMs).[1] Moreover, this type of hysteresis represents the plasticity of the resistance with respect to the applied voltage/current so that the hysteresis can form the basis for artificial synapses following the Hebbian learning rule.[2] Regarding the technological importance of I-V hysteresis in TMOs, the understanding of migration of point defects in TMOs is of great importance. In this study, we chose TiO_2 as an example of non-stoichiometric, to be precise, hypo-stoichiometric, TMOs.

TiO_2 is a prototypical TMO for the application to RRAMs, and thus this material has been widely and globally investigated in the past decade.[3,4] Besides this technological aspect, the defect structure of hypo-stoichiometric TiO_2 was intensively investigated because it was controversial, whether the dominant defect-type is Ti interstitial[5] or oxygen vacancy[6]. This controversy enriched information on the kinetics of defect migration in TiO_2, e.g. diffusion coefficients of oxygen vacancy and Ti interstitial. However, one can find that most of available information was extracted from high temperature measurement at thermodynamic equilibrium. Obviously, we lack data on TiO_2 at low temperature and non-equilibrium.

The aim of our investigation is to understand defect migration kinetics in sputter-grown TiO_2 films at room temperature by means of electrical characterization. To rule out faradaic reactions at the interface between a TiO_2 film and an electrode layer we chose Pt, which is inert, as the electrode material of TiO_2-based metal-insulator-metal (MIM) capacitors, i.e. Pt/TiO₂/Pt capacitors. The fabricated MIM cells were characterized using cyclic voltammetry (CV) with a moderate range of the applied voltage. This voltage range was carefully chosen to rule out the contribution of the dc electronic current to the overall current in the CV measurement.

Measured CV data at different voltage scan rates are plotted in Fig. 1(a). These CV curves represent typical non-fradaic and double-layer-type behavior, implying that no electrochemical reactions take place in the cell. Figure 1(b) shows a CV curve measured in a wider range of the applied voltage, which includes a dc current contribution at high voltages. It can be noted that the CV hysteresis in the high voltage region is counter-clockwise unlike the clockwise hysteresis in the low voltage region [see Figure 1(a)]. The counter-clockwise hysteresis appears to arise from non-static galvani potential distribution in the TiO_2 layer, which is most likely attributed to the sluggish migration of point defects at the given voltage scan rate (0.1 V/s). The non-static galvani potential leads to electromotive force increasing dc electronic current in the given voltage region. In fact, the counter-clockwise hysteresis was found to disappear as decreasing a voltage scan rate, which obviously supports our hypothesis.

For quantitative analysis, we also conducted CV calculations on a Pt/TiO₂/Pt capacitor with fully taking into account the migration of charged point defects as well as electronic carriers. The calculations were done by numerically solving the time-dependent drift-diffusion equation. By comparing the calculation results with the experimental results, important information, e.g. defect distribution in the capacitor and the defects' diffusion coefficient, was obtained.

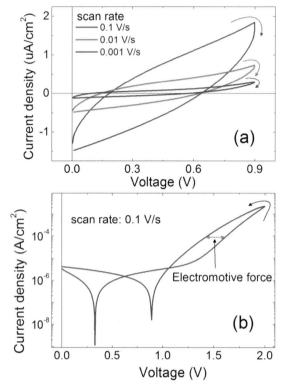

Figure 1:

(a) CV curves of a Pt/TiO$_2$/Pt capacitor. The thickness of all the three layers is 50 nm. The TiO$_2$ layer and the Pt electrodes were grown using sputtering and electron-beam evaporation, respectively, at room temperature. The measurement was performed using a CHI700 potentiostat in a voltage range of 0-0.9 V at three different voltage scan rates, 0.1, 0.01, and 0.001 V/s.

(b) CV curve of the same cell, measured in the wide voltage range 0-2V at a voltage scan rate of 0.1 V/s. Current in the high voltage region dominantly arises from a dc electronic contribution. The CV hysteresis in the high voltage region is most likely due to the evolution of electromotive force, i.e. non-static galvani potential, which accelerates the electron flow through the capacitor. This electromotive force was found to disappear as decreasing a voltage scan rate.

[1] R. Waser et al., Adv. Mater. **21**, 2632 (2009).
[2] S. H. Jo et al., Nano Lett. **10**, 1297 (2010).
[3] D. S. Jeong et al., Electrochem. Solid-State Lett. **10**, G51 (2007).
[4] B. J. Choi et al., J. Appl. Phys. **98**, 033715 (2005).
[5] J. F. Baumard et al., J. Solid State Chem. **20**, 43 (1977).
[6] P. Kofstad, J. Phys. Chem. Solids **23**, 1579 (1962).

TANTALUM OXIDE ULTRA-THIN FILMS BY METAL OXIDATION FOR APPLICATION IN RESISTIVE RANDOM ACCESS MEMORY (RRAM)

Sebastian Schmelzer[1,2], Ulrich Böttger[1,2], Rainer Waser[1,2]

[1]Institute for Materials in Electrical Engineering and Information Technology (IWE2), RWTH Aachen University of Technology, Aachen, Germany;
[2]JARA-Fundamentals of Future Information Technology, Research Center Jülich, Jülich, Germany

The interests in a successor to charge-based memories such as dynamic random access DRAM and flash are increasing with each technology node in the ITRS [1]. Among the large number of concepts, resistance-based memories on the basis of valency-change (VCM) are a promising candidate for the replacement of flash memory [2]. In this work, tantalum oxide (TaOx) based devices are presented with respect to their endurance and the underlying conduction mechanisms. The results will be critically compared with literature [3].

The devices are prepared by low-rate sputtering of tantalum in a reducing atmosphere, followed by surface oxidation at elevated temperatures in an oxygen partial pressure of 1 mbar. The resulting oxide thickness strongly depends on process temperature and ranged from 7 nm to 25 nm; the very smooth TaOx surface is shown in Fig.1. Finally, platinum top electrodes are patterned by photo-lithography and ion beam etching. A principal schematic of the devices is shown in the inset of Fig.2.

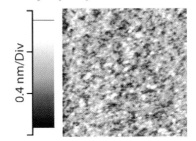

Figure 1: Atomic force microscopy measurement of the TaOx surface, showing an area of 1x1 µm² with a surface roughness of below 0.2 nm RMS.

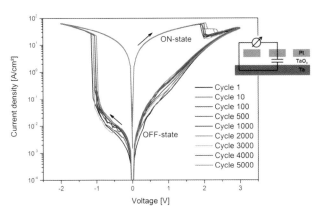

Figure 2: Switching endurance of TaOx device with an area of 0.01 mm². The inset shows the schematic of the devices and the measurement setup. IV-measurements were performed with a Keithley source meter by sweeping the applied voltage referred to the Pt electrode, the sweep speed is in the range of 20 seconds per cycle. The arrows designate the sweep direction. Statistics on the switching behavior are shown in Fig.3.

The outstanding switching endurance of TaOx reported in literature [4,5] are supported by the IV-measurements in Fig.2. Statistics on the switching behavior are shown in Fig.3 and Fig.4. The high and low resistance states (HRS, LRS) in Fig. 3 show Gaussian type distribution with maxima at 460 kΩ and 310 Ω, respectively. The value of the LRS is particularly well defined showing almost no deviation. Statistics on the applied voltages necessary for switching in the particular state are shown in Fig.4. For both switching polarities, around 90 % of the applied voltages are located in a voltage window of 150 mV, though showing slight deviations to higher values for OFF-switching. This stable switching behavior is eminently important for memory applications.

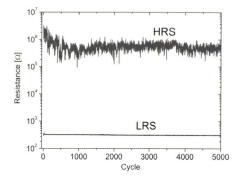
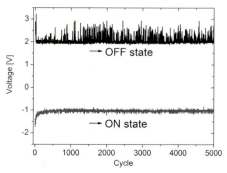

Figure 3: Resistance states for each cycle, showing a medium OFF/ON ratio of about 1500.

Figure 4: Statistics on applied voltages for switching in the particular state.

Compared to pulsed measurements in the nanosecond range, which are also feasible using the cells in this work, IV-measurements enable the investigation of conduction mechanisms. Fig.5 shows the analysis of conduction mechanisms in the pristine state, i.e. the virgin cell before electroforming, and the OFF state. The pristine state for negative applied voltage is mostly determined by a Schottky diode in inverse direction. For positive voltages, Fowler-Nordheim tunneling becomes the dominant mechanism above 1.5 V. This tunneling barrier remains unaffected by the forming event.

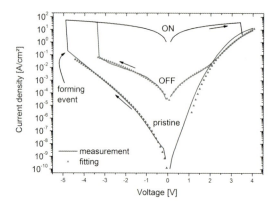

Figure 5: Analysis of conduction mechanisms in the pristine and OFF state. For better analysis, a device with higher switching voltages is used. The solid lines represent the values determined by IV-measurements; the triangles are calculated values using the Schottky barrier and Fowler-Nordheim tunneling equation, respectively.

The OFF state is dominated by two similar, back-to-back Schottky diodes. The barrier height of these diodes determines the OFF/ON ratio to a large extent and can be further reduced by certain current treatment. Transferring the acquired findings on cell preparation, design-rules and in consequence tailor-made devices for different applications should be feasible.

[1] ITRS, International Technology Roadmap for Semiconductors, www.itrs.net.
[2] R. Waser, R. Dittmann, G. Staikov, and K. Szot, Advanced Materials, **21**, 2632-2663 (2009).
[3] J. H. Hur, M.-J. Lee, C. B. Lee, Y.-B. Kim, and C.-J. Kim, Phys. Rev. B, **82**, 155321- (2010).
[4] A. C. Torrezan, J. P. Strachan, G. Medeiros-Ribeiro, and R. S. Williams, Nanotechnology, **22**, 485203 (2011).
[5] M. J. Lee, C. B. Lee, D. Lee, S. R. Lee, M. Chang, J. H. Hur, Y. B. Kim, C. J. Kim, D. H. Seo, S. Seo, U. I. Chung, I. K. Yoo, and K. Kim, Nature Materials, **10**, 625-630 (2011).

RESISTIVE SWITCHING PHENOMENA IN Li_xCoO_2 THIN FILMS

Olivier Schneegans[1], **Van Huy Mai**[1], **Alec Moradpour**[2], **Pascale Auban-Senzier**[2], **Claude Pasquier**[2], **Kang Wang**[2], **Sylvain Franger**[3], **Alexandre Revcolevschi**[3], **Efthymios Svoukis**[4], **John Giapintzakis**[4], **Philippe Lecoeur**[5], **Pascal Aubert**[5], **Guillaume Agnus**[5], **Thomas Maroutian**[5], **Raphaël Salot**[6], **Pascal Chrétien**[1]

[1]Laboratoire de Génie Electrique de Paris, UMR 8507 of CNRS, UPMC and Paris-Sud Universities, Supélec, 91192 Gif-sur-Yvette Cedex, France
[2]Laboratoire de Physique des Solides, UMR C8502 of CNRS, Paris-Sud University, 91405 Orsay, France
[3]Institut de Chimie Moléculaire et des Matériaux d'Orsay, Laboratoire de Physico-Chimie de l'Etat Solide, UMR 8182 of CNRS, Paris-Sud University, 91405 Orsay, France
[4]Department of Mechanical and Manufacturing Engineering, University of Cyprus, Nicosia, 1678, Cyprus
[5]Institut d'Electronique Fondamentale, UMR 8622 of CNRS, Paris-Sud University, 91405 Orsay, France
[6]CEA/LITEN of Grenoble, 38054, Grenoble, France

We have demonstrated, for the first time, a "resistive switching" phenomenon occurring in some mixed valence cobalt oxides (in single-crystal Na_xCoO_2 [1] and more recently, in thin films of Li_xCoO_2 [2]). It is widely accepted that this material, used today in rechargeable lithium batteries, exhibits a bulk-type electrical conductivity, involving cobalt redox reactions coupled to lithium intercalation/deintercalation processes. These materials may represent a possible alternative to the currently studied oxides involving, by contrast, filamentary electronic conductivities mediated by diffusions of oxygen vacancies.

We have investigated this switching phenomenon both on M-I-M devices (I stands for Li_xCoO_2 thin films), and by CP-AFM tips in direct contact with the thin films surfaces. Two drastically different behaviors have been unexpectedly observed in these two approaches (see Figure 1 below). Thus, the corresponding "eightwise" current-voltage curves, obtained with M-I-M devices as compared to CP-AFM, involve reverse directions, indicating opposite electrochemical modifications mediated by surprisingly the same polarity, and potential range.

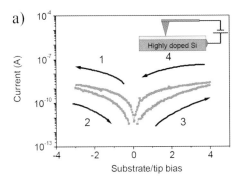
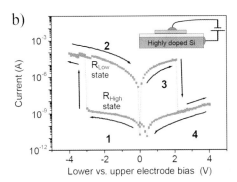

Figure 1: a) I (V) curve of a CP-AFM tip / Li_xCoO_2 thin-film contact; b) I (V) curve of a (100x400μm²) $Au/Li_xCoO_2/Si$ stack, exhibiting a resistive switching between the R_{High} and R_{Low} states, with $R_{High} / R_{Low} > 10^4$

These results will be discussed and the electrochemical mechanisms involved in both M-I-M and CP-AFM approaches will be specified, highlighting the significance of the thin-film/electrode interfaces. The experimental confirmation of the non-filamentary character of the conductivity, as well as the first results of the device's downscaling, will be reported.

[1] O. Schneegans et al., J. Am. Chem. Soc. **129**, 7482 (2007).
[2] A. Moradpour et al., Adv. Mater. **23**, 4141 (2011).

NANASCALE ANALYSIS OF FORMING AND RESISTIVE SWITCHING IN Fe:STO THIN FILM DEVICES

R. Dittmann[1], R. Muenstermann[1], I. Krug[2], D. Park[3], F. Kronast[4], A. Besmehn[5], J. Mayer[3], C. M. Schneider[2] and Rainer Waser[1,6]

[1]Peter Grünberg Institut 7 and JARA-FIT, Forschungszentrum Jülich, Jülich, Germany,
[2]Peter Grünberg Institut 7 and JARA-FIT, Forschungszentrum Jülich, Jülich, Germany,
[3]Central Facility for Electron Microscopy (GFE), RWTH Aachen, Germany,
[4]Helmholtz Zentrum Berlin, Berlin, Germany,
[5]Zentralabteilung für Chemische Analysen, Forschungszentrum Juelich, Jülich, Germany,
[6]Institut für Werkstoffe der Elektrotechnik, RWTH Aachen, Aachen, Germany

A large variety of binary and ternary oxides exhibit resistive switching phenomena, or so-called memristive behavior [1]. In the search for promising oxide materials for future non-volatile memories, special attention has to be paid to their scaling capabilities. The issue of scaling is strongly linked to the question of whether the switching current is distributed homogeneously over the device area or localized to one or a few conducting filaments. In this work, we investigated in detail the nanoscale current distribution in single crystalline Fe:doped $SrTiO_3$ (STO) thin films devices and linked it to chemical and microstructural changes.

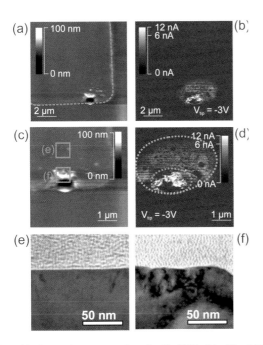

Fig. 1: (a),(c) Atomic force microscope (AFM) topography scans of a formed Fe-STO sample after electrode delamination with different scan sizes; The position of the former electrode is marked in blue; (b),(d) The simultaneously recorded current scans. Green: crater region; orange: conducting halo region near the carter (e) TEM image of the region far outside the crater; (f) TEM images of the crater region (e).

Fig. 1(a) shows the topography of a Fe-STO thin film MIM structure after forming and delamination of the top electrode. A small region of the junction shows a crater-like structure, which is caused by the applied forming step. Considering the current image in Figure 1(b), only the junction area in the vicinity of the forming crater shows good conductivity. It can be clearly deduced that most of the current during electroforming was flown across this channel (green). The Joule heating associated with this current flow has induced the reduction of the channel and the observed morphological changes. Figure 1(f) and (e) show magnified TEM bright field images obtained at the channel region and a region next to it, respectively (positions are indicated in Figure 1(c)). The defect density in the channel region is strongly increased compared to the film next to it.

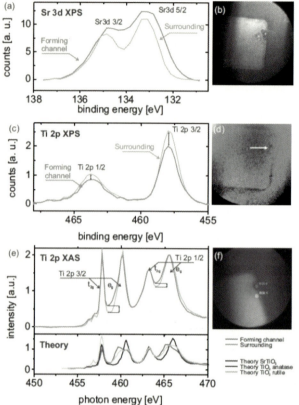

Fig. 2: Photoemission electron microscopy (PEEM) performed on at the BESSY UE49 PGM beamline;
(a) Sr3d XPS spectrum of the forming crater (red) and a region well outside the crater (green) performed at 270eV;
(b) PEEM image showing the position of both measurement sites.
(c) Ti2p XPS spectrum of the forming crater (red) and a region well outside the crater (green) performed at 600eV;
(d) PEEM image showing reduced brightness at the crater site.
(e) Upper part: Ti2p XAS spectra of the forming crater (red) and the outside region (green). Lower part: Reference spectra extracted from [4].
(f) PEEM image showing the position of both measurement sites.

Fig. 2 summarizes the spectromicroscopic analysis of the delaminated Fe-STO samples performed by photoemission electron microscopy (PEEM) in order to elucidate the chemical state of the current carrying filament region. Figure 2(a) shows Sr3d XPS spectra measured at the crater region (red) and outside the crater (green), respectively. The height of the Sr signal within the crater region is roughly 30% larger than outside the crater. In addition to the higher Sr signal we observed a chemical shift to lower binding energies, indicating the formation of new Sr containg phases on the surface in the crater region. XPS spectra of the Ti2p lines in the different regions are shown in Figure 2(c). Although no change of the peak shape could be detected, the Ti content is decreased about 15% in the crater region within the XPS information depth of about 1nm beneath the surface. A larger information depth is yielded by the XAS analysis of the Ti L-edge determined in the total electron yield (Figure 2 (e)). The spectrum in the crater region (red) contains additional states between their t2g and eg peaks compared to the spectrum recorded outside the crater (green). According to a comparison with the literature [4], this observation could be explained by the existence of Ti rich phases in the lower regions of the forming channel. This would be furthermore consistent with the Sr enrichment observed in the upper regions by XPS (Fig. 2(a)).

[1] R. Waser, R. Dittmann, G. Staikov, and K. Szot, Adv. Mater. **21**, 2632 (2009).
[2] R. Münstermann, T. Menke, R. Dittmann, R. Waser, Adv. Mat. 22, 4819 (2010)
[3] R. Dittmann, R. Muenstermann, I. Krug, D. Park, T. Menke, J. Mayer, A. Besmehn, F. Kronast, C.M.Schneider, and Rainer Waser, Proceedings of the IEEE" in press.
[4] P. Krueger, 14th International Conference on X-Ray Absorption Fine Structure (XAFS14), ser. Journal of Physics: Conference Series, vol. 190, 2009, p. 012006.

IN-OPERANDO HAXPES ANALYSIS OF THE RESISTIVE SWITCHING PHENOMENON IN Ti/HfO$_2$-BASED SYSTEMS

<u>Malgorzata Sowinska</u>[1], Thomas Bertaud[1], Damian Walczyk[1], Sebastian Thiess[2], Christian Walczyk[1], and Thomas Schroeder[1,3]

[1]IHP, Im Technologiepark 25, 15236 Frankfurt (Oder), Germany,
[2]Deutsches Elektronen-Synchrotron DESY, Notkestrasse 85, 22607 Hamburg, Germany,
[3]Brandenburgische Technische Universität, Konrad-Zuse-Strasse 1, 03046 Cottbus, Germany

Low-cost embedded nonvolatile memories (eNVMs) with high density, high speed and low-power are attractive for a growing number of applications. Among all the alternative technologies currently investigated, Resistance change Random Access Memory (RRAM) is a very promising and worldwide studied candidate for eNVM [1-3]. Despite the fact that the resistive switching phenomenon was almost proven for all the transition metal oxides, a tendency is emerging focusing on HfO$_2$-based RRAM. Besides, the insertion of a thin Ti layer between the TiN top electrode and the HfO$_2$ has shown an improvement of the performance. Nonetheless, despite numerous efforts, the driving mechanism for the resistive switching effect of HfO$_2$-based RRAM devices is still under debate. Interestingly, many models considering resistive switching [4-5] highlight the importance of the top interface as the localization of the switching event, as well as the possible role of the oxygen vacancies ($V_O^{\cdot\cdot}$) [6]. Regrettably, oxygen vacancies are particularly unstable and thus cannot be investigated by destructive techniques. However, hard X-ray photoelectron spectroscopy (HAXPES) is a powerful method to study deeply buried interfaces in a non-destructive way.

In this contribution, we demonstrate *in-operando* HAXPES studies of the resistive switching phenomenon observed in Ti/HfO$_2$-based systems. Thanks to a setup developed at IHP, we performed an *in-situ* current-voltage cycling of one and the same MIM cell with simultaneous HAXPES measurements. The Ti/HfO$_2$/TiN system, in size of 0.7×0.7 mm^2, was dynamically monitored in order to highlight the modifications between virgin-, ON-, and OFF-states. We have successfully switched the device several times at a chamber pressure of around 10^{-7} mbar (Fig. 1) and recorded HAXPES spectra for different resistance states at an excitation energy of 8 keV [7]. Our investigations revealed chemical and electronic differences in the HAXPES spectra between the investigated resistive states of the sample.

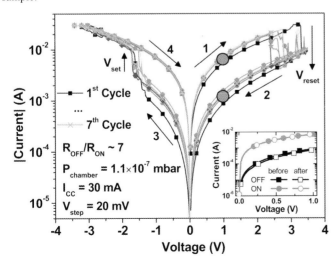

Figure 1: I-V characteristics of the resistive switching obtained *in-situ* at a pressure of 10^{-7} mbar. 7 cycles were performed with a R_{OFF}/R_{ON} ratio of about 7. Inset shows the current levels for both ON- and OFF-states before and after the HAXPES experiments.

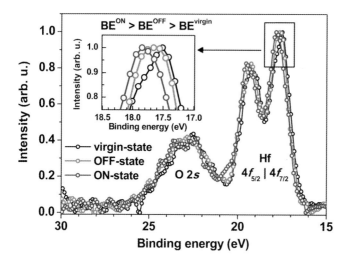

Figure 2: Normalized Hf $4f$ and O $2s$ HAXPES spectra of the virgin-, OFF- and ON-state of the sample recorded at an excitation energy of 8 keV. Inset presents the observed peak shift order.

The chemical changes especially occurred in the Ti $2p$ spectra. An increase of the Ti/HfO$_2$ interface oxidation is underlined for both ON- and OFF-states, compared to the virgin-state. The electronic modifications are characterized by peak shifts in the order of: $BE^{virgin} < BE^{OFF} < BE^{ON}$. This shift order is attributed to different concentrations of n-type oxygen vacancy defects ($[V_O^{\cdot\cdot}]^{virgin} < [V_O^{\cdot\cdot}]^{OFF} < [V_O^{\cdot\cdot}]^{ON}$) in HfO$_2$. As an example of these shifts, we present in Fig. 2 the Hf $4f$ and O $2s$ XPS emission lines collected for all the investigated resistive states. The aforementioned peak shifts are clearly visible in the zoom on the Hf $4f_{7/2}$ line.

These observed chemical and electronic changes point to the possibility to describe the resistive switching event in our Ti/HfO$_2$-based system by a push-pull model coupled with a Schottky interface-like system. We propose that the electroforming process induces a Ti/HfO$_2$ interface oxidation by oxygen scavenging from the HfO$_2$ and the creation of n-type defects (oxygen vacancies) in the HfO$_2$ layer [6]. Furthermore, a positive voltage applied to the top electrode repels these oxygen vacancies and resets the device to the OFF-state, while a negative voltage attracts them and sets the RRAM cell to the ON-state [7]. Detailed results will be presented.

[1] Ch. Walczyk et al., IEEE T. Electron Dev. **58**, 3124 (2011).
[2] Ch. Walczyk et al., J. Vac. Sci. Technol. B **29**, 01AD01 (2011).
[3] D. Walczyk et al., Microelectron. Eng. **88**, 1133 (2011).
[4] A. Sawa, Mater. Today **11**, 28 (2008).
[5] R. Waser et al., Adv. Mater. **21**, 2632 (2009).
[6] M. Sowinska et al., submitted in Appl. Phys. Lett.
[7] T. Bertaud et al., to be submitted.

THE OXYGEN VACANCY DISTRIBUTION IN RESISTIVE SWITCHING Fe-SrTiO$_3$ MIM STRUCTURES BY µXAFS

<u>Christian Lenser</u>[1], Alexei Kuzmin[2], Aleksandr Kalinko[2], Juris Purans[2], Rainer Waser[1,3] and Regina Dittmann[1]

[1]Peter Grünberg Institut 7 and JARA-FIT, Forschungszentrum Jülich, Jülich, Germany
[2]Institute of Solid State Physics, University of Latvia, Riga, Latvia
[3]Institut für Werkstoffe der Elektrotechnik, RWTH Aachen, Aachen, Germany

Binary and ternary metal oxides as emergent materials for non-volatile memory applications are receiving an increasing amount of scientific attention due to the promising scalability, retention and switching characteristics of this material class [1]. The key role of oxygen non-stoichiometry and oxygen-deficient oxide-phases as the underlying mechanism of the resistance change has been recognized for many different oxide systems (e.g. TiO$_2$, Ta$_2$O$_5$, SrTiO$_3$). It is becoming widely accepted that the resistance switching process in SrTiO$_3$ is related to the movement and creation of oxygen vacancies and the associated electron doping. However, direct experimental reports of the redox-reaction induced by an electric field are rare [2].

In this contribution, the distribution of oxygen vacancies in a switched memristor fabricated from epitaxial Fe-doped SrTiO$_3$ will be investigated by spatially resolved x-ray absorption near-edge structure (XANES). SrTi$_{0.95}$Fe$_{0.05}$O$_3$ was grown epitaxially by pulsed laser deposition (PLD) on a conducting Nb:SrTiO$_3$ substrate, and Pt top electrodes were sputter-deposited and structured via optical lithography. XANES measurements at the Fe K-edge with a 7 µm beam spot on the sample were done at beamline ID03 (ESRF, France). Figure 1 compares the Fe K-edge XANES of the virgin thin film to that recorded in the anodic and cathodic regions of an electrocolored single crystal. The cathode was found to contain only Fe^{3+} ions and Fe^{3+}-V$_O$ complexes in the ratio ~ 70/30, and notably no Fe^{4+} [3]. The almost perfect coincidence of the thin film XANES with that of the reduced cathode implies that the thin film is already oxygen deficient after growth, and that the Fe^{3+}/Fe^{4+} redox pair does not serve as an indicator for local resistance changes. The shoulder at 7122 eV excitation energy that marks Fe^{3+}-V$_O$ complexes is indicated by "S" in figure 1.

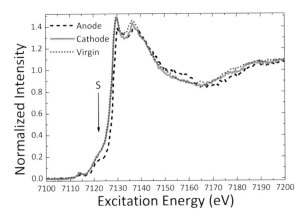

Figure 1: Fe K-edge XANES recorded on the virgin thin film (dots), the cathode region (solid) and the anode region (dash) of an electrocolored Fe-doped SrTiO$_3$ single crystal. The intensity of the shoulder at 7122 eV excitation energy (marked S by the arrow) is a fingerprint of the presence of an oxygen vacancy in the first coordination shell of Fe^3. The chemical state of the virgin film – according to the XANES – is similar to that of the cathode, and the Fe centers in the film are primarily cubic Fe^{3+} centers, with a significant percentage of axial Fe^{3+}-V$_O$ centers.

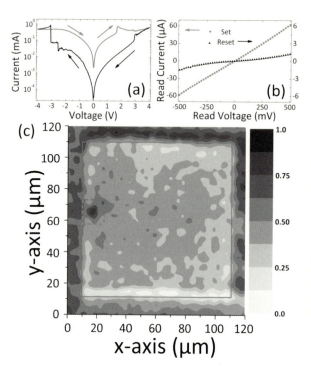

Figure 2 (a): I(V)-characteristics of the investigated memristor after electroforming. (b): Low voltage, non-destructive read-out sweeps in the low resistance "Set"-state (dots) and the high resistance "Reset"-state (triangles). (c): Fe Kα fluorescence map of the switched memristor, recorded at 7122 eV excitation energy (shoulder S). The intensity variations under the electrode correspond to the oxygen vacancy content, the maximum in the dashed box indicates the presence of the conducting filament. The absolute intensity outside the electrode area (solid lines) is higher since the fluorescence radiation is not attenuated, but cannot be directly compared to the intensity under the electrode.

After an electroforming step with a +7V DC voltage applied to the top electrode, the formed memristor can be switched between different resistance states with a bipolar voltage sweep. The current-voltage hysteresis is shown in figure 2(a), the "Set"-state is reached with a negative voltage sweep polarity. A low voltage readout (figure 2(b)) reveals the "Set"-state to show ohmic behavior, while the "Reset"-state shows distinctly non-linear behavior.

A Fe-K edge fluorescence map recorded at 7122 eV excitation energy to maximize the sensitivity to Fe^{3+}-V_O reveals one location on the electrode pad with increased intensity, corresponding to the filament. Moreover, the Fe^{3+}-V_O concentration as measured by the intensity of the shoulder S is increased under the whole electrode area as compared to the virgin film. The important consequence is that before breakdown is achieved during the electroforming, a homogeneous front of vacancy enrichment propagates into the material. Furthermore, the Fe K-edge XANES at the filament location is interpreted via full multiple-scattering calculations and indicates oxygen vacancy clustering in the first shell of Fe [4].

[1] R. Waser, R. Dittmann, G. Staikov, and K. Szot, Adv. Mater. **21**, 2632 (2009).
[2] M. Janousch, G. I. Meijer, U. Staub, B. Delley, S. F. Karg, and B. P. Andreasson, Adv. Materials **19**, 2232 (2007).
[3] C. Lenser, A. Kalinko, A. Kuzmin, D. Berzins, J. Purans, K. Szot, R. Waser, and R. Dittmann, Phys. Chem. Chem. Phys. **13**, 20779 (2011).
[4] C. Lenser, A. Kuzmin, A. Kalinko, J. Purans, R. Waser and R. Dittmann, accepted for publication in JAP

MULTILEVEL RESISTIVE SWITCHING AND METAL –INSULATOR TRANSITION IN SOLUTION-DERIVED La$_{1-x}$Sr$_x$MnO$_3$ THIN FILMS

C. Moreno, J. Zabaleta, A. Palau, J. Gázquez, N. Mestres, T. Puig, C. Ocal, X. Obradors

Institut de Ciència de Materials de Barcelona, ICMAB-CSIC
Campus UAB, 08193 Bellaterra, Spain

Chemical solution deposition (CSD) is a powerful methodology to grow in large areas self-assembled oxide nanostructures and thin or ultrathin epitaxial films. In recent years we have widely investigated the unique microstructural and physical properties of different sorts of CSD-grown functional oxide nanostructures and thin films, including CeO_2, ferromagnetic La$_{1-x}$Sr$_x$MnO$_3$ and $YBa_2Cu_3O_7$ - derived superconductors [1-8]. Here we will report on the interfacial microstructure and the magnetic, transport and nanoscale switching phenomena of La$_{1-x}$Sr$_x$MnO$_3$ (LSMO) films grown by CSD on $SrTiO_3$ (STO) and $LaAlO_3$ (LAO) single crystalline substrates.

We will first show from HRTEM images that highly crystalline and coherently strained ultrathin LSMO films (t~2.4 nm) can be indeed grown on STO (ε~0.9 %) while films with similar thickness on LAO (ε~-2.3 %) appear to be completely relaxed through the formation of interfacial misfit dislocations (separated by ~ 17 nm, as expected for the Burgers vector a$_{LAO}$[100]). It is noteworthy, however, that SQUID measurements show that both sorts of films have similar ferromagnetic properties, without any reduction of T$_c$~350 K, a similar saturation magnetization M$_s$ and a slightly modified coercive field H$_c$, thus differing from the typical observed behavior on PLD-grown thin films. The metal – insulator transition and the colossal magnetoresistance instead, is found to be indeed modified by the film thickness and the strain state of the films.

We report, on the other hand, the observation of reversible transitions from low resistive (LR) to high resistivity (HR) states at the nanoscale in LSMO films, as investigated by Conductive-Scanning Force Microscopy (C-SFM) [9]. The transitions are induced under the application of a bias voltage by means of the conducting tip and the mechanisms at the origin of the modifications induced are investigated in the complete writing and erasing process (Figure 1). Topography, local conductance and contact potential difference (CPD) have been analysed by combining different scanning probe techniques, namely Scanning Force Microscopy (SFM), Conductive-Scanning Force Microscopy (C-SFM) and Kelvin Probe Microscopy (KPM). The transition can be induced through the full thickness of the studied films (up to 25nm) using different geometries such that regions of different resistivity can be topologically isolated (Figure 1). Depending on the magnitude of the applied voltage different HR states can be reached, which can be identified by C-SFM. Moreover, KPM has been revealed as a valuable reading tool which permits a multilevel adscription of the different resistive states by means of work function measurements (Figure 1). Our capability to fabricate multilevel and switchable resistive nanostructures in CSD-grown oxide thin films can be easily envisaged as a suitable approach towards high-density data storage applications [9].

[1] M. Gibert, T. Puig, X. Obradors, A. Benedetti, F. Sandiumenge, R. Hühne; Adv. Mater. 19, 3937 (2007)
[2] C. Moreno, P. Abellán, A. Hassini, A. Ruyter, A. Pérez del Pino, F. Sandiumenge, M.J. Casanove, J. Santiso, T. Puig, X. Obradors; Adv. Funct. Mater. 19, 2139 (2009)
[3] C. Moreno, C. Munuera, A. Pérez del Pino, J. Gutiérrez, T. Puig, C. Ocal, X. Obradors, A. Ruyter; Phys. Rev. B 80, 094412 (2009)
[4] M. Gibert, P. Abellán, A. Benedetti, T. Puig, F. Sandiumenge, A. García, X. Obradors; Small 23, 2716 (2010)
[5] A. Carretero-Genevrier, J. Gázquez, J.C. Idrobo, J. Oró, J. Arbiol, M. Varela, E. Ferain, J. Rodríguez-Carvajal, T. Puig, N. Mestres, X. Obradors; J. Am. Chem Soc. 133, 4053 (2011)
[6] P. Abellán, C. Moreno, F. Sandiumenge, X. Obradors, M.-J. Casanove; Appl. Phys. Lett. 98, 041903 (2011)
[7] C. Moreno, P. Abellán, F. Sandiumenge, M.J. Casanove, X. Obradors; Appl. Phys. Lett. 100, 023103 (2012)
[8] J. Zabaleta, M. Jaafar, P. Abellán, C. Montón, O. Iglesias-Freire, F. Sandiumenge, C.A. Ramos, R.D. Zysler, T. Puig, A. Asenjo, N. Mestres, X. Obradors; J. Appl. Phys. 111, 024307 (2012)
[9] C. Moreno, C. Munuera, S. Valencia, F. Kronast, X. Obradors, C. Ocal; Nano Lett. 10, 3828 (2010)

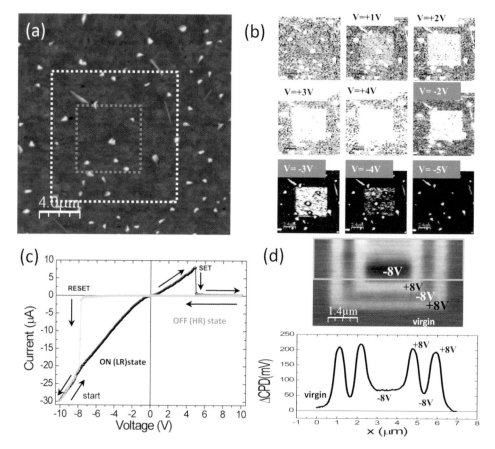

Figure 1.- (a) AFM Topography image of LSMO films; (b) Resistive switching movie in a concentric square at different voltages; (c) Hysteretic bipolar switching I-V curves; (d) KPM image and scan of the corresponding ΔCPD of a sequence of writing and erasing processes.

PUMP AND RELEASE SCENARIO FOR THE BIPOLAR RESISTIVE SWITCHING OF MEMRISTIVE MANGANITE-METAL INTERFACES

Pablo Levy[1], N.Ghenzi[1], M. J. Sanchez[2], M. J. Rozenberg[3], P. Stoliar[1,4], F. G. Marlasca[1], and D. Rubi[1]

[1]GAIANN, CAC - CNEA, Buenos Aires, Argentina ;
[2]GAIANN, CAB-CNEA, Bariloche, Argentina;
[3]LPS-UPS, Orsay, France and FCEN-UBA, Buenos Aires, Argentina;
[4]ECyT-UNSAM, San Martín, Argentina.

We explore different resistance states of metal (Ag, Ti) - manganite ($La_{0.325}Pr_{0.300}Ca_{0.375}MnO_3$) interfaces as prototypes for non-volatile memristive memory devices. In addition to High and Low resistance states accessible through bipolar pulsing with a single pulse, higher (lower) resistance states can be obtained by repeatedly applying Reset (Set) pulses. This accumulative action drives the resistance towards saturation. Our experiments reveal that the pulsing amplitude and the number of applied pulses necessary to reach a predefined target High resistance value have an exponential relationship. Simulations using a phenomenological approach (drift of oxygen vacancies induced by local electric field) confirm obtained results and provide the vacancies profiles associated to these states.

Applying *several* Reset pulses followed by a *single* Set stimulus, a "pump and release" scenario is described and analyzed in terms of the local electric field developed at the interface. While sudden effects (i.e. release) can only be produced in the presence of a strong local electric field, a gradual process (i.e. pumping and accumulation) is necessary to built it up.

By varying the Set amplitude we explore the dependences of the abrupt releasing process. A power law $\Delta R = (I - I_{thr})^p$ is obtained for the response – stimulus dependence, where I_{thr} is the threshold "cohercive amplitude" for the Set switching.

This value is in agreement with the cohercive amplitude obtained by performing major and minor Hysteresis Switching Loops. Based on these findings we propose a pulsing strategy to enhance the performance of existing memristive devices.

RESISTIVE SWITCHING IN NiO BASED NANOWIRE ARRAY FOR LOW POWER RERAM

Sabina Spiga[1], Stefano Brivio[1], Grazia Tallarida[1], Daniele Perego[2], Silvia Franz[2], Damien Deleruyelle[3], Christophe Muller[3]

[1]Laboratorio MDM, IMM-CNR, via C. Olivetti 2, 20864 Agrate Brianza (MB), Italy
[2]Dipartimento di Chimica, Materiali ed Ing. Chimica "G.Natta", Politecnico di Milano, Via Mancinelli, 20131 Milano
[3]IM2NP, UMR CNRS 7334, Aix-Marseille Université, IMT Technopôle de Château-Gombert, 13451 Marseille Cedex 20, France

Metal/oxide/metal (MOM) heterostructure nanowires could represent the building blocks of future ultra-scaled electronics devices. In particular, MOM structures, exhibiting resistance switching between high (reset) and low (set) conductive states under electric pulse application, are currently receiving increasing interest for non-volatile Resistive Random Access Memories (ReRAM) envisaged for high-density 3D cross-bar architectures. The resistance switching in MOM structures is generally based on reduction/oxidation-related electrochemical effects in oxides, and could be related to the formation of conducting filaments in the oxide. Depending on the system the switching behavior could be unipolar, where the resistance change is accomplished using the same voltage polarity, or bipolar. Among the investigated materials for ReRAMs, transition metal binary oxides offer high potential scalability, thermal stability, and can be easily integrated in fabrication processes at industrial level. NiO-based systems are widely studied as unipolar fuse-antifuse type of ReRAM, and could represent a model system for further investigation of operational power scaling. Up to now most of the studies have been performed on thin films [1,2] and only few studies on large diameter nanowires are reported [3-5].

This work reports on the fabrication and resistive switching properties of ordered arrays of segmented Au-NiO-Au nanowires embedded in an alumina matrix (Figure 1a); the resistive switching properties of individual nanowires are addressed using the conductive probe of an Atomic Force Microscope (CP-AFM).

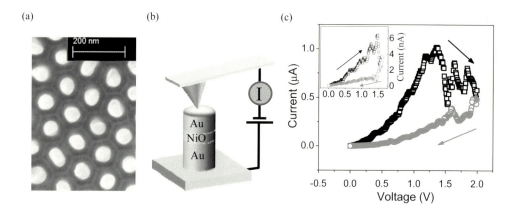

Figure 1: (a) Scanning electron microscopy image of the top view of the AAO matrix with embedded Au-NiO-Au nanowires. (b) CP-AFM set-up used to address the single nanowire embedded in the AAO matrix. (c) Examples of reset operations performed by CP-AFM measurements.

Samples are fabricated by electrodeposition of Au/Ni/Au multilayers into suspended Anodic Aluminum Oxide (AAO) templates with 50 nm pore diameter, followed by mechanical polishing of the AAO membrane and thermal oxidation of the Ni segment. The length of the Ni segment was fixed to 20 nm, while the Au electrodes are several microns long. Oxidation of the Ni segment is achieved without dissolving the matrix, thus allowing the fabrication of large arrays of segmented nanowires with metal/oxide buried interfaces, vertically oriented, and with the hexagonal order derived from that of the AAO nanopores. This approach, which has not been investigated so far in literature, could in principle allow the vertical stacking of several Au/NiO/Au nanowire-based memory cells. Each of the segments would have a controlled length, which is relevant for 3D applications. Although in this work we have been working on free standing AAO matrix, the developed template assisted fabrication method could be transferred on substrates compatible with microelectronic industry.

A scheme of the experimental set-up for CP-AFM measurements of individual Au-NiO-Au nanowire is reported in Figure 1b. Although the spatial resolution of CP-AFM is necessary to contact single nanowires, in this work the AFM probe is used to contact the Au electrode and is not intended to act as a top electrode itself, thus avoiding any possible tip-sample artifact. After a forming procedure, performed with a series load resistor to reduce the flowing current, the Au/NiO/Au nanowire exhibits unipolar switching between two different resistance states, with the possibility to control the operation power from 2 µW down to few nW (examples of reset operations are reported in Figure 1c, and inset). The obtained results extend previous literature data on NiO-based systems, and are related to a combined effect of device area scaling and current control during the forming and set procedures. In order to shed more light on this point, further research effort is currently devoted to the downscaling toward 20 nm-diameter nanowires.

This work was partially supported by Fondazione Cariplo (MORE Project n°2009-2711, http://more.mdm.imm.cnr.it/) and by CNR through the Short Term Mobility Program.

[1] D. Deleruyelle, C. Dumas, M. Carmona, C. Muller, S. Spiga, M. Fanciulli, Applied Physics Express **4**, 051101 (2011).
[2] D. Ielmini, S. Spiga, F. Nardi, C. Cagli, A. Lamperti, E. Cianci, M. Fanciulli, J. Appl. Phys. **109**, 034506 (2011).
[3] S. I. Kim, J. H. Lee, Y. W. Chang, S. S. Hwang, K.-H. Yoo, Appl. Phys. Lett. **93**, 033503 (2008).
[4] E. D. Herderick, K. M. Reddy, R. N. Sample, T. Draskovic, N. P. Padture, Appl. Phys. Lett. **95**, 203505 (2009).
[5] C. Cagli, F. Nardi, B. Harteneck, Z. Tan, Y. Zhang, D. Ielmini, Small **7**, 2899 (2011).

EXPERIMENTAL EVALUATION OF THE TEMPERATURE IN CONDUCTIVE FILAMENTS CREATED IN RESISTIVE SWITCHING MATERIALS

Eilam Yalon, Shimon Cohen, Arkadi Gavrilov, Boris Meyler, Joseph Salzman, and Dan Ritter

Technion – Israel Institute of Technology, Haifa 3200, Israel

Resistive switching (RS) materials are among the most promising candidates for next-generation non-volatile memory technology. The implementation of the technology is hampered, however, by the lack of profound understanding of the switching and conduction mechanisms [1]. The switching effect is widely attributed to the formation and rupture of conductive filaments inside the insulating matrix due to nano-ionic and thermal effects. The extremely high current densities heat the filament significantly, and these thermal effects are believed to be strongly correlated with the switching events in both uni-polar [2] and bipolar switching materials [3]. Clearly, experimental evaluation of the filament temperature is critical for better understanding the switching effects, and for optimizing the endurance of resistive memory devices. However, to the best of our knowledge, measurements of the filament temperature were not yet reported, and only model calculations are available [2,3].

We have recently reported that a metal-insulator-semiconductor bipolar transistor provides information on the energy of electrons injected through insulating materials [4,5]. Here, we demonstrate that this technique can also be used to study thermal effects in the conductive filaments. We have shown that in the low resistance state of the device a filament extends through most of the film, leaving a small tunneling gap adjacent to the semiconductor electrode, and conduction takes place by direct tunneling through the ultrathin (~1nm) gap [5]. Most of the applied voltage drops across the tunneling gap, and as a result most of the heat is dissipated in the electrodes rather than in the filament.

The metal-insulator-semiconductor bipolar transistor monitors minority carrier injection into the semiconductor electrode. As outlined in Fig. 1, at low voltage injection of minority carriers is possible only by thermal excitation. The minority carrier current is proportional to $\exp(qV/\eta kT)$, where V is the applied voltage, T the varying temperature, and η the so called "ideality factor" which may be larger than unity if the tunneling probability is voltage dependent. Assuming that η=1 the temperature at the tip of the filament can be obtained. If η>1 then the actual temperature is somewhat lower. Our method thus yields an upper limit of the filament tip temperature.

The extracted upper limit of the filament temperatures in three 5 nm thick HfO_2 based metal-insulator-semiconductor bipolar transistors are shown in Fig. 2 as a function of the total current squared. The heating effect is quite different in the three devices, reflecting the random nature of filament formation. The obtained temperature range is in agreement with previous modeling [2,3]. Monitoring of the filament temperatures in differently prepared devices is thus a new and valuable tool for studying device physics and reliability.

Nanosession: Valence Change Memories - a look inside

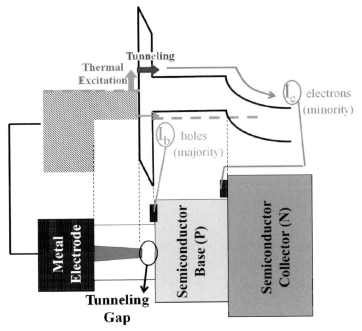

Fig. 1 – Device structure and schematic energy band diagram of a metal-insulator-semiconductor bipolar transistor. Minority carriers (electrons) are injected into the conduction band by thermal exitation, and detected at the collector terminal of the device.

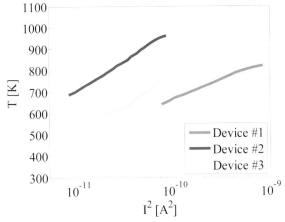

Fig. 2 – Extracted upper limit of the temperature of the conductive fillmat tip in three 5nm thick HfO_2 based metal-insulator-semiconductor bipolar transiostors versus the total current squared.

[1] R. Waser et al. *Adv. Mater.*, 21 (25/26), Jul. 2009.
[2] U. Russo et al., *IEEE Trans. Electron Device*, 56 (2), Feb. 2009.
[3] S. Menzel et al., *Adv. Func. Mater.*, 21 (23), 2011.
[4] E. Yalon et al., *ECS Transactions*, 41 (3), 2011.
[5] E. Yalon et al., *IEEE Electron Device Lett.*, 33 (1), 2012.

FERROELECTRIC RESISTIVE SWITCHING AT SCHOTTKY-LIKE BiFeO₃ INTERFACES

Akihito Sawa, Atsushi Tsurumaki-Fukuchi, Hiroyuki Yamada

National Institute of Advanced Industrial Science and Technology (AIST), Tsukuba, Ibaraki 305-8562, Japan

Resistive switching phenomena in transition-metal oxides have been intensively studied in recent years, because of the potential for nonvolatile memory application, i.e. resistance random access memory (ReRAM). The main mechanism of the resistive switching phenomena is a nanoionic redox reaction triggered by Joule heating or electrochemical migration of oxygen vacancies. Since chemical alterations of materials are inevitably induced in both mechanisms, there is concern for the reliability, such as the data retention and endurance. As a solution of this problem, resistive switching based on an electronic mechanism is being considered.

Ferroelectric resistive switching effects based on polarization reversal are practically attractive, because polarization reversal does not induce a chemical alteration. In this study, we have explored a resistive switching phenomenon in a multiferroic $BiFeO_3$. Capacitor-like structures of $Pt/Bi_{1-\delta}FeO_3$ (BFO)/$SrRuO_3$ were fabricated on $SrTiO_3$ single crystal substrates. Since Bi deficiencies provide hole carriers in BFO, Bi-deficient BFO layers act as a p-type semiconductor. The devices showed rectifying and hysteretic current-voltage (I-V) characteristics. Moreover, in I–V characteristics measured at a voltage-sweep frequency of 1 kHz, positive and negative current peaks originating from ferroelectric displacement current were observed under forward and reverse bias prior to set and reset switching processes, respectively. We also succeeded in resistive switching by pulsed-voltage applications. The devices showed endurance of $>10^5$ cycles and data retention of $>10^5$ s. These results demonstrate promising prospects for application of the ferroelectric resistive switching effect at BFO interfaces to nonvolatile memory.

This work was supported in part by the Japan Society for the Promotion of Science (JSPS) through the "Funding Program for World-Leading Innovative R&D on Science and Technology (FIRST Program)", initiated by the Council for Science and Technology Policy (CSTP), and a Grant-in-Aid for Scientific Research.

HOW CAN WE SWITCH THE RESISTIVITY OF A METALLIC PEROVSKITE OXIDE (SrTiO$_3$:Nb) BY ELECTRICAL STIMULI?

Christian Rodenbücher[1], Krzysztof Szot[1,2], Rainer Waser[1,3]

[1] Peter Grünberg Institut 7, Forschungszentrum Jülich, Jülich, Germany;
[2] Institute of Physics, University of Silesia, Katowice, Poland;
[3] Institut für Werkstoffe der Elektrotechnik 2, RWTH Aachen, Aachen, Germany

Recently the resistive switching phenomenon has been found in a variety of materials [1]. In SrTiO$_3$, which is intensively studied as a model material of perovskites, switching occurs along a network of dislocations which serve as an easy diffusion path for oxygen [2]. In general the change of the resistivity in oxides is connected with a redox process leading to an insulator to metal transition where the transformation of a "d"-electron from d^0 to d^1 plays the key role. This "d"-electron can be generated by donor doping, chemical gradients, electrical gradients or a convolution of these gradients. If we now investigate a donor doped material like Nb- doped SrTiO$_3$ one would expect to find metallic properties since the insulator to metal transition already should have taken place. But the resistive switching effect occurs in Nb-doped thin films [3] as well as in single crystals [4] which immediately raises the question "How is it possible to switch a metallic system which transfer number should be zero?"

To unravel this paradoxical situation a multidisciplinary approach is needed providing information about all significant properties of the material like crystallographic structure, electronic structure, chemical composition and transport phenomena especially between bulk and surface on macro- and nano-scale.

On the basis of surface sensitive measurements we present that on SrTiO$_3$:Nb single crystals a high resistive **native** surface layer exists that has fundamentally different properties than the conductive bulk. Regarding the resistive switching phenomenon we found out that it takes place only in the surface layer whereas the bulk is metallic all the time.

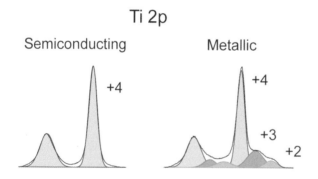

Figure 1: Valences of the Ti 2p line measured by XPS. The native semiconducting surface of the sample shows only one doublet corresponding to the valence +4, whereas the metallic state obtained by reduction or electroreduction shows the two additional valences +3 and +2.

Furthermore we investigated how the surface layer behaves under different gradients and we will show that the electronic structure and the chemical composition can be changed dramatically which implies that the present species, especially Sr and O are highly mobile in the surface layer of the crystal. We could prove by X-ray photoelectron spectroscopy (XPS) measurements shown in Figure 1 that the switching of the surface layer from semiconducting to metallic properties is connected with a redox process changing the valence of Ti from Ti^{4+} to Ti^{3+} and Ti^{2+}.

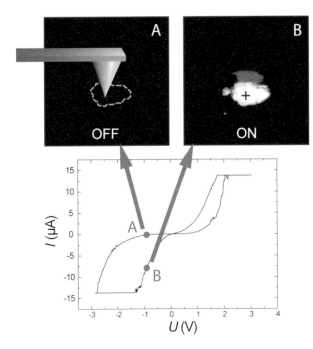

Figure 2: Switching of a conducting cluster from the OFF to the ON state by applying a voltage to one point of the surface via the AFM tip.
The corresponding dynamic IV curve shows that the switching process takes place at relatively low voltages between 1 and 3 V.

To get insight in the basic processes being responsible for the resistive switching it is necessary to investigate the surface layer with nano-scale methods like local conductivity atomic force microscopy (LC-AFM). These measurements reveal that the surface layer is highly inhomogeneous and that the resistive switching effect is a highly local phenomenon showing a cluster like conductivity. These conducting clusters with a size of several tens of nanometers can be switched independently using the AFM tip as top electrode as illustrated in Figure 2.

Finally ab initio calculations give us the opportunity to compare the present measurements with the theory and provide an essential impression of the origin of the electronic structure.

In summary we present that on single crystal $SrTiO_3$:Nb a native surface layer exists that can be switched from semiconducting to metallic behavior by external gradients.

[1] R. Waser et al., Adv. Mater. **21**, 25-26 (2009).
[2] K. Szot et al., Nat. Mater. **5**, 4 (2006).
[3] R. Münstermann et al., Appl. Phys. Lett. **93**, 023110 (2008).
[4] C. Park et al., J. Appl. Phys. **103**, 054106 (2008).

PHYAICAL MECHANISM OF OXYGEN VACANCY MIGRATION IN Pt/Nb:SrTiO$_3$ INTERFACES

Shin Buhm Lee[1], Jong-Bong Park[2], Myoung-Jae Lee[2], Tae Won Noh[1]

[1] ReCFI, Department of Physics and Astronomy, Seoul National University, Seoul 151-747, Korea;
[2] Samsung Advanced Institute of Technology, Yongin, Gyeonggi-do 446-712, Korea

Nanoionics-based bipolar resistance switching (BRS) has attracted emerging attention as a frontier research filed of 21st century. About a decade ago, the BRS was suggested to be applicable for non-volatile memory devices known as resistance random access memory (RRAM). The BRS nanodevices possess excellent device performances, including fast switching, high scalability, long retention, and high endurance. According to many researches, the oxygen vacancies are known to play important roles to determine such device properties.

We investigated the physical mechanism of oxygen vacancy migration in Pt/Nb:SrTiO$_3$ interfaces, which shows the BRS phenomena. By using transmission electron microscopy equipped with electron energy loss spectroscopy, we confirmed the existence of oxygen vacancy clusters in Pt/Nb:SrTiO$_3$ interfaces. We measured the migration velocity of oxygen vacancy under wide temperature (100−400 K) and bias voltage (0.1−10 V) ranges. We found that the velocity is strongly nonlinear with bias voltage and this nonlinearity become larger at lower temperature. Referring the fractal concepts in a disorder system, oxygen vacancy migration in their clusters could be classified into the creep and flow motion according to the bias voltages.

MECHANISM OF RESISTIVE SWITCHING IN BIPOLAR TRANSITION METAL OXIDES

Marcelo J. Rozenberg[1], **María J. Sánchez**[2], **Pablo Stoliar**[1,3], **Ruben Weht**[4,5], **Carlos Acha**[6], **Fernando Gomez-Marlasca**[4] and **Pablo Levy**[4]

[1]CNRS - Laboratoire de Physique des Solides, UMR8502 Université Paris-Sud, Orsay 91405, France; [2]Centro Atómico Bariloche and Instituto Balseiro, CNEA, 8400 - San Carlos de Bariloche, Argentina; [3]ECyT, Universidad Nacional de San Martín, Campus Miguelete, Martín de Irigoyen 3100, 1650 San Martín, Argentina; [4]Gerencia de Investigación y Aplicaciones, CNEA, 1650 - San Martín, Argentina; [5]Instituto Sábato, Universidad Nacional de San Martín -CNEA, Argentina; [6]Departamento de Física J. J. Giambiagi, FCEN, Universidad de Buenos Aires, Ciudad Universitaria Pab. I, 1428 Buenos Aires, Argentina.

Resistive random access memories (RRAM or ReRAM) composed of a transtition metal oxide dielectric in a capacitor-like structure is a candidate technology for next generation non-volatile memory devices. They are based on the physical phenomenon of resistive switching [1,2], which is a sudden and non-volatile change of the resistance under the action of strong electrical stress. Here, we introduce a simple theoretical model [3] which incorporates the main physical ingredients that have emerged from intensive experimental research in the last decade.

Our model accounts for the bipolar resistive switching effect observed in large variety of perovskite transition metal oxides. The numerical study of the model predicts that under the action of initial forming voltage cycles, strong electric fields develop in the highly resistive dielectric-electrode interfaces, leading to a spatially inhomogeneous distribution of oxygen vacancies. After the first few cycles, the distribution approaches a stable profile and the repetitive non-volatile resistance memory effect is observed.

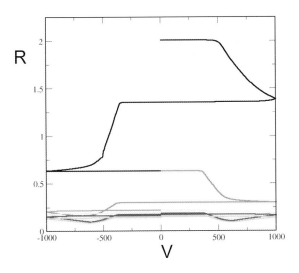

Figure 1: Resistance as a function of voltage (in a.u.) for the first 7 voltage switching loops. The voltage protocol is a triangular ramp following the sequence 0 -> Vmax ->0 -> -Vmax -> 0. The colors indicate the loop, starting from black.

After the first two loops the system is "formed" and the "table-with-legs" resistance switching loop is observed

The theoretical results of the model are validated by successful comparison with the non-trivial "table-with-legs" resistance hysteresis switching loops measured in cuprate (YBCO) and manganite (LPCMO) samples.

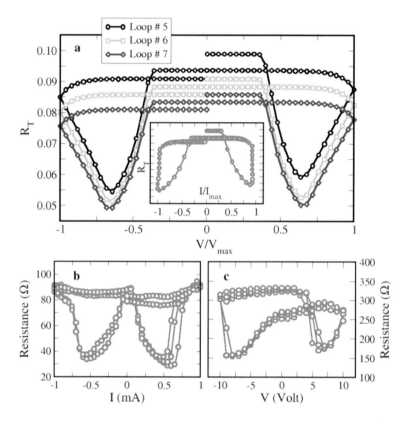

Figure 2: (a) Detail of the resistance as a function of V/Vmax for loops # 5, 6 and 7 of Fig.1. voltage switching loops. The "table with legs" switching loop is observed. Inset: Data of loop #5 as function of current I/Imax. (b) Experimental "table with legs" for the resistance switching loop of a manganite (LPCMO) sample as a function of current. (c) Exprimental "table with legs" for the resistance switching loop of a cuprate superconductor (YBCO) sample as a function of current.

Insights from the model simulations are used to propose a method to precisely control the resistance state. This is further applied in the implementation of a novel multi-level non-volatile memory cell. We shall present results for a working 6-bit (ie 54 resistance states) multi-level memory cell device [4].

[1] M.J. Rozenberg (2011), Scholarpedia, 6(4):11414.
[2] R. Waser et al, Adv. Mater. 2009, 21, 2632–2663.
[3] M.J. Rozenberg et al, Phys. Rev. B **81**, 115101 (2010).
[4] P. Stoliar et al, (submitted).

ELECTRIC FIELD INDUCED RESISTIVE SWITCHING IN A FAMILY OF MOTT INSULATORS: TOWARDS A MOTT-MEMRISTOR?

L. Cario[1], B. Corraze[1], V. Guiot[1], S. Salmon[1], J. Tranchant[1], M.-P. Besland[1], V. Ta Phuoc[2], M. Rozenberg[3], T. Cren[4], D. Roditchev[4], E. Janod[1]

[1] Institut de Matériaux Jean Rouxel (IMN), Université de Nantes, CNRS, 2 rue de la Houssinière, BP3229, 44322 Nantes, France – laurent.cario@cnrs-imn.fr,
[2] GREMAN CNRS UMR 7347, Université F. Rabelais, UFR Sciences - Parc de Grandmont 37200 Tours- France
[3] Laboratoire de Physique des Solides, UMR 8502, Université Paris Sud, Bât 510 91405 Orsay cedex, France
[4] Institut des Nanosciences de Paris (INSP), CNRS UMR 75-88, Université Paris 6 (UPMC), 4 place Jussieu, 75252 PARIS cedex 05, France

The AM_4Q_8 (A = Ga, Ge; M = V, Nb, Ta, Mo; Q = S, Se) compounds represent a new family of narrow gap Mott insulators with very interesting electronic properties. These compounds exhibit a lacunar spinel structure with tetrahedral transition metal clusters M_4 (figure 1(a)) [1]. Compare to most other inorganic Mott insulators, the AM_4Q_8 compounds show very small Mott-Hubbard gap (0.1-0.3 eV) as the electronic repulsion occurs on the scale of these tetrahedral clusters and not on the scale of single atoms [2]. A direct consequence of this low gap value is the high sensitivity of the AM_4Q_8 compounds to doping or pressure. Doping of the GaV_4S_8 leads, for example, to a filling control Metal Insulator Transition and to the emergence of a half ferromagnetic metal [3] or to Colossal Magnetoresistance [4]. On the other hand, application of external pressure induces a bandwidth control Metal Insulator Transition and superconductivity in $GaTa_4Se_8$ and $GaNb_4Se_8$ [2, 5].

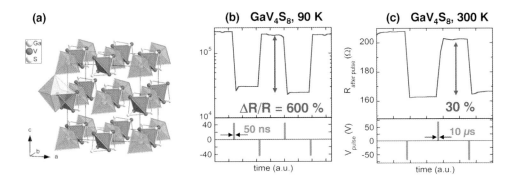

Figure 1: (a) representation of the clustered lacunar spinel structure of GaV_4S_8. Non volatile resistive switching observed on a crystal of GaV_4S_8 at (b) 90K and (c) 300K.

Recently we have discovered that the AM_4Q_8 compounds are sensible to electric field. The application of short electrical pulses on these compounds induces a new phenomenon of volatile or non volatile resistive switching [6-8]. The volatile transition appears above a threshold electric field of a few kV/cm, while the resistive switching becomes non-volatile for higher electric fields. The application of successive electric pulses enables to go back and forth between the high and low resistance states even at room temperature (Figure 1 (b) and (c)). This functionality makes the AM_4Q_8 compounds of interest for RRAM and Memristors applications.

All our results indicate that the resistive switching found in the AM_4Q_8 compounds does not match with any previously described mechanisms based on phase or chemical changes [6, 7, 9]. In contrast, this new resistive switching is related to the formation of an electronic phase separation at the nanometer scale with the creation of some metallic/superconducting domains (see Figure 2) [6, 7]. These results, and the very unusual electrostrictive phenomenon revealed by our STM study [10], highlight the similarity between applying a pressure and applying an electric field on the AM_4Q_8 compounds. The new resistive switching mechanism observed in the AM_4Q_8 compounds is not yet elucidated. However our work suggests that it might be related to an electro-mechanical coupling that leads to a Mott transition at the nanoscale.

Our recent work shows that it is possible to deposit a thin layer of GaV_4S_8 and to retrieve the reversible resistive switching on a metal-insulator-metal (MIM) device [11]. These results therefore lay the foundations of a new type of RRAM or Memristor non-volatile memory using as mechanism of resistive switching a Mott transition at the nanoscale [7].

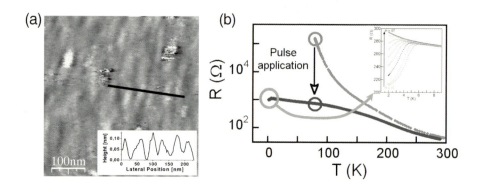

Figure 2: (a) STM and STS study of a $GaTa_4Se_8$ crystal revealing that the electric pulse has generated an electronic phase separation with the appearance of metallic and insulating regions. (b) Resistance of a $GaTa_4Se_8$ crystal measured before and after an electric pulse that generated a non volatile resistive switching. Around 6 K a transition to a granular superconducting state is observed which is reminiscent of the superconducting transition observed under pressure at the same temperature and with the same critical magnetic field (Hc_2) for this compound.

[1] H. Ben Yaich et al., J. Less-Common Met. 102, 9 (1984).
[2] R. Pocha et al., Journal of the American Chemical Society 127, 8732 (2005)
[3] E. Dorolti et al., J. Am. Chem. Soc. 132, 5704 (2010); C. Vaju et al., Chemistry of Materials 20, 2382 (2008)
[4] E. Janod et al. submitted (2012)
[5] M. M. Abd-Elmeguid et al., Physical Review Letters 93, 126403 (2004).
[6] C. Vaju et al., Advanced Materials 20, 2760 (2008).
[7] L. Cario et al., Advanced Materials 22, 5193 (2010).
[8] C. Vaju et al., Microelectronics Engineering 85, 2430 (2008)
[9] R. Waser et al., Advanced Materials 21, 2632 (2009); A. Sawa, Materials Today 11, 28 (2008)
[10] V. Dubost et al., Advanced Functional Materials 19, 2800 (2009)
[11] E. Souchier et al., physica status solidi (RRL) – Rapid Research Letters 5, 53 (2011)

INTRINSIC DEFECTS IN TiO$_2$ TO EXPLAIN RESISTIVE SWITCHING DEVICES

Dieter Schmeißer, Matthias Richter, Massimo Tallarida

BTU Cottbus, Angewandte Physik – Sensorik, Konrad Wachsmann-Allee 17, 03046 Cottbus, Germany.

We use resonant photoemission spectroscopy to study the valence and conduction band partial density of states in TiO$_2$ films and single crystals. We distinguish between covalent contributions of O2p and Ti3d4s states (Fig.1, upper panel) and localized defect states (Fig.1, lower panel). The latter are not related to oxygen vacancies but attributed to polaronic, oxygen based intrinsic defects. They form a continuous band which appears throughout the electronic band gap and which is identified by Fano type resonant behaviour of the valence electrons when passing the O1s resonance, see CIS(-4.3eV) in Figure 1.

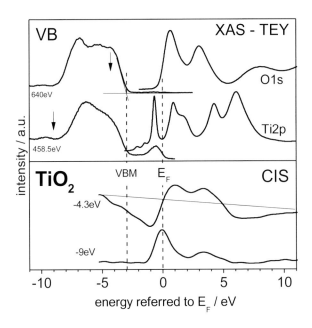

Figure 1: The combined partial density of states of the valence band and conduction band as determined from the XPS and XAS data, respectively. In the lower part we compare to the CIS data of the valence feature and the CIS spectrum at the onset of the Auger process at the O1s threshold as marked by arrows. The common energy is referred to E$_F$ by shifting the absorption spectra by the binding energy of the corresponding core levels. The positions of the VBM and of E$_F$ are indicated by dashed lines.

In addition we identify O2p-Ti3d4s charge transfer states. These are localized around the Fermi energy and are attributed to cause the n-type electronic behaviour. They show up as a peak in the CIS (-9eV) as shown in Fig.1. They are associated to a particular Auger decay at the resonance, which involves these localized O2p-Ti3d4s CT-states. Thereby, the states at the lowest oxygen resonance show a different Auger profile, which proceeds with a slope of 67.5° in a 2-dim plot of all valence band data vs. the excitation energies around the O1s threshold [1-3]. These findings contrast the common Auger process which is also evident in our data but can be observed only at photon energies above 535eV and proceeds under 45°. We explain such behaviour by a multiple Auger process, which causes a 3h final state. It is shown schematically in Figure 2. The starting point for this resonant mechanism is the existence of the localized O2p^5 state of the CT complex (O in Fig.2).

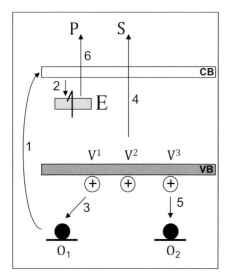

Figure 2: Schematic presentation of the 3h-Auger process. The first step is the excitation into the empty CB from the O1s level (1) which causes the creation of the primary core hole (1). This primary electron becomes trapped (O_1) into an empty localized state O and is stabilized by a Coulomb energy. Consequently its life time increases and while it is in the localized state, its primary core hole can be filled from any valence band state. This creates a pair of valence holes (V^1, V^2) by a common Auger decay (3+4). The three holes process now involves the primary electron in its localized state to become emitted as a participator Auger electron. This step creates the third valence hole V^3 (5). The electron from V^3 will fill an additional core (O_2) hole (available at resonant conditions) and transfers its energy to the primary electron to escape (6). The resulting final state may be denoted as ($O1s$-V^1-V^2-V^3).

In our model we describe the role of these two different intrinsic defect mechanisms to account for the resistive switching mechanism observed in titania based devices. In the normal (high resistivity) state the transport properties are based on the energy difference between the VBM and the CT-band (2eV). In the low resistivity state the band gap is filled by field assisted charged polarons. In this state the transport is by hopping conductivity and is enabled by the fluctuating charge density waves of the polaronic states. They also enable a field assisted reversible phase separation. The polaronic states are originally neutral and intrinsic states with a very short life time. However, an increase of their life time and dissociation into charged polaronic states can be achieved by both, the introduction of structural defects (preconditioning) and the application of external fields.

[1] M. Richter et al., BioNanoScience **2**, 59 (2012).
[2] M. Michling et al., IOP C. Ser. Mater. Sci. Eng., submitted (2011).
[3] S. Schmidt et al., Solid State Ionics, submitted (2011).

CONDUCTANCE QUANTIZATION IN RESISTIVE SWITCHING

Shibing Long[1,2], Carlo Gagli[3], Xavier Cartoixà[1], Riccardo Rurali[4], Enrique Miranda[1], David Jiménez[1], Julien Buckley[3], Ming Liu[2] and Jordi Suñé[1]

[1]Departament d'Enginyeria Electrònica, Universitat Autònoma, Bellaterra, Spain; [2]Lab of nanofabrication and Novel Device Integration, Institute of Microelectronics, Chinese Academy of Sciences, Beijing, China; [3]CEA-LETI, Grenoble, France; [4]Institut de Ciència de Materials de Barcelona, CSIC, Bellaterra, Spain.

Many different transition metal oxides show resistive switching properties which have been attributed to the formation and reversible partial rupture of a nanoscale conducting filament (CF) [1]. In this work we focus on the conduction properties of these CFs in Pt/HfO$_2$/Pt structures operated in the unipolar mode. In these samples, the RESET transition is progressive until the CF conductance decreases to a value close to the quantum of conductance, $G_o=2e^2/h$, and it finalizes with an abrupt conductance drop of several orders of magnitude. This final drop corresponds to the opening of a spatial gap (potential barrier) in the CF. Abrupt conductance transitions of the order of G_o between well-defined discrete states are found in the final stages of the RESET transient (Fig. 1). The behavior is completely analogous to that reported in metallic point contacts using mechanically controllable break junctions, STM contact-retraction experiments and the current induced local oxidation of nanoscale constrictions [2]. These results are evidence of conductance quantization or of structural changes at the atomic scale (i.e. single atom displacements from/to the CF). The temperature dependence is of metallic type for low-resistance CFs, while it is thermally activated when the CFs is in the high-resistance state. The transition from one regime to the other takes place for a CF conductance of the order of G_o. A compact phenomenological model for the CF conduction based on the Quantum Point Contact concept is also shown to nicely fit the current-voltage characteristics both in the low and high resistance states. All these experimental results suggest that a CF with conductance of the order of G_o is the natural boundary between the ON and OFF resistive switching states. For conductance above this limit, the CF supports at least one quasi-1D extended quantum state. Below this boundary, a spatial gap exists in the CF, the conduction takes place by tunneling or hopping and hence, it is strongly non-linear. We have also carried out first principles calculations of the transport properties of metal-HfO$_2$-metal structures where filaments composed of oxygen vacancies are present, in order to determine whether theory supports transport through vacancy paths. We have found that vacancy paths in both the monoclinic (crystalline) and amorphous phase of HfO$_2$ are capable of sustaining conductive channels in the HfO$_2$ energy gap region, though the gap is not completely closed at the path diameters that we have addressed. In particular, Fig. 2 shows the transmission coefficient of a metal/amorphous-HfO$_2$/metal structure for vacancy paths of different thicknesses. We can observe that the stoichiometric oxide still presents a transport gap even in its amorphous phase. As expected, transmission is more favored with increasing path width. Note that full transport calculations are needed in order to obtain transmission curves: the density of states (DOS) alone is not sufficient for qualitative predicting of the transmission properties, since it does not contain information about the spatial extent of the states (cf. inset in Fig. 2).

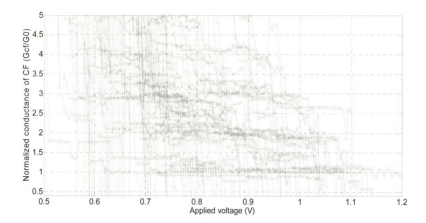

Fig. 1. Evolution of CF conductance in the last stage of 100 switching successive RESET cycles of Pt/HfO$_2$/Pt capacitors. Discrete levels at Go, 2Go are revealed. If plotted in a conductance histogram, peaks at 3Go and 4Go also become evident.

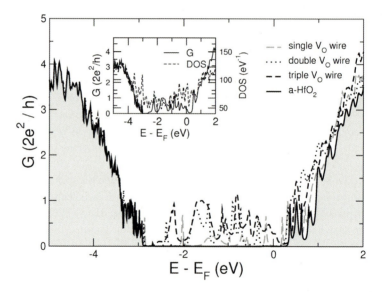

Fig. 2. Transmission coefficient for a metal/amorphous-HfO$_2$/metal structure for vacancy paths of different diameters. The inset overlaps the transmission curve and the density of states (DOS) for the structure with the widest vacancy path.

[1] R. Waser, R. Dittmann, G. Staikov and K. Szot, Adv. Mater., **21**, 2632 (2009).
[2] N. Agraït, A. Levy Yeyati and J.M. van Ruitenbeek, Phys. Rep. **377**, 81 (2003).

MAG 1

HIGHLY SPIN-POLARIZED CONDUCTING STATE AT THE INTERFACE BETWEEN NON-MAGNETIC BAND INSULATORS: $LaAlO_3/FeS_2$ (001)

J. D. Burton and E. Y. Tsymbal

University of Nebraska, Department of Physics and Astronomy and Nebraska Center for Materials and Nanoscience, Lincoln, NE, USA

Interface engineering of complex oxide heterostructures has led to the exploration of interfaces with properties and functionalities distinct from those typical for the respective bulk constituents. In the spirit of the well-known conducting $LaAlO_3/SrTiO_3$ interface [1] we study a similar interface with the added functionality of being unambiguously ferromagnetic. Our first-principles density functional calculations demonstrate that such a spin-polarized two-dimensional conducting state can be realized at the (001) interface between the two non-magnetic band insulators FeS_2 and $LaAlO_3$.[2] The (001) surface of FeS_2 (pyrite), a diamagnetic insulator, supports a localized surface state deriving from the Fe 3d-orbitals near the conduction band minimum. We find that, similar to the $LaAlO_3/SrTiO_3$ system, the deposition of a few unit cells of the polar perovskite oxide $LaAlO_3$ leads to electron transfer into these surface bands, thereby creating a conducting interface (see Fig. 1). The occupation of these narrow bands leads to an exchange splitting between the spin sub-bands, yielding a highly spin-polarized conducting state quite distinct from the rest of the non-magnetic, insulating bulk. We show that the ferromagnetism in the occupied surface bands is consistent with the Stoner model for itinerant magnetism. Such an interface supporting a nearly half-metallic, spin-polarized two-dimensional electron gas presents intriguing possibilities for spintronics applications.

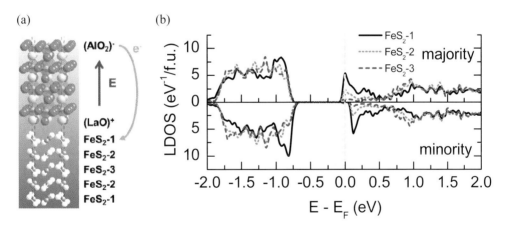

Figure 1: (a) Atomic structure of an $FeS_2/LaAlO_3$ interface consisting of five (001) atomic layers of FeS_2 and 4 u.c. of $LaAlO_3$. (b) Spin polarized local density of states (LDOS) projected onto layers FeS_2-1 through FeS_2-3. The vertical dashed line indicates the Fermi level, E_F.

[1] A. Ohtomo and H. Y. Hwang, *Nature* **427**, 423 (2004).
[2] J. D. Burton and E. Y. Tsymbal, *Phys. Rev. Lett.* **107**, 166601 (2011).

NEW INSIGHTS INTO NANOMAGNETISM BY SPIN-POLARIZED SCANNING TUNNELING MICROSCOPY AND SPECTROSCOPY

<u>Dirk Sander,</u> Hirofumi Oka, Safia Ouazi, Sebastian Wedekind, Guillemin Rodary, Pavel Ignatiev, Larissa Niebergall, Valeri Stepanyuk, and Jürgen Kirschner

Max-Planck-Institute of Microstructure Physics, Weinberg 2, D-06120 Halle, Germany

The study of the magnetic properties of *individual* nm small nanostructures requires dedicated experimental techniques, which provide the required spatial resolution and sensitivity on the nm scale. Spin-polarized scanning tunnelling microscopy (SP-STM) is a method of choice to image and to characterize spin-dependent electronic properties of nanostructures with unsurpassed spatial resolution on the atomic scale. Its working principle exploits the dependence of the tunnel current on the relative magnetization orientation of a sample and the magnetic STM tip.

We present results by SP-STM, where we investigate the correlation between structural, electronic, and magnetic properties of individual nm small, two atomic layers thin Co islands with several hundred to thousands of atoms. These islands are formed upon Co deposition on Cu(111) at 300 K, and they are predominantly of triangular shape.

We use external magnetic fields of up to 7 T with a liquid He cooled SP-STM working at 8 K to tune the magnetic state of both tip and sample. The magnetic field is normal to the sample surface. It allows a switching of the relative magnetization orientation between tip and Co islands from anti-parallel to parallel. This leads to corresponding changes of the transport properties of the tunnel junction between tip and sample. From this we extract magnetic switching fields of individual Co islands, which reach 2.5 T at 8 K [1]. Our quantitative analysis reveals a crossover of the magnetization reversal with increasing island size from an exchange-spring behaviour with vanishing magnetic anisotropy of rim atoms to reversal by domain wall nucleation and growth for larger islands at an Co islands size of around 5500 atoms (base length of islands: 12 nm) [1].

We exploit the high spatial resolution of the STM to obtain maps of the tunnel current and of the differential conductance on single Co islands. We investigate the variation of these properties upon a change of the magnetic state of the system. We derive maps of the tunnel magneto resistance ratio (TMR) of the Co islands [2]. These maps reveal a pronounced spatial modulation of the TMR ratio of up to 20% on a nm scale. In conjunction with density functional theory we ascribe this to a spatially modulated spin-polarization, which is induced by spin-dependent electron confinement within the nanoscale Co islands [3].

[1] Ouazi, Wedekind, Rodary, Oka, Sander, Kirschner, Phys. Rev. Lett. **108** (2012) 107206.
[2] Oka, Tao, Wedekind, Rodary, Stepanyuk, Sander, Kirschner, Phys. Rev. Lett **107** (2011) 187201.
[3] Oka, Ignatiev, Wedekind, Rodary, Niebergall, Stepanyuk, Sander, Kirschner, Science **327** (2010) 843.

SELECTIVE ORBITAL OCCUPATION AT MANGANITE INTERFACES INDUCED BY CRYSTAL SYMMETRY BREAKING

B. Martínez[1], S. Valencia[2], L. Peña[1], Z. Konstantinovic[1], Ll. Balcells[1], R. Galceran[1], D. Schmitz[2], F. Sandiumenge[1], M. Casanove[3]

[1]Instituto de Ciencia de Materiales de Barcelona – CSIC. Campus UAB, Bellaterra 08193. SPAIN
[2]Helmholtz-Zentrum-Berlin für Materialien und Energie, Albert-Einstein Str. 15 D-12489 Berlin, Germany
[3]CNRS - CEMES (Centre d´Elaboration de Matériaux et d´Etudes Structurales) ; BP 94347, 29 rue Jeanne Marvig, F- 31055 Toulouse, France.

Complex oxides have revealed to be one of the most interesting classes of materials due to their amazing variety of properties of strong theoretical and technological interest, including superconductivity, ferromagnetism, ferroelectricity, piezoelectricity and more. In addition, the recent discovery of unexpected properties, going from the formation of a two-dimensional (2D) electron gas[1,2], at the interface between two insulators, to the appearance of interfacial ferromagnetism adjoining two nonmagnetic oxides[3], have further boosted the interest in these materials. In this work interfacial effects between $La_{2/3}Sr_{1/3}MnO_3$ (LSMO) thin films and different capping layers are analyzed by using x-ray absorption techniques, x-ray diffraction and transport measurements. LSMO samples used in this work were prepared by using radio frequency (RF) magnetron sputtering on top of (001)-oriented STO substrates from stoichiometric ceramic targets. The thickness of the samples was about 40nm and that of the capping layers was t_c ~1.6 ±0.2 nm) and it was determined by controlling the evaporation time after a careful calibration of the growth rate of each of the different materials used. High resolution transmission electron microscopy (HRTEM) imaging confirmed nominal values. We make use of x-ray linear dichroism (XLD) to show that, independently of the capping layer, LSMO films exhibit a preferential occupation of Mn 3d $3z^2$-r^2 e_g orbitals at the interface.

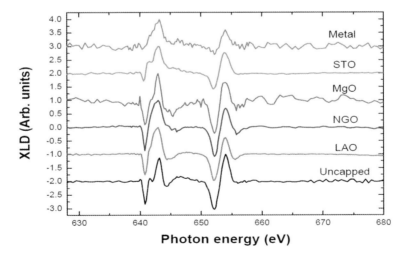

Figure 1: XLD spectra obtained at $L_{3,2}$-edge for LSMO samples with different capping layers. The XLD has been normalized to its maximum value and an offset has been artificially applied for better comparison. All samples, independently on the capping show XLD spectra alike to that of the uncapped sample. The XLD spectroscopic shape reflects the presence of an interfacial phase with preferential occupation of the $3z^2$-r^2 eg orbital.

The presence and extension of such a phase does not depend on the capping layer material. It is neither related to the previously observed degradation of the Mn oxidation state nor to interface magnetic properties. Most remarkably, the XLD amplitude on uncapped films with bulk-like Mn oxidation state at the interface is similar to that observed in capped films. This fact allows identifying the symmetry breaking at the LSMO surface as the major factor behind the observed depressed magnetotransport properties in manganite-based tunnel junctions. Transport measurements across LSMO/capping layer interfaces allow estimating the scale length of this effect showing that the disruption of the double exchange ferromagnetic (DE-FM) phase occurs only by the interface, in fact around 2 to 4 unit cells. Our results highlight the role played by interfacial symmetry over strain and/or chemical segregation on the tunneling process and allow envisaging the correct strategy to obtain manganite-based tunnel junctions working to temperatures close and above room temperature.

[1] A. Ohtomo and H. Hwang, Nature (London) 427, 423 (2004).
[2] M. Huijben, G. Rijnders, D. H. A. Blank, S. Bals, S. Van Aert, J. Verbeeck, G. Van Tendeloo, A. Brinkman and H. Hilgenkamp, Nat. Mater. 5, 556 (2006).
[3] J. Garcia-Barriocanal, J. C. Cezar, F. Y. Bruno, P. Thakur, N. B. Brookes, C. Utfeld, A. Rivera-Calzada, S. R. Giblin, J. W. Taylor, J. A. Duffy, S. B. Dugdale, T. Nakamura, K. Kodama, C. Leon, S. Okamoto and J. Santamaria, Nature Communications 1, 1 (2010).

SCALABLE EXCHANGE BIAS IN LSMO/STO THIN FILMS

Daniel Schumacher[1], Alexandra Steffen[1], Jörg Voigt[1], Jürgen Schubert[1], Hailemariam Ambaye[2], Valeria Lauter[2], John Freeland[2], Thomas Brückel[1]

[1]Jülich Center for Neutron Science JCNS and Peter Grünberg Institut PGI, JARA-FIT, Forschungszentrum Jülich GmbH, 52425 Jülich, Germany;
[2]Spallation Neutron Source, Oak Ridge National Laboratory, Oak Ridge, TN 37831, USA;
[3]Advanced Photon Source, Argonne National Laboratory, Argonne, IL 60439, USA

Even though the Exchange Bias (EB) effect is known for more than 50 years [1], there is still no comprehensive understanding of all different occurring variations. But in general there is one requirement for the presence of the EB: the sample needs to contain an antiferromagnetic (AFM) region which is in contact to a region having a net magnetic moment (ferro- or ferrimagnet). It is therefore very surprising that an EB effect has been reported by Zhu et al. [2] in multilayers of ferromagnetic (FM) $La_{0.66}Sr_{0.33}MnO_3$ (LSMO) and nonmagnetic $SrTiO_3$ (STO). In order to understand the unusual occurrence of the effect, it was our aim to analyse the structural and magnetic depth profile by Polarized Neutron Reflectometry (PNR) and X-ray Resonant Magnetic Scattering (XRMS). This is not only of importance from the scientific point of view to reveal another unusual facet of the EB effect. In addition to the numerous already realized technological applications of the EB effect (e.g. MRAM), the usage in oxide thin film systems opens up many new possibilities of combining the EB effect with the properties of other strongly correlated oxides. As an example, switching between the two distinct EB states by reversing the electric polarization of $BiFeO_3$ has been reported in $BiFeO_3/La_{0.66}Sr_{0.33}MnO_3$ thin films [3], showing the importance of the EB effect for possible device applications and the necessity of a better understanding of the effect.

We have prepared epitaxial LSMO single and LSMO/STO bilayers on (001) oriented single-crystalline STO substrates by both High Oxygen Pressure Sputter Deposition (HSD) and Pulsed Laser Deposition. It was possible to reproduce the Exchange Bias effect in the samples grown by HSD by reducing the oxygen pressure during the layer growth. In fact, the size of the effect can be increased by further reduction of the oxygen pressure. The macroscopic sample analysis by X-ray Diffraction and Vibrating Sample Magnetometry suggests that the occurrence of the Exchange Bias effect is linked to oxygen deficiencies in the LSMO layer. By combining X-ray Reflectometry (XRR), PNR and XRMS, the magnetic depth profile of the samples has been determined. By this, a region in LSMO at the interface to STO has been detected, where the magnetic moment is strongly suppressed (figure 1). By putting together the results of the macroscopic sample analysis and the scattering experiments, an explanation for the occurrence of the effect can be given: It is proposed, that a combination of strain and oxygen deficiencies shifts the LSMO at the interface in the AFM phase of the LSMO strain vs. doping phase diagram. This interface region thus couples to the FM part of the LSMO causing the Exchange Bias effect. In this case, the EB effect is purely due to the LSMO layer and the STO substrate is only needed to mediate the strain. By such a system one is not limited to having to chose an antiferromagnet on the one side and a ferromagnet on the other side of the chemical interface to create an EB effect. This opens up the opportunity of choosing the second material independently as long as it sets the right strain to the magnetic layer. As in our case, one layer need not to be magnetic at all, but could have other properties, which possibly create completely new and more versatile functionalities.

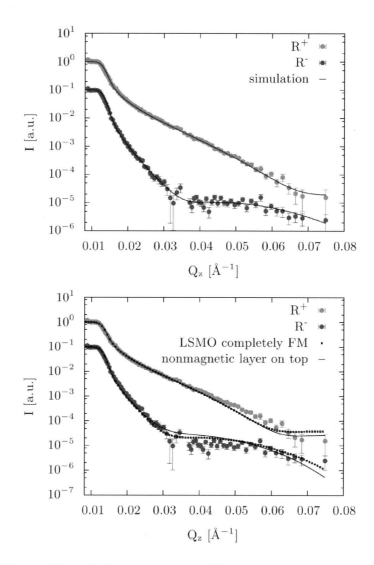

Figure 1: PNR data at 5K of a LSMO single layer grown by HSD at 0.8mbar. Since the sample has been saturated in a field of 1T, spin-flip scattering was not expected. Therefore, in order to obtain higher detected intensities, no polarization analysis was performed. The reflected intensities of a polarized neutron beam with moments aligned parallel (R^+) and antiparallel (R^-) to the applied field are shown. The R^- data is divided by 10 for a better visibility. Due to the small contrast in the nuclear scattering length densities of LSMO and STO, the oscillations are purely due to the magnetic scattering length density in the LSMO layer. This allows for an accurate determination of the thickness of the FM ordered part of the LSMO layer. The best fit (top picture) yields a FM layer thickness of 95Å, which is about 10Å smaller than the thickness determined by XRR. The second plot shows the best fit for a completely FM ordered LSMO layer. In addition, the macroscopically non-magnetic layer needs to be at the interface to STO and not on top of the LSMO layer, which also results in a significantly different simulation. Additionally, the XRMS data also revealed that the thickness of the region with no or at least drastically reduced magnetic moment increases with increasing EB effect.

[1] S. M. Wu et al., Nat. Mater. **9**, 756 (2010).
[2] W. H. Meiklejohn and C. P. Bean, Phys. Rev. **102**, 1413 (1956).
[3] S. M. Wu et al., Nat. Mater. **9**, 756 (2010).

STRUCTURAL AND MAGNETIC PROPERTIES OF NANOPARTICLE SUPERLATTICES

O. Petracic[1], D. Greving[2], D. Mishra[2], M. J. Benitez[2], P. Szary[2], G. Badini Confalonieri[3], A. Ludwig[2], M. Ewerlin[2], L. Agudo[4], G. Eggeler[4], B.P. Toperverg[2], and H. Zabel[2]

[1] Jülich Centre for Neutron Science JCNS and Peter Grünberg Institute PGI, Forschungszentrum Jülich GmbH, 52425 Jülich, Germany
[2] Institut für Experimentalphysik/Festkörperphysik, Ruhr-Universität Bochum, 44780 Bochum, Germany
[3] Instituto de Ciencia de Materiales de Madrid, CSIC, 28049 Madrid, Spain
[4] Institut für Werkstoffe, Ruhr-Universität Bochum, 44780 Bochum, Germany

Nanoparticle superlattices can be considered as novel type of materials with controllable electronic, optical, magnetic or mechanical properties. Their building blocks are nanoparticles (or 'nanocrystals') from a metallic, metal-oxide, or semiconducting material or a hybrid between different materials. Moreover, magnetic and electronic interactions between the particles can be strongly modified by choosing different matrix materials. I will report about the structural and magnetic properties of two types of systems: (a) chemically prepared iron oxide nanoparticles and (b) metal cobalt nanoparticles prepared by sputtering. In the first case we fabricate regular 2- or 3-dimensional colloidal 'superlattices' of iron oxide nanoparticles. Depending on the self-assembly technique we can fabricate 'superlattice films' showing all three growth modes known from thin films, i.e. Stranski-Krastanov, Frank-van-der-Merwe or Volmer-Weber growth. Depending on the thermal treatment we can tune the internal structure and composition of the particles and hence the overall behavior of the superlattice [1]. The lateral ordering is quantified using scanning electron microscopy (SEM) and grazing incidence small angle X-ray scattering (GISAXS) [4]. The magnetic behavior and correlations are investigated by superconducting quantum interference device (SQUID) magnetometry and polarized neutron reflectometry (PNR). We also demonstrate the combination of self-assembly and templating via electron beam lithography or using electrochemically etched alumina pore membranes [2]. In the second case, metallic Co nanoparticles are formed spontaneously by extreme Volmer-Weber growth during deposition of metals on an alumina buffer layer. The magnetic interaction and the transport properties between the nanoparticles can significantly be altered by deposition of a Pt [3] or Cr layer on top of the particles.

[1] M. J. Benitez, D. Mishra, P. Szary, G.A. Badini Confalonieri, M. Feyen, A. H. Lu, L. Agudo, G. Eggeler, O. Petracic, and H. Zabel, J. Phys.: Condens. Matter **23**, 126003 (2011).
[2] G. A. Badini Confalonieri, V. Vega, A. Ebbing, D. Mishra, P. Szary, V. M. Prida, O. Petracic, and H. Zabel, Nanotechnology **22**, 285608 (2011).
[3] A. Ebbing, O. Hellwig, L. Agudo, G. Eggeler, and O. Petracic, Phys. Rev. B. **84**, 012405 (2011).
[4] D. Mishra, M. J. Benitez, O. Petracic, G. A. Badini Confalonieri, P. Szary, F. Brüssing, K. Theis-Bröhl, A. Devishvili, A. Vorobiev, O. Konovalov, M. Paulus, C. Sternemann, B. P. Toperverg, and H. Zabel, Nanotechnology **23**, 055707 (2012).

NANOPARTICLES OF ANTIFERROMAGNETIC AND FERRIMAGNETIC OXIDES AS MAGNETIC HETEROSTRUCTURES

Veronica Salgueiriño,[1] Nerio Fontaíña-Troitiño,[1] Ruth Otero-Lorenzo,[1] Sara Liébaña-Viñas[1]

[1]Departamento de Física Aplicada, Universidade de Vigo, 36310, Vigo, Spain

Interfaces between different oxides in hybrid systems offer the perfect environment for manipulating the complex interplay between the electronic and lattice degrees of freedom and drawing out new functionalities by exploiting epitaxial strain, local symmetry breaking, frustration or charge transfer between the materials. Magnetic interfaces are highly relevant for technological applications and in most of them, exchange bias plays a key role.[1]

We intend to exploit oxide interfaces established in composite nanostructures synthesized by colloidal chemistry methods. The great advantage of using different types of inorganic nanostructures as building blocks comes from the fact that permits the design and fabrication of colloidal and supracolloidal assemblies knowing first their magnetic characteristics. As a proof of concept we have developed mixed systems, driving on the surface of AFM substrates (goethite nanorods or cobalt oxide octahedrons), cobalt ferrite nanoparticles or magnetite shells (the study of bimagnetic systems opens new degrees of freedom to tailor the overall properties and offers the Meiklejohn-Bean paradigm).[2],[3] Opposite structures driving the antiferromagnetic material on a ferrimagnetic substrates is also possible to attain.[4].

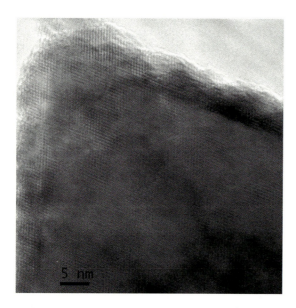

Figure 1: HRTEM image of an octahedron of antiferromagnetic CoO coated with a very thin outer shell of ferrimagnetic Fe_3O_4, epitaxially grown by means a colloidal chemistry method of thermal decomposition of precursors [3].

[1] M. Gibert, P. Zubko, R. Scherwitzl, J. Iñiguez, J. M. Triscone, Nature Materials, **11**, 195 (2012).
[2] R. Mariño-Fernández, S. H. Matsunaga, N. Fontaíña-Troitiño, M. P. Morales, J. Rivas, V. Salgueiriño, J. Phys. Chem. C **115**, 13991 (2011).
[3] Nerio Fontaíña-Troitiño, Beatriz Rivas, V. Salgueiriño, in preparation (2012).
[4] Sara Liébana-Viñas, R. Otero-Lorenzo, V. Salgueiriño, in preparation (2012).

SELF ASSEMBLED IRON OXIDE NANOPARTICLES – FROM A 2D POWDER TO A SINGLE MESOCRYSTAL

Elisabeth Josten[1], Erik Wetterskog[2], Doris Meertens[3], Ulrich Rücker[1], German Salazar-Alvarez[2], Oliver Seeck[4], Peter Boesecke[5], Tobias Schulli[5,] Manuel Angst[1], Raphael Hermann[1], Lennart Bergström[2], Thomas Brückel[1]

[1]JCNS-2 and PGI-4, Forschungszentrum Jülich, Jülich, Germany;
[2]Stockholm Universitet, Department of Materials and Environmental Chemistry, Stockholm, Sweden;
[3]ER-C and PGI-5, Forschungszentrum Jülich, Jülich, Germany;
[4]Desy, Hamburg, Germany;
[5]European Synchrotron Radiation Facility, Grenoble, France

Fundamental research on magnetic nanostructures is an important part of today's scientific effort in information technology [1]. Self assembled structures of nanoparticles are candidates for a new generation of magnetic storage media [2] and other applications include e.g. magnetic field sensors with superparamagnetic particle clusters to reach high sensitivities on extremely small sensing volumes. Understanding and optimizing the structures, at all length scales, of nanocrystal arrays or superlattices is an important step towards controlled design of novel devices.

The investigated nanoparticles are monodisperse gamma-Fe_2O_3 nanocubes and nanospheres with a diameter ≈10 nm. The nanoparticles themselves have been characterized for their well-defined shape, good monodispersity (5-6%) and crystalline structure [3]. The particles have been deposited on a substrate to form highly ordered superstructures using a drop casting method. In a first step, structural characterization has been carried out by SEM, AFM, TEM and GISAXS. Depending on the shape of the particles, they order in either bcc [4] or fcc with long correlation length of about 2-10μm.

In the presented work, nanoparticle assemblies have been investigated in-situ during the self-assembling process using an optimized GISAXS setup to investigate the kinetic during deposition. In addition a new approach, diffraction on a single isolated mesocrystal with a microfocused x-ray beam, has been applied to extract the order beyond the 2D powder average present in normal GISAXS studies of such assemblies.

The time dependent GISAXS study of the self-assembling kinetics, carried out at the ID01 beamline at ESRF, resulted in an understanding how the structures evolve with time and how the evaluation can be controlled with external parameters. The in-situ cell, which was developed to monitor the structure as well as the height and shape of the droplet, was employed for additional control of the process parameters. We observed that the nucleation of the mesocrystal always started after the droplet of nanoparticle solution was already collapsed (height<10μm), and that the organization takes place near the substrate (in contrast to earlier publications [5,6,7]). Within the time resolution of the experiment, several seconds in an experiment of many hours duration, even the just-nucleated mesocrystals were already long range ordered. An example of the GISAXS patterns collected is shown in Fig 1.

Another aspect was to understand the exact structure formed with the already established deposition methods under well-defined parameters. Due to arbitrary orientation of the mesocrystals on the substrate, the grazing incidence scattering experiments cannot yield the full supercrystal structure information, but only a 2D powder average. For this study we have used a completely new approach performing a small angle Bragg diffraction experiment on a single isolated mesocrystal (see Fig 2), detached from a sample using focused ion beam preparation techniques. This analysis gave us the structure information of a single mesocrystal. This challenging experiment proved the feasibility on the investigation on a single, small mesocrystal of nanoparticles and opens a new field for further investigations of the mesocrystal structure.

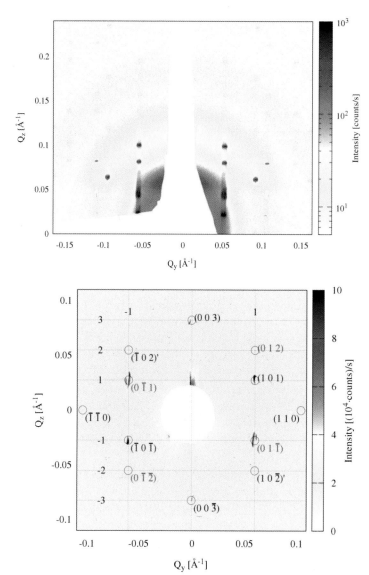

Fig. 1 Scattering from an ensemble of mesocrystal a few seconds after the nucleation started.

Fig. 2 Scattering from a single mesocrystal - 500nm height, 2000nm diameter. The labels indicate the index of the structure.

[1] A. Moser et al., J.Phys.. **35**, R157-R167 (2002).
[2] F.M. Fowkes et al., Colloids and Surf. **29**, 243 (1988).
[3] S. Disch , Phd thesis, RWTH Aachen (2010).
[4] S. Disch et al., Nano Lett. **11**, 1651 (2011).
[5] Z. Jiang et al., Nano Lett. **10**, 799 (2010).
[6] M. J. Campolongo et al., ACSNano. **5**, 7978 (2011).
[7] K. Bian et al., ACSNano.**5**, 2815 (2011).

CHARACTERIZATION OF VACANCY-RELATED DEFECTS IN Fe-DOPED SrTiO$_3$ THIN FILMS USING POSITRON ANNIHILATION LIFETIME SPECTROSCOPY

D. J. Keeble[1], S. Wicklein[2], G.S. Kanda[1], W. Egger[3], and R. Dittmann[2]

[1]University of Dundee, Dundee, Scotland; [2]Institut für Festköperforschung, Forschungszentrum Jülich, 52425 Jülich, Germany; [3]Universität Bundeswehr München, D-85577 Neubiberg, Germany.

Controlled doping of perovskite oxide thin films will provide a key technology for the further development of oxide electronics. Thin films are normally deposited far from equilibrium; this hinders the detailed understanding of the growth mechanisms which, in turn, influences point defect formation and dopant atom incorporation. Advanced characterization methods capable of detecting and identifying point defects, and the local environment of incorporated dopants are required. Conventional electron or ion beam methods typically have sensitivities in the 0.1 - 10 at. % range, and often provide only limited local structure information. Local probe spectroscopies, such as positron annihilation and electron magnetic resonance methods are orders of magnitude more sensitive and can provide direct information on the local environment, but are restricted in the types of defects and impurities that can be detected. Variable energy positron annihilation lifetime spectroscopy performed using a high intensity positron beam can identify cation vacancy defects in perovskite oxides, and provides depth profiling [1]. In this work VE-PALS measurements are reported on a series of Fe-doped SrTiO$_3$ thin films grown by pulsed laser deposition (PLD). Fe-doped SrTiO$_3$ thin films provide an insightful model system for the study of resistive switching memory device mechanisms [2].

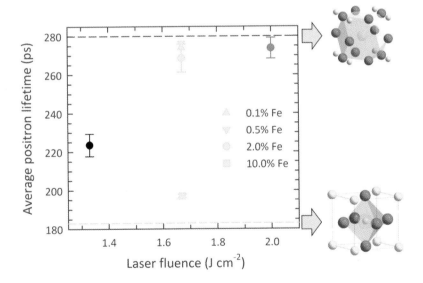

Figure 1: The average positron lifetime, τ_{ave}, determined from the thin film is shown for samples grown with different laser fluences, and as a function of Fe concentration. τ_{ave}, is weighted average of individual positron state lifetime components. The lifetime values for the Ti and the Sr vacancies are shown as dashed lines at 183 ps and 280 ps, respectively. For example, if only V_{Ti} defects were present then $\tau_{ave} \leq 183$ ps. The positron lifetime of a defect increases with increasing open volume size.

A series of 200 nm Fe-doped SrTiO$_3$ thin films were grown by PLD using a laser fluence of 1.67 Jcm^{-2} with [Fe] of 0.1, 0.5, 2.0, and 10.0 at. %, additional 2.0 % films were grown using fluencies of 1.33 and 2.0 Jcm^{-2}. Variable energy positron lifetime measurements were performed at the high intensity NEPOMUC beamline at the Munich research reactor FRMII.

The behavior of the average positron lifetime, obtained from the bulk region of each film, is shown in Fig. 1. While the high number of counts collecting in each spectrum allows de-convolution into individual positron states insight is already available from τ_{ave}. If the films contained positron trapping vacancy-related defects at concentrations less than ~ 0.1 ppm all positrons would annihilate from essentially perfect lattice and a single lifetime of ~155 ps is predicted. If only V_{Ti} defects where present, τ_{ave} would increase with increasing [V_{Ti}] until saturation trapping occurs where every positron is trapped by a V_{Ti}, at this point τ_{ave} will equal the lifetime for the V_{Ti} localized state at ~183 ps. If a τ_{ave} greater than this value is observed, it can be concluded that vacancy defects with an open volume size larger than V_{Ti} must be present, for example Sr vacancies which are characterized by a trapped state lifetime of 280 ps [1].

From Fig. 1 it can be concluded that all the films studied contained positron trapping vacancy-related defects and that vacancies larger that Ti vacancies must also be present. The observed τ_{ave} values lie between the lifetime values for V_{Ti} at ~183 ps and V_{Sr} at 280 ps. The results of deconvolution fitting of individual lifetime spectra provide evidence that the dominant positron trapping defects were V_{Ti} and V_{Sr}, but also showed in some films a low concentration of larger vacancy cluster defects. In the [Fe] = 10 % film trapping is dominated by V_{Ti} defects, while the other 1.67 Jcm^{-2} films and the 2.0 Jcm^{-2} film trapping is dominated by Sr vacancies. Increased trapping to B-site vacancies was observed either for a high concentration of Fe or when using a lower fluence for the growth.

[1] D. J. Keeble et al., Phys. Rev. Lett. **105**, 226102 (2010).
[2] R. Muenstermann et al., Adv. Mater. **22**, 4819 (2010).

AB INITIO CALCULATIONS OF DEFECTS IN GALLIUM OXIDE

T. Zacherle[1], P.C. Schmidt[2], M. Martin[1]

[1]Institute of Physical Chemistry, SFB 917, RWTH Aachen University, Germany;
[2]Institute of Physical Chemistry, Technical University of Darmstadt, Germany

Gallium oxide, β-Ga_2O_3, is a wide-band-gap (4.9 eV) semiconductor and has raised considerable interest due to its possible application as TCO-material [1]. It is already used as gas-sensor with its characteristic p_{O2}-dependence of the conductivity of $\sigma \sim p_{O2}^{-1/4}$ [2]. Nevertheless, the defect structure of the system, which is essential for understanding its electrical and optical properties, is still controversial.

In this study we perform *ab initio* calculations (using the VASP-code [3]) to address the defect structure of β-Ga_2O_3. We consider all intrinsic defects like oxygen and gallium vacancies and oxygen and gallium interstitials. In a first step we calculate defect formation energies [4] with the GGA-functional for different super-cell sizes. We correct our energies for the spurious image-charge interactions with a method proposed by Freysoldt *et al.* [5] and then extrapolate to infinite dilution. The remaining elastic scaling is proportional to $1/L_{sc}^3$, where L_{sc} is the average size of the supercell, i.e. the defect-separation distance (see Figure 1).

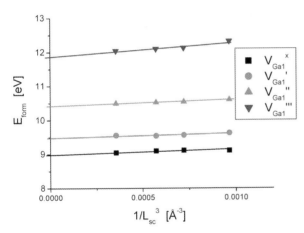

Figure 1: Elastic scaling of the defect formation energies of differently charged gallium vacancies V_{Ga1}. The image-charge corrected formation energies are plotted against $1/L_{sc}^3$. The extrapolated energy is the intersection of the fitting lines with the energy-axis. The plot shows the energies of all the different charge states for supercell sizes with 80, 120, 160 and 240 atoms.

In addition, we perform hybrid-functional-calculations (HSE06) for a smaller cell in order to correct for the well-known band gap error of the GGA-functional. We then apply the found GGA-scaling of the elastic effects to our HSE06-energies, resulting in fully band-gap and finite-size corrected defect formation energies.

In our analysis we focus on quantities, which are directly comparable to experiment, trying to bridge the gap between theoretical investigations and experimental data. Therefore Schottky-, Frenkel- and Anti-Frenkel-energies are presented. Defect concentrations are also calculated. Comparing our theoretical results to experiment we are able to propose a defect-model of effective donor-doping, because the purely intrinsic defect concentrations are very low and already small amounts of unintentional dopants dominate the defect-structure. We also find that entropic effects, i.e. formation entropies due to the change of vibrational frequencies upon introduction of a defect into the system, are very important to account for the experimental behavior. We calculate these contributions for the dominating defects. Figure 2 shows the calculated concentrations for $T = 1273$ K. We also compute migration barriers for the dominant defect species in our system using the NEB-method.

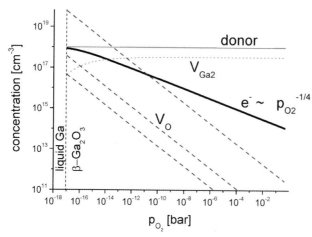

Figure 2: Calculated defect-concentrations for gallium oxide at a temperature of $T = 1273$ K. The concentrations are calculated numerically for each p_{O2} using the charge-neutrality condition. The plot shows that differently charged oxygen vacancies V_O (dashed lines) are only important at low p_{O2}. The donor is compensated over nearly the whole p_{O2}-range by gallium vacancies V_{Ga2}''' (dotted line) turning the electrons to minority charge carriers with a p_{O2}-dependence $c_e \sim p_{O2}^{-1/4}$, as observed experimentally.

[1] J. B. Varley, J. R. Weber, A. Janotti, C. G. Van de Walle, Appl. Phys. Lett. **97**, 142106 (2010).
[2] M. Fleischer, H. Meixner, Sensors Actuat. B- Chem **4**, 437 (1991).
[3] G. Kresse, J. Furthmüller, Phys. Rev. B **54**, 11169 (1996).
[4] C. G. Van de Walle, J. Neugebauer, J. Appl. Phys. **95**, 3851 (2004).
[5] C. Freysoldt, J. Neugebauer, C. G. Van de Walle, Phys. Rev. Lett. **102**, 016402 (2009).

CATION DEFECT ENGINEERING IN STO THIN FILMS BY PLD - VERIFICATION AND IMPLICATIONS ON MEMRISTIVE PROPERTIES

S. Wicklein[1], C. Xu[1], A. Sambri[2], S. Amoruso[2], D.J. Keeble[3], R.A. Mackie[3], W. Egger[4], R. Dittmann[1]

[1]Forschungszentrum Jülich, Peter Grünberg Institut, Germany;
[2]Università degli Studi di Napoli Federico II, Dipartimento di Scienze Fisiche & CNR-SPIN, I-80125 Napoli, Italy;
[3]University of Dundee, School of Engineering, Physics and Mathematics, Dundee DD1 4HN, Scotland;
[4]University Bundeswehr, D-85577 Munich

The realm of electronic properties of a complex oxide is defined by its defect structure. Therefore the control of defects in an oxide is key to influence its electronic properties. Homoepitaxial $SrTiO_3$ (STO) thin films are a suitable oxide model system to investigate the effect of defects on the electrical properties. Ohnishi et.al [1] showed that the variation of the laser energy leads to non-stoichiometric STO thin films which exhibit an elongation of the c-axis parameter.

Therefore, homoepitaxial STO thin films were investigated by positron annihilation lifetime spectroscopy (PALS) to investigate the origin of the c-axis expansion (Fig 1a) and the defect composition of various STO thin films. Furthermore, the cause of the shift in thin film stoichiometry was investigated by OES of the PLD plasma plume.

PALS measurements in general reveal that both types of cation vacancies V_{Ti} and V_{Sr} are present in all STO thin films regardless of its stoichiometry. Extreme non-stoichiometry leads to an extended defect structure like nano-voids deduced by the increase of the average positron lifetime and displayed by TEM. Specifically, PALS also showed that the variation of the pulsed laser energy shifts the ratio of the cation vacancy concentration in the STO thin film (Fig 1b). For stoichiometric films the ratio of $[V_{Ti}] / [V_{Sr}] = 1$. It was possible to detect that non-stoichiometric thin films e.g. A-site rich films show the incorporation of B-site vacancies and vice versa. In essence, the excess of one type of cation vacancy causes the STO unit cell to expand.

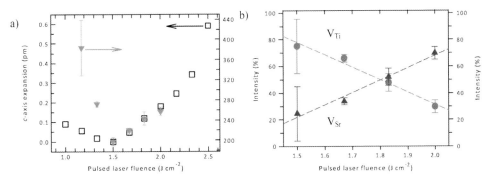

Figure: 1a) c-lattice parameter and defect concentration; b) V_{Sr} & V_{Ti} intensity signal shift with varying laser energy. Both graphs retrieved from [3]

Moreover, the laser energy for a stoichiometric film shifts when the background pressure or the target substrate distance is varied. That indicates a time of flight effect of the plume species causing a non-stoichiometric film.

Therefore, OES and fast imaging investigations of the plume propagation, was utilized to probe for a time of flight (ToF) pattern for various species in the plume as well as their kinetic energy.

The OES data of the plume in its early stage (2400ns) shows evidence that Sr is preferentially ablated and a change in laser energy results in a shift of the plume composition (Fig. 2a right). The Ti content increases with increasing laser energy as reported before by Dam et al. [2]. The initial plume composition can then be further disturbed by the interaction of the species with the background gas where species are oxidized and scattered during the plume propagation. The ToF data indicate a preferred scattering of Ti. For high pressures the loss of lighter Ti-species can be compensated by an increase of the laser energy. For lower background pressures <5e^{-2}mbar the scattering of species becomes less pronounced, however, the kinetic energy of the species increases (Fig 2a left). This energy if sufficient to back-sputter the growing STO film, whereas Sr is preferentially sputtered due to its lower binding energy causing the incorporation of V_{Sr} into the STO thin film.

The effect of the different defects on the electronic properties and the memristive properties respectively, were also investigated. Ti-rich or Sr-rich STO films show different current-voltage (I(V)) characteristics. The ratio of cation vacancies influences the ON/OFF ratio, the reliability of the switching effect and the switching voltage. Sr-rich and stoichiometric films exhibit in general a more stable and pronounced switching mechanism compared to Ti-rich films (Fig. 2b).

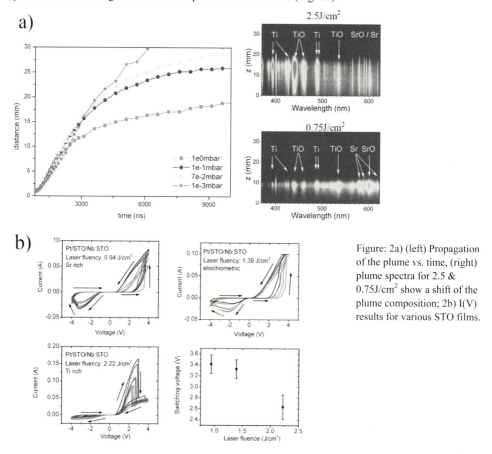

Figure: 2a) (left) Propagation of the plume vs. time, (right) plume spectra for 2.5 & 0.75J/cm^2 show a shift of the plume composition; 2b) I(V) results for various STO films.

[1] T. Ohnishi, K. Shibuya, T. Yamamoto, and M. Lippmaa, Defects and transport in complex oxide thin films, *J. Appl. Phys.* **103**, 103703/1-6 (2008)

[2] B. Dam, J. H. Rector, J. Johansson, J. Hujibregtse, and D.G. De Groot, Mechanism of incongruent ablation of STO, *J. Appl. Phys.* **83**, 3386 - 3389 (1998)

[3] Keeble, Wicklein, Dittmann, Ravelli, Mackie, and Eder, Identification of A- and B-Site Cation Vacancy Defects in Perovskite Oxide Thin Films, *PRL* **105**, 4 (2010)

STRUCTURAL RESPONSE OF SINGLE CRYSTAL SrTiO$_3$ ON O-VACANCY MIGRATION IN THERMAL AND ELECTRICAL FIELDS

<u>Barbara Abendroth</u>[1], Juliane Hanzig[1], Hartmut Stöcker[1], Florian Hanzig[1], Ralph Strohmeyer[1], Solveig Rentrop[1], Uwe Mühle[2], Dirk C. Meyer[1]

[1]Institute for Experimental Physics, TU Bergakademie Freiberg, Freiberg, Germany
[2]Fraunhofer Institute for Nondestructive Testing, Dresden, Germany

SrTiO$_3$ is a model transition metal oxide with perovskite structure. Its electronic properties can be tuned from insulating to metallic conductive by doping. For this reason SrTiO$_3$ is intensively studied for different applications including high k materials and in metal- insulator-metal structures with resistive switching characteristics.

The initial conductivity of SrTiO$_3$ can be enhanced by introduction of donor or acceptor dopants or intrinsic point defects by increasing the oxygen vacancy concentration. To achieve the latter, oxygen has to be removed from the material. In research, a common method is to reduce SrTiO$_3$ single crystals by vacuum annealing above 900 °C. At elevated temperatures, oxygen can be desorbed from the SrTiO$_3$ surface and is extracted from the system by the vacuum pump. Since the oxygen in the bulk becomes mobile above 700 °C the concentration gradient between the bulk and the surface can be balanced and an equilibrium is established quickly between O Vacancy concentration and the Oxygen partial pressure in the surrounding atmosphere [1]. The defect equilibrium is described by $O_O^x \leftrightarrow V_O^{\bullet\bullet} + \frac{1}{2} O_2(g) + 2e'$. In consequence, a corresponding free electron concentration can be observed [2].

The switching between high resistance and low resistance states and also the initial electroforming are, as well, closely connected with the redistribution of positively charged oxygen vacancies in the oxide. In both cases, existing extended defects, such as vacancy clustering, dislocations, stacking faults and interfaces (including the surface) affect the oxygen vacancy migration. On the other hand, the formation or annealing of these extended defects is an answer of the crystal lattice to accommodate non-equilibrium stoichiometry changes.

For conventional reducing annealing with an isotropic furnace heating, we observe increasing free carrier concentrations and electrical conductivities with increasing annealing times, being evidence of a continuous increase of the vacancy concentration, only up to a certain maximum. For prolonged annealing, a reduction of the conductivity is then observed. At the same time, the analysis of etch pit distributions on the SrTiO$_3$ (100) surface and transmission electron microscopy indicate, that the dislocation density decreases and dislocations are rearranged into clusters.

Anisotropic annealing, heating only one surface of the single crystal specimen leads to a much more effective loss of oxygen even at moderate temperatures of 750° C. This severe loss of oxygen is accommodated by the formation of a secondary phase with edge sharing Ti-O octahedra.

Such strong reactions on increased oxygen vacancy concentrations are not observed for electrical field induced vacancy migration. Conductivity measurements and in situ XRD analysis during electroforming give evidence that the redistribution of oxygen vacancies in the bulk in an external electric field and the connected deformation of the lattice are relatively week and reversible.

Our results are discussed in context with the non-volatile, i.e. persisting without electrical field, nature of the resistance changes, which are consentaneously attributed to local pathways of fast electron conductivity or vacancy migration.

[1] N. H. Cha et al., J. Electrochem. Soc. **128**, 1762 (1981).
[2] R. Waser, J. Am. Ceram. Soc. **74**, 1934 (1991).

CRYSTAL- AND DEFECT- CHEMISTRY OF REDUCTION RESISTANT FINE GRAINED THERMISTOR CERAMICS ON BaTiO$_3$-BASIS

Christian Pithan[1], Hayato Katsu[2], Rainer Waser[1,3], Hiroshi Takagi[2]

[1]Peter Grünberg Institute – Electronic Materials (PGI 7), Forschungszentrum Jülich GmbH, Jülich, North Rhine-Westphalia, Germany;
[2]Materials Development Department, Murata Manufacturing Corporation Ltd., Yasu, Shiga, Japan;
[3]Institute for Electronic Materials, RWTH Aachen University, Aachen, North Rhine-Westphalia, Germany

Bulk polycrystalline thermistor ceramics on BaTiO$_3$-basis with positive temperature coefficient (PTC) posses an increasing current limiting electric insulation resistance when they are heated above the Curie temperature T_C of the ferroelectric compound BaTiO$_3$. This behavior relies on the developement of resistive back-to-back Schottky barriers formed at the grain boundaries, when the ceramics are sintered and cooled usually under oxidizing conditions. The mechanism responsible for the resistivity jump of typically several orders of magnitude is well understood, particularly in terms of point defect chemistry [1, 2]. Here the formation of the insulating grain boundaries depends on the absorption of oxygen combined with the segregation of acceptor-type metal vacancies, enriched at these interfaces.

Only in the rather recent past multilayer PTC thermistors [3-5] consisting of laminated metallic (Ni- or Ti-based alloys) and ceramic layers on the basis of donor-doped BaTiO$_3$ formulations with BaO-excess that are configured in a consecutive but alternating serial order have been developed. These devices have to be sintered under reducing atmosphere at a partial pressure of oxygen $p(O_2)$ below 10^{-7} MPa in order to prevent the oxidation of the internal electrodes. They only achieve their distinct PTC-characteristics after a gentle post-sintering re-oxidation process. The exact physico-chemical mechanism for this behavior has practically not been clarified so far and therefore the present study reports on the crystal- and defect-chemistry of La-doped BaO-rich BaTiO$_3$ ceramics. The communicated results are based on impedance spectroscopic experiments, high temperature DC-conductivity measurements carried out at partial pressures of oxygen $p(O_2)$ in the range from 10^{-20} MPa up to 0.1 MPa, the determination of ^{18}O-tracer diffusion profiles and the analytical, crystallographic and microstructural characterization using X-ray diffraction and electron microscopy.

Figure 1 shows for an example of a BaTiO$_3$-ceramic with BaO-excess processed in the way described before that the total PTC-response (R_{Total} versus temperature in fig.1b) depends on several contributions. According to this the microstructure consists of semiconducting grains that are separated by more insulating grain boundaries, showing a distinct positive temperature coefficient of resistance. The absolute values of resistance at high temperatures seem however to be governed by external surface layers, that disappear after grinding and reappear after a second re-oxidation treatment. Interestingly the BaO-excess does not result in any wetting of the grain boundaries by a secondary phase or in the formation of extended defects in the lattice (fig.1c). Crystallographic data show that the lattice constants of the BaTiO$_3$ matrix are practically not affected by the BaO-content, suggesting that the bulk solubility is very low. It is assumed that BaO is preferentially enriched in the grain boundary regions.

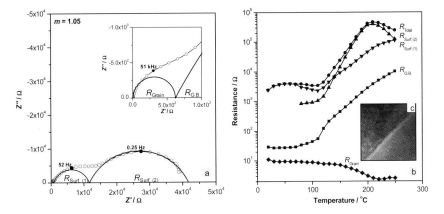

Figure 1: (a) Cole-Cole diagram at 160 °C showing the frequency response locus of $(Ba_{1.048}La_{0.002})TiO_{3+\delta}$ with a Ba-excess of 5 at.-%, sintered at 1300 °C at $p(O_2)=10^{-9}$ MPa and re-oxidized in air for two hours. The small inset represents the high frequency end of the impedance spectrum. (b) Temperature dependence of the total resistance R_{Total} and different resistive components arising from the response of the grains (R_{Grain}), of the grain boundaries ($R_{G.B.}$) and the ceramics surfaces ($R_{Surf.}$) for this material. (c) High resolution TEM-image of a grain boundary.

SIMS investigations after re-oxidation with the isotope ^{18}O prove that the resistive surface layers are due to enhanced oxygen diffusion along grain boundaries. This is only the case in hyper-stoichiometric compositions with BaO-excess. Ceramics with a Ba:Ti ratio of exactly 1 do not show this behaviour. A more detailed study of the defect chemical relations in BaO-rich donor doped Ba-TiO$_3$ supports the idea that excessive BaO dissolves mainly in the grain boundary region resulting in the local formation of both Ti- and O-vacancies upon sintering in reducing atmosphere. During re-oxidation O-vacancies are filled again preferentially by grain boundary diffusion and the acceptor type Ti-vacancies finally remain to form the insulating barrier layers on the grain boundaries. This interpretation is supported by the results of figure 2.

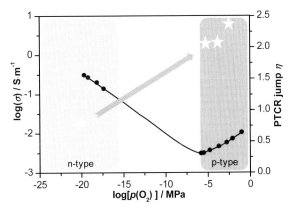

Figure 2: DC conductivity and the PTC-parameter η (white symbols) quantifying the height of the jump in resistivity upon heating in measured in situ in dependence of $p(O_2)$ at 700°C for the same material described in figure 1. The PTC effect is efficiently enhanced when post sintering annealing is performed in the region of overall p-type semiconduction (dark gray area) above $p(O_2)=10^{-6}$ MPa at this temperature. The grain boundaries become insulating, while the grain interior still shows n-type semiconduction.

[1] W. Heywang, Solid-State Electron. **3**, 51 (1961).
[2] J. Daniels et al., Philips Res. Rep. **31**, 544 (1976).
[3] K. Mihara et al., US-Patent 7,075,408 B2 (2006).
[4] H. Niimi et al., Jap. J. Appl. Phys. **46**, 675 (2007).
[5] H. Niimi et al., Jap. J. Appl. Phys. **46**, 6715 (2007).

RED-OX DRIVEN POINT DEFECT EQUILIBRIA, ANISOTROPIC CHEMICAL AND THERMAL EXPANSION AND FERROELASTICITY OF ACCEPTOR DOPED LaMO$_3$ PEROVSKITE OXIDE AT THE NANO-SCALE

Xinzhi Chen[1], Julian R. Tolchard[1], Sverre M. Selbach[1], <u>Tor Grande</u>[1]

[1]Department of Materials Science and Engineering, Norwegian University of Science and Technology, NO-7491 Trondheim, Norway

Acceptor doped LaMO$_3$ materials (M=Mn,Co,Fe) have attractive magnetic and electrical properties and possess ferroelasticity. Acceptor doping in LaMnO$_3$ is charge compensated either by oxygen vacancies ($V_O^{\bullet\bullet}$) or oxidation of Mn, which strongly influences the charge carrier density and the magnetism, as shown for La$_{1-x}$Sr$_x$MnO$_{3+\delta}$ (LSMO) [1]. The valance of M is also influenced by Schottky *point defects*; nil $\rightarrow V_{La}''' + V_M''' + 3V_O^{\bullet\bullet}$, where V_{La}''', V_M''' represent vacant cation sites. Cation defects can be dominating in LSMO at low acceptor levels and intermediate temperatures

$$2\delta M_{Mn} + \delta/2 O_2(g) \rightarrow \delta O_O + 2\delta Mn_{Mn}^{\bullet} + \delta/3 V_{La}''' + \delta/3 V_{Mn}''' \qquad (1)$$

where $Mn_{Mn}^{\bullet}$ and Mn_{Mn} correspond to Mn^{4+} and Mn^{3+}, while oxygen vacancies is dominating in LSMO at high acceptor levels and elevated temperatures [2]

$$2\delta M_{Mn}^{\bullet} + \delta O_O \rightarrow \delta 2 M_{Mn} + \delta V_O^{\bullet\bullet} + \delta/2 O_2(g) \qquad (2)$$

The change in oxidation state is associated with *chemical expansion* due to the increasing ionic radii with decreasing charge [3]. The sensitivity of the unit cell volume by changes in oxidation state has pawed the way to study point defect equilbria in nano-scale LSMO [4]. The unit cell volume of La$_{0.8}$Sr$_{0.2}$MnO$_{3+\delta}$ in O$_2$ and N$_2$ atm. is shown in Fig. 1. Despite the kinetics of eq. (1) is dependent on cation diffusion [4], oxidation/reduction of Mn with an associated volume change take place from ~400 °C at the nano-scale and will influence on the functional properties.

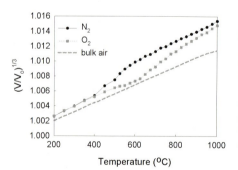

Figure 1: a) In situ high temperature X-ray diffraction (HTXRD) of nano-crystalline La$_{0.8}$Sr$_{0.2}$MnO$_{3+\delta}$ during heating in O$_2$ or N$_2$ atm [4]. The materials were annealed and cooled down in air prior to the measurement. Chemical expansion due to reduction of Mn in N$_2$ and chemical contraction due to oxidation of Mn in O$_2$ are evident. The dotted line show the corresponding behavior of bulk LSMO by dilatometry showing only linear thermal expansion due to the *freeze in* of defect eq. (1) at relatively high temperature in bulk.

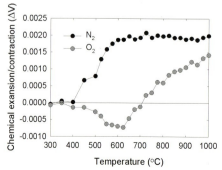

b) The volumetric effect of change in the valance state of Mn calculated by subtracting a linear thermal expansion term from the unit cell volume in the two atm. In N$_2$ an expansion of about 2 vol% is evident from about 400 to 600 °C, where the material becomes essentially stoichiometric with respect to oxygen. In O$_2$ a completely different behavior is observed, where first a contraction of ~0.7 vol% occurs due to oxidation of Mn, followed by an expansion above 600 °C due to reduction of Mn also in O$_2$ atm.

The defect equilibria (1) and (2) will *freeze in* during cooling due to the associated diffusion of vacancies [4]. The characteristic temperature for the *broken ergodicity* depends on the *length scale*, and equilibration of the point defects continues to take place at considerable lower temperatures at the nano-scale relative to the bulk. The second order ferroelastic phase transition of rhombohedral $LaMO_3$ is strongly related to the unit cell volume, and red-ox equilbria similar to eq. (2) will influence on the spontaneous strain during cooling. An example is shown for Sr doped $LaCoO_3$ in Fig. 2. The spontaneous strain occurs along the pseudo cubic c_{pc}, due to antiferrodistortive tilting ($a^-a^-a^-$ in the Glazer tilt system). As a consequence the thermal expansion is anisotropic, and is considerably larger along c_{pc} relative to a_{pc}.

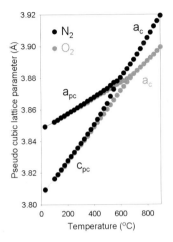
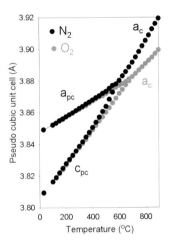

Figure 2: In situ high temperature X-ray diffraction of $La_{0.7}Sr_{0.3}CoO_{3-\delta}$ during heating in O_2 or N_2 atm. The materials were annealed and cooled down in air prior to the measurement. The hexagonal a and c unit cell parameters are normalized to pseudo cubic lattice parameters $a_{pc} = a/2^{1/2}$ and $c_{pc} = c/12^{1/2}$. The second order phase transition from the ferroelastic (rhombohedral) to paraelastic (cubic) state is strongly dependent on the partial pressure of O_2. Chemical expansion due to reduction of Co is significantly more pronounced in N_2 relative to O_2, which reflects the considerable chemical expansion of Sr doped $LaCoO_3$ materials [3]. The chemical expansion is highly anisotropic in the ferroelastic state, recognized by the pronounced sensitivity of the c_{pc} lattice parameter to changes in the partial pressure of O_2 relative to the a_{pc} lattice parameter.

The chemical expansion due to reduction of Co affects the ferroelastic phase transition, and the transition is shifted to lower temperatures by lowering the partial pressure of O_2. The chemical expansion is highly anisotropic and is stronger along the direction of ferroelastic twinning (c_p). The onset of chemical expansion due to (2) is observed form ~300 °C, but may take place at lower temperatures since the minor changes in oxygen non-stoichiometry at these conditions is hard to detect.

Anisotropic chemical/thermal expansion and red-ox driven point defect equilbria have been documented in Sr-doped $LaMO_3$ (M=Mn,Fe,Co) [4,5]. Since these processes takes place below the typical temperatures for thin film growth, we propose that the valance state and charge carrier density in oxides with mixed valence state transition metals such as Fe, Mn, Co or even Ti, will be strongly influenced by the thermal history. Finally, we would like to address the possibility for chemical relaxation of strain by chemical expansion/contraction, which also will be strongly anisotropic.

[1] M.B. Salamon, M. Jaime, Rev. Mod. Phys. **73**, 583 (2001).
[2] L. Rørmark, K. Wiik, S. Stølen, T. Grande, J. Mater. Chem. **12**, 1058 (2002).
[3] S.B. Adler, J. Am. Ceram. Soc. **84**, 2117 (2001).
[4] T. Grande, J.R. Tolchard, S.M. Selbach, Chem Mater. **24**, 338 (2012).
[5] X. Chen et al., in preparation

OXYGEN EXCHANGE KINETICS ON PEROVSKITE SURFACES: IMPORTANCE OF ELECTRONIC AND IONIC DEFECTS

R. Merkle[1], L. Wang[2], Y. A. Mastrikov[3], E. A. Kotomin[1], J. Maier[1]

[1]Max Planck Institute for Solid State Research, Stuttgart, Germany
[2]present address: Massachusetts Institute of Technology, Cambridge, MA, USA
[3]present address: Insitute for Solid State Physics, University of Latvia, Riga, Latvia

At sufficiently high temperature, any oxide will adjust its point defect concentrations to equilibrate with the surrounding oxygen partial pressure. This is at least a twofold process, starting with the oxygen exchange surface reaction (converting O_2 to oxide ions in the first layer of the material) followed by bulk chemical diffusion. Further complications may arise due to impeded or accelerated transport at grain boundaries, or from extended defects such as dislocations.

The kinetics of the oxygen exchange surface reaction is of fundamental interest (one of the "simplest" gas-solid reactions), but also important for the functioning of electrochemical devices such as gas sensors and solid oxide fuel cells (SOFC) as well as resistive switching memories (oxygen loss in the initial "forming" step). The role of electronic and ionic defects for the oxygen exchange reaction can nicely be studied on perovskite structured oxides ranging from semiconductors such as $SrTiO_3$ to good electronic conductors such as $(La,Sr)CoO_{3-\delta}$.

For slightly Fe-doped $SrTiO_3$ single crystals, conduction electrons were found to be involved before or in the rate-determining step of the surface reaction [1]. In the perovskite solid solution series $SrTi_{1-x}Fe_xO_{3-\delta}$, an increase of the Fe content above $\approx 10\%$ leads to a pronounced decrease of the band gap and corresponding increase of the electronic conductivity. For such "electron-rich" $SrTi_{1-x}Fe_xO_{3-\delta}$ perovskites, a change of the reaction mechanism can be recognized, and the effective rate constant k^δ increases strongly (Fig. 1) [2,3].

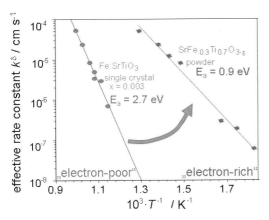

Figure 1: Effective rate constants k for oxygen exchange on slightly Fe-doped $SrTiO_3$ crystals and $SrTi_{0.7}Fe_{0.3}O_{3-\delta}$ powder samples. The different activation energy and pO_2 dependence of k indicate a change in the reaction mechanism at Fe contents $\geq 10\%$. The strong increase in k for $SrTi_{0.7}Fe_{0.3}O_{3-\delta}$ occurs in parallel to the much higher availability of electronic carriers for charge transfer to adsorbed oxygen species, suggesting that electron transfer is not limiting any more.

Turning to $(La,Sr)(Mn,Fe,Co)O_{3-\delta}$ and $(Ba,Sr)(Co,Fe)O_{3-\delta}$ perovskites which are investigated as ionically and electronically conducting SOFC cathode materials, the presence of the transition metal in a mixed oxidation state provides a sufficiently high electronic carrier concentration so that electron transfer is not expected to be limiting. The strong increase of the oxygen exchange rate between $(La,Sr)MnO_{3-\delta}$ and $(La,Sr)(Fe,Co)O_{3-\delta}$ perovskites (see e.g. [4]) suggests that oxygen vacancies are involved in the rate-determining step [5]. This interpretation of the reaction mechanism is supported by DFT calculations for $LaMnO_3$, which indicate that the approach of a surface oxygen vacancy to adsorbed O^- is the bottleneck of the oxygen incorporation reaction [6]. Thus the vacancy mobility is not only important for the later bulk diffusion step, but also for the actual surface reaction.

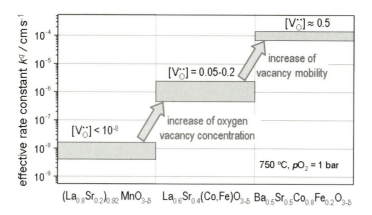

Figure 2: Effective rate constants k for oxygen exchange for various perovskites. Since the mobility of oxygen vacancies does not change significantly between (La,Sr)MnO$_{3-\delta}$ and (La,Sr)(Fe,Co)O$_{3-\delta}$, it is essentially the increased vacancy concentration which increases k. Moving from (La,Sr)(Fe,Co)O$_{3-\delta}$ to (Ba,Sr)(Co,Fe)O$_{3-\delta}$, the increase in vacancy mobility makes the larger contribution to the accelerated reaction rate.

The importance of the vacancy mobility became particularly obvious in the family of (Ba,Sr)(Co,Fe)O$_{3-\delta}$ perovskites [7], where the vacancy diffusivity is remarkably enhanced for Co- and Ba-rich materials. The contributions of vacancy concentration and vacancy mobility for the increase of the oxygen exchange rate constant are summarized in Fig. 2. Having identified the the importance of these quantities for surface kinetics, materials such as (Bi,Sr)(Co,Fe)O$_{3-\delta}$ perovskites are currently explored [8].

[1] R. Merkle et al., Phys. Chem. Chem. Phys **4**, 4140 (2002).
[2] R. Merkle et al., Topics in Catalysis **38**, 141 (2006)
[3] R. Merkle et al., Angew. Chemie Int. Ed. **47**, 3874 (2008)
[4] R.A. De Souza et al., Solid State Ionics **126**, 153 (1999)
[5] R. Merkle et al., Angew. Chemie Int. Ed. **43**, 5069 (2004)
[6] Y.A. Mastrikov et al., J. Phys. Chem. C **114**, 3017 (2010)
[7] L. Wang et al., J. Electrochem. Soc. **157**, B1802 (2010)
[8] A. Wedig et al., Phys. Chem. Chem. Phys. **13**, 16530 (2011)

ORDERS OF MAGNITUDE VARIATIONS IN THE ELECTRICAL CONDUCTION PROPERTIES OF ACCEPTOR AND DONOR DOPED STRONTIUM TITANATE ON DOWNSIZING

Giuliano Gregori,[1] Piero Lupetin[1], Joachim Maier[1]

[1] Max Planck Institute for Solid State Research, Stuttgart, Germany

The study of the conduction properties of nanosized materials has revealed in the last years a large number of exciting size-induced phenomena, which take place when interfaces are so close to each other that they prevail over the bulk. Many of these effects have been understood in the context of nanoionics, which enables elucidating the defect chemistry not only in well separated boundary regions but also in the mesoscopic regime where boundary layers overlap, giving origin to novel conductors (Figure 1).[1]

Here we consider $SrTiO_3$ as prototypical mixed conductor for the investigation of the conduction properties at the nanoscale, since the defect chemistry of this perovskite has been deeply studied for the macroscopic situation and gives the possibility of a better understanding of the properties at the nanoscale.

In this context, the oxygen non-stoichiometry plays a role not only for the ionic conductivity but also with respect to the n- and p-type contributions. In this way, ionic and electronic conduction properties are studied over a broad range of oxygen partial pressures and temperatures as a function of size for donor as well as for acceptor doping by means of impedance spectroscopy.

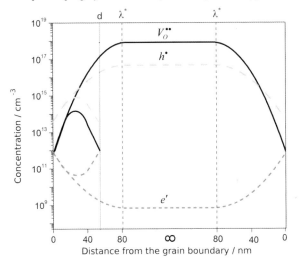

Figure 3 Charge carrier concentration profiles calculated in pure oxygen for an 0.01at.% acceptor doped sample at T=544 °C and $\Delta\phi_0 = 0.5$ V . λ^* defines the space charge width, whereas d the effective grain size.

As already observed in the case of undoped $SrTiO_3$ material (characterized by intrinsic acceptor impurities),[2] the investigation of the electrical conductivity with respect to the variation of the oxygen stoichiometry reveals enormous size effects due to the mesoscopic situation. In particular, the depression of p-type conductivity and the consequent increase of the n-type contribution generate a giant shift of the conductivity minimum (at which the conduction mechanism switches from p-type to n-type) by several orders of magnitude in terms of oxygen partial pressure. The differences between the acceptor doped nanocrystalline SrTiO₃ and the nominally pure material can be explained by the higher doping level.

All the results can be explained in the light of space charge effects occurring as a consequence of a positive charge excess in the grain boundary core.

Another aspect of great interest in the field of mesoscopic conducting materials concerns the possibility of tuning the grain boundary properties and, consequently, the overall conduction of the materials. With respect to iron doped SrTiO₃, it has been shown that the spatial distribution of the dopant (homogeneously dispersed throughout the grain or selectively added at the GBs) can crucially affect the space charge properties and therefore the overall electrical transport (Figure 2a).

Particularly intriguing are the studies on donor (niobium) doped SrTiO₃, which is a well known n-type conductor at relatively high temperatures in the high oxygen partial pressure range. Surprisingly, the nanocrystalline material showed p-type conductivity in oxidizing conditions at 550°C and a blocking effect of the grain boundaries with respect of the electron transport when the material switches to the n-type regime (Figure 2b). These results can be ascribed to the presence of a negative space charge potential, which generates the enrichment of holes and the depletion of electrons in the space charge region.

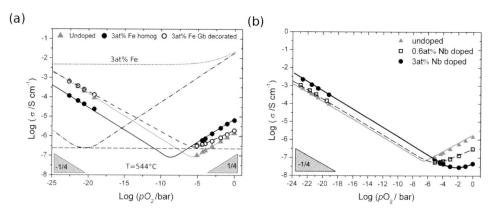

Figure 4 Oxygen partial pressure dependence of the conductivity measured at 544°C for nanocrystalline (a) Fe-doped and GB decorated SrTiO₃ (b) Nb-doped SrTiO₃. In (a) the dotted line refers to the theoretical conductivity of 3at% Fe doped SrTiO₃ according to Denk et al..[3]

In conclusion, these investigations on the electrical properties of the grain boundaries of the SrTiO₃ model system demonstrates the impact that grain size has on the overall ionic and electronic conduction properties of the materials and confirmed the possibility of further adjusting the transport properties by controlling grain size and grain boundary charge.

[1] J. Maier, *Nature Materials*, **4**, 805 (2005).
[2] P. Lupetin, G. Gregori and J. Maier, *Angewandte Chemie International Edition*, **49**, 10123 (2010).
[3] I. Denk, W. Munch and J. Maier, *Journal of the American Ceramic Society*, **78**, 3265 (1995).

DYNAMIC SIMULATION OF OXYGEN MIGRATION IN TiO$_2$

Jan M. Knaup[1], Michael Wehlau[1], Thomas Frauenheim[1]

[1]Bremen Center for Compuational Materials Science, Universität Bremen, Bremen, Germany

Oxygen deficient titania has gained considerable technological interest as material for resistive-switching nonvolatile memory applications. Experimental evidence shows that the dynamics of oxygen vacancies and oxygen stoichiomentry related phase transitions play a crucial role in the resistive switching effect. In order to achieve fundamental understanding of the migration dynamics of oxygen vacancy defects we perform molecular dynamics simulations of oxygen vacancies in rutile titania under external driving forces using Density Functional Based Tight Binding (DFTB). Our results show that such DFTB-simulations of the oxygen diffusion dynamics in rutile is of comparable reliability to Density Functional Theory (DFT) simulations at much reduced computational cost, thus allowing extensive statistical sampling. By simulating the oxygen vacancy diffusion under the influences of external electric fields and temperature gradients, we determine how these driving forces influence the oxygen diffusion mechanisms. Understanding the relative influences of different driving forces is crucial for further knowledge based materials and device engineering.

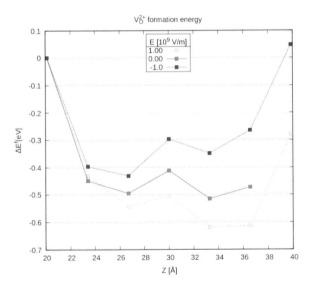

Figure 1: Comparison of relative formation energies of positively charged V$_O$ defects, depending oh their position along an external electric field through a slab of rutile TiO$_2$.

Figure 1 shows the variation of static formation energies of oxygen vacancies in rutile with shift along an external electric field, perpendicular to a (110) surface slab model. The slope of about -1/80 eV/V is consistent with the dielectric constant of rutile $\varepsilon a = \sim 110$ [1]. The resulting forces on neighboring atoms are minute, necessitating the use of metadynamics[2] simulation to determine the diffusion barriers and their response to external electric fields.

[1] R.A. Parker, Phys. Rev., **124** (1961), 1719
[2] A. Laio, M. Parrinello; Proc. Natl. Acad. Sci. USA **99** (2002), 12562.

ATOMISTIC SIMULATION STUDY ON OXYGEN DEFICIENT STRONTIUM TITANATE

Marcel Schie[1], Astrid Marchewka[1], Roger A. De Souza[2], Thomas Müller[3], Rainer Waser[1,4]

[1]Institute of Materials in Electrical Engineering and Information Technology, RWTH Aachen University, 52074 Aachen, Germany;
[2]Institute of Physical Chemistry, RWTH Aachen University, 52056 Aachen, Germany;
[3]Jülich Supercomputer Centre, Institute of Advanced Simulation, Forschungszentrum Jülich, 52425 Jülich, Germany; [4]Peter Grünberg Institute 7, Forschungszentrum Jülich, 52425 Jülich, Germany

Strontium titanate ($SrTiO_3$) is a perovskite-type transition metal oxide which exhibits insulating behavior, but can be reversibly switched into a conducting state by applying external voltage or current pulses [1]. This qualifies $SrTiO_3$ as a model material for future non-volatile redox-based random access memory (ReRAM) cells. The understanding of the switching phenomena requires detailed information about the physical and chemical conditions on the atomic scale. Especially the diffusion of donor-type oxygen vacancies needs to be examined further because it induces structural modifications and a subsequent reduction of the Ti sublattice which leads to a drastic change in conductivity. For this purpose classical force field simulations provide a powerful tool to calculate structures, energies and many other static physical properties as well as dynamical characteristics.

In this work, we present lattice dynamics and molecular dynamic simulations of $SrTiO_3$ based on the classical force field by Thomas *et al.* [2], which proved to be adequate for simulations of simple point defects like strontium, titanium and oxygen vacancies and interstitials [3]. The force field is a partial charge model consisting of three Born-Mayer potentials for Sr-O, Ti-O and O-O interaction, but without cation-cation pair potentials and dispersion contributions. Using this potential we calculated the heat capacity to check the characteristics of the force field in the harmonic region beyond the energetic minimum. The thermal expansion was simulated to investigate the anharmonic region of the potential. Both properties were compared to experimental values by de Ligny *et al.* [4] and revealed good qantitative congruence. Hence, defect simulations which depend on non-equilibrium positions in the potential landscape are expected to be reliable. We studied the static migration and interaction of oxygen vacancies using the Mott-Littleton [5] approach in GULP [6] by scanning the potential surface in the vicinity of the expected migration path. Further we optimized configurations of two oxygen vacancies with respect to the defect energy and calculated the related association energies. The dynamical behavior of vacancies in the bulk phase of $SrTiO_3$ was investigated via molecular dynamic simulations using DL_POLY [7]. For this purpose 15 to 250 oxygen vacancies were randomly distributed in a supercell consisting of 10x10x10 unit cells with a side length of 3.905 nm. We compensated surplus charges through a constant charge shift of all titanium and oxygen ions in the simulation box. The simulations are analyzed with respect to the character of vacancy-vacancy interaction, the possibility of clustering and diffusion properties in relation to temperature and vacancy concentration.

[1] R. Waser, R. Dittmann, G. Staikov, K. Szot, Adv. Mater., **21**, 2632 (2009)
[2] N. A. Marks, B. D. Begg and B. S. Thomas, Nucl. Instrum. Methods B, **228**, 288-92 (2005)
[3] N. A. Marks, B. D. Begg and B. S. Thomas, Nucl. Instrum. Methods B, **254**, 211-18 (2007)
[4] D. De Ligny and P. Richet, Phys. Rev. B, **53**, 3013 (1996)
[5] N. F. Mott and M. J. Littleton, Trans. Faraday Soc., **34**, 485 (1938)
[6] J. D. Gale, J. Chem. Soc., Farady Trans., **93**, 629 (1997)
[7] W. Smith, Molecular Simulation, **32**, 933-1121 (2006)

IOR 5

FIRST PRINCIPLE STUDY AND MODELING OF STRAIN-DEPENDENT IONIC MIGRATION IN ZIRCONIA

Julian A. Hirschfeld[1], Hans Lustfeld[2]

[1]Forschungszentrum Jülich – Institute for Advanced Simulation (IAS-1), Jülich, Germany;
[2]Forschungszentrum Jülich – Peter Grünberg Institute (PGI-1), Jülich, Germany

Electrolytes with high ionic conductivity at lower temperatures are the prerequisite for the success of Solid Oxide Fuel Cells (SOFC). One promising candidate is doped zirconia. In the past the ionic resistance of the electrolyte has mainly been reduced by decreasing its thickness. However, the influence of the thickness is only linear, whereas the impact of migration barriers is exponential. Therefore understanding the oxygen transport in doped zirconia is of fundamental importance.

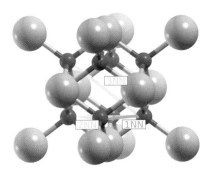

Figure 1: The cubic fluorite phase has the lowest migration barrier for oxygen ions in zirconia. Possible pathways for oxygen ion migration are shown as arrows. The three arrows indicate the possible jump processes to the nearest [1NN (purple)] location and to the next nearest ones [2NN (teal)], [3NN (green)]. A vacancy has to be present at the final position for any of these jumps to happen.

In this work [1] we pursue the approach of the strain dependent ionic migration in zirconia. We investigate how the migration barriers for oxygen ions respond to a change of the atomic strain. We employ the method of Density Functional Theory (DFT) calculations to relax the atomic configurations to the ground state. In connection with the Nudged Elastic Band (NEB) method we obtain the migration barrier of the oxygen ion jumps in zirconia for a given lattice constant. Similar to other publications (e.g. [2]) we observe a decrease in the migration barrier for expansive strain, but in addition we also find a migration barrier decrease for high compressive strains beyond a maximal height of the migration barrier at an intermediate compressive strain. We present a simple analytic model that, by using interactions of the Lennard-Jones type, gives an explanation for this behavior.

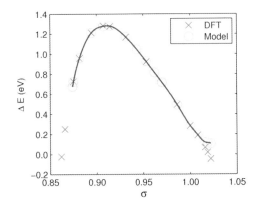

Figure 2: The height of the oxygen ion migration barrier ΔE in zirconia is shown as a function of the strain σ. The (blue) solid line shows a spline interpolation of the model data. The model agrees over a wide strain range with the DFT results. For $\sigma < 0.87$ the model is not valid any more and deviates for $\sigma > 1.008$.

[1] J. H. Hirschfeld and H. Lustfeld, Phys. Rev. B **84**, 224308 (2011).
[2] C. Korte, A. Peters, J. Janek, D. Hesse, and N. Zakharov, Phys. Chem. Chem. Phys. **10**, 4623 (2008).

INVESTIGATIONS ON THE INTEGRATED CATHODES FOR HIGH ENERGY DENSITY LITHIUM RECHARGEABLE BATTERIES

S.B. Majumder[1], C. Ghanty [1], R.N. Basu[2]

[1]Materials Science Centre, Indian Institute of Technology, Kharagpur, West Bengal, India;
[2]Fuel Cell and Battery Division, Central Glass and Ceramic Research Institute, Kolkata, West Bengal, India

Unlike traditional layered oxide based cathode materials for lithium-ion batteries, lithium and manganese rich oxide cathode materials form a structurally integrated nanocomposite of Li_2MnO_3 and $LiMO_2$ (where M=Co,Ni and Mn) by sharing a common oxide lattice. These cathodes, represented in two component composite notation as xLi_2MnO_3-$(1-x)LiMO_2$, can yield a reversible capacity of as high as ~300 mAhg^{-1}. The discharge capacity is almost double than most of the traditional cathode materials. We have demonstrated that the lithium ion intercalation mechanism in these integrated cathodes is a complex process and the electrochemical performance of these cathode materials depends on various interrelated factors. Through extensive structural and electrochemical characterization, we have found that (i) a significant structural change occurs during first cycle, (ii) a relatively slower and continuous structural change occurs during repeated cycling and (iii) an electronically insulating solid electrolyte interface layer grows with repeated charge-discharge cycling. All these factors significantly influence the electrochemical performance of the cathode material. In the present work, various approaches have been adopted to improve the electrochemical performances of the nanocomposite xLi_2MnO_3-$(1-x)Li(Mn_{0.375}Ni_{0.375}Co_{0.25})O_2$ ($0.0 \leq x \leq 1.0$) cathode materials. First, the composition of the integrated cathode and process parameters are optimized to yield high discharge capacity. Thus a discharge capacity of ~300 mAhg^{-1} is obtained in x=0.5 cathode. A reversible layer to spinel conversion is identified in cathodes with higher Li_2MnO_3 content. By stabilizing with small amount of $LiMO_2$ (M=Ni, Co, Mn), it is demonstrated that the so called inactive Li_2MnO_3 can be made electrochemically active to yield high capacity. Second, we have modified these integrated cathodes by partially dope oxygen with fluorine. Fluorine modified $Li(Li_{0.111}MnNi_{0.249}Co_{0.166})O_{2-z}F_z$ ($0.02 \leq z \leq 0.2$) cathode materials are synthesized by a low temperature fluorination method using NH_4HF_2 as fluorine source. A systematic change in lattice parameters with varying fluorine content indicates the incorporation of fluorine in the lattice. No significant change in the surface morphology and particle size is observed with the variation in fluorine content. Although the discharge capacity is found to be systematically reduced with the increase in fluorine content, both the cycleability as well as the rate capability is found to be improved in 0.02 mole fluorine doped integrated cathodes. Third, much better improvement in electrochemical properties is achieved in zirconium oxide coated integrated cathode materials. The lattice parameter of the integrated cathodes remains unchanged with the increase in zirconia contents. This indicates that the surface modification does not cause any structural change to the integrated cathode particles. Transmission electron micrographs of these cathodes show a discrete nano-crystalline porous particulate coating on the surface of the integrated cathode particles. Though initial discharge capacity decreases with the increase in ZrO_2 content, both the capacity retention with cycling and rate performance have been significantly improved in zirconia modified samples as compared to their unmodified counterpart. By estimating the lithium ion diffusion co-efficient from the Warburg impedance of the Nyquist plots, we have argued that the bulk of the cathode particles are marginally changed with repeated charge-discharge cycling. It is found that the capacity fading of these integrated cathodes correlates well with the systematic increase of interfacial resistance of the cathode particles. We have demonstrated that the porous particulate ZrO_2 coating improved the capacity retention of these integrated cathodes by suppressing the impedance growth at the electrodes-electrolyte interface. Finally, we have studied the effect of particle size on the electrochemical performance of the integrated cathode materials and found that among several interrelated factors (viz. cathode composition, activation of Li_2MnO_3 component, crystallinity of the cathode particles etc.) an optimum particle size is very much crucial for the improved performance of the synthesized cathode materials.

Acknowledgement: CSIR, Govt. of India and Alexander von Humboldt Foundation, Germany

SDY 1

VORTEX DOMAIN WALL DYNAMICS IN MAGNETIC NANOTUBES

Attila Kákay[1], **Ming Yan**[1], **Christian Andreas**[1,2], **Felipe García-Sánchez**[1], **Riccardo Hertel**[2]

[1]Peter Grünberg Institut (PGI-6), Forschungszentrum Jülich, Jülich, Germany;
[2]Institut de Physique et Chime des Matériaux de Strasbourg, Strasbourg, France

Recently, domain wall (DW) propagation in magnetic nanostructures has been intensively studied because of the potential application in novel devices, such as the race-track memory [1] and magnetic logic gates [2]. Here we present a micromagnetic study on DW dynamics in ferromagnetic nanotubes. A typical vortex-type DW formed in the tube shows advantageous properties regarding the DW speed and stability. For topological reasons, the Walker breakdown (WB), i.e., the collapse of the DW structure above a critical velocity - is effectively suppressed in tubes. This allows for higher non-turbulent DW speeds compared to thin strips. In addition, the curvature of the tube causes an unusual effect: the left-right (or chiral) symmetry the DW propagation is broken.

The inset of Fig. 1 shows the configuration of a head-to-head (h2h) DW formed in a 4 μm long Permalloy tube with 60 nm outer diameter and 10 nm thickness. In the wall region, the magnetization circles around the tube and forms a core-less vortex structure. Given the two possible vorticities, there are two energetically degenerate wall configurations. When an external field is applied along the tube to drive the DW, one can define the chirality of the DW by combining its vorticity with the field direction, as displayed in the inset of Fig. 1.

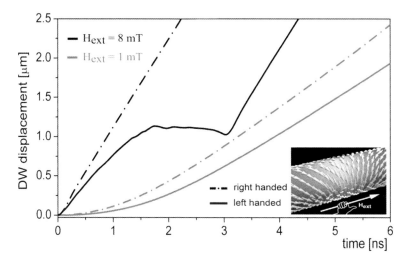

Figure 1: Field-driven DW displacement versus time in a 4 μm long Permalloy tube with 60 nm outer diameter and 10 nm thickness. The solid and dashed lines indicate left and right handed chiralities, as defined in the inset. The propagation of a DW driven by field values of 1 mT and 8 mT is plotted in grey and in black, respectively. Inset: A h2h vortex-type DW formed in a nanotube. The combination of the DW vorticity and the external field direction defines the chirality.

The field-drivenDW propagation in the tube was systematically studied by numerically solving the Landau-Lifshitz-Gilbert equation using a finite element algorithm [3]. It is found that the DW displays different dynamic behavior depending on the chiralities. This is illustrated in Fig. 1, which shows the DW displacement as a function of time for both types of handedness. Driven by a small field of 1 mT, the DW speed already shows a notable difference with a higher speed in the right handed chirality than in the left one. This chiral symmetry break originates from the static DW configuration in tubes, which possesses a non-zero radial component of the magnetization [4]. The radial component of the DW, resulting from magnetostatic considerations, is a purely geometric effect. It therefore does not occur in flat geometries, where the rotational symmetry is broken. Because the precessional torque exerted by an external field is in the radial direction, with the sign determined by the chirality, the initial non-zero radial component of the DW results in a chiral symmetry break of the dynamics.

Contrary to the case of lower fields, the DW in the left handed chirality driven by an 8 mT field experiences a period of halt. The DW resumes its motion after switching to the right handed chirality. This process is a WB involving the nucleation and annihilation of a vortex-antivortex pair. Once the pair is created, the vortex and anti-vortex move away from each other. Eventually, they meet on the other side of the tube and annihilate, thereby switching both, the vorticity and the chirality of the wall. This process is different from the WB in flat strips, where only a single (anti)vortex is nucleated. In tubes, a pair has to be created due to the lack of a lateral boundary and the conservation of the winding number.

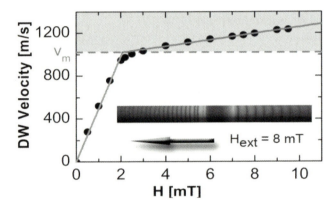

Figure 2: DW velocity as a function of field in the case of right handed chirality for the h2h DW in a 4 μm long Permalloy tube with 60 nm outer diameter and 10 nm thickness. The two segments of lines are linear fits to the data at two distinct regions, showing a clear drop of the DW mobility above a critical velocity v_m. Inset: A snapshot of the DW propagation driven by an 8 mT external field. SW tails with well-defined yet different wave lengths are emitted both in front of and behind the moving DW.

In the right handed case, for this particular tube, the WB is completely suppressed. Instead, another speed limit of the DW motion is encountered: the magnonic limit [5]. Fig. 2 shows the DW speed as a function of field. The DW mobility displays a significant drop above a critical velocity v_m. This is found to be related to the spontaneous emission of spin waves (SWs) by the moving DW. As shown in the inset, SW tails with well-defined yet different wave lengths are attached both in front of and behind the DW. The attached SW tails increases the effective mass of the DW, thereby causing the mobility drop. This effect only occurs when the DW velocity exceeds the minimum SW phase velocity. The bichromatic emission of SWs results from the two-fold degeneracy of a given SW phase velocity. The calculation of the SW dispersion in the tube confirms the SW emission mechanism by the moving DW.

[1] S.S.P. Parkin, M. Hayashi, and L. Thomas, *Science* **320**, 190 (2008).
[2] D. A. Allwood, G. Xiong, C. C. Faulkner, D. Atkinson, D. Petit, and R. P. Cowburn, *Science* **309**, 1688 (2005).
[3] A. Kákay, E. Westphal, and R. Hertel, *IEEE Trans. Magn.* **46**, 2303 (2010).
[4] M. Yan, C. Andreas, A. Kákay, F. García-Sánchez, and R. Hertel, submitted.
[5] M. Yan, C. Andreas, A. Kákay, F. García-Sánchez, and R. Hertel, *Appl. Phys. Lett.* **99**, 122505 (2011).

SPIN-TORQUE DYNAMICS OF STACKED VORTICES IN MAGNETIC NANOPILLARS

Daniel E. Bürgler[1], Volker Sluka[1,2], Alina Deac[2], Attila Kakay[1], Riccardo Hertel[3], Claus M. Schneider[1]

[1]Peter Grünberg Institute, Electronic Properties (PGI-6), Research Center Jülich, D-52425 Jülich, Germany;
[2]Institute of Ion-Beam Physics and Materials Research, Helmholtz-Zentrum Dresden-Rossendorf, D-01328 Dresden (Rossendorf), Germany;
[3]Institut de Physique et Chimie des Matériaux de Strasbourg, Université de Strasbourg, CNRS UMR 7504, F-67034 Strasbourg Cedex 2, France

Since the first experimental evidence for spin-torque generated magnetization dynamics the field has seen years of intense research. The major attention has been attracted by devices exhibiting homogeneously magnetized layers, which under the action of a spin-polarized current can emit microwave signals [1]. But recently, the inhomogeneously magnetized vortex structure has gained interest, because it has been shown that the so-called gyrotropic mode can be excited by spin-torque [2,3]. The signal associated with this mode is of higher power at lower line widths than for single domain spin-torque oscillators, which makes vortices interesting candidates for integrated microwave sources.

We investigate nanopillars consisting of two Fe disks separated by a nonmagnetic spacer (Ag, 6 nm thickness), see Fig. 1. Top and bottom Fe disks have thicknesses of 15 and 30 nm, respectively, while the pillar diameter is roughly 150 nm. The layers are MBE grown and preparation of the nanostructures is accomplished using electron beam and optical lithography.

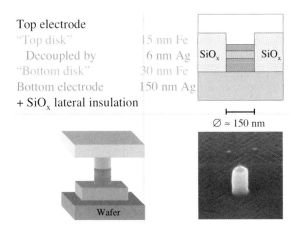

Figure 1: Sequence, materials, layer thicknesses of the multilayer structure, from which nanopillars with a diameter of 150 nm are fabricated by e-beam and optical lithography. The 3D representation shows the connection of the nanopillar to the top and bottom electrodes, which are integral parts of the final structure. The SEM image shows a freestanding nanopillar after ion-beam etching before the lateral insulation (SiO$_x$) and the Au top electrode are applied.

At suitable applied fields vortices enter the sample resulting in states with one vortex and one homogeneous layer (single-vortex states [2,3]) and states, where each layer contains a vortex (double-vortex states [4]). These configurations are investigated under the influence of spin-polarized currents. We find that not only the single-vortex states can be excited by spin-torque [5]. The double-vortex state can also show current-induced dynamics depending on the configuration in terms of the vortex chiralities and current direction. Our observations for a variety of configurations are in agreement with theoretical predictions [4]. In addition, we find that the vortices' motions are coupled via a rather strong dipolar core-core interaction, which lifts the degeneracy of double-vortex states with parallel and antiparallel core polarities. Micromagnetic simulations confirm this picture. A comparison of simulated and measured mode frequencies of different double-vortex configurations is shown in Fig. 2.

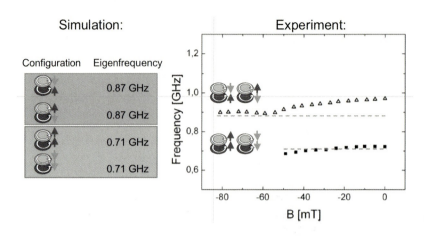

Figure 2: Comparison of simulated and measured mode frequencies of different double-vortex configurations. Dashed lines in the right-hand part indicate the simulated eigenfrequencies.

The results demonstrate means to reproducibly create and read-out double-vortex states characterized by different chirality and polarity combinations, which are prerequisites for applications of vortex-type nanopillars for storage or oscillator devices.

[1] S. I. Kiselev et al., Nature **425**, 380 (2003).
[2] V. S. Pribiag et al., Nat. Phys. **3**, 498 (2007).
[3] R. Lehndorff et al., Phys. Rev. B **80**, 054412 (2009).
[4] A. V. Khvalkovskiy et al., Appl. Phys. Lett. **96**, 212507 (2010).
[5] V. Sluka et al., J. Phys. D: Appl. Phys. **44**, 384002 (2011).

PURE SPIN CURRENTS IN FERROMAGNETIC INSULATOR/NORMAL METAL HYBRID STRUCTURES

Matthias Althammer[1], Mathias Weiler[1], Franz D. Czeschka[1], Johannes Lotze[1], Georg Woltersdorf[2], Michael Schreier[1], Stephan Gepraegs[1], Hans Huebl[1], Matthias Opel[1], Rudolf Gross[1,3], Sebastian T.B. Goennenwein[1]

[1]Walther-Meißner-Institut, Bayerische Akademie der Wissenschaften, Garching, Germany;
[2]Department of Physics, Universität Regensburg, Regensburg, Germany;
[3]Physik Department, Technische Universität München, 85748 Garching, Germany

In analogy to the well-established charge currents, a pure spin current is the directed flow of spin angular momentum. In spite of the conceptual analogy, however, the properties of charge and spin currents are very different: spin currents can flow not only in electrical conductors, but also in electrical insulators, since no charge motion is required for the propagation of angular momentum.

We experimentally study pure spin currents in magnetic insulators, using two complementary approaches. The samples consist of an epitaxial yttrium iron garnet (YIG) thin film – a ferrimagnetic insulator – covered with a thin metallic platinum layer. On the one hand, we use spin pumping in combination with the inverse spin Hall effect [1] to generate and detect pure spin currents across the YIG/Pt interface. The spin mixing conductance derived from these experiments is quantitatively comparable to that of conductive ferromagnet/Pt hybrids. On the other hand, we investigate the local magneto-thermo-galvanic voltages induced in YIG/Pt by a focused, scanning laser beam, and discuss the contribution of spin currents generated by the spin Seebeck effect [2].

The YIG/Pt heterostructures have been grown via laser-MBE and electron beam evaporation on (111)-oriented $Gd_3Ga_5O_{12}$ (GGG) substrates. The Pt was deposited in situ, without breaking the vacuum, after the laser-MBE YIG film growth. The structural properties of the samples were investigated by high-resolution X-ray diffractometry. No secondary phases were detected. The YIG thin film grows (111)-oriented with a low mosaic spread. The full width at half maximum of the rocking curve for the YIG (444) reflection is as low as 0.03°. The magnetic properties of the YIG thin film were investigated by superconducting quantum interference device (SQUID) magnetometry with the magnetic field applied in-plane. We extract a saturation magnetization of 125 kA/m, which is 90% of the bulk value. Ferromagnetic resonance (FMR) measurements at 10.3 GHz on our YIG thin films yield a peak to peak linewidth of 0.78 mT for the external magnetic field applied parallel to the surface normal. The structural and magnetic properties of our YIG/Pt heterostructures thus are state-of-the-art.

In our spin pumping experiments we generate a spin current by driving the magnetization of YIG into resonant precession (FMR). The magnetization then relaxes by emitting a spin current into the Pt layer. The efficiency of the spin pumping process is dependent on the spin-mixing conductance, which is in the order of $1 \cdot 10^{19}$ m^{-2} [1]. In our spin pumping experiments we simultaneously record the FMR signal and the DC voltage generated by the inverse spin Hall effect in the Pt layer at different temperatures between room temperature and 3 K (Fig. 1). From these experiments we extract a spin-mixing conductance in the range $1.2 \cdot 10^{18}$ m^{-2} to $8 \cdot 10^{19}$ m^{-2}, which is consistent with values measured on conductive ferromagnet/Pt interfaces [1].

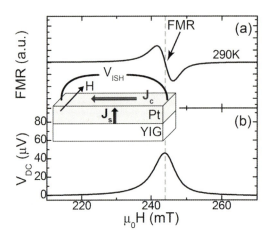

Figure 1: Spin pumping experiment using a YIG (20 nm)/Pt (7 nm) heterostructure on a GGG substrate at room temperature. (a) FMR signal observed for the external magnetic field H along the short side of the sample (see inset). (b) DC voltage drop along the long side of the sample detected during the FMR measurement. The FMR resonance field and the maximum in DC voltage occur at the same field values. The detected voltage V_{DC} is due to a pure spin current $\mathbf{J}_S$ pumped from the YIG into Pt, converted into a charge current $\mathbf{J}_C$ due to the inverse spin Hall effect.

In a second set of experiments we use a focused, scanning laser beam to generate a local thermal gradient along the surface normal in our FM/Pt heterostructures. This thermal gradient gives rise to a local spin current via the spin Seebeck effect that we electrically detect by means of the inverse spin Hall effect in the Pt thin film. From these experiments we determine a spin Seebeck coefficient of 600 nV/K [2]. In addition with this technique we demonstrate all-electrical detection of the YIG magnetic texture in our experiments [2].

Taken together, our findings show that it is possible to generate a pure spin current in a ferromagnetic insulator/normal metal hybrid structure using either spin pumping or the spin Seebeck effect. These hybrid structures thus can be used as a convenient spin current source and/or detector.

Financial support by DFG SPP 1538 is gratefully acknowledged.

[1] F. D. Czeschka et al., Phys. Rev. Lett. **107**, 046601 (2011).
[2] M. Weiler et al., Phys. Rev. Lett. **108**, 106602 (2012).

FEMTOSECOND SPIN DYNAMICS AND NANOMETER IMAGING WITH LASER-BASED EXTREME ULTRAVIOLET SOURCE

Roman Adam,[1] Dennis Rudolf,[1] Alexander Bauer,[1] Christian Weier,[1] Moritz Plötzing,[1] Patrik Grychtol,[1,2] Chan La-O-Vorakiat,[2] Emrah Turgut,[2] Henry C. Kapteyn,[2] Margaret M. Murnane,[2] Justin M. Shaw,[3] Hans T. Nembach,[3] Thomas J. Silva,[3] Stefan Mathias,[2,4] Martin Aeschlimann[4] and Claus M. Schneider[1]

[1]Peter Grünberg Institut PGI-6 & JARA-FIT, Research Centre Jülich, 52425 Jülich, Germany
[2]Department of Physics and JILA, University of Colorado and NIST, Boulder, CO, USA
[3]Electromagnetics Division, National Institute of Standards and Technology, Boulder CO, USA
[4]University of Kaiserslautern and Research Center OPTIMAS, 67663 Kaiserslautern, Germany

Further understanding of complex materials, such as agnetic alloys and metallic multilayers or multiferroics, requires an experimental technique that allows testing the electronic and magnetic properties of a targeted material element-selectively on femtosecond time scales and with nanometer lateral resolution. The above requirements were recently achieved using a laser-based extreme ultraviolet (XUV) light source generating photon energies ranging from 22 eV up to 72 eV. This energy range covers absorption edges of elements used in technologically significant application, such as Fe (52.7eV), Ni (68eV), Co (59.9 eV), Cr (42 eV) in magnetic multilayers and alloys, or Gd (28eV), Sb(32 eV), Te(41eV) in phase change materials. If the photon energy is tuned to an absorption edge of a targeted material, transmission or reflectivity is resonantly enhanced allowing element selective analysis. At the same time, as the XUV radiation is composed of harmonics of an intense laser pulse, and thus, maintains the time and repetition rate of the fundamental laser light and supports temporal resolution in femtosecond time scale.

Combining both the element- and time-resolution we studied the magnetic response of thin ferromagnetic trilayers containing Ni and Fe layers separated by a thin layer of Ru. By exciting the Ni/Ru/Fe multilayer with a femtosecond laser pulse, we could observe the resulting magnetization response in the Ni and Fe layers simultaneously but separately by tuning XUV probe to the 3p absorption edges of Ni and Fe (Fig.1), detecting magnetization quenching in Ni and Fe layers within 300 fs after laser excitation. This experimental approach has been successfully demonstrated recently also in ferromagnetic alloys [1-3].

Another aspect of laser-generated XUV light is the short wavelength, in range of few nm, more than one order of magnitude shorter compared to visible light. Taking advantage of this fact, we demonstrated a capability to image structures below lateral size of 1 μm in a proof-of-principle experiment using zone plates (Fig.2). Further improvement in Fourier optics is expected to push the lateral resolution well below 1 μm. New analytical possibilities can then be accessible by combining this high lateral resolution with the element selectivity and femtosecond temporal resolution.

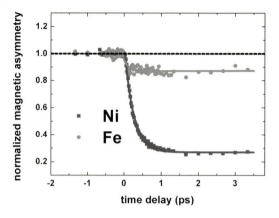

Fig.1. Time- and element-resolved magnetization of Ni and Fe at the corresponding M absorption edges of 66 eV and 52 eV. The magnetization is quenched within 300 fs after laser excitation.

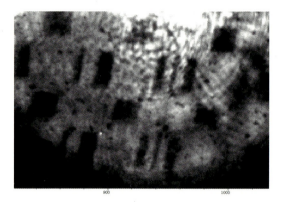

Fig.2 Imaging of a test structures using XUV light with l= 30nm and 5 mm focal distance zone plate demonstrating resolution below 1 mm in the proof-of-principle experiment.

[1] La-O-Vorakiat, C., Siemens, M., Murnane, M., Kapteyn, H., Mathias, S., Aeschlimann, M., Grychtol, P., Adam, R., Schneider, C. M., Shaw, J., Nembach, H. & Silva, T.; *Ultrafast Soft X-Ray Magneto-Optics at the M-edge Using a Tabletop High-Harmonic Source* Phys. Rev. Lett., **2009**, Vol. 103, pp. 257402
[2] La-O-Vorakiat, C., Turgut, E., Teale, C.A., Kapteyn, H.C., Murnane, M.M., Mathias, S., Aeschlimann, M., Schneider, C.M., Shaw, J., Nembach, H. & Silva, T. *Ultrafast Demagnetization Measurements Using Extreme Ultraviolet Light: Comparison of Electronic and Magnetic Contributions* Physical Review X, **2012** Vol. 2(1), p. 011005
[3] Mathias, S., La-O-Vorakiat, C., Grychtol, P., Granitzka, P., Turgut, E., Shaw, J.M., Adam, R., Nembach, H.T., Siemens, M., Eich, S., Schneider, C. M., Silva, T., Aeschlimann, M., Murnane, M.M. & Kapteyn, H. *Probing the timescale of the exchange interaction in a ferromagnetic alloy* PNAS, 2012, 1201371109v1-6

THEORETICAL STUDY OF ULTRAFAST LASER INDUCED MAGNETIC PRECESSIONS.

Daria Popova, Andreas Bringer, Stefan Blügel

Peter Grünberg Institute and Institute for Advanced Simulation, Forschungszentrum Jülich and JARA, 52425 Jülich.

Ultrafast optical control of a magnetic state of a medium is one of the most attracting subjects in the modern magnetism due to its great potential for the development of novel concepts of data storage and information processing. Direct optical control of a magnetic state of a material on subpicosecond time-scales can be realized via the ultrafast inverse Faraday effect (IFE) [1]. In these experiments circularly polarized high-intensity laser pulses with the length of several tens of femtoseconds are used to excite the magnetic system of a sample. The principles of the ultrafast IFE are still unclear.

In order to get insight into this process, we study the interaction of a circularly polarized laser pulse with the length of 100 femtoseconds with a magnetic system. We describe the ultrafast IFE via the stimulated Raman scattering process by the solution of the time-dependent Schrödinger equation [2,3]. We consider a system of atoms coupled with the exchange interaction in a crystal field environment. We study the time evolution of the magnetization during the action of light and magnetic oscillations, which are excited in the system.

We are thankful for the support by the FANTOMAS project.

[1] A. Kimel et al., Nature **435**, 655 (2005).
[2] D. Popova, A. Bringer, S. Blügel, Phys. Rev. B **84**, 214421 (2011).
[3] D. Popova, A. Bringer, S. Blügel, Phys. Rev. B **85**, 094419 (2012).

SPIN RELAXATION INDUCED BY THE ELLIOTT-YAFET MECHANISM IN 5d TRANSITION-METAL THIN FILMS

N. H. Long, Ph. Mavropoulos, S. Heers, B. Zimmermann, Y. Mokrousov and S. Blügel

Peter Grünberg Institut and Institute for Advanced Simulation, Forschungszentrum Jülich and JARA, D-52425 Jülich, Germany

The spin relaxation in non-magnetic metallic thin films is an important effect in spintronics, especially due to the decreasing dimensionality of magnetic multilayers [1]. In systems with space-inversion symmetry, the spin relaxation is mainly due to the Elliott-Yafet mechanism [2-3]: momentum scattering ($\mathbf{k} \to \mathbf{k'}$) at impurities or phonons can cause a spin-flip due to the presence of spin-orbit coupling that entails a $\mathbf{k}$-dependent superposition of spin-up and spin-down in degenerate Bloch states. If one neglects the spin-orbit coupling of the impurity, the spin relaxation time is in a first approximation proportional to the momentum relaxation time. The proportionality factor is the Elliott-Yafet parameter b^2, which quantifies the degree of admixture of states with different spin character in a crystal.

In this work, using the Korringa-Kohn-Rostoker Green function method, the Elliott-Yafet parameter as well as the spin relaxation time by adding adatom impurities are calculated for 5d transition-metal ultrathin films. Our results show that the spin relaxation induced by the Elliott-Yafet mechanism in these materials strongly depends on the film thickness, the film orientation, the shape of Fermi surface as well as the surface states. Moreover, a significant anisotropy effect of the Elliott-Yafet parameter and the spin relaxation rate depending on the spin quantization axis observed.

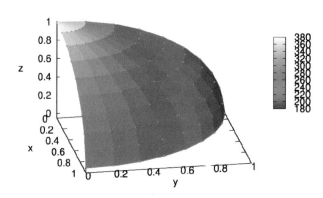

Figure 1: Spin relaxation rate (in units of 1/ps) in bcc W(110) film of 10 layers thickness by adding 1% adatom impurities on top of the film. The position on the surface denotes the direction of the spin-quantization axis, while the relaxation time is given in a colour-code. The maximum value here is for the case of the spin quantization axis in the [001] direction, while the minimum value is for the spin quantization axis in the [010] direction. The anisotropy, which is defined as (max. value – min. value) / (min. value), is 111%.

[1] I. Zutic, J. Fabian, S. Das Sarma, Rev. Mod. Phys. **76**, 323 (2004).
[2] R. J. Elliott, Phys. Rev. **96**, 266 (1954).
[3] Y. Yafet, Solid State Physics, Vol. **14**, 2 (1963).

N-TYPE ELECTRON-INDUCED FERROMAGNETIC SEMICONDUCTOR (In,Fe)As

Pham Nam Hai, Le Duc Anh, Daisuke Sakaki, Masaaki Tanaka

Department of Electrical Engineering and Information Systems, The University of Tokyo, 7-3-1 Hongo, Bunkyo-ku, Tokyo 113-8656, Japan.

Carrier-induced ferromagnetic semiconductors (FMSs) have been intensively studied for decades as they have novel functionalities that cannot be achieved with conventional metallic materials. These include the ability to control magnetism by electrical gating or light irradiation, while fully inheriting the advantages of semiconductor materials such as band engineering. Prototype FMSs such as (In,Mn)As or (Ga,Mn)As, however, are always p-type, making it difficult to be used in real spin devices. This is because manganese (Mn) atoms in those materials work as local magnetic moments and acceptors that provide holes for carrier-mediated ferromagnetism. Here we show that by introducing iron (Fe) and donor atoms into InAs, it is possible to fabricate a new FMS with the ability to control ferromagnetism by both Fe and independent carrier doping. Despite the general belief that the tetrahedral Fe-As bonding is antiferromagnetic, we demonstrate that (In,Fe)As doped with electrons behaves as an n-type electron-induced FMS, a missing piece of semiconductor spintronics for decades [1]. This achievement opens the way to realise novel spin-devices such as spin light-emitting diodes or spin field-effect transistors, as well as helps understand the mechanism of carrier-mediated ferromagnetism in FMSs.

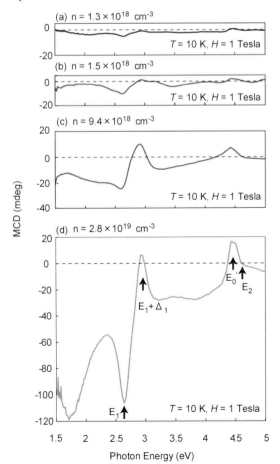

Figure 1: Magnetic circular dichroism (MCD) spectra of 100 nm-thick $(In_{0.82},Fe_{0.08})As$ layers with different electron concentrations. Huge spin-split band was observed when the electron concentration n is larger than 10^{19} cm^{-3}.

Figure 1 shows the magnetic circular dichroism (MCD) spectra of 100 nm-thick $(In_{0.82},Fe_{0.08})As$ layers with different electron concentrations, grown by low-temperature molecular beam epitaxy on semi-insulating GaAs substrates. With increasing the electron concentration (1.3×10^{18} cm^{-3} to 2.8×10^{19} cm^{-3}), the MCD spectra show strong enhancements at optical critical point energies E_1 (2.61 eV), $E_1 + \Delta_1$ (2.88 eV), E_0' (4.39 eV) and E_2 (4.74 eV) of InAs, proving that the band structure of (In,Fe)As is spin-split due to the sp-d exchange interaction between the localized d states of Fe and the electron sea. We demonstrate two phenomena with our new (In,Fe)As FMS: a novel crystalline anisotropic magnetoresistance (AMR) effect with two fold and eight fold symmetry [2], and quantum size effect in ultra thin (In,Fe)As quantum wells in which MCD peaks show systematic blue-shift with decreasing the (In,Fe)As quantum well thickness.

[1] P.N. Hai, L. D. Anh and M. Tanaka, cond-mat, arXiv:1106.0561v3 (2011).
[2] P. N. Hai, D. Sasaki, L. D. Anh, M. Tanaka, cond-mat, arXiv:1202.5874 (2012).

ELECTRICAL SPIN INJECTION AND SPIN TRANSPORT IN ZINC OXIDE

Matthias Althammer[1], Eva-Maria Karrer-Müller[1], Sebastian T.B. Goennenwein[1], Matthias Opel[1], Rudolf Gross[1,2]

[1]Walther-Meissner-Institut, Bayerische Akademie der Wissenschaften, 85748 Garching, Germany;
[2]Technische Universität München, 85748 Garching, Germany

The wide bandgap semiconductor ZnO is interesting for semiconductor spintronics because of its small spin-orbit coupling implying a large spin coherence length. This is a prerequisite for the successful creation, transport, and detection of spin-polarized currents over hundreds of nanometers in typical spintronic devices. In this context, the spin dephasing time of mobile charge carriers - and the associated length scale for coherent spin transport - are fundamental parameters. While other semiconductors like GaAs and related III-V compounds have been studied extensively, only very few reports on the spin coherence in ZnO exist. Using time-resolved *optical* techniques, electron spin coherence at 30 K was observed in epitaxial ZnO thin films with a spin dephasing time of 20 ns [1]. Reports on *electrical* spin injection are rare [2,3] and focus on technical aspects like the improvement of GMR read heads rather than the fundamental spin-dependent properties of ZnO.

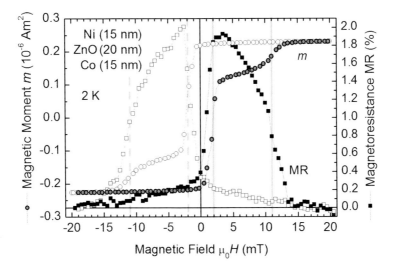

Figure 1: Magnetic moment m (circles, left axis) and magnetoresistance MR (squares, right axis) of a Ni/ZnO/Co spin valve device at 2 K as a function of the magnetic field H applied in-plane. Closed symbols represent data taken with increasing field, open symbols with decreasing field. The $m(H)$ and MR(H) hystereses nicely correspond to each other with regard to their coercive fields (vertical grey lines), evidencing a spin valve behavior of the MR.

We investigate the injection, transport, and detection of spin-polarized charge carriers in ZnO utilizing all-electrical, vertical spin valve devices with ferromagnetic electrodes. Using pulsed laser deposition and electron-beam evaporation, we fabricated epitaxial multilayers of TiN/Co/ZnO/Ni/Au on (0001)-oriented Al_2O_3 substrates with different thicknesses of the ZnO spacer layer ranging from 5 nm to 100 nm. The multilayers were patterned into vertical mesa structures with junction areas between 100 μm^2 and 400 μm^2. Magnetotransport (MR) measurements with the magnetic field and the current applied in and perpendicular to the plane, respectively, show a clear spin valve behavior (Fig. 1). The switching fields correspond to the coercive fields of the ferromagnetic layers as determined by SQUID magnetometry (Fig. 1). For a ZnO thickness of 20 nm, the magnetoresistance (MR) increases from 0.8% at 200 K to 8.5% at 2 K (Fig. 1). We systematically analyze the maximum MR values as a function of the ZnO thickness in the framework of a two spin channel model with a spin-dependent interface resistance [4,5]. From our fits, we obtain spin diffusion lengths for ZnO of 12.3 nm (2 K), 9.2 nm (10 K) and 8.3 nm (200 K). This corresponds to a spin dephasing time of 110 ns at 2 K which exceeds previously published data determined from optical experiments. Here, we electrically create and detect a spin-polarized ensemble of electrons and demonstrate the transport of the spin information across several nanometers in ZnO.

This work was supported by the Deutsche Forschungsgemeinschaft via SPP 1285 (project no. GR 1132/14).

[1] S. Ghosh et al., Appl. Phys. Lett. **86**, 232507 (2005).
[2] Y. Chen et al., Phys. Lett. A **303**, 91 (2002).
[3] K. Shimazawa et al., IEEE Trans. Mag. **46**, 1487 (2010).
[4] T. Valet and A. Fert, Phys. Rev. B **48**, 7099 (1993).
[5] A. Fert and H. Jaffres, Phys. Rev. B **64**, 184420 (2001).

SIT 3

SPIN RELAXATION BY IMPURITY SCATTERING: IMPORTANCE OF RESONANT SCATTERING

<u>Phivos Mavropoulos</u>, Swantje Heers, Rudolf Zeller, and Stefan Blügel

Peter Grünberg Institut and Institute for Advanced Simulation, Forschungszentrum Jülich and JARA, D-52425 Jülich

We present calculated results on spin relaxation in Cu, Ag and Au due to scattering off impurities of the 4th, 5th and 6th row of the periodic table. We consider the processes of spin-flip scattering by spin-orbit coupling (SOC) at the impurity as well as SOC at the host atoms (Elliott-Yafet mechanism). Our results for Cu show that the SOC in the host is not so important, especially in the case of heavy impurities where the impurity SOC dominates. On the contrary, SOC in the host plays a major role in Au. Especially critical, however, is the case of resonant scattering off the impurity, in which the ratio of spin-flip to spin-conserving transition probability can reach very high values (see Figure 1).

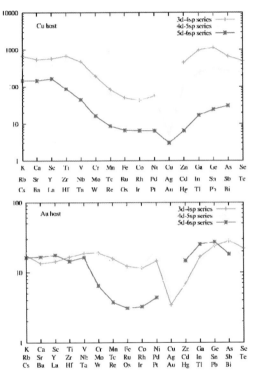

Figure 1: Ratio of total scattering rate, $1/\tau$, to the spin-flip rate, $1/T_1$, for defects in Cu (upper panel) and Au (lower panel) host. Note the logarithmic scale in the ordinate. The ratio T_1/τ shows how many scattering events are needed on the average before a spin-flip event occurs. The effect of the strong spin-orbit coupling in the Au host, as well as in the 5d and 6p impurities, is evident. In addition it can be seen that the d impurities, characterized by resonant scattering, cause an increase of the spin-flip rate.

In our approach we employ density-functional theory and the full-potential Korringa-Kohn-Rostoker Green-function method (KKR) for the calculation of the electronic structure and the spin-flip transition probability $P_{kk'}^{ss'}$, where k is the Bloch wavevector and s the spin-index. A Fermi-surface integral of P over k and k' yields then $1/T_1$.

In order to go beyond the approximation of independent scattering events, implied by this approach, we calculate the scattering by impurity dimers at increasing distance from nearest to third-nearest neighbours. In this way we simulate possible effects of impurity clustering in the sample, caused *e.g.* by impurity attraction or increasing concentration. We find that the presence of nearest-neighbour dimers has a considerable effect on the spin-relaxation rate $1/T_1$, while the results for dimers with distance beyond second-nearest neighbours are practically the same as in the independent scattering approximation.

EXPERIMENTAL AND THEORETICAL ANALYSIS OF OXYGEN-DEFICIENT EuO THIN FILMS

A. Ionescu[1], M. Barbagallo[1], P.M.D.S. Monteiro[1], N.D.M. Hine[1,4], J.F.K. Cooper[1], N.-J. Steinke[1], J.-Y. Kim[1], K.R.A. Ziebeck[1], C.H.W. Barnes[1], C. J. Kinane[2], B.R.M. Dalgliesh[2], T.R. Charlton[2], S. Langridge[2], T. Stollenwerk[3] and J. Kroha[3]

[1] Cavendish Laboratory, Physics Department, University of Cambridge, Cambridge CB3 0HE, U.K.
[2] ISIS, Harwell Science and Innovation Campus, STFC, Oxon OX11 0QX, U.K.
[3] Physikalisches Institut and Bethe Center for Theoretical Physics, Bonn University, D-53115, BRD
[4] Thomas Young Centre, Department of Materials and Department of Physics, Imperial College London, Exhibition Road SW7 2AZ, U.K.

Stoichiometric EuO is a fully spin-polarised insulator with a Curie temperature, T_c=70 K, rendering it one of the most promising materials for spintronics applications. We have carried out both experimental and theoretical studies of the magnetic properties of thin films of oxygen-deficient EuO. Accurate control of the oxygen vacancy concentration in these films was achieved by sputter co-deposition of Eu and Eu_2O_3. The films were characterized by superconducting quantum interference device, x-ray reflectometry and polarized neutron reflectometry (PNR) and the magnetic moment was found to increase monotonically with oxygen vacancy concentration. The electronic structure of EuO_{1-x} was calculated using density-functional theory (DFT+U). In agreement with previous studies, these calculations show that oxygen vacancies act as n-type dopant in EuO and that the excess electrons preferentially populate the majority spin branch of the spin-polarized conduction band. The observed increase in the magnetic moment originating from these excess electrons was accurately determined experimentally and found to be in good agreement with our quantitative predictions [1]. Furthermore, we have also studied how the magnetic properties of oxygen-deficient EuO thin films vary as a function of thickness. The magnetic moment, measured by PNR, and T_c are found to decrease with reducing thickness. Our results indicate that the reduced number of nearest neighbours, band bending and the partial depopulation of the electronic states that carry the spins associated with the 4f orbitals of Eu are all contributing factors in the surface-induced change of the magnetic properties of EuO_{1-x} [2].

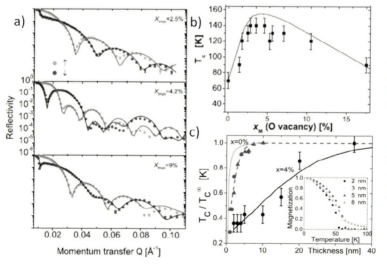

Figure 1: a) PNR measurements and fits to the data of EuO_{1-x} films. b) T_c versus O vacancy concentration and Mauger fit. c) Normalised T_c versus film thickness and (inset) saturation magnetisation versus temperature.

[1] M. Barbagallo et al., PRB **81**, 235216 (2010).
[2] M. Barbagallo et al., PRB **84**, 075219 (2011).

DELTA DOPED ANTIFERROMAGNETIC MANGANITES

T. S. Santos,[1] B. J. Kirby,[2] S. Kumar,[3] S. J. May,[4,5] J. A. Borchers,[2] B. B. Maranville,[2] J. Zarestky,[6] S. G. E. te Velthuis,[4] J. van den Brink,[3] B. Nelson-Cheeseman[4] and A. Bhattacharya[1,4]

1. Center for Nanoscale Materials, Argonne National Laboratory
2. NIST Center for Neutron Research.
3. Institute for Theoretical Solid State Physics, IFW Dresden.
4. Materials Science Division, Argonne National Laboratory.
5. Department of Materials Science and Engineering, Drexel University.
6. Ames Laboratory and Department of Physics and Astronomy, Iowa State University.

In the seminal work of de Gennes, it was recognized that when carriers are added to the bottom of the conduction band in an insulating antiferromagnetic (AF) manganite, they want to delocalize via the double-exchange mechanism.[1] In the simplest single-orbital model this should cause the AF spins to cant, leading to a metallic state with a net ferromagnetic moment. This transformation is thwarted in real materials by various localizing effects such as Jahn-Teller polarons and orbital-ordering, and tendencies towards phase-separation. Using digital synthesis techniques, we demonstrate that delta-doping can be used to realize de Gennes' model system. Using this approach, we create a dimensionally confined region of double-exchange metallic ferromagnetism in an AF manganite host, without introducing any explicit disorder due to dopants or frustration of spins.[2] The delta-doped carriers cause a local enhancement of double-exchange with respect to superexchange, resulting in local canting of the AF spins and enhanced metallicity. This leads to a highly modulated magnetization, as measured by polarized neutron reflectometry and neutron diffraction. The spatial modulation of the canting is related to the spreading of charge from the doped layer, and establishes a fundamental length scale for charge transfer, transformation of orbital occupancy and magnetic order in these manganites. Time permitting I will also discuss some of our more recent results with delta-doping of Ruddlesden-Popper analogs.

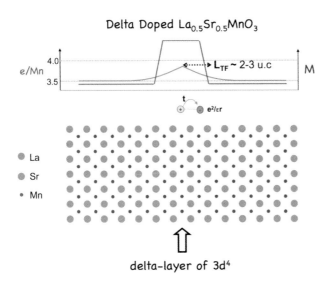

Fig. 1. Schematic of delta-doping in a $LaMnO_3/SrMnO_3$ superlattice.

[1] P.-G. de Gennes, *Phys. Rev.* **118**, 141 (1960).
[2] T. Santos et al., *Phys. Rev. Lett.* **107**, 167202 (2011).

SPIN-ORBIT MEDIATED TORQUES IN HETEROSTRUCTURES WITH STRUCTURAL INVERSION ASYMMETRY

Frank Freimuth[1], Yuriy Mokrousov[1], Stefan Blügel[1]

[1]Peter Grünberg Institut & Institute for Advanced Simulation, Forschungszentrum Jülich and JARA, 52425 Jülich, Germany

Several recent experiments [1,2,3,4] have shown that an in-plane current can switch magnetization in thin ferromagnetic metallic layers asymmetrically sandwiched between an oxide layer on one side and a paramagnetic metal layer on the other side. Two geometrically distinct torques have been identified experimentally. One allows to switch the magnetization direction from perpendicular to in-plane [1,5,6], the other one to switch between the two perpendicular configurations [2,3,4]. The first kind is usually attributed to the Rashba effective magnetic field [7]. Its underlying magnetoelectric effect gives rise to the current-induced spin-polarization (CISP) in paramagnets with structural inversion asymmetry (SIA) [8]. In order to explain torques of the second geometry, both the spin Hall effect (SHE) [3,4] and the perpendicular spin accumulation [2,9] in Rashba systems in the presence of an additional exchange interaction have been considered. In the SHE scenario, the spin current flowing in the perpendicular direction due to the SHE of the paramagnetic metallic layer is absorbed by the ferromagnetic layer, which thereby experiences a torque.

The microscopic origin of the current induced torques in these systems is currently only pooly understood, especially because the magnitude of the effective Rashba interaction has not been measured directly in the experiments, making estimations based on the Rashba model difficult. Interestingly, several experimental works [1,5,6] report a sizable torque due to the Rashba effect, while several others [3,4] find only the torque expected from the SHE, suggesting a strong dependency of the torques on the details of the interface between the magnetic and the oxide layers. Moreover, models which consider contributions both due to SHE and due to the Rashba effect on an equal footing are difficult to construct. Determination of the current-induced torques from density-functional theory (DFT) calculations of the realistic heterostructure is therefore highly needed.

In this presentation we discuss DFT calculations of the current-induced torques in Pt/Co/X heterostructures (X=Vacuum, or Al, or O, or AlO) based on linear response calculations. Our computational linear response formalism is structurally analogous to the one used for magnetoresistance [10,11], anomalous Hall [12] and spin Hall effect [13]. In particular, the linear response theory for the magnetoresistance is formally analogous to the one for the in-plane effective magnetic fields due to the Rashba effect, while the linear response theory for the Hall effects is formally analogous to the one for the perpendicular switching. Thus, the two geometrically different torques differ also with respect to their behavior on the longitudinal resistivity and under time-reversal. We find that while the current-induced in-plane Rashba field is almost negligible in Pt/Co/Vacuum, it becomes sizeable if Al, O, or AlO are deposited on Co, reaching the magnitude measured by Pi et al. [5] in Pt/Co/AlO. This suggests that the origin of the discrepancy between the experimental findings [1,3] on the size of the Rashba field in this system might lie in the details of device fabrication. Moreover, our calculation allows to determine the spin-current in the transverse direction layer resolved and to switch off spin-orbit interaction in the Co layer. This allows to disentangle contributions from the Rashba effect and contributions due to SHE. It is found that both mechanisms can contribute in realistic systems and can generally be of the same size of magnitude. However, in Pt/Co/Vacuum the Rashba contributions to both kinds of torques are comparatively small and only the SHE contribution is substantial, a scenario suggested for Pt/Co/AlO by Liu et al. [3]. On the other hand the Rashba effect contributes substantially to both torques if Al, O, or AlO are deposited on Co, showing that also the scenario of combined SHE and Rashba effects proposed by Miron et al. [1,2] is possible in Pt/Co/X systems depending on the details of the Co/X interface.

[1] I. M. Miron et al., Nature Mat. **9**, 230 (2010)
[2] I. M. Miron et al., Nature **476**, 189 (2011)
[3] L. Liu et al., arXiv: 1110.6846
[4] L. Liu et al., arXiv: 1203.2875
[5] Pi et al., Appl. Phys. Lett. **97**, 162507 (2010)
[6] T. Suzuki et al., Appl. Phys. Lett. **98**, 142505 (2011)
[7] A. Manchon et al., Phys. Rev. B **79**, 094422 (2009)
[8] V. M. Edelstein, Solid State Communications **73**, 233 (1990)
[9] H.-A. Engel et al., Phys. Rev. Lett. **98**, 036602 (2007)
[10] K. M. Seemann et al., Phys. Rev. Lett. **108**, 077201 (2012)
[11] K. M. Seemann et al., Phys. Rev. Lett. **107**, 086603 (2011)
[12] J. Weischenberg et al., Phys. Rev. Lett. **107**, 106601 (2011)
[13] F. Freimuth et al., Phys. Rev. Lett. **105**, 246602 (2010)

MAGNETICALLY ENHANCED MEMRISTOR

<u>Mirko Prezioso</u>[1], Alberto Riminucci[1], Ilaria Bergenti[1], Patrizio Graziosi[1] and Valentin A. Dediu[1]

[1]Consiglio Nazionale delle Ricerche - Istituto per lo Studio dei Materiali Nanostrutturati (CNR-ISMN) via P. Gobetti 101, 40129 Bologna, Italy

The unrelenting demand for increased density of processing and storage elements in the field of Information and Communication Technology (ICT) is clashing against fundamental physical limitations. This calls for conceptually new solutions leading to innovative devices capable of both storage and processing. Recently, architectures capable to meet these requirements have been proposed which have the memristor as the sole building element[1].

We describe here an experimental achievement which adds conceptually new features to a standard memristor principle[2]. We show that electrically controlled magnetoresistance[3] can be achieved in organic devices by combining magnetic bistability, GMR effect[4], and memristive effects. Moreover, this behavior leads to innovative and useful features for information processing. This Magnetically Enhanced Memristor (MEM) is distinguished by a strong intermixture of the electrical and the magnetic degrees of freedom leading to a magnetic modulation of memristance states multitude.

In more details, the devices presented in this work consist of a 20 nm thick bottom LSMO electrode, on which a layer of Alq_3 (with thickness ranging between 3 nm and 250 nm) is evaporated. The top electrode consists of a 20 nm thick Co film, separated from the Alq_3 by a 2 nm thick AlO_x tunnel barrier.

These devices show the typical memristor fingerprint known as the pinched I-V hysteresis. The electrical switching is driven by the applied voltage and are in the 1-3 V range. In the negative branch the I-V features a well controllable NDR region which allows the selection of the memristance state of the device by reaching the necessary voltage and then come back to zero.

By the electrical point of view the R_{off}/R_{on} ratio can reach up to 10^4 while the retention time is in the order of days (at 100K). We also performed Write/Read/Erase/Read cycles for checking the repeatability up to 14000 times.

The same devices, when operated at biases as low as -0.1 V, behaves also as a spin-valve and can show a GMR effect of up to 22% (at 100K).

The most interesting effect is the dependence of the GMR effect on the memresistance state of the memristor. Indeed, the higher is the resistance state the lower is the GMR amplitude. This, added to the possibility of selecting the memristance state, leads to a controllable amount of GMR effect. We present experimental results demonstrating 32 different resistance states with their corresponding GMR amplitude.

Moreover we show that by using the programming voltage and the external magnetic field as inputs, it is possible to obtain an AND logic gate with only a single MEM device.

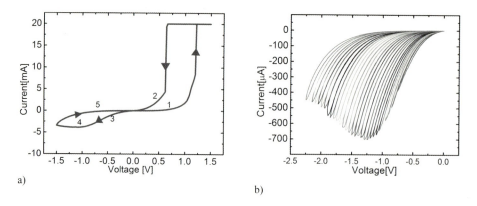

Figure 1:

a) Typical I-V found in our devices. The numbers and the arrows show the sequence. The NDR region correspond to number 4. b) These I-Vs in the negative branch are successively taken to show how to obtain 32 different resistance state by reaching increasingly higher negative programming voltages

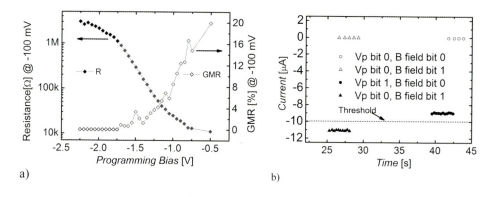

Figure 2:

a) This graphs shows the resistance and the GMR of the device in every of the 32 resistance states resulted by the I-Vs of figure1b. The value are recorded at -100mV the standard bias for the MR curves, which maximize the GMR effect.

b) Here the AND logic gate operation is experimentally demonstrated. Vp (programming bias) represent bit 1 when used to turn ON the device and 0 when it turns it OFF. B field corresponding to antiparallel (lower resistance) state is coded to 1 while it is 0 when it saturates and aligns the magnetization of both electrodes. In this way the current is lower than the threshold only when both inputs are 1. If we code this current level to 1, inputs and outputs maps to the AND truth table.

[1] J. Borghetti, et al., Nature, **464**, 873,(2010).
[2] L. Chua, Applied Physics a-Materials Science & Processing, **102**, 765,(2011).
[3] M. Prezioso, et al., Adv. Mater., **23**, 1371,(2011).
[4] V. A. Dediu, et al., Nat. Mater., **8**, 707,(2009).

NOVEL FUNCTIONALITIES AT INTERFACES IN $La_{0.7}Ca_{0.3}MnO_3$ /$PrBa_2Cu_3O_7$/$La_{0.7}Ca_{0.3}MnO_3$ MAGNETIC TUNNEL JUNCTIONS

Fabián A. Cuellar[1], Yaohua Liu[2], Norbert M. Nemes[1], Mar Garcia Hernandez[3], John Freeland[4], Juan Salafranca[5], Satoshi Okamoto[5], Suzanne G. E. te Velthuis[2], María Varela[5], Stephen J. Pennycook[5], Manuel Bibes[6], Agnes Barthélémy[6], Zouhair Sefrioui[1], Carlos Leon[1], Jacobo Santamaria[1]

[1] GFMC, Dpto. Fisica Aplicada III, Univ. Complutense Madrid, 28040 Madrid, Spain.
[2] Materials Science Division, Argonne National Laboratory, Argonne, IL 60439, USA.
[3] Instituto de Ciencia de Materiales de Madrid, 28049 Cantoblanco, Spain.
[4] Advanced Photon Source, Argonne National Laboratory, Argonne, IL 60439, USA.
[5] Materials Sci. & Technology Div., Oak Ridge National Lab, Oak Ridge, TN 37831, USA.
[6] Unité Mixte de Physique CNRS/Thales, 91767 Palaiseau, France.

Novel emergent phenomena at interfaces between correlated transition metal oxides have attracted great interest due to both new fundamental physics and technological opportunities [1,2]. In particular, it has been reported that charge transfer and orbital reconstruction at the interface in cuprate-manganite heterostructures may provide new superexchange paths which induce magnetism in otherwise non magnetic compounds [3], and this effect seems to be a quite general phenomenon [4,5]. Here we report novel functionalities of magnetic tunnel junctions ensuing from this interfacial magnetism. We have studied $La_{0.7}Ca_{0.3}MnO_3$/ $PrBa_2Cu_3O_7$/ $La_{0.7}Ca_{0.3}MnO_3$ (LCMO/ PBCO/ LCMO) magnetic tunnel junctions (MTJs) where the non superconducting cuprate PBCO serves as a barrier. High quality epitaxial layers have been grown by high pressure sputtering technique in pure oxygen and MTJs have been patterned as square pillars (4 x 4 µm) using standard lithography and ion milling techniques. The LCMO top and bottom electrodes were 8 and 50 nm thick respectively, while the thickness of the PBCO barrier was varied between 2 and 6 nm. Electron microscopy observations evidence that the barrier is continuous and free of pinholes. Studies of magnetic switching of top and bottom layers have been carried out by fits of both the non spin flip and the spin flip components of polarized neutron reflectometry (PNR). Also from these measurements, the saturation moment is found to be consistent with the optimal hole doping of both layers. X-ray magnetic circular dichroism (XMCD) measurements show a large induced Cu moment at both interfaces that disappears when temperature is increased beyond the Curie temperature of the neighboring manganite electrode.

Magnetic field was applied in plane along the [110] direction of the $SrTiO_3$ substrate and swept in a hysteresis loop sequence in order to measure the tunnelling magnetoresistance (TMR). Below the Curie temperature of both electrodes, (magneto) resistance (MR) exhibited clear plateaus between the coercive fields of top and bottom layers, characteristic of a misaligned magnetization state of the two ferromagnetic electrodes with a positive spin polarization. For MTJs with a thin enough barrier, we have found that TMR at the highest temperatures (between 90 and 100 K) shows an anomalous negative value at remanence, in a narrow magnetic field range before the coercivity of the bottom layer (see Fig.1(a)). This negative TMR shows a strong dependence on the applied bias. In fact, it is only observed in a limited bias range (see Fig.1(b)).

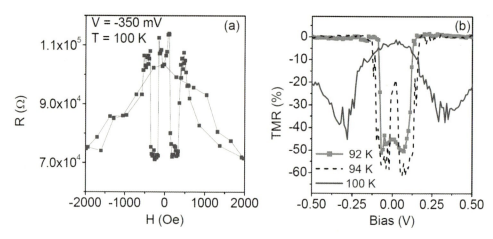

Figure 1: (a) R(H) at T=100 K showing negative TMR values at low magnetic fields. (b) Bias dependence of the negative TMR at several temperatures.

We believe that this behaviour can be accounted for by two competing magnetic interactions. Biaxial anisotropy with top and bottom layers having different easy axes which tends to keep the moments of top and bottom electrodes pointing in different directions (misaligned) in low applied fields, and a (ferro) magnetic coupling mediated by the induced Cu moments which tends to align moments [6].

Our results show that for sufficiently thin PBCO barriers the moment induced on the interfacial Cu atoms may yield a novel mechanism of magnetic coupling between the LCMO electrodes, mediated by the localized (spin-polarized) Cu electrons in the CuO_2 planes. This finding will stimulate the search of novel functionalities of MTJs for future spintronic devices based on the use of emergent properties of oxide interfaces.

Acknowledgement: Work supported by Spanish MICINN Grants MAT 2008 06517 and CSD2009-00013. Work at ANL supported by US-DOE, Office of Science, BES, No. DE-AC02-06CH11357. Research at ORNL (SJP and MV) was sponsored by the Materials Sciences and Engineering Division of the U.S. Department of Energy (DOE).

[1] J. Mannhart and D. G. Schlom, Science **327**, 1607-1611 (2010).
[2] H. Y. Hwang, et al., Nature Materials **11**, 103–113 (2012).
[3] J. Chakhalian et al., Nature Physics **2**, 244-248 (2006); Science **318**, 1114-1117 (2007).
[4] J. Garcia-Barriocanal et al., Nature Comm. **1**:82 doi: 10.1038/ncomms1080 (2010).
[5] S. Valencia, et al., Nature Materials **10**, 753-758 (2011).
[6] J. Salafranca and S. Okamoto, Phys. Rev. Lett. **105**, 256804 (2010).

INTEGRATION OF A MAGNETIC OXIDE DIRECTLY WITH SILICON

Martina Müller[1], C. Caspers[1], A. X. Gray[2], A. M. Kaiser[1,2], A. Gloskovskii[3], W. Drube[4], M. Gorgoi[5], C. S. Fadley[2], and C. M. Schneider[1]

[1]Peter Grünberg Institut (PGI-6), Forschungszentrum Jülich, 52428 Jülich, Germany;
[2]Department of Physics, University of California Davis, CA, USA;
[3]Institut für Anorganische Chemie, Johannes Gutenberg Universität, 55128 Mainz, Germany;
[4]DESY Photon Science, Deutsches Elektronen-Synchrotron, 22603 Hamburg, Germany;
[5]Helmholtz-Zentrum für Materialien und Energie, BESSY II, Berlin, Germany

One major challenge in present-day spintronics research is the efficient electrical injection of spins in semiconductors. Owing to its ability to generate almost fully spin-polarized currents, the magnetic oxide europium oxide (EuO) has recently been revisited as a material with outstanding potential for spintronics. The intriguing coexistence of magnetic and insulating properties - although at low temperatures - makes EuO "spin filter" tunnel barriers highly interesting for solving the long-standing conductivity mismatch problem of spin injection into SCs. Besides, the structural compatibility and theoretically predicted chemical stability with silicon should principally allow for a seamless integration of EuO with this mainstay of semiconductors.

We present a comprehensive study on the growth and electronic structure of ultrathin EuO films on Si(001) [Fig.1]. We succeeded in epitaxially integrating the binary magnetic oxide EuO with silicon without any buffer layer via reactive molecular beam epitaxy (MBE). Given the high reactivity of both constituents, this presents a major challenge and requires to prepare a structurally and chemically sharp interface.

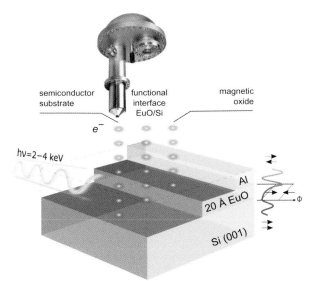

Figure 1: Schematic view of a magnetic oxide/silicon heterostructure as investigated by hard x-ray photoemission spectroscopy (HAXPES) experiments at BESSY (Berlin, Germany) and PETRAIII (Hamburg, Germany). The information depth of the HAXPES experiment can be sensitively tuned to probe (i) the silicon substrate, (ii) the EuO/Si interface, or (iii) the EuO thin film. Functionalized EuO/Si heterostructures will serve as spin-selective magnetic tunnel barriers for silicon spintronic uses [2,3].

Our pathway towards a functional EuO /Si heterostructure is based on the sensitive choice of EuO synthesis parameters and Si(001) surface treatment [1]. By providing a hydrogen-terminated surface, we observe fcc EuO layer-by-layer growth on Si(001) up to 10 nm thickness in a reflection high-energy electron diffraction (RHEED) study. The magnetic properties of such ultrathin EuO/Si heterostructures are close to that of bulk EuO.

Using hard X-ray photoemission spectroscopy (HAXPES), we demonstrate the successful chemical stabilization of stoichiometric EuO thin films grown directly on Si without any buffer layer [2, 3]. Due to its large probing depth of several nm, HAXPES allows direct access to the bulk electronic structure of buried films and interfaces, and enables a depth profiling of their chemical homogeneity, which is not possible with conventional PES. Experiments have been carried out at the beamlines HIKE (BESSY, Berlin) and P09 (PETRAIII, Hamburg). We carefully determine the initial state Eu valency via a quantitative analysis of core-level and valence spectra. We performed depth-sensitive measurements and extracted the Eu chemical state in the bulk and interface regions, confirming that nearly ideal, homogeneous and stoichiometric thin films of the magnetic oxide EuO can be grown on silicon. Furthermore, our depth-sensitive HAXPES study reveals that an interfacial silicide ($EuSi_2$) and/or silicon dioxide (SiO_x) formation can be minimzed down to 0.3 nm, respectively.

Our study explicitly demonstrates the successful structural and chemical stabilization of EuO thin films directly on silicon, and points encouragingly towards the future integration of this functional magnetic oxide into silicon-based spintronic devices.

M. M. acknowledges financial support by DFG under grant MU 3160/1-1.

[1] C. Caspers, M. Müller, A. Gloskovskii, W. Drube, M. Gorgoi, C. S. Fadley, and C. M. Schneider, submitted (2012).
[2] C. Caspers, M. Müller, A. X. Gray, A. M. Kaiser, A. Gloskovskii, W. Drube, C. S. Fadley, and C. M. Schneider, Physical Review B **84**, 205217 (2011).
[3] C. Caspers, M. Müller, A. X. Gray, A. M. Kaiser, A. Gloskovskii, W. Drube, C. S. Fadley, and C. M. Schneider, Phys. Status Solidi RRL **5**, Vol. 12, 441 (2011).

A SPINTRONIC MEMRISTOR

J. Grollier[1], A. Chanthbouala[1], J. Sampaio[1], P. Metaxas[1], R. Matsumoto[1], A. Anane[1], A. V. Khvalkovskiy[1], V. Cros[1], A. Fert[1], K. A. Zvezdin[2], A. Fukushima[3], H. Kubota[3], K. Yakushiji[3], S. Yuasa[3]

[1]Unité Mixte de Physique CNRS/Thales, Palaiseau, France
[2]Istituto P.M., Turin, Italy [3]National Institute of Advanced Industrial Science and Technology, Tsukuba, Japan

We will present a spintronic memristor, with resistance changes based on purely electronic phenomena. Basically, a memristor is a tiny non-volatile analog tuneable resistance. The more intense is the current through the structure, and the longer it is injected, the more the resistance changes [1]. In spintronics, the "spin transfer" effect allows to manipulate a magnetic domain wall by current injection [2]. The domain wall displacement is precisely proportional to the amplitude and width of the current pulse [3]. A spintronic memristor can therefore be obtained by converting the domain wall displacements in resistance variations, through the use of a magneto-resistive structure [4].

Most experiments on domain wall motion by spin transfer have been performed on single magnetic layer or metallic spin valves, leading to very small resistance variations of the order of 1-10% (Anisotropic or Giant Magneto-Resistance) [5]. In addition, the usual configuration is "lateral", i.e. the current is injected directly through the wire in which the domain wall propagates. In this geometry, the critical current densities are of the order of 10^8 A.cm^{-2}, too large for applications [6].

Here we show that by combining the use of a sub-micron magnetic tunnel junction together with a vertical current injection, we can obtain large resistance variations (~ 100 %), low current densities (~ 5 10^6 A.cm^{-2}), and fast switching (< 1 ns, corresponding to a domain wall speed of ~ 500 m/s) [7,8]. Our device stack is similar to the one that will be used in the next generation of STT-MRAMs, and is therefore CMOS compatible. This work paves the way towards the implementation of reliable, purely electronic memristors based on spintronics effects.

We thank Canon ANELVA for the preparation of magnetic films. Financial support by the ERC 2010 Stg 259068 is acknowledged.

[1] L. O. Chua, IEEE Trans. Circuit Theory 18, 507–519 (1971)
[2] L. Berger, J. Appl. Phys. 55, 1954 (1984); 71, 2721 (1992)
[3] A. Yamaguchi et al., Phys. Rev. Lett. **92**, 077205 (2004)
[4] X. Wang et al., **30** (3), 294 (2009)
[5] J. Grollier et al., J. Appl. Phys **92**, 4825 (2002) & Appl. Phys. Lett **83**, 509 (2003)
[6] Stuart S. P. Parkin, et al. Science **320**, 190 (2008)
[7] A.V. Khvalkovskiy et al. , Phys. Rev. Lett. **102**, 067206 (2009)
[8] A. Chanthbouala et al., Nature Phys. **7**, 626 (2011)

MEMRISTIVE MAGNETIC TUNNEL JUNCTIONS AND THEIR APPLICATIONS

Andy Thomas[1], Patryk Krzysteczko[1], Günter Reiss[1], Jana Münchenberger[1], Markus Schäfers[1]

[1]Bielefeld University, Thin films and physics of nanostructures, Bielefeld, Germany

We demonstrate that tunnel magnetoresistance and resistive switching can be observed simultaneously in nanoscale magnetic tunnel junctions with MgO barriers. The devices show bipolar resistive switching of 6% and tunnel magnetoresistance ratios of about 100%. For each magnetic state, multiple resistive states are created depending on the bias history. The electronic transport measurements are discussed in the framework of a memristive system, where we show that the flux is a good variable for describing voltage-induced resistance variations. Several systems are compared to gain insight into the switching mechanism [1].

The simultaneous observation of the resistive switching and magnetoresistance is promising to improve already established technologies. We use the memristance in magnesium-oxide-based magnetic tunnel junctions to improve the error tolerance in magnetic random access memory and magnetic field programmable logic arrays [2].

We used the same memristive magnetic tunnel junctions to demonstrate that the synaptic functionality is complemented by neuron-like behavior in these nanoscopic devices. The synaptic functionality originates in a resistance change caused by a voltage-driven oxygen vacancy motion within the MgO layer. The additional functionality provided by magnetic electrodes enabled a current-driven resistance modulation due to spin-transfer torque. Since resistivity provides a natural measure of the synaptic strength, and because of the bipolar nature of the resistance change, long term potentiation and long term depression were mirrored. Furthermore, it provides the scope for the emulation of spike timing dependent plasticity as well [3].

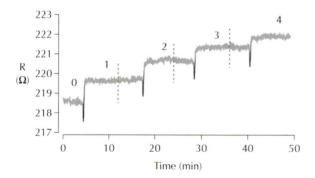

Figure 1: Resistance change in an MgO based magnetic tunnel junction after repeated treatment with spike trains. A good stability of the resistive states (1,2,3 and 4) is observed between the spike trains. The tunnel magnetoresistance ratio at every state is close to 100%.

[1] P. Krzysteczko, G. Reiss, A. Thomas: Memristive switching of MgO based magnetic tunnel junctions, Appl. Phys. Lett. **95**, 112508 (2009).
[2] J. Münchenberger, P. Krzysteczko, G. Reiss, A. Thomas: Improved reliability of magnetic field programmable gate arrays through the use of memristive tunnel junctions, J. Appl. Phys. **110**, 096105 (2011).
[3] P. Krzysteczko, J. Münchenberger, M. Schäfers, G. Reiss, A. Thomas: The memristive magnetic tunnel junction as a nanoscopic synapse-neuron-system, Adv. Materials **24**, 761 (2012).

STS 6

ANTIFERROMAGNETIC COUPLING ACROSS SILICON REGULATED BY TUNNELING CURRENTS

Rashid Gareev[1], MaximilianSchmid[1], Johann Vancea[1], Christian Back[1], Reinert Schreiber[2], Daniel Bürgler[2], Claus Schneider[2], Frank Stromberg[3], Heiko Wende[3]

[1] Institute of Experimental and Applied Physics, University of Regensburg, 93040 Regensburg, Germany;
[2] Institute of Solid State Research (IFF), Research Center Jülich, 52428 Jülich, Germany;
[3] Faculty of Physics and Center for Nanointegration Duisburg-Essen (CeNIDE), University of Duisburg-Essen, 47048 Duisburg, Germany

We report on the enhancement of antiferromagnetic coupling in Fe/Si/Fe structures prepared by molecular beam epitaxy. Using the ballistic electron magnetic microscopy [1] we established that the hot-electron collector current reflects magnetization alignment and the magnetocurrent exceeds 200 % at room temperature (Figure 1) [2]. The saturation magnetic field for the collector current corresponding to the parallel alignment of magnetizations rises up with the tunneling current, thus demonstrating stabilization of the antiparallel alignment and increasing antiferromagnetic coupling (Figure 2). We connect the enhancement of antiferromagnetic coupling with local dynamic spin torques mediated by voltage-driven spin-polarized tunneling electrons with high current densities [2].

This work is supported by the project DFG 9209379.

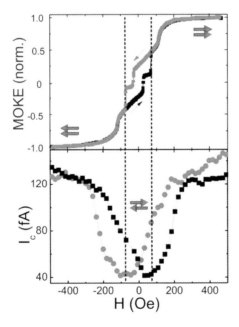

Figure 1: Longitudinal MOKE hysteresis and collector current I_c hysteresis taken at $U_{bias} = -2.5$ V. The in-plane magnetic field is aligned along the easy-axis [110] direction. The $I_c(H)$ hysteresis loops are averaged over 20 cycles. Thin arrows indicate the sweep direction of the magnetic field, thick arrows show magnetization alignment.

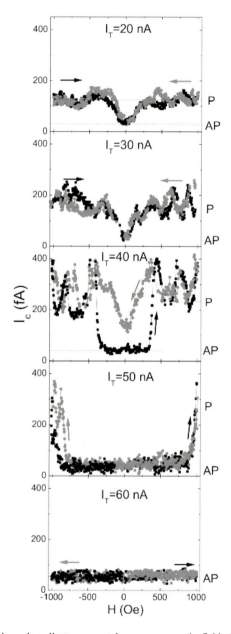

Figure 2: Single-cycle collector current I_c versus magnetic field aligned along the easy-axis [110] direction for different values of the tunneling current I_T measured at $U_{bias} = -2.5$ V [2]. The quadrats and circles correspond to positive and negative magnetic field sweep directions indicated by arrows, correspondingly. Parallel and antiparallel alignment of magnetizations is marked as P and AP, accordingly. Details of the measurement techniques and the material parameters are given in Ref. [2].

[1] E. Heindl et al., J. Magn. Magn. Mater. **321**, 3693 (2009).
[2] R.R. Gareev et al., Appl. Phys. Lett. **100**, 022406 (2012).

NON-VOLATILE ELECTRICAL CONTROL OF MAGNETISM IN MANGANESE-DOPED ZINC OXIDE

Antonio Ruotolo[1], Xiao Lei Wang[1], Chi Wah Leung[2], Rolf Lortz[3]

[1]City University of Hong Kong, Kowloon, Hong Kong, China; Hong Kong Polytechnic University, Kowloon, Hong Kong, China; The Hong Kong University of Science and Technology, Kowloon, Hong Kong, China

Semiconductor oxides can be functionalized to show magnetic properties. In particular, Manganese-doped Zinc Oxide (Mn-ZnO) shows ferromagnetism at room temperature [1]. Ferromagnetism in ZnO is believed to arise from oxygen vacancies [1,2]. Oxygen vacancies are also believed to be responsible for resistive switching in semiconductor oxides [3,4], an effect that could be applied for solid-state memories.

We here show that resistive switching in Mn-ZnO coexists with a switching of the magnetic phase. Schottky junctions were fabricated that use Mn-ZnO as a semiconductor and show reliable resistive switching with good retention time. We found that a change of the resistance corresponds to a change of the magnetic moment in the film. In other words, the magnetic phase can be altered in a reversible and non-volatile manner by applying an electric command.

The studied system consists of $Zn_{0.98}Mn_{0.02}O$ film grown on a highly conductive Niobium-doped Strontium Titanate (NSTO) substrate to allow the current to flow uniformly perpendicular to the film after the deposition of a Platinum (Pt) top electrode. After an irreversible forming step, the *I-V* curve becomes hysteretic (see Figure 1), and the device can reliably be switched between a high-resistive state (HRS) and a low-resistive state (LRS). A similar behavior has been reported in several Schottky contacts to semiconductor oxides [5,6] and ascribed to a change of carrier concentration in the space charge due to a redox process.

Figure 2 shows the full magnetization loop of the film-device at room temperature in the two states. The loops are measured by using a SQUID magnetometer after switching the film-device in either state. We found a reduction of the magnetic moment by 40% in samples with a thickness of ZMO of 120 nm. This change is consistent with the change of the depletion layer width in the two resistive states as estimated by capacitance versus voltage measurements.

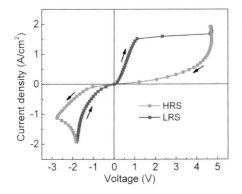
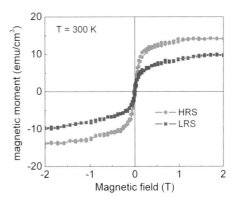

Figure 1: Bipolar d.c. I-V switching of a 5 mm²-wide, 120 nm-thick $Zn_{0.98}Mn_{0.02}O$ film device. The blue curve represents a lower resistive state, while the red one shows a higher resistive state.

Figure 2: Magnetization versus in-plane applied field in the HRS (red curve) and LRS (blue curve) of a 120 nm-thick $Zn_{0.98}Mn_{0.02}O$ film device at T = 300 K.

The experiment was carried out on films with different thickness. We could conclude that the effect does not arise from a formation of filamentary channels in the depletion layer but it is rather a homogeneous process, confined to the space-charge region of the Schottky junction.

[1] P. Sharma et al. Nat. Mater. **2**, 673 (2003).
[2] D. C. Kundaliya et al., Nat. Mater. **3**, 709 (2004).
[3] D. B. Strukov et al., Nature **453**, 80 (2008).
[4] R. Waser and M. Aono, Nat. Mater. **6**, 833 (2007).
[5] T. Fujii et al., Phys. Rev. B **75**, 165101 (2007).
[6] A. Ruotolo et al., Phys. Rev. B **77**, 233103 (2008).

MFH 1

MAGNETOELECTRICALLY INDUCED GIANT TUNNELING ELECTRORESISTANCE EFFECT

J. D. Burton,[1] Yuewei Yin,[2] X. G. Li,[3] Young-Min Kim,[4] Albina Y. Borisevich,[4] Qi Li[2] and Evgeny Y. Tsymbal[1]

[1]University of Nebraska, Department of Physics and Astronomy and Nebraska Center for Materials and Nanoscience, Lincoln, NE, USA.
[2]Penn State, Department of Physics, University Park, PA, USA.
[3]University of Science and Technology of China, Hefei, China.
[4]Oak Ridge National Laboratory, Oak Ridge, TN, USA.

The control and utilization of the magnetic, ferroelectric, and transport properties in complex oxide heterostructures is one of the most promising avenues for electronic devices. Here we present theoretical and experimental evidence of a novel magnetoelectric mechanism producing a giant resistive switching effect at the interface between a ferroelectric perovskite oxide, $BaTiO_3$, and a complex oxide manganite electrode, $La_{1-x}Sr_xMnO_3$. First principles density functional calculations predict a cross-coupling between ferroelectric polarization of the ferroelectric $BaTiO_3$ and the magnetic order of the manganite $La_{1-x}Sr_xMnO_3$.[1] The hole doped La-manganites like $La_{1-x}Sr_xMnO_3$ possess a rich magnetic phase diagram as a function of carrier concentration, x. In addition to chemical doping, carrier concentration can be modulated electrostatically at the interface with a ferroelectric material. By choosing x to reside near the ferromagnetic-antiferromagnetic phase boundary around $x \sim 0.5$, the $La_{1-x}Sr_xMnO_3/BaTiO_3$ interface exhibits a transition from ferromagnetically ordered Mn spins to antiferromagnetic order. This leads to a net change in interface magnetization of $\Delta M = 7.05\mu_B$ per interface Mn, constituting a giant interfacial magnetoelectric effect.

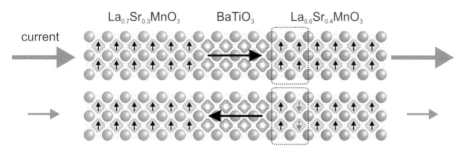

Figure 1: The structure and functionality of an $La_{0.7}Sr_{0.3}MnO_3/BaTiO_3/La_{0.6}Sr_{0.4}MnO_3$ ferroelectric tunnel junction. When the ferroelectric polarization (large arrow) of the $BaTiO_3$ tunneling barrier is reversed the ordering of the magnetic moments (small arrows) near the interface with the right $La_{0.6}Sr_{0.4}MnO_3$ electrode changes. This change leads to a significant decrease in the tunneling current, and therefore a large change in the resistance.

This magnetoelectric effect provides additional functionality when it is incorporated into a Ferroelectric Tunnel Junction (FTJ). FTJs consist of two metal electrodes separated by a thin ferroelectric barrier through which electrons can tunnel.[2] FTJs, in general, exhibit the fundamental characteristic of a large change in tunneling resistance with reversal of the ferroelectric polarization, or Tunneling ElectroResistance (TER). Using first principles density functional calculations we predict that FTJs incorporating a magnetoelectrically active $La_{1-x}Sr_xMnO_3/BaTiO_3$ interface exhibit a giant TER effect. [3] We show that this effect is due to a few atomic layers of the highly spin-polarized $La_{1-x}Sr_xMnO_3$ layer near the interface acting as an electrically controlled atomic scale spin-valve in series with the ferroelectric tunnel barrier creating a switch for the tunneling conductance.

Stimulated by these theoretical predictions, we have experimentally fabricated epitaxial all oxide FTJs with La-manganite electrodes sandwiching a $BaTiO_3$ tunneling barrier via pulsed laser deposition on $SrTiO_3$ substrates.[7] We find that $La_{0.7}Sr_{0.3}MnO_3/BaTiO_3/La_{0.7}Sr_{0.3}MnO_3$ FTJs display a modest TER of ~30%. When, however, one interface in the FTJ is modified to incorporate a thin layer (2 unit-cells) of $La_{0.5}Ca_{0.5}MnO_3$ between the $BaTiO_3$ layer and the $La_{0.7}Sr_{0.3}MnO_3$ electrode, the TER effect is enhanced to as large as ~10,000%. This enhancement can be attributed to the extreme sensitivity of the thin $La_{0.5}Ca_{0.5}MnO_3$ interfacial layer to the electrostatic doping induced by the $BaTiO_3$ ferroelectric polarization. This giant effect is reproducible across several samples and provides a strong indication that this novel effect can serve as a viable route to future all oxide electronic and spintronics applications.

[1] J. D. Burton and E. Y. Tsymbal, Phys. Rev. B **80**, 174406 (2009).
[2] E. Y. Tsymbal and H. Kohlstedt, Science **313**, 181 (2006).
[3] J. D. Burton and E. Y. Tsymbal, Phys. Rev. Lett. **106**, 157203 (2011).
[4] Y. W. Yin, J. D. Burton, X. G. Li, Y.-M. Kim, A. Y. Borisevich, E. Y. Tsymbal and Q. Li (unpublished)

MFH 2

REVERSIBLE ELECTRICAL SWITCHING OF SPIN POLARIZATION IN MULTIFERROIC Co/Pb(Zr$_{0.2}$Ti$_{0.8}$)O$_3$/La$_{0.7}$Sr$_{0.3}$MnO$_3$ TUNNEL JUNCTIONS

Daniel Pantel, Silvana Goetze, Marin Alexe, and <u>Dietrich Hesse</u>

Max Planck Institute of Microstructure Physics, D-06120 Halle, Germany

Spin polarized transport in ferromagnetic tunnel junctions, characterized by tunnel magnetoresistance (TMR), has already proven a high application potential in the field of spintronics and in magnetic random access memories. Until recently, in such a junction the insulating barrier played only a passive role keeping apart the ferromagnetic electrodes in order to allow electron tunneling. However, a new dimension was added to these devices by replacing the insulator with a ferroelectric, which possesses permanent dielectric polarization switchable between two stable states. The obtained multiferroic tunnel junction (MFTJ) is a non-volatile memory device with four states, given by a combination of two possible ferroelectric polarization directions in the barrier and two different magnetization alignments of the electrodes.

Here, we show that due to the coupling between magnetization and ferroelectric polarization at the interface between a macroscopic magnetic cobalt electrode and the ferroelectric Pb(Zr$_{0.2}$Ti$_{0.8}$)O$_3$ barrier of a Co/PbZr$_{0.2}$Ti$_{0.8}$O$_3$/La$_{0.7}$Sr$_{0.3}$MnO$_3$ (Co/PZT/LSMO) MFTJ, the spin polarization of the tunneling electrons can be reversibly and remanently *inverted* by switching the ferroelectric polarization of the barrier.

Co/PZT/LSMO tunnel junctions were grown on (100)-oriented SrTiO$_3$ (STO) crystals. PZT and LSMO films were epitaxially grown by pulsed laser deposition at temperatures between 550 °C and 600 °C in an oxygen atmosphere of 0.2 mbar by ablating stoichiometric ceramic targets (with 10% Pb excess for the PZT target). The 0.1° off-cut STO (100)-oriented substrates were etched and annealed before deposition to obtain an atomically flat, single TiO$_2$-terminated surface. The energy fluence of the KrF excimer laser (λ = 248 nm) at the target was about 300 mJ/cm^2. The ferromagnetic top Co electrodes (as large as 4000 μm^2) were prepared by thermal evaporation from a tungsten coil through a shadow mask and subsequently capped by a protective Au layer. The pressure during the room temperature deposition process was kept below 7x10^{-6} mbar to prevent oxidation of the Co layer.

The oxide heterostructures are fully strained to the STO substrate as shown by high-resolution transmission electron microscopy (HRTEM) and X-ray diffraction. The ferroelectricity of the ultrathin PZT layers was proven by piezoresponse force microscopy (PFM). An a.c. probing voltage (f = 24.5 kHz) with an amplitude of 0.5 V (below the effective coercive voltage of the PZT films) was applied to the tip. A lock-in amplifier was used to detect the piezoresponse signal. For remnant hysteresis measurements a poling voltage pulse (200 ms) was applied to the tip to switch the polarization. Afterwards the PFM signal was measured. The poling voltage was varied between +/−3 V.

Temperature-dependent magnetoresistance measurements were performed using a low-temperature probe station equipped with a superconducting solenoid. The measurements were performed in two-probe geometry in constant voltage mode (10 mV). The resistance of the bottom electrode (about 30 Ω at 10 K) was estimated from the LSMO resistivity to be well below the junction resistance not influencing the measurement. The ferroelectric polarization was switched by rectangle voltage pulses (±3 V) applied to the top Co electrode generated by an arbitrary function generator.

Figure 1a shows a HRTEM image of part of a Co/PZT/LSMO MFTJ on STO, whereas Figures 1b-d show an AFM topography image of the 50 nm thick as-grown LSMO layer, a PFM image of the 3.2 nm thick as-grown PZT layer, and PFM phase (upper) and amplitude (lower) curves proving the ferroelectric nature of the PZT layer sandwiched between Co and LSMO layers.

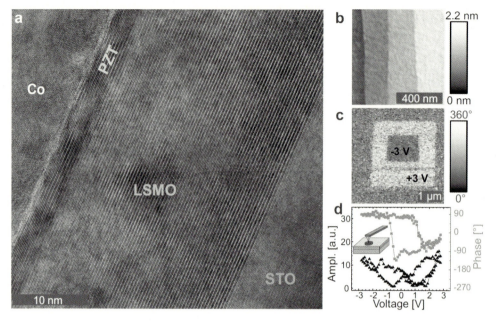

Fig. 1 (from D. Pantel et al., Nature Materials 2012)

Figure 2a shows the resistance vs. magnetic field curves measured at 50 K in the as-grown state of the MFTJ (lower curve) and after ferroelectric polarization switching with +3V applied electrical bias (upper curve). The polarization state of the barrier as well as the magnetization directions in each magnetic layer are schematically shown for each non-volatile state. Figure 2b shows the resistance (squares) and TMR (dots) after repeated switching with ± 3V voltage pulses for a different junction at 10 K.

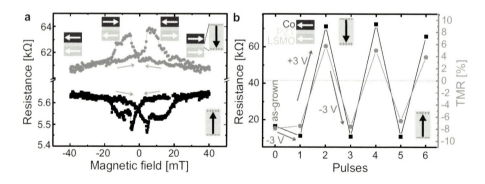

Fig. 2 (from D. Pantel et al., Nature Materials 2012)

Most remarkably, the sign of the TMR is reversibly and remanently switched from inverse to normal and back by the polarization switching of the PZT layer. This is most probably due to an inversion of the spin polarization at the Co/PZT interface. Thus the spin direction of carriers injected into, e.g., a spin-based device can be selected without any applied magnetic field, only by the ferroelectric polarization direction. The microscopic mechanism of this effect is most probably given by a strong magnetoelectric coupling between ferroelectric polarization and magnetization at the Co/PZT interface.

MFH 3

MAGNETOELASTIC AND MAGNETOELECTRIC EFFECTS IN COMPOSITE MULTIFERROIC HYBRID STRUCTURES

Stephan Geprägs[1], Matthias Opel[1], Sebastian T.B. Goennenwein[1], Rudolf Gross[1,2]

[1]Walther-Meissner-Institut, Bayerische Akademie der Wissenschaften, 85748 Garching, Germany; [2]Technische Universität München, 85748 Garching, Germany

Multiferroic materials, which simultaneously combine at least two long-range ordering phenomena, have attracted widespread interest over the last years due to their rich physics and large variety of potential applications [1]. In particular, cross-coupling effects between the different ferroic phases pave the way to novel phenomena as well as enhanced and improved features in future engineered devices [2]. For example, the control of the magnetic properties by local electric fields in extrinsic multiferroic composites, in which ferromagnetic and ferroelectric compounds are artificially assembled, enable large and robust converse magnetoelectric effects at room temperature by exploiting the elastic coupling between the two constituents [3]. This extrinsic converse magnetoelectric coupling relies on mechanical deformations of the ferroelectric layer caused by the converse piezoelectric effect, or by ferroelectric/ferroelastic domain reconfigurations. These strain changes are then transferred into the ferromagnetic thin film clamped onto the ferroelectric layer, modifying its magnetic properties due to magnetoelastic effects [4]. Thus, a detailed understanding of the piezoelectric and magnetoelastic effects as well as the elastic coupling across the interface is mandatory to predict magnetoelectric effects in novel multiferroic hybrid structures.

Here, we present a detailed study of the magnetic properties of multiferroic hybrid systems consisting of $BaTiO_3$ (BTO) as the ferroelectric and Ni or Fe_3O_4 as the ferromagnetic layers.

As a first step, we investigate the magnetization changes caused by magnetoelastic effects exploiting the structural phase transitions of BTO. In these experiments, it is mandatory to control the ferroelectric/ferroelastic domains in BTO. We show that for well-defined ferroelectric/ferroelastic domain states, the magnetization changes in BTO-based multiferroic hybrid structures can indeed be predicted on the basis of first-principles effective Hamiltonian simulations. In the corresponding calculations, we first determine the strain state of the BTO crystal as a function of temperature by means of molecular dynamics simulations. Since the ferromagnetic layer is elastically coupled to the BTO crystal, each change of the in-plane strain state of the BTO modifies the strain state of the ferromagnetic thin film. Thus, second, by using a phenomenological thermodynamical model, the strain state of the ferromagnetic thin film is calculated. With the knowledge of the elastic behavior of the ferromagnetic thin film, the magnetization changes can now be simulated (Fig. 1).

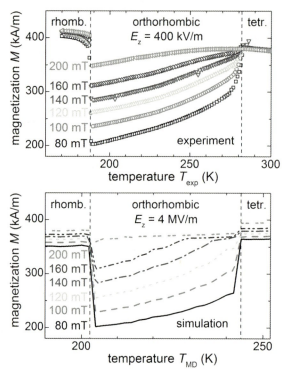

Figure 1: Experimental data (upper panel) and theoretical simulations (lower panel) of the temperature-dependent magnetization $M(T)$ for different magnetic field strengths applied in the film plane of a Ni/BTO multiferroic hybrid structure. The experiment was performed by SQUID magnetometry while applying an electric field of 400 kV/m along the out-of-plane direction of the Ni/BTO hybrid structure. The calculation is based on molecular dynamic simulations using the FERAM code, which is described in detail in Ref. [5]. In contrast to the experiment, an electric field strength of 4 MV/m is needed in the simulation to ensure a ferroelectric/ferroelastic single domain state in the tetragonal (tetr.), orthorhombic, and rhombohedral (rhomb.) phases of BTO. The temperature of the phase transitions are indicated by vertical dashed lines. For the simulation of the magnetic behavior of the Ni thin film, a phenomenological thermodynamic model based on a macrospin model was used [4].

As a second step, we simulate converse magnetoelectric effects at room temperature in BTO-based multiferroic hybrid structures. There, the magnetization behavior as a function of the applied electric field can be explained in first order by the magnetic behavior of those parts of the ferromagnetic thin film clamped to ferroelastic a-domains of the BTO substrate, while the magnetization of regions of the ferromagnetic thin film located on top of ferroelastic c-domains stays nearly unaffected. Using Fe_3O_4 thin films epitaxially grown on BTO substrates by pulsed laser deposition, we show that the experimentally obtained converse magnetoelectric effects at room temperature in multiferroic hybrid are well described within this approach.

The excellent agreement between experiment and simulation demonstrates that this approach allows to predict the magnetic behavior of existing and novel strain-mediated ferromagnetic/ferroelectric multiferroic hybrid structures not only as a function of temperature but also as a function of the applied electric field.

Financial support by the German Excellence Initiative via the Nanosystems Initiative Munich (NIM) is gratefully acknowledged.

[1] W. Eerenstein *et al.*, Nature **442**, 759 (2006); R. Ramesh *et al.*, Nat. Mater. **6**, 21 (2007).
[2] N. Spaldin *et al.*, Science **309**, 391 (2006); S. Geprägs *et al.*, Philos. Mag. Lett. **87**, 141 (2007).
[3] S. Geprägs *et al.*, Appl. Phys. Lett. **96**, 142509 (2010).
[4] A. Brandlmaier *et al.*, Phys. Rev. B **77**, 104445 (2008); M. Weiler *et al.*, New J. Phys. **11**, 013021 (2009); M. Opel *et al.*, Phys. Status Solidi A **208**, 232 (2011).
[5] T. Nishimatsu *et al.*, Phys. Rev. B **78**, 104104 (2008).

ELECTRIC CONTROL OF THE MAGNETIZATION IN BiFeO$_3$/LaFeO$_3$ SUPERLATTICES.

Zeila Zanolli[1], Jacek C. Wojdel[2], Jorge Iniguez[2], Philippe Ghosez[1]

[1]Université de Liège, Institut de Physique, Sart Tilman, Liège, Belgium;
[2]ICMAB-CSIC, Bellaterra, Spain

Transition-metal oxides of perovskite structure can present a wide variety of physical properties such as ferroelectricity, piezoelectricity, colossal magnetoresistance, spin-dependent transport and superconductivity, which can be exploited in various technological applications. In particular, there is nowadays a strong interest in multiferroic materials that are simultaneously ferroelectric and magnetic, since the so-called 'magnetoelectric' coupling between electrical polarization and magnetism could permit electrically writable and magnetically readable data storage. Due to the scarcity of natural magnetoelectric multiferroics and thanks to the recent advances in the epitaxial growth techniques, the idea of designing new magnetoelectric multiferroic heterostructures seems the most promising approach to succeed in this quest.

We will discuss the possibility of achieving electric control of the magnetization, possibly at room temperature, through a specific (trilinear) coupling of the polarization with two other non-polar lattice instabilities which occurs in the so-called hybrid improper ferroelectrics [1-3]. First-principles modelling techniques are used to investigate a promising system: BiFeO$_3$/LaFeO$_3$ superlattice. We found this system to exhibit magnetism at room temperature, trilinear coupling of structural instabilities, improper ferroelectricity and magneto-electric coupling. Electric switching of magnetization in this material will be discussed.

We acknowledge financial support of the European project OxIDes.

[1] E. Bousquet et al., *Nature* **452**, 732 (2008).
[2] N.A. Benedek, C.J. Fennie, *Phys. Rev. Lett.* **106**, 107204 (2011).
[3] Ph. Ghosez and J. M. Triscone, *Nature Materials* **10**, 269 (2011).

THE NEXT STEP ON THE SPIRAL – TbMnO$_3$ THIN FILMS

Artur Glavic[1], Jörg Voigt[1], Enrico Schierle[2], Eugen Weschke[2], Thomas Brückel[1]

[1]Jülich Center for Neutron Science JCNS and Peter Grünberg Center PG, JARA-FIT, Forschungszentrum Jülich GmbH, 52425 Jülich, Germany
[2]Helmholtz-Zentrum Berlin für Materialien und Energie, Elektronenspeicherring BESSY II, Albert-Einstein-Str.15, 12489 Berlin, Germany

Among the improper multiferroics TbMnO$_3$ is the system studied most extensively in bulk. Thin film preparation of this compound, however, has concentrated on strained TbMnO$_3$ films so far [1-6]. These films show an emergent ferromagnetism, which contradicts the cycloidal magnetic order responsible for ferroelectricity. For future application in devices, the persistence of the multiferroicity of such spiral magnets in thin films is crucial. We have investigated sputtered TbMnO$_3$ on YAlO$_3$ (1 0 0) substrates leading to untwinned, almost unstrained epitaxial layers [7].

The films of 6, 11, 16 and 100 nm have been characterized with x-ray reflectometry, x-ray diffraction and SQUID magnetometry, proving low roughness (<1.5 nm), epitaxial quality and a magnetic transition between 40 and 43 K (bulk transition to the spin density wave phase).

We will present the investigation of the multiferroic properties in these films with soft x-ray resonant magnetic scattering XRMS performed at the UE46-PGM1 beamline of BESSY II. This method allows the element specific investigation of the magnetic order in such systems including the direction of the magnetic moments and the chirality of the magnetic structure even in thin films [8-13].

In all investigated films the development of a magnetic reflection at (0 τ_{Mn} 0) was observed on the Mn L absorption edge (Figure 1). Besides the influence of the film thickness on the peak width, the temperature dependence of this peak was found to be very similar for all films and was comparable to bulk. Extraction of the site specific magnetic order parameters m$_i$ (see Figure 2 extracted as described in [14]) shows two transitions at T$_{SDW}$=43 K and T$_C$=27 K, which is comparable to the bulk behavior (spin density wave below 46 K and cycloidal order below 28 K [15]).

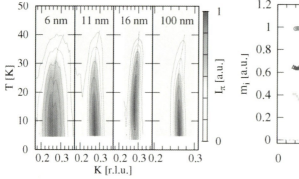
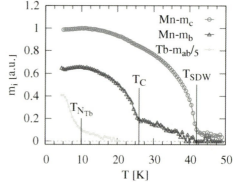

Figure 1: Temperature dependence of the resonant magnetic scattering from TbMnO$_3$ films with different thicknesses. The (0 τ_{Mn} 0)-peak width is dominated by the coherence limiting film thicknesses.

Figure 2: Temperature dependent magnetic order parameters of the 11 nm film extracted from σ and π channel measurements on the (0 τ_{Mn} 0)-reflection.

The Tb order measured at (0 τ_{Tb} 0) on the Tb M-edge showed an enhanced ordering temperature of T$_N$=10 K (bulk 7 K).

Cycloidal domains, prepared with a point charge created with the photoelectric effect while cooling into the low temperature phase as described in [12], could be measured with circular polarized photons. A switch in the sign of the circular dichroism (and thus the chirality of the magnetic structure) at the point where the charge was created could be observed for all films (Figure 3). The switching point could be changed by cooling with the beam at a different position proving the multiferroic properties (ferroelectric order and coupling to the magnetic structure) of the films.

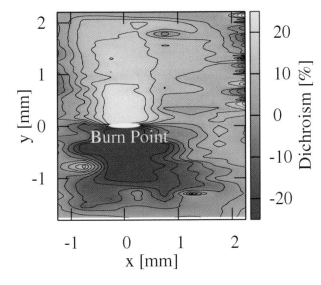

Figure 3: Circular dichroism of the magnetic peak intensity mapped over the sample surface of the 100 nm TbMnO$_3$ film.

[1] Cui, Y. M.; Wang, C. C. & Cao, B. S. , *Solid State Communication,* **2005**, *133*, 641-645
[2] Daumont, C. J. M.; Mannix, D.; Venkatesan, S.; Catalan, G.; Rubi, D.; Kooi, B. J.; Hosson, J. T. M. D. & Noheda1, B. , *Journal of Physics: Condensed Matter,* **2009**, *21*, 182001
[3] Kirby, B. J.; Kan, D.; Luykx, A.; Murakami, M.; Kundaliya, D. & Takeuchi, I. , *Journal of Applied Physics,* **2009**, *105*, 7D917
[4] Marti, X.; Skumryev, V.; Ferrater, C.; Garcia-Cuenca, M. V.; Varela, M.; Sanchez, F. & Fontcuberta, J. , *Applied Physics Letters,* **2010**, *96*, 222505
[5] Rubi, D.; de Graaf, C.; Daumont, C. J. M.; Mannix, D.; Broer, R. & Noheda, B. , *Physical Review B,* **2009**, *79*, 14416
[6] Venkatesan, S.; Daumont, C.; Kooi, B. J.; Noheda, B. & De Hosson, J. T. M. , *Physical Review B,* **2009**, *80*, 214111
[7] Glavic, A.; Voigt, J.; Persson, J.; Su, Y. X.; Schubert, J.; de Groot, J.; Zande, W. & Brückel, T. , *Journal of Alloys and Compounds,* **2011**, *509*, 5061-5063
[8] Fabrizi, F.; Walker, H. C.; Paolasini, L.; de Bergevin, F.; Boothroyd, A. T.; Prabhakaran, D. & Mcmorrow, D. F. , *Physical Review Lettetters,* **2009**, *102*, 237205
[9] Feyerherm, R.; Dudzik, E.; Aliouane, N. & Argyriou, D. N. , *Physical Review B,* **2006**, *73*, 180401
[10] Forrest, T. R.; Bland, S. R.; Wilkins, S. B.; Walker, H. C.; Beale, T. A. W.; Hatton, P. D.; Prabhakaran, D.; Boothroyd, A. T.; Mannix, D.; Yakhou, F. & Mcmorrow, D. F. , *Journal of Physics: Condensed Matter,* **2008**, *20*, 422205
[11] Schüßler-Langeheine, C.; Schlappa, J.; Tanaka, A.; Hu, Z.; Chang, C. F.; Schierle, E.; Benomar, M.; Ott, H.; Weschke, E.; Kaindl, G.; Friedt, O.; Sawatzky, G. A.; Lin, H.-J.; Chen, C. T.; Braden, M. & Tjeng, L. H. , *Physical Review Letters,* **2005**, *95*, 156402
[12] Schierle, E.; Soltwisch, V.; Schmitz, D.; Feyerherm, R.; Maljuk, A.; Yokaichiya, F.; Argyriou, D. N. & Weschke, E. , *Physical Review Letters,* **2010**, *105*, 167207
[13] Weschke, E.; Ott, H.; Schierle, E.; Schüßler-Langeheine, C.; Vyalikh, D. V.; Kaindl, G.; Leiner, V.; Ay, M.; Schmitte, T.; Zabel, H. & Jensen, P. J. , *Physical Review Letters,* **2004**, *93*, 157204
[14] Jang, H.; Lee, J.-S.; Ko, K.-T.; Noh, W.-S.; Koo, T. Y.; Kim, J.-Y.; Lee, K.-B.; Park, J.-H.; Zhang, C. L.; Kim, S. B. & Cheong, S.-W. , *Physical Review Letters,* **2011**, *106*, 47203
[15] Kajimoto, R.; Yoshizawa, H.; Shintani, H.; Kimura, T. & Tokura, Y. , *Physical Review B,* **2004**, *70*, 12401

MAGNETIC CHIRAL DOMAINS IN MULTIFERROIC THIN FILMS KEEP MEMORY

Josep Fontcuberta, I. Fina, L. Fàbrega, X. Martí and F. Sánchez

Institut de Ciència de Materials de Barcelona (ICMAB-CSIC)
Campus UAB, Bellaterra, 08193. Catalonia. Spain

Antiferromagnetic perovskites with cycloidal magnetic order are known to display ferroelectric polarization (P). The sense of P is dictated by the chirality of the cycloid. Upon cooling the material through its Néel temperature, domains of distinct chirality and ferroelectric polarization, are expected to be formed. It is also known that the cycloidal plane can be flopped from bc(ac) to ac(bc) by an appropriate magnetic field.

As in any ferroic material, ferroic domains lead to hysteresis and coercivity, the antiferromagnetic chiral domains, upon flopping under a magnetic field and field-retreating, may preserve memory of the applied field(s) history. Here we explore in detail the flopping process in orthorhombic $YMnO_3$ thin films.

Orthorhombic $YMnO_3$ films (about 100 nm thick) were on $Nb:SrTiO_3$ (110) substrates. It turns out that when measuring the polarization along the a-axis (Pbnm), under a magnetic field (H//c) a pronounced rising of polarization ($P_a \approx 80$ nC/cm^2) occurs at $\mu_0 H > 4$ T, signalling the polarization flop from the basal bc-plane to the ab plane [1], indicating cycloidal magnetic arrangement.

From these results, some questions arise: a) if $YMnO_3$ is cooled through its Néel temperature without any bias filed, which are the relative populations cycloidal domains (clockwise or anticlockwise) and b) if a bias electric field is used to define a single domain polar state and then H is used to induce a flop and subsequently zeroed to recover the initial state, is the single-domain chiral state recovered?

To address these issues the magnetic-field dependent polarization P(H) of a-textured $YMnO_3$ films has been measured 5K under distinct cooling conditions.

In Fig.1, main panel we show a schematic view of the polarization results. The measured ferroelectric currents, obtained by using an adapted PUND protocol that allows to discriminate the fraction of P_{a+}/P_{a-} domains, are shown in panels (a-e). Data in (a) recorded by applying 6 T field after a (magnetic) zero-field-cooling process (blue arrow) shows that identical fractions of P_{a+} and P_{a-} are obtained. More interesting, when the sample is polarized by an E-field (E//a+) to obtain a single domain ferroelectric and chiral state (b) and then the magnetic field is zeroed (red arrow) and subsequently increased again up to 6 T (green arrow), the measured polarization (c) shows that the polarization is majority along a+ and thus the chiral order is preserved. This implies that, upon flopping, the system has memory of its previous chiral state [3].

Detailed inspection of (c) and data in Fig. 1(a-d) indicates that: 1) only a fraction of the initial chiral order is preserved (~2/3) and thus multiferroic domain walls are formed and 2) the sense of the magnetic field (d) does not modify this memory effect. Finally, changing the sense of the pooling produces the corresponding change of the remnant polarization upon flopping.

In summary, chiral order in magnetic ferroelectric is preserved upon magnetic-field flopping but this memory effect is limited ($\approx 2/3$ in the present case). These results show that chirality keeps memory of the initial polarization state after subsequent flop processes. It thus follows that antiferromagnetic materials, is spite of the null magnetization, can be used to store magnetic information.

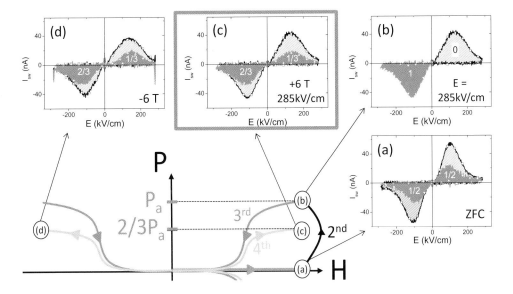

Figure 5. Polarization dependence on magnetic field. Blue line (1st) shows that P_a after electric and magnetic zero field cool process. Red (3rd) and green (4th) line show that after applying a electric pulse to pole the samples (2nd), the polar state is partially retained after flopping process. (a-d) Switching current versus electric field measured using PUND (squares), ND (down-triangles) and PU (up-triangles). Stripped area represents the total amount of switchable charge (PUND) and the solid areas the switched charge during PU or ND measurements.

1 I. Fina et al. Appl. Phys. Lett. 97, 232905 (2010); J. Fontcuberta et al. Phase Transitions, 84, 555 (2011)
2 I. Fina et al, Phys. Rev. Lett. 107, 257601 (2011)

EXCHANGE BIAS AND MAGNETOELECTRIC COUPLING EFFECTS IN $ZnFe_2O_4$ – $BaTiO_3$ COMPOSITE THIN FILMS

Michael Lorenz[1], Michael Ziese[1], Gerald Wagner[2], Pablo Esquinazi[1], Marius Grundmann[1]

[1]Universität Leipzig, Institut für Experimentelle Physik II, D-04103 Leipzig, Germany;
[2]Universität Leipzig, Institut für Mineralogie, Kristallographie und Materialwissenschaften, D-04275 Leipzig, Germany

The combination of different ferroic ordering phenomena such as ferroelectricity, ferroelasticity and ferromagnetism in one and the same multiferroic material is currently a challenging topic of condensed matter physics and materials science. However, the number of intrinsic, single phase multiferroics is rather limited to several perovskite, boracite, and delafossite compounds. Therefore, two-phase, strain-coupled multilayers and composites gain more and more attention because of the flexibility concerning materials selection and geometrical arrangement.

Multiferroic composite thin films prepared from ferrimagnetic zinc ferrite ($ZnFe_2O_4$, ZFO) and ferroelectric barium titanate ($BaTiO_3$, BTO) show sizable magnetic exchange bias and magnetoelectric coupling effects. After field cooling in +3 T, an exchange-bias field at 10 K of about -37 mT for a 65% ZFO / 35% BTO and of about -34 mT for a 35% ZFO / 65% BTO composite is observed. Exchange biasing is accompanied by a significant vertical loop shift of about 10% of the total saturation magnetization after field cooling in 3 T. Clear indication for magnetoelectric coupling is found by magnetocapacitance measurements. Depending on the ZFO to BTO ratio of the target for pulsed laser deposition, the composite films show either preferential spinel- or perovskite-like X-ray diffraction patterns. The composite films show nm-size amorphous precipitates in a ZFO-like matrix which are most probably responsible for the observed exchange bias effects.

MFO 1

CHARGE ORDER IN LUTETIUM IRON OXIDE: AN UNLIKELY ROUTE TO FERROELECTRICITY

J. de Groot[1], T. Mueller[1], R.A. Rosenberg[2], D.J. Keavney[2], Z. Islam[2], J.-W. Kim[2], and M. Angst[1]

[1]Peter Grünberg Institut PGI and Jülich Centre for Neutron Science JCNS, JARA-FIT, Forschungszentrum Jülich GmbH, 52425 Jülich, Germany;
[2]Argonne National Laboratory, 9700 S. Cass Avenue, Argonne, IL 60439, USA

Magnetic ferroelectric ("multiferroic") materials with a strong magnetoelectric coupling are of high interest for potential information technology applications, e.g. non-volatile memories [1]. Of the different mechanisms proposed to achieve this, ferroelectricity (FE) originating from charge (or valence state) ordering (CO) is particularly intriguing because it potentially combines the necessary large electric polarizations and magnetoelectric couplings – however examples of materials where this mechanism is in effect are exceedingly rare, with none ever unambiguously demonstrated experimentally [2].

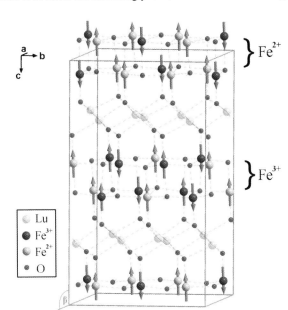

Figure 1: The $LuFe_2O_4$ monoclinic crystal structure for one particular charge order domain [4]. The Fe atoms are arranged in Fe/O bilayers separated by Lu single layers. The different Fe valances are indicated by different colors. The Brackets beside the unit cell indicate the majority charge configuration for each bilayers present in the cell.

The red arrows indicate the spin direction form the ferrimenatic structure reported in [10]; also compatible with our local spin configurations for Fe^{2+} and Fe^{3+} determined by XMCD.

For $LuFe_2O_4$ the initial proposal was in 2005 that below T_{CO}=320K a peculiar arrangement of the Fe^{2+} and Fe^{3+} valences gives rise to a configuration where each Fe/O bilayers develops an electric dipole, resulting in macroscopic ferroelectricity [3]. This material was generally considered to be the best (and only) example of ferroelectricity from charge order [1], attracting an increasing amount of attention, see e.g. [4-11].

For a long time the complex charge order domain structure of this material [6] prevented a direct crystal reconstruction by single crystal diffraction measurements. We were able to overcome this problem by screening different samples and selecting an almost mono domain specimen, alleviating the structure determination in a new monoclinic cell (see Fig.1) [4]. For these refinement results, the arrangement of the Fe^{2+} and Fe^{3+} valences was readily determined by the Bond-Valence-Sum method (which shows an almost ionic valence character). Unexpectedly, the different valences are arranged into charged and thus intrinsically non-polar bilayers (see Fig.1). Such a charge configuration is in strong contrast to all previously suggested charge orders and it is highly incompatible with ferroelectricity. Our CO result is also confirmed by additional x-ray magnetic circular dichroism (XMCD) measurements, which determine the local magnetic environment of the two different valence states (all spins of Fe^{2+} and only one third of the Fe^{3+} point in field direction, the rest point opposite to the field). This XMCD result is remarkably similar to previously published XMCD work [8,9] and from the local spin configurations, the new CO configuration inevitably follows from our recently published [10] ferrimagnetic spin structure. This independent verification is further supporting the presented non ferroelectric CO result for this material.

Additionally, the possibility of polarizing the charged bilayers by external electric and magnetic fields is considered by in-situ high energy x-ray diffraction experiments; here a robust charge order configuration is found [4] as a negative result for the idea of induced ferroelectricity [11]. At this point the question of how to explain the reported ferroelectric behavior remains; however, in recent published and unpublished work it is shown that the observed macroscopic indications for ferroelectricity are most likely due to contact [12,13] and conductance [14] effects.

All these results mean that a good example material with ferroelectricty originating from charge order has yet to be identified experimentally, i.e. research in this field has to take a step back.

[1] K. F. Wang, J.-M. Liu and Z. F. Ren, Advances in Physics **58**, 321 (2009).
[2] J. van den Brink and D. I. Khomskii, J. Phys.: Condens. Matter **20**, 434217 (2008).
[3] N. Ikeda et al., Nature **436**, 1136 (2005).
[4] J. de Groot et al., arXiv:1112.0978; PRL accepted (2012).
[5] J. Bourgeois et al., Phys. Rev. B **85**, 064102 (2012).
[6] C. Lee et al., Phys. Rev. B **85**, 014303 (2012).
[7] J. Rouquette et al., Phys. Rev. Lett. 105, 237203 (2010).
[8] K.-T. Ko et al., Phys. Rev. Lett. **103**, 207202 (2009).
[9] K. Kuepper et al., Phys. Rev. B. **80**, 220409(R) (2009).
[10] J. de Groot et al., Phys. Rev. Lett. **108**, 037206 (2012).
[11] M. Angst et al., Phys. Rev. Lett. **101**, 227601 (2008).
[12] P. Ren et al., J. Appl. Phys. **109**, 074109 (2011).
[13] D.Niermann et al., arXiv:1203.1200v1.
[14] M. Maglione and M. A. Subramanian, Appl. Phys. Let. **93**, 032902 (2008).

CRYSTAL STRUCTURE, PHASE TRANSITION, CHEMICAL EXPANSION AND DEFECT CHEMISTRY OF HEXAGONAL HoMnO$_3$

Sverre M. Selbach[1], **Kristin Bergum**[1], **Amund Nordli Løvik**[1], **Julian R. Tolchard**[1], **Mari-Ann Einarsrud**[1], **Tor Grande**[1]

[1]Department of Materials Science and Engineering, Norwegian University of Science and Technology, NO-7491 Trondheim, Norway

HoMnO$_3$ is a hexagonal manganite possessing both ferroelectric and frustrated antiferromagnetic order, and is hence termed a multiferroic material. Hitherto the hexagonal manganites have been presumed to be oxygen stoichiometric, but this depends strongly on the thermal history of the materials both during processing and during cooling to ambient. As the valence of Mn is controlled by the oxygen stoichiometry, it should not be neglected in the preparation and characterization of these compounds.

Anisotropic thermal and chemical expansion [1] of hexagonal HoMnO$_3$ was investigated by in situ high temperature X-ray diffraction in inert (N$_2$) and oxidizing (air) atmospheres up to 1623 K [2]. A 2nd order structural phase transition directly from $P6_3cm$ to $P6_3/mmc$ was found at 1298 ± 4 K in N$_2$, and 1318 ± 4 K in air. For the low temperature polymorph $P6_3cm$ the contraction of the c-axis was more rapid in inert than in oxidizing atmosphere, as shown in Fig. 1. The c-axis of the $P6_3/mmc$ polymorph of HoMnO$_3$ displayed anomalously high expansion above 1400 K, which is attributed to chemical expansion caused by increasing oxygen vacancy population and reduction of Mn. This can be described with Kröger-Vink notation by eq. (1) where Mn^x and Mn' corresponds to Mn^{3+} and Mn^{2+}, respectively:

$$O_O^x + 2Mn_{Mn}^x \rightleftarrows V_O^{\bullet\bullet} + 2Mn_{Mn}' + 1/2\,O_2(g) \qquad (1)$$

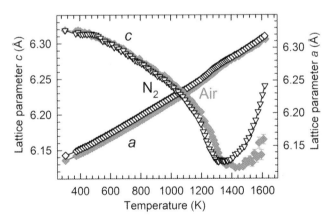

Figure 1: Lattice parameters of HoMnO$_3$ obtained by Rietveld refinement of HTXRD patterns collected in oxidizing (air) and reducing (N$_2$) atmospheres [2]. The tri-merization transition from $P6_3cm$ to $P6_3/mmc$ is evident from the change of slope in the thermal expansion of the c–axis at 1300-1320 K. Chemical expansion is more pronounced for lattice parameter c than a.

The orthorhombic perovskite polymorph of HoMnO$_3$ appeared gradually in air above 1200 K at the expense of the hexagonal phase. A rapid, reconstructive transition from the perovskite back to the hexagonal polymorph was observed at 1623 K upon *in situ* reduction of the partial pressure of oxygen. This *in situ* observation confirmed previous discussions of the stability of the competing phases and the strong influence of the chemical potential of oxygen [3,4]. Distinctly non-linear electrical conductivity was observed for both HoMnO$_3$ and YMnO$_3$ in oxidizing atmosphere between 555 and 630 K, and is associated with oxygen hyper-stoichiometry. Excess oxygen in hexagonal manganites is expected to occupy interstitial lattice sites. This can be described with Kröger-Vink notation by eq. (2) where Mn$^{\bullet}$ and O$_i''$ represent Mn^{4+} and interstitial oxygen, respectively:

$$\delta/2\,O_2(g) + 2\delta Mn_{Mn}^{x} \rightleftarrows \delta O_i^{''} + +2\delta Mn_{Mn}^{\bullet} \qquad (2)$$

This situation differs from perovskites, where excess oxygen only can be accommodated by the creation of cation vacancies [5]. The understanding of defect chemistry in perovskites can not be directly transferred to the hexagonal manganites, and the study of point defects in hexagonal manganites is still in its infancy.

Finally, we have also recently demonstrated that the oxygen non-stoichiometry and oxygen partial pressure are important for competing kinetics and stability of nano-crystalline orthorhombic and hexagonal $YMnO_3$ [3]. Careful annealing of amorphous oxide precursors has enabled the exploration of finite size effects in hexagonal $YMnO_3$ [3]. Reduction of the c-axis with decreasing crystallite size was evident and the ferrielectric displacements of Y^{3+} cations along the polar c-axis decays progressively with decreasing size below 100 nm.

[1] S. B. Adler, J. Am. Ceram. Soc. **84**, 2117 (2001).
[2] S. M Selbach et al. submitted.
[3] K. Bergum et al., Dalton Trans. **40**, 7583 (2011).
[4] H. W. Brinks, H. Fjellvåg, A. Kjekshus, J. Solid State Chem. **129**, 334 (1997).
[5] J.A.M. van Roosmalen, E.H.P. Cordfunke, J. Solid State Chem. **110**, 109 (1994).

COLLECTIVE MAGNETISM AT FERROELECTRIC DOMAIN WALLS

<u>Weida Wu</u>[1], Yanan Geng[1], Y.J. Choi[1,2], N. Lee[1] and Sang-Wook Cheong[1]

[1]Rutgers Center for Emergent Materials and Department of Physics and Astronomy, Piscataway, NJ, USA;
[2]Department of Physics and IPAP, Yonsei University, Seoul 120-749, South Korea.

Topological defects are pervasive in complex matter such as superfluids, liquid crystals, earth atmosphere, and early universe [1, 2]. They have been fruitful playgrounds for many emergent phenomena [3, 4]. Recently, vortex-like topological defects with six interlocked structural antiphase and ferroelectric domains merging into a vortex core were revealed in ferroelectric hexagonal manganites [5]. Numerous vortices are found to form an intriguing self-organized network, where charged domain walls with emergent properties are protected [6-8]. Thus, it is imperative to find out the magnetic nature of the multiferroic vortices. Using cryogenic magnetic force microscopy (MFM) in applied magnetic fields and ambient piezo-response force microscopy (PFM) [9], we discovered that the alternating domain wall magnetizations around vortices can correlate over the entire vortex network [10]. The collective nature of the domain wall magnetism appears originated from the uncompensated Er^{3+} moments at domain walls, the field-controllable chirality of antiferromagnetic domains, and the correlated organization of the vortex network. Our results demonstrate a new route for nanoscale magnetoelectric coupling in single-phase multiferroics, and open the possibility of detecting spin chirality by harnessing domain wall magnetism.

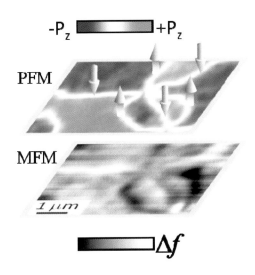

Figure 1: RT-PFM image and MFM image (5.5 K, 0.2 T) taken at the *same* location on the (001) surface of multiferroic hexagonal $ErMnO_3$. The arrows in PFM image represent the net magnetic moments in ferroelectric domain walls. $\Delta f = 0.8$ Hz.

[1] P. M. Chaikin, and T. C. Lubensky, *Principles of Condensed Matter Physics* (Cambridge University Press, Cambridge, UK, 2000).
[2] A. A. Fraisse, C. Ringeval, D. N. Spergel, and F. R. Bouchet, Phys. Rev. D **78**, 043535 (2008).
[3] J. Seidel *et al.*, Nat. Mater. **8**, 229 (2009).
[4] A. Mesaros *et al.*, Science **333**, 426 (2011).
[5] T. Choi *et al.*, Nature Materials **9**, 253 (2010).
[6] S. C. Chae *et al.*, Proc. Natl. Acad. Sci. USA **107**, 21366 (2010).
[7] E. B. Lochocki, S. Park, N. Lee, S.-W. Cheong, and W. Wu, Appl. Phys. Lett. **99**, 232901 (2011).
[8] W. Wu, Y. Horibe, N. Lee, S.-W. Cheong, and J. R. Guest, Phys. Rev. Lett. **108**, 077203 (2012).
[9] S. Park *et al.*, Appl. Phys. Lett. **95**, 072508 (2009).
[10] Y. Geng, N. Lee, Y. J. Choi, S.-W. Cheong, and W. Wu, arXiv:1201.0694 (2012).

HARD X-RAY NANOSCALE STRUCTURAL IMAGING OF MULTIFERROIC THIN FILMS

Martin Holt[1], Stephan Hruszkewycz[2], Chad Folkman[2], Robert Winarski[1], Volker Rose[1,3], Paul Fuoss[2], Ian McNulty[1,3]

[1]Center for Nanoscale Materials, Argonne National Laboratory, Argonne, IL, USA;
[2]Materials Science Division, Argonne National Laboratory, Argonne, IL, USA;
[3]Advanced Photon Source, Argonne National Laboratory, Argonne, IL, USA;

Nondestructive visualization of lattice strain, rotation, and distortion in confined ferroelectric and multiferroic materials under applied fields is of key importance to understanding the complex interplay between fundamental material physics and nanoscale device functionality and design. Much recent progress has been made in both experimental demonstration and theoretical understanding of the broad range of electronic and magnetoelectric behavior that can be generated in multiferroic thin film materials subject to complex stresses – the optimal use of these behaviors in potential microelectronic devices will require an equally quantitative understanding of *in operando* nanoscale material response to applied fields and lithographic patterning.

Nanoscale scanning probe hard x-ray diffraction microscopy is ideally suited to answering these and related questions of internal structural ordering in complex materials without sectioning or otherwise modifying the sample. This technique has been developed at the Argonne Center for Nanoscale Materials Hard X-ray Nanoprobe located at the Advanced Photon Source to apply to a wide range of nanoscale active material systems – previous results include structural measurements of semiconductor heterostructures [1], and Ti:NiO RRAM devices [2]. We report here recent results related to thin-film multiferroics utilizing scanning probe nanodiffraction microscopy at a 40nm spatial resolution quantifying i) elastic relaxation within lithographically patterned $BiFeO_3$ nanostructures[3] (Fig. 1), and ii) formation and organization of $BiFeO_3$ lattice tilt domains due to anisotropic electrical field poling [4] (Fig. 2). Use of coherent Bragg diffraction ptychography imaging techniques [5] to resolve structural contrast within ferroelectric domains at a sub-10nm spatial resolution will also be presented.

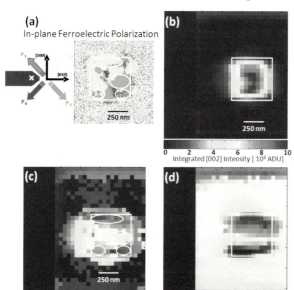

Figure 1: Results of nanofocused x-ray diffraction lattice mapping in a single 500nm BFO nanostructure [(b)-(d)] compared to the ferroelectric domain structure observed via PFM (a). Repeated 2D lateral scans were taken while varying the sample angle across the BFO (002) rocking curve, from which the integrated intensity (b), out-of-plane lattice strain (c), and out-of-plane lattice (C-axis) rotation (d) were extracted. The lattice constant of the film in the nanostructure is relaxed relative to the planar film by a strain value of as much as -1.8% dc/c, with a strain distribution that generally corresponds to the ferroelectric domain structure [3].

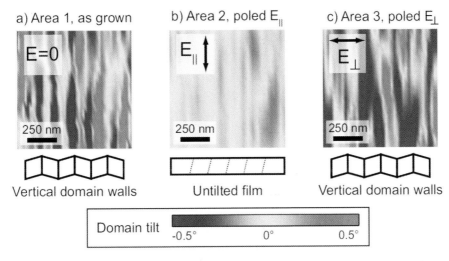

Figure 2: Focused beam nanodiffraction tilt maps of the as-grown and poled BiFeO3 film are shown. The as-grown and poled E⊥ areas exhibit out-of plane tilting indicative of vertical domain walls, while the E∥ poled film has been switched to an untilted state consistent with a monodomain film or with slanted domain walls (dashed lines) [4].

[1] O. Heinonen, M. Siegert, A. Roelofs, A. K. Petford-Long, M. V. Holt, K. d'Aquila, and W. Li, Appl. Phys. Lett. **96**, 103103 (2010).
[2] C. E. Murray, A. Ying, S. M. Polvino, I. C. Noyan, M.V. Holt, and J. Maser, Journ. Appl. Phys. **109**, 083543 (2011).
[3] J. A. Klug, M.V. Holt, R. N. Premnath, A. Joshi-Imre, S. Hong, R. S. Katiyar, M. J. Bedzyk, and O. Auciello, Appl. Phys. Lett. **99**, 052902 (2011).
[4] S. O. Hruszkewycz, C. M. Folkman, M. J. Highland, M. V. Holt, S. H. Baek, S. K. Streiffer, P. Baldo, C. B. Eom, and P. H. Fuoss Appl. Phys. Lett. **99**, 232903 (2011).
[5] S. O. Hruszkewycz, M. V. Holt, A. Tripathi, J. Maser, and P. H. Fuoss Optics Letters **36**, 2227 (2011).

THE STRUCTURE OF THE MULTIFERROIC BaTiO$_3$/Fe(001) INTERFACE

H.L. Meyerheim,[1] F. Klimenta[1], A. Ernst[1], K. Mohseni[1], S. Ostanin[1], M. Fechner[1], S.S. Parihar[1], I.V. Maznichenko[1,2], I. Mertig[1,2], and J. Kirschner[1]

[1]Max-Planck-Institut für Mikrostrukturphysik, D-06120 Halle, Germany, [2]Institut für Physik, Martin-Luther-Universität Halle-Wittenberg, D-06099 Halle, Germany

We present a combined surface x-ray diffraction (SXRD) and theoretical study of the multiferroic BaTiO$_3$/Fe(001) junction [1]. Multiferroic tunnel junctions have become a central topic in solid state research because of their potential for developing new device architectures on the nanometer scale. Theoretical work has established the importance of the interface bonding between the FE barrier and the ferromagnetic (FM) electrode [2-5]. In this context, the BaTiO$_3$/Fe(001) junction represents an archetype system in which the classical FE film is combined with the FM electrode in an almost perfect lattice match (misfit 1.4 %). Surprisingly, current knowledge is limited to theoretical predictions, while quantitative structure information on growth, film and interface structure is not available so far.

Ultrathin BaTiO$_3$ films were grown on Fe(001) by pulsed laser deposition. The SXRD measurements were carried out by using Co-Kα radiation generated by a rotating anode generator. In contrast to common assumptions we find that BaTiO$_3$ films are characterized by the presence of complete unit cells (m) terminated by a BaO-layer, schematically written as: Fe/(TiO$_2$-BaO)$_m$. Onset of polarization is observed at a minimum thickness of two unit cells. Figs (a) and (b) schematically show the m=1 and m=2 structures in side view, respectively.

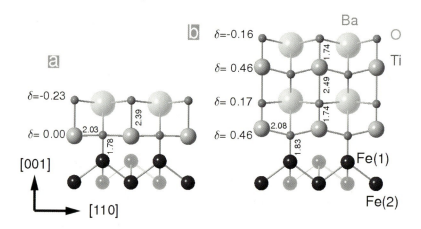

Figure 1: Model of the BaTiO3(001)/Fe(001) structure: (a) and (b) show side views for films with 1 and 2 unit cells thickness. Large and small spheres correspond to Ba- and O-atoms, respectively, while medium sized spheres represent Ti and Fe-atoms as labeled. Numbers represent interatomic distances.

Distances are given in Ångstrom units. The parameter δ represents the height of the cation above ($\delta>0$) or below ($\delta<0$) the plane of oxygen atoms. For the m=1 sample, we find (δ=0.00 at the first TiO_2 layer, while (δ=-0.23 for the top BaO layer to ensure a flat surface charge profile. Large values for δ (0.46 Å) are found for the m=2 structure at the TiO_2 layers. Based on the quantitative atomic positions first principles calculations indicate that the $BaTiO_3$/Fe(001) junction is multiferroic in nature. We find a significant multiferroic effect as a result of the substantial polarizations given by the relative Ti-O displacements in the 0.4 Å range.

We thank F. Weiss for technical support. This work is supported by the DFG through SFB 762

[1] H.L. Meyerheim, F. Klimenta, A. Ernst, K. Mohseni, S. Ostanin, M. Fechner, S.S. Parihar, I.V. Maznichenko, I. Mertig, and J. Kirschner, Phys. Rev. Lett. 106, 087203 (2011)
[2] E. Y. Tsymbal and H. Kohlstedt, Science 313, 181 (2006).
[3] C.-G. Duan, S. S. Jaswal, and E. Y. Tsymbal, Phys. Rev. Lett. 97, 047201 (2006).
[4] M. Fechner, I.V. Mazhnichenko, S. Ostanin, A. Ernst, J. Henk, P. Bruno, and I. Mertig, Phys. Rev. B 78, 212406 (2008)
[5] S. Valencia, A. Crassous, L. Blocher, V. Garcia, X. Moya, R.O. Cherifi, C. Deranlot, K. Bouzehouane, S. Fusil, A. Zobelli, A. Gloter, N.D. Mathur, A. Gaupp, R. Abrudan, F. Radu, A. Barthélémy and M. Bibes, Nat. Mat. 10, 753 (2011)

INVESTIGATION OF TWO MECHANISMS FOR MULTIFERROICITY IN PbCrO$_3$ BY DFT

Martin Schlipf[1], Marjana Lezaic[1]

[1]Forschungszentrum Jülich, Peter Grünberg Institut and JARA, 52425 Jülich, Germany

The cubic perovskite PbCrO$_3$ is a rare example of a semiconducting compound of Cr in oxidation state 4+ [1,2]. It has gained renewed experimental interest in the recent years. However, so far the theoretical calculations based on DFT have not established the origin of the semiconducting state and have found metallic ground state instead.[3]

In this contribution, we employ DFT to investigate the perovskite PbCrO$_3$ including an on-site Hubbard U like correction for the d orbitals of Cr. We analyze in detail all possible Jahn-Teller distortions and tilts of the oxygen octahedra and find semiconducting solutions. The ground state solution depends on the particular choice of the Hubbard U parameter. If we allow for polar distortions, PbCrO$_3$ establishes a ferroelectric polarization. We describe two competing mechanisms that lead to the multiferroic behaviour, whose relative stability depends on the size of the U parameter.

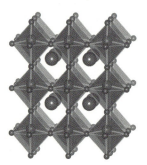

Figure 1: Stable PbCrO$_3$ structure at small values of U. Pb, Cr, and O ions are shown in gray, blue, and red, respectively. The largest distortions are a large c/a ratio and an off-centering of the Cr ions along the z-axis. A smaller Jahn-Teller distortion in plane leads to a splitting of the t_{2g} states into three separate levels at each Cr site.

In Fig. 1, we depict the structure associated with the first mechanism, which is energetically favorable for small U. Similar to PbVO$_3$ [4,5], the combination of a large c/a ratio and an off-centering of the transition metal cation, stabilizes the occupation of the d_{xy} orbital. For PbCrO$_3$, we obtain a small Jahn-Teller distortion in plane lifting the degeneracy of the d_{xz} and the d_{yz} orbital. As a consequence, the ground state exhibits an orbital order of a checkerboard arrangement of the d_{xz} and the d_{yz} orbital in plane. We find a semiconducting state in this structural arrangement for values of U between 2 and 5 eV. For larger values of U a different, metallic structure in the same spacegroup is energetically favorable.

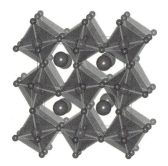

Figure 2: Stable PbCrO$_3$ structure at large values of U. Pb, Cr, and O ions are shown in gray, blue, and red, respectively. The largest distortions are tilts of the oxygen octahedra in all three axes and a displacement of the Pb ions. The size of the shift is different for all of the Pb ions.

In Fig. 2, the stable structure for large values of U is shown. The tilts of the oxygen octahedra in all three coordinate axes break the symmetry of the Pb sites. Hence, a charge order on these sites is allowed by symmetry. In our calculations, the charge order of the Pb ions can be formally associated with a half-and-half mixing of Pb^{2+} and Pb^{4+} ions. To maintain overall charge neutrality, the Cr ions exhibit a formal charge of 3+, so that all t_{2g} states are occupied. As a consequence, the structure becomes insulating for values of U larger than 3 eV.

[1] W. L. Roth and R. C. DeVries, J. Appl. Phys. **38**, 951 (1967).
[2] B. L. Chamberland and C. W. Moeller, J. Solid State Chem. **5**, 39 (1972).
[3] P. Ganesh and R. E. Cohen, Phys. Rev. B **83**, 172102 (2011).
[4] R. V. Shpanchenko, *et al.*, Chem. Mater. **16**, 3267 (2004).
[5] K. Oka, *et al.*, Inorg. Chem. **47**, 7355 (2008).

REALIZATION OF FULL MAGNETOELECTRIC CONTROL AT ROOM TEMPERATURE

Kee Hoon Kim[1], Sae Hwan Chun[1], Yi Sheng Chai[1], Byung-Gu Jeon[1], Kwang Woo Shin[1], Hyung Joon Kim[1], Yoon Seok Oh[1], Ingyu Kim[1], Ju-Young Park[1], Suk Ho Lee[1], Jae-Ho Chung[2], Jae-Hoon Park[3]

[1]CeNSCMR, Dept. of Physics and Astronomy, Seoul National University, Seoul 151-747, Rep. of Korea; [2]Dept. of Physics, Korea University, Seoul 136-713, Rep. of Korea; [3]Dept. of Physics and Div. of Advanced Materials Science, POSTECH, Pohang 790-784, Rep. of Korea

The control of magnetization by an electric field at room temperature remains as one of great challenges in materials science. Multiferroics, in which magnetism and ferroelectricity coexist and couple to each other, could be the most plausible candidate to realize this long-sought capability. While recent intensive research on the multiferroics has made significant progress in sensitive, magnetic control of electric polarization, the electrical control of magnetization, the converse effect, has been observed only in a limited range far below room temperature. Here, we demonstrate at room temperature the control of both electric polarization by a magnetic field and magnetization by an electric field in a Co_2Z-type hexaferrite, $Ba_{0.52}Sr_{2.48}Co_2Fe_{24}O_{41}$ single crystal. The electric polarization rapidly increases in a magnetic field as low as 5 mT and the magnetoelectric susceptibility reaches up to 3200 ps/m, the highest value in single phase materials. The magnetization is also modulated up to 0.34 μ_B per formula unit in an electric field of 1.14 MV/m. Furthermore, four magnetoelectric states induced by different magnetoelectric poling exhibit unique, non-volatile magnetization versus electric field curves, offering an unprecedented opportunity for multi-bit memory or spintronic device applications.

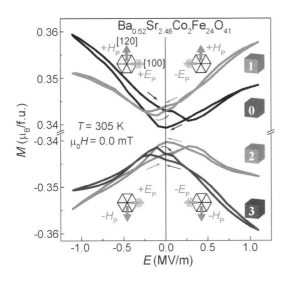

Figure 1: The $M(E)$ curves at zero H-bias obtained in $Ba_{0.52}Sr_{2.48}Co_2Fe_{24}O_{41}$ single crystal at 305 K after applying four different ME poling (*states* **0**, **1**, **2**, and **3**) as indicated in the inset [1].

[1] S. H. Chun et al., arXiv: 1111.4525; Phys. Rev. Lett. in press (2012).

SPIN WAVES AND LATTICE ANOMALY OF BiFeO$_3$ MEASURED BY NEUTRON SCATTERING

Jaehong Jeong[1], E. A. Goremychkin[2], T. Guidi[2], K. Nakajima[3], Gun Sang Jeon[4], Shin-Ae Kim[5], S. Furukawa[6], Yong Baek Kim[6], Seongsu Lee[5], V. Kiryukhin[7], S-W. Cheong[7], and <u>Je-Geun Park</u>[1,8]

[1]FPRD Department of Physics & Astronomy, Center for Strongly Correlated Materials Research, Seoul National University, Seoul 151-747, Korea
[2]ISIS Facility, STFC Rutherford Appleton Laboratory, Oxfordshire OX11 0QX, UK
[3]Neutron Science Section, MLF Division, J-PARC Center, Tokai, Ibaraki 319-1106, Japan
[4]Department of Physics, Ewha Womans University, Seoul 120-750, Korea
[5]Neutron Science Division, Korea Atomic Energy Research Institute, Daejeon 305-353, Korea
[6]Department of Physics, University of Toronto, Toronto M5S 1A7, Canada
[7]Rutgers Center for Emergent Materials and Department of Physics and Astronomy, Rutgers University, Piscataway NJ 08854, USA
[8]Center for Korean J-PARC Users, Seoul National University, Seoul 151-747, Korea

Multiferroic materials having a coexistence of otherwise seemingly incompatible phases of magnetic and ferroelectric ground states have been the focus of intensive materials researches recently. BiFeO$_3$ is arguably one of the most interesting such materials with both magnetic and ferroelectric transitions occurring above room temperature: T_N=650 K and T_C=1050 K. Moreover, it has an unusual incommensurate magnetic transition with an extremely long period of 650 Å.

Despite the interesting and attractive physical properties as well as numerous studies carried out on this particular compound, some fundamental questions about the underlying mechanism of both transitions remain largely unanswered. There are two main technical reasons for that. First, the transition temperatures are unfortunately little bits too high for most experimental set-ups. Second, it is notoriously difficult to grow sufficiently large single crystals although some breakthrough has been achieved recently in obtaining single crystals of high quality for bulk measurements.

By using 10 single crystals co-aligned within 3° of one another, we have recently measured the spin waves of the antiferromagnetic phase at two state-of-the-art inelastic neutron scattering instruments: one is AMATERA of J-PARC and another MERLIN of ISIS. These two experiments allowed us to map out the full dispersion curve, for the first time, over the entire Brillouin zone and succeeded in determining the essential magnetic exchange interactions: two antiferromagnetic interactions and one Dzyaloshinskii-Moriya term [1]. We will also discuss our latest experimental studies using high-resolution neutron and synchrotron diffraction studies [2] as well as high field studies [3] to find that there are clear anomaly in the temperature dependence of lattice constants. We will discuss our findings in the context of more broad physical properties of this interesting material.

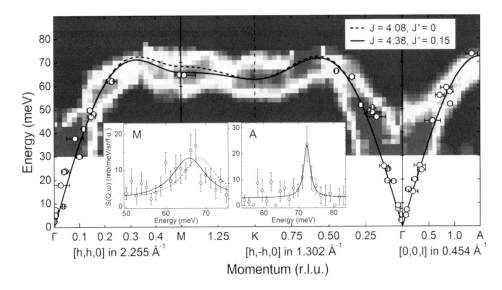

Figure 1: Experimental spin waves measured at AMATERAS beamline of the J-PARC (circles) and MERLIN beamline of the ISIS (contour plot) together with the theoretical spin waves (solid line) calculated with J=4.38 and J'=0.15 meV: the dashed line is for the theoretical spin waves calculated with the Hamiltonian having the NN interaction alone. Inserts are for the momentum cut at the M and A points.

[1] Jaehong Jeong et al., Phys. Rev. Lett. **108**, 077202 (2012).
[2] Junghwan Park et al., J. Phys. Soc. Jpn. **80**, 114714 (2011).
[3] Kenji Ohoyama et al., J. Phys. Soc. Jpn. **80**, 125001 (2011).

TUNING THE MULTIFERROIC PHASE OF CuO WITH IMPURITIES

J. Hellsvik[1], M. Balistieri[2], A. Stroppa[1], A. Bergman[3], L. Bergqvist[4], O. Eriksson[3], S. Picozzi[1], J. Lorenzana[2,5]

[1]CNR-SPIN, Superconducting and Innovative Materials and Devices, L'Aquila, Italy
[2]Dipartimento di Fisica, Università di Roma "Sapienza", Roma, Italy
[3]Department of Physics and Astronomy, Uppsala University, Uppsala, Sweden
[4]Department of Materials Science and Engineering, KTH, Stockholm, Sweden
[5]Istituto dei Sistemi Complessi, CNR, Roma, Italy

The recent discovery that cupric oxide (CuO) is a type-II multiferroic with a high antiferromagnetic (AF) transition temperature T_N of 230 K opened a possible route to room-temperature multiferroicity with a strong magnetoelectric coupling [1]. The type-II behavior of CuO is only present at finite temperatures, between T=210 K and T=230 K, disappearing above and below. As shown by Giovannetti et al. [2], CuO belongs to a new class of multiferroic materials where the so called "order by disorder mechanism" [3] plays a crucial role. A prerequisite for this mechanism is the presence of two or more antiferromagnetic sublattices. At the classical Heisenberg level the exchange field due to spins on one sublattice acting on the other cancels by symmetry so that the angle between the magnetization in the two sublattices remains undetermined and the system has a degenerate zero temperature ground-state. The degeneracy is lifted by quantum and thermal fluctuations rendering collinear alignment of the sublattices.

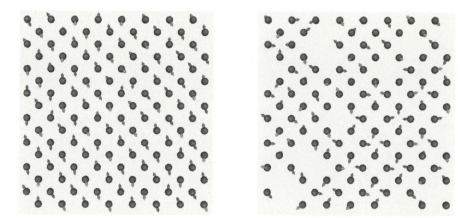

Figure 1: Order-by-disorder in a 2D square J1-J2 antiferromagnet, with or without vacancies, at finite temperature below the Neel temperature. In the left panel, thermal fluctuations couple the two sublattices to yield an essentially collinear state. In the right panel, vacancies cause the net exchange field acting between the two sublattices to be finite, with a preferred perpendicular orientation between the sublattices as result.

Quenched chemical disorder can also act to lift the degeneracy, but in contrast to the quantum and thermal fluctuations, the "chemical fluctuations" favors a canted or perpendicular arrangement of the sublattices, to the benefit of ferroelectric polarization. In this work we study the effect of different impurities on the phase diagram of CuO aiming at engineering the multiferroic properties. As a first step, extensive density functional theory (DFT) calculations were performed for a large number of fixed spin configurations in pure CuO and CuO doped with a small (up to 6%) fraction of the Cu atoms substituted with the nonmagnetic elements Zn or Cd. This enabled, to obtain Heisenberg exchange, anisotropies and biquadratic terms for a parametrized magnetic Hamiltonian. In addition our computations established that the energy difference between the low-temperature collinear AF1 phase and the intermediate temperature multiferroic AF2 phase decreased monotonously with increasing doping level confirming that impurities favor the multiferroic phase.

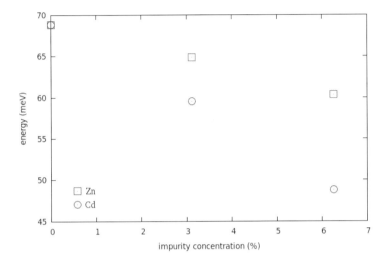

Figure 2: The energy difference between the low-temperature collinear AF1 phase and the intermediate temperature AF2 phase for pure CuO and for CuO doped with 3% or 6% Zn or Cd. The energy is given per cell of 64 atoms.

In subsequent Monte Carlo simulations, the magnetic phase diagram has been mapped out in simulations for classical Heisenberg spins, augmented by investigation into the dynamical properties by means of Langevin spin dynamics simulations. The susceptibility of the spin current has a peak at finite temperature. This peak is connected to the emergence of a spontaneous polarization in the AF2 phase. The inclusion of vacancies or impurities is shown to widen the susceptibility peak and thus the range in temperature in which the AF2 phase can exist.

[1] T. Kimura et al., Nature Mat. **7**, 291 (2008)
[2] G. Giovannetti et al., Phys. Rev. Lett. **106**, 026401 (2011)
[3] C. L. Henley, Phys. Rev. Lett. **62**, 2056 (1989)

MULTIFERROICITY AND MAGNETOELECTRICITY IN A DOPED TOPOLOGICAL FERROELECTRIC

Marco Scarrozza, Maria Barbara Maccioni, Alessio Filippetti, and Vincenzo Fiorentini

CNR-IOM, UOS Cagliari, and Dipartimento di Fisica, Università di Cagliari

Among the family of the layered perovskites, La2Ti2O7 (LTO) is a wide gap ferroelectric (Tc=1770 K, Ps=0.05 C/m2) that has recently attracted special attention for its unconventional mechanism of ferroelectricity. It is termed "topological" ferroelectric [1], as the macroscopic polarization (P) is here produced by antiferrodistortive octahedra rotations failing to compensate due to the layered structure. In the search for multiferroicity, i.e. to add magnetic order in addition to the polarization, we investigated magnetic doping of LTO from first-principles within density-functional theory.

The isovalent substitution of Mn for Ti produces weak multiferroicity at all dopings, as expected due to superexchange between Mn 3d ions. In particular, in the fully-substituted compound La2Mn2O7, many ordering patterns compete, the lowest being a variant of G-type antiferromagnetism. The AF spins are actually canted, giving rise to weak ferromagnetism (FM). The same system is also magnetoelectric, because the rotations are involved in both magnetic and ferroelectric order: as a coercive field undoes the rotations causing ferroelectricity, magnetic coupling doubles in intensity. The magnetoelectric tensor has a single off-diagonal non-zero component, producing changes in M (P) orthogonal to electric (magnetic) field. Notably, the ionic contribution is already 10-fold that of Cr2O3 [2].

Heterovalent substitution of Ti with Sc, Cr and V does yield ferromagnetism. Sc doping induces resonant electronic states at the top valence band (TVB) causing metallicity, it is ferromagnetically coupled, but oxygen vacancies form very easily, spoiling the effect entirely. Cr doping produces deep gap states (~1.8 eV from the TVB).

V doping is the most promising case, resulting in a small gap insulating material. The computed density of states (DOS) is shown in Fig. 1. The valence band (VB) is mostly composed of O p states, the conduction band (CB) is formed by Ti d states hybridized with O p ones. The key role here is played by the V impurity, inducing two peaks of states in proximity of the conduction band (CB) edge, in the majority spin channel: the filled impurity level contains the extra electron provided to the host by the donor, separated by a small gap (0.28 eV) from the other peak, at the CB edge. In the minority spin channel, the V related states are located at ~ 0.80 eV from the Fermi level.

The V dopants align along a direction orthogonal to P, with strong ferromagnetic order (an estimated stabilization energy of ~78 meV/atom), hence forming magnetic chains in the ferroelectric host. Supercell calculations at lower doping concentration indicate a tendency of V to clustering along chains, preserving both the insulating nature and the ferromagnetic unidirectional order.

The analysis of the electronic properties of the V-doped LTO shows that the origin of the robust ferromagnetic order lies in the peculiar layered structure of the host, whose structural anisotropy favours directional orbital overlap, resulting in the splitting of the $t2g$ triplet levels. Specifically, the dxy and dxz orbitals form two quite dispersed impurity bands separated by a small gap (bonding/antibonding splitting [3]), overlapping with the O p states. On the other hand, the dyz projected states are higher in energy, localized on a single peak and degenerate with the eg doublet. This is an indication of the strongly directional V-impurity interaction mediated by O, and involving exclusively dxy and dxz orbitals.

Even more intriguingly, it turns out that the V-based magnetic chains are an example of orbital degenerate system where superexchange is exceptionally ferromagnetic, exhibiting both spin ordering (ferro) and orbital ordering (antiferro) [4]. Unfortunately, the magnetoelectric tensor lattice contribution appears to be negligible.

In summary, our results indicate that V-doped LTO is properly multiferroic. Moreover, we showed that the layered structure of this material, which is at the origin of its ferroelectricity, is also a key to enhance ferromagnetic order when doped with vanadium.

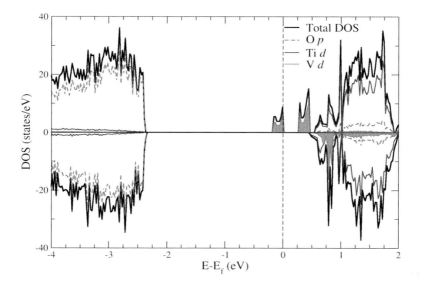

Fig.1 Computed density of states (DOS) of the V-doped LTO, and projected contributions resolved per chemical species. Upper and lower panels are the majority and minority spin channels, respectively.

[1] J. Lopez-Perez and J. Iniguez, Phys. Rev. B **84**, 075121 (2011)
[2] J. Iniguez, Phys. Rev. Lett. **101**, 117201 (2008)
[3] P. Mahadevan, A. Zunger, and D. D. Sarma, Phys. Rev. Lett. **93**, 177201 (2004)
[4] K. I. Kugel and D. I. Khomskii, Sov. Phys. Usp. **25**, 231 (1982).

SEARCH FOR NEW STRAIN-INDUCED MULTIFERROICS WITH HIGH CRITICAL TEMPERATURES

<u>Stanislav Kamba</u>[1], Veronica Goian[1], Přemysl Vaněk[1], Carolina Adamo[2], Charles M. Brook[2], Alexander Melville[2], Nicole A. Benedek,[3] Craig. J. Fennie,[3] June Hee Lee[4], Karin M. Rabe[4], Alexei A. Belik[5], Darrell G. Schlom[2,6]

[1] Institute of Physics, Academy of Sciences of the Czech Republic, Prague, Czech Republic;
[2] Department of Materials Science and Engineering, Cornell University, Ithaca, New York, USA;
[3] School of Applied and Engineering Physics, Cornell University, Ithaca, New York, USA;
[4] Department of Physics and Astronomy, Rutgers University, Piscataway, New Jersey, USA;
[5] National Institute for Materials Science, Tsukuba, Japan;
[6] Kavli Institute at Cornell for Nanoscale Science, Ithaca, New York, USA

Recently, we have experimentally verified that biaxial strain can induce the ferroelectric and ferromagnetic state in epitaxial thin film of $EuTiO_3$ which is paraelectric and antiferromagnetic in the bulk form. We have found that 1% tensile strain in $EuTiO_3/DyScO_3$ films induces ferroelectricity at 250 K and ferromagnetism at 4.2 K.[1] This paper is devoted to the experimental search for other strain-induced multiferroics with higher critical temperatures than $EuTiO_3$. **We will focus on the study of EuO, $Ca_3Mn_2O_7$ and $SrMnO_3$ thin films**, which were recently predicted to be multiferroic using first principles calculations.[2-4] All the films were grown using reactive molecular-beam epitaxy. Direct low-frequency dielectric measurements are mostly impossible due to the leakage current present in the strained thin films. Therefore we have used infrared (IR) spectroscopy, which is not influenced by the conductivity.

We have investigated three kinds of **EuO thin films** deposited on yttrium-stabilized ZrO_2 (YSZ), $LuAlO_3$ and $YAlO_3$ substrates. The films exhibited tensile strain from 0 up to +2.2%. Theoretical strain for induction of ferroelectric state is approx. 4%,[2] so we could not reach the ferroelectric phase. Nevertheless, no significant shift of the phonon frequency with rising strain in EuO films was observed. It indicates no tendency to lattice instability with the strain, i.e. the theoretical predictions from Ref. [2] were not confirmed.

IR reflectivity spectra of $SrMnO_3$ ceramics show strong (17%) hardening of the lowest frequency phonon on cooling below $T_N \approx 230$ K resulting in a 32% decrease of the static permittivity below T_N (see Fig. 1). Exactly the same temperature dependence of the TO1 phonon frequency we theoretically obtained from first principle calculations. **The observed phonon anomaly near T_N gives evidence for extremely strong spin-phonon coupling in $SrMnO_3$**. It is very promising for the possibility of stabilizing the ferroelectric phase in the strained films of $SrMnO_3$.[3] Measurements of the $SrMnO_3$ films grown on $DyScO_3$ and $TbScO_3$ substrates, (strain of 3.4 - 3.9%) are in progress.

Thin film of $Ca_3Mn_2O_7$ with 1.5% compressive strain should theoretically allow 180° switching of magnetization and simultaneous switching of the ferroelectric polarization in an external electric field.[4] We have measured an epitaxial film of $Ca_3Mn_2O_7/YAlO_3$ as well as $Ca_3Mn_2O_7$ ceramics (Néel temperature $T_N \approx 130$ K). **Signature of an improper ferroelectric phase transition between 600 and 700 K was revealed in the IR reflectivity spectra of $Ca_3Mn_2O_7$ ceramics**. Other experiments like XRD, calorimetry and Raman scattering for the confirmation of the phase transition are in progress.

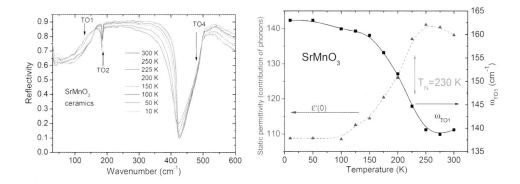

Fig. 1. (a) Infrared reflectivity spectra of SrMnO$_3$ ceramics taken at various temperatures. Phonon frequencies are marked by arrows. (b) Temperature dependence of static permittivity calculated from phonon contributions (left scale). The change of permittivity is caused by the shift of the TO1 phonon frequency near T_N (right scale).

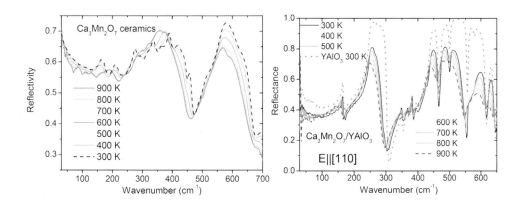

Fig. 2. (a) Infrared reflectivity spectra of Ca$_3$Mn$_2$O$_7$ ceramics showing a gradual shift of phonon frequencies typical for improper ferroelectric phase transition. (b) Infrared reflectance of Ca$_3$Mn$_2$O$_7$ thin film deposited on YAlO$_3$ substrate taken at various temperatures and compared with the room-temperature reflectivity spectra of YAlO$_3$ substrate.

[1] J.H. Lee et al. *Nature*, **466**, 954 (2010); ibid **476**, 114 (2011).
[2] E. Bousquet, N.A. Spaldin et al., *Phys. Rev. Lett.* **104**, 037601 (2010).
[3] J.H. Lee and K.M. Rabe, *Phys. Rev. Lett.* **104**, 207204 (2010).
[4] N.A. Benedek and C.J. Fennie, *Phys. Rev. Lett.* **106**, 107204 (2011).

Nanosession: Qubit systems

QUB 1

Quantum Electronic Materials

Andrew Briggs

Department of Materials, University of Oxford, Parks Road, Oxford OX1 3PH, UK

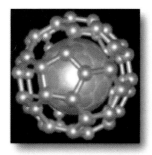
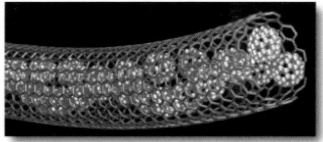

Courtesy of Dr Simon Benjamin

Quantum superposition and entanglement offer deep resources which are ripe for harnessing in practical devices. Superposition incorporates a phase with information content surpassing any classical mixture. Entanglement offers correlations stronger than any which would be possible classically. Together these give quantum computing its spectacular potential, but earlier applications may be found in metrology and sensing. Fundamental progress is being made in the development of quantum devices incorporating electron and nuclear spins which can be controlled with high precision.[1]

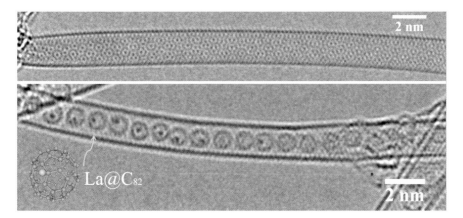

Courtesy of Dr Jamie Warner

Fullerene molecules offer remarkable electron spin properties.[2] They can be assembled in single walled carbon nanotubes, and the resulting atomic structures can be imaged using low voltage aberration corrected transmission electron microscopy.[3] $N@C_{60}$ contains a single nitrogen atom in a cage of sixty carbon atoms, whose spin superposition states are coherent for hundreds of microseconds,[4] and other endohedral fullerenes can be almost as good.[5] Information can be transferred from electron to nuclear spins and back again to give even longer memory times,[6] and can be stored and retrieved holographically in collective spin states.[7] Small tip-angle excitations can be used to demonstrate many of the fundamental principles.[8] Correlated spins can be used for magnetic field sensors that surpass the standard quantum limit.[9] Devices can be made in which the active materials can be imaged with atomic resolution,[10] and whose transport properties can detect a single electron spin.[11] These results open the way for new technologies using the remarkable resources of quantum superposition and entanglement. This kind of quantum nanotechnology also enables fundamental concepts such as reality to be tested experimentally, stimulating new philosophical insights.[12] Somewhat remarkably, these basic studies serve in turn to push the limits of technology, by extending the range of 'quantumness' which can be embodied in practical systems.

1   Ardavan, A. & Briggs, G. A. D. Quantum control in spintronics. *Phil. Trans R. Soc. A* **369**, 3229-3248 (2011)
2   Benjamin, S. C. *et al.* Towards a fullerene-based quantum computer. *Journal of Physics-Condensed Matter* **18**, S867-S883 (2006)
3   Warner, J. H. *et al.* Capturing the motion of molecular nanomaterials encapsulated within carbon nanotubes with ultrahigh temporal resolution. *ACS Nano* **3**, 3037-3044 (2009)
4   Morton, J. J. L. *et al.* Environmental effects on electron spin relaxation in N@C_{60}. *Phys. Rev. B* **76**, 085418 (2007)
5   Brown, R. M. *et al.* Electron spin coherence in metallofullerenes: Y, Sc, and La@C_{82}. *Phys. Rev. B* **82**, 033410 (2010)
6   Brown, R. M. *et al.* Coherent state transfer between an electron and nuclear spin in ^{15}N@C_{60}. *Phys. Rev. Lett.* **106**, 110504 (2011)
7   Wesenberg, J. H. *et al.* Quantum computing with an electron spin ensemble. *Phys. Rev. Lett.* **103**, 070502 (2009)
8   Wu, H. *et al.* Storage of multiple coherent microwave excitations in an electron spin ensemble. *Phys. Rev. Lett.* **105**, 140503 (2010)
9   Jones, J. A. *et al.* Magnetic field sensing beyond the standard quantum limit using 10-spin NOON states. *Science* **324**, 1166-1168 (2009)
10 Warner, J. H. *et al.* Resolving strain in carbon nanotubes at the atomic level. *Nature Mater.* **10**, 958-962 (2011)
11 Chorley, S. J. *et al.* Transport spectroscopy of an impurity spin in a carbon nanotube double quantum dot. *Phys. Rev. Lett.* **106**, 206801 (2011)
12 Knee, G. C. *et al.* Violation of a Leggett-Garg inequality with ideal non-invasive measurements. *Nature Commun.* **3**, 606 (2012)

QUB 2

IDENTIFYING CAPACITIVE AND INDUCTIVE LOSS IN LUMPED ELEMENT SUPERCONDUCTING RESONATORS

Martin P. Weides*, Michael R. Vissers, Jeffrey S. Kline, Martin O. Sandberg, and David P. Pappas

National Institute of Standards and Technology, Boulder, Colorado 80305, USA
*Present address: Karlsruher Institut für Technologie, Karlsruhe, Germany

Superconducting resonators with low loss are of great interest for photon detection [1] and quantum computation [2]. Hybrid devices, using different materials in specific parts of the resonant circuit, allow the alteration of intrinsic materials properties such as the superconducting gap, radiation cross-section, and capacitive and inductive loss within the resonators to optimize the entire circuit. Locating whether the residual loss sources are inductive or capacitive in nature is of considerable relevance. So far, most resonator studies have been done on single films in either coplanar waveguide resonators (CPWs) or lumped element circuits [3,4].

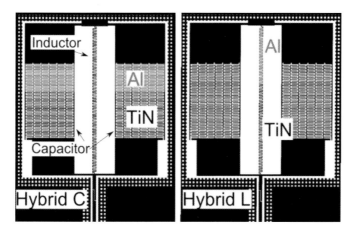

Figure 1. Drawn resonator designs hybrid capacitor (left) and hybrid inductor (right) with 3 : 2 TiN : Al fraction. The TiN (black) interdigitated capacitors and meandered inductors are progressively replaced with Al (red). The frequency multiplexed resonators are capacitively coupled to the feedline (not shown).

We present a method to systematically locate and extract capacitive and inductive losses at microwave frequencies by use of mixed-material, lumped element resonators, see Fig. 1. In frequency multiplexed resonators, ultra-low loss titanium nitride [5,6] was progressively replaced with aluminum in the interdigitated capacitor and meandered inductor elements. By measuring the loss as the Al/TiN fraction in each element is increased (Fig. 2), we find that at low field, i.e., in the low photon limit, the loss is correlated with the amount of Al capacitance, rather than the inductance. By changing geometry we show that the power (i.e., photon number) dependent loss is located in the capacitive region and is two-level system in nature. Even with small contacts, fits to the measured loss at different Al-TiN fractions indicates that any loss at the Al-TiN material interface is $<1 \times 10^{-6}$.

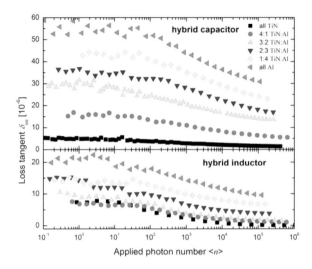

Figure 2. Loss for (a) hybrid capacitor and (b) hybrid inductor resonators as a function of the photon number in the capacitor. When Al is added to the capacitor, there is a greater increase in loss than if Al is added to the inductor.

The lack of interface loss does not preclude future hybrid devices from having low losses. This work suggests that future devices can be designed with their individual circuit elements optimally tuned for their purpose e.g. low TLS loss, superconducting gap, kinetic inductance, detection efficiency etc. specifically tuned. [7-9]

[1] P. K. Day, H. G. LeDuc, B. A. Mazin, A. Vayonakis, and J. Zmuidzinas, Nature (London) **425**, 817 (2003).
[2] M. Mariantoni, H. Wang, T. Yamamoto, M. Neeley, R. C. Bialczak, Y. Chen, M. Lenander, E. Lucero, A. D. O'Connell, D. Sank, M. Weides, J. Wenner, Y. Yin, J. Zhao, A. N. Korotkov, A. N. Cleland, and J. M. Martinis, Science **334**, 61 (2011).
[3] J. Gao, M. Daal, J. M. Martinis, A. Vayonakis, J. Zmuidzinas, B. Sadoulet, B. A. Mazin, P. K. Day, and H. G. Leduc, Appl. Phys. Lett. **92**, 212504 (2008).
[4] A. D. O'Connell, M. Ansmann, R. C. Bialczak, M. Hofheinz, N. Katz, E. Lucero, C.McKenney, M. Neeley, H. Wang, E. M. Weig, A. N. Cleland, and J. M. Martinis, Appl. Phys. Lett. **92**, 112903 (2008).
[5] M. R. Vissers, J. Gao, D. S. Wisbey, D. A. Hite, C. C. Tsuei, A. D. Corcoles, M. Steffen, and D. P. Pappas, Appl. Phys. Lett. **97**, 232509 (2010).
[6] M. Sandberg, M. R. Vissers, J. Kline, M. Weides, J. Gao, D. Wisbey, and D. Pappas, in preparation (2012).
[7] J. M. Martinis, K. B. Cooper, R. McDermott, M. Steffen, M. Ansmann, K. D. Osborn, K. Cicak, S. Oh, D. P. Pappas, R. W. Simmonds, and C. C. Yu, Phys. Rev. Lett. **95**, 210503 (2005).
[8] M. Weides, R. C. Bialczak, M. Lenander, E. Lucero, M. Mariantoni, M. Neeley, A. D. O'Connell, D. Sank, H. Wang, J. Wenner, T. Yamamoto, Y. Yin, A. N. Cleland, and J. Martinis, Supercond. Sci. Technol. **24**, 055005 (2011).
[9] M. P. Weides, J. S. Kline, M. R. Vissers, M. O. Sandberg, D. S. Wisbey, B. R. Johnson, T. A. Ohki, and D. P. Pappas, Appl. Phys. Lett. **99**, 262502 (2011).

SUPERCONDUCTIVITY IN QUASI-1D LaAlO$_3$/SrTiO$_3$ NANOSTRUCTURES

Joshua Veazey[1], Guanglei Cheng[1], Patrick Irvin[1], Shicheng Lu[1], Mengchen Huang[1], Chung Wung Bark[2], Sangwoo Ryu[2], Chang-Beom Eom[2], Jeremy Levy[1]

[1]University of Pittsburgh, Pittsburgh, PA, USA; [2]University of Wisconsin-Madison, Madison, WI, USA

Superconductivity in quasi-one-dimensional geometries poses fundamental challenges while providing several pathways for solid-state quantum information preservation and processing. Theoretically, superconducting long-range order should be suppressed in one dimension due to quantum fluctuations; in real systems, it may be possible to sustain or even enhance long-range order through interactions with 3D surroundings or via other mechanisms.

The 2D electron gas at the LaAlO$_3$/SrTiO$_3$ interface becomes superconducting below a critical temperature T_c~200 mK [1]. Here, we discuss transport characteristics of superconductivity observed in quasi-1D LaAlO$_3$/SrTiO$_3$ nanostructures.

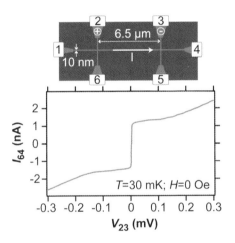

Figure 1: Superconductivity in q-1D LAO/STO nanostructures. A six-terminal Hall cross is sketched with c-AFM [2]. The structure has line widths, w~10 nm, and a main channel length, L=6.5 μm. Here, the resistance in the superconducting phase is vanishing below a critical current I_c~1.5 nA. To perform these measurements, a current is sourced as noted in the schematic, while the voltage is measured independently by leads with ~TΩ input-impedances.

Quasi-1D nanostructures are created using a conductive atomic-force microscopy lithography technique [2]. Figure 1 shows four-terminal transport measurements for a six-terminal Hall cross (line widths w~10 nm and a main channel length L=6.5 μm) at temperature T=30 mK. In the superconducting phase, the resistance is vanishing. However, similar devices exhibit finite superconducting resistances of $R_C \sim h/4e^2$, while the normal-state resistances are approximately $R_N \sim h/e^2$ above T_c. The critical current magnitude scales as $I_c \sim \Delta/\Phi_0$, where Δ is the superconducting gap energy and Φ_0 is the flux quantum. This is characteristic of a single superconducting quantum channel.

The existence of full superconductivity in LaAlO$_3$/SrTiO$_3$ nanostructures represents an important milestone for the creation of new families of quantum devices at extreme length scales. The presence of strong spin-orbit coupling as well as ferromagnetism in this system may lead to novel properties that are relevant for superconducting phases that support Majorana fermions.

[1] N. Reyren et al., Science **31**, 1196 (2007).
[2] C. Cen et al., Nature Mater. **7**, 298 (2008).

ON-DEMAND SINGLE ELECTRON TRANSFER BETWEEN DISTANT QUANTUM DOTS — ELECTRON "PING-PONG" IN A SINGLE ELECTRON CIRCUIT

R. P. G. McNeil[1,*], M. Kataoka[1,2], C. J. B. Ford[1], C. H. W. Barnes[1], D. Anderson[1], G. A. C. Jones[1], I. Farrer[1] and D. A. Ritchie[1].

[1] Cavendish Laboratory, University of Cambridge, J. J. Thomson Avenue, Cambridge CB3 0HE, UK.
[2] National Physical Laboratory, Hampton Road, Teddington TW11 0LW, UK.

Electrons in a quantum dot (QD) may be used to represent a quantum bit of information (qubit). Such qubits may be transferred between tunnel-coupled quantum dots [1, 2] but over micron distances a tunnel barrier is not suitable. We demonstrate the repeated transfer of a single electron over a distance of four microns back and forth from one quantum dot to another through a depleted one-dimensional channel using a short surface acoustic wave (SAW) pulse. This technique may allow the movement of quantum information between distant quantum dots.

The device consists of two QDs (RQD & LQD) connected by a 4 μm long channel, all defined by surface gates (Fig. 1(a)), on a GaAs/AlGaAs heterostructure with a 2DEG 90nm below the surface. Transducers are placed 1 mm to the left and right of the channel. The number of electrons a dot can hold is controlled by the barrier and plunger gates (Fig. 1(b)). Changes in the occupation of the QDs are monitored by quantum point contacts (Fig. 1(c)). Electrons trapped in the LQD are raised above the Fermi energy by stepping plunger and barrier gates. These trapped non-equilibrium electrons can then be held for many seconds [3]. Having set the LQD to contain one electron and the RQD to be empty a SAW pulse is sent. The SAW potential lifts an electron from the left dot and carries it to the right dot, where it is captured by a large voltage on the barrier gate. The RQD potential can then be raised and a left-going SAW pulse sent to return the electron. [4]

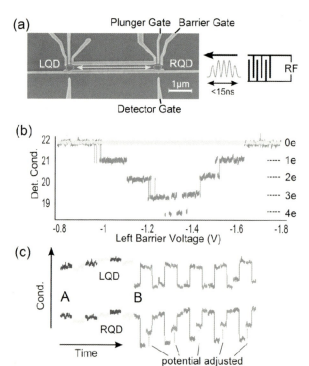

Figure 1: (a) SEM of device. Electrons transferred between QDs (outlined) by SAW pulses. (b) QDs may be initialised empty (grey/blue) or occupied with 0 – 4 electrons (red/black) by stepping barrier and plunger gates. (c) Alternating right- and left-going SAW pulses are sent through device. Sec. A, no electrons in QDs as a control, Sec. B single electron in LQD transferred back and forth. Electrons leaving (entering) QD cause step increase (decrease) in conductance. Potential of RQD is adjusted after receiving electron giving change is height.

The pulse width dependence of depopulating the dots suggests that transfer occurs during the first few nanoseconds of the pulse. A single electron may be transferred reliably back and forth between the QDs (Fig. 1(c) section B) over a hundred times.

[1] T. Hayashi et al., Phys. Rev. Lett. **91**, 226804 (2003).
[2] J. R. Petta et al., Science **309**, 2180 (2005).
[3] M. Kataoka et al., Phys. Rev. Lett. **98**, 046801 (2007).
[4] R. McNeil et al., Nature **477**, 439 (2011).
*current address: Physikalisches Institut IIC, RWTH-Aachen, 52074 Aachen, Germany.

ULTRAFAST ENTANGLING GATES BETWEEN NUCLEAR SPINS USING PHOTO-EXCITED TRIPLET STATES

Vasileia Filidou,[1] Stephanie Simmons,[1] Steven D. Karlen,[2] Feliciano Giustino,[1] Harry L. Anderson,[2] and John J. L. Morton,[1,3]

[1] Department of Materials, University of Oxford, Parks Road, Oxford OX1 3PH, UK
[2] Department of Chemistry, University of Oxford, Oxford OX1 3TA, UK
[3] CAESR, The Clarendon Laboratory, Department of Physics, University of Oxford, OX1 3PU, UK

Nuclear spins despite their long decoherence times suffer from slow interactions and weak polarization, An electron spin can be used efficiently to polarize nuclear spins, however its permanent presence is a source of decoherence. Here we show how a transient electron spin, arising from the optically excited triplet state of C_{60}, can be used to hyperpolarise, couple and measure two nearby nuclear spins. We use electron spin resonance (ESR) methods in combination with laser photoex-citation and DFT to characterise our system and determine the interactions. We use two methods to implement entangling operations within the lifetime of the triplet state and by exploiting the spinor nature of the electron we show that an entangling gate can be performed in hundreds of nanoseconds, five orders of magnitude faster than the liquid state J-coupling.

NOISE SPECTROSCOPY USING CORRELATIONS OF SINGLE-SHOT QUBIT READOUT

Thomas Fink, Hendrik Bluhm

2nd Institute of Physics C, RWTH Aachen University, D-52074 Aachen, Germany

Qubits that fulfill the DiVincenzo criteria are an essential prerequisite for the realization of a scalable quantum computer. For fault-tolerant quantum computation, qubits with sufficiently long coherence times are required. Understanding the noise causing decoherence is therefore crucial for improving qubit performance.

In the past, pulse sequences have gained attention for effectively decoupling the qubit from its noisy environment and therefore reducing decoherence. Besides enhancing dephasing times, they can also be used for noise spectroscopy by observing the dependence of qubit coherence on the evolution time under the application of suitable pulse sequences. However, this is subject to certain limitations. For a fixed pulse sequence, the frequency region over which the pulse sequence is sensitive shifts with the inverse evolution time while longer durations increase the overall sensitivity. This relation makes it hard to probe low frequency noise: by the time the frequency region of interest is accessible, the qubit may be fully dephased leaving no measurement contrast. This problem can be circumvented somewhat by adding pulses and clever choice of their timing. However, this strategy may be limited by π-pulse error and energy relaxation time of the qubit.

We propose an alternative method to determine the noise spectrum filling the gap between direct and pulse sequence-based spectroscopy and not necessitating any π-pulses. It is based on correlating single-shot measurements of free induction decay experiments. The measurement cycle is depicted in Fig. 1 and consists of initializing a state $|I\rangle$, letting it evolve under the influence of the noise process for time τ, projective measurement of the final state and repeating this process after a delay time Δt. Averaging over many such measurements, one can compute the correlation between consecutive measurements as a function of delay time Δt.

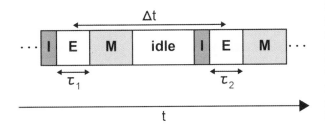

Figure 1: Measurement cycle: initialization (I) of state $|I\rangle$, evolution (E) for time and measurement (M) of the outcome. The delay time between two pulses is Δt. While the measurement time sets a lower limit for Δt, the idle block can be used to enhance the delay between pulses.

We consider pure dephasing leading to a Hamiltonian $H = \frac{\hbar}{2}[\Omega + \beta(t)]\hat{\sigma}_z$ where β represents a classical noise process. The autocorrelation of the qubit is then approximately given by $\langle P_I(t)P_I(t+\Delta t)\rangle \approx e^{-\frac{1}{2}\chi_-}$ where χ_- contains the spectrum and filter function [1] of our measurement cycle and depends on the evolution time τ and the delay time Δt. The fundamental advantage of our method over pulse sequence-based spectroscopy is the possibility to maintain a small χ_- and thus a good measurement contrast over a wide range by independently tuning the evolution and delay time.

In order to use our method for noise spectroscopy in a concrete application, we calculate χ_- for a Singlet-Triplett spin qubit based on a gate-defined GaAs double quantum dot. In these devices, the different Overhauser fields in the two dots are the dominant source of dephasing [2]. We consider a spectrum $S_\beta(\omega) \propto \exp[-(\omega/\omega_E)^\gamma]$ introduced in [3] to mimic the nuclear spin diffusion processes and their expected suppression at frequency ω_E. Fig. 2 (a) shows that different γ can be identified due to their slope with a constant evolution time. Simultaneously varying evolution and delay time allows for maintaining a good measurement contrast over an even wider range, as seen in Fig. 2 (b).

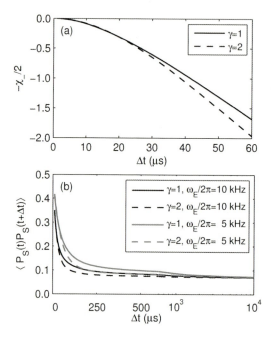

Figure 2: (a) χ_- for evolution time 5 μs, exponential (Gaussian) cutoff with $\gamma = 1$ (2), and $\omega_E/2\pi = 10$ kHz. Note the difference for $\Delta t > 1/\omega_E \sim 16$ μs that makes it possible to distinguish different cutoff behaviors. For long delay times the two curves merge as in this regime, χ_- is independent of the form of the spectrum. (b) Adjusting the evolution time allows for detection of the crossover and slope of curves with different cutoff behavior, thus providing well distinguishable characteristics on an experimentally measurable scale.

We can thus use our method to extract characteristics of the underlying noise spectrum acting on the qubit. Simulations indicate that our technique can be used similarly to identify different α investigating the ubiquitously observed $1/f^\alpha$ noise in superconducting qubits.

[1] Ł. Cywiński et al., Phys. Rev. B **77**, 174509 (2008).
[2] J. R. Petta et al., Science **309**, 2180 (2005).
[3] M. Biercuk and H. Bluhm, Phys. Rev. B **83**, 235316 (2011).

INVESTIGATING THE IRON BASED SUPERCONDUCTOR ($FeSe_{0.4}Te_{0.6}$) WITH SPECTROCOPIC-IMAGING SCANNING TUNNELING MICROSCOPE

Stefan Schmaus[1], Udai Raj Singh[1], Seth White[1], Joachim Deisenhofer[2], Vladimir Tsurkan[2], Alois Loidl[2], Peter Wahl[1]

[1]Max-Planck-Institut für Festkörperforschung, Stuttgart, Germany;
[2]Lehrstuhl für Experimentalphysik V, Universität Augsburg, Germany

The recent discovery of iron-based superconductors attracted strong interest in the hope that understanding superconductivity in these compounds might facilitate to solve the riddle of high temperature superconductivity [1]. The ability to characterize electronic excitations simultaneously in real and in momentum space has made Spectroscopic-Imaging Scanning Tunneling Microscopy (SI-STM) a valuable tool to study correlated electrons systems and to contribute to an understanding of unconventional superconductivity [2,3].

Here, we present an SI-STM study of the iron chalcogenide $FeSe_{0.4}Te_{0.6}$. The layered crystal structure of this simplest iron-based superconductor with its natural cleavage plane is predestined for investigations by SI-STM. Topographic images of the *in situ* cleaved sample surface allow for a stoichiometric analysis of the composition of the material which yields excellent agreement with energy-dispersive X-ray spectroscopy (EDX) measurements. By spatially mapping the differential conductance as a function of bias voltage of the *in situ* cleaved sample surface, we have studied the electronic properties of $FeSe_{0.4}Te_{0.6}$ in the temperature range from above the critical temperature ($T_C \approx 14$ K) down to 1.8 K. The temperature dependent measurements clearly reveal the emergence of the superconducting gap, consistent with previous measurements [4]. A detailed analysis of the gap parameters shows significant spatial inhomogeneity of the superconducting gap. We discuss possible origins of the gap inhomogeneity in relation to the sample composition. In the autocorrelation and Fourier transformation of spectroscopic maps we find evidence for dispersing electronic states which break the C_4 symmetry of the underlying crystal lattice. This symmetry breaking persists even in measurements performed above T_C. Similar to recent observations of a nematic state in the non-superconducting parent compound of 122 iron pnictides [5], the electronic structure of iron chalcogenides apparently shows a nematic state, which is not expected from the crystal structure. Our measurements clearly show coexistence of this state with superconductivity.

[1] Y. Kamihara et al., J. Am. Chem. Soc. **128**, 10012 (2006).
[2] J. E. Hoffman et al., Science **297**, 1148 (2002).
[3] K. McElroy et al., Nature **422**, 592 (2003).
[4] T. Hanaguri et al., Science **328**, 474 (2010).
[5] T.-M. Chuang et al., Science **327**, 181 (2010).

NANOSCALE LAYERING OF ANTIFERROMAGNETIC AND SUPERCONDUCTING PHASES IN $Rb_2Fe_4Se_5$

Aliaksei Charnukha[1], Antonija Cvitkovic[2], Thomas Prokscha[3], Daniel Proepper[1], Nenand Ocelic[2], Andreas Suter[3], Zaher Salman[3], Elvezio Morenzoni[3], Joachim Deisenhofer[4], Vladimir Tsurkan[4,5], Alois Loidl[4], Bernhard Keimer[1], and Alexander Boris[1]

[1]Max Planck Institute for Solid-State Research, Heisenbergstrasse 1, D-70569 Stuttgart, Germany
[2]Neaspec GmbH, D-82152 Martinsried (Munich), Germany
[3]Laboratory for Muon Spin Spectroscopy, Paul Scherrer Institute (PSI), CH-5232 Villigen PSI, Switzerland
[4]Experimental Physics V, Center for Electronic Correlations and Magnetism, Institute of Physics, University of Augsburg, D-86159 Augsburg, Germany
[5]Institute of Applied Physics, Academy of Sciences of Moldova, MD-2028 Chisinau, R. Moldova

The recent discovery of intercalated iron-selenide superconductors has stirred up the condensed-matter community accustomed to the proximity of the superconducting and magnetic phases in various cuprate and pnictide superconductors. Never before has a superconducting state with a transition temperature as high as 30 K been found to coexist with such exceptionally strong antiferromagnetism with Neel temperatures up to 550 K as in this new family of iron-selenide materials. Significant experimental evidence suggests, however, that the superconducting and antiferromagnetic phases are spatially separated. All the experimental effort notwithstanding, the microscopic scale of the phase separation and the domain order with a possible coupling between the phase order parameters remain unclear.

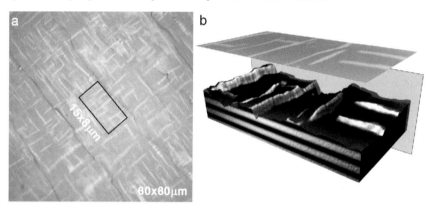

Figure 1: Topography and optical properties of the RFS surface. a, Microscope image of a 60x60 mm² surface patch of a freshly-cleaved superconducting RFS single crystal. Typical rectangular 15x8 mm² area studied via near-field microscopy. b, Superposition of the topography of a 15x8 mm² rectangular area (terrain) and the optical contrast (brightness) normalized to that of silicon. Glossy areas indicate high silicon-RFS contrast and thus metallicity, while the matt areas are insulating.

Here we use a unique combination of scattering-type scanning near-field optical microscopy (s-SNOM) and low-energy muon spin rotation (LE-mSR) to shed light on microscopic character of the phase separation in superconducting Rb2Fe4Se5 (RFS) single crystals. We demonstrate that the phases segregate on the nanoscale out of plane, reminiscent of a quasiregular heterostructure, while the characteristic size of the paramagnetic domains in plane reaches 10 mm (see Fig.1). By means of LE-mSR we further show that the antiferromagnetic semiconducting phase is strongly weakened near the sample surface. Self-organization in a chemically homogeneous structure indicates an intimate connection between the modulated superconducting and antiferromagnetic phases.

RESONANCE MODE IN RARE-EARTH SYSTEMS WITH VALENCE INSTABILITY

Kirill Nemkovski[1], Pavel Alekseev[2,3], Jean-Michel Mignot[4]

[1]Forschungszentrum Jülich GmbH, Jülich Centre for Neutron Science, Außenstelle am FRM II, Garching, Germany;
[2]National Research Centre "Kurchatov Institute", Moscow, Russia;
[3]National Research Nuclear University "MEPhI", Moscow, Russia;
[4]Laboratoire Léon Brillouin, CEA-CNRS, CEA/Saclay, Gif sur Yvette, France

An extended study of the spin dynamics in the Kondo-insulator YbB_{12} [1,2], as well as in the Sm- and Eu-based unstable-valence systems SmB_6 [3], $Sm_{1-x}Y_xS$ [4] and $EuCu_2Si_xGe_{1-x}$ [5], has been performed by means of inelastic neutron scattering (INS) spectroscopy, including neutron polarization analysis.

The results of [1-5], along with new experimental data, clearly show that rare-earth (RE) compounds with a valence instability may demonstrate an exciton-like in-gap excitation, generally similar (though different in some details) to the so-called *resonance mode* in HTSC. For the illustration, the characteristic behavior of the resonance excitation in YbB_{12} is shown in Fig. 1.

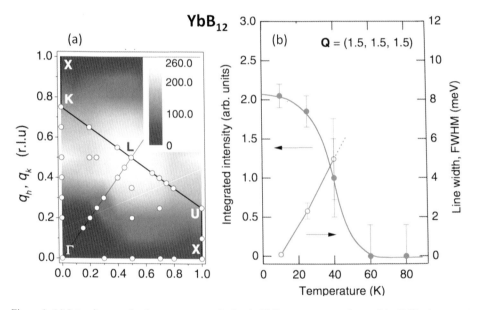

Figure 1: (a) Intensity map for the resonance excitation in YbB_{12} over one quadrant of the Brillouin zone at T = 5 K; circles denote Q vectors at which energy spectra have been measured [2]. (b) Temperature dependence of the integrated intensity and line width for the resonance peak in the INS spectra m⎯ ⎯d at Q = (1.5, 1.5, 1.5) [1]. Lines are guides to the eye.

The observed excitations can be classified into two types: (1) excitations base⎯ with intermediate radius in systems with "strong" intermediate valence; (2) ⎯ tions arising from dynamical antiferromagnetic correlations between the loc⎯ of the RE ions. The possible role of an interplay with lattice degrees of free⎯

The results obtained along with similar studies of HTSC [6] point to s⎯ tions being characteristic for strongly correlated electron systems with the presence of competing interactions.

[1] J.-M. Mignot, P. A. Alekseev, K. S. Nemkovski *et.al.*, Phys. Rev. Lett., **94**, 247204, 2005.
[2] K. S. Nemkovski, J.-M.Mignot, P.A.Alekseev *et al.*, Phys.Rev. Lett. **99**, 137204 (2007).
[3] P. A. Alekseev, J.-M. Mignot, J. Rossat-Mignod *et al.*, J. Phys.: Condens. Matter **7**, 289 (1995).
[4] P. A. Alekseev, J.-M. Mignot, E. V. Nefeodova, *et al.*, Phys. Rev. B **74**, 035114 (2006)
[5] P. A. Alekseev, K. S. Nemkovski, J.-M. Mignot et al., Magnetic excitations in $EuCu_2(Si_xGe_{1-x})_2$: from mixed valence towards magnetism, to be published.
[6] V. Hinkov, P. Bourges, S. Pailhès *et.al.*, Nature Physics **3**, 780 (2007).

RESONANCE MODE IN RARE-EARTH SYSTEMS WITH VALENCE INSTABILITY

Kirill Nemkovski[1], Pavel Alekseev[2,3], Jean-Michel Mignot[4]

[1]Forschungszentrum Jülich GmbH, Jülich Centre for Neutron Science, Außenstelle am FRM II, Garching, Germany;
[2]National Research Centre "Kurchatov Institute", Moscow, Russia;
[3]National Research Nuclear University "MEPhI", Moscow, Russia;
[4]Laboratoire Léon Brillouin, CEA-CNRS, CEA/Saclay, Gif sur Yvette, France

An extended study of the spin dynamics in the Kondo-insulator YbB_{12} [1,2], as well as in the Sm- and Eu-based unstable-valence systems SmB_6 [3], $Sm_{1-x}Y_xS$ [4] and $EuCu_2Si_xGe_{1-x}$ [5], has been performed by means of inelastic neutron scattering (INS) spectroscopy, including neutron polarization analysis.

The results of [1-5], along with new experimental data, clearly show that rare-earth (RE) compounds with a valence instability may demonstrate an exciton-like in-gap excitation, generally similar (though different in some details) to the so-called *resonance mode* in HTSC. For the illustration, the characteristic behavior of the resonance excitation in YbB_{12} is shown in Fig. 1.

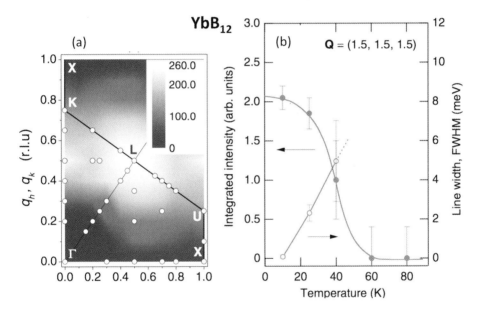

Figure 1: (a) Intensity map for the resonance excitation in YbB_{12} over one quadrant of the Brillouin zone at T = 5 K; circles denote Q vectors at which energy spectra have been measured [2]. (b) Temperature dependence of the integrated intensity and line width for the resonance peak in the INS spectra measured at **Q** = (1.5, 1.5, 1.5) [1]. Lines are guides to the eye.

The observed excitations can be classified into two types: (1) excitations based on a charge exciton with intermediate radius in systems with "strong" intermediate valence; (2) spin-exciton-like excitations arising from dynamical antiferromagnetic correlations between the localized magnetic moments of the RE ions. The possible role of an interplay with lattice degrees of freedom is also discussed.

The results obtained along with similar studies of HTSC [6] point to such type of resonance excitations being characteristic for strongly correlated electron systems with gap-like dynamical response in the presence of competing interactions.

[1] J.-M. Mignot, P. A. Alekseev, K. S. Nemkovski *et.al.*, Phys. Rev. Lett., **94**, 247204, 2005.
[2] K. S. Nemkovski, J.-M.Mignot, P.A.Alekseev *et al.*, Phys.Rev. Lett. **99**, 137204 (2007).
[3] P. A. Alekseev, J.-M. Mignot, J. Rossat-Mignod *et al.*, J. Phys.: Condens. Matter **7**, 289 (1995).
[4] P. A. Alekseev, J.-M. Mignot, E. V. Nefeodova, *et al.*, Phys. Rev. B **74**, 035114 (2006)
[5] P. A. Alekseev, K. S. Nemkovski, J.-M. Mignot et al., Magnetic excitations in EuCu$_2$(Si$_x$Ge$_{1-x}$)$_2$: from mixed valence towards magnetism, to be published.
[6] V. Hinkov, P. Bourges, S. Pailhès *et.al.*, Nature Physics **3**, 780 (2007).

MULTI-BAND EFFECTS ON SUPERCONDUCTING INSTABILITIES DRIVEN BY ELECTRON-ELECTRON INTERACTIONS

Stefan Uebelacker[1], Carsten Honerkamp[1]

[1]RWTH Aachen University, Institue for Theoretical Solid State Physics, Aachen, Germany

We explore multi-band effects on d-wave superconducting instabilities driven by electron-electron interactions. Our models on the two-dimensional square lattice consist of a main band with an extended Fermi surface and predominant weight from d (x^2-y^2) -orbitals, whose orbital character is influenced by the admixture of other energetically neighbored orbitals. Using a functional renormalization group description of the superconducting instabilities of the system and different levels of approximations, we study how the energy scale for pairing and hence the critical temperature is affected by the band structure. We find that a reduction of orbital admixture as function of the orbital energies can cause a Tc-enhancement although the Fermi surface becomes more curved and hence less favorable for antiferromagnetic spin-fluctuations. While our study does not allow a quantitative understanding of the T_c-differences in realistic high- T_c cuprate systems, it may reveal an underlying mechanism contributing to the actual material trends.

FIELD-INDUCED SUPERCONDUCTIVITY IN A LAYERED TRANSITION METAL DICHALCOGENIDE

Jianting Ye, Yijin Zhang, Yoshihiro Iwasa

Department of Applied Physics, The University of Tokyo, 7-3-1 Hongo, Bunkyo-ku, Tokyo 113-8656, Japan

Recent developments in electric double layer transistors (EDLTs) are attracting growing interests because of its stronger field effect than other transistor techniques. This orders of magnitude larger gating effect provides unique abilities to reach the high carrier densities required for inducing superconductivity in several kinds of materials. Among them, layered materials are convenient examples to work with since high quality surface suitable for transistor channel could be easily obtained after mechanical cleavage. Especially, after the introduction of graphene techniques, high quality atomically flat surface can be routinely fabricated on a broad range of layered materials.

Combining double layer gating method with novel materials processing techniques on layered materials provides new opportunities in manipulating their electronic properties. We can achieve high carrier density up to 10^{14} cm^{-2} electrostatically in a layered transition metal dichalcogenide and induce metal insulator transitions. Superconductivity, similar as that shown in ZrNCl [1], could be observed when we cool down this dichalcogenide system to low temperature after reaching a metal insulator transition with large amount of accumulated carriers. The versatility of this combination shows its potential as a protocol to study varieties of layered materials for broader scope of possibilities in accessing superconducting properties. And hopefully, this method could also facilitate to induce superconductivity in new materials.

[1] J. T. Ye et al., Nature Materials **21**, 4487 (2011).

TOWARDS IDEAL HIGH-T_c JOSEPHSON JUNCTIONS

Yuriy Divin[1], Irina Gundareva[1,2], Matvei Lyatti[1,2], Ulrich Poppe[1]

[1]Peter Gruenberg Institute, Research Center Jülich, Germany;
[2]Kotel'nikov Institute of Radio Engineering and Electronics, RAS, Moscow, Russian Federation

Electrical transport in high-T_c superconductors is limited by grain boundaries and, at high misorientation angles, these boundaries even demonstrate Josephson behavior [1]. Bicrystal technology is used to fabricate grain boundaries with a large range of misorientations. Previously, the main development was focused on the [001]-tilt bicrystal high-T_c junctions [1]. Much better structural and electrical parameters have been found for [100]-tilt high-T_c junctions [2]-[4]. Starting from consistency and reproducibility of I-V curves and high-frequency behaviour for low-T_c Josephson junctions, a concept of an "ideal" junction was introduced in late 70th of last century [5]. In a similar way, here, we present our approach to the "ideal" high-T_c Josephson junction, based on our study of [100]-tilt $YBa_2Cu_3O_{7-x}$ bicrystal junctions.

TEM image of a cross section for one of our [100]-tilt $YBa_2Cu_3O_{7-x}$ bicrystal junctions are presented in Fig. 1 (inset). The junction electrodes consist of thin films with mutually tilted c-axes. The amplitude of meandering of the $YBa_2Cu_3O_{7-x}$ bicrystal boundary with respect to the substrate bicrystal boundary is less than 20 nm, which is more than one order lower compared with that of the [001]-tilted $YBa_2Cu_3O_{7-x}$ bicrystal junctions. The improved nanostructure of [100]-tilt junctions was reflected in higher spatial homogeneity of electrical transport through these junctions and high values of the characteristic voltages I_cR_n up to 8 - 10 mV at liquid-helium temperatures (Fig.1). The highest I_cR_n-values obtained for our [100]-tilt junctions are still 4.5 times lower than the Ambegaokar-Baratoff value $\pi\Delta(0)/2$ [6], when the energy gap value $\Delta_X = 29$ meV for $YBa_2Cu_3O_{7-x}$ [7] is taken into account. A suppression of the order parameter of high-T_c materials at the grain boundaries may be responsible for lower values of I_cR_n in high-T_c Josephson junctions.

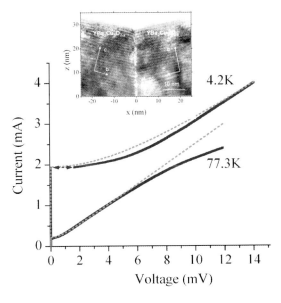

Figure 1: I-V curves of a [100]-tilt $YBa_2Cu_3O_{7-x}$ junction at temperatures of 4.2 K and 77.3 K (solid lines). The corresponding I-V curves in RSJ model with the same values of the critical currents I_c and normal-state resistances R_n as for experimental data are shown by dotted lines. The values of I_cR_n-product are equal to 1 mV and 8 mV at 77.3 K and 4.2 K, correspondingly. The inset is an image of a cross section of a [100]-tilt $YBa_2Cu_3O_{7-x}$ bicrystal junction obtained by transmission electron microscopy [3].

Low-frequency fluctuations δI_c of critical current and δR_n of normal-state resistance in the [100]-tilted $YBa_2Cu_3O_{7-x}$ bicrystal junctions for the first time demonstrated the same relative intensities and complete anticorrelation [8]. These circumstances should hold when both quasiparticle current and supercurrent are uniformly distributed across the barrier area or, at least, they are tunnelling through the same parts of the barrier.

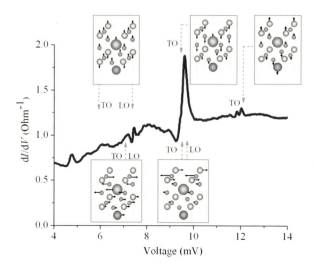

Figure 2: Differential conductance dI/dV vs. voltage V for [100]-tilt YBa$_2$Cu$_3$O$_{7-x}$ bicrystal junction, where arrows show Josephson voltages V_i =hf_i/2e, calculated from frequencies f_i of phonon modes in YBa$_2$Cu$_3$O$_{7-x}$ (insets). Phonon modes of both polarizations, with atomic displacements along the c-axis and perpendicular to the c-axis, are presented and their frequencies are close to the Josephson frequencies of the junction, at which strong interaction with environment takes place.

Josephson behavior of [100]-tilt YBa$_2$Cu$_3$O$_{7-x}$ junctions has been estimated by the interaction of Josephson oscillations with the environment of the junction and external terahertz monochromatic radiation. It was found, that I-V curves of [100]-tilt YBa$_2$Cu$_3$O$_{7-x}$ junctions at low temperatures contain a set of reproducible features at specific voltages. Fine structure on the I-V curve of one of our junctions at 5K is presented in Fig. 2. Most pronounced structure on the voltage dependence of the differential conductance dI/dV is observed at the voltages centered near 9.4 mV, smaller ones are situated at one half of this voltage, near 4.7 mV, at voltages near 7.4 mV and close to 12 mV. According to Josephson spectroscopy [9], these features on the IV curve localized at particular voltages V_i correspond to an increase of the shunting admittance $Y_{ext}(f)$ of junction environment at several frequencies f_i = 2eV_i/h. Among various reasons for frequency-selective shunting of Josephson oscillation, optical phonon modes in junction electrodes look most probable. In Fig. 2, optical phonon modes with atomic displacements in YBa$_2$Cu$_3$O$_{7-x}$ are presented as insets and arrows indicate the voltage positions V_i, corresponding to the frequencies f_i of these phonon modes, calculated from YBa$_2$Cu$_3$O$_7$ lattice dynamics [10].

A single crystallographic facet for each of high-T_c electrodes at the bicrystal boundary, optimal oxygen content inside the boundary, I_cR_n-values up to 10 mV, phonon-induced structures on the I-V curves at the voltages from 4 to 12 mV and the ac Josephson effect up to 5 THz are the main features of an "ideal" high-T_c junction.

[1] H. Hilgenkamp, J. Mannhart, Rev. Mod. Phys. **74**, 485 (2002)
[2] P.M.Shadrin, Y.Y.Divin, S.Keil, J.Martin, R.P.Huebner, IEEE Tr. Appl. Supercond. **9**, 3925(1999)
[3] Y.Y. Divin, U. Poppe, C.-L. Jia, P.M. Shadrin, K. Urban, Physica C **372**, 115 (2002).
[4] Y.Y. Divin et al., In: Applied Superconductivity 2003 (Ed. A. Andreone, G.P. Pepe, R. Cristiano, G. Masulo, IOP Conf. Ser. No.181, Bristol: IOP Publishing Ltd; 2004) p. 3112-3118
[5] D.A. Weitz, W.J. Skocpol, M.Tinkham, Phys. Rev. Lett. **40**, 253 (1978)
[6] V. Ambegaokar, A. Baratoff, Phys. Rev. Lett. **10**, 486(1963)
[7] A. Damascelli, Z. Hussain, Z.-X. Shen, Rev. Mod. Phys. **75**, 473 (2003)
[8] M.V. Liatti, U. Poppe, Y.Y. Divin, Appl. Phys. Lett. **88**,152504 (2006)
[9] A.F. Volkov, Radiotekhnika i Elektronika **12**, 2581(1972)
[10]J. Humilíček et al., Physica C **206**, 345 (1993)

TUNED EPITAXY OF OXIDE HETEROSTRUCTURES

M. I. Faley[1], U. Poppe[1], C. L. Jia[1], O. M. Faley[2], R. E. Dunin-Borkowski[1]

[1]Institute for Microstructure Research, Jülich Research Centre, Jülich, Germany; [2]RWTH Aachen University, Aachen, Germany

Epitaxial growth of high-T_c superconducting $YBa_2Cu_3O_7$ (YBCO) films and heterostructures on MgO substrates and buffer layers is used for making highly sensitive superconducting quantum interference devices (SQUIDs), high-Q microwave resonators, electrical power transmission cables, current leads, fault current limiters, transformers, generators, motors and energy storage devices. For such applications, it is usually important to align the c-axis of the YBCO film normal to the substrate surface and to ensure an absence of in-plane-misoriented grains in the films. However, degradation of the MgO surface due to exposure to air leads to the growth of 45° in-plane-misoriented YBCO grains, reducing the critical current and increasing noise and the spread of parameters in oxide heterostructures and devices made with such films. For tilted MgO substrates, surface degradation also leads to a deviation of the c-axis of the YBCO film from the substrate crystallographic direction and its alignment with the substrate surface normal. Here we demonstrate that ion beam etching (IBE) of the MgO surface eliminates the growth of misoriented grains and aligns the c-axis of the film with the substrate crystallographic direction (Fig.1).

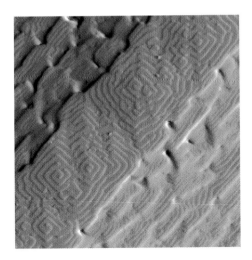

Figure 1: AFM image of a 140 nm thick YBCO film deposited on IBE-cleaned trapezoidal feature on a (100) MgO substrate with a 600 nm wide top and about 3 degree slopes. The AFM scan area is 2 μm x 2 μm. The growth spirals on the YBCO film show the orientation of its crystallographic directions: the direction of the c-axis of the YBCO film is normal to the terraces of the growth spirals. Here, we show that the c-axis of the YBCO film on an MgO surface refreshed by IBE follows the substrate crystallographic direction, which deviates from the substrate surface normal on the slopes. Exposure of the MgO surface to air before deposition of the YBCO film changes the growth mode of the YBCO film from epitaxial to graphoepitaxial.

By controlled exposure of the IBE-refreshed surface of the MgO substrate to air or hydro-carbons, it is possible to tune the coupling strength between the substrate and the YBCO and to change the growth of the YBCO film from epitaxial to graphoepitaxial. This approach can be used to achieve a controlled deviation of the film c-axis from the substrate crystallographic direction and its alignment with the substrate normal, while avoiding the formation of in-plane-misoriented YBCO grains.

STRAIN EFFECTS ON THE ELECTRONIC SUBBAND STRUCTURE OF SrTiO$_3$

Vladimir Laukhin[1,2], Olivier Copie[3], Marcelo Rozenberg[4], Karim Bouzehouane[3], Éric Jacquet[3], Manuel Bibes[3], Agnès Barthélémy[3], <u>Gervasi Herranz</u>[1]

[1] Institut de Ciència de Materials de Barcelona ICMAB-CSIC, Campus de la UAB, 08193 Bellaterra, Catalonia, Spain;
[2] Institució Catalana de Recerca i Estudis Avançats (ICREA), 08010 Barcelona, Catalonia, Spain;
[3] Unité Mixte de Physique CNRS/Thales, 1 avenue Fresnel, Campus de l'Ecole Polytechnique, 91767 Palaiseau, France and Université Paris-Sud, 91405 Orsay, France;
[4] Laboratoire de Physique des Solides, Université Paris-Sud, Bâtiment 510, 91405 Orsay, France

The perovskite SrTiO$_3$ (STO) is nowadays the cornerstone element for the emerging field of oxide electronics [1]. The interest in STO has been boosted after the recent discovery of high-mobility conduction at the STO-LaAlO$_3$ and STO-LaGaO$_3$ heterointerfaces [2, 3] and of a two-dimensional electron gas at the STO surface [4]. These findings have spurred a flurry of intensive research towards the exploitation of these interface and surface phenomena for electronics applications, as well as to understand the basic mechanisms governing the physics of transport in STO [5-7]. One of the endeavors has been to find ways to boost the electronic properties of these systems and, in particular, of the electronic mobility. Very interestingly, recent reports have shown that strain can be exploited to attain very large enhancements of the electronic mobility in STO thin films, by factors up to ~ 300% [8]. Although this represents an important advancement to achieve better electronic responses, the underlying microscopic physical mechanisms leading to such an outstanding transport enhancement are not completely elucidated. This is a point of paramount relevance, since understanding the evolution of the electronic band structure with strain –that has not been disclosed so far– and its relationship with transport parameters, like the mobility and carrier density, are essential to apply appropriate strategies to boost the electronic properties of systems such as high-mobility 2DEGs in LaAlO$_3$/STO or LaGaO$_3$/STO interfaces.

To identify the fundamental mechanisms that determine the evolution of the electronic structure and properties of STO under strain, we have carried out an extensive and systematic analysis of the magnetotransport under different quasi-hydrostatic pressures (see Figure 1).

Interestingly, due to the small Fermi energy of doped STO – of only a few meV [7] – pressure is expected to be an extremely sensitive probe to detect the subtle effects of moderate values of strain on the electronic structure. Our study has unveiled a multiple conduction subband structure that implies transport of electrons with different masses, with an energy subband hierarchy that depends on the applied pressure [9]. We have uncovered that the reason for the extraordinary mobility enhancement lies on a very subtle rearrangement of electrons among these bands. Although the associated energy scales of this electronic reconfiguration are rather small –on the scale of the meV– they generate a conspicuous mobility enhancement up to ≈ 300% for quite moderate pressures [9]. Therefore, our unprecedented discovery of an extremely fine modulation of the electronic structure shows that such remarkable effects are accomplished with moderate strain, compatible with epitaxial engineering, opening novel unexplored fascinating avenues for the 2DEG oxide interfaces.

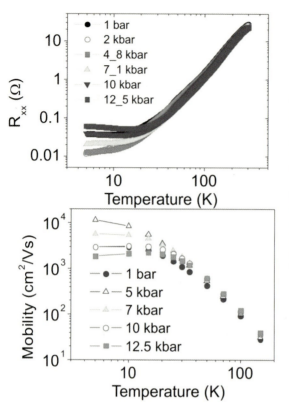

Figure 1
(Upper panel) Temperature dependence of the longitudinal resistance R_{xx} at different quasi-hydrostatic pressures. The values indicated in the panels are those corresponding to the pressure applied at room temperature; at lower temperatures, the pressure value was corrected according to the calibration curves of the pressure transmitting medium.
(Lower panel). Temperature dependence of the electronic mobility for each applied pressure (corrected for temperature).

[1] H.Y. Hwang et al., Nature Materials **11**, 103 (2012).
[2] A. Ohtomo and H.Y. Hwang, Nature (London) 427, 423 (2004).
[3] P. Perna et al., Appl. Phys. Lett. 97, 152111 (2010).
[4] A. F. Santander-Syro et al., Nature **469**, 189 (2011).
[5] S. Thiel et al., Science **313**, 1942 (2006)..
[6] N. Reyren et al., Science **317**, 1196 (2007).
[7] G. Herranz et al. Phys. Rev. Lett. **98**, 216803 (2007).
[8] Bharat Jalan et al., Appl. Phys. Lett. **98**, 132102 (2011).
[9] V. Laukhin et al., submitted.

ELECTRON OCCUPANCY OF 3D-ORBITALS IN MANGANITE THIN FILMS

D. Pesquera[1], A. Barla[2], E. Pellegrin[2], F. Sánchez[1], F. Bondino[3], E. Magnano[3] and J. Fontcuberta[1]

[1]Institut de Ciència de Materials de Barcelona (ICMAB-CSIC). Campus UAB, Bellaterra, Catalonia 08193, Spain; [2]ALBA Synchrotron Light Source, Carretera BP-1413 de Cerdanyola a Sant Cugat, Km 3.3. E-08290 Cerdanyola del Vallès, Barcelona, Spain; [3]Laboratorio TASC, IOM CNR, S.S. 14 km 163.5, Area Science Park Basovizza (Ts), I-34149, Italy

Electron occupancy of 3d-orbitals of transition metal is determined by the total number of electrons in the 3d shell and by the energies of the five 3d orbitals and the resulting spin state. It is well known that at interfaces between dissimilar oxides in epitaxial heterostructures, charge redistribution, strain effects and symmetry breaking may lead to emerging new properties [1]. It has been shown, for instance, that interface-mediated tensile strain acting on the 3d-orbitals of regular octahedrally-coordinated metal ions, such as $3d^4$-Mn^{3+}, breaks the degeneracy of x^2-y^2 and $3z^2-r^2$ states favoring electron filling of in-plane x^2-y^2 orbitals subsequently determining the orbital and magnetic ordering [2,3].

On the other hand, the interface with the vacuum or more generally, the free-surface of transition metal oxides, is a particular interface of major interest in areas like green energy and catalysis. From this point of view, bulk transition metal oxides are receiving much attention. For instance, it has been recently shown that the oxygen evolution reaction, of high relevance in many applications including water splitting, is determined by the degree of filling of the 3d-orbitals, and thus depending on the electron number *and* spin state of the transition metal at the oxide surface [4].

Interestingly, despite modern techniques of oxide thin film growth allowing for the control of oxides structure and surface composition almost at the atomic scale, the role of symmetry breaking at free surfaces has not received much attention.

Here, we will use of X-ray absorption linear dichroism (XLD) at Mn L-edge in $La_{2/3}Sr_{1/3}MnO_3$ (LSMO) to disentangle the free surface contribution to the orbital degeneracy x^2-y^2 and $3z^2-r^2$ occupancy from its strain-induced counterpart. LSMO films of various thicknesses (from 4 unit cells up to 150 uc) where grown on single crystalline (001)ABO_3 perovskites substrates. The substrate selection: $SrTiO_3$, LSAT, $NdGaO_3$ and $LaAlO_3$, having distinct mismatch with LSMO, allows to explore distinct strain states of the films. Moreover, LSMO on (001)$SrTiO_3$ substrates, either TiO_2 or SrO terminated are also grown and used to explore surface termination effects. RHEED was used to monitor the growth of the thinnest films on (001)$SrTiO_3$.

In Figure 1 we show one of the crucial results, that is the XLD spectra of a 27 nm thick LSMO films grown on (001)$SrTiO_3$ and LSAT substrates. We note that $SrTiO_3$ produces a tensile in-plane strain on LSMO due to the slightly larger cell parameter of the cubic STO when compared to that of LSMO. In contrast to this, the LSAT almost perfectly matches LSMO thus producing virtually strain-free LSMO. The XLD data of the LSMO//STO film in Fig. 1 show a negative dichroism in the 648-660 eV region. According to the common understanding of XLD, this indicates a preferential occupancy of in-plane x^2-y^2 orbitals. On the other hand, the XLD spectrum of LSMO//LSAT shows a positive XLD in the same energy region, thus indicating a preferential $3z^2-r^2$ electron occupation, which is at odds with the vanishing strain in LSMO//LSAT. The systematic study of all films as a function of substrate, chemical termination, thickness and orientation ((001) or (110)) conclusively shows that there is a constant $3z^2-r^2$ contribution in all XLD spectra that adds to the strain-induced orbital occupation.

We will argue that this net $3z^2-r^2$ contribution arises due to the symmetry-breaking at the free surface which stabilizes the $3z^2-r^2$ orbitals with respect to the x^2-y^2. This finding, supported by theoretical calculations, shows that electronic occupation at the free surface orbitals in transition metal oxides, can be tailored at wish by balancing strain and surface contributions, and thus opening the way to active control of surface properties.

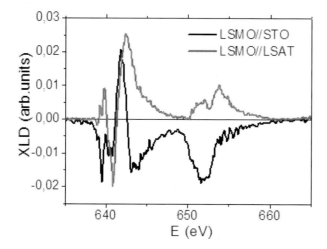

Figure 1. X-ray linear dichroism on the Mn L-edge obtained a 27 nm LSMO film grown on STO (black line) and 60 nm LSMO film grown on LSAT (red line)

[1] H. Y. Hwang et al. Nature Materials 11, 103 (2012)
[2] E. Benckiser et al. Nature Materials 10, 189 (2011)
[3] A.Tebano, et al, Phys. Rev. Lett. **100,** 137401 (2008); M. Huijben et al, Phys. Rev. B **78**, 094413 (2008); A. Tebano et al, Phys. Rev. B **82**, 214407 (2010)
[4] J. Suntivich et al. Nature Chemistry 3, 546 (2011)

SUBSTRATE COHERENCY DRIVEN PHASE SEPERATION AND INTRINSIC ANISOTROPY IN EPITAXIAL $La_{0.67}Ca_{0.33}MnO_3/NdGaO_3(001)$ EPITAXIAL FILMS

Lingfei Wang[1], Wenbin Wu[1]

[1]Hefei National Laboratory for Physical Sciences at Microscale, University of Science and Technology of China, Hefei, Anhui, People's Republic of China.

$La_{0.67}Ca_{0.33}MnO_3$ (LCMO) films, of which the bulk material is a typical manganite with ferromagnetic metal (FMM) ground state, were epitaxially grown on $NdGaO_3(001)$ single crystal substrates. After ex-situ annealing, the anisotropic strained LCMO epitaxial films show remarkable antiferromagnetic insulator (AFI) phase transition and phase separation (PS) due to the coexistence of AFI and FMM phases [1] as shown in Figure 1. To probe the origin lies behind, atomic force microscopy topographic and magnetotransport measurement were correspondingly established. The results revealed the strong correlation between surface morphology and electronic properties. Based on this observation as well as the distinct octahedral distortion patterns of substrate and film, we propose that the intrinsic coupling of octahedral distortion at heterointerface, namely the substrate coherency, plays critical role in the formation of AFI phase. More strikingly, pronounced intrinsic anisotropy in magnetization, transport and magnetoresistance [2] are observed in the LCMO/NGO(001) films. Based on the temperature dependent measurements of these anisotropic properties, we conclude that the PS and resultant strong phase completion could be responsible for the intrinsic anisotropy of the films.

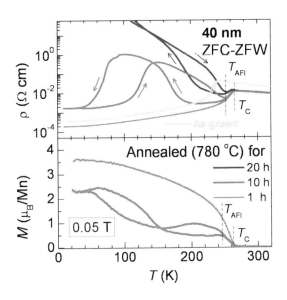

Figure 1: the ρ-T and M-T curves of 40 nm LCMO/NGO(001) films after various post-annealing process. For the as-grown and 1 h-annealed samples show bulk-like FM transition, while the 10 h and 20 h-annealed samples show clear AFI phase transition below the FMM phase transition. And the complex hysteresis low-temperature plateau observed in both ρ-T and M-T curves indicate that the PS and strong phase competition between FMM and AFI are induced.

[1] Z. Huang et al., J. Appl. Phys. **105**, 113919 (2009).
[2] L. F. Wang et al., Appl. Phys. Lett. **97**, 242507 (2010).

THICKNESS DEPENDENCE OF LATTICE DISTORTIONS IN EPITAXIAL FRAMEWORK STRUCTURES OF STRONGLY CORRELATED OXIDES: $La_{2/3}Sr_{1/3}MnO_3/SrTiO_3$

Felip Sandiumenge[1], Jose Santiso[2], Lluís Balcells[1], Z. Konstantinovic[1], Jaume Roqueta[2], Alberto Pomar[1], Benjamin Martínez[1]

[1]ICMAB-CSIC, Campus de la UAB, 08193 Bellaterra, Catalonia, Spain; [2]CIN2 (CSIC-ICN), Campus de la UAB, 08193 Bellaterra, Catalonia, Spain

Epitaxial strongly correlated oxides constitute an ideal arena to explore the effects of biaxial strain on the delicate balance between lattice, spin, charge and orbital degrees of freedom. Such materials typically exhibit ABO_3 perovskite-type corner sharing octahedral framework structures whose elastic behavior is governed by the relative strength between B - O bonds and B - O - B bond angles bridging adjacent BO_6 octahedra. When the strength of the former is larger, strain easily couples with octahedral tilts at a lower energy cost than plastic (misfit dislocation) relaxation, independently of film thickness. Conversely, charge and orbital degrees of freedom may favor the coupling of strain with B - O bonds. The intricate interplay between both mechanisms define a rather complex misfit strain relaxation scenario with unexpected consequences on the physical properties of epitaxial films. To address this issue, here we investigate the film thickness dependence of lattice distortions, transport and magnetic properties in the canonical ferroelastic, ferromagnetic and metallic $La_{0.7}Sr_{0.3}MnO_3/(001)SrTiO_3$ (LSMO/STO) heterostructure for thicknesses comprised between 1.8 nm and 500 nm [1].

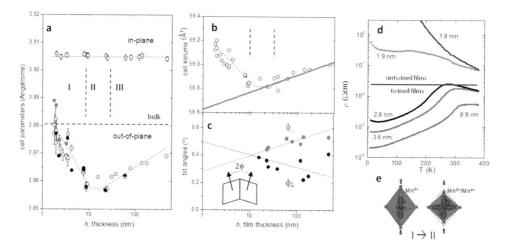

Figure 1: Thickness dependence of lattice distortions in LSMO/STO films [1]. (a) Evolution of the in-plane and out-of-plane lattice parameters with film thickness (h). Labels I, II, III indicate three deformation regimes. (b) *Pseudo*cubic unit cell volume. Red line is a calculation based on pure octahedral tilting mechanisms. (c) Evolution of the in-plane ($\square \parallel$) and out-of-plane ($\square \perp$) twin angles. (d) Resistivity *versus* temperature curves. Below h_{shear} films are insulating. (e) Schematics of the transition from regime I to II involving a progressive e_g degeneracy towards a pure elastic strain state of the MnO_6 octahedra. Below h_{shear} charge enrichment and preferential occupation of $3z^2 - r^2$ orbitals induces octahedral dilation and stretching along the c-axis.

Figure 1 summarizes main results. The out-of-plane c-axis parameter follows a progressive decrease from the thinnest film ($h \sim 1.8$ nm) down to a minimum value at thicknesses comprised between 10 nm and 25 nm followed by a smooth increase (a). Notably, this trajectory is accompanied by a constant in-plane fully strained a-axis parameter throughout the whole thickness range (a). The resulting evolution of the unit cell volume, $V(h)$, is depicted in Figure 1b. The trend followed above the c-axis minium ($h > 25$ nm), is well described on a pure octahedral tilting basis as indicated by the red (gray) straight line. Figure 1c shows the evolution of the in-plane ($\phi_\parallel$) and out-of-plane ($\phi_\perp$) components of the twin angle, $2(\alpha_{rh} - 90°)$. According to the changing thickness dependence of these structural parameters, we identify in Figure 1a three different deformation regimes labelled I, II and III in Figure 1a. The calculated value of the Poisson's ratio, $v = \varepsilon_\perp/(\varepsilon_\perp - 2\varepsilon_\parallel) \sim 0.33$, and the convergence of $\phi_\parallel$ and $\phi_\perp$, indicate that in regime II the film is elastically deformed without invoking significant octahedral tilting perturbations. In this regime, films exhibit bulk like metallic and ferromagnetic properties [1]. In regime I we also detected a shear transition at a thickness $h_{shear} \sim 2.5$ nm defined by the formation of (100) and (010) twins. Below h_{shear}, clamping to the substrate is mediated by a monoclinic distortion which suppresses the in-plane shear contribution to the misfit. Figure 1d shows that this interfacial layer, $h<h_{shear}$, is insulating without any signature of insulating-to-metal transition between 10K and 380K. Moreover, we observed that this behavior is accompanied by a sudden drop of the magnetic moment and the loss of ferromagnetic order [1]. Altogether, these results allow as to draw the following conclusions:

(1) For $h<h_{shear}$, the structure of the film is determined by the interaction between the electronic structures of film and substrate. As also recently reported by other authors [2], the large $Mn(d_z^2)$ - $O(2p_z)$ - $Ti(d_z^2)$ orbital overlap across the interface plane favours a preferential occupation of the $Mn(3z^2 - r^2)$ orbitals causing an elongation of the c-axis parameter and the Jahn-Teller distortion of the MnO_6 octahedra.

(2) As growth proceeds beyond h_{shear}, regime I, the electronic energy gained by removing the orbital degeneracy within the $Mn^{3+/4+}O_6$ coordination environment becomes progressively exceeded by the elastic energy opposing a similar expansion of the equatorial Mn - O distances of the octahedra, until a tensilely strained rhombohedral phase exhibiting bulk transport and magnetic properties condenses at a thickness of ~ 10 nm (regime II, see Figure 1e).

(3) Regime III is characterized by pure octahedral tilting mechanisms which do not affect the transport and magnetic behavior of the films.

[1] F. Sandiumenge, J. Santiso, Ll.Balcells, Z. Konstantinovic, J. Roqueta, A. Pomar, B. Martínez, submitted for publication.
[2] M.-B. Lepetit, B. Mercey, Ch.Simon, Phys. Rev. Lett. **108**, 087202 (2012).

ORBITAL ENGINEERING BY STRAIN IN THIN FILMS OF $La_{1-x}Sr_{1+x}MnO_4$ GROWN BY PULSED LASER DEPOSITION

Mehran Vafaee Khanjani[1], Philipp Komissinskiy[1], Mehrdad Baghaie Yazdi[1], Roberto Krauss[2], Valentina Bisogni[2], Jochen Geck[2], and Lambert Alff[1]

[1]Institute of Materials Science, Technische Universität Darmstadt, 64287 Darmstadt, Germany;
[2]Leibniz Institute for Solid State and Materials Research Dresden, 01171 Dresden, Germany

It is well known that charge and orbital ordering (COO) occurs in a variety of manganese oxide compounds. For the single layered Ruddlesden-Popper manganite $La_{1-x}Sr_{1+x}MnO_4$ (LSMO), by doping holes and thereby changing the manganese valance from Mn^{3+} to $Mn^{3.5+}$, COO has been observed. This phenomenon (for $x=0.5$) is believed to be strongly related to the lattice degrees of freedom ($a = 3.86$, $c = 12.42$ Å) and Mn-O bond length [1-3]. The charge localization and orbital ordering is thought to be caused by electron-lattice interactions [4]. Thus, a change in lattice constants instead of hole-doping should also affect the COO. For the first time, we report on the thin film growth of LSMO ($x=0.0$, $x=0.5$) using pulsed laser deposition. X-ray diffraction reveals the growth of either fully strained or totally relaxed thin films on (001) ($La_{0.3}$, $Sr_{0.7}$) ($Al_{0.65}$, $Ta_{0.35}$) O_3 (LSAT) and (001) LaSrAlO$_4$ (LSAO) substrates. Strain in these thin films changes the structural parameters and hence alters the electron-lattice interactions. Such thin films have provided the foundation to study the strain dependence of orbital ordering by X-ray absorption techniques. In bulk LSMO ($x = 0.5$), electrons reside on in- and out-of-plane orientated orbitals symmetrically as evidenced by X-ray absorption spectra [5]. The data presented in Figure 1 show that for the $x = 0.5$ compound compressive strain (elongated out-of-plane lattice constant) breaks the symmetry of the electron distribution and changes the orbital occupation. This result suggests that strain can be used to control orbital occupation and, thereby, change e.g. interfacial conductivity.

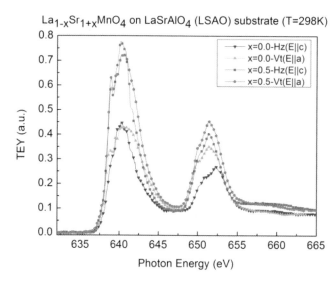

Figure 1: X-ray Absorption Spectra for horizontal and vertical polarization conducted at the BESSY-beamline UE52-PGM, probing the $L_{2,3}$ edges of Mn for $La_{1-x}Sr_{1+x}MnO_4$ ($x=0.0$, $x=0.5$) thin films on LaSrAlO$_4$ substrate.

[1] D. Senff et al., Phys. Rev. B **71**, 024425 (2005).
[2] R. Mahesh et al., J. Solid State Chem. **122**, 448 (1996).
[3] C. S. Hong et al., Chem. Mater. **13**, 945 (2001).
[4] J.H. Jung et al., Phys. Rev. B **61**, 6902 (2000).
[5] M. Merz et al., Eur. Phys. J. B **51**, 315 (2006).

STRUCTURE AND TRANSPORT PROPERTIES OF SmNiO$_3$ THIN FILMS

Flavio Y. Bruno[1], Konstantin Rushchanskii[2], Cécile Carretero[1], Yves Dumont[3], Marjana Lezaic[2], Stefan Blügel[2], Manuel Bibes[1] and Agnès Barthélémy[1]

[1]Unité Mixte de Physique CNRS/Thales, Campus de l'Ecole Polytechnique, 1 Av. A. Fresnel, 91767 Palaiseau, France and Université Paris-Sud 11, 91405 Orsay, France; [2]Peter Grünberg Institut, Forschungszentrum Jülich GmbH, 52425 Jülich and JARA-FIT, Germany; [3]Groupe d'Etude de la Matière Condensée, Université de Versailles St-Quentin en Yvelines/CNRS, 45 avenu des Etats-Unis, 78035 Versailles

The RENiO$_3$ (RE=rare earth) family is a prototypical strongly correlated electron system exhibiting a Mott metal-insulator transition (MIT), unusual magnetic order and possibly charge and orbital order. The MIT transition temperature T_{MI} depends on the ionic radius of the RE ion. For RE=La, the compound is paramagnetic and metallic at all temperatures while for smaller RE atoms (RE= Nd and Pr) nickelates exhibit a first order MIT accompanied by a transition to an antiferromagnetic ground state (with a Néel temperature T_N). Interestingly for even smaller RE atoms (Re=Sm, Eu, etc) the antiferromagnetic ordering temperature T_N and the T_{MI} are different, emphasizing the influence of structural properties on the interplay between magnetic order and transport [1]. Here we report on thin films of SmNiO$_3$, a compound that presents bulk ordering temperatures of T_N=220K and T_{MI}=400K[2]. The fact that T_N differs from T_{MI} is interesting because it allows for the study of thickness (dimensionality) and strain (octahedral distortions - rotations) on both ordering temperatures. SmNiO$_3$ (SNO) crystallizes in the orthorhombic distorted perovskite structure with Pbnm symmetry and unit cell dimensions $\sqrt{2}a_p \times \sqrt{2}a_p \times 2a_p$, where a_p=0.379 nm is the simple perovskite lattice parameter.

We have grown high quality epitaxial SNO thin films by pulsed laser deposition under different strain states on (001) oriented substrates: LaSrAlO$_4$ (1.3% compressive strain) LaAlO$_3$ (~0% strain), and LaSrGaO$_4$ (1.2% tensile strain). In Figure 1a X-ray diffraction spectra of a 25 u.c. thin film deposited on LaAlO$_3$ (LAO) is shown. Due to the close lattice matching with the substrate the Bragg peak of the SNO thin film is not observed, however the finite size oscillations indicate its presence and the high quality of the thin films. The clear streak patterns in the reflection high energy electron diffraction pattern observed in the inset of Figure 1a indicate flat surface and epitaxial growth of the film.

Transport measurements were performed on films of different thickness and grown on different substrates. In Figure 1b we show resistivity as a function of temperature for a series of thin films grown on LAO with thickness: 60, 25, 9, 5 u.c. The 60 u.c. thick sample display T_{MI}=390K close to the bulk value, as thickness is reduced the sharpness of the MIT decreases and for the thin film of 5 u.c. it is completely suppressed. Transport data unexpectedly shows that T_{MI} decreases as compared to strain-free films independently on whether the strain is compressive or tensile. We will explain these results based on first principle calculations.

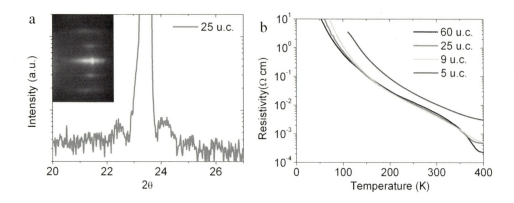

Figure 1. (a) X-ray diffraction spectra of a SmNiO$_3$ 25u.c. thick film grown on LaAlO$_3$ substrate, (Inset) reflection high energy electron diffraction pattern after deposition of 7 u.c. SmNiO3 on LaAlO$_3$. (b) Resistivity as a function of temperature for SmNiO3 thin films grown on LaAlO$_3$ with thickness: 60 (black), 25 (red), 9 (green), and 5 u.c.(blue).

We acknowledge financial support from the European Research Council (ERC advanced grant no. 267579).

[1] G Catalan, Phase Transitions, 81 (2008) 729-749.
[2] J Perez-Cacho, *et al*, J. Phys. Condens. Matter 11 (1999) 405–415.

METAL-INSULATOR TRANSITION AND INTERFACE PHENOMENA IN NICKELATE HETEROSTRUCTURES

Raoul Scherwitzl[1], Marta Gibert[1], Pavlo Zubko[1], Stefano Gariglio[1], Gustau Catalan[2], Jorge Iniguez[3], Marc Gabay[4], Alberto Morpurgo[1] and Jean-Marc Triscone[1]

[1]University of Geneva, Geneva, Switzerland;
[2]ICREA and CIN2, Barcelona, Spain;
[3]ICMAB, Barcelona, Spain;
[4]University of Paris Sud, Paris, France

Transition metal oxides display a wide range of physical properties arising from the complex interactions between their spin, charge, orbital and lattice degree of freedom. Rare earth perovskite nickelates are particularly exciting materials as they display a temperature-driven metal-insulator transition, charge disproportionation and a unique antiferromagnetic ordering.

We will demonstrate the ability to control the metal-insulator transition in $NdNiO_3$ using strain, field-effect and light [1,2]. The implications on the underlying mechanisms behind the transition will be discussed.

Then, we will focus on $LaNiO_3$, which is the only nickelate that, in bulk, does not display a metal-insulator transition or any other ordering phenomena. We will show that a metal-insulator transition can be induced as the film thickness is reduced to only a few unit cells [3]. Building on these results, superlattices of ultrathin $LaNiO_3$ and ferromagnetic $LaMnO_3$ were found to display exchange bias, indicating that a complex magnetic order develops in $LaNiO_3$. [4].

[1] R. Scherwitzl et al., Adv. Mater. **22**, 5517 (2010)
[2] A.D. Caviglia. R. Scherwitzl et al., Phys. Rev. Lett. (2012)
[3] R. Scherwitzl et al., Phys. Rev. Lett. **106**, 246403 (2011)
[4] M. Gibert et al., Nat. Mater. **11**, 195 (2012)

PHOTOVOLTAIC ENERGY CONVERSION BASED ON STRONGLY CORRELATED OXIDES

Christian Jooss[1], Gesine Saucke[1], Jonas Norpoth[1], Dong Su[2] and Yimei Zhu[2]

[1]Institute of Materials Physics, University of Goettingen, [2]Brookhaven National Laboratory, Upton NY 11973, USA

The main constraint for photovoltaic energy conversion in conventional semiconductor pn junctions is that the chemical potential $\Delta\mu$ of photo-excited charge carriers is limited by the presence of an indirect bandgap of energy E_g, i.e. $\Delta\mu < E_g$. Transmission losses of photons with $E < E_g$ and thermalization losses of photons with $E > E_g$ determine the so-called Shockley-Queisser limit of photovoltaic energy conversion [1]. It is proposed that the conversion efficiency can be strongly improved by quenching the rapid thermalization process of photo-excited electron-hole pairs and converting such "hot carriers" into a photo-voltage [2]. Consequently, finding mechanisms, where large lifetimes of photo-excited carriers are established without the presence of a bandgap in the optical absorption spectrum would be highly desirable. Pn-junctions formed by complex oxides offer new opportunities for controlling lifetimes of optical excitations and charge separation by tunable electron-electron or electron-phonon correlation interactions. Systematic studies of photovoltaic effects in such systems require the fabrication of interfaces with atomic level control of electronic and chemical structure.

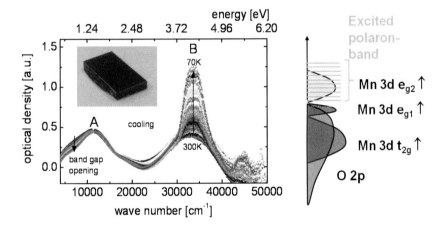

Fig. 1: Absorption spectra of $Pr_{0.67}Ca_{0.33}MnO_3$ as a function of temperature showing a polaron absorption band (A) and an O2p→Mn3d interband transition (B).

Here, we present a case study on the relation between atomic level structure, electric transport and photovoltaic effect in a manganite-titanite pn-heterojunction [3]. For the p-doped material we have selected $Pr_{0.64}Ca_{0.36}MnO_3$ (PCMO) which exhibit strong electron-electron as well as strong electron-phonon interactions. In the visible range, its absorption spectrum is dominated by a polaron band (Fig. 1) with lifetimes of the optically excited "hot polarons" up to ns. At room temperature an optical bandgap is absent whereas a charge bandgap is opening below the charge ordering temperature of T_{CO} ~240 K. The n-doped material is $SrTi_{1-y}Nb_yO_3$ (STNO, y = 0:002 and 0.01) which has a bandgap of 3.2 eV. High-resolution electron- microscopy and -spectroscopy reveal a nearly dislocation-free, epitaxial interface and gives insight into the local atomic and electronic structure (Fig. 2a). The presence of a photovoltaic effect under visible light at room temperature suggests the existence of mobile excited polarons within the band gap-free PCMO absorber (Fig. 2b). More than 90% of the photovoltage is generated by photons with hv < 2.3 eV. The temperature-dependent rectifying current-voltage characteristics prove to be mainly determined by the presence of an interfacial energy spike in the conduction band and are strongly affected by the colossal electro-resistance (CER) effect. The open circuit voltage of 10-20 mV at room temperature is strongly increased below TCO to values above 700 mV. The photocurrent can be strongly modified by the CER effect. From the comparison of photo-currents and spatiotemporal distributions of photo-generated carriers (deduced from optical absorption spectroscopy) we obtain an estimate of the excited polaron diffusion length. The results are confirmed by an independent study based on electron beam induced current (EBIC) which reveals a polaron diffusion length of about 20 nm.

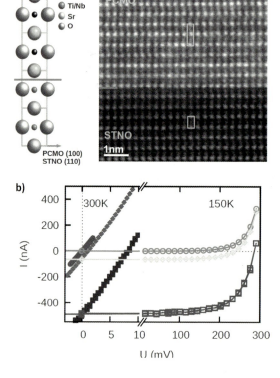

Fig. 2: (a) Structure of a PCMO-STNO junction (film thickness d = 200 nm, y = 0:01, T = 80 K).The viewgraph shows a HAADF-STEM image of a cross-section of the interface along the [110]-STNO zone axis. The projected structure and stacking order are schematically depicted on the left and the sub-unit cells are marked in the STEM image. (b) Temperature-dependent photovoltaic effect of the PCMO-STNO junction. (a) Current-voltage characteristics in the dark (red circles) and under illumination with Hg lamp (blue squares) or additional 2.3 eV high-pass filter (green diamonds) respectively of the 380 nm PCMO sample with y = 0:002. Dark colored filled symbols represent measurements at room temperature, light colored open symbols mark measurements at T = 150 K below the charge order transition.

Combined atomic column resolved Electron Energy Loss Spectroscopy (EELS) and DFT calculation using GGA and PBE0 functions give detailed insights into the electron structure of the involved materials and the interface and confirm the presence of an energy spike. We discuss the results in the view of the single particle bandstructure. In addition, since the optical excitations are beyond the capabilities of DFT, we use semi-empirical models for the analysis of the small polaron absorption.

The temperature, electric field and spectral dependence of the photovoltaic energy conversion of light into electric power in both materials systems sheds light onto the role of electron-phonon and magnetic interactions on the conversion mechanism. It may offer new insights in general pathways for light harvesting based on strongly correlated electron devices.

[1] W. Shockley and H. J. Queisser, J. Appl. Phys. 32, 510 (1961)
[2] W. A. Tisdale, K. J. Williams, B. A. Timp, D. J. Norris, E. S. Aydil, and X.-Y. Zhu, Science 328, 1543 (2010).
[3] G. Saucke, J. Norpoth, D. Su, Y. Zhu and Ch. Jooss, Phys. Rev. B, accepted

TERAHERTZ AND INFRARED BEHAVIOR OF STRAINED $Sr_{n+1}Ti_nO_{3n+1}$ THIN FILMS WITH RUDDLESDEN-POPPER STRUCTURE

Veronica Goian[1], Stanislav Kamba[1], Nathan D. Orloff[2], Che-Hui Lee[3], Viktor Bovtun[1], Martin Kempa[1], Dmitry Nuzhnyy[1] and Darrell G. Schlom[3,4]

[1] Institute of Physics, Academy of Sciences of the CR, Na Slovance 2, 182 21 Prague 8;
[2] National Institute of Standards and Technology, Boulder, Colorado 80305, USA;
[3] Department of Materials Science and Engineering, Cornell University, Ithaca, New York 14853 USA;
[4] Kavli Institute at Cornell for Nanoscale Science, Ithaca, New York 14853 USA;

Ruddlesden-Popper (RP) compounds have the general formula $A_{n+1}B_nO_{3n+1}$, n=1,2,3...∞ and are consisting of n perovskite blocks separated and sheared by AO layers along [001].[1] One of the interesting examples of RP-n compounds is $Sr_{n+1}Ti_nO_{3n+1}$. The number of perovskite layers controls the value of permittivity ε' in bulk $Sr_{n+1}Ti_nO_{3n+1}$ ceramics. At room temperature the ε' increases from 37 (for n=1) to 100 (for n=4).[2] The n = ∞ end member of the RP series, bulk $SrTiO_3$, has ε' = 290 at 300 K and it is the most known quantum paraelectric material, in which the ε' increases on cooling and saturates at ~25000 below 10 K. Electric tunability of materials increases with their rising ε'. Thanks to this $Sr_{n+1}Ti_nO_{3n+1}$ ceramics with high n are very well tunable, but their tunability is the best at low temperatures. Simultaneously, the $Sr_{n+1}Ti_nO_{3n+1}$ ceramics exhibit low dielectric loss, which makes them perspective for low-temperature microwave (MW) devices.[2] The question arises, how the operating temperature can be shifted to room temperature.

It is well known that ε' can be enhanced by suitable strain in $SrTiO_3$ thin films and that the 1% tensile strain can even induce the ferroelectricity near 270 K in $SrTiO_3$ thin films deposited on $DyScO_3$ substrate.[3] The permittivity of such films can be well tuned by electric field not only in kHz and MW regions, but also in the THz region, because the soft mode frequency is sensitive to external electric field.[4]

Here we will show that strained $Sr_{n+1}Ti_nO_{3n+1}$ thin films deposited on (110)-oriented $DyScO_3$ substrates by reactive molecular beam epitaxy exhibit a ferroelectric phase transition for $n \geq 3$. THz transmission and infrared reflectivity spectra reveal the ferroelectric soft mode, which drives the ferroelectric phase transition. The T_c increases with rising number of perovskite layers - see Fig.2. In $Sr_{n+1}Ti_nO_{3n+1}$ thin films with n≤2, no ferroelectric phase transition was found. Near room temperature the lowest phonon frequency (e.g. the soft mode frequency) decreases with increasing number of layers. Due to this fact, the static ε' increases with the number of layers – see Fig.1(a). Also, the permittivity is larger in strained $Sr_{n+1}Ti_nO_{3n+1}$ films than in the bulk ceramics.[5,6] In bulk $Sr_{n+1}Ti_nO_{3n+1}$ no ferroelectric phase transitions have been observed down to liquid He temperature. In all films a dielectric relaxation was observed below the soft mode which is responsible for several times higher permittivity in radio-frequency region than in the THz region - see e.g. Fig.1(b). In spite of it ε' is well tunable, which makes strained $Sr_{n+1}Ti_nO_{3n+1}$ thin films a promising candidate for applications in microwave integrated circuits.

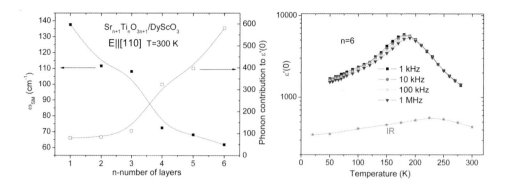

Fig.1.a) Dependence of the soft mode frequency and total phonon contribution to permittivity on the number of perovskite layers in $Sr_{n+1}Ti_nO_{3n+1}$; b) Comparison of temperature dependence of radio-frequency permittivity with sum of phonon contributions to permittivity. Dielectric dispersion typical for relaxor ferroelectrics is seen.

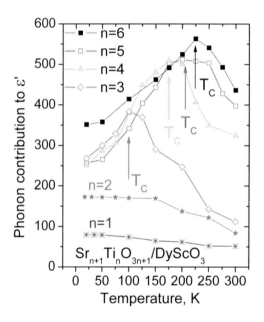

Fig. 2. Temperature dependence of phonon contributions to the permittivity in strained $Sr_{n+1}Ti_nO_{3n+1}$ (n=1 - 6) thin films. The radio-frequency permittivity of these films is several times higher due to dielectric relaxation below the phonon frequencies.

[1] S. N. Ruddlesden and P. Popper, Acta Crystallogr., **10**, 538 (1957);
[2] P.L. Wise et al., J. Eur. Ceram. Soc., **21**, 1723(2001);
[3] J.H. Haeni et al., Nature, **430**, 7589(2008);
[4] C. Kadlec, V. Skorometz, F. Kadlec et al. Phys. Rev. B **80**, 174116 (2009).
[5] D. Noujni et al., Int. Ferroelectrics **62**, 199(2004);
[6] S. Kamba et al., J. Eur. Ceram. Soc., **23**, 2639(2003);

FERROELECTRIC ENHANCED CHARGE GENERATION IN SOLAR ENERGY HARVESTING

Yeng Ming Lam[1,2], Teddy Salim[2], Theo Schneller[1]

[1]Institut für Werkstoffe der Elektrotechnik II, RWTH Aachen, Sommerfeldstraße 24, D-52074 Aachen, Germany
[2]School of Materials Science and Engineering, Nanyang Technological University, 50 Nanyang Avenue, 639798, Singapore

Organic solar cells (OSCs) have attracted significant interest from both scientific and industrial communities due to their promising properties, such as low-cost, flexibility and large-area processibility. The power conversion efficiencies of OSCs can reach as high as 8-9 % [1][2][3][4]. One of key difference between OSCs and the inorganic solar cells, such as silicon and thin film solar cells, is the absorption of photons in OSCs leads to the generation of bound charges instead of free charges. These bound charges, also known as excitons, have to dissociate to result in free charges. Exciton dissociation happens at the interface between the donor and acceptor due to the presence of an internal electric field. Increasing the probability of dissociation and reducing the probability of recombination can improve the efficiencies of OSCs.

The driving energy for the charge transfer is the energy offset between the lowest unoccupied orbitals (LUMOs) of the donor and the acceptor which has to be larger than the exciton binding energy. In some organic donor-acceptor heterojunctions, these energy offsets can be lower than 0.3 eV at room temperature, which is insufficient to overcome the binding forces, and hence unfavorable for the charge transfer to occur. To circumvent this restriction, an ultrathin organic or inorganic interlayer can be introduced between cathode and active layer with a preferred dipole orientation. The exciton dissociation could be enhanced when the recombination loss is reduced by the enhanced internal electric field induced by the interlayer. At the same time, the induced electric field would aid the carriers drift length and hence improve the charge collection.

Some polymeric materials have shown ferroelectric behavior and one such material is the copolymer of vinylidene fluoride and trifluoroethylene (PVDF-TrFE). This material is chemically inert, required low processing temperature and can generate large polarization of the order of 100 mC m^{-2} with a film as thin as 1 nm [5]. PVDF-TrFE has shown to improve the performance of organic photovoltaic cells because this film induced an internal field that both drive exciton separation and suppress charge recombination.[6][7] These films were deposited using Langmuir-Blodgett (LB) deposition method. The deposition method is less suitable for large-scale production but it is able to produce highly ordered films. In this work, the formulation of PVDF-TrFE has been modified in order to generate ordered organization without the need of LB deposition method.

One issue with the use of organic ferroelectric films is the orthogonality of the solvents used. As the ferroelectric films have to be either deposited on the active film or vice versa, the choice of solvents is very limited. One way to circumvent this problem is with the use of inorganic films such as lead zirconate titanate(PZT) which is also compatible with solution processes. This work will explore the considerations and effectiveness of both approaches to the issue of the "supposedly" mismatched energy levels. The thickness/morphology/structure of the interlayer has an important impact on the effectiveness of the charge dissociation. Therefore, the effect of the interlayer properties on the performance of organic solar cells will also be demonstrated.

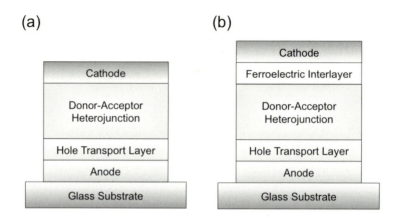

Figure 1: (a) Conventional device structure of an organic solar cell (OSC). The device consists of a donor-acceptor heterojunction layer deposited on an anode-coated coated glass substrate. Hole transport layer can be inserted in between to improve hole collection. Cathode is deposited on the organic layer to collect the electrons. (b) OSC device structure with insertion of the ferroelectric interlayer.

[1] Z. He, C. Zhong, X Huang, W.Y. Wong, H. Wu, L. Chen, S. Su, Y. Cao, Adv. Mater. **23**, 4636 (2011).
[2] C.M. Amb, S. Chen, K.R. Graham, J. Subbiah, C.E. Small, F. So, J.R. Reynolds, J. Am. Chem. Soc. **133**, 10062 (2011).
[3] N.C. Giebink, G.P. Wiederrecht, M.R. Wasielewski, S.R. Forrest, Phys. Rev. B **83**, 195326 (2011).
[4] H.Y. Chen, J. Hou, S. Zhang, Y. Liang, G. Yang, Y. Yang, L. Yu, Y. Wu, G. Li, Nat. Photonics 3, **649** (2009).
[5] A.V. Bune, V. M. Fridkin, S. Ducharme, L. M. Blinov, S. P. Palto, A. V. Sorokin, S. G. Yudin, A. Zlatkin, Nature **391**, 874 (1998)
[6] Y. Yuan, T.J. Reece, P. Sharma, S. Poddar, S. Ducharme, A. Gruverman, Y. Yang, J. Huang, Nature Mater. 10, 296 (2011).
[7] B. Yang, Y. Yuan, P. Sharma, S. Poddar, R. Korlacki, S. Ducharme, A. Gruverman, R. Saraf, J. Huang, Adv. Mater. **24**, 1455 (2012).

A FACILE PREPARATION AND EXTREMILY FAST PHOTOCATALYTIC PROPERTIES OF OXIDE SEMICONDUCTOR/FERROELECTRIC NANO HETEROSTRUCTURE

Huiqing Fan, Pengrong Ren, Xin Wang

State Key Laboratory of Solidification Processing, School of Materials Science and Engineering, Northwestern Polytechnical University, Xi'an 710072, China

Oxide semiconductor / ferroelectric nano heterostructure were prepared through a facile milling and annealing process. Powder X-ray diffraction (XRD), transmission electron microscopy (TEM), X-ray photoelectron spectroscopy (XPS) and ultralviolet-visible (UV-visible) absorption spectra were used to characterize the as-prepared nano powders. Furthermore, UV-induced catalytic activities of semiconductor/ferroelectric nano heterostructure was studied by a degradation reaction of methyl orange (MO) dye. An extremely fast performance of the photocatalytic property is ascribed to the inhibited electron-hole recombination due to the built-in electric filed at the interface and in the oxide semiconductor Bi_2O_3 and ferroelectric $BaTiO_3$ heterojunction.

LIGHT CONTROLLED AMORPHOUS-Al_2O_3 MEMRISTIVE DEVICES

M. Ungureanu[1], R. Zazpe[1], F. Golmar[1,2], Pablo Stoliar[1,3,4], R. Llopis[1], F. Casanova[1,5], L.E. Hueso[1,5]

[1] CIC nanoGUNE Consolider, Tolosa Hiribidea 76, 20018 San Sebastian, Spain;
[2] I.N.T.I. - CONICET, Av. Gral. Paz 5445, San Martín, Bs As, Argentina;
[3] GIA, CAC - CNEA, Av. Gral. Paz 1499, San Martín, Bs As, Argentina;
[4] ECyT, UNSAM, Martín de Irigoyen 3100, San Martín, Bs As, Argentina;
[5] IKERBASQUE, Basque Foundation for Science, 48011 Bilbao, Spain

In this work we present a novel concept consisting on light-controlled metal-oxide-semiconductor memory devices, with optically active Si as bottom electrode. The current through the device depends on the illumination conditions as a result of electron photogeneration in silicon. With this architecture we follow the path towards multifunctional memory devices, in this specific case with light-sensing capabilities included.

In detail, we investigate resistive switching in 20 nm-thick amorphous Al_2O_3 films prepared by atomic layer deposition on p-Si/SiO_2 substrates. Pd top metal contacts are deposited by sputtering after photo-lithography patterning, while Si is used as bottom electrode.

The current (I)-voltage (V) curves show a highly reproducible hysteresis when the samples are illuminated with radiation in the ultraviolet (UV)-infrared (IR) range. However, when external illumination is switched off, the samples present a weak non-hysteretic response to applied voltages (Figure 1).

We prove that our devices behave as memristors by analyzing the remnant current hysteresis switching loops, HSL (Figure 2). To construct the HSL, we apply voltage pulses of 5 ms following the sequence 0 V, +10 V, -10 V, 0 V, in steps of 0.1 V. After each of these pulse steps we wait 100 ms in short-circuit conditions, then we measure the remnant current (I_{rem}) for a fixed voltage of 6 V. This specific measurement of the remnant current reflects an authentic memristive behavior, since capacitive effects are excluded by the long 100 ms time in short-circuit conditions.

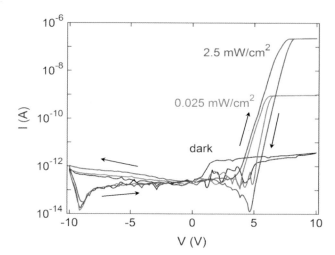

Figure 1: Characteristic I-V curves for a Pd/Al_2O_3/SiO_2/Si memristive device illuminated with UV light of different irradiances. When illuminated with a chosen irradiance, the device can be switched between two states, corresponding to the increasing and decreasing branches of the hysteresis loop at positive applied voltages.

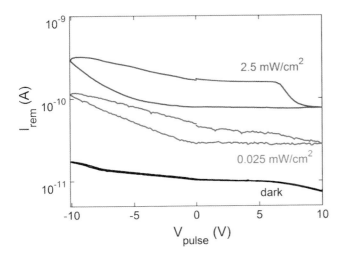

Figure 2: Remnant current hysteresis switching loops for a Pd/Al$_2$O$_3$/SiO$_2$/Si memristive device illuminated with UV light of different irradiances.

From the electrical characterization, the resistive switching can be attributed to electrons from Si combining with traps in the Al$_2$O$_3$ film, thus changing the oxide resistance state. Our system presents interesting applications, for example in codified data storing or in mimicking learning processes for light sensitive bio-systems.

M. Ungureanu, R. Zazpe, F. Golmar, P. Stoliar, R. Llopis, F. Casanova, L. Hueso, Advanced Materials, accepted for publication.

FERROELECTRIC SWITCHING DYNAMICS AT THE NANOSCALE WITH HIGH SPEED SPM

Bryan D. Huey[1,2]

[1] University of Connecticut, Institute of Materials Science, Storrs, CT, USA; [2] Aarhus University, Interdisciplinary Nanoscience Center, Aarhus, Denmark.

Optimizing the performance of data storage devices requires a thorough understanding of the dynamics of the switching process, particularly at the nanoscale. High Speed SPM imaging is employed here, in which the local device state is simultaneously detected and manipulated *in situ* by biasing through a conducting AFM tip [1]. Applied to ferroelectric thin films, this enables domain nucleation times, growth velocities, growth directions, and/or activation energies to be mapped and statistically analyzed as a function of position, defects, processing, switching parameters, etc.

As an example, Figure 1 displays a switching movie acquired by polarizing a 3 μm x 3 μm region of epitaxial (001) PZT, first positively, then negatively. A profound difference in the switching mechanism, from growth dominated to nucleation controlled, is apparent for this film depending on the bias polarity. Correspondingly, the film exhibits a strong polarization preference according to both macroscopic and local hysteresis measurements. Average defect densities can easily be extracted from such data, both spatially and energetically, providing insight into their overall distribution. More importantly, by tracking the evolution of distinct domains, local domain dynamics can be correlated to specific features. Leveraging complementary measurements, for instance topography (standard AFM), local mechanical compliance (AFAM or nanoindentation), or chemical content (*ex situ* x-ray microanalysis), switching dynamics can further be correlated to step edges, ferroelastic domain walls or intersections, concentration, etc.

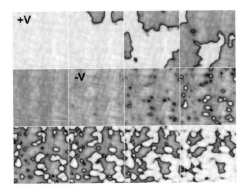
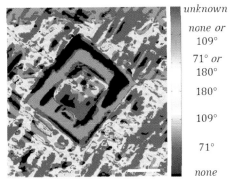

Figure 1: 3 μm x 3 μm HSSPM images extracted from a sequence of 256 consecutive images [2], with steps of ±55 meV$_{dc}$ for each to cycle through an entire hysteresis loop for a (001) epitaxial PZT thin film. 3 seconds/image frame.

Figure 2: Switching movies for a 5 μm x 5 μm region of epitaxial (001) BiFeO$_3$, resolving both out-of-plane and in-plane polarization, reveals <111> domain variants and hence the initial switching angle based on 5 μsec poling pulses.

Extending HSSPM to epitaxial (001) BiFeO$_3$, with eight possible <111> domain variants, requires simultaneous resolution of both out-of-plane and in-plane piezoactuation. The resulting switching movies therefore enable individual steps during polarization reversal to be visualized, including the direction and especially the angle of polarization change. The number of switches per location is also easily determined, with certain regions switching just once, several undergoing multiple discrete steps, and many re-orienting in response to adjacent domain reorientations, evidencing the important of local strain on domain stability. Domain wall statistics are accessible as well, such as the domain wall density, interfacial charging, and angle between adjacent domains [3].

HSSPM is therefore a powerful technique for considering nanoscale ferroelectric domain dynamics, stability, and reliability. The concept is additionally applicable to a range of other data storage systems, particularly phase change materials, the subject of ongoing research.

Acknowledgements: This work is sponsored by NSF:MWN and DOE:BES:Electron and Scanning Probe Microscopies. Specimens are provided by Y. H. Chu and R. Ramesh, UC Berkeley, as well as J. Ihlefeld and D. Schlom, Cornell University.

[1] R. Nath, Y.H. Chu, N. Polomoff, R. Ramesh, B. D. Huey, APL, **93**, 072905, 2008.
[2] B. D. Huey, R. Nath, S. Lee, N. A. Polomoff, J. American Ceramic Society, **95** (4), *Feature Article and Cover Image* (2012).
[3] J. Desmarais, J. F. Ihlefeld, T. Heeg, J. Schubert, D. G. Schlom, B. D. Huey, Applied Physics Letters, **93**, 162902, (2011).

IONIC CHARGE INTERACTIONS WITH FERROELECTRIC SURFACES: POLARIZATION OF ULTRATHIN PbTiO$_3$ WITH CONTROLLED SURFACE COMPENSATION

S.K. Streiffer[1], M.J. Highland[2], T.T. Fister[3], D.D. Fong[2], P.H. Fuoss[2], Carol Thompson[4], J.A. Eastman[2], and G.B. Stephenson[5]

[1]Physical Sciences and Engineering Directorate, Argonne National Laboratory, Argonne, Illinois 60439, USA; [2]Materials Science Division, Argonne National Laboratory, Argonne, Illinois 60439, USA; [3]Chemical Sciences and Engineering Division, Argonne National Laboratory, Argonne, Illinois 60439, USA; [4]Department of Physics, Northern Illinois University, DeKalb, Illinois 60115, USA; [5]Advanced Photon Source, Argonne National Laboratory, Argonne, Illinois 60439, USA

Studies of the behavior and stability of ultrathin ferroelectrics have found that the extent of compensation of the polarization by free charge at interfaces strongly affects the phase transition and switching properties of these materials [1–4]. Furthermore, competition between energy terms that scale with surface area and those that scale with volume largely dictate the behavior of these systems as their size decreases. It has also been shown that sufficient ionic compensation of bare (unelectroded) ferroelectric surfaces can occur such that ultrathin films remain polar [5, 6, 7]. For this case, the polarization orientation can be altered by changing the chemistry of the environment [8, 9], for example by changing pO$_2$ in a manner that alters the balance of positively and negatively charged ionic species on the ferroelectric surface. Recent synchrotron x-ray scattering experiments on ultrathin (001) PbTiO$_3$ films [10] have found that, depending upon film thickness and temperature, this polarization reorientation can occur either through the nucleation and growth of oppositely oriented domains or by a continuous mechanism, in which the polarization uniformly decreases to zero and reverses without the formation of domains. These observations have motivated efforts to determine the equilibrium polarization phase diagram of ultrathin ferroelectric PbTiO$_3$ films as a function of temperature and the external chemical potential controlling their ionic surface compensation [11].

Predicted equilibrium phase diagrams [12] are qualitatively similar to those measured experimentally using x-ray scattering. At lower temperatures, the polarization has opposite signs for high and low pO$_2$, respectively. Most interestingly, the observed T$_C$ is minimum at the intermediate pO$_2$ value separating regions of the phase diagram in which the polarization orientation below T$_C$ reverses. The degree to which T$_C$ is suppressed at these intermediate pO$_2$ values becomes greater as film thickness decreases. Additionally below T$_C$ in this pO$_2$ region, diffuse scattering satellite peaks in the in-plane directions around PbTiO$_3$ Bragg peaks indicate that a mixed 180° domain structure separates regions of the phase diagram of opposite polarization orientation.

These results provide an explanation for polarization switching without domain formation at the intrinsic coercive field in sufficiently thin PbTiO$_3$ films [10]. For a thickness-dependent range of temperatures above the minimum T$_C$, but at sufficiently low temperature such that stable polarizations of opposite signs exist for lower or higher pO$_2$'s, continuous switching induced by changing pO$_2$ occurs because the polar phase loses stability with respect to the nonpolar phase, giving rise to a situation during switching in which the local polarization magnitude passes through zero. The nonpolar phase is stabilized and T$_C$ is suppressed at intermediate pO$_2$ because the concentrations of charged surface ions of either sign are insufficient to compensate either orientation of the polar phase. In contrast to this, at lower temperatures below the minimum T$_C$, the regions of equilibrium 180° stripe domains in the phase diagram match to the traditional nucleated switching mechanism in which a net zero polarization at the midpoint of switching is a result of an equal mixture of domains of opposite polarization orientation.

These studies show that the chemical environment at the surface of an ultrathin ferroelectric film has strong effects on its phase transition. Such ionic compensation offers a new tool for tuning ferroelectric properties and a new avenue for devices, and must be considered when comparing studies carried out in different environments, e.g. ambient air or vacuum.

This work was supported by the U.S. Department of Energy, Office of Science, Office of Basic Energy Sciences under Contract DE-AC02-06CH11357. The PI's and scientific program were supported by the Materials Sciences and Engineering Division, while use of the Advanced Photon Source was supported by the Scientific User Facilities Division.

[1] M. F. Chisholm, W. Luo, M. P. Oxley, S. T. Pantelides, and H. N. Lee, Phys. Rev. Lett. **105**, 197602 (2010).
[2] A. M. Kolpak et al., Phys. Rev. Lett. **105**, 217601 (2010).
[3] P. Zubko, N. Stucki, C. Lichtensteiger, and J.-M. Triscone, Phys. Rev. Lett. **104**, 187601 (2010).
[4] M. G. Stachiotti and M. Sepliarsky, Phys. Rev. Lett. **106**, 137601 (2011).
[5] D. D. Fong et al., Phys. Rev. Lett. **96**, 127601 (2006).
[6] R. Takahashi et al., Appl. Phys. Lett. **92**, 112901 (2008).
[7] C. Lichtensteiger, J.-M. Triscone, J. Junquera, and P. Ghosez, Phys. Rev. Lett. **94**, 047603 (2005).
[8] R. V. Wang et al., Phys. Rev. Lett. **102**, 047601 (2009).
[9] Y. Kim, I. Vrejoiu, D. Hesse, and M. Alexe, Appl. Phys. Lett. **96**, 202902 (2010).
[10] M. J. Highland et al., Phys. Rev. Lett. **105**, 167601 (2010).
[11] M. J. Highland et al. Phys. Rev. Lett. **107**, 187602 (2011).
[12] G. B. Stephenson and M. J. Highland, Phys. Rev. B **84**, 064107 (2011).

STRAIN TUNNING OF FERROELECTRIC-ANTIFERRODISTORTIVE COUPLING IN PbTiO$_3$/SrTiO$_3$ SUPERLATTICES

Pablo García-Fernández[1], **Pablo Aguado-Puente**[1], **Javier Junquera**[1]

[1]Departamento de Ciencias de la Tierra y Física de la Materia Condensada, Universidad de Cantabria, Santander, Spain

The interest in ultra-short period PbTiO$_3$/SrTiO$_3$ (PTO/STO) superlattices has been fueled during the last years, mostly due to the appearance of an improper ferroelectric (FE) polarization from the coupling of two rotational antiferrodistortive (AFD) modes of the oxygen octahedra cages [1]. Using first-principles simulations, based on the density functional theory, we have explored the phase diagram of the PTO/STO superlattices to gain further insight about the coupling between the FE, AFD and the strain degrees of freedom in monodomain phases, and to find new paths to engineer functional properties in this system [2]. In a second step we have focused on the formation of polarization domains and the structure of domain walls in PTO/STO superlattices, considering the influence of the lateral domain size, orientation and energy of domain walls, and the influence of AFD instabilities [3]. These are challenging simulations because they require accurate computations on systems with up to almost 1000 atoms in the simulation box.

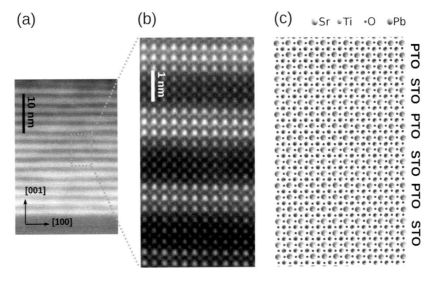

Figure 1: High-resolution transmission electron microscopy (TEM) image of a PTO/STO superlattice experimentally grown and the simulation box theoretically computed, showing how realistic our simulations are (overlap in size with experiment).

The main results of this work are: (i) In the monodomain phase the presence of the paraelectric STO layer imposes an electrostatic penalty on the out of plane polarization of the PTO, favouring the rotation of the polarization out of the normal direction and giving rise to an enhanced piezoelectric response at very low strain values. (ii) An enhanced FE-AFD coupling is observed at the PTO/STO interface that cannot be accounted by typical steric effects. Instead a covalent model is invoked to explain this effect. (iii) The dependence of the FE-AFD coupling (known to be responsible of the improper-ferroelectric behaviour in this system) with respect to the periodicity of the superlattice reveals that this effect is dominant for the shortest periodicity (1|1) and contributes to stabilize the FE instability. This contribution, however, decreases in magnitude rapidly with the periodicity and changes sign (thus contributing to destabilize the FE distortion) for periodicities larger than (3|3). (iv) Results on polydomain phases reveal structures departing from the typical 180° domains typically assumed, forming vortices at domain walls. (v) We have found a progressive transition, as a function of the periodicity of the superlattice, from a strongly electrostatic coupled regime (where the ground-state is a monodomain configuration with a constant out-of-plane component of the polarization preserved throughout the structure), to a weakly coupled regime (where the polarization is confined within the PTO layers forming domains). (vi) Our results are in good agreement with x-ray diffraction, transmission electron microscopy, and ultra-high resolution electron energy loss spectroscopy experimental measurements [4].

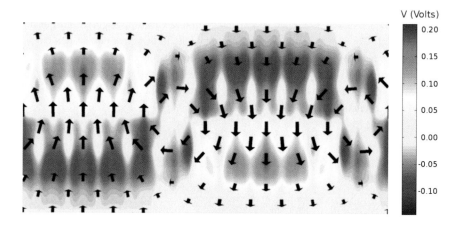

Figure 2: Electrostatic potential inside a $(PTO)_3/(STO)_3$ superlattice in a polydomain configuration. The arrows represent the values of the local polarization, which displays the structure of a domain of closure.

[1] E. Bousquet et al., Nature **452**, 732 (2008).
[2] P. Aguado-Puente, P. García-Fernández, and J. Junquera, Phys. Rev. Lett. 107, 207601 (2011).
[3] P. Aguado-Puente and Javier Junquera, submitted to Phys. Rev. B. Pre-print available at http://arxiv.org/abs/1202.5481
[4] P. Zubko, N. Jecklin, A. Torres-Pardo, P. Aguado-Puente, A. Gloter, C. Lichtensteifer, J. Junquera, O. Stéphan, and J.-M. Triscone, submitted to Nano Letters.

INTERFACE CONTROL OF BULK FERROELECTRIC POLARIZATION

D. Yi[1], P. Yu[1], W. Luo[2,3], J. X. Zhang[1], M. D. Rossell[4], C. –H. Yang[5], G. Singh-Bhalla[6], S. Y. Yang[1], Q. He[1], Q. M. Ramasse[4], R. Erni[4], L. W. Martin[7], Y. H. Chu[8], S. T. Pantelides[3,2], S. J. Pennycook[2,3] and R. Ramesh[1,6]

[1]Department of Physics and department of materials science and engineering, University of California, Berkeley, CA, USA, 94720
[2]Materials Science and Technology Division, ORNL, Oak Ridge, TN, USA, 37831
[3]Department of Physics and Astronomy, Vanderbilt University, Nashville, TN, USA, 37235 [4]National Center for Electron Microscopy, LBNL, Berkeley, CA, USA, 94720
[5]Department of Physics, KAIST, Daejeon, Republic of Korea, 305-701
[6]Materials Science Division, LBNL, Berkeley, CA, USA, 94720
[7]Department of Materials Science and Engineering, UIUC, Urbana, IL, USA, 61801
[8]Department of Materials Science and Engineering, National Chiao Tung University, Taiwan, 300100

Recent advance in deposition technics provides us method to precisely construct atomically sharp perovskite oxide heterointerfaces, which show novel electronic properties and functionalities that are different from the bulk. Although different novel states are well studied at the interface in atomic scale, it remains intriguing to use the interfacial structure to control the order parameter and properties of the bulk components.

In this talk, we demonstrate the ability to deterministically control a bulk property, namely the ferroelectric polarization, of a heteroepitaxial bilayer by precise atomic-scale interface engineering. More specifically, the control is achieved by exploiting the interfacial valence mismatch (polar discontinuity) to influence the electrostatic potential step across the interface, which further manifests itself as the internal field in ferroelectric hysteresis loops and determines the ferroelectric state. The generality of this experimental approach has been further extended to different ferroelectrics and conducting oxides, which indicates that electrostatic potential step is formed as a consequence of the interface valence mismatch and controls the polarization state of the ferroelectric. Clearly, such a coupling effect between interface properties and the bulk order parameters of the thin films is, in principle, not limited to the electrostatic degree of freedom.

ANALYZING POLARIZATION AND LATTICE STRAINS AT THE INTERFACE OF FERROELECTRIC HETEROSTRUCTURES ON ATOMIC SCAL VIA CS-CORRECTED SCANNING TRANSMISSION ELECTRON M IXROSCOPY (STEM)

D. Park[1], A. Herpers[2], T. Menke[2], R. Dittmann[2] and J. Mayer[1]

[1]Central Facility for Electron Microscopy, RWTH Aachen, Ahornstr. 55, D-52074 Aachen, Germany; [2]Institute of Solid State Research and JARA-FIT, Jülich Aachen Research Alliance, Fundamentals of Future Information Technology, Research Center Jülich, Germany

Ferroelectric thin films are attractive candidates for capacitors in random access memory (FeRAM) devices, in which a reversible spontaneous polarization is utilized to store information. However, below a critical thickness, the ferroelectric polarization usually disappears, since imperfections at the interface obstruct the spontaneous polarization of ferroelectric films. Therefore, an in-depth understanding of the chemical and structural information at the interface on the atomic scale is indispensable for processing high-quality epitaxial thin films.

The investigated thin film heterostructures were grown by RHEED-assisted pulsed laser deposition (PLD). The ultrathin epitaxial $BaTiO_3$ was grown on niobium doped $SrTiO_3$ and a top electrode $SrRuO_3$ was then deposited. One unit cell of $BaRuO_3$ is embedded between $BaTiO_3$ and $SrRuO_3$. Cross-section specimens were prepared by focused ion beam (FIB). To remove artifacts during FIB preparation, the specimen was finally thinned with a low acceleration voltage (2 kV).

To analyze the interface of ferroelectric heterostructures, the HAADF imaging technique is an ideal tool due to its high sensitivity to the atomic number Z. The use of the HAADF detector in STEM allows nearly a direct interpretation of the images unlike the ambiguities in high-resolution transmission electron microscopy (HRTEM).

To minimize the sample drift in STEM mode, a series of 30 HAADF images is acquired every second and the sample drift is corrected by displacement correction with the *imtools* software package [1]. The drift corrected HAADF image shown in Figure 1 proves perfect epitaxial growth of the thin film heterostructures. The $BaTiO_3$ thin film is fully strained by the $SrTiO_3$ substrate because of the lattice parameter difference. The in-plane lattice parameters of the bulk form of $SrTiO_3$ and $BaTiO_3$ are 3.90 Å and 3.99 Å respectively [2]. For the identification of the tetragonal distortion of the $BaTiO_3$ film, the in-plane and out-of-plane lattice parameters are determined by analyzing the intensity profile (Table 1). As a result, the out-of-plane lattice parameter of 4.17 Å of the $BaTiO_3$ thin film exceeds considerably the value of 4.03 Å of the bulk form. The lattice parameters measured by HAADF are in good agreement with reported values from synchrotron x-ray diffraction and theoretical prediction [3].

The results shown in Figure 1 and Table 1 demonstrate the presence of the tetragonal distortion of the thin $BaTiO_3$ film strained by the $SrTiO_3$ substrate. Moreover, the small shift of Ti atomic column was visible in the HAADF image. However, local mistilting of $BaTiO_3$ film could lead to a misinterpretation. Therefore, to elucidate ferroelectric polarization the relative shift of Ti and O columns along the [110] direction should be studied by HRTEM in negative Cs imaging condition.

[1] L. Houben, http://www.er-c.org/centre/software/imtools.html
[2] A. Visinoiu et al. jpn. Jppl. Phys. 41 (2002), p 6633-6638
[3] A. Petraru et al. J. Appl. Phys. 101, (2007), p114106.
[4] The authors gratefully acknowledge funding from Deutsche. Forschungsqemeinschaft (DFG) under grant number Ma 1280/32-1.

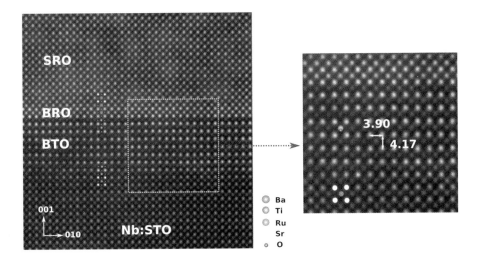

Figure 1: The drift corrected HAADF image of ferroelectric heterostructures along the [100] direction. The image was acquired an aberration-corrected STEM at an acceleration voltage of 300 kV. The atomic column positions are indicated by different colours. The in-plane and out-of-plane lattice parameter of the $BaTiO_3$ are indicated and the small shift of Ti column is visible on the right image.

	in-plane (a) [Å]	out-of-plane (c) [Å]	c/a
SRO	3.90 ± 0.05	3.90 ± 0.05	1.00 ± 0.03
BTO	3.90 ± 0.05	**4.17 ± 0.05**	**1.07 ± 0.03**
STO	3.90 ± 0.05	3.93 ± 0.05	1.01 ± 0.03

Table 1: The lattice parameters of ferroelectric heterostructures by analyzing the intensity profile.

CONTROL OF CONDUCTION THROUGH DOMAINS AND DOMAIN WALLS IN BiFeO$_3$ THIN FILMS

Saeedeh Farokhipoor and Beatriz Noheda

Zernike Institute for Advanced Materials, University of Groningen, The Netherlands

Because BiFeO$_3$ (BFO) is, at room temperature, a rhombohedrally distorted, ferroelectric, perovskite, there are eight possible polarization directions (or domains) which give rise to three different types of domain walls, namely, 180° (ferroelectric) and 109° and 71° (ferroelectric and ferroelastic) domain walls (DWs). These DWs have raised recent interest because of the distinct characteristics they reveal and the novel physical properties they give rise to. 109° DWs have been reported to be responsible for the photo-currents observed in BiFeO$_3$[1]. Moreover, when antiferromagnetically ordered BFO is put in contact with a soft ferromagnetic (permalloy) layer, the exchange bias observed is reported to be directly related to the amount of 109° DWs in the BFO layer[2]. In the same type of DWs, magnetoresistance as large as 60% has been measured[3]. In addition, recent conducting atomic force microscopy (C-AFM)-based studies on artificially-written, both 109° and 180° DWs[4] and those on 71° DWs in the as-grown state[5], reveal enhanced conduction in the walls compared to the domains in this interesting material. In order to get insight into the origin of domain wall/domain conductivity, the underlying conduction mechanisms have been extensively studied. The large current regime turns out to be determined by Schottky emission from the tip. Migration of oxygen vacancies to the domain walls lower the Schottky barrier heights at the interface with the metallic tip compare to those in the domains. This results in the observed difference of conduction levels between domains and domain walls[5]. It also implies that the domain wall conductivity is extrinsically tunable by changing the defect chemistry [6,7]. Moreover, in ferroelectrics the Schottky barriers with the metal electrodes can also be tuned by switching the polarization state and the combination of these two effects determine the conductivity in BiFeO$_3$[8,9]. Here we present the effects of this interplay between defect migration and polarization direction and the degree of control that can be attained in the conductivity of BFO thin films.

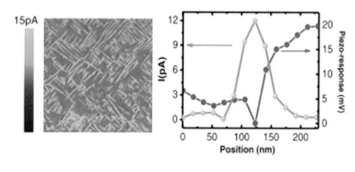

Fig. 1 Left) Room temperature C-AFM image of a 50nm BiFeO$_3$ film deposited on a SrRuO$_3$-buffered SrTiO$_3$ substrate, with a bias voltage of 2.75 V (low current regime). Right) Line scan across one domain wall for both the piezo-response amplitude and the current images. Ref. [5].

[1] S.Y. Yang, Nature Nanotech. 5, 143 (2010)
[2] L.W. Martin et al. Nano Letters 8, 2050 (2008)
[3] Q. He et al., PRL 108, 067203 (2012)
[4] J. Seidel et al., Nature Mat. 8, 229 (2009)
[5] S. Farokhipoor, B. Noheda. PRL 107, 127601 (2011)
[6] J. Seidel et al., PRL 105, 197603 (2010)
[7] S. Farokhipoor, B. Noheda, arXiv:1201.0144
[8] S. Farokhipoor and B. Noheda (to be submitted); Oral presentation at the Fall MRS meeting (symposium P14.3), Boston December 2011.
[9] H.T.Yi et al., Adv Mater 23, 3403 (2011)

CMOS COMPATIBLE FERROELECTRIC MATERIALS BASED ON HAFNIUM OXIDE

Thomas Mikolajick[1)2)], Uwe Schroeder[1)], Johannes Müller[3)], Stefan Müller[1)] and Stefan Slesazeck[1)]

[1)] NaMLab GmbH, Nöthnitzer Str. 64, 01187 Dresden
[2)] Institut für Halbleiter und Mikrosystemtechnik, TU Dresden, Nöthnitzer Str. 64, 01187 Dresden
[3)] Fraunhofer Center Nanoelectronic Technologies, Königsbrücker Str. 180, 01099 Dresden

Ferroelectrics are very interesting materials for nonvolatile data storage. However, the progress in this field is mainly limited by the low compatibility of conventional ferroelectrics like lead zirconate titanate (PZT) and the required electrodes with CMOS processing. Therefore conventional 1T/1C ferroelectric memories are not available with features sizes below 130 nm [1] and 1T ferroelectric FETs are still struggling with retention and very thick memory stacks [2]. In other words ferroelectric memories face serious scaling limitations and can only explore a very small fraction of their potential. Recently it was shown, that hafnium oxide can show ferroelectric hysteresis with very promising characteristics. Since hafnium oxide is a standard material in sub 45 nm CMOS processes [3], this discovery could open the path towards further exploiting the potential of ferroelectric memories.

By adding a few percent of silicon and annealing the films in a mechanically confined manner Boescke et al. demonstrated ferroelectric properties in hafnium oxide for the first time [4]. Recently also hafnium-zirconium oxide [5] as well as yttrium- [6] and aluminum- [7] additions to hafniumoxide have shown to enable the occurrence of a ferroelectric phase in layers as thin as 10 nm. The effect is attributed to the fact that the doping stabilizes a non-centrosymmetric phase near the monoclinic to tetragonal/cubic phase boundary. An example for the ferroelectric hysteresis of a metal/ferroelectric/metal capacitor using yttrium doped hafnium oxide is shown in fig.1. Fig. 2 illustrates the memory window achieved in a ferroelectric field effect transistor using silicon doped hafnium oxide.

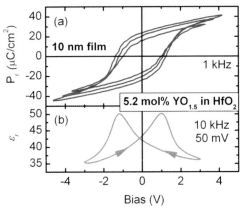
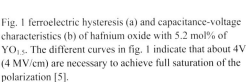
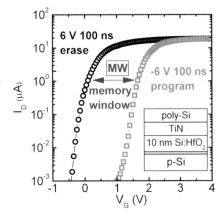

Fig. 1 ferroelectric hysteresis (a) and capacitance-voltage characteristics (b) of hafnium oxide with 5.2 mol% of $YO_{1.5}$. The different curves in fig. 1 indicate that about 4V (4 MV/cm) are necessary to achieve full saturation of the polarization [5].

Fig. 2 Id–Vg measurements of a Si : HfO_2-based MFIS-FET after a 100-ns +6-V program and −6-V erase pulse [7].

In this paper the recent status of ferroelectric hafnium oxide will be reviewed. In the first part the different doping elements that have been shown to enable ferroelectricity will be compared. The second part will focus on the memory relevant characterization data as well as possible memory concepts leading to a discussion of the applicability of the material towards 1T/1C ferroelectric memories and 1T ferroelectric field effect transistors. Finally an outlook on further developments will be given.

[1] H. P. McAdams at al., IEEE Journal of Solid State Circuits, Vol. 39, No. 4, p. 667, 2004
[2] L.V.Hai et al., Prodeedings of IEEE International Memory Workshop, p. 1, 2011
[3] T. S. Böscke et. al., Appl. Phys. Lett., vol. 99, p. 102903, 2011.
[4] J. Müller et al., Appl. Phys. Lett., vol. 99, p. 112901, 2011.
[5] J. Müller et al., J. Appl. Phys., vol. 110, p. 114113, 2011.
[6] S. Mueller et al., Adv. Func. Mater., 2012, 10.1002/adfm.201103119.
[7] J. Müller at al., IEEE Electron Device Letters, vol. 33, 2, p. 185, 2012

ATOMIC LAYER DEPOSITED Gd-DOPED HfO_2 THIN FILMS: FROM HIGH-K DIELECTRICS TO FERROELECTRICS

Christoph Adelmann[1], Lars-Åke Ragnarsson[1], Alain Moussa[1], Joseph A. Woicik[2], Stefan Müller[3], Uwe Schroeder[3], Valeri V. Afanas'ev[4], and Sven Van Elshocht[1]

[1]Imec, B-3001 Leuven, Belgium;
[2]National Institute of Standards and Technology, Gaithersburg, Maryland 20899, USA;
[3]Namlab gGmbH, 01187 Dresden, Germany;
[4]Department of Physics, University of Leuven, B-3001 Leuven, Belgium

In the past, the continuous scaling of complementary metal oxide semiconductor (CMOS) transistors has allowed for both increasing their performance and decreasing their size. To address the ever increasing leakage current densities for scaled CMOS transistors, alternative (to SiON) gate dielectrics have received intense attention in the last decade. For the 45 nm node and beyond, Hf-based dielectrics, such as HfO_2, have been integrated into commercial devices [1]. However, while HfO_2 remains the material of choice for the next technology nodes, further scaling of the gate may eventually require the replacement of HfO_2 by a dielectric with an even higher dielectric constant. The cubic/tetragonal phases of HfO_2 have recently received increasing interest because it offers a dielectric constant of the order of 30 or higher. Although the cubic/tetragonal phases are stable only at high temperature, they can be stabilized by alloying HfO_2 with a small concentration (cation concentration typically ~8-10%) of other metal oxides, such as Gd_2O_3 [2], Y_2O_3 [3,4], Dy_2O_3 [5], or SiO_2 [6].

To assess the scalability of Gd-doped HfO_2 to the relevant thicknesses for gate dielectrics in scaled CMOS devices, which are of the order of 2 nm, one has to understand the relation between the crystallographic phase of the films and their dielectric properties as a function of the thermal budget. Because of the increased crystallization temperatures for ultrathin films [7], it is difficult to extrapolate the properties of "thicker" films (e.g. of the order of 10 nm) down to the target thickness range. Because conventional XRD becomes elusive for 2 nm thin films, we have studied the crystallographic phase of ultrathin Gd_2O_3-doped HfO_2 by extended x-ray absorption

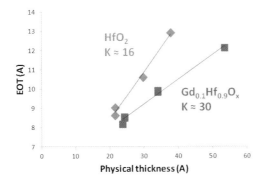

Figure 1: Equivalent oxide thickness (EOT) vs. physical thickness for HfO_2 and $Gd_{0.1}Hf_{0.9}O_x$. Dielectric constants of 16 and 30 were obtained for HfO_2 and $Gd_{0.1}Hf_{0.9}O_x$, respectively.

fine structure (EXAFS) measurements. The EXAFS measurements indicated that 2 nm thin films can be crystallized into the cubic/tetragonal phase by spike annealing at 1050°C. Indeed, scaled capacitors with physical dielectric thicknesses down to 2.2 nm show a dielectric constant of ~30 for $Gd_{0.1}Hf_{0.9}O_x$, comparable to that measured for thicker films [2]. As a consequence, an equivalent oxide thickness (EOT) of 0.8 nm has been demonstrated for $Gd_{0.1}Hf_{0.9}O_x$ deposited on a nominally 1 nm thick chemical SiO_2/Si starting surface using TiN as metal gates. Although the band gap and band offsets were not measurably influenced by the alloying with Gd_2O_3, an increase in the leakage current with respect to

HfO$_2$ was however observed for films of the order of 10 nm, which could be related to an increased Poole-Frenkel conduction across the films [5]. For scaled films of 2-3 nm, the measured leakage currents across the gate dielectric were also found to be an order of magnitude higher than for HfO$_2$ for identical EOT values. The origins of this leakage current increase for ultrathin films are currently under study. The results indicate that the assessment of the dielectric constant alone is thus insufficient to fully determine the scaling properties of Gd$_2$O$_3$-doped HfO$_2$.

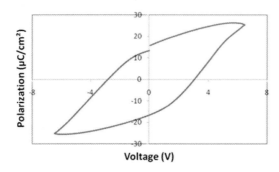

Figure 2: Polarization *vs.* applied voltage for a TiN/Gd$_{0.05}$Hf$_{0.95}$O$_x$/TiN MIM stack indicating ferroelectricity of the Gd$_{0.05}$Hf$_{0.95}$O$_x$ layer.

In addition to the stabilization of the cubic/tetragonal phase of HfO$_2$ by doping, an orthorhombic phase has also recently been observed for doping concentration around 5% after high-temperature annealing and quenching in presence of a TiN layer [8]. We have also studied the behaviour of Gd$_2$O$_3$-doped HfO$_2$ with 5% Gd/(Gd+Hf) in metal-insulator-metal (MIM) structures in combination with TiN. Similar to the results for SiO$_2$-doped HfO$_2$ [8], the orthorhombic phase was indeed stabilized after annealing at 1000°C. Ferroelectricity was observed in such films, making them promising for the integration into ferroelectric transistor or memory devices.

[1] M. T. Bohr, R. S. Chau, T. Ghani, and K. Mistry, IEEE Spectrum **44**, 29 (2007).
[2] C. Adelmann et al., J. Electrochem. Soc. **157**, G105 (2010).
[3] K. Kita, K. Kyuno, and A. Toriumi, Appl. Phys. Lett. **86**, 102906 (2005).
[4] E. Rauwel et al., Appl. Phys. Lett. **89**, 012902 (2006).
[5] C. Adelmann, et al., Appl. Phys. Lett. **91**, 162902 (2007).
[6] K. Tomida, K. Kita, and A. Toriumi, Appl. Phys. Lett. **89**, 142902 (2006).
[7] C. Adelmann et al., Appl. Phys. Lett. **95**, 091911 (2009).
[8] T. S. Böscke et al., Appl. Phys. Lett. 99, 102903 (2011).

FER 3

TEMPERATURE-DEPENDENT ELECTRICAL CHARACTERIZATION OF HAFNIUM OXIDE BASED FERROELECTRIC ULTRA-THIN FILMS

U. Böttger[1], I. Müller[1], J. Müller[2], U. Schröder[3]

[1] IWE 2, RWTH Aachen, Sommerfeldstr. 24, 52074 Aachen, Germany,
[2] Fraunhofer CNT, Königsbrücker Str. 180, 01099 Dresden, Germany
[3] NaMLab GmbH, Nöthnitzer Str. 64, 01187 Dresden, Germany

Stable ferroelectric crystalline phases were recently found in thin hafnium oxide films with suitable dopants and processing [1]. Due to the excellent and already demonstra-ted CMOS compatibility, this class of materials represents promising candidates for functional layers in ferroelectric random access memories as well as in ferroelectric field effect transistors [2].

In this paper, the temperature dependence of the large signal $P(V)$ and the small signal $C(V)$ response of hafnium oxide films with different additives (3.6 mol% Y and 5.6 mol% Si) is investigated in a range between 80 K and 400 K. Y:HfO_2 and Si:HfO_2 are prepared by metal organic ALD based on metal organic precursors tetrakis-(ethylmethylamino)-hafnium (TEMA-Hf), tetrakis-dimethylamino-silane (4DMAS) / tris(methylcyclopentadienyl)yttrium (Y(MeCp)3), and ozone. The MIM structures are realized by 10 nm thick layers of these materials and TiN electrodes which are deposited by CVD as well as by PVD. Details of the fabrication are given in [1, 3].

The $P(V)$ hysteresis curve of Y:HfO_2 and Si:HfO_2 thin films at different temperatures are illustrated in Figs. 1 and 2. With increasing temperature, the yttrium doped sample shows a reduction of the mean value and of the asymmetry of the coercive field, while the switched polarization $\Delta P_{sw} = P_+ - P_-$ remains constant. In addition, the tilt of hysteresis decreases. There is evidence that regions with non-switchable ("frozen") polarization in the bulk or at the interfaces exist similar to acceptor-oxygen vacancy defect dipoles which are responsible for the imprint behavior in "classical" ferroelectrics [4]. By thermal activation the dipoles become mobile and contribute to the electrical polarization.

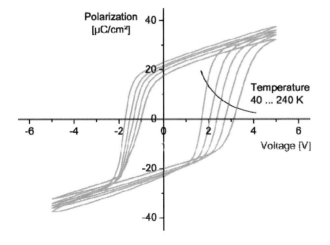

Figure 1: $P(V)$ hysteresis loops of Y:HfO_2 thin films at different temperatures.

In contrast to Y:HfO_2 the Si:HfO_2 thin films undergo abrupt changes at 150 K with respect to the remanent polarization and the shape of the $P(V)$ curve, see Fig. 2. The behavior is correlated with the appearance of a fourfold peak in the small signal $C(V)$, see Fig. 3.

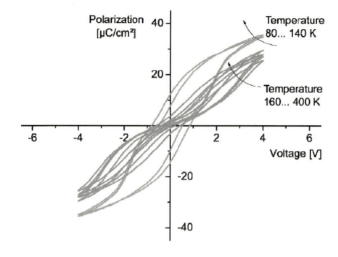

Figure 2: $P(V)$ hysteresis loops of Si:HfO$_2$ thin films at different temperatures.

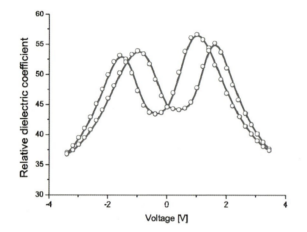

Figure 2: Dielectric coefficient from $C(V)$ measurements of Si:HfO$_2$ thin films at room temperature.

The abrupt change at 150 K is assumed to be linked with a phase transition from a ferroelectric to an antiferroelectric or relaxor-like phase. The final identification of the phase needs further investigations.

[1] T. Boescke, J. Mueller, D. Braeuhaus, U. Schroeder, and U. Boettger, Appl. Phys. Lett. 99, 102903 (2011).
[2] T. Boescke; J. Mueller; D. Braeuhaus; U. Schroeder; U. Boettger, IEDM Techn. Dig., 547 (2011).
[3] J. Müller, U. Schröder, T. Böscke, I. Müller, U. Böttger, L. Wilde, J. Sundqvist, M. Lemberger, P. Kücher, T. Mikolajick, and L. Frey, J. Appl. Phys. 110, 114113 (2011).
[4] R. Lohkamper, H. Neumann, G. Arlt, J. Appl. Phys. 68, 4220 (1990).

CORRELATION BETWEEN COMPOSITION AND ELASTIC PROPERTIES OF $Ca_xBa_{1-x}Nb_2O_6$ RELAXOR FERROELECTRICS

Chandra Shekhar Pandey[1], **Jürgen Schreuer**[1], **Manfred Buranek**[2] **and Manfred Mühlberg**[2]

[1]Ruhr University Bochum, Bochum, Germany; [2]University of Cologne, Cologne, Germany

$Ca_xBa_{1-x}Nb_2O_6$ (CBN-x), a relaxor ferroelectric material crystallizing in the partially filled tetragonal tungsten bronze (TTB) structure type, provides an excellent alternative to strontium barium niobate (SBN) for device applications because of its relatively high Curie temperature ($T_C \approx 264°C$ for $x \approx 0.28$).

Here we report on the anomalous behaviour of thermal expansion and elastic properties of varying Ca/Ba ratio in CBN-x single crystals (congruently grown by the Czochralski method) in between 100 K and 1423 K, employing high-resolution dilatometry and resonant ultrasound spectroscopy (RUS), respectively.

Like in CBN-28 [1] the temperature evolution of the elastic constants c_{ij} in the paraelectric phase (point symmetry group 4/mm) of CBN-x shows pronounced anomalies. All independent elastic constants evolved differently, with temperature reflecting their coupling to different types of the reorientational motion of polar nanoregions (PNRs) through their interaction with acoustic waves. Interestingly, the elastic properties show a linear increase with a decreasing Ca content. This was attributed to the enhanced coupling between the polarization and the strain due to the formation of quasistatic PNRs in the paraelectric phase. The onset of elastic softening is frequency dependent. In the ferroelectric phase strong ultrasound dissipation effects appear which are probably related to interactions between sound waves and ferroelectric domain walls. This study of CBN-x in its paraelectric phase enables us to understand the relaxor phenomenon.

[1] C. S. Pandey et al., Physical Review B. **84**, 174102 (2011).

LONE PAIR-INDUCED COVALENCY AS THE CAUSE OF TEMPERATURE AND FIELD-INDUCED INSTABILITIES IN BISMUTH SODIUM TITANATE

Denis Schütz[1*], Marco Deluca[2,3], Werner Krauss[1], Antonio Feteira[4], Klaus Reichmann[1]

[1]Christian Doppler Laboratory for Advanced Ferroic Oxides, Institute for Chemistry and Technology of Materials, Graz University of Technology, Stremayrgasse 9/3, A-8010 Graz, Austria
[2]Institut für Struktur- und Funktionskeramik, Montanuniversitaet Leoben, Peter Tunner Straße 5, A-8700 Leoben, Austria
[3]Materials Center Leoben Forschung GmbH, Roseggerstraße 12, A-8700 Leoben, Austria
[4]Christian Doppler Laboratory for Advanced Ferroic Oxides, Materials Engineering Research institute, Sheffield Hallem University, Howard Street Sheffield S1 1WB, United Kingdom
*Author to whom all correspondence should be addressed (denis.schuetz@tugraz.at).

One of the primary challenges today within the field of piezoelectric ceramics is the necessity of replacing Lead with alternative (non-toxic) elements. Since the combination of valence, ionic radius, polarizability and its electronic structure is unique, the replacement(s) for PZT will most likely not be able to include all of the varied properties that PZT possesses. Bismuth sodium titanate derived ceramics are emerging as the prime contender to replace PZT in actuator. The reasons and mechanisms leading to the extraordinary strain under high fields are poorly understood, especially at its physico-chemical origins[1]. The temperature dependent properties are also puzzling, presenting an intermediate transition, the depolarization temperature (Td), above which no low-signal ferroelectricity is visible but a gradual transition to relaxor ferroelectric behavior is observed. Crystallographically multiple symmetries have been reported for the same nominal compositions. Multiple symmetries have been found to exist within the same sample or even the same grain without macroscopic separation or noticeable chemical differences[1].

Recently we manufactured a prototype consisting of 50 layers (35μm) of a BNT-BKT solid solution using industry standard processes (water as a solvent; Ag/Pd as an electrode) exhibiting temperature stable properties, and strains (at 200V; 7 kV/mm) exceeding those of state of the art PZT actuators (0,22%).[2] The low thickness of the ceramic layers allows us to safely apply high electric fields on the stacks within a Raman spectrometer, and by this route not only acquire temperature dependent spectra but also field dependent ones. [3]

We propose here a comprehensive explanation combining the short-range chemical and structural sensitivity of in situ Raman spectroscopy (under applied electric field and temperature) with macroscopic electrical measurements. Our results clarify the causes for extended strain as well as the peculiar temperature-dependent properties encountered in this system. Special attention was given to the A-O vibrations below 100 cm^{-1} which are indicative of the bonding between A-site and oxygen. The underlying cause is determined to be mediated by the complex-like bonding of the octahedra at the center of the perovskite. Namely, a loss of hybridization of the 6s^2 Bismuth lone pair interacting with oxygen p-orbitals occurs, which triggers both the field-induced phase transition and the loss of macroscopic ferroelectric order at the depolarization temperature. Additionally it serves as an explanation of the unstable room temperature phases which are close in cell constants but differ in symmetry[3]. The details of the field dependent behavior of the polar nanoregions which are argued to be the reason for the development of relaxor properties under an applied field and temperature will also be shown.

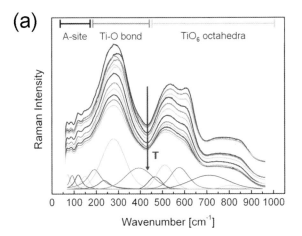

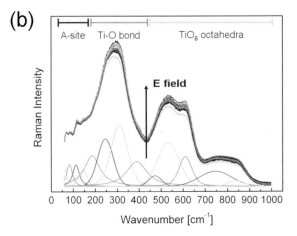

Figure 1: Depolarized Raman spectrum of (Li, Nd)-doped BNT-BKT in function of (a) temperature and (b) applied electric field. The spectrum is deconvoluted according to 13 Gaussian-Lorentzian peak functions, and the fitting is displayed for high values of both temperature and field. The assignment of spectral modes to particular lattice vibrations is also indicated.

[1] Schmitt, L. A. *et al.* Structural investigations on lead-free Bi1/2Na1/2TiO3-based piezoceramics. *Journal of Materials Science* **46**, 4368-4376, doi:10.1007/s10853-011-5427-6 (2011).
[2] Krauss, W., Schütz, D., Naderer, M., Orosel, D. & Reichmann, K. BNT-based multilayer device with large and temperature independent strain made by a water-based preparation process. *Journal of the European Ceramic Society* **31**, 1857-1860, doi:10.1016/j.jeurceramsoc.2011.02.032 (2011).
[3] Denis Schütz, Marco Deluca, Werner Krauss, Antonio Feteira, Tim Jackson, Klaus Reichmann *Advanced Functional Materials*; Article first published online: 13 MAR 2012
DOI: 10.1002/adfm.201102758

INTERACTION OF POINT DEFECTS AND FERROELECTRIC POLARIZATION IN A LEAD-FREE PIEZOELECTRIC MATERIAL

Sabine Körbel[1], **Christian Elsässer**[1,2]

[1]Fraunhofer Institute for Mechanics of Materials IWM, Wöhlerstraße 11, 79108 Freiburg, Germany;
[2]Karlsruhe Institute of Technology (KIT), Institute of Applied Materials (IAM), 76131 Karlsruhe, Germany

The ferroelectric perovskite-type compound $KNbO_3$ is an end member of the solid-solution system $(K,Na)NbO_3$ (KNN), which is a possible lead-free substitute for today's standard material $Pb(Zr,Ti)O_3$ (PZT) in piezoelectric applications. The piezoelectric properties of PZT and KNN can be optimized by doping. Depending on the application, ferroelectric hardness or large piezoelectric strain is desired. Replacing part of the A or B element in the ABO_3 perovskite structure by a different element can result in a morphotropic phase transition, at which the piezoelectric strain can increase. Defect complexes consisting of aliovalent dopants and charge-compensating oxygen vacancies can impede domain wall motion and contribute to ferroelectric hardness. In aged ferroelectrics containing defect complexes extraordinarily large piezoelectric strains were observed [1]. Combining density-functional theory and classical empirical interatomic potentials like in [2], we studied isolated Cu substitutionals and defect complexes consisting of Cu substitutionals and oxygen vacancies in $KNbO_3$. In this way we found a morphotropic phase transition in Cu-doped $KNbO_3$ [2] similar to the one in Li-doped $(K,Na)NbO_3$ [3] and obtained the energy needed for switching the ferroelectric polarization in a crystal region with a defect complex. This energy determines whether the defect complex can pin domain walls and contribute to ferroelectric hardness and/or large piezoelectric strain.

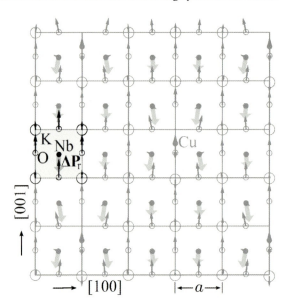

Figure 1: Ionic displacements and dipole moments in a $KNbO_3$ supercell where some of the K atoms are replaced by Cu atoms [2].

[1] X. Ren, Nature Materials **3**, 91 (2004).
[2] S. Körbel and C. Elsässer, Phys. Rev. B **84**, 014109 (2011).
[3] Y. Guo, K. Kakimoto, and H. Ohsato, Appl. Phys. Lett. **85**, 4121 (2004).

ATOMIC LAYER DEPOSITION FOR MICROELECTRONIC DEVICES

Cheol Seong Hwang[1]*

[1] Department of Materials Science & Engineering and Inter-university Semiconductor Research Center, Seoul National University, Seoul 151-744, Korea

The enormous improvement in information technology, accomplished mostly by the better computers, has been one of the most important impetuses that have driven the development of modern civilization over the last ~ 50 years. Improving computers has been accomplished by scaling of semiconductor devices. As the scaling of semiconductor devices proceeds, the demands for a thin film deposition with low thermal budget, higher accuracy in thickness control, better conformality over three dimensional (3D) structures are increased. This is mainly due to the highly scaled metal oxide field effect transistors (MOSFET), high-k gate dielectric/metal gate technology, and capacitors for dynamic random access memory (DRAM). The atomic layer deposition (ALD) often adopts a lower process temperature compared to the chemical vapor deposition in an atomically controlled manner making it a more desirable method for the low thermal budget and atomic thickness control, which are the key ingredients of thin film deposition process for future microelectronic devices.

There are several other emerging memories which will subsidize the present mass-production memories (DRAM and Flash) in near future. These emerging memories include the phase change RAM (PcRAM), ferroelectric RAM (FeRAM), magnetic RAM (MRAM, or spin transfer torque RAM (STTRAM)), and resistance switching RAM (ReRAM). Although any of these emerging memories may not replace the present mass-production memories, they will find their position in the wide spectrum of memory hierarchy for modern computer systems soon. As these devices will commonly adopt design rule < 30 nm, they will also need ALD processes in their fabrication.

Certainly, there are very diverse logic devices ranging from the state-of-the-art multi-core main processors to fairly simple microprocessors. However, all those logic chips are combinations of high performance transistors without special capacitors or junction devices involved. Therefore, from the ALD point of view, the logic can be treated as a device composed of the front-end and back-end of the line steps. In addition, great attention has been paid recently in the adoption of new substrates with high-mobility, such as Ge, compound semiconductors (GaAs and InP-based materials), or even carbon-based materials. These substrates have very distinctive physical and chemical interactions with the growing high-k films.

On the other hand, the ALD layers work not only as the functional materials in the final products but also as the sacrificial layers that enhance several process capabilities. For example, the double patterning process is a typical example of sacrificial ALD, which largely improves the photolithographic capability without astronomical cost.

In this talk, the author will briefly review the diverse application fields of ALD for the modern semiconductor devices, present status, and near-/long-term prospects.

ATOMIC LAYER DEPOSITION OF SrTiO₃ FILMS WITH Cp-BASED PRECURSORS FOR DRAM CAPACITORS

Woongkyu Lee[1], Jeong Hwan Han[1], Woojin Jeon[1], Yeon Woo Yoo[1], Changhee Ko[2], Julien Gatineau[2] and Cheol Seong Hwang[1*]

[1]WCU Hybrid Materials Program, Department of Materials Science and Engineering and Inter-university Semiconductor Research Center, Seoul National University, Seoul 151-744, Korea; [2]Air Liquide, 28, Wadai, Tsukuba-Shi, Ibaraki Pref., 300-4247, Japan *cheolsh@snu.ac.kr

Currently, the dielectric material used in mass production of dynamic random access memory (DRAM) capacitor is ZrO_2 with an interposed Al_2O_3 layer.[1] However, in order for a further scaling of DRAM down to the design rule of <20nm, application of materials with higher dielectric constants is necessary. Strontium titanate ($SrTiO_3$, STO) is one of the promising next generation dielectric materials for DRAM capacitor. Although numerous studies on atomic layer deposition of STO films with its excellent conformality and promising electrical property have been reported, there still remains a difficulty that arise during the STO deposition such as the formation of $SrCO_3$ due to the violent reactivity of Cp-based Sr precursor with Ru(O) substrate. To overcome this problem, a TiO_2 layer has been interposed as a barrier layer to suppress the unwanted growth of Sr-rich STO films.[2]

In this study, $Sr(iPr_3Cp)_2$ and $Cp^*Ti(OMe)_3$ ($Cp^*= C_5(CH_3)_5$) were employed as the Sr- and Ti-precursor, respectively, for ALD of STO films. Since both Sr-precursor and Ti-precursor have strong reactivity with the Ru(O) substrate, the composition of STO could be controlled stoichiometric without the formation of Sr-excess layer or $SrCO_3$ through the all thicknesses even without the TiO_2 barrier layer, only by changing the sub-cycle ratio of $SrO:TiO_2$ (Left figure below). By this approach an equivalent oxide thickness of as small as 4.3Å was achieved with an acceptable leakage current density (right figure below).

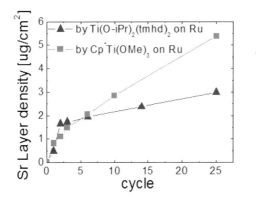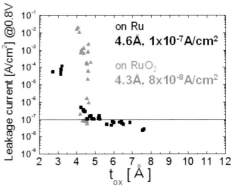

Figure (left) Comparison of Sr layer density as a function of STO cycle with different Ti precursors; (right) Leakage current density vs. equivalent oxide thickness plot of STO films

[1] Seong Keun Kim et al., Adv. Fuct. Mater. 20, 2989 (2010).
[2] Sang Woon Lee et al., Chem. Mater. 23, 2227 (2011).

DEPOSITION OF INNOVATIVE MATERIALS BY GAS PHASE TECHNOLOGIES FOR THE SEMICONDUCTOR INDUSTRY

B. Gouat[1], U. Weber[1], Peter K. Baumann[1], Michael Heuken[1], and B. Lu[2]

[1]AIXTRON SE, Herzogenrath, Germany; [2]AIXTRON Inc., Sunnyvale, CA, USA

AIXTRON is a leading provider of gas phase deposition equipment for the semiconductor industry. Those equipments enable the building of advanced components for electronic and opto-electronic applications based on compound, silicon, or organic semiconductor materials and more recently carbon nanotubes (CNT), graphene and other nanomaterials. Such components are used in display technology, signal and lighting technology, fiber communication networks, wireless and cell telephony applications, optical and electronic data storage, computer technology as well as a wide range of other high-tech applications.

Semiconductor device cell design changes and shrinking became the major factors for every new generation of highly integrated commodity logic or memory devices during the past decades. Technology-wise, this development was significantly enabled by the thickness scaling of the functional film layers applied in these components. As outlined in the International Technology Roadmap for Semiconductors (ITRS), a broad range of alternative materials have to be introduced and integrated for current and future device nodes and further advances will be necessary to allow continued scaling of nanoelectronic device technologies. This includes logic, capacitive and memory applications. The deposition of such material systems with high uniformity, high purity, precisely controlled impurities as well as well defined interfaces sets high standards for possible enabling deposition technologies.

In this study high-k, metal nitride and phase change materials have been deposited by AVD® (atomic vapor deposition) and ALD (atomic layer deposition). AVD® is a pulsed MOCVD (metal organic vapor deposition) method combining basic operation of conventional MOCVD and ALD processes. For our AVD® and ALD processes we have used a vaporizer with several independent injectors for pulsed direct liquid injection of various metal-organic precursors. However for conventional MOCVD and ALD typically heated bubblers are used for the precursors. This causes precursor deterioration and a change in deposition quality over time. In comparison in our study the precursors are stored at room temperature right until injection into the vaporizer. High precursor gas phase saturation is achieved by precise pulse dosing and flash evaporation. This allows deposition of multi-component materials with high throughput and precise composition control even for metal-organic precursors with low volatility, low thermal stability and instability in air.

High-k dielectrics are widely used in DRAMs and MIM structures for memory and communication applications. For those applications, high-k values and low leakage currents are required. Perovskites, and especially Barium Strontium Titanate (BST), are excellent candidates for high-k materials. However, there is some challenge in low-temperature deposition needed for integration purposes as well as in combining high-k value and low leakage. Another issue is the control of the interfaces between the electrode and the high-k oxide. BST was deposited by AVD® at a low-temperature of 400°C on various electrodes to study the influence of the substrate on the growth and morphology. The film was partially crystalline as deposited with the presence of many crystallites on the surface. The crystallites were grown from nucleation centers to form domes in the layer and hemispheres on the surface. Thus the crystallites increased the roughness of the surface. Film roughness could be decreased by reducing the number of nucleation centers. After annealing the layer at 625°C in argon, the films were completely crystalline. Compared to as-deposited layers, the annealed film showed increased dielectric constants up to 200 and increased leakage current by several orders of magnitude. Further experiments showed that the leakage could be reduced by adding other doping elements to the high-k layer.

Another example is the deposition of germanium antimony tellure (GST) materials for phase-change applications. An important issue is the control of the composition along with the control of the morphology. The composition of the deposited GST films was determined by XRF and could be adjusted in a wide range of suitable choices including $Ge_2Sb_2Te_5$. The morphology was found to be strongly dependent on the composition and on the electrode substrate. GST materials were successfully deposited in trench structures for future device applications.

Metal nitrides like tantalum nitride (TaN) were also deposited for electrode applications with state-of-the-art properties.

ATOMIC LAYER DEPOSITION OF TRANSITION METAL OXIDE THIN FILMS FOR RESISTIVE MEMORY APPLICATIONS

S. Hoffmann-Eifert[1], M. Reiners[1], N. Aslam[1], I. Kärkkänen[1], J. H. Kim[1] and R. Waser[1]

[1]Peter Grünberg Institut (PGI-7), Forschungszentrum Jülich and JARA-FIT, 52425 Jülich, Germany

The resistance switching (RS) effect was re-discovered about a decade ago and since then it emphasis the promise for the realization of an alternative type of fast and non-volatile memory [1]. What it makes so special is the fact, that the materials can adopt at least two different resistance states - a High Resistive State (HRS) and a Low Resistive State (LRS), which can be interpreted as binary information representing a logical '0' or '1'. Because the change of the resistance state is combined with a change in material structure, it is a non-volatile effect, which means, that the memory cell keeps the information even if the power is turned off. The logic states can be toggled by exceeding a threshold voltage, whereby, depending on the material system, this can be either a unipolar or a bipolar switching mechanism with different underlying physical mechanisms [2]. With increasing integration density the lateral area of the MIM device is decreasing down to 1000 - 100 nm^2. The reduction of the operation voltage affords a reduction of the thickness of the switching layer down to a few to 10 nm. For stable RS device operation, the extremely thin transition metal oxide films have to exhibit a homogeneous microstructure, either amorphous or nano crystalline, with a controlled defect density. In addition, the films may have to be deposited conformal over structured bottom electrodes. Therefore a deposition technique is needed which supports conformal and dense film growth. Atomic layer deposition (ALD) fulfills these requirements by its unique surface-reaction controlled self-limiting growth behavior supporting an excellent control of the thin film thickness and composition even for complex geometries.

The talk will comprise recent results on ALD transition metal oxide thin films integrated into cross point MIM structures for future resistive switching applications.

The functional oxide layers were sandwiched between Pt bottom and Ti/Pt top electrodes in a nano crosspoint configuration. In a first step, the Pt bottom electrode was deposited on a Si/SiO$_2$/TiO$_2$ substrate using sputter deposition in combination with nano imprint lithography (NIL) and plasma etching (see Fig. 1) [3]. Atomic layer deposition was utilized for a conformal growth of the oxide layer with thickness between 8 and 25 nm over the bottom electrode. Finally, the top electrodes were fabricated in a perpendicular arrangement with a two layer electron beam lithography process, electron beam evaporation of Ti/Pt and lift-off technique [4].

For the ALD TiO$_2$ process a liquid delivery injection technique was chosen which also enabled the use of a solid Ti-precursor with high thermal stability. Different ALD TiO$_2$ processes were utilized using different alkoxide and amide based Ti sources. Nanocrystalline TiO$_2$ films with anatase-type structure down to a film thickness of 8 nm were obtained at a growth temperature of about 365°C using Ti(O-iPr)$_2$(tmhd)$_2$ and H$_2$O.[5] ALD TiO$_2$ films from Ti(O-iPr)$_4$ and H$_2$O show a transition in growth behavior related to the deposition of either 'amorphous' films or films with a nano crystalline structure.[6] Dense ALD TiO$_2$ films conformal covering three dimensional device structures were also obtained from Ti(N(Me)$_2$)$_4$ and H$_2$O.

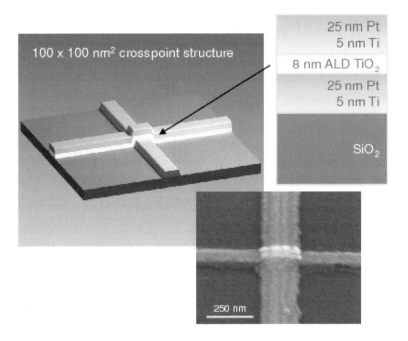

Fig. 1: Schematic and layer structure of a single crosspoint test structure. Each test device consists of a Si/SiO$_2$/Ti/Pt/TiO$_2$/Ti/Pt stack. The SEM picture shows the top view of a single 100x100 nm² crosspoint device.

For the cross point structures the thickness of the TiO$_2$ layer could be reduced to about 8 nm without an increase in the device current due to leakage effects. [7]

In addition, results of ALD ZrO$_2$ thin films obtained from an amide type precursor in combination with ozone or water and selected studies on ALD Nb$_2$O$_5$ and Ta$_2$O$_5$ thin films will be subsumed. Results on the effect of the film material, thickness, structure and annealing conditions on the resistive switching behavior of the above mentioned cross point cells will be discussed in an exemplary manner.

The examples highlight the important role of atomic layer deposition for the concept of RRAM devices which requires the integration of dense oxide films with thickness in the nm-regime onto metal layers which already exhibit a certain surface roughness. In addition, the ALD technique enables a control of the film defect structure even for thicknesses in the nm-regime and for low growth temperatures.

This work was founded in part by the Deutsche Forschungs Gemeinschaft (DFG, HO2480/2-1) and by the European Community's Seventh Framework Programme (FP7/2007-2013) under grant agreement number ENHANCE-238409.

[1] A. Beck et. al, Appl. Phys. Lett. **77**, 139 (2000).
[2] R. Waser et.al., Adv. Mat. 21, 2632 (2009).
[3] C. Kügeler et. al, Solid State Elec., **53**, 1287 (2009).
[4] C. Nauenheim et. al., Appl. Phys. Lett. **96**, 122902 (2010).
[5] S. K. Kim, S. Hoffmann-Eifert, M. Reiners, R. Waser, J. Electrochemical Soc. **158**, D6 (2011).
[6] S. K. Kim S. Hoffmann-Eifert, R. Waser, Electrochem. and Solid-State Lett. **14**, H146 (2011).
[7] C. Kuegeler, J. Zhang, S. Hoffmann-Eifert, S. K. Kim, R. Waser, J. Vac. Sci. Technol. B **29**, 01AD01 (2011).

ALD PROCESS CONTROL FOR TAILORING THE NANOSTRUCTURE OF TiO$_2$ FILMS FOR RESISTIVE SWITCHING APPLICATIONS

M. Reiners[1], N. Aslam[1], S. Hoffmann-Eifert[1], R. Waser[1]

[1]Peter Grünberg-Institut and JARA-FIT, Forschungszentrum Jülich, 52425 Jülich, Germany

The fast growing demand for non volatile memory with high speed data access and good reliability pushes the research on new memory concepts. One promising concept is the resistive random access memory (RRAM). It relies on the change of the resistance of a functional oxide layer sandwiched between two metallic conducting electrodes by applying voltage pulses. For the resistive switching oxide layer TiO$_2$ is one of the most auspicious materials. The working principle is based on the valence change mechanism (VCM) where the reversible change between Ti^{4+} and Ti^{3+} states leads to the high resistance state and low resistance state, respectively, within a so called conducting filament (CF). One drawback for the integration into memory architectures is the initially necessary forming step to generate this CF. For circumventing the forming it has been reported that an downscaling of the TiO$_2$ thickness is desirable [1]. But this can be accompanied by an insufficient coverage of the devices which ends in shorting the functional layer. Here we utilized a well controlled atomic layer deposition (ALD) process to grow highly dense films with an excellent control of the thickness. For the TiO$_2$ growth we applied a liquid injection type ALD method. The deployed Ti source was tetra-dimethylamine-titanium (TDMAT) in combination with water vapor as oxygen source. Growth experiments were conducted at different temperatures with different thicknesses to analyze the growth behavior of TiO$_2$ as well the resulting thin film properties. The thickness and the Ti mass layer content investigation were performed by x-ray reflectance and x-ray fluorescence spectroscopy. The ALD growth saturation is shown in figure 1 for temperatures ranging from 150 °C up to 300 °C with growth rates of 0.040 nm/cycle to 0.054 nm/cycle. The ALD growth mode was demonstrated by the excellent step coverage of the films grown onto deep pinhole silicon structures with an aspect ratio of 1:32 as depicted in figure 2. Grazing incidence x-ray diffraction revealed the formation of anatase, rutile and brookite type phases of TiO$_2$ depending on the thickness as well as on the growth temperature of the thin films. Films which were grown at 200 °C and below did not show any diffraction peaks. A low roughness is essential for further integration of the films. From atomic force microscopy analysis RMS values below 0.7 nm were determined for all ALD TiO$_2$ films except those grown at 300 °C. Thin films processed at 300 °C showed a linear dependency of the roughness on the thickness. Nevertheless below about 15 nm the roughness was lower than at maximum 1 nm. The observation of surface objects revealed polycrystalline structures at a thin film thickness of 25 nm. Films deposited at 300 °C showed solely nanocrystallites from early stages of growth. The residual carbon and nitrogen content was analyzed by x-ray photo emission spectroscopy with no more residual content than 2 %. TiO$_2$ thin films with different thicknesses and from different growth temperatures were integrated into micro- as well nano structured cross point devices. As bottom electrode Pt was used while the top electrode consists of a Ti|Pt stack. The resistive switching properties were analyzed by I-V sweeps to examine the dependence on the thicknesses as well on the nanostructure. Low current compliances for the setting of the ON state were achieved as depicted in figure 3.

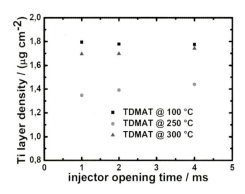

Figure 1: Ti mass layer concentration vs. precursor injector opening time: For temperatures ranging from 100 °C up to 300 °C the deposition of TiO_2 is saturated. Interestingly the growth shows a minimum at 250 °C.

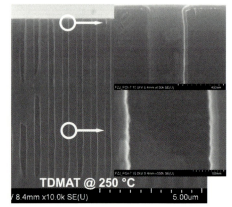

Figure 2: Scanning electron microscopy cross section view of a miro pinhole test structure. The ALD TiO_2 thin film grown by TDMAT and H_2O is visible as the bright layer. The insets show the sidewalls at different vertical positions at higher magnification. The TiO_2 thin film is conformal within the hole with a thickness of 30 nm.

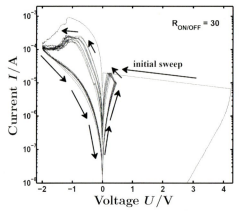

Figure 3: IV characteristics of an 8 nm quasi-amorphous TiO_2 thin film grown by ALD. It is integrated into a nano cross point structure. The stack consists of a 30 nm thick Pt bottom electrode, the functional TiO_2 layer and 10 nm Ti top electrode, covered by Pt. The device needs a SET current of 20 µA with an average voltage of around 0.5 V. For the RESET operation a voltage of -2.0 V is necessary.

[1] J. J. Yang, F. Miao, M. D. Pickett, D. A. A. Ohlberg, D.R. Stewart, C. N. Lau, and R. S. Williams, Nanotechnology **20**, 215201 (2009).

This work was founded in part by the Deutsche Forschungs Gemeinschaft (DFG, HO2480/2-1) and by the European Community's Seventh Framework Programme (FP7/2007-2013) under grant agreement number ENHANCE-238409.

INVESTIGATION OF ATOMIC LAYER DEPOSITION PROPERTIES OF $(GeTe_2)_{1-x}(Sb_2Te_3)_x$ PSEUDO-BINARY COMPOUND FOR PHASE CHANGE MEMORY APPLICATION

Taeyong Eom[1], Taehong Gwon[1], Si Jung Yoo[1], Moo-Sung Kim[2], Manchao Xiao[3], Iain Buchanan[3], and Cheol Seong Hwang[1*]

[1]Department of Materials Science & Engineering and Inter-university Semiconductor Research Center, Seoul National University, Seoul 151-744, Republic of Korea.
[2]Air Products Korea, 15 Nongseo-dong, Giheung-gu, Yongin-si, Gyeonggi-do, 446-920, Republic of Korea.
[3]Air Products and Chemicals, Inc., 1969 Palomar Oaks Way, Carlsbad, CA 92011, USA

Phase change random access memory (PCRAM) is one of the most probable next generation non-volatile memory devices. [1] However, a high current level for the transition from the crystalline to the amorphous state has been an obstacle for the scaling down of PCRAM. The modified cell structure that phase changing materials is confined in the hole plug, can significantly reduce the current level due to the large reduction of the heat dissipation. [2] For the fabrication of this type of the structure, forming the phase changing materials in the contact hole structure with an excellent conformality is indispensable. Therefore, the growth behavior and characteristics of Ge-Sb-Te films by an atomic layer deposition (ALD) method is investigated.

$GeTe_2$, Sb_2Te_3, and their solid solution films were grown at the temperatures between 50 and 250°C using $Ge(OC_2H_5)_4$, (or $Ge(OCH_3)_4$), $Sb(OC_2H_5)_3$, and $((CH_3)_3Si)_2Te$ precursors. The films were grown by the ligand exchange reaction between methyl-silyl ligands in $((CH_3)_3Si)_2Te$ and alkoxy ligands in $Ge(OC_2H_5)_4$, $Ge(OCH_3)_4$, and $Sb(OC_2H_5)_3$, according to following reaction route.

$$Ge(OC_2H_5)_4 + 2[(CH_3)_3Si]_2Te \rightarrow GeTe_2 + 4(CH_3)_3Si\text{-}OC_2H_5 \quad (1)$$

$$Ge(OCH_3)_4 + 2[(CH_3)_3Si]_2Te \rightarrow GeTe_2 + 4(CH_3)_3Si\text{-}OC_2H_5 \quad (2)$$

$$2Sb(OC_2H_5)_3 + 3[(CH_3)_3Si]_2Te \rightarrow Sb_2Te_3 + 6(CH_3)_3Si\text{-}OC_2H_5 \quad (3)$$

The growth of Sb and Te in Sb-Te film showed the ALD-specific saturation behavior. However, Ge in Ge-Te film was not saturated with the precursor purge processes, as shown in Fig. 1, suggesting that the chemical interaction between the Ge-precursors and substrate is generally quite weak. It was observed that the $Ge(OC_2H_5)_4$ precursor adsorbed chemically with quite weak chemical binding, and $Ge(OCH_3)_4$ precursors adsorbed physically on the substrate. The growth properties were substantially influenced by the composition of film or the thickness of sub-layer i.e.; Sb-Te sub-layer thickness for Ge-Te and vice versa.

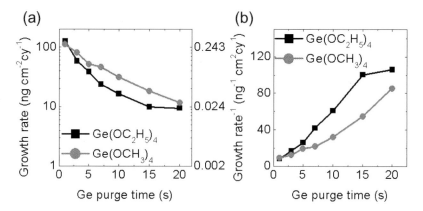

Figure 1: Growth rate change as a function of Ge precursor purge time. (a) Semi-log growth rate change graph. The Ge(OCH$_3$)$_4$ precursor show linear change, it means the Ge(OCH$_3$)$_4$ precursor desorption mechanism is physisorption. (b) Semi-reciprocal growth rate change graph. The Ge(OC$_2$H$_5$)$_4$ precursor show linear change, it means the Ge(OC$_2$H$_5$)$_4$ precursor desorption mechanism is chemisorption.

The conformal (GeTe$_2$)$_{0.66}$(Sb$_2$Te$_3$)$_{0.33}$ film was formed in the 62-nm-diameter, 390-nm-deep slightly tapered cylindrical hole. The energy dispersive spectroscopy composition analysis results along the line indicated in Fig. 2a shows that the (GeTe$_2$)$_{0.66}$(Sb$_2$Te$_3$)$_{0.33}$ layer has very uniform thickness and chemical composition along the depth direction in the hole

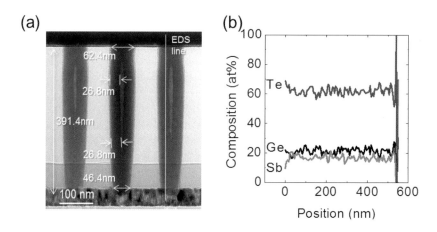

Figure 2: (a) Cross sectional image of (GeTe$_2$)$_{0.66}$(Sb$_2$Te$_3$)$_{0.33}$ film deposited hole structure. (b) Composition profile of Ge, Sb, and Te along the EDS line in (a).

[1] Ovshinsky, S. R., Phys. Rev. Lett, **21**, (20), 1450 (1968)
[2] Kim, Y.-T. et al., Jpn. J. Appl. Phys., **44**, (No. 4B), 2701 (2005)

DIRECT PATTERNING OF OXIDE INTERFACE WITH HIGH MOBILITY 2DEG WITHOUT PHYSICAL ETCHING

Nirupam Banerjee[1], Mark Huijben[1], Gertjan Koster[1] and Guus Rijnders[1]

[1]Faculty of Science & Technology and MESA+ Institute for Nanotechnology, University of Twente, P.O. Box 217, 7500 AE, Enschede, The Netherlands

Discovery of highly mobile two dimensional electron gas (2DEG) at the atomically engineered interface between two wide band-gap perovskite insulators, $SrTiO_3$ (STO) and $LaAlO_3$ (LAO) has stimulated the research to apply oxide materials in electronic devices such as high mobility electron transistors (HMET). In spite of excellent interfacial transport properties manifested, challenges remained in structuring these heterointerfaces without damaging the STO single crystal. Top-down physical etching process was an unsuitable choice to serve the purpose since it induces substrate conductivity through creation of oxygen vacancies.

Here, we will demonstrate development of a novel procedure for fabricating patterned functional interfaces based on epitaxial-lift-off technique. With its help devices incorporating patterned interfaces of LAO-STO was fabricated devoid of any physical etching process performed and temperature dependent magneto transport properties were investigated. The results demonstrated conservation of the high-quality interface properties in the patterned structures enabling future studies of low-dimensional confinement on high mobility interface conductivity as well as interfacial magnetism.

EUV INTERFERENCE LITHOGRAPHY WITH LABORATORY SOURCES

Serhiy Danylyuk[1], Larissa Juschkin[1], Sascha Brose[1], Hyun-Su Kim[1], Jürgen Moers[2], Peter Loosen[1], Detlev Grützmacher[2]

[1]Chair for the Technology of Optical Systems, RWTH Aachen University and JARA - Fundamentals of Future Information Technology (FIT), 52074 Aachen, Germany;
[2]Peter Grünberg Institute 9 (PGI-9): Semiconductor Nanoelectronics, Research Center Jülich and JARA-FIT, 52425 Jülich, Germany;

Recent progress in nanotechnology and a constant increase in requirements on nano-structures based devices in terms of integration density and energy efficiency is rapidly pushing feature sizes towards sub-10 nm dimensions. At the moment, high density structures on this scale are not only beyond what is possible with available mass production technology, but also not reachable with low-volume research oriented techniques, such as electron beam or ion beam writing [1]. Interference lithography (IL) with extreme ultraviolet (EUV) radiation is the most promising technique to achieve the sub-20 nm resolution for large and dense arrays of periodic nanostructures [2]. Thanks to the short wavelength (typically between 10 nm – 15 nm) structuring with EUV light on a scale of a few tens of nanometers is not limited by diffraction. And, in contrast to high energy techniques, proximity and charging effects are negligible due to the very strong interaction of EUV radiation with matter leading to short (<100 nm) absorption distances.

Furthermore, the combination of top-down pre-patterning via EUV-IL with bottom-up self-assembly is of special interest as it is not only an innovative research instrument but also highly suitable for large-scale production [3].

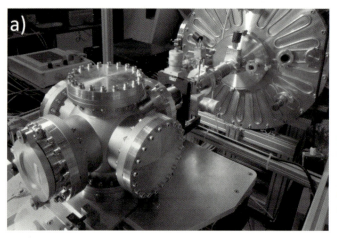

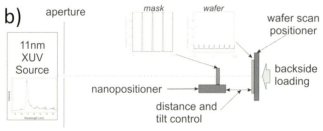

Figure 1: XUV-IL illumination set-up realised at RWTH-TOS with a wafer loading chamber and a source behind (a). The principal scheme of the XUV-IL nano-patterning system (b).

A laboratory EUV interference lithography setup optimized for partially coherent discharge produced plasma EUV light sources is presented (see Figure 1). Interference lithography with transmission gratings at a wavelength around 11 nm is used for producing nanoscale patterns in a photoresist. Source emission was optimized to achieve highest possible intensity within the necessary bandwidth [4]. Mask fabrication technology based on free standing thin Nb membranes with areas reaching 2000 x 2000 μm^2 is developed.

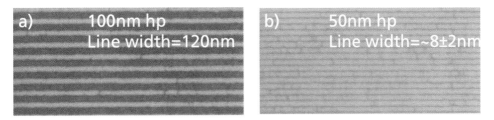

Figure 2: SEM images of line/space pattern produced by proximity printing (a) and Talbot interference lithography (b) with the same mask. The period reduction is demonstrated.

Analysis of the efficiency of possible interference schemes is highlighting advantages of an achromatic Talbot self-imaging approach [5, 6], making use of a spatially extended source with spectral bandwidth around 1 nm. Non monochromatic XUV light with a finite bandwidth creates a continuous self-image with half of the grating period. Analytical estimations are supported by wavefront interference simulations and numerical diffraction calculations. Spatial coherence requirement of illumination is limited to several tens of micrometers which is available from the laser or discharge based sources. The beam power is used very efficiently as all of the diffraction orders contribute to the exposure of the pattern. With this technology lines or dots as small as 10 nm with 10 nm spacing can be patterned thus approaching the ultimate limits of device scaling.

Test exposure results address a number of perspective applications of the technique including crossbar arrays manufacturing for phase change memory modules and substrate pre-patterning for self-organized growth of ordered quantum dots. The 2:1 reduction in periods of arrays from mask to wafer is demonstrated, confirming results of analytical and numerical calculations (see Figure 2). Single exposure areas with up to 4 mm^2 are achieved with feature sizes reaching down to 10 nm.

[1] C. Vieu et al., Appl. Surface Science **164**, 111–117 (2000)
[2] H.H. Solak et al., J. Vac. Sci. Technol. B **25** (1), 91-95 (2007)
[3] D. Grützmacher et al., Nano Lett. **7** (10), 3150–3156 (2007)
[4] K. Bergmann et al., J. Appl. Phys. **106**, 073309–073309-5 (2009)
[5] N. Guerineau at al., Optical Comm. **180**, 199 (2000)
[6] H. H. Solak et al., J. Vac. Sci. Technol. B **23** (6), 2705 – 2710, (2005)

HIGH EFFICIENCY TRANSMISSION MASKS FOR EUV INTERFERENCE LITHOGRAPHY

S. Brose[1], S. Danylyuk[1], L. Juschkin[1], K. Bergmann[2], J. Moers[3], G. Panaitov[3], S. Trellenkamp[4], P. Loosen[1], D. Grützmacher[3]

[1]Chair for the Technology of Optical Systems, RWTH Aachen University and JARA - Fundamentals of Future Information Technology (FIT), 52074 Aachen, Germany;
[2]Fraunhofer Institute for Laser Technology (ILT), 52074 Aachen, Germany;
[3]Peter Grünberg Institute 9 (PGI-9): Semiconductor Nanoelectronics, Forschungszentrum Jülich and JARA-FIT, 52425 Jülich, Germany;
[4]Peter Grünberg Institute 8 (PGI-8): Bioelectronics, Forschungszentrum Jülich and JARA-FIT, 52425 Jülich, Germany

For the second generation of extreme ultraviolet (EUV) lithography at wavelengths below the Si L-absorption edge (12.4 nm) conventional silicon nitride based transmission masks are no longer suitable due to high absorption. To overcome this limitation we developed a new fabrication process of high efficiency transmission masks [1]. In this design 100 nm – 300 nm thin niobium films are used as a support membrane that additionally serves as a build-in filter for radiation above 16 nm. The absorbing part of the masks is comprised from 80 nm thick nickel layer, which is structured with periodic patterns by means of e-beam lithography and ion beam etching. The transmission masks are equipped with a set of fiducial markers allowing implementation in process chains of different applications (see Figure 1).

Figure 1: Scanning electron microscopy image of mask layout including fiducial markers. The field size is 1000 x 1000 µm² (scale bar = 200 µm).

High quality free-standing masks with areas up to 1000 x 1000 µm² are realized with mask periods reaching 100 nm. An example of lines and spaces mask with 150 nm period is shown in Figure 2. Good uniformity of the pattern was achieved over the whole mask area.

The performance of the masks was characterized in the EUV interference lithography setup constructed at the RWTH Aachen [2]. The setup utilizes a xenon gas discharge produced plasma EUV source with 3.2% bandwidth at 11 nm wavelength [3]. Investigated masks were fixed in a mask holder designed to control the distance between a mask and a wafer with sub-10 nm resolution. This allows for both proximity lithography of the mask's structures as well as for interference lithography with spatial frequency multiplication. The patterns with down to 100 nm periods were successfully reproduced with minimal feature sizes reaching sub-10 nm scale.

The patterning by this method with 1000 x 1000 µm² mask offers ten times higher throughput than a comparable e-beam lithography process.

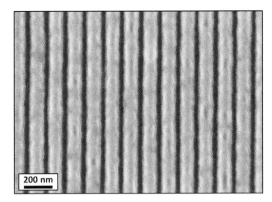

Figure 2: Scanning electron microscopy image of 150 nm period transmission mask with 35 nm gap size and nickel absorber layer thickness of 80 nm (scale bar = 200 nm).

Further improvements of the fabrication process should allow mask periods below 50 nm. Main limitation currently is caused by the limited resolution of the absorber pattern transfer process, thus the implementation of nickel electroplating is currently under investigation. Another alternative might be the creation of pure Nb-based phase-shift masks without an additional absorber layer.

[1] S. Brose et al., Thin Solid Films (accepted) (2012).
[2] S. Danylyuk et al., JARA-FIT annual report (2008).
[3] K. Bergmann et al., J. Appl. Phys. **106**, 073309 (2009).

SELF-PATTERNED ABO₃(001) SUBSTRATES: A PLAYGROUND FOR FUNCTIONAL NANOSTRUCTURES

Romain Bachelet, Florencio Sánchez, Carmen Ocal, Josep Fontcuberta

Institut de Ciencia de Materials de Barcelona (ICMAB-CSIC), Campus UAB, 08193 Bellaterra, Spain

Cost-effective fabrication of ordered nanostructures and novel devices with increased functionalities are nowadays required. These last years, the chemical terminations (CT) of functional $ABO_3(001)$ perovskites (i.e. AO and BO_2) have appeared critical to control interfaces (properties) of heterostructures. More recently, the possibility to self-assemble at the nanoscale the CT with lateral order has paved the way towards nanostructure fabrication by selective growth exploiting the CT-dependent interface energy [1-4]. Figure 1 is an illustrative example using a self-patterned LSAT single-crystal. We will show here that several atomically-flat $ABO_3(001)$ substrates with self-ordered (charged and neutral) CT can be used as template to realize distinct nanostructures of epitaxial oxides [1-2] and hybrid organic-inorganic [4].

The epitaxial oxide nanostructures have been grown by pulsed-laser deposition monitored by *in-situ* reflection high-energy electron diffraction and all the nanostructures have been characterized mainly by *ex-situ* atomic force microscopy (AFM) techniques (amplitude-modulation, friction, and conductive modes).

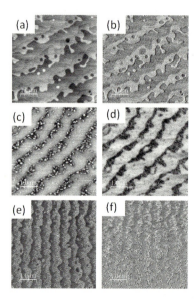

Figure 1. Atomic force micrographs of (a-b) self-patterned chemical-terminations of LSAT(001) substrates after thermal treatment at 1300°C [2], (c-d) after selective growth of $SrRuO_3$ dots, and (e-f) after selective adsorption of organic matter from ambient atmosphere. (a,c,e) Topographic images and (b,d,f) phase-lag images in amplitude-modulation AFM mode.

Particularly, we will show that the morphology of the deposited ordered-arrays of oxide can be controlled by the epitaxial strain. Selected results will be presented as the fabrication of one-dimensional array of dots (e.g. $SrRuO_3$/LSAT, cf. figure 1 c-d) [2] and of conducting nanostripes (e.g. $SrRuO_3$/$SrTiO_3$) [1].

We will then show that these self-patterned substrates can also provide selective adsorption of inorganic and organic matter (water, molecules, etc..., cf. figure 1 e-f) [4]. These results point-out the critical importance of CT of functional perovskites for sensing applications in micro-fluidic environment.

[1] R. Bachelet, F. Sánchez, J. Santiso, C. Munuera, C. Ocal, and J. Fontcuberta, Chem. Mater. 21, 2494 (2009).
[2] R. Bachelet, C. Ocal, L. Garzón, J. Fontcuberta, and F. Sánchez, Appl. Phys. Lett. 99, 051914 (2011).
[3] J. E. Kleibeuker, G. Koster, W. Siemons, D. Dubbink, B. Kuiper, J. L. Blok, C.-H. Yang, J. Ravichandran, R. Ramesh, J. E. ten Elshof, D. H. A. Blank, and G. Rijnders, Adv. Funct. Mat. 20, 3490 (2010)
[4] M. Paradinas, L. Garzón, F. Sánchez, R. Bachelet, D. B. Amabilino, J. Fontcuberta, and C. Ocal, Phys. Chem. Chem. Phys. 12, 4452 (2010)

ANOMALOUS GAS SENSING CHARACTERISTICS OF EMBEDDED AND ISOLATED MAGNESIUM ZINC FERRITE NANO-TUBES

S.B. Majumder[1], K. Mukherjee[1], A. Maity[1], S. Basu[2], C. Lang[3], and M. Topic[4]

[1] Materials Science Centre, Indian Institute of Technology, Kharagpur, India;
[2] IC Design & Fabrication Centre, Dept. of Electronics & Telecom. Engg. Jadavpur University, Kolkata 700032, India;
[3] Centre for Materials Engineering, University of Cape Town, Private Bag, Rondebosch 7701 South Africa;
[4] Materials Research Department, iThemba LABS, Somerset West 7129, South Africa

We have investigated the gas sensing characteristics of wet chemical synthesized $Mg_{0.5}Zn_{0.5}Fe_2O_4$ (MZFO) nano-tubes embedded into porous alumina template and their isolated counterpart coated on quartz substrate. Figure 1(a) shows the electron microscopy image of synthesized MZFO tubes embedded in alumina template. The electron microscopy image of entangled MZFO tubes and the enlarged view of hollow MZFO tubes (coated on quartz substrate) are shown in Fig. 1 (b) and (c) respectively. As shown in the Fig. 1(c) each single nano-tube is composed with innumerous tiny crystallites. The indexing of the selected area electron diffraction (SAED) pattern (Fig. 1(d)) confirms the cubic spinel structure of the synthesized MZFO nano-tubes.

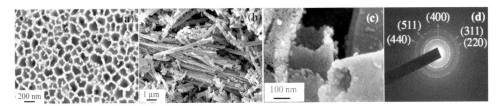

Figure 1. FESEM image of $Mg_{0.5}Zn_{0.5}Fe_2O_4$ nano-tubes (a) embedded into porous alumina template (b) coated on the quartz substrate. (c) Magnified FESEM image of the isolated $Mg_{0.5}Zn_{0.5}Fe_2O_4$ nano-tubes (d) SAED pattern (acquired during transmission electron microscopy imaging) of the $Mg_{0.5}Zn_{0.5}Fe_2O_4$ nano-tube.

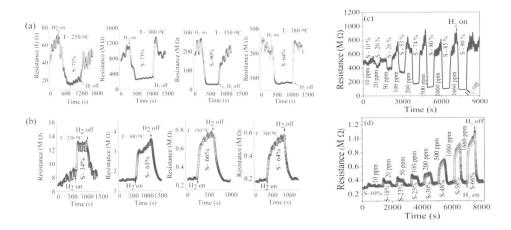

Figure. 2. Resistance transients of (a) embedded and (b) isolated $Mg_{0.5}Zn_{0.5}Fe_2O_4$ nano-tubes for the detection of H_2 (~ 1660 ppm) gas measured by varying the sensor operating temperature. Resistance transients of (c) embedded and (d) isolated $Mg_{0.5}Zn_{0.5}Fe_2O_4$ nano-tubes for the detection of various concentrations (10-1660 ppm) of H_2 gas measured at ~350°C. The respective response (S) (%) and H_2 on-off points are marked in the figure.

The resistance transients of both the embedded and isolated MZFO tubes are recorded by varying the sensor operating temperature and test gas (H_2, CO, N_2O) concentration using a dynamic flow gas sensing set-up. For H_2 sensing, Fig. 2 (a) and (b) show the respective response and recovery transients of embedded and isolated MZFO tubes by varying the sensor operating temperature (250-380 °C). At the optimized operating temperature (T_{opt} ~350 °C), Fig. 2 (c) and (d) exhibit the response and recovery transients of embedded and isolated MZFO nano-tubes for sensing H_2 gas in a wide concentration range (10-1660 ppm). Both the embedded and isolated MZFO sensing elements are capable to detect 10 ppm H_2 gas. Reviewing Fig. 2, it is interesting to note that irrespective of the operating temperature (250-380°C) and H_2 gas concentration (10-1660 ppm) the embedded tubes exhibit 'n' type (i.e. resistance decreases when the sensing elements are exposed to reducing gas) whereas isolated nano-tubes exhibit 'p' type (i.e. resistance increases when the sensing elements are exposed to reducing gas) gas sensing characteristics. Similar phenomenon has also been observed when the resistance transients for embedded and isolated tubes are recorded in presence of other test gases (e.g. CO, N_2O). Such inversion of dominant charge carriers (from n to p type in isolated MZFO tubes) are related to the higher chemi-adsorption of oxygen content over isolated tubes during their ageing in air prior to the exposure in test gas. When the embedded and isolated tubes are exposed in air prior to the exposure of test gas, the embedded tubes are depleted electronically from the inner surfaces only whereas the isolated tubes are depleted from both inner as well as outer surfaces leading to the higher chemi-adsorbed oxygen contents over isolated tubes. For a semiconducting oxide, the nature of conduction behavior (n or p type) is determined by the availability of free electrons in the lattice and usually modulated by their intrinsic oxygen vacancies. At elevated temperature, the formation of oxygen vacancies in MZFO lattice can be represented as follows

$$O_o^x \leftrightarrow \tfrac{1}{2} O_2 (g)\uparrow + V_{\ddot{O}} + 2e^- \qquad (1)$$

where, O_o^x is the lattice oxygen, O_2 (g) is the lost oxygen, $V_{\ddot{O}}$ is the created oxygen vacancies and e^- is the trapped electron. Due to the low concentration of chemi-adsorbed oxygen the trapped electrons act as donor level resulting 'n' type conduction behaviour of the embedded tubes. However in presence of high concentration of chemi-adsorbed oxygen the oxygen vacancies generate holes (Eqn. 2) in the valence band of the MZFO lattice.

$$V_{\ddot{O}} + \tfrac{1}{2} O_2 \leftrightarrow O_o + 2 p^- \qquad (2)$$

The generated holes are the major charge carriers for isolated nano-tubes. The inversion of the dominant charge carrier (from n to p type) is solely dependent on the surface morphology of nano-structured $Mg_{0.5}Zn_{0.5}Fe_2O_4$ in the form of isolated nano-tube. Thus, in addition to their central role in gas sensing, engineering of the surface to volume ratio of these spinel ferrite may open up interesting possibilities to tailor their electronic conduction.

Acknowledgement: IBSA-DST, CSIR, NPMASS-ADA, Govt. of India.

ORDERED MESOPOROUS METAL OXIDES BY STRUCUTRE REPLICATION: STRUCTURE-PROPERTY-RELATIONSHIPS AND APPLICATION IN GAS SENSING

Thorsten Wagner[1], Michael Tiemann[1]

[1]University of Paderborn, Faculty of Science, Department of Chemistry, Paderborn, Germany

Structure replication of ordered mesoporous matrices (nanocasting) offers a new way of controlling structure and composition of materials in the nanometer region [1]. Ordered mesoporous metal oxides offer a high specific surface area (e.g. SnO_2 up to 200 m^2/g), pores with diameters of typically 10 nm with narrow size distributions and high thermal stability [2]. The utilization of a rigid structure matrix (see figure 1) allows for high conversion temperatures and offers the possibility to control the crystallinity of the products. In the field of semiconducting gas sensors (chemiresistors) improvements of sensing properties by applying this structuring concept to the functional layer material have been shown [3-5].

Here we present some results of a structure-property relationship found for ordered mesoporous In_2O_3 synthesized by nanocasting. The mesoporous In_2O_3 shows sensing properties different from those in non-structured materials. Specific photoreduction properties occur which make light-assisted low-temperature sensing possible. Nanocomposites prepared by nanocasting are also suitable for application as battery electrodes and in fuel cells.

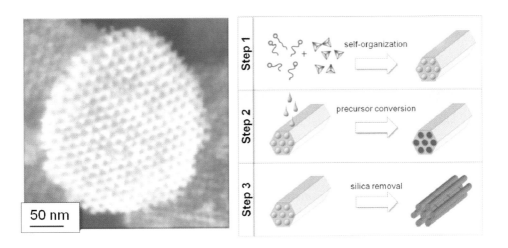

Figure 1: SEM of ordered mesoporous In_2O_3 with cubic pore system (left) and schematic of the nanocasting procedure used for its synthesis. 1st step: synthesis of an ordered mesoporous silica utilizing a self-organization process; 2nd step: impregnation of the porous silica matrix with a metal oxide precursor and conversion of the precursor to the oxide; 3rd step: removal of the silica matrix e.g. by etching with NaOH.

[1] M. Tiemann, Chem. Mater. **20** 961 (2008).
[2] T. Waitz et al., Sens. Actuators B **150** 788-793 (2010).
[3] T. Wagner et al., Sensors **6** 318 (2006).
[4] T. Wagner et al., Thin Solid Films **515** 8360 (2007).
[5] T. Wagner et al., Thin Solid Films **517** 6170 (2009).

WIDTH CONTROL AND OPTICAL NONLINEARITY OF PLATINUM NANOWIRES

<u>Yoichi Ogata</u>, and Goro Mizutani

School of Materials Science, Japan Advanced Institute of Science and Technology, Nomi, Ishikawa, JAPAN

We have fabricated arrays of Pt nanowires with around 2 *nm* width on the MgO(210) faceted substrates by using shadow deposition method in a UHV chamber with base pressure of 3.7×10^{-7} Pa, and observed their optical second-harmonic generation (SHG) response. The faceted MgO(210) templates were prepared by self-organization with periodicity depending on the homo-epitaxial growth condition. The platinum of 2 *nm* nominal thickness was deposited from the shorter face side on the MgO faceted template. In the plan-view TEM image, well-aligned nanowires with widths of as small as 2 *nm* were confirmed (see Fig.1). A big advantage of this shadow deposition method is that the entire fabricated nanowire array with ultra-small width is in a macroscopic size as large as 10 *mm* x 10 *mm*.

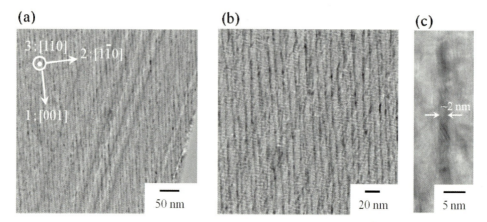

Fig. 1 (a) Plan-view and its (b,c) expanded TEM images of an ultra-small-width Pt nanowire array on a MgO(210) faceted template. Pt nanowires seen in dark contrast run along direction 1.

The SHG intensity was measured from a frequency-doubled mode-locked Nd:YAG laser at a photon energy of 2.33 *eV* with a pulse duration of 30 *ps* and a repetition rate of 10 *Hz*. The pulse energy was set at 120 *μJ*. Figure 2 indicates the plots of the SHG intensity for *p*-in/*p*-out polarization configuration from the ultra-small-width Pt nanowires as a function of the rotation angle φ around the substrate normal. The SHG intensity pattern shows elliptic shape with maxima at φ=0, 180°. Here, we found that the maximum SHG intensity from the fabricated 2 *nm* width Pt nanowires was much higher than that of thicker nanowires. One candidate origin of this enhancement of the SHG intensity at the 2 *nm* width nanowires may be a quantum confinement effect of electrons. In the phenomenological analysis of the angular SHG intensity pattern as seen in Fig.2, we confirmed contribution from five nonlinear susceptibility elements χ_{113}, χ_{223}, χ_{311}, χ_{322}, and χ_{333} originating from the broken symmetry in the direction 3; [110] of the MgO substrate. The suffices 1 and 2 denotes the [001] and [1$\bar{1}$0] directions, respectively. Among them, the contribution of the nonlinear susceptibility element χ_{113} was the highest. The absolute value of the χ_{113} bears comparison with that of well-known nonlinear optical crystals. Development of new optical nonlinear materials consisting of multi-layer stack of nanowires is expected in the near future.

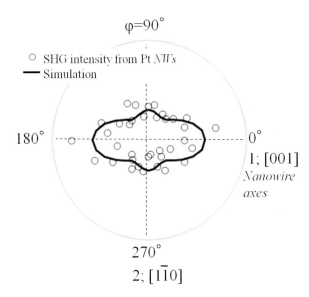

Fig. 2 Angular SHG intensity pattern of the Pt nanowires for *p*-in/*p*-out polarization configuration

NON-LINEAR PROPERTIES OF BALLISTIC ELECTRON FOCUSING DEVICES

Arkadius Ganczarczyk[1], Martin Geller[1], Axel Lorke[1], Dirk Reuter[2], Andreas D. Wieck[2]

[1]Experimental Physics and CeNIDE, Universität Duisburg-Essen, Duisburg, Germany; [2]Chair of Solid State Physics, Ruhr-Universität Bochum, Bochum, Germany

Ballistic electron focusing in nanostructured, two-dimensional electron gases was first demonstrated more than 20 years ago [1]. While the linear transport characteristics of such devices were examined in detail [2], nonlinear effects were so far investigated to a lesser extent. However, non-linear transport may be of great interest for possible new functional devices such as ballistic rectifiers and transistors.

In this work detailed studies of non-linear (and linear) transport properties of ballistic electron focusing devices are presented. Such a device is shown in the inset of fig. 1. It consists of a two-dimensional electron gas (2DEG) embedded in a AlGaAs heterostructure. The carrier density n_0 and mobility μ at a temperature of 4.2 K are 4.2 10^{15} m^{-2} and 89 m^2/Vs, respectively, resulting in a ballistic free path [3] of 2 µm. The 2DEG (green) is patterned by surface gate electrodes (yellow) which define electrostatically two quantum point contacts (QPC) with a separation of $d = 3.5$ µm. One QPC acts as an electron injector, the other as an electron collector. With an applied voltage at the injector and an applied magnetic field B perpendicular to the 2DEG, electrons are ballistically guided on circular paths. For a resonant magnetic field, when a multiple of the cyclotron radius matches d, peaks in the collector voltage are observed. The width of the injector or collector can be changed by applying a suitable negative gate voltage to the left or right gate, while the bias of the center gate remains unchanged.

This can be seen in fig. 1, where the focusing efficiency V_c/I_i (collector voltage divided by the injector current) as a function of B is shown for different gate bias situations: (i) wide open injector and collector (magenta), (ii) nearly closed point contacts (orange) or (iii) a combination of both (cyan and dark blue). With decreasing injector or collector width, the focusing efficiency increases and the position of the peaks shifts to higher magnetic fields. The latter is attributed to the fact that the focusing distance between the collector and injector decreases and higher magnetic fields are needed to decrease the cyclotron diameter accordingly.

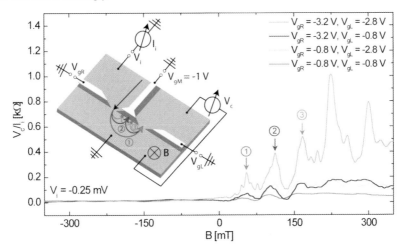

Figure 1: Focusing efficiency spectra $V_c/I_i(B)$ for different collector and injector widths. A gate voltage of -0.8 V corresponds to a widely opened injector or collector, -3.2 V and -2.8 V correspond to a nearly closed injector and collector, respectively. Inset: Schematic picture of the electron focusing device.

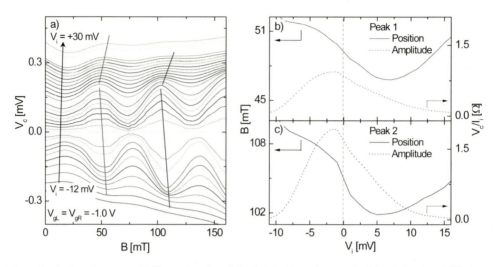

Figure 2: a): Focusing spectra $V_c(B)$ as a function of the the injection voltage V_i. Position (left axis, solid line) and amplitude (right axis, dotted line) as a function of the injection bias of the first (b) and second (c) focusing peak shown in a).

Figure 2 a) shows the focusing spectra for different injection voltages varying from $V_i = 30$ mV to -12 mV. The thorough analysis of the amplitude and the position of the focusing peaks is given in fig.2 b) and c). The plots show the position (left axis, solid line) and the amplitude (right axis, dotted line) of the first (b) and second (c) peak in dependence on the injection energy. For increasing negative injection bias (emission of electrons above the Fermi energy E_F), we observe an increase in the resonant magnetic field in agreement with the increased velocity of the injected electrons. For small positive bias, a similar electron focusing pattern is observed as for negative bias, which can be understood in the framework of electron-hole-symmetry. For increasing positive bias up to $V_i \approx 7$ mV, the resonances are shifted towards smaller magnetic fields, indicating that the transport can also be probed for (missing) carriers below E_F [4]. However, for even higher positive bias, the trend changes and the resonances start to shift to higher magnetic fields. At present, we do not have an explanation for this striking observation.

Additionally, the behavior of the peak amplitude shows also an unexpected anomaly. The maximum focusing efficiency, which is visible for $V_i \approx -2$ mV in fig. 2 b) and c), was expected for a vanishing injection bias, as cool electrons have a limited number of scattering possibilities. For increasingly higher positive or negative bias the amount of possible scattering possibilities for hot electrons/cool holes increases, leading to a decreasing ballistic free path [5].

[1] C.W.J. Beenakker et al., Europhys. Lett. **7**, 359 (1988).
[2] H. van Houten et al., Phys. Rev. B **39**, 8556 (1989).
[3] F. Nihey et al., Appl. Phys. Lett. **57**, 1218 (1990); J. Spector et al., Surf. Sci. **263**, 240 (1992).
[4] J.G. Williamson et al., Phys. Rev. B **41**, 1207 (1990); Surf. Sci. **229**, 303 (1990); R.I. Hornsey et al., Phys. Rev. B **48**, 14679 (1993).
[5] U. Sivan et al., Phys. Rev. Lett. **63**, 992 (1989).

HALL EFFECT IN AN ASYMMETRIC BALLISTIC CROSS JUNCTION

M. Szelong[1], U. Wieser[1], M. Knop[1], U. Kunze[1], D. Reuter[2], A. D. Wieck[2]

[1]Werkstoffe und Nanoelektronik, Ruhr-Universität Bochum, D-44780 Bochum, Germany;
[2]Angewandte Festkörperphysik, Ruhr-Universität Bochum, D-44780 Bochum, Germany

In the linear transport regime, the Hall resistance of ballistic cross junctions is known to exhibit deviations from classical results like quenched, negative, or enhanced values [1–3]. Recently we have studied the Hall effect in asymmetric ballistic cross junctions in the nonlinear transport regime and observed a current-direction-dependent Hall voltage violating the classical commutation relation $V_H(I,B) = -V_H(-I,B)$ [4]. In the present work we study this effect for the transition from linear to nonlinear transport regime.

Our 4-terminal cross junction consists of a $w_S = 200$ nm wide straight stem between an upper and lower contact "U" and "L" while two $w_B = 140$ nm wide branches (Hall probes "1" and "2") merge into the stem under an angle of 30° [see inset of Figure 1 (a)] [5]. A large-area top gate is covering junction and leads. The device is processed on a high-mobility GaAs/AlGaAs heterostructure with a two-dimensional electron density and mobility at a temperature of $T = 4.2$ K of $n_{2D} = 3.6 \cdot 10^{11}$ cm^{-2} and $\mu_{2D} = 7.5 \cdot 10^5$ cm^{2}V^{-1}s^{-1}, respectively, resulting in a mean free path of about 8 µm. Detailed information about device processing is given in [6].

DC transport measurements have been performed in perpendicular magnetic field $|B| \leq 500$ mT at $T = 4.2$ K. The classical cyclotron radius for $B = \pm 500$ mT corresponds to $r_C \sim w_S$. Therefore, Landau quantization is negligible and our device is operating in classical ballistic regime. In Hall configuration a current I_L is injected into contact "L" and extracted from contact "U" by push-pull voltage sources which are transformed into current sources by means of two 1-MΩ resistances (R_U, R_L). In this way the voltages at contacts "U" and "L" obey $V_U = -V_L$ and the channel potential is close to zero. The gate voltage V_G is held constant at about 220 mV above threshold V_T at negligible leakage current.

The result is shown in Figure 1. It can be clearly observed that for large currents $|V_{21}(I,B)| < |V_{21}(-I,B)|$ independently of B-field direction, an increase that amounts to about 21 % (6 %) for negative (positive) fields. We interpret the difference as caused by ballistic electrons whose transmission probability $T_{L\rightarrow 1/2}$ from contact "L" to "1" or "2" ($I_L < 0$) is large as the angular deviation of the Hall probe from the electron trajectory is small and vice versa for $T_{U\rightarrow 1/2}$ from contact "U" to "1" or "2" ($I_L > 0$). So the absolute value of V_{21} is increased for $I_L < 0$. As $|I_L|$ decreases the difference between the current directions vanishes, and for $|I_L| \leq 300$ nA both curves collapse into one line [Figure 1 (b)], which indicates the transition into the linear regime.

For further investigation we extracted the deviation $V_\pm(|I_L|,B) = V_{21}(I_L,B) + V_{21}(-I_L,B)$ from experimental data. In linear regime $V_\pm$ is expected to be equal to zero for all I_L and B. Any nonzero value is caused by nonlinearities. In addition, all antisymmetric terms of $V_{21}(I_L,B)$ (mode-controlled density variation [7], classical Hall voltage, geometrical lateral offset between Hall probes) vanish and only symmetric terms (ballistic component, density variation due to $R_U \neq R_L$) are left in $V_\pm$. We distinguish two different dependencies of $V_\pm$ on B. First, considering an influence solely by a ballistic component, resulting in $T_{L\rightarrow 1/2} > T_{U\rightarrow 1/2}$, we expect a strictly monotonic increasing function $V_\pm(B)$. Second, in our experimental setup R_U exceeds R_L by about 0.8 %. This tiny difference causes a shift of the effective gate voltage as current is increased. If the corresponding shift in density causes a deviation $V_\pm(B)$ its slope would be opposite to that observed in Figure 2 (a) from which we infer that the dependence is dominated by ballistic electrons.

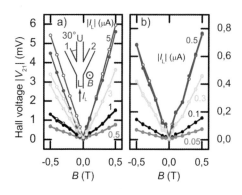 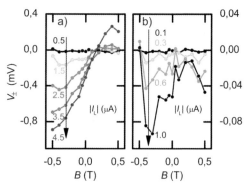

Figure 1: $|V_{21}|$ vs. B with parameter $|I_L|$, solid circles: $I_L > 0$, open circles: $I_L < 0$. Parameters in (b) are one order of magnitude smaller than in (a).

Figure 2: Deviation $V_\pm$ vs. B with parameter $|I_L|$ for large (a) and small currents (b).

In Figure 2 (b) we again observe a vanishing ballistic influence for smaller currents approaching the linear transport regime where the absolute value of any measurable voltage is independent on current direction.

To conclude, we have investigated the influence of a symmetry-breaking geometry on Hall voltage V_H in the classical ballistic transport regime and observed an enhancement of V_H for negative and a reduction of V_H for positive current. This influence has been found only in nonlinear transport regime whereas in linear transport regime the effect disappears and the commutation relation $V_H(I,B) = -V_H(-I,B)$ still holds. This is in contrast to hitherto known magnetotransport effects which are inherent to linear transport regime.

[1] Roukes et al., Phys. Rev. Lett. **59**, 3011 (1987)
[2] Ford et al., Phys. Rev. Lett. **62**, 2724 (1989)
[3] Baranger et al., Phys. Rev. B **44**, 10637 (1991)
[4] Wieser et al., Physica E **40**, 2179 (2008)
[5] Knop et al., Appl. Phys. Lett. **88**, 082110 (2006)
[6] Knop et al., Semicond. Sci. Technol. **20**, 814 (2005)
[7] Wiemann et al., Appl. Phys. Lett. **97**, 062112 (2010).

ELIMINATION OF HOT-ELECTRON THERMOPOWER FROM BALLISTIC RECTIFICATION USING A DUAL-CROSS DEVICE

J. F. von Pock[1], D. Salloch[1], U. Wieser[1], U. Kunze[1], T. Hackbarth[2]

[1]Werkstoffe und Nanoelektronik, Ruhr-Universität Bochum, D-44780 Bochum, Germany;
[2]DaimlerChrysler Forschungszentrum Ulm, D-89081 Ulm, Germany

If the critical dimensions of a device structure are smaller than the mean free path of the charge carriers the electronic transport is carried by ballistic motion. Hence the trajectories of electrons can be guided into directions which are determined by the device geometry and which do not necessarily follow the direction of the electric field lines. Such behaviour has been observed from bend resistance [1] and is applicable to a ballistic full-wave rectifier, which has been realized as orthogonal cross junction with symmetry-breaking central scatterer [2] or as an asymmetric cross junction [3]. The latter structure consists of a straight voltage stem and oblique injecting branches which are tilted with respect to the stem by an angle of less than 90° (Figure 1 (a) upper part). While the injectors (1,2) are supplied with current the electrons stationary charge the lower stem (L) because of their inertial ballistic motion. This causes a voltage V_{UL} between upper (U) and lower stem, which is independent of the current polarity. Additionally, Joule heating of the electron system in the injectors leads to a non-equilibrium distribution of carriers which diffuse into the stem leads U and L. Even a small difference in the lateral confining potentials in (U) and (L) imposes a voltage upon the ballistic signal, which is called hot-electron thermopower (HET) [4]. Hence, V_{UL} contains both the ballistic response and the HET, where the latter is particularly large close to pinch-off and at low temperatures [5].

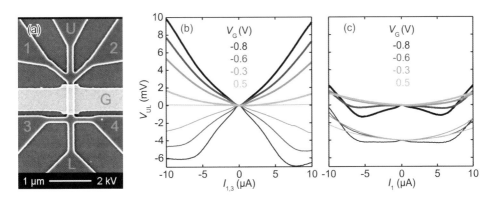

Figure 1 (a): Scanning electron micrograph top-view of the dual-cross device. (1,2) denote the injector branches, (3,4) the heating branches, (U,L) the voltage probes, and (G) the gate.
(b): Characteristics of V_{UL}-vs-I_1 (thick lines) via injector (1,2) and V_{UL}-vs-I_3 (thin lines) via heating branches (3,4) at different V_G.
(c): Characteristics of V_{UL}-vs-$(I_1 = I_3)$ (thick lines) via (1,2) and (3,4) with V_G as parameter. Thin lines (shifted by −4 mV) indicate numerically added curves V_{UL}-vs-I_1 (1,2) and V_{UL}-vs-0.8·I_3 (3,4).

In the present work we aim to eliminate the HET from V_{UL} by removing the gradient of the electron temperature. This is done by heating the stem on the opposite side of the gate by an orthogonal cross-junction (OCJ) (Figure 1 (a) lower part). The remaining signal should only be due to inertial ballistics. Our preliminary work introducing this method [6] was restricted to a single gate voltage $V_G = -0.2$ V which is far above threshold voltage V_T. Here we will focus on the V_G dependence of this device until pinch-off.

The channels of our device (Figure 1) are 120 nm wide and the injectors are tilted by 45°. Between the two cross-junctions an 840 nm wide Schottky gate is located. The Si/SiGe heterostructure, which is used for the device, has a two-dimensional electron gas (2DEG) 60 nm below the surface. At a temperature of $T = 1.5$ K the 2DEG mobility is $\mu = 1.8 \cdot 10^5$ cm^{2}V^{-1}s^{-1} and the density $n = 6.3 \cdot 10^{11}$ cm^{-2}. Details of the fabrication process can be found elsewhere [6].

The device was measured at 4.2 K. The threshold was detected as $V_T \sim -0.75$ V. The rectifier characteristics of V_{UL} versus injector current I_1 at different V_G are shown in Figure 1 (b) as thick lines. We consider the strong increase of V_{UL}, as the gate voltage V_G is lowered towards threshold voltage, to reflect the increase in HET and not a ballistic effect [5]. Conversely, we assume that the small remaining V_{UL} at $V_G = 0.5$ V mainly represents the ballistic component. The V_{UL}-vs-I_3 characteristics at different V_G of the current-carrying OCJ (Figure 1 (c), thin lines) represent pure HET. The saturation of the signal at $|I_3| > 7$ µA probably is the result of a mode-controlled potential correction at the centre of the cross junction [7].

If both cross junctions are supplied with the same current ($I_1 = I_3$) the influence of HET is widely removed (Figure 1 (c), thick lines). However, a closer look on the curvature at low current indicates a transition from ballistics-dominated at $V_G = -0.3$ V to HET-dominated at $V_G = -0.6$ V. Obviously, the HET is not fully eliminated near pinch-off. We suppose that the width of the heating channels (1,2) and (3,4) as well as the separation of the cross junctions from the gate centre are not identical. A possible way to achieve equal electron temperature is different heating currents $I_1 > I_3$, which is illustrated in thin lines for a numerical addition of $V_{UL}(I_1) + V_{UL}(0.8 \cdot I_3)$. We see, that at a ratio of $I_1/I_3 = 0.8$ the curvatures of the parabolas vanish for $V_G \sim V_T$. This has still to be verified experimentally.

This work was supported by the DFG and by the Ruhr-University Research School funded by Germany's Excellence Initiative (Grant No. DFG GSC 98/1).

[1] U. Wieser et al., Appl. Phys. Lett. **87**, 252114 (2005).
[2] A. M. Song et al., Phys. Rev. Lett. **80**, 3831 (1998).
[3] M. Knop et al., Appl. Phys. Lett. **88**, 082110 (2006).
[4] L. W. Molenkamp et al., Phys. Rev. Lett. **65**, 1052 (1990).
[5] D. Salloch et al., Appl. Phys. Lett. **94**, 203503 (2009).
[6] D. Salloch et al., American Institute of Physics Conference Proceedings **1399**, 321 (2011).
[7] M. Wiemann et al., Appl. Phys. Lett. **97**, 062112 (2010).

STRUCTURAL INFLUENCES ON ELECTRONIC TRANSPORT IN NANOSTRUCTURES

Robert Frielinghaus[1,5], K. Sladek[1,5], K. Flöhr[2,5], L. Houben[1,3,5], St. Trellenkamp[1,5], T.E. Weirich[4,5], M. Morgenstern[2,5], H. Hardtdegen[1,5], Th. Schäpers[1,2,5], C.M. Schneider[1,5], C. Meyer[1,5]

[1]Peter Grünberg Institut, Forschungszentrum Jülich, 52425 Jülich, Germany;
[2]II. Physikalisches Institut, RWTH Aachen University, 52056 Aachen, Germany;
[3]Ernst Ruska-Center for Microscopy and Spectroscopy with Electrons, Forschungszentrum Jülich, 52425 Jülich, Germany;
[4]Central Facility for Electron Microscopy GFE, RWTH Aachen University, 52074 Aachen, Germany;
[5]JARA – Fundamentals of Future Information Technologies

Self-assembled nanostructures such as carbon nanotubes (CNTs) or InAs nanowires (NWs) are suitable candidates for future nanoscale electronic devices. It is possible to functionalize these structures to tailor fundamental properties. However, the extent of functionalization and atomic arrangements usually differ from device to device leading to variations in the electric behavior and thus impeding the reproducibility of measurement results.

We present an approach based on fabricating suspended nanodevices on a Si_3N_4 membrane substrate that enables us to perform (quantum) transport measurements, Raman spectroscopy, and transmission electron microscopy (TEM) all on one individual nanostructure [1].

To fabricate suspended transport devices, electron beam lithography patterning and reactive ion beam etching are used to open holes in a Si_3N_4 TEM membrane. Transport devices are fabricated in a successive step, placing the nanostructures across these holes [2]. The free suspension of the devices does not only enable various characterization techniques to work on the very same structure, but also inhibits interactions of the nanostructure with a substrate that inevitably produces artifacts in transport measurements.

In the case of InAs NWs (cf. Figure 1) the measurements nicely resolve universal conductance fluctuations (UCFs) as shown in Figure 2. The electron phase-coherence length measured in these experiments varies between different devices. The UCFs and the variation of the phase-coherence length are compared to the atomically-resolved structure of the devices measured by TEM.

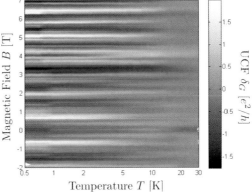
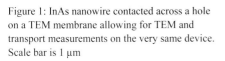

Figure 1: InAs nanowire contacted across a hole on a TEM membrane allowing for TEM and transport measurements on the very same device. Scale bar is 1 μm

Figure 2: Universal conductance fluctuation patterns at various temperatures measured in a three-probe configuration on the device in Figure 1.

CNTs are a versatile material for functionalization, but due to their chirality even pristine CNTs split into many families whose properties can be accurately described by theory. We show combined measurements for a contacted triple-walled CNT. High-Resolution TEM imaging as in Figure 3 was used to measure refined shell diameters while Raman spectroscopy was applied to determine the metallic or, respectively, semiconducting character of the CNT shells (cf. Figure 4). Room-temperature transport measurement on the same CNT confirmed the assignments.

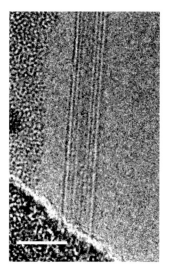
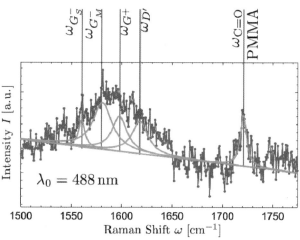

Figure 3: TEM image of a triple-walled CNT. The setup is similar to the InAs nanowire in Figure 1. A contact can be seen on the lower left corner. Scale bar is 5 nm

Figure 4: Raman measurement of the triple-walled CNT shown in Figure 3. High-energy modes that can be assigned to a semiconducting innermost and a metallic outer shell can be distinguished.

The results presented here show the necessity of monitoring the structural and chemical properties of individual devices for gaining a deeper understanding of electronic transport in nanoelectronic devices.

[1] R. Frielinghaus et al., Phys. Stat. Sol. (b) **248**, 2660 (2011)
[2] K. Flöhr, R. Frielinghaus et al., Rev. Sci. Instrum. **82**, 113705 (2011)

TUNNEL-INDUCED SPIN-ANISOTROPY IN QUANTUM DOT SPIN VALVES

<u>Maciej Misiorny</u>[1], Michael Hell[1], Maarten Wegewijs[1,2]

[1]Peter Grünberg Institut, Forschungszentrum Jülich & JARA Jülich
Aachen Research Alliance, 52425 Jülich, Germany;
[2]Institute for Theory of Statistical Physics, RWTH Aachen, 52056 Aachen, Germany

Atomic-scale spintronic systems, such as single-molecule magnets (SMMs) and magnetic adatoms, have recently been studied intensely mainly because of their large spin-anisotropy arising from strong spin-orbit and ligand fields. We show that spin-anisotropy can also be generated in spin-isotropic systems by spin-dependent transport of electrons. For a generic spin-1 quantum dot tunnel-coupled to two metallic ferromagnetic electrodes we show that quantum fluctuations induce a quadrupolar exchange field, generalizing the well established (dipolar) exchange field. This fields generates a uniaxial spin-anisotropy barrier that increases with the tunnel coupling, achieving values comparable to that of SMMs, but with the added flexibility of electric and magnetic tuneability. Besides inducing it, the transport can also be used to directly read out the quadrupolar field, utilizing its competition with Kondo spin-exchange processes with the ferromagnets. In this regime the proximity-induced quadrupolar exchange field is found to dominate over the dipolar exchange field, strongly enhancing the low-temperature spin-filtering as compared to spin-½ quantum dot spin-valves. Consequently, not only do spin-quadrupole effects in spin-polarized transport seem inevitable [1,2] in high-spin nanosystems, but they also offer new prospects for spintronic applications.

[1] M. Baumgärtel, M. Hell, S. Das and M. R. Wegewijs, Phys. Rev. Lett. **107**, 087202 (2011).
[2] B. Sothmann and J. König, Phys. Rev. B **82**, 245319 (2010)

CHARGING EFFECTS AND ELECTRON TRANSPORT PHENOMENA ASSOCIATED WITH THE REDOX PROPERTIES OF SELF-ASSEMBLED POLYOXOMETALATE MOLECULES

Angeliki Balliou[1,2], Antonios Douvas[1], Dimitrios Velessiotis[1], Vassilis Ioannou-Sougleridis[1], Pascal Normand[1], Panagiotis Argitis[1], Nikos Glezos[1]

[1]Institute of Microelectronics, NCSR Demokritos, Aghia Paraskevi, Athens 15310, Greece; [2] NTUA, Department of Chemical Engineering, Zographou Campus, Athens 15773, Greece

The combination of molecular systems with semiconductors opens the way to the realization of hybrid systems with a potential for new applications. This involves sensors, switches, memories, or quantum devices based on discrete molecular levels. In this work we investigate the possibility of implementing inorganic tungsten polyoxometalate (POM) molecules as charge-trapping medium in novel hybrid molecular/silicondevices for memory applications.

Capacitor devices containing molecular layers of a Keggin-structure polyoxometalate (POM: 12-tungstophosphoric acid) as trapping material and Isopentylamine (IPA: $C_5H_{13}N$) as cap dielectric on 3-aminopropyl triethoxysilane (APTES)-modified silicon surface, were fabricated via the layer-by-layer (LBL) bottom-up nanofabrication technique [1] (Fig. 1). Our intention is to isolate the POM molecular layer by means of two insulating layers, i.e the SiO_2 layer from below and the IPA on top, allowing thus, charge retention within the molecular layer.

The fabrication process parameters as well as the effect of the precursor solution composition were monitored by means of UV/vis reflection spectroscopy, FT-IR spectroscopy, multi wavelength variable angle ellipsometry, AFM and SEM. The technique resulted in an average density of $5*10^{12}$ cm^{-2} active trap centers. The conduction and charging mechanisms in the composite Metal-Insulator-Semiconductor (MIS) structures were elucidated by electrical characterization of quasi-static, dynamic and transient type in a wide range of temperatures (80-300 K) and the appropriate theoretical analysis. The special features rising in J-V characteristics yielded to the extraction of electrical parameters as well as to identification of the electronic structure of the functional molecules. More specifically, we determined the tunneling capacitance [2] ($C_t=4.59*10^{-11}$ F) associated with the transfer process to the available POM states via determination of the displacement current under gate injection and assumption of linear dielectric medium. The two transient peaks (Q1 and Q2) observed in the experimental current versus voltage ramp rate curve (Fig. 2) are manifestations of POM molecular states and are attributed to the filling of the ground and first excited quantum states of POM molecules respectively.

The effective carrier confinement in the states of POM nanostructures and the relatively large separation of their molecular levels, allows for single-electron effects to be observed even at RT.

The carrier transport under gate injection in mesoscopic scale consists of coupled processes of thermionic emission over the Schottky barrier induced by the HOMO-LUMO/ Al Fermi level mismatch and Poole-Frenkel transport along with resonant tunneling through molecular states at relative high and intermediate electric fields. Conduction is mediated by hopping, coherent and incoherent tunneling in the low field regime. The mechanisms encountered are in reasonable correlation with the structural characteristics of the film. Each mechanism dominates the current in a different bias and/or temperature regime.

Transient capacitance measurements indicate that presence of POM molecules influences strongly the effective generation lifetimes of the substrate and supply information, via a modified Zerbst equation, for the effective generation lifetimes within the molecular epilayers and the rate of change in the space charge region generated charge.

Moreover, the device exhibits discrete resistive states under gate injection, possibly due to strong correlation effects originating from POM charging in that field regime, indicating the potential of the POM molecules as sensitive resistive switches or memory elements.

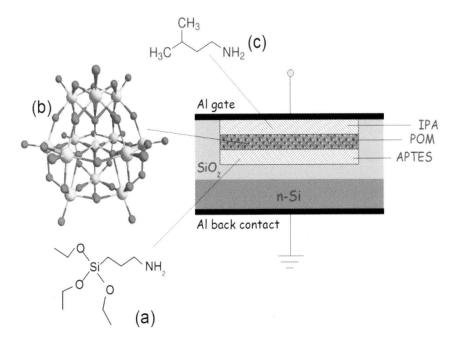

Figure 1: Planar MIS and incorporated functional layers. (a) APTES molecules used for surface chemical modification of the 5.7-nm-thick thermally grown SiO$_2$. (b) POM molecules self organized via electrostatic forces on top of the APTES monolayer. (c) IPA molecules electrosta-tically linked on POM structures, used as capping/passivating layer.

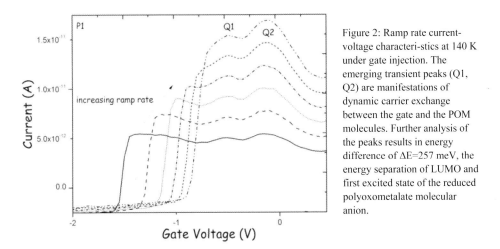

Figure 2: Ramp rate current-voltage characteri-stics at 140 K under gate injection. The emerging transient peaks (Q1, Q2) are manifestations of dynamic carrier exchange between the gate and the POM molecules. Further analysis of the peaks results in energy difference of $\Delta E=257$ meV, the energy separation of LUMO and first excited state of the reduced polyoxometalate molecular anion.

[1] A.M. Douvas, E. Makarona, N. Glezos, J.A. Mielczarski, E. Mielczarski, ACS Nano **2**, 733-742 (2008).
[2] V. Ioannou-Sougleridis and A. Nassiopoulou, Journal of Applied Physics **106**, 054508 (2009).

MOLECULAR ELECTRONICS MEETS SPINTRONICS: AN *AB INITIO* EXPLORATION

Nicolae Atodiresei[1], Vasile Caciuc[1], Predrag Lazic[2], Stefan Blügel[1]

[1]Peter Grünberg Institut and Institute for Advanced Simulation, Forschungszentrum Jülich, Germany;
[2]Massachusetts Institute of Technology, Cambridge-Massachusetts, USA

Merging the concepts of molecular electronics with spintronics opens a very exciting avenue in designing and building future nanoelectronic devices. The ability to reliably describe the electronic properties of molecules adsorbed on magnetic surfaces is essential to understand and design the functionality of hybrid organic molecular devices. This strongly depends on the accuracy of the state-of-the-art theoretical methods used to assess the interaction between a molecule or a molecular layer and a substrate of choice. Our theoretical studies [1-6] show that, with its predictive power, the density functional theory (DFT) provides a framework where a realistic description of these hybrid systems can be expected and can be further used to tune both the electronic and magnetic properties at hybrid organometallic interfaces. Besides this, our studies demonstrate the decisive role played by the van der Waals forces to correctly describe the interaction between aromatic molecules and metallic surfaces.

The π-electrons of the cobalt-phthalocyanine molecule hybridize strong with the *d*-electrons of a magnetic metal and, as a consequence, a complex energy dependent magnetic structure is formed. Therefore, near the Fermi level, at the molecular site an *inversion* of the spin-polarization with respect to the ferromagnetic surface occurs. Our studies demonstrate that electrons of different spin [i.e. up (↑) and down (↓)] can selectively be injected from the same ferromagnetic surface by locally controlling the inversion of the spin-polarization.

We will present conceptual studies performed to understand how to tailor the magnetic properties at a hybrid organic-ferromagnetic interface by adsorbing organic molecules containing π-electrons onto a magnetic substrate. For such hybrid systems, the magnetic properties like molecular magnetic moments and their spatial orientation can be specifically tuned by an appropriate choice of the chemical substituents. Our *ab initio* calculations demonstrate that, by employing an appropriate chemical functionalization of organic molecules adsorbed on a ferromagnetic surface, a fine tuning of the spin-unbalanced electronic structure can be achieved. For example, by using molecular substituents with different electronegativities attached to π-conjugated systems adsorbed on a ferromagnetic surface, the electrons with a specific spin [i.e. up (↑) and down (↓)] can selectively be injected at the molecular site from the same ferromagnetic substrate. Even more important, we show that there is a direct correspondence between the substituent's electronegativity and the size of the induced molecular magnetic moment. As regarding the stability of the magnetization direction of the hybrid organic-ferromagnetic system, we demonstrate that the adsorbed hydrogenated molecules destabilize more the out-of-plane magnetization of the ferromagnetic surface as compared to molecules containing more electronegative atoms as Cl and F which could also enhance it. Ultimately, this knowledge allows us to precisely engineer the magnetic properties of the hybrid organic-ferromagnetic interfaces which can be further exploited to design more efficient spintronic devices based on organic molecules.

[1] N. Atodiresei, P. H. Dederichs, Y. Mokrousov, L. Bergqvist, G. Bihlmayer, S. Blügel, *Phys. Rev. Lett.* **100**, 117207 (2008);
[2] N. Atodiresei, V. Caciuc, P. Lazic, S. Blügel, *Phys. Rev. Lett.* **102**, 136809 (2009);
[3] N. Atodiresei, J. Brede, P. Lazic, V. Caciuc, G. Hoffmann, R. Wiesendanger, S. Blügel, *Phys. Rev. Lett.* **105**, 066601 (2010);
[4] J. Brede, N. Atodiresei, S. Kuck, P. Lazic, V. Caciuc, Y. Morikawa, G. Hoffmann, S. Blügel, R. Wiesendanger, *Phys. Rev. Lett.* **105**, 047204 (2010);
[5] C. Busse, P. Lazic, R. Djemour, J. Coraux, T. Gerber, N. Atodiresei, V. Caciuc, R. Brako, A. T. N'Diaye, S. Blügel, J. Zegenhagen, T. Michely, *Phys. Rev. Lett.* **107**, 036101 (2011).
[6] N. Atodiresei, V. Caciuc, P. Lazic, S. Blügel, *Phys. Rev. B* **84**, 172402 (2011).

MOL 2

CHARGE TRAPPING AND ELECTROFORMING IN METAL OXIDE /POLYMER RESISITVE SWITCHING MEMORY DIODES

<u>Stefan C. J. Meskers</u>[1], Benjamin F. Bory[1], Henrique L. Gomes[2], René A. J. Janssen[1], Dago M. de Leeuw[3]

[1] Molecular Materials and Nanosystems, Eindhoven University of Technology, P.O. Box 513, 5600 MB Eindhoven, The Netherlands,
[2] Center of Electronics Optoelectronics and Telecommunications (CEOT), Universidade do Algarve, Campus de Gambelas, 8005-139 Faro, Portugal,
[3] Philips Research Laboratories, Professor Holstlaan 4, 5656 AA Eindhoven, The Netherlands

Al/Al$_2$O$_3$/poly(spirofluorene)/Ba/Al diodes show resistive switching after electroforming. The electroforming process is found to be initiated by trapping of electrons at the polymer/oxide interface. This leads to an increase of the potential difference over the oxide layer. This enhances tunneling of electrons over the oxide. Upon increasing the potential further, soft breakdown in the oxide occurs. In case of ZnO as oxide, it can be shown that electroforming is associated with hole injection into the oxide.

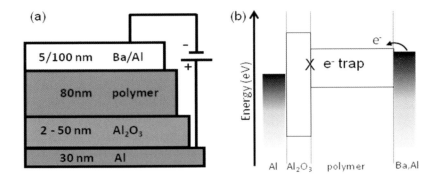

Figure: (a) layout of the Al/Al$_2$O$_3$ /poly(spirofluorene)/Ba/Al diode with respective thickness studied. (b) band diagram of the diode.

The resistive switching processes itself may be described using mean field theory for the metal-to-insulator transition in the metal oxide layer. This involves formation and breaking of conducting filaments of the metallic phase in the formed metal oxide.

MOL 3

ELECTROFORMING IN LIF/POLYMER RESISITVE SWITCHING MEMORY DIODES AND HOLE INJECTION

Benjamin F. Bory[1], Stefan C. J. Meskers[1], Henrique L. Gomes[2], René A. J. Janssen[1], Dago M. de Leeuw[3]

[1]Molecular Materials and Nanosystems, Eindhoven University of Technology, P.O. Box 513, 5600 MB Eindhoven, The Netherlands;
[2]Center of Electronics Optoelectronics and Telecommunications (CEOT), Universidade do Algarve, Campus de Gambelas, 8005-139 Faro, Portugal;
[3]Philips Research Laboratories, Professor Holstlaan 4, 5656 AA Eindhoven, The Netherlands

Al/LiF/poly(spirofluorene)/Ba/Al diodes can be electroformed, and show resistive switching after electroforming. Presently electroforming in diodes for resistive switching is only partially understood. However, full control over electroforming would be required for successful application in solid state memory.

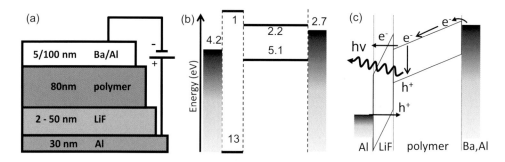

Figure: (a) layout of the Al/LiF/poly(spirofluorene)/Ba/Al diode with respective thickness studied. (b) band diagram of the diode. (c) schematic band diagram at electroforming voltage showing injection of holes and consequent light emission.

The electroforming process in Al/LiF/poly(spirofluorene)/Ba/Al diodes starts with injection of electrons into the polymer via the quasi-ohmic Ba contact. The poly(spirofluorene) is a p-conjugated polymer capable of electron transport. After injection, the electrons get trapped at the LiF/polymer interface. This causes the potential drop over the LiF layer to increase. At sufficiently high potential difference over LiF, hole injection into LiF and the polymer occurs, as evidenced by light emission from the p-conjugated polymer. We find that the light emission coincides with electroforming suggesting that electroforming is induced by injected holes.

EFFECT OF HETEROMETALLIC CONTACTS ON CHARGE TRANsPORT

N. Babajani[1], C. Kaulen[2], M. Homberger[2], U. Simon[2], R. Waser[1], S. Karthäuser[1]

[1]Peter Grünberg Institut (PGI-7) and JARA-FIT, Forschungszentrum Jülich GmbH, D-52425 Jülich, Germany;
[2]Inorganic Chemistry (IAC) and JARA-FIT, RWTH Aachen University, D-52056 Aachen, Germany

Molecular electronics focuses on the understanding of electronic transport properties of organic molecules with the aim to use them as functional units in information technology. A challenge in electronic transport measurements consists in connecting reliably molecules or molecular capped nanoparticles (NPs) to the macroscopic outer world [1].

Here we present a method to achieve this goal by immobilizing single AuNPs (*13.3 ± 0.8 nm*) capped with 1,8-mercaptooctanoic acid (MOS) in between heterometallic nanoelectrodes (gap size: *15 ± 2 nm*) by dielectrophoretic trapping (DEPT) [2] (see fig. 1). The AuNPs are prepared from citrate stabilized precursors in a ligand exchange reaction with MOS. The MOS stabilized nanoparticles exhibit pH-dependent aggregation behavior, which we examined by UV-vis, DLS and Zeta-Potential measurements. Based on these results we adjusted the pH in DEPT experiments in order to achieve trapping of individual particles.

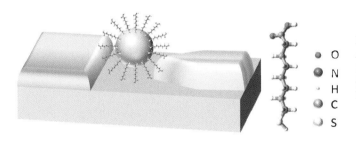

Figure 1: Schematic of a single Au-nanoparticle capped with 1,8-mercaptooctanoic acid (MOS) in between heterometallic nanoelectrodes.

The process of controllable fabrication of heterometallic nanoelectrodes is achieved by combining two methods, the electron-beam lithography (EBL) and a self-alignment procedure. The nanoelectrodes fabricated for these measurements exhibit a gap size of 15 nm and consist of a platinum electrode with a thickness of 13 nm and an AuPd electrode with a thickness of 11 nm. The fabrication of the nanoelectrodes starts with the formation of the platinum electrode by e-beam lithography in a lift off process. On top of the platinum electrode aluminum is evaporated and oxidized in air. Thus, an overhang of aluminum on top of the platinum is formed which can be used as a shadow mask for the next evaporation step. In this second step the fabrication of the AuPd electrodes with a separation from the Pt-electrode defined by the aluminumoxide mask is performed. In the last step contact pads are formed, to enable the electric access to the nanoelectrodes.

Subsequently, the immobilization of a single MOS capped NP in between heterometallic nanoelectrodes is realized. From a HEPES-buffer solution at pH = 6.5 as well as from a phosphate buffer solution at pH = 5 the MOS capped AuNPs were immobilized. Successful trapping has been achieved applying a DC electrical field to immobilize single MOS capped AuNPs from the HEPES-buffer and an AC electrical field for the MOS-AuNPs in phosphate buffer solution. The yield for DC-DEPT is 1%, while 30% could be obtained in case of AC-DEPT. At pH = 5, the carboxyl-terminated ligand shell of the MOS capped AuNPs is partly deprotonated, thus the MOS capped AuNPs are charged, which explains the high yield upon applying an AC electrical field.

I/U measurements performed on heterometallic electrodes in vacuum showing tunneling characteristics exhibit an asymmetry for voltages larger than the work function of the electrodes. The same is true, if the space between the two different metallic electrodes is filled with an insulating material, like Al_2O_3 [3]. In contrast, according to the theoretical paper published by Simmons [4], a material with an electronic state exhibiting an energy smaller than the work-function of the electrode, deposited between the heterometallic nanoelectrodes should lead to symmetrical *I/U* characteristics. With respect to these considerations we expect for immobilized MOS-AuNPs in between heterometallic nanoelectrodes symmetrical *I/U* curves as well.

The *I/U* measurements across these "electrode-molecule-NP-molecule-electrode" junctions reveal a detailed fingerprint of the molecule capped nanoparticle in the nanogap. By fitting the Simmons tunneling curve to the *I/U* measurements the characteristic parameter of MOS forming the tunneling barrier can be determined, i. e., the tunneling barrier height and the length of the remaining vacuum gap. In addition, the *I/U* curves show symmetrical curves. The obtained tunneling barrier heights are found to correspond very well to literature data for the respective molecular moieties. Supplementary, transition voltage spectroscopy is used to confirm these results.

After performing the electrical measurements the device is characterized by scanning electron microscopy to confirm the device geometry. Thus, transport properties of single nanoparticle devices revealing molecular tunneling barriers are obtained.

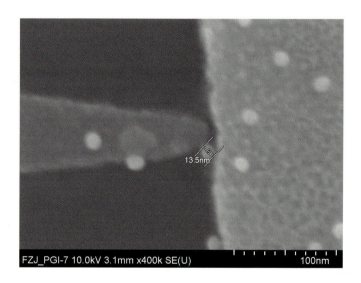

[1] H. B. Akkerman, B. de Boer, "Electrical conduction through single molecules and self-assembled monolayers," J. Phys. Cond. Mat. **20**, 013001 (2008)
[2] M. Manheller, S. Karthäuser, K. Blech, U. Simon, R. Waser, "Electrical Characterization of single Biphenyl-propanethiol capped 4 nm Au Nanoparticles," Proceedings of 10th IEEE International Conference on Nanotechnology (IEEE-Nano), 919 – 923, (2010)
[3] E. W. Cowell III *et al.* "Advancing MIM Electronics: Amorphous Metal Electrodes," Adv. Mater. **23**, 74-78, 2011
[4] J. G. Simmons, "Electric Tunnel Effect between Dissimilar Electrodes Separated by a Thin Insulating Film," J. Appl. Phys. **34**, 9, 2581-2590 (1963)

SEMI-EMPIRICAL VS. AB-INITIO CORRELATION EFFECTS: DFT STUDY OF THIOPHENE ADSORBED ON THE Cu(111) SURFACE

Martin Callsen[1], Nicolae Atodiresei[1], Vasile Caciuc[1], Stefan Blügel[1]

[1]Peter Grünberg Institut (PGI-1) and Institute for Advanced Simulation (IAS-1), Forschungszentrum Jülich and JARA, 52425 Jülich, Germany

In the emerging field of nano-electronics organic molecules adsorbed on metal surfaces are of high scientific interest as possible candidates for future electronic devices. Thiophene and its oligomers represent prototypes of such molecular electronic components and are in consequence intensely studied.

The adsorption mechanism of thiophene (C_4H_4S) on the Cu(111) surface has been studied by density functional theory (DFT) calculations as implemented in the VASP code. As approximation for the xc-functional we have used GGA-PBE and the PAW-approach has been applied. To account for long range van der Waals effects and to investigate their importance on the adsorption geometry and the corresponding binding energy a first principles method (vdW-DF [1]) implemented in the JuNoLo code [2] and two semi-empirical (DFT-D2 [3] and DFT-D3 [4]) have been compared.

While the adsorption structure predicted by DFT agrees well with experimental results [5] the calculated adsorption energy of -137 meV underestimates the experimental value -0.59 eV without including dispersion interactions. Interestingly this physisorption character of the thiophene bonding on the Cu(111) surface is changed to weak chemisorption even for the DFT ground-state geometry, when a non-local correlation energy functional [1] is used.

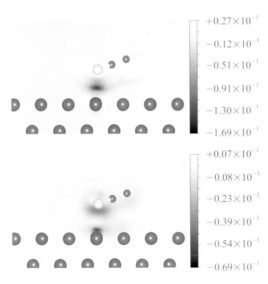

Figure 1: Sideview of thiophene adsorbed on Cu(111). In the ground-state the S atom adsorbs on top of a Cu atom and forms a weak chemical bond. The data given in grayscale is the correlation binding energy density in eV/Å^3 for the semi-local contribution to PBE correlation (upper panel) and for the non-local vdW-DF (lower panel). The difference in adsorption energy from physisorption to weak chemisorption arises just due to the different spatial distribution of these two quantities.

This difference in the predicted adsorption energy arises from the substantially different spatial distribution of the corresponding correlation energy densities (see Fig. 1). While the correlation energy density of the semi-local PBE functional is confined to regions where charge transfer takes place the non-local binding energy density has also contributions from regions, which are not relevant for the former.

[1] M. Dion et al., Phys. Rev. Lett. **92**, 246401 (2004).
[2] P. Lazić et al., Comput. Phys. Commun. **181**, 371 (2010).
[3] S. Grimme J. Comput. Chem **27**, 1787 (2006)
[4] S. Grimme et al., J. Chem. Phys. **132**, 154104 (2010).
[5] P. K. Milligan et al., J. Chem. Phys. B **105**, 140 (2001)

INTERFACE DIPOLE FORMATION IN DITIOCARBAMATE BASED SURFACE FUNCTIONALIZATIONS

Philip Schulz[1,2], Tobias Schäfer[1], Christopher Zangmeister[2], Dominik Meyer[1], Christian Effertz[1], Riccardo Mazzarello[3], Roger Van Zee[2] and Matthias Wuttig[1]

[1] I. Institute of Physics (IA) and JARA-FIT, RWTH Aachen University, Aachen, Germany
[2] National Institute of Standards and Technology, Gaithersburg, Maryland 20899, United States
[3] Institute for Theoretical Solid State Physics, RWTH Aachen University, Aachen, Germany

Organic light emitting diodes (OLED), organic photovoltaic devices (OPV) and organic field effect transistors (OFET) are well-known for their potential in the field of optoelectronics. However, this class of organic electronics share a common set of challenges in meeting the ideal charge transport requirements within the organic layer. Charge transport models try to reflect the multiple effects that conspire in the fundamental electronic transport processes ranging from band transport theory to correlation and disorder limited regimes [1]. The interfaces between the organic layers and especially between organic layer and inorganic electrode are identified as natural sinks for charge transfer rates. Significant improvements in the device functionality are expected if the electronic coupling of the organic/inorganic interface at the contact area of the metal or transparent conductive oxide electrode to the functional organic layer can be controlled and optimized [2]. In the past decades the application of self-assembled monolayers has been proposed to achieve such surface modifications [3]. Recently, first successes of dithiocarbamate (DTC) monolayer formation yielding promising electronic properties have been reported [4]. Compared to their well investigated thiolate counterparts, DTC molecules promise to exhibit ideal geometric properties of the carbodithiolate binding group, which strongly adheres to the gold surface.

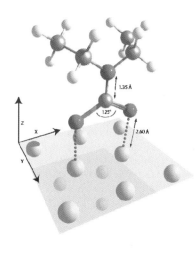

Figure 1: Diethyldithiocarbamate molecule adsorbed on a Au (111) surface.

Here, we present a study on dithiocarbamate (DTC) monolayers on pristine gold surfaces which were investigated by Photoemission Spectroscopy and Fourier-Transform Infrared Spectroscopy. While the growth of well oriented and densely packed monolayers was confirmed in the x-ray photoemission and infrared spectra, the valence band structure was probed in the ultra violet photoemission experiments. We were able to achieve very low work functions down to 3.2 eV for the SAM covered gold substrates which can be explained by the formation of strong bond dipoles combined with a high packaging density of DTC molecules which carry an exceptionally high intrinsic dipole moment. The alignment of the electronic structure at the metal/SAM interface has also been reproduced by density functional theory. Subsequently, first OFET devices comprising DTC functionalized gold electrodes were fabricated, which show superior contact properties compared to unmodified devices. Thus, we present a promising and reliable route towards low work function metal contact interfaces for acceptor-type materials and ambipolar devices.

[1] V. Coropceanu, J. Cornil, D.A. da Silva Filho, Y. Olivier, R. Silbey, and J.L. Bredas, Chemical Reviews 2007, Vol. 107, 926-952
[2] S.R. Forrest and M.E. Thompson, Chemical Reviews, 2007, Vol. 107, 923-925
[3] A. Aviram and M.A. Ratner, Chemical Physics Letters, 1974, Vol 29, 277;
G.Heimel, F. Rissner and E.Zojer, Adv. Mater. 2010, 22, 2494-2513
[4] Y.Zhao, W. Pérez-Segarra, Q. Shi and A. Wie, J. Am. Chem. Soc., 2005, 127 (20), pp 7328-7329

HYBRIDIZATION OF PARALLEL CARBON NANOTUBE QUANTUM DOTS

Carola Meyer[1], K. Goß[1], N. Peica[2], S. Smerat[3], M. Leijnse[4], M. R. Wegewijs[1], C. Thomsen[2], J. Maultzsch[2], C. M. Schneider[1]

[1]Peter Grünberg Institute, Research Centre Jülich & JARA Fundamentals of Future Information Technologies, Jülich, Germany;
[2]Institut für Festkörperphysik, Technische Universität Berlin, Berlin, Germany;
[3]Physics Department, Arnold Sommerfeld Center for Theoretical Physics, Ludwig-Maximilians-Universität München, München, Germany;
[4]Niels Bohr Institute & Nano-Science Center, University of Copenhagen, Copenhagen, Denmark

Carbon nanotubes (CNTs) are ideal one-dimensional conductors and show exceptional transport properties as for instance ballistic transport. Furthermore, like other molecular systems CNTs can strongly interact with their environment. For instance interaction with neighbouring molecules can lead to strong changes in the transport behaviour [1]. Here, we show that the molecular hybridization of CNTs bundled together in a rope is clearly reflected in the quantum transport behaviour.

To interpret the transport measurements correctly, first the structure of the actual transport device has to be characterized. Tip-enhanced Raman spectroscopy (TERS) offers an increased spatial resolution with a high signal enhancement and can thus be used as a probe for single molecules. We identify several carbon nanotubes with different diameters (see Figure 1(a)) and electronic character from the diameter-dependent Raman modes. Besides chiral index assignment, the resonance effect of the measurements done with different laser energies reveals a red shift of the resonances, which can be attributed to re-hybridization within the rope [2].

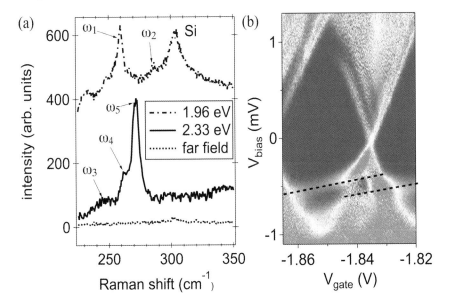

Figure 1: (a) TERS measurements at two laser energies on the carbon nanotube rope, which was used for quantum transport spectroscopy (b). Different radial breathings modes ω_i indicate CNTs with different diameters. Signatures of five different tubes within the rope are indicated in these measurements. Only a very weak silicon signal is visible in the far field when the tip is not approached. (b) Quantum transport measurement of the CNT device shows clear additional resonances (dashed lines) within the coulomb blockade indicating a parallel quantum dot formed on a different strand on the CNT rope.

We show that the number of tubes found in the TERS measurements, as well as their hybridization, is reflected in quantum transport measurements on the same device. Parallel quantum dots are formed on the different CNT strands of the rope (see Figure 1(b)). Hybridization of electronic states between parallel molecular quantum dots shows up as anti-crossings in the transport pattern [3]. We extract both the magnitude and the sign of the hybridization and find that current transport occurs via the bonding states of the coupled QD system. Differential gating is used to add charges to individual stands of the CNT rope and to control the coupling between electronic states. Furthermore, by applying a magnetic field the electronic hybridization is selectively suppressed due to spin effects. This offers prospects for accessing individual charge and spin degrees of freedom in coupled carbon-based molecular systems.

[1] M. Urdampilleta et al., Nature Materials **10**, 502 (2011).
[2] K. Goß et al., Phys. Status Solidi B **248**, 2577 (2011).
[3] K. Goß et al., Phys. Rev. B **83**, 201403(R) (2011).

LOW-TEMPERATURE SCANNING PROBE MICROSCOPY EXPERIMENTS ON ATOMICALLY WELL-DEFINED GRAPHENE NANORIBBONS

Peter Liljeroth[1], Joost van der Lit[2], Mark P. Boneschanscher[2], Mari Ijäs[1], Andreas Uppstu[1], Ari Harju[1], Daniël Vanmaekelbergh[2]

[1]Department of Applied Physics, Aalto University School of Science, Finland; [2]Condensed Matter and Interfaces, Debye Institute for Nanomaterials Science, Utrecht University, the Netherlands

Despite the availability of good quality, large scale graphene layers, the realization of both the room-temperature graphene transistor as well as the more advanced theoretical ideas require well-defined samples, in particular in terms of the graphene edge structure. This level of control is currently not available through conventional lithographic techniques and there is a lack of experimental data on atomically well-defined graphene nanostructures. For example, opening a sufficient gap for room-temperature operation through quantum confinement requires structures in the size range of 10 nm. Furthermore, the electronic structure of graphene nanostructures is sensitive to the structure of the edges (e.g. zig-zag vs. armchair) [1]. There is a possibility of edge reconstructions and attachment of various functional groups, which further complicate the comparison between theory and experiment [2].

We use on-surface polymerization from molecular precursors to form narrow graphene nanoribbons (GNRs) [3]. The GNRs have a fixed width and edge termination as determined by the choice of chemical precursor, resulting in armchair edges along the long axis of the ribbon and zigzag ends along the short axis. Graphene interacts only weakly with the underlying Au(111) substrate and retains the electronic structure of isolated graphene.

We explore the size-dependent electronic properties of these atomically well-defined graphene nanostructures using low-temperature scanning tunneling microscopy (STM) and atomic force microscopy (AFM). We measured the atomic structure of individual GNRs and spatially resolved the local density of states at different energy values. We find that the electronic states of the GNRs close to the Dirac point are located at the zigzag ends of the nanoribbons, whereas the states away from the Dirac point show increased amplitude along the armchair edges (Figure 1). Comparison of our experimental results with density functional theory and tight-binding calculations allows us to unravel the nature of these findings in detail.

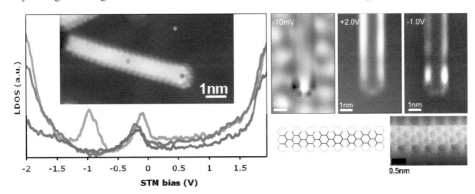

Figure 1: Energy spectroscopy on a GNR (left). Mapping the local density of states shows that states around the Dirac point are localized on the zigzag ends of the ribbon, while states far away from the Dirac point are more localized along the armchair edges (right top). The model for the GNR indicating the armchair edges in blue and the zigzag ends in red shown together with an AFM image of the atomic backbone of one of the GNRs (right bottom).

[1] A.H. Castro Neto et al., Rev. Mod. Phys. 81, 109 (2009).
[2] P. Koskinen, S. Malola, H. Häkkinen, Phys. Rev. Lett. 101, 115502 (2008).
[5] J. Cai et al., Nature 466, 470 (2010).

CORRELATIONS BETWEEN SWITCHING OF CONDUCTIVITY AND OPTICAL RADIATION OBSERVED IN THIN CARBON FILMS

Sergey G.Lebedev

Institute for Nuclear Research of Russian Academy of Sciences

At the study of conductive properties of some carbon condensates, produced by methods of the carbon arc (CA) and chemical vapor deposition (CVD), the phenomenon of spasmodic increase in electrical resistance on ~4-5 orders of magnitude under electric current increase up to some critical value is revealed (see Fig.1). The critical current decreases with temperature and at a room temperature has the values of 5 - 500 m (depending on deposition condition of a carbon film and its further processing) at the applied DC voltage in the range of 5-50 V. This work presents the results on registration of IR radiation caused by sharp change of conductivity in thin graphite-like carbon films. The IR radiation with the wavelength of about 1 micron has been registered with the slow and fast silicon avalanche photo diodes (see Fig.2). During the moment close to switching an oscilloscope registered the optical radiation consisting the series of optical pulses which amplitudes considerably exceeds a level of "substrate" so enters the photo diode into a condition of saturation. The possible explanation of the results observed is presented.

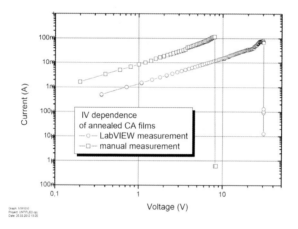

Figure 1. Switching of conductivity in carbon films.

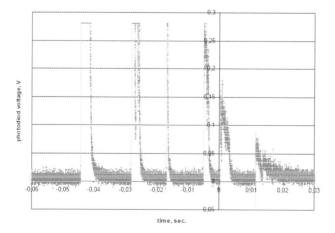

Figure 2. Optical radiation detected by fast photodiode in the moment of switching of conductivity.

FIRST-PRINCIPLES AND SEMI-EMPIRICAL VAN DER WAALS STUDY OF π-CONJUGATED MOLECULES PHYSISORBED ON GRAPHENE AND A BORON NITRIDE LAYER

Caciuc Vasile[1], Nicolae Atodiresei[1], Martin Callsen[1], Predrag Lazic[2], and Stefan Blügel[1]

[1]Peter Grünberg Institut (PGI-1) and Institute for Advanced Simulation (IAS-1), Forschungszentrum Jülich, Germany; [2]Massachusetts Institute of Technology, Cambridge, Massachusetts, USA

In the last decade, graphene has emerged as a material of choice in the field of nanoelectronics. In particular, recent experiments [1] suggested that the hexagonal boron nitride (BN) could be used to decouple graphene from a reactive substrate like SiO_2 to preserve its unique electronic structure.

Therefore, in the present contribution we analyze in detail the crucial role played by the van der Waals (vdW) interactions on the adsorption of molecular π-systems like benzene (C_6H_6), triazine ($C_3N_3H_3$) and borazine ($B_3N_3H_6$) on graphene and a BN sheet to gain a fundamental understanding on how to tune the properties of graphene-based electronic devices. Note that benzene and borazine are the molecular counterparts of graphene and BN, respectively.

To determine the proper ground-state adsorption geometry, the vdW interactions are mandatory and they were included in our first-principles density functional theory (DFT) simulations by using the semi-empirical DFT-D2 [2] and DFT-D2 [3] methods as well as the *ab initio* vdW-DF non-local correlation functional [4], the latter as implemented in our real-space code JuNoLo [5].

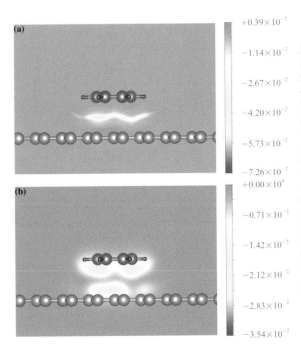

Figure 1: (a) Semi-local versus (b) non-local correlation binding energy density for benzene on graphene. The semi-local correlation binding energy density is localized at the interface between the molecule and substrate where the Pauli repulsion effects are important. On the contrary, the non-local correlation binding energy density is delocalized over the molecular plane and the graphene underneath reflecting the non-local character of the vdW interactions. More importantly, this striking qualitative difference is solely responsible for a significant quantitative difference between the adsorption energy evaluated with DFT without vdW corrections (~0.03 eV) and that obtained when including them via the vdW-DF functional (~0.7 eV).

The semi-empirical vdW calculations suggest that the strength of the molecule-surface interaction corresponds to a weak "chemisorption" although the bonding mechanism is physisorption since no net charge transfer between the molecules and the graphene and BN substrates takes place. As a general trend, the DFT-D2 approach predicts lower adsorption energies and molecule-surface equilibrium distances than the DFT-D3 method. Interestingly, the results obtained with the vdW-DF functional lie between those obtained with the semi-empirical approaches.

Beside this, starting from the DFT-D2 geometries, our vdW-DF calculations allowed us to reveal the importance of the non-local correction effects on the molecule-surface bonding mechanism which can not be assessed from the semi-empirical vdW methods. As an example, as shown in Figure 1, the visualization of the non-local versus semi-local correlation binding energy density of the benzene-graphene system provides a clear understanding of the qualitative difference between a non-local and a semi-local description of the correlations effects present in molecular systems with π-π stacking. Importantly, only this qualitative difference is responsible for a significant increase of the adsorption energy evaluated with DFT without vdW corrections (~0.03 eV) with respect to that obtained when including them via the vdW-DF functional (~0.7 eV).

[1] C. R. Dean et al., Nat. Nanotech. **5**, 722 (2010).
[2] S. Grimme, J. Comput. Chem. **27**, 1787 (2006).
[3] S. Grimme et al., J. Chem. Phys. **132**, 154104 (2010).
[4] M. Dion et al., Phys. Rev. Lett. **92**, 246401 (2004).
[5] P. Lazic et al., Comput. Phys. Commun. **181**, 371 (2010).

FINITE-TEMPERATURE EXACT DIAGONALIZATION STUDY OF THE HUBBARD MOLECULES IN HETEROSTRUCTURES

H. Ishida[1], A. Liebsch[2]

[1]College of Humanities and Sciences, Nihon University, Tokyo 156-8550, Japan;
[2]Peter Grünberg Institut and Institute of Advanced Simulations, Forschungszentrum Jülich, D-52425 Jülich, Germany

A key issue in developing single molecular devices is to clarify the effects of strong electron correlations on the properties of finite interacting systems. A model suitable for this purpose is the Hubbard-type Hamiltonian made out of N atomic sites with onsite repulsive interactions and linked to non-interacting leads at both ends. The model with a single interaction site ($N=1$) was used to discuss the Kondo prolem in quantum dots. More recently, various techniques such as numerical renormalization group theory were applied to the systems with $N>1$ to study the Kondo effect and transport properties of atomic chains and single molecules linked between leads.

In the present work, we adopt the finite-temperature exact diagonalization (ED) method to study the equilibrium (zero-bias limit) electronic structure of the Hubbard molecules ($N=2$ to 5) as functions of chemical potential, Coulomb repulsion energy, and the hopping integral between lead and molecular sites. In doing so, the semi-infinite cubic lattice representing the two leads is mimiced by a 5-site cluster, where the tight-binding parameters of the cluster are chosen such that the difference between the surface-site Green function of the cluster and that of the semi-infinite lead is minimized on Matsubara frequencies. Then, the finite-temperature Green function of the total system made out of the molecule plus two clusters (N+10 sites) is evaluated by applying the ED technique and the Lanczos procedure for the Green function.

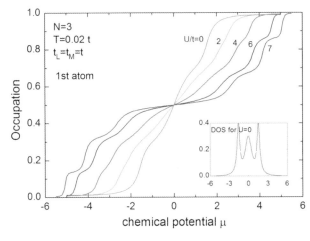

Figure 1: Electron occupation of the outer atomic site of the $N=3$ chain molecule between two leads. Energy is measured in units of the hopping integral between nearest-neighbor (NN) sites in the molecule, t. The hopping integral between the NN sites in metal leads (t_M) and that between the lead and molecular sites (t_L) are chosen as t, and temperature T is $0.02t$. The inset shows the partial DOS of the outer site when $U=0$.

As an example, we show in Figure 1 the occupation per spin of the outer atomic site of the $N=3$ chain molecule as a function of chemical potential μ, where the system is electron-hole symmetric at $\mu=0$. It is seen that a region with a small slope appears in the occupation curves around $\mu=0$ with increasing Coulomb repulsive energy U, indicating that electron correlation effects reduce the quasiparticle density of states (DOS) at μ most significantly for the half-filled system. Figure 2 shows $-G_{11}(\beta/2)$, the quasiparticle density of states (DOS) integrated around μ, of the outer atomic site of the $N=3$ chain molecule as a function of chemical potential μ. For small U values, it possesses 3-peak structure as in the case of the DOS of the non-interacting chain. On the other hand, for larger U values, $-G_{11}(\beta/2)$ exhibits a minimum for the electron-hole symmetric case, and three peaks appear in both hole and electron doping sides.

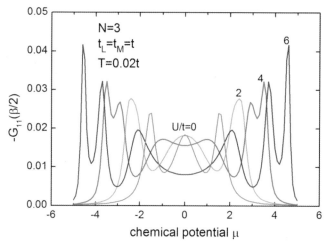

Figure 2: $-G_{11}(\beta/2)$, Integrated quasiparticle DOS around μ, as a function of chemical potential for the $N=3$ chain molecule between two leads. The parameters are the same as those in Figure 1.

DFT+CI CALCULATIONS OF QUANTUM DOTS IN GRAPHENE NANORIBBONS

<u>Tobias Burnus</u>[1], Gustav Bihlmayer[1], Daniel Wortmann[1], Ersoy Şaşıoğlu[1],
Yuriy Mokrousov[1], Stefan Blügel[1], Klaus Michael Indlekofer[2]

[1]Peter Grünberg Institut & Institute for Advanced Simulation, Forschungszentrum Jülich and JARA, 52425 Jülich, Germany; [2]Hochschule RheinMain, Unter den Eichen 5, 65195 Wiesbaden, Germany

Graphene nanoribbons (GNR) hold great future promise for field-effect transistors and quantum-dot devices. The gate electrodes and the electric field distribution play a crucial role. With gate electrodes an in-plane electric field can be generated, which localizes quantum-dot states in the bandgap of armchair GNR. Density-functional theory (DFT) calculations [1] have been used to calculate GNR under an in-plane gate electric field, taking correctly the edge termination of the ribbon into account. The results obtained via DFT have been combined with the screened Coulomb interaction calculated within the random-phase approximation [2] to setup a configuration interaction (CI) calculation [3] for the quantum dot, which properly describes the multiplet states of the few electrons in the quantum dot. Results of the DFT+CI calculations will be shown.

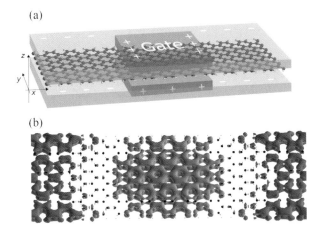

Figure 1: (a) Free-standing, hydrogen-terminated armchair graphene nanoribbon of 13 atoms width. The inner rectangles indicate the location of the positively charged top and bottom gate electrodes. Their charge is compensated by the negatively charged top and bottom electrodes.

(b) Total charge density difference between a calculation with gate electric field and without; the density is reduced at outside and increased in the centre.

The work is supported by the DFG Research Unit 912 "Coherence and Relaxation Properties of Electron Spins."

[1] FLEUR, http://flapw.de.
[2] Ersoy Şaşıoğlu, Christoph Friedrich, and Stefan Blügel, Phys. Rev. B **83**, 121101(R) (2011).
[3] Newbpotcoul, http://sf.net/projects/newbpotcoul

Poster Sessions

Electronic structure, lattice dynamics, and transport (ELT)	473
Memristive systems (MEM)	523
Spin-related phenomena (SRP)	589
Polar dielectrics, optics, and ionics (POL)	633
Advances in technology and characterization (ATC)	665

FUNDAMENTAL PROPERTIES OF THE SUPERCONDUCTING STATE AT THE LaAlO$_3$/SrTiO$_3$ INTERFACE

Christoph Richter, Hans Boschker, Werner Dietsche, Jochen Mannhart

Max Planck Institute for Solid State Research, Stuttgart, Germany

The LaAlO$_3$/SrTiO$_3$ interface provides an intriguing 2-dimensional electron system in which the coexistence of superconductivity and magnetism has been observed. It remains unclear whether the same electrons are responsible for both superconductivity and magnetism. It has been suggested that the superconductivity and the magnetism are spatially separated: scanning SQUID microscopy measurements showed ferromagnetic domains surrounded by superconducting regions. A separation of the two states between different bands has been proposed as well: the magnetism is due to localized electrons in the d_{xy} derived band, while the superconductivity occurs in the d_{xz} and d_{yz} derived bands. Nevertheless, it is also possible that the groundstate of the LaAlO$_3$/SrTiO$_3$ interface is an unconventional superconductor which is intimately connected to the magnetism. We will present detailed transport measurements of the superconducting state in order to clarify this issue.

TORQUE MAGNETOMETRY ON LaAlO$_3$-SrTiO$_3$ HETEROSTRUCTURES

M. Brasse[1], R. Jany[2], Ch. Heyn[3], J. Mannhart[4], M.A. Wilde[1], D. Grundler[1]

[1]Lehrstuhl für Physik funktionaler Schichtsysteme, Technische Universität München, Physik Department, James-Franck-Str. 1, D-85747 Garching, Germany;
[2]Experimentalphysik VI, Institut für Physik, Universität Augsburg, Universitätsstraße 1, D-86135 Augsburg, Germany;
[3]Institut für Angewandte Physik, Universität Hamburg, Jungiusstraße 11, D-20355 Hamburg, Germany;
[4]Max Planck Institut für Festkörperforschung, D-70569 Stuttgart, Germany

Two-dimensional electron systems (2-DESs) have been found to form at the interface between the otherwise insulating oxides LaAlO$_3$ and SrTiO$_3$ [1]. The strongly correlated electron system shows a metallic phase and coexistence of superconductivity [2] and magnetism at low temperatures [3,4,5]. To explore the nature of the magnetic phase we use highly sensitive micromechanical torque magnetometry [6]. This technique allows us to address different phenomena such as the de Haas-van Alphen effect [7], dia- and paramagnetism in superconducting states as well as magnetic hysteresis and anisotropy [8] in case of correlated magnetism.

Here we present preliminary results of ongoing angular dependent torque magnetometry measurements on SrTiO$_3$-LaAlO$_3$ heterostructures at different magnetic fields and temperatures. The heterostructures consist of a 100 μm thick SrTiO$_3$ substrate on top of which four unit cells of LaAlO$_3$ were grown by pulsed laser deposition [9]. We find that the magnetic anisotropy is dominated by a fourfold symmetry. These findings suggest either a cubic anisotropy or two crossed uniaxial anisotropies. Surprisingly, the magnetic easy axes seem to be oriented at ~45° with respect to the interface plane. In addition, we observe signatures of rotational hysteresis, indicating ferromagnetism in the interface. These results hint at a highly nontrivial anisotropic magnetic behavior of the strongly correlated electron systems in SrTiO$_3$-LaAlO$_3$ heterostructures.

This work is supported by the DFG via TRR 80 "From Electronic Correlations to Functionality" and by the German Excellence Cluster Nanosystems Initiative Munich (NIM).

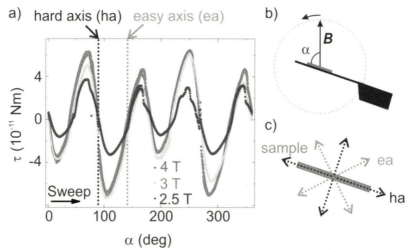

Figure 1: a) Torque as a function of angle between magnetic field and sample. The data implies a fourfold symmetry in the magnetic anisotropy with position of easy and hard axis as denoted. b) Orientation of magnetic field with respect to the SrTiO$_3$-LaAlO$_3$ heterostructure and torque magnetometer (side view). c) Orientation of easy and hard magnetic axes (side view).

[1] A. Ohtomo, H.Y. Hwang, Nature **427**, 423 (2004).
[2] N. Reyren *et al.*, Science **317**, 1196 (2007).
[3] L. Li *et al.*, Nature Physics **7**, 762 (2011).
[4] D.A. Dikin *et al.*, Phys. Rev. Lett. **107**, 5 (2011).
[5] J.A. Bert *et al.*, Nature Physics **7**, 767 (2011).
[6] M.A. Wilde, D. Grundler, D. Heitmann in *Quantum Materials,* edited by D. Heitmann (Springer, 2010) pp. 245-275.
[7] M.A. Wilde *et al.*, Phys, Rev. B. **73**, 125325 (2006).
[8] B. Rupprecht *et al.*, J. Appl. Phys. **107**, 093711 (2010).
[9] S. Thiel *et al.*, Science 313, 1942-1945 (2006).

BUILDING UP A STRATEGY TO CONTROL LATTICE THERMAL CONDUCTIVITY IN LAYERED COBALT OXIDES

Masahiro TADA[1,2], Yohei MIYAUCHI[1], Masato YOSHIYA[1,3], Hideyuki YASUDA[1]

[1]Department of Adaptive Machine Systems, Osaka University, Suita, Osaka, Japan;
[2]Research Fellow of the Japan Society for Promotion of Science, Chiyoda, Tokyo, Japan;
[3]Nanostructures Research Laboratory, Japan Fine Ceramics Center, Nagoya, Aichi, Japan.

Newly emerged thermoelectric(TE) materials such as layered cobalt oxides including Na_xCoO_2, which can directly convert wasted heat to electricity, have attracted much attention since they have a higher figure of merit, originating in part from the layered structure, at elevated temperature and lower materials cost than conventional Bi-Te based alloy. However, there is still no clear pathway toward controlling their lattice thermal conductivity, one of three factors to determine a figure of merit, since conventional theories of thermal conduction developed for metals often fail to be applied for oxides. In this study, aside from the theories, we calculated the lattice thermal conductivity in layered oxide TE materials using perturbed molecular dynamics methods and tried to reveal the mechanisms of thermal conduction, thereby building up a strategy to control it.

It is found that Na vacancies are responsible for supressing in-plane lattice thermal conductivity through disturbing vibrations of CoO_2 layers which mainly conduct thermal energy. For deeper understanding, we have compared Na_xCoO_2 with Li_xCoO_2, K_xCoO_2 and Ca_xCoO_2 to reveal what property of the cations at M sites in M_xCoO_2 dominates the decrease in lattice thermal conductivity. As we expected, when M vacancies were introduced, lattice thermal conductivity was decreased in all these cobaltites. However, against our simple expectations, the magnitude of the decrease was lower in the M_xCoO_2 when M ions are larger. Further analyses indicate that two-dimensionality of each layer affects the magnitude of the decrease in thermal conductivity. Besides, the lattice thermal conductivity can also be controlled by valences of M sites via distortion of CoO_6 octahedra.

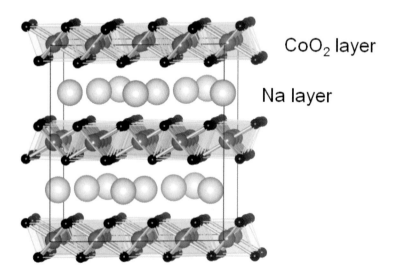

Figure 1: A crystal structure of $Na_{0.5}CoO_2$. Na_xCoO_2 consists of Na layers and CoO_2 layers stacked alternatively. The Na layers have Na vacancies when $x < 1$. Co and O compose edge-sharing CoO_6 octahedra in CoO_2 layers.

THE EFFECT OF AMMONOLYSIS ON THE STRUCTURE AND THERMOELECTRIC PROPERTIES OF $EuTiO_3$ AND $EuTi_{0.98}Nb_{0.02}O_3$

Leyre Sagarna[1], Alexandra Maegli[1], Songhak Yoon[1], Sascha Populoh[1], Andrey Shkabko[1], Anke Weidenkaff[1]

[1]Empa – Swiss Federal Laboratories for Materials Science and Technology, Solid State Chemistry and Catalysis, Überlandstrasse. 129, CH-8600 Dübendorf, Switzerland

Transition metal oxides have shown good performance in the field of thermoelectricity. Perovskite-type $EuTiO_3$ is one of the promising thermoelectric n-type oxides, with a figure of merit (ZT) which can be enhanced by Nb substitution on the Ti position.

Additionally, oxynitrides have drawn attention due to their reduced band gap in comparison with the corresponding oxides [1]. XPS analyses have been performed in order to investigate the core level and valence band electronic structures of polycrystalline $EuTiO_xN_y$ and $EuTi_{0.98}Nb_{0.02}O_xN_y$. The nitrogen content has been determined by hotgas extraction and the optical band gaps have been estimated from UV-Vis reflectance measurements. The crystal structure of oxynitrides, studied by X-ray-, electron- and neutron-diffraction, has been compared with the structure of oxides.

The thermoelectric properties of the oxynitrides have been investigated in the temperature range of $200K < T < 973$ K. The electrical resistivity the oxynitrides is up to 3 orders of magnitude higher than for the oxides. A morphological characterization by SEM and TEM illustrated weak grain-interconnections indicating inefficient intergrain electron transport. To improve the grain connectivity of oxynitrides, more advanced sintering techniques like spark plasma sintering would be needed.

[1] Ebbinghaus, S. G. *et al.* Perovskite-related oxynitrides – Recent developments in synthesis, characterisation and investigations of physical properties. *Progress in Solid State Chemistry* **37**, 173–205 (2009).

ELECTRICAL TRANSPORT INVESTIGATIONS ON NANOPARTICLE TEST STRUCTURES

Silvia Karthäuser[1], Marcel Manheller[1], Rainer Waser[1], Kerstin Blech[2], Ulrich Simon[2]

[1]Peter Grünberg Institut (PGI-7) and JARA-FIT, Forschungszentrum Jülich GmbH, D-52425 Jülich, Germany;
[2]Inorganic Chemistry (IAC) and JARA-FIT, RWTH Aachen University, D-52056 Aachen, Germany

In order to achieve the next generation of nanometer sized electronic devices a detailed understanding and control of electrical transport is essential. One approach to fabricate nanodevices based on functional components is to assemble a 3D-array of nanoparticles on electrode structures, while another method is to bridge the gap between two nanoelectrodes by a single nanoparticle. Here we report on electronic transport measurements of biphenylpropanethiol (BP3) capped gold nanoparticles (AuNPs) with a diameter of 4 nm used as functional units studied in both set-ups.

Possible transport mechanisms in such "nanoelectrode-(molecule-NP-molecule)n-nanoelectrode" systems, i.e. in arrays of NPs and single NPs between nanoelectrodes (Fig. 1), resp., are explored by temperature dependant complex impedance or current/voltage measurements. Complex impedance measurements were performed on IDE structures (interdigitated electrode structures) on which the cluster solution was drop-casted to form a densely packed film. The electrical transport properties of single BP3 capped AuNPs were investigated after immobilization between gold nanoelectrodes with a separation of approximately 3 - 5 nm fabricated by electron-beam lithography [1].

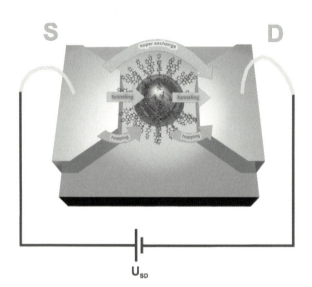

Figure 1: Schematic showing a single BP3 capped AuNP in between nanoelectrodes and possible charge transport mechanisms.

Differing from the expected behavior for ligand-stabilized AuNPs known from literature [2] the Arrhenius-Plot of the BP3-AuNPs array exhibits three different transport regimes (upper curve in Fig. 2): I) 300 K – 340 K linear dependence of the conductance according to the Arrhenius relation; II) 230 K – 290 K: discontinuity in the curve, III) 100 K – 220 K linear dependence of the conductance according to the Arrhenius relation. The electrical transport properties of the AuNP array can be explained by an simply activated, Arrhenius-like transport based on electron hopping between nearest neighbors at high temperatures (I) while at lower temperatures (III) a differentiation between the granular metal model and the superexchange coupling model based on these data is not possible.

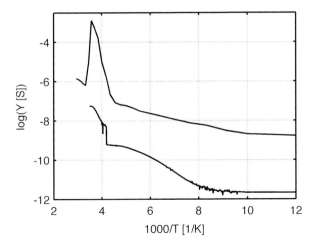

Figure 2: Arrhenius-type diagram of the electric conductance (Y) through a NP-array (upper curve) and the nano-electrode/BP3-NP-BP3/nanoelectrode device (lower curve) while applying a constant source-drain voltage, $U_{SD} = 1$ V.

The temperature dependent conductance of a single BP3 capped AuNP is given in Fig. 2 as well (lower curve). Four different transport regimes are identified clearly and can be attributed according to their temperature dependence to different transport mechanisms: (I) (265 K – 245 K) with an exponential temperature dependence, (II) (245 K – 200 K) intermediate region with significant singularity in conductance, (III) (200 – 110 K) with a slope indicating a weaker temperature dependence characteristic for superexchange coupling and (IV) (T < 110 K) with no significant temperature dependence corresponding to a tunnelling mechanism.

In conclusion, the charge transport behavior of BP3 AuNPs has been studied as well in a single NP device as in a 3D array and the complex temperature dependence of the conductance has been characterized in detail. However, most interestingly in temperature region (II) a reproducible discontinuity in conductance is measured for both experimental set-ups. It clearly shows up as a sudden change in conductance for a single NP while it is broadened up to 60 K for the NP array.

[1] M. Manheller, S. Trellenkamp, R. Waser, S. Karthäuser, Nanotechnology **23**, 125302 (2012)
[2] K. Blech, M. Homberger, U. Simon in G. Schmid, G. (ed.) Nanoparticles – From Theory to Application, Wiley-VCH, Weinheim, Germany, 401-454 (2010).

FIRST-PRINCIPLES CALCULATION OF MAGNETISM OF SUBSTITUTIONAL TRANSITION IMPURITIES IN BINARY IRON-SELENIUM SYSTEM

T. Pengpan, A. Boonthummo

Department of Physics, Faculty of Science, Prince of Songkla University, Hatyai, Thailand

An iron-based pnictide has been found to be a new type of superconductors [1,2,3]. In this work, a binary iron-selenium system, which is one of the iron-based pnictide compounds, is studied by substituting a transition element X (Cu or V) at a corner as shown in Figure 1.

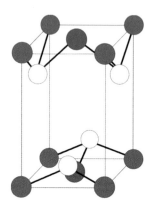

Figure 1: A binary FeSe tetragonal structure, where the Fe atoms are situated in the planar square planes, and the Se atoms are above and below the Fe-based planes and tetrahedral-coordinated to the Fe atoms. Within virtual crystal approximation, the pseudopotential of the Fe atom at all corners is alchemically mixed to the pseudopotential of a transition X(= Cu or V).

The $Fe_{1-x}X_xSe$ disordered compound is virtual and defined as a spatial average of the real one by an approach so called virtual crystal approximation (VCA) [4]. From first-principles based on density functional theory, a pseudopotential of the virtual atom can be constructed with a composition-weighted average of the real atoms. The Fe and X pseudopotentials are generated with equal cut-off radii of $2.1 a_0$ in $3d$-orbitals. Their fully optimized structures are calculated by using ABINIT code [5]. The lattice parameters of the $Fe_{1-x}X_xSe$, where x is increased at a step of 0.01 from 0.00 to 0.20, can be fitted by a modified Vegard law [6]: $a(x) = 4.430x + 3.777(1-x) - 0.462x(1-x)$ and $c(x) = 5.664x + 5.533(1-x) - 0.688x(1-x)$ for $Fe_{1-x}Cu_xSe$, and $a(x) = 4.018x + 3.777(1-x) - 0.100x(1-x)$ and $c(x) = 5.095x + 5.532(1-x) - 0.194x(1-x)$ for $Fe_{1-x}V_xSe$. However, it is noted that the experimental lattice parameters of $Fe_{1.01-x}Cu_xSe$ in Ref. [3] can be fitted by just the linearized Vegard law: $a_{exp}(x) = 4.149x + 3.769(1-x)$ and $c_{exp}(x) = 5.241x + 5.521(1-x)$.

Their band structures are calculated and show a similar profile, possessing multi-band crossing their Fermi energy (E_F) levels. However, as x increases, E_F shifts lower, resulting in variation of density of states at E_F and deforming in topology of Fermi surfaces similar to other electron-doped pnictide compounds. Lastly, their local magnetic moments at the virtual atoms are calculated. As shown in Figure 2, they vary dependent upon a proportion of the transition element X. For x less than 0.08, both compounds show no local magnetic moment. But for x above 0.09 their local magnetic moments drastically change, jumping to be about $2.5\mu_B$ and then gradually decline. Magnetism in the compounds possibly plays a key role in metal-insulator transition [3,7].

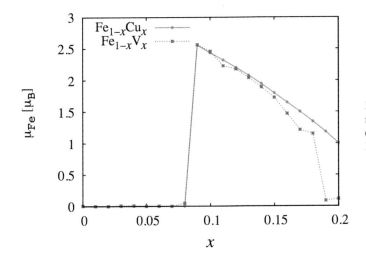

Figure 2: Local magnetic moments of the pseudo-atom ($Fe_{1-x}X_x$) at corners, solid for $Fe_{1-x}Cu_x$ and dash for $Fe_{1-x}V_x$.

[1] F.-C. Hsu et al., PNAS. **105**, 14262 (2008).
[2] A. Subedi, L. Zhang, D. J. Singh, M. H. Du, Phys. Rev. B **78** 134514 (2008).
[3] A.J. Williams et al., J. Phys. Condens. Matter. **21**, 305701 (2009).
[4] R. Poloni et al., J. Phys.: Condens. Matter. **22**, 415401 (2010).
[5] X. Gonze et al., Computer Phys. Commun. **180**, 2582 (2009).
[6] Y.-K. Kuo et al., Optics Commun. **237**, 363 (2004).
[7] S. Chadov, D.Schärf, G.H. Fecher, C. Felser, Phys. Rev. B. **81**, 104523 (2010).

FIRST-PRINCIPLES CALCULATIONS OF PHONON-PHONON INTERACTION IN ROCK-SALT TYPE CRYSTALS

Atsushi Togo[1], Laurent Chaput[2], Isao Tanaka[1,3]

[1]Department of Materials Science and Engineering, Kyoto University, Kyoto, Japan; [2]Institut Jean Lamour, UMR CNRS 7198, Nancy Université, Nancy, France; [3]Nanostructures Research Laboratory, Japan Fine Ceramics Center, Nagoya, Japan

Anharmonic properties of rock-salt type crystals, MgO, LiF, NaF, KCl, NaCl, and PbTe, are investigated using first-principles calculations. The lowest order imaginary parts of phonon self energies $\Gamma(\omega)$ for those crystals were calculated from the third-order force constants. The supercell approach with the finite displacement method [1] as implemented in phonopy code [2] was employed to obtain the third-order force constants. The forces on atoms were calculated using VASP code [3-5] with GGA-PBE. In this abstract, only the results of MgO are presented.

The imaginary part of self energy of LO-mode of MgO at Γ-point is shown in Fig. 1. The two phonon density of states (two phonon DOS) is the DOS of phonon triplets that satisfy the sum rule among three wave vectors and the conservation of energy [1]. Therefore the shape of the two phonon DOS becomes similar to that of the harmonic phonon DOS. When we consider a decay of an optical phonon at Γ-point, a simple model is often employed. In this model, two acoustic phonons are created symmetrically in the Brillouin zone under the conditions among the frequencies, $\omega_O = 2\omega_A' = 2\omega_A''$, and the wave vectors, $\mathbf{q}_A' + \mathbf{q}_A'' = \mathbf{0}$, and therefore the shapes of $\Gamma(\omega)$ and the two phonon DOS should also be similar. However $\Gamma(\omega)$ shows that this simple model is inappropriate even for MgO since the peak position of $\Gamma(\omega)$ deviates from that of the two phonon DOS. It is considered that this deviation is originated from the phonon-phonon interaction coefficients since $\Gamma(\omega)$ is the product of two phonon DOS and the phonon-phonon interaction coefficients. Indeed, the calculated decay channels show that these decay processes are asymmetric; 58 % of the LO-mode phonon decays into TA- and TO-modes and 42 % of that decays into TA- and LA-modes.

$2\Gamma(\omega_{LO})$ of the LO-mode and $2\Gamma(\omega_{TO})$ of the TO-mode of MgO as a function of temperature are presented in Fig. 2. The calculations approximately agree with the experiments by Hisano and Toda [6]. The two experiments for the LO-mode are inconsistent to each other. The values of the experiments by Jasperse et al. [7] and Hisano and Toda were obtained by analyzing infrared reflection spectra and thermal emission spectra, respectively. Their results largely depend on the models that they employed. Therefore the calculations and the experiments may not be directly compared. If linewidth of the spectrum is measured, it can be directly compared with the calculated 2Γ at the frequency of the phonon mode. However to our knowledge, such an experiment is not available for MgO.

This study was supported by the Grant-in-Aid for Scientific Research on Priority Areas of Nano Materials Science for Atomic Scale Modification (No. 474) from Ministry of Education, Culture, Sports, Science and Technology (MEXT) of Japan.

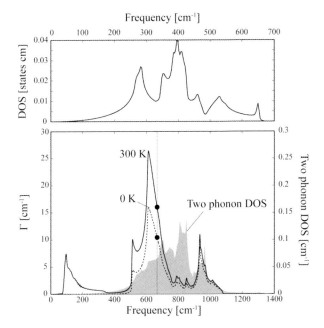

Figure 1: Imaginary part of phonon self energy $\Gamma(\omega)$ of LO-mode of MgO. The top figure shows the harmonic phonon DOS. The bottom figure depicts $\Gamma(\omega)$ at 0 K (dashed curve) and 300 K (solid curve) and two phonon DOS. The vertical line at around 650 cm^{-1} represents the LO-mode phonon frequency. The circles on the vertical line give $\Gamma(\omega_{LO})$ at 0 K and 300 K.

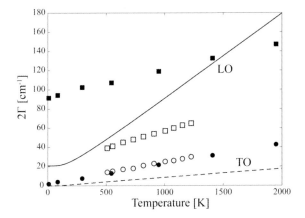

Figure 2: $2\Gamma(\omega_{LO})$ of LO- and TO-modes of MgO as a function of temperature. The solid and dashed curves show the calculations of the LO- and TO-modes, respectively. The squares and circles depict the experiments of the LO- and TO-modes, respectively. The open and filled symbols give the experiments by Jasperse et al. [6] and Hisano and Toda [7], respectively.

[1] L. Chaput et al., Rhys. Rev. B **84**, 094302 (2011).
[2] A. Togo, F. Oba, and I. Tanaka, Phys. rev. B **78**, 134106 (2008).
[3] G. Kresse, J. Non-Cryst. Solids **193**, 222 (1995).
[4] G. Kresse and J. Furthmüller, Comput. Mater. Sci. **6**, 15 (1996).
[5] G. Kresse and D. Joubert, Phys. Rev. B **59**, 1758 (1999).
[6] K. Hisano and K Toda, J. Phys. C: Solid State Phys. **15**, 1111 (1982).
[7] J. R. Jasperse et al., Phys. Rev. **146**, 526 (1966).

MICROSTRUCTURAL CHARACTERIZATION OF VARISTOR CERAMICS AFTER ACCELERATED AGEING WITH DC VOLTAGE.

M. A Ramírez[1], A.Z, Simões[1], E. Longo[2], J. A Varela[2]

(1) Universidade Estadual Paulista, UNESP, Faculdade de Engenharia de Guaratinguetá, Av. Dr. Ariberto Pereira da Cunha, 333, CEP 12516-410, Guaratinguetá, SP, Brasil.
(2) Universidade Estadual Paulista, UNESP, Instituto de Química, Rua Prof. Francisco Degni, 55, CEP 14801-970, Araraquara, SP, Brasil

In this study, the structural and microstructural changes of ZnO and SnO_2 based varistors after the degradation process with continuous tension were analyzed. The SnO_2-based varistors showed no changes on electrical and microstructural after the degradation process (fixed dc bias voltage at different temperatures). However, the ZnO-based varistors showed degradation of the non-Ohmic properties due the phase transformation. Before the degradation process the ZnO- based varistors showed the presence of grains of zincite phase (n-type semiconductor behavior) and grain boundary and triple points with δ-Bi_2O_3 enriched phase (p-type semiconductor behavior) leading to *n-p-n* contacts explaining the non-Ohmic behavior. After degradation, it occurs a phase transformation of δ-Bi_2O_3 to BiO_{2-x} (n-type semiconductor behavior) damaging the *n-p-n* contacts which explains the degradation of the non-Ohmic properties. This phase transformation was detected by the X-ray diffraction (DRX) technique and confirmed by using transmission electron microscopy (TEM) with electron diffraction. Additionally, phase changes could be detected by scanning electron microscopy (SEM) which evidenced increase in porosity of the ZnO-based varistors after the degradation phenomenon.

FIRST-PRINCIPLES ELECTRONIC STRUCTURE OF β-FeSi$_2$ AND FeS$_2$ SURFACES

Pengxiang Xu[1], Timo Schena[1], Gustav Bihlmayer[1], Stefan Blügel[1]

[1]Peter Grünberg Institut & Institute for Advanced Simulation, Forschungszentrum Jülich and JARA, 52425 Jülich, Germany

Applying density functional theory in the framework of the full potential linearized augmented plane-wave (FLAPW) method FLEUR [1], we investigate the electronic structure of potential future photovoltaic materials, β-FeSi$_2$ and FeS$_2$, for selected surface orientations and terminations. Surface passivation has become an essential factor for translating high-efficiency solar cell concepts into industrial production schemes due to trapping of charge carriers in surface states at the passivated surface layer.

We study the atomic and electronic structure of β-FeSi$_2$ and FeS$_2$ thin films for (001) and (100) orientations with different terminations. The most stable orientations are determined by comparing their cohesive energy. Detailed electronic structure calculations show that surface states originating from Fe play an important role and might determine their photovoltaic properties. The effects of passivation on the electronic structure are also presented.

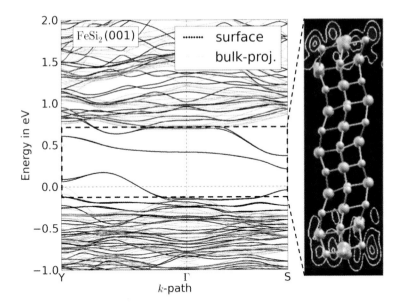

Figure 1: Left panel shows the band structure for a 3-layer FeSi$_2$ (001) film with the energetically favorable Si-terminated surface on both sides. The bulk projected band structures are marked by grey lines as a reference to distinguish surface states clearly. In right panel, integrated charge density over energy window from E_F-0.2eV to E_F+0.7eV is plotted and shows surface states originating from top/bottom Fe.

This work is supported by BMBF under project Nr. 03SF0402A (NADNuM).

[1] www.flapw.de.

ACCURATE BAND GAPS OF TRANSPARENT CONDUCTING OXIDES WITH A SEMILOCAL EXCHANGE-CORRELATION POTENTIAL

A. Thatribud, T. Tungsurat, T. Pengpan

Department of Physics, Faculty of Science, Prince of Songkla University, Hatyai, 90112, Thailand

Transparent conducting oxides (TCOs) are widely used as transparent electrodes for flat-panel displays and photovoltaic devices [1, 2]. They are materials that combine electrical conductivity and optical transparency, having carrier concentration of at least 10^{20} cm^{-3} and optical band gap more than 3 eV [3]. To study the electronic properties of a family of the *p*-type TCOs such as $CuBO_2$, $CuAlO_2$, $CuGaO_2$, and $CuInO_2$, we used first-principles calculations based on density functional theory (DFT), DFT with on-site Coulomb interaction (DFT+U) and density functional perturbation theory with quasi-particle approximation (GWA). Firstly, we calculated by using DFT and DFT+U for fully optimized structure and electronic band gaps of $CuBO_2$. It was found that errors of both direct and indirect band gaps from DFT+U were improved from DFT. It is also known that band gaps can be corrected by using GWA which treats strongly interacting electrons as weakly interacting quasi-particles. A quasi-particle energy is contributed by Kohn-Sham energy and a self-energy term (Σ) which composes of a single-particle Green's function (G) and a dynamically screened Coulomb interaction (W), shortly $\Sigma = iGW$. Theoretically, we need to solve Σ from Hedin's equation self-consistently [4]. However, Σ can be simply estimated to first-order correction by $\Sigma^{(1)} = iG_0W_0$. Our calculated results are shown in Figure 1. We compared results of $CuBO_2$ band gaps from different approaches by using ABINIT code [5]. In all our calculations, we used norm-conserving pseudo-potentials for G_0W_0 and TB09, the projected augmented-wave (PAW) for DFT and DFT+U. It is clearly seen that the Tran and Blaha (TB09) [6] and the DFT with hybrid functional PBE0 approaches yield the indirect and the direct band gaps close to its experimental values, respectively.

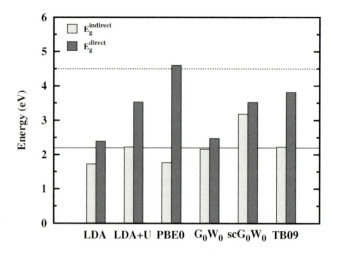

Figure 1: Band gaps of $CuBO_2$ using LDA, LDA+U, hybrid functional PBE0, G_0W_0 and scG_0W_0 (Trani *et al*'s G_0W_0@scCOHSEX [7]). The horizontal solid and dot lines at 2.20 and 4.50 eV are the experimental indirect and direct gaps, respectively [8].

Since, the quasi-particle self-consistent *GW* (QPsc*GW*) calculation according to Faleev *et al*'s scheme [9] is quite time- and memory-consuming. So, we are back to do the electronic band structure calculation as suggested by Tran and Blaha [6], where an exchange potential of Becke-Johnson (BJ) type is modified by including a semi-local one which can reproduce the shape of the exact exchange optimized effective potential of atoms.

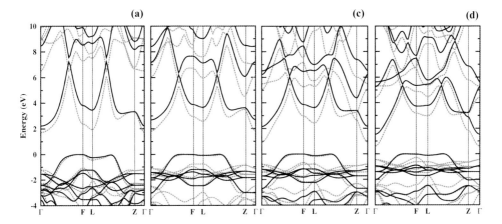

Figure 2: Band structures of (a) CuBO$_2$, (b) CuAlO$_2$, (c) CuGaO$_2$, and (d) CuInO$_2$, where the dot and solid lines for LDA and TB09, respectively. In each figure, the Fermi energy level is shifted to top of valence band.

Finally, we compared the calculated band structures of LDA and TB09 for CuBO$_2$, CuAlO$_2$, CuGaO$_2$, and CuInO$_2$ as shown in Figure 2. From our TB09 calculations for CuBO$_2$, CuAlO$_2$, CuGaO$_2$, and CuInO$_2$, the indirect band gaps are 2.22, 2.69, 2.15, and 1.53 eV, and direct ones are 3.81, 3.90, 3.86, and 3.75 eV, respectively. These results are close to their experimental values [10, 11, 12].

[1] T. Minami, Semicond. Sci. Technol. **20**, S35 (2005).
[2] H. Hosono, Int.J. Appl. Ceram. Technol. **1**, 106 (2004).
[3] R. G. Gordon, MRSBull. **25**, 52 (2000).
[4] L. Hedin, Phys. Rev. **139**, A796 (1965).
[5] X. Gonze et al., Comput, Phys. Commun. **180**, 2582 (2009).
[6] F. Tran and P. Blaha, Phys. Rev. Lett. **102**, 226401 (2009).
[7] F. Trani et al., Phys. Rev. B **82**, 085115 (2010).
[8] M. Snure and A. Tiwari, Appl. Phys. Lett. **91**, 092123 (2007).
[9] S. V. Faleev et al., Phys. Rev. Lett. **93**, 126406 (2004).
[10] H. Yanagi et al., Appl. Phys. **88**, 4159 (2000).
[11] H. Yanagi et al., Appl. Phys. Lett. **78**, 1583 (2001).
[12] K. Ueda et al., Appl. Phys. **89**, 1790 (2001).

OCTAHEDRAL-TILTING-DEPENDENT STRUCTURE DISTORTION IN EPITAXIAL PEROVSKITE OXIDE FILMS

X. L. Tan, P. F. Chen, L. F. Wang, B. W. Zhi, and W. B. Wu[1]

Hefei National Laboratory for Physical Sciences at Microscale, University of Science and Technology of China, Hefei 230026, People's Republic of China

We investigated the octahedral tilts/rotations mismatch in epitaxial orthorhombic $SmFeO_3/NdGaO_3$ (110) system by using high resolution reciprocal space mapping to characterize the structure distortion evolution with layer thickness. The tilts patterns can propagate across the layer/substrate interface, accompanying oxygen octahedral deformation in ultra-thin films and a symmetry-lowing monoclinic structure distortion with increasing layer thickness. Besides, we designed to grow heteroepitaxial bi-layers $La_{0.67}Ca_{0.33}MnO_3/SmFeO_3/NdGaO_3$ (110) with both sharp interfaces, confirming that octahedral-tilting continuity takes in charge of the growth mechanism of $GdFeO_3$-type perovskite oxide films. These results indicate that tuning the octahedral tilts can enrich strain-engineering to explore artificial oxide interface with novel properties.

THE γ→α CHANGE IN CERIUM IS HIDDEN STRUCTURAL PHASE TRANSITION: THEORY AND EXPERIMENT

A.V. Nikolaev[1,4], K.H. Michel[2], A.V. Tsvyashchenko[3,4], A.I. Velichkov[5], A.V. Salamatin[5], L.N. Fomicheva[3], G.K. Ryasny[4], A.A. Sorokin[4], O.I. Kochetov[5] and M. Budzynski[6]

[1]Institute of Physical Chemistry and Electrochemistry of RAS, Leninskii pr. 31, Moscow 119071, Russia;
[2]University of Antwerp, Department of Physics, Groenenborgerlaan 171, Antwerp 2020, Belgium;
[3]Vereshchagin Institute for High Pressure Physics, RAS, Troitsk 142190, Russia;
[4]Skobeltsyn Institute of Nuclear Physics, Moscow State University, Moscow 119991, Russia;
[5]Joint Institute for Nuclear Research, PO Box 79, Moscow, Russia;
[6]Institute of Physics, M. Curie-Sklodowska University, 20-031 Lublin, Poland

Usually different crystallographic forms of a same element are distinguished by their crystal space symmetry. In some cases, for example in metallic cerium, there are two well defined thermodynamic phases: γ and α, which apparently have the same face centred cubic lattice. The existence of such twin phases in cerium has become a challenge to the theory and prompted the search of a main factor responsible for the difference, Ref. [1]. Here we show both theoretically [2-3] and experimentally [4,5] (results of Tsvyashenko et al.) that the well known γ-α phase transition in cerium is not really isostructural.

Figure 1: 3-q-AFQ ($Pa\bar{3}$) structure of α-Ce proposed by Nikolaev and Michel[3,4]. Quadrupoles represent the $l=2$ valence electron ($4f+5d6s^2$) charge density distributions. Trigonal site symmetry (C_3) is clearly visible.

A mechanism of symmetry lowering was provided by a new scenario [2,3] of quadrupolar ordering, which contrary to the previous approaches predicted a hidden and very peculiar symmetry change at the γ→α transition, Fig. 1. The transition γ→α is of the first order and is driven by the minimization of the electron repulsion between the neighboring cerium sites. The active electronic mode belongs to the X point of the Brillouin zone and involves its three arms, q_X^x, q_X^y, q_X^z, that causes the quadrupolar order parameter components alternate sign along x, y, and z-axis. The space group symmetry lowering, $Fm\bar{3}m$ (γ-Ce) → $Pa\bar{3}$ (α-Ce), is accompanied by a uniform lattice contraction so that the fcc structure of the atomic centers of mass (cerium nuclei) is fully conserved. The long range order is due to the orientation of the local $Y_{l=2}^{m=0}$ quadrupolar charge density component of the valence electrons of cerium ($4f+5d6s^2$) on four different sublattices, Fig. 1.

The quadrupolar electron densities induce electric field gradient at the nuclear positions. Through the electric quadrupole hyperfine interaction electric field gradient at a lattice site is directly experienced by a probe nucleus. Such nuclear quadrupole interactions in solids are exploited by many famous methods (for example, by nuclear quadrupole resonance, Mössbauer effect, etc.). Below we present results obtained with a less used technique of time-differential perturbed angular correlations[5] (TDPAC) with the ^{111}In/^{111}Cd probe atoms introduced in the cerium lattice.

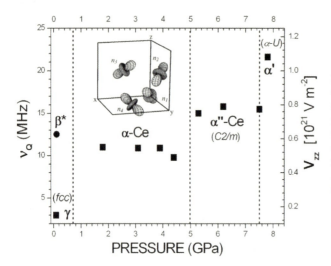

Figure 2: Pressure dependence of electric field gradients (V_{zz}, right scale) and the nuclear quadrupole frequency (v_Q, left scale) measured at the ^{111}Cd probe nuclei implanted in cerium lattice. *Data for β-Ce are from Forker et al., J. Phys. F: Met. Phys. **18**, 823-832 (1988).

The TDPAC pressure experiments of Tsvashchenko et al. [4,5] clearly indicate that the γ-α phase transition is not isostructural. An appreciable electric field gradient in α-Ce is comparable with the electric field gradients of the noncubic phases (β, α″), which border α-Ce in the pressure-temperature phase diagram, Fig. 2.

The only theory predicting such a symmetry change for cerium was the quadrupolar model [2,3]. The local site symmetry is trigonal (C_3) with a quadrupolar electron charge density component $Y_{l=2}^{m=0}$ oriented along one of the main cube diagonals.

Additional support is provided by phonon dispersion measurements on elemental Ce by inelastic x-ray scattering [6]. A pronounced softening of certain branches is found toward the X point of the Brillouin zone.

This work is partially supported by the RFFI project № 11-02-00029.

[1] G. Elisashberg and H. Capellmann, JETP **67**, 125 (1998).
[2] A.V. Nikolaev and K.H. Michel, Eur. Phys. J. B **9**, 619-634 (1999).
[3] A.V. Nikolaev and K.H. Michel, Phys. Rev. B **66**, 054103 (2002).
[4] A.V. Tsvyashchenko et al., JETP **111**, 627 (2010).
[5] A.V. Tsvyashchenko et al., Phys. Rev. B **82**, 092102 (2010).
[6] M. Krisch et al., PNAS **108**, 9342 (2011)

HETEROSTRUCTURES BASED ON EPITAXIAL OXIDE THIN FILMS

P. Prieto[1], M.E. Gómez[2]

[1] Excellence Center for novel Materials CENM, (www.cenm.org) Physics departmentUniversidad del Valle, A. A. 25360, Cali, Colombia
[2] Thin Film Group, (http://calima.univalle.edu.co) Physics department, Universidad del Valle, A. A. 25360, Cali, Colombia

Metal oxides show fascinating physical properties such as high temperature superconductivity, ferro- and antiferromagnetism, ferroelectricity or even multiferroicity. Progress in oxide thin film technology allows us to integrate these materials with semiconducting, normal conducting, dielectric, in complex oxide heterostructures. The combination of magnetic properties with dielectric, semiconducting, or ferroelectric materials in one and the same material (e.g. magnetic semiconductors (MS) or intrinsic multiferroics) as well as in artificial heterostructures (e.g. ferromagnetic/dielectric heterostructures for magnetic tunnel junctions (MTJs) or artificial multiferroic heterostructures) allows for the design of materials with novel functionalities.

For many possible electronic applications as well as fundamental studies, it is essential to fabricate epitaxial layered structures of insulators, semiconductors or normal metals as well as magnetic layers together with high temperature superconductors (HTS). However, HTS materials have complex lattice structures and this makes difficult to grow multilayers with sharp interfaces, preserving epitaxiallity through the whole structure. Here we describe transport measurements and microstructural analysis of $YBa_2Cu_3O_{7-\delta}$/ $PrBa_2Cu_3O_{7-d}$/$YBa_2Cu_3O_{7-\delta}$ (YBCO/PBCO/YBCO), $Bi_2Sr_2CaCu_2O_{8+\delta}$/$Bi_2Sr_2YCu_2O_{8+d}$/ $Bi_2Sr_2CaCu_2O_{8+\delta}$ (BSCCO/BSYCO/BSCCO) and $La_xCa_{1-x}MnO_3$/$YBa_2Cu_3O_{7-\delta}$ (FM/HTS) heterostructures deposited on (001) $SrTiO_3$ substrates by using an *in situ* DC sputtering technique at high oxygen pressures. Conductance measurements on this type of multilayers showed a clear quasiparticle tunneling indicating a gap structure around 25 mV in the case of YBCO compounds, 30 – 35 mV and a zero bias anomaly in the case of BSCCO materials. We will discuss also the Josephson behavior in heterostructures based on BSCCO compounds as well as the interplay between superconductivity and magnetism in superconductor/ferromagnetic heterostructures and superlattices. We report the study of the temperature dependence of magnetization and magnetotransport properties in [AF-LCMO(t_{AF})/F-LCMO(t_F)]$_N$ superlattices grown"*in situ*" on (001)-oriented $SrTiO_3$ substrates via a high-pressure dc sputtering process.

Recent results on the electric and magnetic properties of multiferroic $BiFeO_3$ and $YMnO_3$ thin films exhibiting both ferromagnetic and ferroelectric polarizations along with coupling between them will also be presented.

This research was supported by "El patrimonio Autónomo Fondo Nacional de Financiamiento para la Ciencia, la Tecnología y la Innovación Francisco José de Caldas" under contract RC-No. 275-2011 with the Excellence Center for Novel Materials (CENM)

[1] **P. Prieto**, M. Chacón, M. E. Gómez, O. Morán and D. Oyola; "*Superconducting properties of HoBa2Cu3O7/SrTiO3/HoBa2Cu3O7 heterostructures*"; Solid State Communications; Vol. **83** (1992), 195 - 198.
[2] E. Baca, M. Chacón, L. F. Castro, W. Lopera, M. E. Gómez, **P. Prieto** and J. Heiras; "*Superconducting and structural properties of epitaxial $Bi_2Sr_2CaCu_2O_{8-\delta}$/ $Bi_2Sr_2YCu_2O_{8-\delta}$/ $Bi_2Sr_2CaCu_2O_{8-\delta}$ heterostructures*". Physica C: Superconductivity; Vol. **235-240** (1994), 727 - 728
[3] A. M. Cucolo, R. Di Leo, P. Romano, E. Baca, M. E. Gómez, W. Lopera, **P. Prieto**, and J. Heiras. "*Epitaxial deposition and properties of $Bi_2Sr_2CaCu_2O_{8+\delta}$/$Bi_2Sr_2YCu_2O8+\delta$/ $Bi_2Sr_2YCu_2O_{8+\delta}$/ $Bi_2Sr_2CaCu_2O_{8+\delta}$ trilayers*"; Applied Physics Letters; Vol. **68** (3) (1996), 253-255.
[4] W. Lopera, D. Giratá, J. Osorio and **P. Prieto**. "*Structural and electrical properties of grain boundary Josephson Junctions based on $Bi_2Sr_2CaCu_2O_{8+?}$*"; Physica Status Solidi B; Vol. **220** (1) (2000), 483 – 487.

[5] **P. Prieto**, P. Vivas, G. Campillo, E. Baca, L. F. Castro, M. Varela, C. Ballesteros, J. E. Villegas, D. Arias, C. León, and J. Santamaría "*Magnetism and Superconductivity in $La_{0.7}Ca_{0.3}MnO_3/YBa_2Cu_3O_{7-\delta}$ superlattices*". Journal of Applied Physics; Vol. **89** (12), (2001), 8026 – 8029

[6] E. Baca, W. Saldarriaga, J. Osorio, G. Campillo, M. E. Gómez, **P. Prieto** "*Quasiparticle-injection in $YBa_2Cu_3O_{7-\delta}/La_{1/3}Ca_{2/3}MnO_3/La_{2/3}Ca_{1/3}$ heterostructures*" Jour. App. Phys. **93** (10) 8206-8208 (2003).

[7] **P Prieto**, M. E. Gómez, G Campillo, A. Berger, E. Baca, R. Escudero, F. Morales "*Exchange-coupling effect and magneto transport properties in epitaxial $La_{2/3}Ca_{1/3}MnO_3/La_{1/3}Ca_{2/3}MnO_3$ superlattices*" A Physica Status Solidi A-Applied Research. **201,10**, p.2343 - 2346, (2004)

[8] A. Berger, G. Campillo, P. Vivas, J. E. Pearson S. D. Bader, E. Baca and **P. Prieto**, "*Critical exponents of inhomogeneous ferromagnets*" Journal of Applied Physics; Vol. **91** (10), (2002) 8393-8395

[9] J. G. Ramírez, F. Pérez, M. E. Gómez and **P. Prieto** " *Statistical Study of AFM-images on Manganite Thin Films"* Physica Status Solidi C., **V1, n.S1**, p.13 - 16, (2004)

[10] G. Campillo, A. Hoffmann M. E. Gomez and **P. Prieto**, "Exchange bias and magnetic structure in modulation-doped manganite superlattices" J. Appl. Phys. **97**, 10K104 s (2005)

[11] G. Campillo, M. E. Gomez, A. Berger, A. Hoffmann, R. Escudero, **P. Prieto**, "Influence of ferromagnetic thickness on structural and magnetic properties of exchange-biased manganite superlattices"; J. Appl. Phys. **99** 08C106 (2006)

[12] J. M. Caicedo, J. A. Zapata, M. E. Gómez, and **P. Prieto** "Magnetoelectric coefficient in $BiFeO_3$ compounds" J. Appl. Phys **103**, 07E306 (2008)

[13] J. Zapata, J Narvaez, W. Lopera M.E. Gómez and **P. Prieto** " *Electric and magnetic propertie of multiferroic $BiMmO_3$ and $YMnO_3$ thin films"* IEEE Transactions on Magnetics Vol **44** No 11 2895-2898 (2008)

INCOHERENT INTERFACES AND LOCAL LATTICE STRAINS IN SOLUTION-DERIVED YBCO NANOCOMPOSITES: A NOVEL VORTEX PINNING MECHANISM

T. Puig[1], X. Obradors[1], A. Palau[1], A. Llordés[1], M. Coll[1], R. Vlad[1], J. Gazquez[1], J. Arbiol[1], R. Guzmán[1], A. Pomar[1], F. Sandiumenge[1], S. Ricart[1], V. Rouco[1], S. Ye[1], G. Deutscher[2], D. Chataigner[3], M. Varela[4], C. Magen[5], J. Vanacken[6], J. Gutierrez[6], V. V. Moshchalkov[6]

[1] Institut de Ciència de Materials de Barcelona, ICMAB-CSIC, Bellaterra, Spain
[2] School of Physics and Astronomy, Tel Aviv University, Tel Aviv, Israel
[3] Laboratoire de Cristallographie et Sciences des Matériaux, CRISMAT, Caen, France
[4] Condensed Matter Sciences Division, Oak Ridge National Laboratory, Oak Ridge, USA
[5] Univ. Zaragoza, Inst. Nanociencia Aragon, Zaragoza, Spain
[6] INPAC-Institute for Nanoscale Physics and Chemistry, K.U.Leuven, Leuven, Belgium

Interfaces in oxides have become one of the most relevant issues to generate, enhance and control new physical phenomena. In many of the cases, interfaces have been promoted by growing nanocomposites where each phase is properly designed to undertake a specific role. Heteroepitaxial growth has therefore become the key process in controlling the strain of the designed semicoherent interfaces. Epitaxial growth of high temperature superconducting nanocomposites has emerged as a solution to control and enhance the vortex pinning landscape. In this work, we demonstrate that a not so often used type of interface, incoherent interfaces, give rise to a new and highly effective vortex pinning mechanism in $YBa_2Cu_3O_7$ nanocomposites, where local lattice strains precludes Cooper pair formation inducing nanoscale regions effective for core pinning of vortices. For that purpose, solution-derived epitaxial nanocomposites with randomly oriented and distributed second phase nanoparticles ($BaZrO_3$, Y_2O_3, $BaCeO_3$ and Ba_2TaYO_6) were grown. X-ray diffraction was used to determine the fraction of randomly oriented nanoparticles and the nanoscale strain from line broadening. Angular dependent transport critical current density measurements underlined the huge improvement of vortex pinning [Fig. 1], which could be correlated with the amount of incoherent interfaces. HRSTEM analysis revealed a 3D ramified network of localized and highly strained nanoscale regions (mainly due to extra Cu-O chains and associated partial dislocations) emanating from the incoherent interfaces. Strain maps were determined by Peak Pairs Analysis of STEM cross-sectional images [Fig 2]. We have demonstrated that this nanostrain is the responsible for the huge quasi-isotropic pinning forces and a vanishing anisotropy of the critical currents, though with preserved YBCO intrinsic mass anisotropy as evidenced by high field $H_{c2}(T)$ measurements. This methodology has become an excellent low cost and scalable processing option, making of this material the highest performance superconductor ever reached and most promising material for coated conductor production.

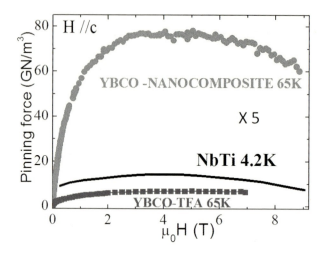

Figure 1: Pinning force of a solution derived YBCO nanocomposite film and a standard YBCO film also grown from solution chemistry at 65 K compared to NbTi material at 4.2 K. Notice that the performance of the nanocomposite at 65 K is 5 times larger than NbTi at 4.2 K.

Figure 2: (left) STEM image showing a BaZrO$_3$ nanoparticle randomly oriented within the epitaxial YBCO matrix with an enormous amount of Cu-O intergrowth surrounding it. (right) HRSTEM image showing a region with two intergrowth and corresponding ε_{xx} strain map determined from Peak Pairs Analysis demonstrating local compressive and tensile nanoregions associated to the intergrowth.

[1] J. Gutierrez et al, Nat. Mat 6, 367 (2007)
[2] A. Llordés et al, Nat. Mat.(2012), DOI: 10.1038/NMAT3247

DENSITY INFLUENCE ON AMORPHOUS HfO$_2$ STRUCTURE: A MOLECULAR DYNAMICS STUDY

Giulia Broglia[1], Monia Montorsi[1], Luca Larcher[2], Andrea Padovani[2].

[1] DIMA Università of Modena and Reggio Emilia, Modena, Italy; [2] DISMI Università of Modena and Reggio Emilia, Reggio Emilia, Italy.

The scaling and power reduction of non-volatile memories (NVM) pose great challenges in the study and development of new materials.

In the last years, hafnium dioxide (HfO$_2$) has been identified as one of the most promising materials for the fabrication of novel memory and logic devices. The interest on HfO$_2$ rises from its properties, such as the high permittivity and the chemical and thermal stability on silicon, that allowed a easy integration into the microelectronic manufacturing process. [Notwithstanding its large diffusion in technological devices, several studies revealed that the structure of HfO$_2$ has still to be ambiguously assessed and so the relationships between its atomic structure and its final properties.

It is well known that HfO$_2$ morphology is strongly dependent on the deposition techniques, the post annealing process, and its thickness. Precisely it was reported that below 10 nm the structure tend to be amorphous. Differently, for thicker layers, the post annealing structure is usually polycrystalline the atomic structure of grains is constituted by different crystalline HfO$_2$ phases (monoclinic, cubic, tetragonal, etc) and residual portion of material between grains is amorphous. In addition, the deposition techniques and post annealing process strongly affect the density. This parameter has a deep effect on the structure and all the correlated properties, such the electronic performances, and therefore its effects on the final electronic properties of the system have to be deeply investigated.

In this scenario, the aim of this work is to analyse systematically the influence of the material density on the structure of amorphous HfO$_2$ (a-HfO$_2$). We will focus on investigating the atomic structure in the short, medium and long range in order to understand which is the preferential atomic structure.

The molecular dynamics technique has been chosen for this analysis because it permits to investigate accurately the short and medium structural order of this material. Furthermore, this method, combined with other computational techniques, can be used to cover the gap between the first-principles calculations and the macrostructure evidences experimentally observed on this oxide. A rigid ionic force-field has been implemented to include the Hf–O interatomic pair parameters. Such model has shown to reproduce the structural properties of HfO$_2$ crystal phases with good accuracy. The following procedure has been developed to simulate the amorphous formation. The system was heated at 4000 K and subsequently cooled continuously from 4000 to 300 K with a nominal cooling rate of 5 K/ps. Configurations at every 0.1 ps were recorded for structural analysis.

It is well known that the amorphous structures have lower density with respect to the crystalline ones (i.e. monoclinic HfO$_2$: 10.19 g/cm^3). In order to investigate the dependence of the structure property on density we have investigated also material systems with higher density values. Both monoclinic and cubic structures have been computed to compare the results with the ones obtained for the a-HfO$_2$ structures. The structural parameters of systems we simulated are very close to the ones derived by previous ab initio molecular dynamics works, demonstrating the physical accuracy of this method. In addition, this study reveals that the material density affects strongly the morphology of the HfO$_2$ film. In particular, the increase of the density induces in amorphous hafnium a change in the Hf coordination number, which passes on average from six to eight, while in monoclinic phase the Hf atom is 7-fold coordinated. The same trend has been observed also for the average oxygen coordination number, which indicates a structural change from trigonal to tetrahedral. This behaviour might be associated with a more compact structure, which is promoted by the higher density, being related to the smaller volume available for the system to relax. The pair distribution functions (PDF) and the bond angle distributions (BAD) well compare and agreed the hypothesis based on coordination confirming that density of HfO_2 induces significant structural modification into the amorphous system.

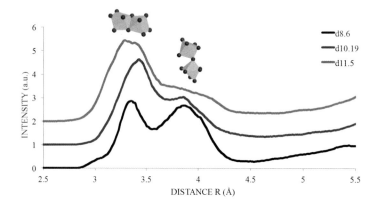

Fig. 1: Hf-Hf Pair Distribution Function at various density values

Information on Hf distribution into the glass structure can be derived considering the Hf-Hf PDF (Fig. 1) as a function of the density values. It is possible to note that decreasing the density, a second peak around 3.9 Å appears and this is consistent with Hf local environment characterized by corner sharing polyhedra. This evidence indicates that at low density the Hf polyhedrons tend to share more corners than edges or faces: there is a less packed structure, corresponding to an increase of the free volume. These strong changes into the structure, induced by a variation of the density might aid to explain why the layer structure is influenced by the deposition technique and post annealing process. deposited layers. In fact, the density might facilitate the formation of structure more or less similar to the crystal. Infact, being the density dependent on the deposition technique, every HfO_2 layer might show a different atom rearrangement and a structure more or less similar to the crystal. The methodology presented in this paper provides an effective tool to correlate the deposition technique to the HfO_2 properties.

Next step in the research will be devoted to find a correlation between the structure and the presence of oxygen vacancies, which are considered to be – at least partially – responsible of hafnia electronic response. For example, a structure having a high free volume might facilitate the diffusion of these species into the system.

ELECTRIC FIELD TUNING OF THE QUASIPARTICLE WEIGHT IN THIN FILMS MADE OF STRONGLY CORRELATED MATERIALS

D. Nasr Esfahani[1], L. Covaci[1] and F. M. Peeters[1]

[1]Departement Fysica, Universiteit Antwerpen, Groenenborgerlaan 171, B-2020Antwerpen, Belgium

The ground state properties of a paramagnetic Mott insulator at both at half-filling and away half-filling are investigated in the presence of an external electric field using the inhomogeneous Gutzwiller approximation for a single band Hubbard model in a slab geometry. For the half filled case we find that the metal insulator transition is shifted towards higher Hubbard repulsions by applying an electric field perpendicular to the slab [1]. The main reason, is the accumulation of charges near the surface and the spatial distribution of site dependent quasiparticle weight shows that it is maximal in few layers beneath the surface while the central sites where the field is screened have a very low quasiparticle weight. For the doped case, similar to half-filling, the quasiparticle weight recovers exponentially its bulk value deep into the slab. Unlike the half-filled case the correlation length is controlled not only by the Hubbard repulsion but also by the charge density. The critical density is n=1.0 for all of the $U>U_c^{bulk-halffilled}$, moreover the correlation length follows a power law as function of filling. In the presence of an electric field, due to charge transfer from one side to the other, the quasiparticle weight is highly inhomogeneous, i.e. suppressed on one side and enhanced on the other side. The field effect presented here is of direct experimental relevance because it could be used to modulate the resistance of a three-terminal device that takes advantage of the modification of surface states. Upon switching the direction of the field, large on/off ratios could be achieved.

[1] D. Nasr Esfahani, L. Covaci, and F. M. Peeters, Phys. Rev. B 85, 085110 (2012).

METALLIC STATE INDUCED BY SPIN-CANTING IN LIGHTLY ELECTRON-DOPED CaMnO$_3$

Hiromasa Ohnishi[1,3], Taichi Kosugi[1], Takashi Miyake[1], Shoji Ishibashi[1,3], and Kiyoyuki Terakura[1,2,3]

[1]National Institute of Advanced Industrial and Science Technology (AIST), Tsukuba, Ibaraki, Japan;
[2]Japan Advanced Institute of Science and Technology (JAIST), Nomi, Ishikawa, Japan;
[3]JST-CREST, Kawaguchi, Saitama, Japan

Perovskite transition-metal oxides exhibit a variety of exotic electronic and magnetic properties by chemical doping, changing temperature, etc. Control of these properties is a topical issue for future device applications, as well as their scientific importance. Among these applications, it is a new frontier in depelopment of field effect transistor (FET), that utilizes the metal-insulator transition (MIT) in strongly correlated electron systems as a resistivity switch. This new kind of FET is sometimes called "Mott transistor", and is expected to overcome the limitation of existing semiconductor devices [1].

CaMnO$_3$ (CMO) is one of promising materials to realize the Mott transistor. The non-doped CMO is a G-type antiferromagnetic (G-AFM) insulator at low temperature. Small amounts of electron-doping to CMO cause the emergence of a weak ferromagnetic (FM) component in the background of the G-AFM order [2]. The experimental study for the Ce-doped CMO [3] has revealed that this lightly electron-doped system shows a metallic behavior. In other words, the phase change from the G-AFM state to the weak FM one is a MIT by the electron-doping. Very recent experiment has shown that this MIT temperature is controllable by the filling change through electric field [4].

So far, the origin of the weak FM component has been studied, in terms of the double-exchange (DE) mechanism, and two scenarios have been proposed. One is the spin-canting in the G-AFM (canted G-AFM), and the other is the AFM-FM phase separation [5]. Although both scenarios are compatible with the experimental magnetic property, we cannot expect metallic behavior naively by the phase separation scenario with small metallic FM domains. On the other hand, in the spin-canting scenario, metallic behavior is naturally expected with the DE hopping of electrons in the entire system.

In this study, we give quantitative arguments of the FM-component in the canted G-AFM as a function of electron-doping, based on the detailed electronic structure calculations with the noncollinear version of local spin density approximation. For its purpose, total energy calculations are performed with our computational code QMAS (Quantum MAterials Simulator) based on the projector augmented-wave (PAW) method and plane wave basis [6].

We show that the canted G-AFM state is stabilized in the lightly electron-doped system, as shown in Fig.1(a), and the weak FM component observed in the Ce-doped CMO [2] is quantitatively well explained by the spin-canting from the G-AFM [7]. As shown in Fig.1(b), density of states (DOS) for the canted G-AFM state shows the broadening of the band width with increase of the doping-amount. This is due to the increase of the hopping integral among neghboring Mn's. As a result, the system shows strong metallic characteristics, as observed in the experiment [3].

Poster: Electronic structure, lattice dynamics, and transport 499

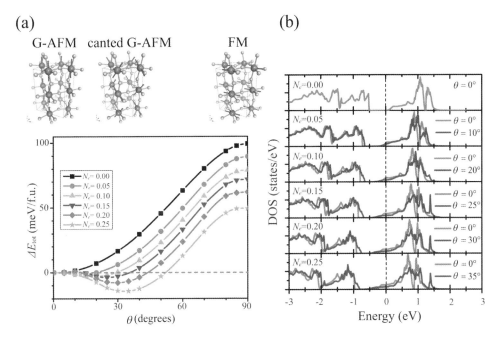

Figure 1: (a) Canting angle θ-dependence of total energy for several doping amounts N_e (electrons/f.u.). $\theta=0$ and 90 correspond to the G-AFM and FM states, respectively. In between them, the canted G-AFM state is realized. (b) Density of states (DOS) for the G-AFM and canted G-AFM state for several N_e.

[1] C. H. Ahn et al., Nature **424**, 6952 (2003).
[2] E. N. Caspi et al., Phys. Rev. B **69**, 104402 (2004).
[3] P. –H. Xiang et al., Appl. Phys. Lett. **94**, 062109 (2009).
[4] P. –H. Xiang et al., Adv. Mater. **23**, 5822 (2011).
[5] E. Dagotto, T. Hotta and A. Moreo, Phys. Rep. 344, 1 (2001).
[6] http://qmas.jp/
[7] H. Ohnishi et al., submitted to Phys. Rev. B.

PRESSURE-INDUCED STRUCTURAL CHANGES AT THE CROSSOVER FROM LOCALIZED TO ITINERANT BEHAVIOUR IN PrNiO$_3$

<u>Marisa Medarde</u>[1], Thierry Straessle[1], Vladimir Pomjakushin[1], María Jesus Martínez-Lope[2] and José Antonio Alonso[2]

[1] Paul Scherrer Institut, CH-5232 Villigen PSI, Switzerland
[2] Instituto de Ciencia de Materiales de Madrid, CSIC, Cantoblanco, E-28049 Madrid, Spain

The RNiO$_3$ perovskite family (R= rare earth and Y) constitutes a particularly interesting, single-valence system where a complete evolution from a Pauli paramagnetic metal to a Curie-Weiss insulator can be achieved by applying chemical pressure (i.e., by decreasing the size of the rare earth ion) [1]. LaNiO$_3$ is paramagnetic, metallic and crystallizes in the rhombohedral R-3c space group. The remaining nickelates display orthorhombic Pbnm symmetry and undergo metal to insulator (MI) transitions with critical temperatures T_{MI} increasing as a function of decreasing size of the lanthanide. The mechanism responsible for the electronic localization involves a subtle symmetry breaking from orthorhombic to monoclinic and a charge disproportionation - or self-doping - $2Ni^{3+} \rightarrow Ni^{3+\delta} + Ni^{3-\delta}$ ($0 < \delta < 1$), which splits the unique Ni site of the high-temperature metallic phase into two inequivalent Ni(I) and Ni(II) sites with slightly different valences below T_{MI}. [2, 3]

The evolution of T_{MI} along the series is controlled by the degree of hybridization between the Ni 3d:e_g and O 2p:σ bands, which is directly related to the bandwith W_σ in the metallic state through the expression $W_\sigma \propto \cos\phi / d_{Ni-O}^{7/2}$. [4] Here ϕ is the average buckling angle of the NiO$_6$ octahedra and d_{Ni-O} the average Ni-O bond length. Experimentally, it has been found that the progressive metallization of the system is primarily achieved by reducing the tilting angle of the NiO$_6$ octahedra, the comparatively modest compression of the nearly perfect Ni coordination polyhedra having only a minor effect on W_σ.

An alternative way to access the crossover region between localized and itinerant behaviour is to apply external pressure. Resistivity and x-ray diffraction measurements on PrNiO$_3$ (T_{MI} = 130K) have shown that both the full metallic state and the orthorhombic-rhombohedral transition can be reached by applying respectively 13 and 50 kbar [5, 6]. Moreover, the existence of two different non-Fermi Liquid (FL) phases below 30 kbar, one with resistivity $\rho \propto T^{4/3}$ and the other with $\rho \propto T^{5/3}$ has been reported [7]. Since a FL power law ($\rho \propto T^2$) is observed in rhombohedral LaNiO$_3$ at ambient pressure [7], the existence of an additional transition into FL behaviour has been suggested for P > 50kbar.

In order to understand the origin this surprising step-like behaviour, a precise knowledge of the evolution of the structural parameters (lattice constants & atomic positions) as a function of pressure and temperature is necessary. Unfortunately, only the behaviour of the cell parameters up to P ≤ 50 kbar has been reported to date [6]. Since both chemical and external pressure lead to an increase of W_σ, it could be tempting to extrapolate the structural behaviour observed by chemical substitution. However, in a previous investigation of PrNiO$_3$ under moderate pressures (5 kbar) [4] we found that, for an equivalent ΔT_{MI}, the variation of ϕ is *smaller* and that of d_{Ni-O} *much larger* than those observed by chemical substitution.

In order to provide reliable structural parameters for further electronic structure calculations we have carried out neutron powder diffraction measurements of PrNiO$_3$ up to 100kbar. In spite of the experimental difficulties related to the small sample mass we have been able to extract valuable information about the evolution of the structure from 0 to 100 kbar. Besides the potential interest of these results for many other perovskite-related systems, they may be also relevant to understand the puzzling behaviour of the transport properties recently reported for RNiO$_3$ multilayers and thin films [8,9].

[1] M.L. Medarde, J. Phys.: Condens. Matter **9**, 1679 (1997).
[2] J.A. Alonso et al., Phys. Rev. Lett. **82**, 3871 (1999).
[3] M. Medarde et al., Phys. Rev. B **80**, 245105 (2009).
[4] M. Medarde et al., Phys. Rev. B **52**, 9248 (1995).
[5] P.C. Canfield et al., Phys. Rev. B 47, 12357 (1993). X. Obradors et al., Phys. Rev. B **47**, 12353 (1993).
[6] J.-S. Zhou, J.B. Googenough and B. Dabrowski, Phys. Rev. B **70**, 081102 (2004).
[7] J.-S. Zhou, J.B. Googenough and B. Dabrowski, Phys. Rev. Lett. **94**, 226602 (2005).
[8] R. Schewitzl et al., PRL **106**, 246403 (2011)
[9] A.V. Boris et al., Science **332**, 937 (2011)

MECHANICS MEETS ELECTRONICS IN NANOSCALE: THE MYSTERY OF CURRENT SPIKE AND NANOSCALE-CONFINEMENT

R. Nowak[1], D. Chrobak[1], W.W. Gerberich[2], K. Niihara[3], T. Wyrobek[4]

[1]Nordic Hysitron Laboratory, Dpt. Materials Science, Aalto University, 00076 Aalto, Finland,
[2]Department of Chemical Engineering & Materials Science, University of Minnesota, Minneapolis MN55455, USA,
[3]Nagaoka University of Technology, Nagaoka 940-2188, Japan,
[4]Hysitron Inc., Minneapolis, MN 55344, USA

Our discovery highlighted for the first time in the Letters to the *Nature Nanotechnology* [1,2] offers an enhanced understanding of the link between nano-scale mechanical deformation and electrical behaviour, and ultimately suggests key advances in pressure-sensing, pressure-switching, and unique phase-change applications in future electronics. It is a very encouraging demonstration of the way in which nanomechanics may contribute to electronics and optoelectronics developments.

One of the fundamental questions in materials science concerns the nature of deformation of solids [3]. The onset of plasticity is traditionally understood in terms of dislocation nucleation and motion. A study of nanoscale deformation has proven that initial displacement transient events occurring in metals are the direct result of dislocation nucleation [4,5]. In this presentation we show that this is not always true: instead of dislocation activity, nanoscale deformation may simply be due to transition from one crystal structure (semiconductor) to another (metal) as confirmed in the case of GaAs (see Fig. 1), while predicted by our earlier atomistic calculations [6].

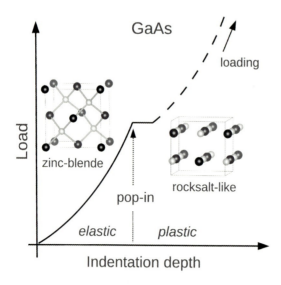

Fig.ure 1: Nanoindentation induced structural changes in GaAs. The initial part of the loading cycle exhibits perfectly elastic deformation followed by the onset of plasticity (pop-in event) caused by zinc-blende to rocksalt-like structure transfor-mation. The displayed scenario contrasts the widely accepted dislocation–nucleation origin of the pop-in event [4,5].

Using a novel conductive nanoindentation technique, which is highly sensitive to micro-structural changes under pressure [1], we discovered the essential link between this electrical phenomenon (current spike) and the mechanical transient (pop-in) exhibited by GaAs exclusively during nanoscale deformation (Fig. 2).

This correlation is solidified by our *ab initio* calculations, leading to the conclusion that a previously unseen phase transformation is the fundamental cause of nanoscale plasticity in GaAs [1]. Indeed, anyone wishing to project to nanoscale, even with such classic phenomena as elastic or plastic deformation, will inevitably look to the atomistic approach for answers. The presented results lead to a major shift in our understanding of elastic-plastic transition as well as inherent Schottky barrier formed in semiconductors under local high pressures.

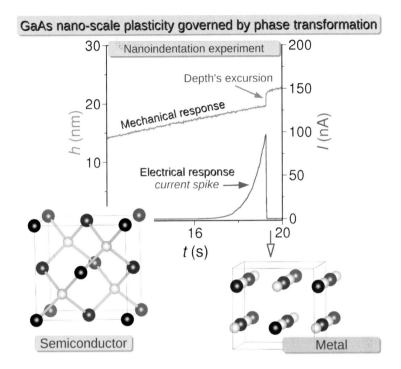

Figure 2: The result of *in situ* nanoscale electrical resistivity measurements of the (100)GaAs crystal demonstrate the peculiar electric current spike (reverse bias) that appears simultaneously with pop-in event during indentation with conducting tip. The Current-Time curve recorded during elastic nano deformation proves leaking junction and immediately after pop-in event –the restoration of a perfect Schottky barrier.

The phenomenon called the "Current Spike" is clearly visible, and its explanation relies heavily on quantum calculations [1].

The results obtained for GaAs and dramatic impact of crystal imperfections on the functional properties of Si nano-volumes motivated studying the onset of incipient plasticity in Si nanoparticles. Molecular Dynamics calculations and supporting experimental results reveal that plasticity onset in Si nano-spheres below 130 nano-meter diameter is governed by dislocation-driven mechanisms, in striking contrast to bulk Si where incipient plasticity is dominated by phase transformations [2]. With the broad implications for nanotechnology, we establish previously unforeseen role of 'nanoscale confinement' governing a transition in mechanical response from "bulk" to 'nanovolume' behaviour. This understanding will benefit processing future nano-structures for electronic, magnetic and optical devices as well as biomedical applications including drug delivery and biosensors as indicated by Cross [7].

[1] R. Nowak, D. Chrobak, S. Nagao, D. Vodnick, M. Berg, A. Tukiainen and M. Pessa, Nature Nanotechnology **4**, 287 (2009).
[2] D. Chrobak, N. Tymiak, A. Beaber, O Ugurlu, W.W. Gerberich and R. Nowak, Nature Nanotechnology **6**, 480 (2011).
[3] R. Nowak, T. Sekino and K. Niihara, Phil. Mag. A74, 171 (1996)
[4] S.G. Corcoran, R.J. Colton, E. Lilleodden and W.W. Gerberich, Phys. Rev. B 55, R16057 (1997)
[5] C.A. Schuh, J.K. Mason and A.C. Lund, Nature Mater. **4**, 617 (2005)
[6] D. Chrobak, K. Nordlund and R. Nowak, Phys. Rev. Lett. **98** (2007) 045502.
[7] G.L.W. Cross, Nature Nanotechnology **6** (2011) 467-468.

EFFECT OF THE CAPPING ON THE MANGANESE OXIDATION STATE IN SrTiO$_3$/La$_{2/3}$Ca$_{1/3}$MnO$_3$ INTERFACES AS A FUNCTION OF ORIENTATION

S. Estradé[1,2], J. M. Rebled[1,3], M. G. Walls[4], F. de la Peña[5], C. Colliex[4], R. Córdoba[6], I. C. Infante[3,7], G. Herranz[3], F. Sánchez[3], J. Fontcuberta[3], F. Peiró[1].

[1]LENS-MIND-IN2UB, Departament d'Electrònica, Universitat de Barcelona, Barcelona, Spain;
[2]TEM-MAT, CCiT- UB, Barcelona, Spain;
[3]Institut de Ciència de Materials de Barcelona - CSIC, Bellaterra, Spain;
[4]Laboratoire de Physique des Solides, Université Paris-Sud, Orsay, France;
[5]CEA, LETI, MINATEC, Grenoble, France;
[6]Instituto Universitario de Investigación en Nanociencia de Aragón (INA), Zaragoza, Spain;
[7]Laboratoire de Structure, Propriétés et Modélisation de Solides, UMR8580 CNRS-École Centrale Paris, Châtenay Malabry, France

It has been reported [1] that epitaxial (001) and (110) La$_{2/3}$Ca$_{1/3}$MnO$_3$ (LCMO) grown on SrTiO$_3$ (STO) substrates (STO$_S$//LCMO) display an intriguing magnetic asymmetry: the magnetization and the Curie temperature of (110) films are repeatedly higher than those of (100) orientation grown simultaneously.

Quite similarly, magnetization and Nuclear Magnetic Resonance experiments on (001) and (110) LCMO films capped with nanometric layers of SrTiO3 (STO$_C$) indicate that whereas the properties of the (110) LCMO are robust and quite insensitive to capping, this is not the case of STO-capped (001) LCMO which displays a significant reduction in magnetization [2, 3].

It follows that some electronic reconstruction at the interface occurs upon capping the (001)LCMO films, which strongly affects their magnetic properties.

In order to shed some light on these dramatic effects, we will address a comparative study of epitaxial(001) and (110) STO$_S$//LCMO/STO$_C$ heterostructures, using electron energy-loss spectroscopy (EELS). Selection of Ca-based manganite for this study is dictated by the fact that its narrower (compared to Sr-based manganites) conduction band weakens the double-exchange coupling while enhancing electron-phonon coupling and sensitivity to disorder effects.

Electron energy-loss spectroscopy allows direct determination of the local Mn oxidation state and elemental quantification at the nanometric scale and is thus one of the most suitable techniques for direct local evaluation of electronic and chemical reconstructions.

It will be shown that STO capping promotes a clear hole-depletion of the (001)LCMO layer that extends a few nanometers deep into the film. In contrast, the (110) robustly remains electronically stable after capping.

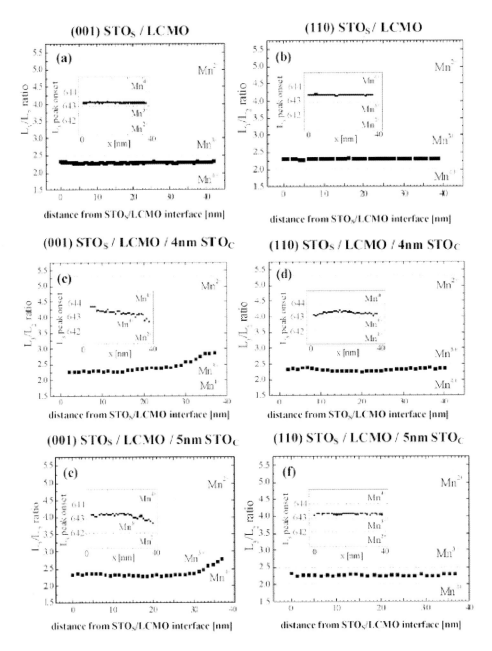

Figure 1: (a), (b) Mn L_3/L_2 intensity ratio and -inset- Mn L_3 peak onset along the bare LCMO electrodes for (001) and (110) LCMO orientations, respectively. (c), (d) Mn L_3/L_2 intensity ratio and -inset- Mn L_3 peak onset in the (001) and (110) 4 nm STO_S//LCMO/STO_C systems, respectively. (e), (f) Mn L_3/L_2 intensity ratio and -inset- Mn L3 peak onset in the (001) and (110) 5 nm STO_S//LCMO/STO_C systems, respectively.

[1] I. C. Infante et al., Phys. Rev. B **76**, 224415 (2007).
[2] I. C. Infante et al., J. Appl. Phys. **103**, 07 E302 (2008).
[3] I. C. Infante et al., J. Appl. Phys. **101**, 093902 (2007).
[4] S. Estradé et al., J. Appl. Phys. **110**, 103903 (2011).

ON THE ELECTRICAL BEHAVIOR OF PLANAR TUNGSTEN POLYOXOMETALATE SELF-ASSEMBLED MONO- AND BI-LAYER JUNCTIONS

D. Velessiotis[1], A. M. Douvas[1], P. Dimitrakis[1], P. Argitis[1], N. Glezos[1]

[1]Institute of Microelectronics, NCSR Demokritos, Athens, 15310, Greece

Polyoxometalates are complex inorganic anions that can be considered as the molecular analogs of transition metals oxides [1]. Their wealthy, redox and photochemical, chemistry and their ability for electron storage and transfer without disintegration, have recently led to an increasing interest in using them for future molecular electronic and photoelectronic applications [2], [3], a goal that we have been already pursuing for quite some time [4]. More recently, we focused on studying the electrical properties of polyoxometalate (POM) self-assembled layers using planar junctions [5-6], vertical capacitors and STM measurements [7], aiming to use this kind of materials in fabrication of hybrid Si/molecular memories, two terminal memory elements or planar molecular nanotransistors. In this work, the lateral conduction properties of single and double POM layers are discussed and compared and the physical mechanisms that govern this conduction are revealed.

To do so, planar Au electrodes were constructed on top of oxidized Si wafers using standard e-beam lithography and lift-off process [6]. In this work we used parallel electrodes of 2μm overlap and 250nm height, so that the molecular layers were placed in a uniform electric field. POM layers were selectively applied on the oxide using the layer-by-layer self-assembly method [8]. Initially, the substrates were functionalized with 3-aminopropyltriethoxy silane (APTES), providing amino (NH_2) groups chemically linked to the SiO_2 covered substrate. Then, anions of the Keggin-structure POM, 12-tungstophosphoric acid ($H_3PW_{12}O_{40}$) were deposited from aqueous solution forming a single POM layer attached to the APTES sub layer by electrostatic interactions between POM anions ($PW_{12}O_{40}^{3-}$) and protonated amino groups (NH^{3+}) of APTES. For the fabrication of double POM layers, on top of the first POM layer, cations of 1,12-diaminododecane (DD) were deposited from a HCl-acidified aqueous solution forming a DD molecular layer, also attached to the POM layer with electrostatic interactions. Finally, POM anions were deposited in the same way on top of the DD layer. Throughout the process, pH was kept constant at 0.5, i.e. in the pH stability range of $H_3PW_{12}O_{40}$ (pH<1) in order to avoid POM structure alterations [5]. After the deposition, the I-V characteristics of 50, 75 and 100nm-distance molecular junctions were measured in the temperature range of 80K to 360K using a Janis low-temperature wafer prober and a HP4155A semiconductor parameter analyzer. The obtained characteristics were tested in various conduction models (namely space charge limited current, thermionic conduction and hopping) in order to reveal the prevailing conduction mechanisms for the POM-layer materials.

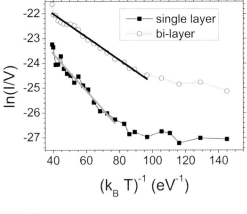

Figure 1: Arrhenius plots of the conductivity for the 50nm junctions and applied voltage of 14V i.e. 2.8MV/cm applied field. The higher position of the bi-layer plot indicates the higher conductivity obtained by the bi-layer material compared to the single layer one. The slope of the plots, indicated by the straight lines, is the activation energy calculated in this example as 46meV for the bi-layer material and 75meV the single layer one.

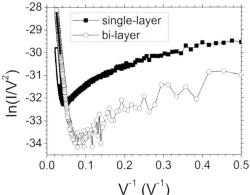

Figure 2: Fowler-Nordheim (FN) plot of the 80K I-Vs for the 50-nm distant junctions along with their respective linear fits. The corresponding barrier heights are 421meV for the bi-layer junction and 458meV for the single-layer counterpart.

The conductivity of the bi-layer samples was typically one order of magnitude higher than that of the single-layer ones (Fig. 1). The obtained I-Vs had an exponential form, with the low-conductance region corresponding to indirect tunneling and the high-conductance region corresponding to hopping or Fowler-Nordheim (FN) tunneling depending on the measurement temperature. Hopping prevailed above 150K for the single-layer junctions and above 120K for the double-layer ones (Fig. 1). The extracted activation energies for the single-layer were 85.5±0.4meV for the 50nm-distant junctions, 76.3±0.7meV for the 75nm-distant junctions and 50.5±0.5eV for the 100nm-distant junctions, while for the double-layer were 55.5±0.4meV, 41.1±0.9meV and 49±1meV, respectively. The effective tunneling barrier (φ) was calculated using the Fowler-Nordheim representation of the I-Vs (Fig. 2), and found to be 480±10meV for the single-layer POM junctions and 335±3meV for the bi-layer ones.

Authors wish to acknowledge MC2 ACCESS Program (Contr. No:026029) for partially funding this work.

[1] D. E. Katsoulis, Chem. Rev. 98 (1998) 359-387.
[2] W. Qi, L.X. Wu, Polym. Inter. 58 (2009) 1217-1225
[3] Y.L. Zhong, W. Ng, J.X. Yang, K.P. Loh, J. Am. Chem. Soc. 131 (2009) 18293-18298
[4] N. Glezos, P. Argitis, D. Velessiotis, P. Koutsolelos, C. D. Diakoumakos, A. Tserepi, K. Beltsios, MRS Symposium Proc. 705 (2002) 49-59
[5] A.M. Douvas, E. Makarona, N. Glezos, J.A. Mielczarski, E. Mielczarski, ACS Nano 2 (2008) 733-742
[6] D. Velessiotis, A. M., Douvas, S. Athanasiou, B. Nilsson, G. Petersson, U. Södervall, G. Alestig, P. Argitis, and N. Glezos, Microel. Engin. 88 (2011) 2775–2777
[7] E. Makarona, E. Kapetanakis, D. Velessiotis, A. M. Douvas, P. Argitis, P. Normand, T. Gotszalk, M. Woszczyna, and N. Glezos, Microel. Engin. 85 (2008) 1399-1402
[8] J. D. Hong, K. Lowack, J. Schmitt, and G. Decher, Progr. in Coll. & Surf. Sci. 93. (1993) 98-102

MECHANICS MEETS ELECTRONICS IN NANOSCALE: NANOSCALE DECONFINEMENT OF SILICON ALTERS ITS PROPERTIES

Dariusz Chrobak[1,2], William W. Gerberich[3], Roman Nowak[1]

[1]Nordic Hysitron Laboratory, School of Chemical Technology, Aalto University, Finland; [2]Institute of Materials Science, Silesia University, Poland; [3]Department of Chemical Engineering and Material Science, Minnesota University, USA

Rapid expansion of nanotechnology is accompanied by growing interest in developing nanoparticles exhibiting mechanical behaviour tailored to optimize their production routes and to satisfy the demands of their intended use as electronic materials [1-3]. Nanoparticle and nano-wire based novel technologies include hydrogen storage, solar energy conversion, plasmonic nano-sensors, drug delivery, medicinal imaging, and tissue engineering. The validity of predicting mechanical behaviour of material in its nanovolume form by "scaling" the relationships established for bulk materials is a subject of a continuing debate. By combining atomistic calculations and nanoscale compression experiments we demonstrate that mechanical behaviour of bulk Si deconfined to isolated nanovolume is governed by the mechanisms distinctively different from those dominant in the bulk Si. Our study is motivated by unique functional properties of silicon nanoparticles offering promise for photovoltaic solar cells, metal oxide silicon capacitors, ultra violet photo detectors, "laser on a chip" devices, distributed floating gate memory devices and biological markers. Given the evidence of the dramatic impact of crystal imperfections on the functional properties of Si nanovolumes, understanding the evolution of lattice defects in Si nanoparticles is essential.

The presented results revealed a hitherto unknown dislocation-driven mechanisms governing deformation in Si nanoparticles. The observed behaviour was examined in the framework of the dilemma concerning dislocation, or phase-transformation origin of the incipient plasticity in nanoscaled semiconductors and ceramics. With the broad implications for the design of nano-sized devices, our study clarifies a unique previously unforeseen role of 'nano-scale deconfinement' and captures conditions governing a transition in mechanical/electrical response from "bulk" to "nanovolume" behaviour.

It is well known that deformation of Silicon is governed by its phase transition from semiconducting diamond to metallic beta-tin structure. We are demonstrating here a new fact - the isolation of the silicon nanovolume leads to drastic alteration of its mechanical properties. We show that deformation of smaller nanoparticles occurs exclusively by dislocation nucleation and movement without contribution of the phase transformation. The molecular dynamics simulated response of Si nanospheres matches the experimental data exhibit striking contrast to mechanical response of bulk Si dominated by amorphization and phase transitions.

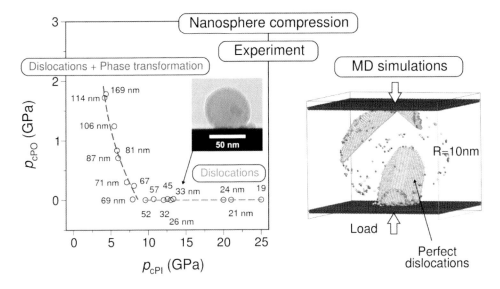

Figure 1: Molecular dynamics calculations and experimental results reveal that plasticity onset in silicon nanospheres below 57 nm radius is governed by dislocation-driven mechanisms. These findings are in contrast to bulk Si where incipient plasticity is dominated by phase transformations. The transition from transition-driven to dislocation-governed incipient plasticity is determined to be a consequence of the increasing role of shear stresses.

We have proved that the behaviour of the silicon nanoparticles is a direct consequence of shear stress increase that occurs when nanoparticle size decreases. Furthermore, the deconfinement of nano-volume, a progression from a state of relative constraint of the bulk to a less constrained state of the nanoparticle or nanowire, results in increasing stabilization of the Si diamond phase and enhanced dislocation activity. The phenomenon in question appears to have a universal appeal, as it contributes to the debate regarding the dislocation versus phase transformation origin of elastic-plastic transition and possible pressure-control of semiconductor-base devices [4-8].

[1] M. H. Nayfeh et al., Silicon nanoparticles ..., ed. V. Kumar, Elsevier, Amsterdam, 1 (2008).
[2] R. R. He et al., Nature Nanotechnology **1**, 42 (2006).
[3] A. L. Hochbaum et al., Nature **451**, 163 (2008).
[4] R. Nowak et al., Nature Nanotechnology **4**, 287 (2009).
[5] D. Chrobak et al., Nature Nanotechnology 6, 480 (2011).
[6] C. A. Schuh et al., Nature Materials **4**, 617 (2005).
[7] J. K. Mason et al., Phys. Rev. B **73**, 054102 (2006).
[8] W. W. Gerberich et al., Nature Materials **4**, 577 (2005).

BREATHING-LIKE MODES IN AN INDIVIDUAL MULTIWALLED CARBON NANOTUBE

Carola Meyer[1,2], C. Spudat[1,2], M. Müller[3], L. Houben[1,2,4], J. Maultzsch[3], K. Goss[1,2], C. Thomsen[3], C. M. Schneider[1,2]

[1]Peter Grünberg Institute, Research Centre Jülich, 52425 Jülich, Germany;
[2]JARA – Fundamentals of Future Information Technologies;
[3]Institut für Festkörperphysik, Technische Universität Berlin, Berlin, Germany;
[4]Ernst Ruska-Center for Microscopy and Spectroscopy with Electrons, Forschungszentrum Jülich, 52425 Jülich, Germany;

Multiwalled carbon nanotubes (MWCNTs) can be imagined as multiple graphene sheets rolled up in tubes, which are embedded into one another (inset FIG. 1). This material is interesting for application in nanoelectromechanical devices [1]. It is crucial for this purpose, however, to understand and measure the coupling between the nanotube shells. This coupling is caused by inter-shell interactions, and Raman spectroscopy is a versatile tool to study the nature of this coupling. Its strength shows up in the movement of the shells against each other. The corresponding vibrational motion has its origin in the radial breathing modes (RBMs) of the nanotube walls, which can be observed with Raman spectroscopy. Thus, carbon nanotubes can be seen as a model system for few-layer graphene, where the corresponding B_{2g} mode is Raman and infrared silent.

Raman spectroscopy, however, provides only indirect information of the atomic structure of CNTs. A correlation of aberration-corrected high-resolution electron-transmission microscopy (HR-TEM) and spectroscopy measurements on the same CNT significantly helps to interpret the low-frequency Raman spectrum experimentally observed [2].

MWCNTs were grown by chemical vapour deposition on a grid suitable for HR-TEM. The grid was prepared with a marker structure in order to identify individual tubes for the spectroscopic and microscopic measurements.

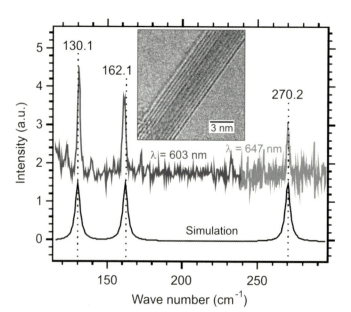

Figure 1: Low-frequency Raman spectrum of the MWCNT measured at two different wavelengths (blue and red line) and its simulation. Inset: HR-TEM micrograph of the same MWCNT. The nanotube shells appear bright.

The HR-TEM image (inset Figure 1) reveals that this MWCNT is composed of six walls. The shell diameters of the tubes range from d = 0.84 nm for the innermost tube to d = 4.34 nm for the outermost tube. The distance between two shells is constant within the error of the measurements and $\Delta r = 0.35$ nm. We compare the diameters obtained in the HR-TEM measurement to the energy of the resonant Raman modes in the low-frequency spectrum. Usually, these resemble the RBMs of single walled carbon nanotubes and depend on the diameter of the tubes.

Frequency dependent Raman measurements of the same MWCNT are shown in Figure 1. They were performed prior to the HR-TEM imaging in order to avoid any influence of defects that might be induced by electron irradiation. We observe three modes, which exhibit a strong resonance effect as expected for RBMs.

The Raman shifts ω_{Raman} of the modes observed can be compared with the diameters obtained in the HR-TEM measurements using the relation $\omega_{RBM} = A/(d_T+B)$ to convert the tube diameters d_T into Raman shifts ω_{RBM} as they would be expected for individual single-walled carbon nanotubes. A and B are sample-dependent constants, which are chosen to be A = 223.75 cm-1 and B = 0 for tubes without any additional interactions as suggested by Bandow et al. [3]. Strong shifts toward higher wave numbers are found in the experiment for the second and especially for the third inner tube, where the mode is shifted by more than 23% with respect to the RBM position expected from the HR-TEM measurement.

We have shown that the stiffening of the Raman modes can be explained by means of a coupling of the phonon modes due to van der Waals interactions between the walls of the MWCNT using a model of coupled harmonic oscillators [2].

In Figure 1 the experimentally obtained spectra are compared with the position of the Raman modes obtained from the simulation. The best agreement between simulated and measured Raman shifts is found at a coupling frequency of $\Omega_c = 1.84$ THz, which is a measure for the coupling strength. It will reach the inter-layer coupling of graphite $\Omega_c = 3.81$ THz for $d_T \rightarrow \infty$ and infinite number of shells. We conclude that we observe coupled breathing-like modes (BLMs) as predicted by Popov et al. [4] rather than individual RBMs.

[1] A. Barreiro, R. Rurali, E. R. Hernández, J. Moser, T. Pichler, L. Forró, A. Bachtold, Science 320, 775 (2008)
[2] C. Spudat, M. Müller, L. Houben, J. Maultzsch, K. Goß, C. Thomsen, C. M. Schneider, C. Meyer, Nano Lett. 10, 4470 (2010)
[3] S. Bandow, S. Asaka, Y. Saito, A. M. Rao, L. Grigorian, E. Richter, P. C. Eklund, Phys. Rev. Lett. 80, 3779 (1998)
[4] V. Popov, L. Henrard, Phys. Rev. B 65, 235415 (2002)
[5] X. Q. He, S. Kitipornchai, K. M. Liew, Nanotechnology 16, 2086 (2005)

EFFECT OF DIFFERENT ACID TREATMENT OF CARBON NANOTUBES (CNT) ON CNT/TiO$_2$ NANOCOMPOSITES VIA SOL-GEL METHOD

Mohammad Reza Golobostanfard[1,*], **Hossein Abdizadeh**[1]

[1]School of Metallurgy & Materials Engineering, University of Tehran, Tehran, Iran

The influence of different acid treatment temperatures and times on CNT/TiO$_2$ nanocomposites prepared with modified sol-gel method was investigated. The composites were characterized using X-ray diffraction (XRD), field emission scanning electron microscopy (FESEM), UV-Vis spectroscopy, differential scanning calorimetry/thermogravimetry (DSC/TG), Raman, and fluorescence spectroscopy. The XRD and Raman analysis confirms the presence of CNTs besides anatase TiO$_2$. The results show that the more severe the acid treatment conditions, the more successful the separation of CNTs, and the less the dispersion of CNTs in ethanol. Furthermore, increasing acid treatment temperature and time causes further fractionation of CNTs. Thus, the most enhanced properties of CNT/TiO$_2$ nanocomposites confirmed by UV-Vis and fluorescence spectroscopies were achieved by medium treatments.

Keywords: CNT/TiO$_2$ nanocomposite, sol-gel method, carbon nanotube acid treatment.

INTERPLAY OF ELECTRONIC CORRELATIONS AND SPIN-ORBIT INTERACTIONS AT THE ENDS OF CARBON NANOTUBES

Manuel J. Schmidt[1]

[1]Institute for Theoretical Solid State Physics, RWTH Aachen, Germany

Due to the vanishing density of states at the charge neutrality point in graphene the Coulomb interaction between electrons is strongly suppressed, so that correlation effects in bulk graphene monolayers are only of minor importance. On the other hand, it is well known that the single-particle local density of states is strongly peaked at a graphene zigzag edge, thus allowing the Coulomb interaction to become active there and drive the zigzag edge into a ferromagnetically correlated state, known as edge magnetism. An analogous mechanism exists at the end of a zigzag carbon nanotube, which is also strongly correlated and shows a large magnetic moment (called superspin) which is proportional to the tube circumference.

The interaction of spin and orbital degrees of freedom of an electron is also rather small in usual graphene. This spin-orbit interaction is suppressed in graphene because of two reasons: (a) carbon is a light element and (b) the mirror symmetry at the graphene plane makes the spin-orbit interaction in the low energy sector an effect of second order in perturbation theory. (a) is a given fact, and one must live with it. However, the symmetry on which (b) is based can be broken. And indeed it is a well-known fact that the spin-orbit interaction is strongly enhanced in carbon nanotubes, since exactly this mirror symmetry is broken by their surface curvature.

In this work [1], the interplay of the two effects described above, the correlations at the edges and the spin-orbit enhancement, is discussed. I will show that the spin-orbit interaction induces an XY magnetic anisotropy in the low-energy sector of the superspin at the carbon nanotube end. The form of this anisotropy is exclusively derived from the curvature-induced spin-orbit interaction of the edge states, derived directly from the $\pi\sigma$-hybridization, and cannot be obtained from effective low-energy theories of the spin-orbit interaction in bulk graphene or carbon nanotubes.

Finally, the interplay of the spin-orbit anisotropy and the tunability of edge magnetism [2] is discussed. It turns out that the anisotropy can be enhanced by more than an order of magnitude via a suppression of the superspin size due to an additional kinetic energy of the edge states.

[1] M. J. Schmidt, Phys. Rev. B **84**, 241403(R) (2011).
[2] M. J. Schmidt and D. Loss, Phys. Rev. B **82**, 085422 (2010).

GRAPHENE CHARGE DETECTOR ON A CARBON NANOTUBE QUANTUM DOT

Stephan Engels[1,2,3], Alexander Epping[1,2,3], Carola Meyer[2,3], Stefan Trellenkamp[2,3], Uwe Wichmann[1,3], and Christoph Stampfer[1,2,3]

[1] II. Institute of Physics B, RWTH Aachen University, 52074 Aachen
[2] Peter Grünberg Institute (PGI-6/8/9), Forschungszentrum Jülich, 52425 Jülich
[3] JARA Fundamentals of Future Information Technologies

Carbon nanomaterials, such as graphene and carbon nanotubes (CNTs) attract increasing interest mainly due to their promises for flexible electronics, high-frequency devices and spin-based quantum circuits. Both materials consist of sp^2-bound carbon and exhibit unique electronic properties. In particular the weak hyperfine interaction makes graphene and CNTs interesting host materials for quantum dots which promise the implementation of spin qubits. Here, we present quantum devices based on both graphene and carbon nanotubes, which combine the advantages of prospectively clean quantum dot systems in CNTs [1] and the ability to pattern graphene into desired geometries [2,3]. In particular, we discuss a carbon nanotube quantum dot (QD) with a capacitively coupled graphene nanoribbon acting as electrostatic gate and charge detector.

The fabrication process is based on chemical vapour deposition (CVD) growth of carbon nanotubes and subsequent deposition of mechanically exfoliated natural graphite. Electron beam lithography and reactive ion etching (RIE) is used to pattern individual graphene flakes. Metallic contacts are fabricated by electron beam lithography and lift-off.

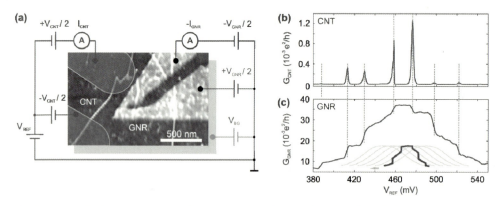

Fig. 1: (a) Scanning force micrograph of the investigated grapheme-carbon nanotube hybrid device. (b), (c) Simultaneously measured conductance through the nanotube (CNT) quantum dot (b) and through the grapheme nanoribbon (GNR) (c).

Fig. 1(a) shows a scanning force micrograph (SFM) of an all-carbon device consisting of a carbon nanotube lying in the close vicinity to an etched graphene nanoribbon (GNR) which acts as a charge detector (CD). From a Raman spectrum the nanoribbon is identified to be bilayer graphene and from a SFM profile the nanotube is determined to be a single-walled carbon nanotube. Both carbon nanostructures are separated by roughly 150 nm, the nanoribbon has a width of around 100 nm and the CNT quantum dot is defined by two metal electrodes (indicated in magenta) which are separated by 350 nm. As illustrated in Fig. 1(a) we apply a symmetric bias voltage (V_{CNT} and V_{GNR} respectively) to both structures and the overall Fermi level can be tuned by the back gate voltage V_{BG} applied to the highly doped Si substrate. Additionally, we can use the CNT as a lateral gate for the GNR by applying a reference potential V_{REF}.

All measurements on the illustrated device were performed in a pumped ^{4}He-cryostat at a base temperature of T≈1.5 K. Figs. 1(b) and 1(c) show the simultaneously measured low-bias V_{CNT}=0.5 meV conductance through the nanotube G_{CNT} and the nanoribbon G_{GNR} as function of V_{REF} respectively. G_{CNT} exhibits Coulomb peaks which are indicating single charging events in the CNT QD. In the simultaneously measured trace of the GNR we observe distinct steps in the conductance at the exact positions of the CNT QD charging events. The steps in conductance can measure up to 20% of the total resonance amplitude and are due to the capacitive coupling of both nanostructures. Increasing the reference potential V_{REF} and decreasing V_{BG} both leads to a higher chemical potential in the QD and subsequently lower occupation numbers at every event. Consequently, the GNR resonance shifts to lower values of V_{REF} giving rise to the unconventional shape of the charge detecting resonance in Fig. 1(c) as illustrated by the inset.

In summary, we present the fabrication and characterization of carbon nanotube graphene hybrid devices. We show an example of a structure where a nanotube quantum dot is capacitively coupled to a graphene nanoribbon. Sharp resonances in the graphene nanoribbon gate characteristics give rise to clear detection signals. The presented results open the road to more sophisticated devices which are entirely fabricated out of carbon nanostructures and exploit the different advantages of these materials.

[1] S.Sapmaz, P. Jarillo-Herrero, J. Kong, L. P. Kouwenhoven, and H. van der Zant, Phys. Rev. B., 71, 153402 (2005).
[2] B. Terrés, J. Dauber, C. Volk, S. Trellenkamp, U.Wichmann, and C. Stampfer, Appl. Phys. Lett. 98, 032109 (2011).
[3] C. Volk, S. Fringes, B. Terrés, J. Dauber, S. Engels, S. Trellenkamp, and C. Stampfer, Nano Lett., 11, 3581 (2011).
[4] S. Engels, P. Weber, B. Terrés, J. Dauber, C. Meyer, C.Volk, S. Trellenkamp, U. Wichmann, and C. Stampfer, submitted (2012).

RESISTIVE SWITCHING ON METAL-OXIDE POLYMER MEMORIES

Paulo R. F. Rocha[1], Qian Chen[1], Asal Kiazadeh[1], **Henrique L. Gomes**[1], Stefan Meskers[2] and Dago de Leeuw[3]

[1]Center of Electronics Optoelectronics and Telecommunications (CEOT), Universidade do Algarve, Campus de Gambelas, 8005-139 Faro, Portugal;
[2]Molecular Materials and Nanosystems, Eindhoven University of Technology, P.O. Box 513, 5600 MB Eindhoven, The Netherlands;
[3]Philips Research Laboratories, Professor Holstlaan 4, 5656 AA Eindhoven, The Netherlands

Metal-oxide polymer memories consisting of an oxide layer in series with a semiconducting polymer exhibit non-volatile resistive states with large retention time and excellent cycle endurance. The resistive switching mechanism is physically located in the oxide layer. The switching is triggered by a soft-breakdown mechanism in the oxide inducing a local conducting path. This process is often referred as electroforming. The role of the adjacent semiconductor polymer layer in electroforming is usually overlooked. It has been suggested that the semiconducting polymer layer only acts as a series distributed parallel resistance preventing catastrophic breakdown. However, the semiconducting polymer layer plays also a role in tuning the memory properties. We have shown that prior to electroforming, electrons are trapped at the oxide/polymer interface and the polymer layer has a crucial role in this trapping process [1, 2]. Trapped electrons establish a dipole layer across the oxide. The associated electric field is ultimately responsible by a soft-breakdown mechanism.

In this contribution, we show that by a careful control of the interfacial charge build-up (speed and magnitude) it is possible to minimize the thermal damage (by Joule heat) and achieve high on/off ratios between resistive states.

The soft-breakdown mechanism is followed in time by monitoring the amount of trapped electrons when the diode is stressed under constant current as illustrated in Fig. 1.

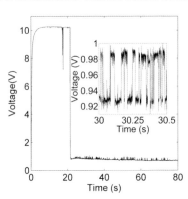

Figure 1: Time dependence of a voltage across a diode when stress with a constant current of 1µA for 80 seconds. After a switching event occurring at t≈20 sec, the diode exhibits RTS noise (see inset).

After a switching event a significant random telegraph noise (RTS) is observed (see inset of Fig.1). By monitoring the evolution of the electrical noise as function of time, we get insight into the traps directly involved in the soft-breakdown mechanism. We extract from the RTS signals an average trap characteristic time in the scale of microseconds. From temperature-dependent measurements we estimate a trap depth of 0.8 eV.

It is also shown that a threshold electrical power is required to induce a non-volatile memory. This threshold power is discussed in terms of the suitable energy to stabilize permanently the defect in the oxide matrix.

[1] Q. Chen e al., Appl. Phys. Lett. **99**, 083305 (2011).
[2] B. F. Bory et al., Appl. Phys. Lett. **97**, 222106 (2010).

THIOLATED (OLIGO)PHENOTHIAZINES AS PROMISING CANDIDATES FOR FUTURE STORAGE ELEMENTS

Michael Paßens[1], Adam Busiakiewicz[1], Adam W. Franz[2], Christa S. Barkschat[2], Dominik Urselmann[2], Thomas J. J. Müller[2], Silvia Karthäuser[1]

[1]Peter Grünberg Institut (PGI-7) and JARA-FIT, Forschungszentrum Jülich, D-52425 Jülich, Germany;
[2]Heinrich-Heine-Universität Düsseldorf, D-40225 Düsseldorf, Germany

For future high-density storage devices the CMOL concept is a promising candidate. The basic idea is to combine the advantages of the CMOS technology with functional molecular elements like molecular switches. Two well known examples of photoswitchable molecular structures are azobenzene and diarylethene, which have been investigated in STM geometry and as capping layer of nanoparticles assembled in an array. Here we are exploring phenothiazines as redox active molecular switches for molecular storage elements.

Phenothiazines are tricyclic nitrogen and sulfur containing heterocycles which offer various advantages. In particular, they show a highly reversible formation of stable radical cations [1], tunable redox and fluorescence properties [2] as well as a tendency to self assembled through π–π interactions [3], and they are electrically and optically addressable. Furthermore phenothiazines exhibit interesting conformational changes. In the neutral electrical state the molecule displays with a folding angle of 158.5° [4] a so called "butterfly conformation", which changes by oxidation into a stable planar radical cation with only minute changes of bond lengths. The origin of this remarkable stability of the radical cation originates from extended charge delocalization [5]. This offers the possibility to change between the neutral and the charged state. In addition the conductivity varies considerably between both states rendering phenothiazines as good candidates for molecular switches.

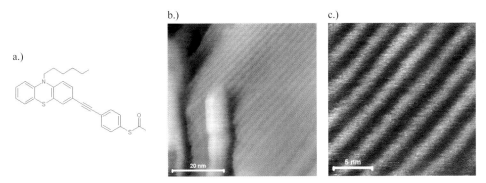

Figure 1: a.) Molecular structure of Phen1Ph, b.) topography of Phen1Ph self assembled on a Au(111) substrate, c.) higher resolution picture showing the striped structure

For application of phenothiazines in electronic devices and for determining charge transport characteristics their ligation to metal surfaces such as Au{111} is required. Here we utilize the reliable thiol-gold bond for the chemisorption of molecules [6].

Three different phenothiazine derivatives (thioacetic acid S-(10-methyl-10H-phenothiazin-3-yl) ester (Phen1), thioacetic acid S-[4-(10-hexyl-10H-phenothiazine-3-ylethynyl)-phenyl] ester (Phen1Ph, Fig. 1a), thioacetic acid S-[4-(10, 10'-dihexyl-10H, 10'H-[3,3']biphenothiazinyl-7-ylethinyl)-phenyl] ester (Phen2Ph)) that were chemisorbed from solution on a Au(111) surface were characterized. The topography obtained by UHV-STM shows a monolayer of Phen1Ph (Fig.1) with a striped structure. The size of the stripes indicates lying down molecules. Furthermore this structure gives strong evidence for the π–π interaction between the phenothiazines. In addition Phen2Ph molecules could be self-assembled onto a gold surface using a matrix isolation method. In this case the molecules are upright (not shown). We obtained reproducible current-voltage and differential conductance curves by UHV-STS investigations. In Figure 2 the results for the Phen1Ph molecules are depicted.

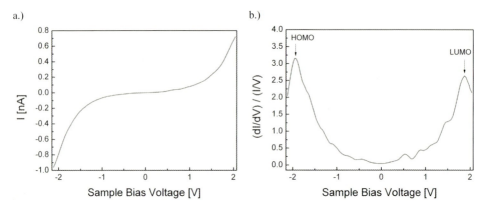

Figure 2: a) current/voltage curve; b) differential conductivity of Phen1Ph showing the HOMO-LUMO gap

The peaks at -1.94 eV and at 1.87 eV can be identified as HOMO and as LUMO, respectively. The data obtained by UV/Vis-measurements given in [7,8] and the results for the HOMO-LUMO gap derived from STS-measurements for the three different phenothiazine derivatives are in good agreement and show clearly the tunability of the gap.

[1] H. J. Oka, J. Mater. Chem. **2008**, 18, 1927-1934
[2] A.W. Franz, L.N. Popa, F. Rominger, T.J.J. Müller, Org. Biomol. Chem. **2009**, 7, 469-475
[3] C.S. Barkschat, R. Guckenberger,T.J.J. Müller, Z. Naturforsch. **2009**, 64b, 707-718
[4] J.J.H. McDowell, Acta Crys. **1976**, B32, 5
[5] T. Uchida, M. Ito, K. Kozawa, Bull. Chem. Soc. Jpn. **1983**, 56, 577-582
[6] F. Maya, A.K. Flatt, M.P. Stewart, D.E. Shen, J.M. Tour, Chem. Mater. **2004**, 16, 2987-2997
[7] C.S. Barkschat, S. Stoycheva, M. Himmelhaus, T.J.J. Müller, Chem. Matter. **2010**, 22, 52-63
[8] A.W. Franz, S. Stoycheva, M. Himmelhaus, T.J.J. Müller, Beilstein J. Org. Chem. **2010,** 6, No. 72

SPM-INVESTIGATIONS OF THE SPIROPYRAN – MEROCYANIN PHOTOISOMERIZATION

A.Soltow, S. Karthäuser, R. Waser

Peter Grünberg Institut (PGI-7) and JARA-FIT, Forschungszentrum Jülich GmbH, Jülich, Germany

One aim of molecular electronics is to exploit organic molecules with optically alterable properties as switching components for nanoscale devices. One interesting family of photoswitchable molecules are the spiropyran derivatives [1]. It is well known that the closed spiropyran form (SP) can be transformed under UV light irradiation into the open merocyanin (MC) form. In solution the MC isomer is unstable at room temperature and can be reverted thermally into the stable SP form on a time scale of seconds or minutes. However, using solvents with a large dipole moment, the MC isomer can be stabilized [2]. The SP – MC transition induced by optical excitation is accompanied by a modification of the dipole moment and the HOMO - LUMO gap. Additionally, due to the widely delocalized π- electron system in the MC- form a significant increase of conductance is expected which makes the SP – MC couple useful as molecular switches.

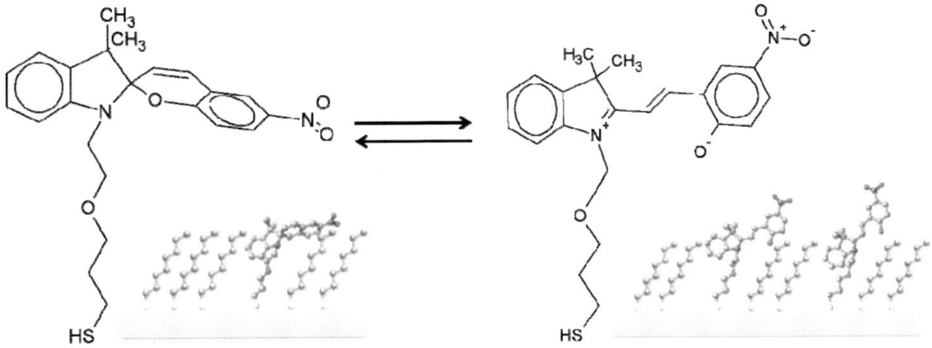

Figure 1. Structure of (1´-(2,3-Mercaptopropyl)ethoxy)-3´,3´-dimethyl-6-nitrospiro[benzopyran-2,2´-indolin]) (SP) synthesized by "emp Biotech GmbH" and merocyanine (MC). Whereas SP is chiral and exhibits two halves with orthogonal planes, the conjugated MC form is planar and prochiral. Schematics of SP and MC embedded in matrices of alkanethiols are shown in the insets.

Crystalline thin gold films with a (111) orientation were prepared on mica substrates by evaporation as described in Ref. [3]. Since our SP and MC derivatives chemisorb on Au (111) with a long mercapto-propyl-ethoxy anchor group, they are not expected to build highly ordered SAM- structures. Therefore, we used an inert matrix to embed them and focused the investigation on isolated molecules. The inert alkanethiol monolayer (SAM) was prepared by depositing the Au(111) substrates for hours in a 1 mM solution of octanethiol or dodecanethiol in ethanol. Small bundles or even single molecules of SP were inserted into the highly ordered alkanethiol matrix by storing the SAM covered Au(111) substrate for 1 to 4 days in a 1 mM solution of SP in acetone. The same procedure was applied for MC, but here a highly polar solvent, i.e. hexafluoro-2-propanol was used [2]. After inserting single molecules into the SAMs the samples were rinsed with the solvent and immediately transferred into vacuum. For MC the storage was in darkness and transfer was performed with light of wavelengths larger than 695 nm. The films were investigate by imaging in constant current mode and electronically characterized in CITS spectroscopic mode using a JEOL STM-4500S UHV-STM at room temperature. The UV/Vis spectra were acquired with a PerkinElmer Lambda 900 spectrometer.

Spectroscopic measurements were carried out on isolated SP derivatives and a HOMO – LUMO gap of 3,8 eV was determined. Furthermore we observed that an applied electrical field is responsible for a shift of about 0,2 eV in the HOMO and LUMO energy levels of SP. This shift is likely caused by a reorientation of the large flexible molecule in the electrical field which in turn changes the relative distances of the molecular moieties with respect to the substrate and the tip and thus, the effective electrical field sensed by the molecular moieties. The thermodynamically unstable MC is embedded into a SAM with longer chain length for stabilization. Spectroscopic measurements on isolated MC derivatives indicate a HOMO – LUMO gap of 2,2 eV, which is in agreement with UV/Vis – investigations. Further spectroscopic measurements reveal also HOMO –LUMO gaps of approx. 1,0 eV, which point to a dimerisation of MC. Most interestingly the MC form can be stabilized by the dodecanethiol matrix for several days without reconversion to SP. However, an electrical field of 2V can trigger this back reaction.

[1] V. I. Minkin, "Photo-, Thermo-, Solvato-, and Electrochromic Spiroheterocyclic Compounds", Chem. Rev. **104**, 2751-2776 (2004)
[2] A. Fissi, 0. Pieroni, F.Ciardelli, Z D. Fabbri, G. Ruggeri, K.Umezawa, "Photoresponsive Polypeptides: Photochromism and Conformation of Poly (L-Glutamic Acid) Containing Spiropyran Units", Biopolymers. 33 1993, 1505–1517
[3] Lüssem, B.; Karthäuser, S.; Haselier, H.; Waser, "The origin of faceting of ultraflat gold films epitaxially grown on mica", R. Appl. Surf. Sci. 2005, 249, 197–202

ELT 30

MEMRISTIVE PHENOMENA OF CONDUCTION POLYMER PEDOT:PSS

Fei Zeng [1], Jing Yang [2], Zhishun Wang [2], Yisong Lin [2] and Sizhao Li [2]

[1,2] Key Laboratory of Advanced Materials (MOE), Department of Materials Science and Engineering, Tsinghua University, Beijing, 100084, People's Republic of China
[1] Presenting author, zengfei@mail.tsinghua.edu.cn

Recently, redox reactions in semiconductors have played important roles in the field of data storage and memory effect. "0-1" storage model, which is prevailed in current computer, have been realized by redox reaction. Furthermore, scientists have found that short term plasticity (STP) and long term potentiation (LTP) of synapse could be mimicked by redox memristor [1,2]. Constructing artificial neural network using memristor on the base of traditional CMOS technique attracts people's attention significantly. We report here that a long-term depression (LTD) like behaviour is observed in a simple metal/polymer junction of Ti/PEDOT:PSS/Ti. The conductance of the device decreases and negative differential resistance appears after training using a DC voltage scanning. Pulse with various temporal values and amplitudes has been used to test the device. The results resemble to LTD relating also to learning and habitation. We analyze roughly that the mechanism of the observed LTD is the combination of de-doping of holes and movement of PSS molecules.

PEDOT:PSS is a complex composed of conductive PEDOT and soluble PSS. The intrinsic asymmetry in conductivity makes PEDOT:PSS sensitive to external factors. We have found that multiple redox mechanisms exist in PEDOT:PSS, such as filaments of PEDOT molecule [3], or filaments of Cu entering into PEDOT:PSS [4]. Electrode has critical selective effect on the redox mechanisms. These properties let PEDOT:PSS be a plausible materials simulating synapse.

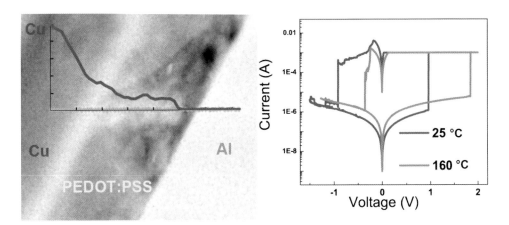

Figure 1. In the structure of Al/PEDOT:PSS/Cu, the switching mechanism is the formation of Cu filament. Cu ions enter into PEDOT:PSS with the aid of redox of Cu and enhance the temperature stability greatly. Ref. [4]

[1] Takeo Ohno et al., Nature Materials **10**, 591 (2011)
[2] Sung Hyun Jo et al., Nano Lett. **10**, 1297 (2010)
[3] Z.S. Wang et al., Appl. Phys. Lett. **97**, 253301 (2010)
[4] Z.S. Wang et al., ACS Appl. Mater. Interfaces **4**, 447 (2012)

A NEW MINIMUM SEARCH METHOD FOR COMPLEX OPTIMIZATION PROBLEMS

Julian A. Hirschfeld[1], Hans Lustfeld[2]

[1]Forschungszentrum Jülich – Institute for Advanced Simulation (IAS-1), Jülich, Germany;
[2]Forschungszentrum Jülich – Peter Grünberg Institute (PGI-1), Jülich, Germany

Optimization is essential in many scientific and economical areas, as well as in the development of products. In many cases the optimization problem is too complex to be tackled by simple straightforward calculations or by trial and error. The reason is the too large phase space with too many local minima. To find the global minimum, or at least a low-lying one more sophisticated methods have to be applied, two well known ones being simulated annealing and the genetic algorithm. In these methods artificial fluctuations control the probability of the system to overcome a local minimum having a certain depth.

Here we present a new method [1], which is complementary to the established ones. The chance to get stuck in a local minimum or to escape is independent of the depth of the minimum but depends on the attractor size of the minimum. Therefore, it can overcome local minima and high barriers equally well.

We successfully applied the method to find the ground states of the phosphorus P_4 and P_8 molecules and the corresponding molecules of As_n, Sb_n, Bi_n ($n = 4, 8$) in the framework of the Density Functional Theory (DFT). In the case of P_8 we have found stable and metastable configurations, some of which are new and have similar energies. As a by-product we obtained an upper bound for the energy barriers between these configurations.

Figure 1: The P_8 isomer (C_{2v} symmetry, "wedge") that has been discovered by Jones and Hohl [2] in 1990 and has been considered as the most stable isomer of P_8 since then. Before 1990 it had been assumed that the high symmetry of the cubic structure would be favored.

Figure 2: The P_8 isomer found by our new optimization scheme is a dimer of two P_4 isomers. The energy of this configuration is slightly lower than the energy of the "wedge" configuration shown in figure 1. This is a result obtained in DFT but MP2 calculations [3] corroborate this finding.

[1] J. H. Hirschfeld and H. Lustfeld, submitted to Phys. Rev. B.
[2] R. O. Jones and D. J. Hohl, Chem. Phys. **92**, 6710 (1990).
[3] R. Janoschek, Chem. Ber. **125**, 2687-2689 (1992).

RESISTIVE SWITCHING CHARACTERISTICS IN HfO_2 THIN FILMS DEPENDING ON THEIR CRYSTALLINE STRUCTURE

Jung Ho Yoon[1], **Hyung-Suk Jung**[1], **Min Hwan Lee**[2], **Gun Hwan Kim**[1], **Seul Ji Song**[1], **Jun Yeong Seok**[1], **Kyung Jean Yoon**[1] and **Cheol Seong Hwang**[1]

[1] Department of Materials Science & Engineering and Inter-university Semiconductor Research Center, Seoul National University, Seoul 151-744, Korea
[2] University of California, Merced5200 North Lake Rd.Merced, CA 95343, US

Devices based on the resistive-switching behavior of transition metal oxides (TMOs) such as TiO_2, HfO_2, ZrO_2, NiO, etc. are the strong contenders for next-generation nonvolatile memory due to their good performance in various aspects such as low-power consumption, rapid switching speed, etc. Among these TMOs, HfO_2 has some advantages as the key material for resistive switching memory (RRAM) because it is already adopted as the high-k gate dielectric in advanced Si based devices so the process maturity is high.

Until now, research concerning HfO_2 based RRAMs was focused on improving the performance and uniformity of RRAM device characteristics. However, studies on the origin of resistive switching behavior in HfO_2 thin film is still lacking although it is very important because it is obvious that maximizing efficiency in any application will be difficult without clarifying the physical principles that govern device operation.

Therefore, in this research, the relationship between the microstructural properties and the resistive switching characteristics of HfO_2 is identified with an emphasis on the influence of grain size and grain boundary area in order to clarify the origin of the resistive switching mechanism. For this purpose, rapid thermal annealing of amorphous ALD HfO_2 films was performed in N_2/H_2 or O_2 environments at temperatures between 400 °C and 600°C.

It was found that stable resistive switching occurred in the crystallized films which were achieved when annealing temperatures exceeded 500°C, irrespective of the annealing environment, while amorphous films did not show any fluent resistive switching behavior. These results suggest that the effect of oxygen exchange during the heat treatment is negligible because the annealing gas environment did not affect the resistive switching behavior. This was further confirmed by XPS results that showed no significant change in the relative ratio of Hf/O in films annealed in different environments. Therefore, crystallization is the major factor that governs resistive switching behavior in HfO_2 thin films.

The presence of local conducting spots (maybe conducting filaments) was investigated by implementing conductive atomic force microscopy (CAFM). These results showed that a clear relationship exists between the current profile and film morphology, especially in terms of grain and grain-boundary distribution. As a result, it was demonstrated that the conducting filaments are formed mainly at locations where multiple grain boundaries are locally merged. Therefore, the geometrical factor of grain and grain boundaries in HfO_2 thin films strongly affect the resistive switching behavior. In this presentation, the authors will report the detailed results on the switching behaviors depending on sample area, serial connection, electrode materials, and others to elucidate the origin of the switching in this material, as well as the optimized device performance.

DEPOSITION OF CHALCOGENIDE THIN LAYERS BY MAGNETRON SPUTTERING FOR RRAM APPLICATIONS

M.-P. Besland, J. Tranchant, E. Souchier, P. Moreau, S. Salmon, B. Corraze, E. Janod, L. Cario

[1] Institut des Matériaux Jean Rouxel (IMN), Université de Nantes, UMR CNRS 6502,
2, rue de la Houssinière, BP 32229, 44322 Nantes Cedex 3, France

In recent works, we have discovered a new type of reversible and non-volatile resistive switching on single crystals of the Mott Insulator compounds AM_4X_8 (A = Ga, Ge; M = V, Nb, Ta; X = S, Se) [1, 2]. These studies indicate that the mechanism of the resistive switching in the AM_4X_8 [1-3] differs from those reported in the literature [4]. Indeed, our results highlight an electric field effect which induces an electronic phase change from a Mott insulating state to a metallic-like state [3]. Among emerging memories [4], Resistive RAM (RRAM) stands as serious candidates to replace flash memories in next information storage devices. RRAM are based on functional materials exhibiting two stable resistive states, i.e. a high resistance state (HRS) and a low resistance state (LRS). The application of electric pulses allows a reversible and non-volatile resistive switching (RS) between these two states. Thus, in that respect, the AM_4X_8 compounds stand as a really promising class of materials.

Prior to envision any development of functional materials towards devices, two major challenges have to be tackled. The first one is to obtain thin layers of active and functional materials. The second challenge is to recover on thin layers the functional properties. RF magnetron sputtering has been chosen as deposition process since it is widely used in the back end of the line (BEOL) of microelectronics processes. Moreover, magnetron sputtering enables to deposit well-crystallized films of insulating or conducting materials, at low temperatures, over large areas, while controlling the film composition and microstructure, even for complex and multi-component materials. Thus, on the basis of well established know-how in deposition process and multi-layered functional structures [5, 6], the deposition of AM_4X_8 material in the form of thin layers has been investigated.

Hence, GaV_4S_8 thin layers (300-500 nm thickness range) were deposited by RF magnetron sputtering, using stoichiometric GaV_4S_8 targets obtained by Spark Plasma Sintering (SPS) technique [7] and starting from polycrystalline GaV_4S_8 powder. Experimentally, the transfer to a thin layer of a complex ternary material is favored by soft deposition conditions [6]. Thin layers of GaV_4S_8 were first deposited at low deposition pressure and RF power, respectively 40 mTorr and 60 W (i.e. 3 W.cm^{-2}). For thin layers deposited in pure Ar, a one hour ex-situ annealing at 600°C in a sulfur-rich atmosphere restore the targeted stoichiometric composition GaV_4S_8 and the expected crystalline structure [8].

We will present here recent results concerning the improvement of the deposition process and the electrical characterizations performed on MIM structures Au/GaV_4S_8/Au/Si (Figure 1). Typically, electrical pulses applied to 50µm x 50µm pads at 300K induce a resistive switching with a $\Delta R/R$ values close to 33% [8]. In particular, the decrease of the sample size (Figure 2) induces a strong increase of the resistive switching amplitude. Moreover, very competitive performances compared to other emerging RRAM technologies have been obtained regarding endurance or writing/erasing time. Our studies therefore demonstrated that the narrow gap Mott Insulator AM_4X_8 compounds could be good candidates for a new type of non-volatile RRAM memory based on a new mechanism of resistive switching, i.e. an electric-field-driven metal insulator transition.

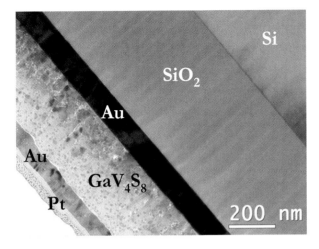

Figure 1: HRTEM image of the cross section of the MIM structure Au/GaV$_4$S$_8$/Au

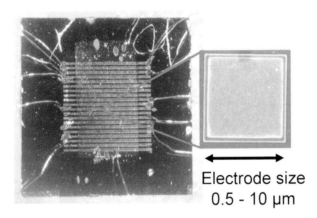

Figure 2: Photography and SEM image of the substrates used for electrical characterizations after deposition of GaV$_4$S$_8$ thin films

[1] C. Vaju et al., Advanced Materials, **20**, 2760 (2008).
[2] C. Vaju et al., Microelectronics Engineering **85** (12), 2430 (2008).
[3] L. Cario , C.Vaju , B.Corraze , V.Guiot E. Janod, Advanced Materials **22** (45), 5193 (2010).
[4] R. Waser, R. Dittman, G. Staikov, K. Szot, Advanced Materials, **21**, 2632 (2009).
[5] C. Duquenne, P-Y. Tessier, M-P. Besland, B. Angleraud , P-Y. Jouan , R. Aubry , S. Delage, M. A. Djouadi, J. Appl. Phys **104** (6), 2008.
[6] M.P.Besland et al. Thin Solid Films, **495**, 86 (2006).
[7] C. Elissalde, M. Maglione, C. Estournes, J. Am. Ceram. Soc. **90**, 973 (2007).
[8] E. Souchier, L. Cario, B. Corraze, P. Moreau, P. Mazoyer, C. Estounes, R. Retoux, E. Janod, M.P. Besland, Phys. Status Solidi RRL, 5, 53 (2011).

DETERMINISTIC RESISTIVE SWITCHING CONTROL IN HfO$_2$-BASED MEMORY DEVICES

Raúl Zazpe[1], Mariana Ungureanu[1], Roger Llopis[1], Federico Golmar[1,2], Pablo Stoliar[1,3,4], Félix Casanova[1,5], Luis Eduardo Hueso[1,5]

[1]Nanodevices Group, CIC nanoGUNE Consolider, Tolosa Hiribidea 76, 20018 Donostia - San Sebastian, Spain

[2]I.N.T.I. - CONICET, Av. Gral. Paz 5445, Ed. 42, B1650JKA, San Martín, Bs As, Argentina
[3]GIA, CAC - CNEA, Av. Gral. Paz 1499, B1650JKA, San Martín, Bs As, Argentina
[4]ECyT, UNSAM, Martín de Irigoyen 3100, B1650JKA, San Martín, Bs As, Argentina
[5]IKERBASQUE, Basque Foundation for Science, 48011 Bilbao, Spain

An intense research effort is currently focused on the development of the next generation of nonvolatile memory devices[1]. Resistance random access memory (RRAM) has emerged as one of the most promising alternatives to current technologies due its inherent properties such as high density, simple structure, low power operation and scalability. RRAM operation is based on the reversible switching between two different resistance states triggered by an electric field or resistive switching (RS). In this work we study in detail a RS HfO$_2$-based memory cell with a metal-insulator-metal (MIM) vertical structure in which the HfO$_2$ was deposited by Atomic Layer Deposition (ALD). We carried out a statistic study on the effect of the HfO$_2$ deposition process conditions on the resistive switching behavior (Figure 1). Different oxygen vacancies concentrations are induced depending on the ALD conditions and its effect on the metal/oxide barrier could explain the diverse RS performances observed[2]. A further correlation found between the RS behavior and the pristine leakage currents (Figure 2) lead us to propose an oxygen vacancy-based model for explaining the switching process[3].

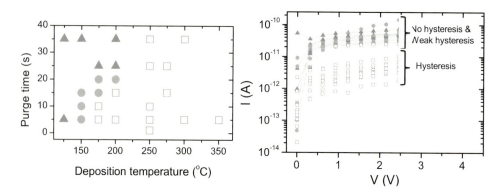

Figure 1: Summary of electrical behavior obtained for samples fabricated at different deposition temperatures and purge times (☐Hysteresis, ●Weak Hysteresis and ▲No Hysteresis, being weak hysteresis devices those which lost the response before 150 cycles). The electrical characterization was carried out at 10 devices per sample in order to get a reliable statistic in the phase diagram.

Figure 2: Current (I) – Voltage (V) monitored from pristine samples. The color indicates the samples with different resistive switching behavior (☐Hysteresis, ●Weak Hysteresis and ▲No Hysteresis).

[1]G.W. Burr et al, IBM J. Res. & Dev. 52, 4/5 (2008)
[2]A. Sawa, Mater.Today, 11, 28 (2008)
[3]M.J. Rozenberg et al, Phys. Rev. B 81, 115101 (2010)

INVESTIGATION OF TRANSIENT CURRENTS DURING ULTRA-FAST DATA OPERATION OF TiO$_2$ BASED RRAM

C. Hermes[1], M. Wimmer[2], S. Menzel[3], K. Fleck[3], V. Rana[1], M. Salinga[2], U. Böttger[3], R. Bruchhaus[4], M. Wuttig[2], R. Waser[1,3]

[1] Peter Grünberg Institut, Forschungszentrum Jülich, Jülich Germany
[2] Institute of Physics, RWTH Aachen University, Aachen, Germany
[3] Institute of Materials in Electrical Engineering and Information Technology II, RWTH Aachen University, Aachen, Germany
all authors mentioned above are with JARA-FIT
[4] JCNS, Forschungszentrum Jülich, Jülich, Germany

Resistive RAM (RRAM) memory elements are the key elements for future non-volatile memory devices. These memory elements have attracted considerable interest due to the prospect of non-volatile data storage combined with excellent scalability and low power consumption [1]. Among the numerous materials proposed so far TiO$_2$ has turned out to be a very attractive and CMOS compatible material. High speed bipolar switching has been reported for the SET as well as the RESET operation underlining the potential for the use in future applications. TiO$_2$ based resistive memory elements exhibit excellent properties in terms of switching speed, retention and stability. However, the analysis of the transient switching currents during the fast switching operation is needed to estimate the switching energy and the transient current densities during the switching event in the metallization lines of the memory arrays. A Requirement for the memory elements in these devices includes very fast data- and readout operations and low power consumption during operation.

In this work we investigated the transient currents during 5-ns resistive switching operations [2]. Transient maximum currents for the SET and RESET processes were as high as 200 and 230 μA (see Fig. 1).

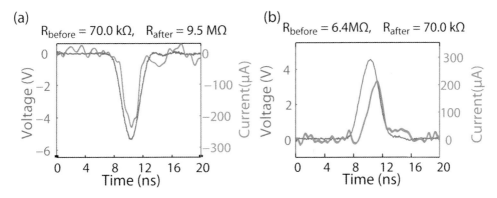

Figure 1: $V(t)$ and $I(t)$ of (a) RESET and (b) SET processes during ultra-fast data operation of Nano-crossbar memory elements. Copyright IEEE. Reuse with permission [2]

To model the observed currents during fast-pulse analysis we used FEM models and simulations. The currents during fast-pulse switching are explained by Joule-heating which enables a high mobility of oxygen vacancies. The measured transient currents enable a further optimization and engineering of resistive switches based on TiO$_2$. Due to the symmetry of the device this equation system is solved in 2D axial symmetry, for which we employed the commercial finite element software COMSOL. The device geometry is shown in Fig. 2 along with the boundary conditions. The actual switching is considered to take place in the disc region in front of a Magnéli phase Ti_4O_7 filament, which has grown during electroforming. A filament diameter of 5 nm is assumed as reported in literature [3]. The electric conductivity of the disc region and the filament are fitted to match the experimental current data of the non-switching events whereas it is assumed to be field-dependent in the disc region to account for the nonlinear $I(V)$-characteristic.

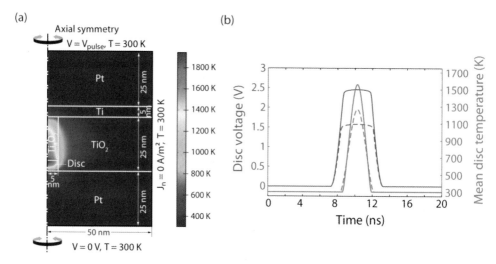

Figure 2 (a) Cross section of the simulated cell. The color represents the temperature distribution 2.5 ns after the beginning of a voltage pulse (define amplitude and shape) applied to the device. (b) Voltage dropping (blue) across the disk, not across the complete device, and (red) mean disk temperature for (dashed lines) ON- and (solid lines) OFF-states as a function of time. Copyright IEEE. Reuse with permission [2]

[1] R. Waser, R. Dittmann, G. Staikov, and K. Szot, Adv. Mater. **21**, 2632 (2009).
[2] C. Hermes et al., IEEE Electron Device Lett. **32**, 1116 (2011).
[3] D.-H. Kwon et al., Nat. Nanotechnol. **5**, 148 (2010).

FAST PULSE FORMING PROCESS FOR TiO$_2$ BASED RRAM NANO-CROSSBAR DEVICES

F. Lentz[1,2], C. Hermes[1,2], B. Rösgen[1,2], T. Selle[1,2], R. Bruchhaus[3], V. Rana[1,2], R. Waser[1,2,4]

[1] Peter Grünberg Institut, Forschungszentrum Jülich GmbH, Jülich, Germany
[2] Jülich-Aachen Research Alliance, Section Fundamentals for Future Information Technology, Jülich, Germany
[3] JCNS, Forschungszentrum Jülich GmbH, Jülich, Germany
[4] Institute of Materials in Electrical Engineering and Information Technology II, RWTH Aachen University, Aachen, Germany

Switchable metal-isolator-metal (MIM) structures play a key role for future non-volatile resistive RAM (RRAM) devices. The low power consumption, excellent scalability and very fast write/read operation in combination with the non-volatile resistance states make these kinds of memory elements a promising candidate to become a successor of current memory technologies [1]. The MIM structures based on TiO$_2$ exhibit excellent properties in terms of data operation speed and retention. However these resistive switching cells have required a quasi-static electroforming step which is a drawback for integration in future memory [2]. In this work, we propose a short voltage pulse induced electroforming (Fig. 4) and demonstrate that this new method is advantageous over the conventional quasi-static current driven electroforming process (Fig. 3).

Fig. 1 shows a SEM image of a Pt/TiO$_2$/Ti/Pt nano-crossbar device. The device fabrication process is mentioned in ref. 3. Here, we describe in short. First, the Pt bottom electrodes were patterned with UV nano-imprinting and reactive ion beam etching. Next, a 25 nm TiO$_2$ layer was sputtered. Finally, a lift-off process with e-beam lithography was used to fabricate the top electrode. This results in the schematic cross section of the device shown in Fig. 2.

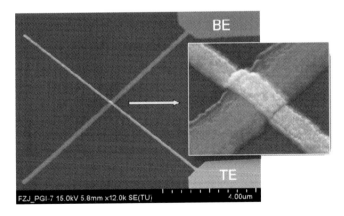
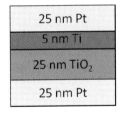

Fig 1: Top view of a 100 nm x 100 nm Pt/TiO$_2$/Ti/Pt nano-crossbar

Fig 2: Schematic cross section of a crossbar device

To enable resistive switching in the Pt/Ti/TiO$_2$/Pt nano-crossbar devices, an electroforming step is needed. A current driven electroforming with negative polarity [2] shown in fig. 3 transforms the device into a stable RESET state. The formed memory device exhibits a resistance of around 10 MΩ.

In a new approach, the electroforming process is carried out with fast voltage pulses. A single voltage pulse of 100 ns, 1 µs or 100 µs or a pulse train of 1000 single pulses of 100 ns voltage pulses of variable amplitude is applied to the memory device. A change in the cell resistance after the pulse application (Fig. 4) hints to a formed state of the memory element. The application of a single 100 µs voltage pulse of -5 V pulse amplitude or the use of a 100 ns pulse repeated 1000 times with the same amplitude of -5 V forms the memory element to a switchable state.

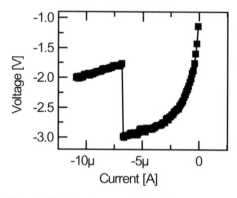
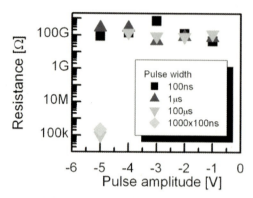

Figure 3: Quasi-static current driven electroforming procedure. The forced current amplitude through the memory element is stepwise increased until a sudden decrease in the necessay voltage is observed.

Figure 4: Negative voltage pulse induced electroforming procedure. The pulse amplitude is raised in 1 V steps until a change in the cell resistance is observed.

After the pulse-forming, the first switching cycle of the pulsed-formed samples starts with the RESET process followed by the SET process by a positive voltage sweep. The switching behaviour of the quasi-static formed sample is the same as in the pulse-formed sample which exhibits the SET process for a positive voltage sweep whereas the RESET is observed for the negative voltage sweep. The difference between the first switching cycles can be explained by the uncontrolled current during the pulse forming application.

In conclusion, the time consuming quasi-static electroforming in the TiO_2 based memory elements could be replaced by the fast pulse electroforming process. During the pulse driven data operation, the peak power consumption in the memory device is about 1 mW.

1. R. Waser, R. Dittmann, G. Staikov, K. Szot, Adv. Mater. **21**, 2632 (2009)
2. C. Nauenheim, C. Kuegeler, A. Ruediger, R. Waser, Appl. Phys. Lett. **96**, 122902 (2010)
3. C. Hermes, F. Lentz, R. Waser, R. Bruchhaus, S. Menzel, K. Fleck, U. Bottger, M. Wimmer, M. Salinga, M. Wuttig, Proceedings NVMTS, 92 (2011)

CURRENT TRANSPORT MODELING IN OXIDE-BASED RESISTIVELY SWITCHING MEMORY CELLS FOR THE INVESTIGATION OF ELECTROFORMATION

Astrid Marchewka[1], Stephan Menzel[1], Ulrich Böttger[1], Rainer Waser[1,2]

[1]Institut für Werkstoffe der Elektrotechnik 2, RWTH Aachen University, 52074 Aachen, Germany
[2]Peter Grünberg Institut (PGI-7), Forschungszentrum Jülich, 52425 Jülich, Germany

Resistively switching memory cells based on transition metal oxides are promising candidates for next-generation nonvolatile memory devices. These cells are simple metal-oxide-metal structures that can be reversibly switched between a high resistive state (HRS) and a low resistive state (LRS) [1]. Usually, the two metal/oxide interfaces are asymmetric, comprising a more resistive Schottky-like interface and a more conductive ohmic-like interface. The switching mechanism in the so-called valence change memory (VCM) cells involves local redox reactions triggered by migration of anions, such as oxygen vacancies. To transform the initially insulating oxide material into a switchable state, an electroforming procedure is typically required. During electroforming, positively charged oxygen vacancies drifting through the bulk film towards the negatively biased electrode form a zone of reduced resistivity, often referred to as the virtual cathode [1,2]. It has been shown that electroforming with opposite polarities results in different final cell resistance states [3,4]. A positive voltage polarity applied to the ohmic-like interface electroforms the sample into the LRS, whereas a negative bias leads to the HRS [4].

In this work, we model the current transport in metal-oxide-metal structures to gain an in-depth insight into the electroforming process. Carrier transport across the metal-oxide interfaces by thermionic emission, thermionic field emission and field emission as well as carrier drift and diffusion in the bulk oxide are accounted for. We show how the profile of the virtual cathode impacts the barrier shape at the Schottky-like interface and the resulting *J-V* characteristics. Polarity dependent electroforming of the cell into the LRS or HRS, respectively, is discussed with respect to the oxygen vacancy distribution in the oxide layer. In case of a positively biased ohmic-like electrode, oxygen vacancies piling up at the Schottky-like interface during the forming step lead to a reduction of the electronic barrier at the Schottky-like contact and leave the cell in the LRS. When forming with the opposite polarity, the oxygen vacancies accumulate at the ohmic-like interface. Thus, the oxygen vacancy concentration at the Schottky-like junction is comparably low resulting in a higher electronic barrier and the HRS as the post-forming state.

[1] R. Waser et al., Adv. Mater. **21**, 2632 (2009).
[2] T. Menke et al., J. Appl. Phys. **105**, 006104 (2009).
[3] J. J. Yang et al, Nanotechnology **20**, 215201 (2009).
[4] C. Nauenheim et al., Appl. Phys. Lett. **96**, 122902 (2010).

MODELING OF SWITCHING DYNAMICS FOR TiO$_{2-x}$ MEMRISTIVE DEVICES

Brian Hoskins[1], Fabien Alibart[1], Dmitri Strukov[1]

[1] University of California at Santa Barbara, Santa Barbara, CA 93106, U.S.A.

Memristive devices are envisioned to provide new functionalities at the nanoscale based on the possibility to tune their resistivity continuously [1]. Important efforts are still needed to understand the switching mechanism and a physical modeling is still lacking. We proposed here a phenomenological approach to gain insight into the memristive behavior and to address (i) the dynamical switching characteristic (i.e. both V and t dependency) and (ii) an evaluation of dispersion between devices. Using our variation tolerant algorithm describe in [1], we performed semi automatized characterization of the switching transition from different intermediate states.

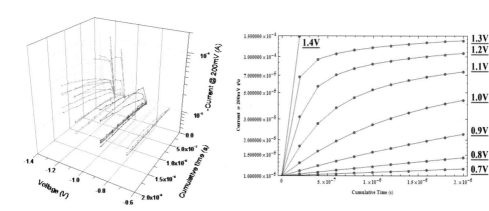

Figure 1: On-state transition dynamics of ten different TiO$_2$ devices. The time is measured in steps of 200ns pulses applied to each device. The resistive state is measured as a function of current at a non-perturbing potential of 200mV.

Figure 2: Compact modeling of Figure 1 based on statistical analysis of the switching dynamics across multiple devices.

Analysis of the transitions (Figure 1) has revealed near universal behavior based on the resistive state and applied voltage independently. A phenomological model based on this behavior has been developed allowing for a reliable prediction of a new switching state in response to an applied pulse of arbitrary voltage (Figure 2). Likewise, the *I-V* characteristics of many devices across many states have also been analyzed revealing non-linear trends in both the forward and reverse biased states of the device.

This work is supported by AFOSR and NSF.

[1] F. Alibart, L. Gao, B. Hoskins, and D. B. Strukov, Nanotechnology **23**, 075201 (2012).

MULTISTATE MEMORY DEVICES BASED ON FREE-STANDING VO$_2$/TiO$_2$ MICROSTRUCTURES DRIVEN BY JOULE SELF-HEATING

Luca Pellegrino [1], Nicola Manca [1,2], Teruo Kanki [3], Hidekazu Tanaka [3], Michele Biasotti [1,2], Emilio Bellingeri [1], Antonio Sergio Siri [1,2], Daniele Marré [1,2]

[1] CNR-SPIN, Genova (Italy); [2] Physics Department, University of Genova, Genova (Italy); [3] ISIR, Osaka University, Osaka (Japan)

Vanadium dioxide (VO$_2$) is a promising material for high speed electronics and optoelectronics due to its fast (sub-ps) thermally driven Metal-Insulator Transition (MIT) occurring above room temperature (68°C) where a decrease of more than 4 orders of magnitudes of electrical resistance associated with variations of optical constants are observed. This MIT presents thermal hysteresis that widens when moving from single crystals to thin films.

In this work, we report the fabrication and electrical characterization of two-terminal multistate memory devices based on VO$_2$/TiO$_2$ thin film microcantilevers [1].

Fabrication of VO$_2$/TiO$_2$ microcantilevers is a multistep process combining Pulsed Laser Deposition (PLD) and optical microlithography. At first, a 200 nm thick rutile-type TiO$_2$ (110) film is deposited on MgO(001) substrate by PLD. Film is then patterned by optical lithography into the shape of double clamped micrometric cantilevers. Free-standing structures are fabricated by selective wet etching and critical point drying in order to avoid stiction. As final step, sample containing microstructures is inserted into the PLD chamber where a 70 nm thick VO$_2$ film is deposited all over the sample. Free-standing cantilevers are about 10 um above the substrate top surface and are thus electrical disconnected from the VO$_2$ film grown on top of the MgO surface, thus different devices can be independently addressed.

VO$_2$ films grown on cantilevers show three orders of magnitude resistance change nearby 340 K and hysteretic behavior (width = 6.5 K) during thermal cycles.

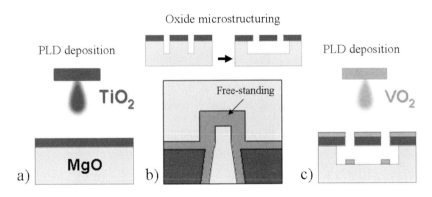

Figure 1: fabrication process for VO$_2$/TiO$_2$ cantilevers. a) PLD deposition of TiO$_2$ film. b) microfabrication by optical lithography and selective wet etching. c) PLD Deposition of the final VO$_2$ layer.

We study the behavior of the electrical resistance of these cantilevers when current pulses are applied. Experiments are made within the thermal hysteresis region, where phase coexistence of metallic and insulating domains exists.

We observe two types of memory effects: **non-volatile** changes of the electrical resistance are observed when a current pulse is applied to the microcantilever. Pulses of the same magnitude produce only one single change of the electrical resistance, while we observe that the resistance decreases with pulses of increasing magnitude. These states persist also if the current is switched-off and can be erased only by cooling the device below the hysteresis region. **Volatile** multilevel resistance states are instead possible by biasing the device with a fixed powering current bias and written with reproducibility by current pulses of different magnitude. These states can be erased by nullifying the bias with a short pulse.

The memory mechanism is based on current-induced creation of metastable metallic clusters by self-heating of micrometric suspended regions and resistive reading *via* percolation. The higher thermal insulation of free-standing structures with respect to patterned thin film devices is a key point of these devices. Hot spots are created at the cantilever center-end where thermal dissipation is lower and efficient Joule heating is possible. The use of current pulses instead of voltage ones prevents catastrophic effects due to negative differential resistance and allow setting the system in selected states with good reproducibility.

We will make experimental comparison between free-standing and non free-standing devices as well as discuss temperature distribution on VO_2 calculated by finite element analysis. Current-programmable memory optical metamaterials [2] based on free-standing VO_2 THz resonators that can be independently addressed are envisaged.

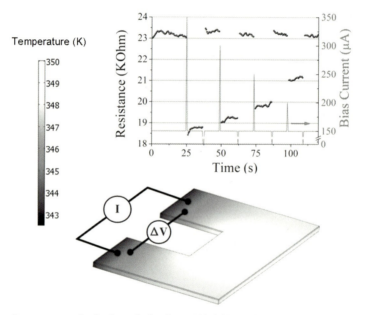

Figure 2: temperature distribution calculated on a VO_2/TiO_2 cantilever under 100 uA bias current. Sample temperature is fixed at 343 K. Hot spot forms at cantilever center-end. Multi resistance states written by current pulses of different magnitude, cantilever is powered under 150 uA current bias. Erasing is possible by a short pulse to zero. *Adapted from Ref.* [1].

[1] L. Pellegrino, N. Manca, T. Kanki, H. Tanaka, M. Biasotti, E. Bellingeri, A. S. Siri, D. Marré, *Adv Mater.* **2012** (*in press*)
[2] T. Driscoll, H.–T. Kim, B.–G. Chae, B.–J. Kim, Y.-W. Lee, N. M. Jokerst, S. Palit, D. R. Smith, M. Di Ventra, D. N. Basov, Science **2009**, 325, 1518.

TANTALUM OXIDE BASED MEMRISTIVE DEVICES AB INITIO ELECTRONIC STRUCTURE CALCULATIONS FOR STABILITY, DEFECTS AND DIFFUSION BARRIERS

Antonio Claudio M. Padilha[1], Gustavo Martini Dalpian[1], Alexandre Reily Rocha[1]

[1]Universidade Federal do ABC, Santo André, São Paulo, Brazil

Memristive devices are attractive candidates to provide a paradigm change in volatile and non-volatile memory devices fabrication. These new devices would be faster, denser and less power consuming than those available today. However, the mechanism through which the memristive property arises is not yet well understood. It is believed that a voltage-driven phase transition occurs in the material which leads to significant changes in the device's conductivity. In the particular case of tantalum oxide-based devices the relevant crystalline phases are still a matter of debate.

In this work we have performed *ab initio* Density Functional Theory based calculations to study the structural and electronic properties of different phases of Ta_2O_5 – the structure which is believed to exist inside Tantalum Oxide dispositives. Preliminary results show that one of the phases which is amongst the most stable is the L one, containing Ta in a pentagonal-bipyramidal environment surrounded by oxygens as well interstitial tantalum atoms. We speculate that these interstitial atoms may act as impurities, possibly doping the system and increasing conductivity.

CONCURRENT RESISTIVE AND CAPACITIVE STATE SWITCHING OF NANOSCALE TiO$_2$ MEMRISTORS

Themistoklis Prodromakis[1], Iulia Salaoru[1], Ali Khiat[1], Christopher Toumazou[1]

[1]Centre for Bio-Inspired Technology, Department of Electrical and Electronics Engineering, Imperial College London, London SW7 2AZ, UK

The interest on nanoscale resistive switching elements has recently shown a rapid increase owing to their potential for non-volatile memory applications, reconfigurable architectures and neuromorphic computing. In turn, this has rendered research on the physical switching mechanisms of distinct materials and architectures that facilitate memristive phenomena[1], with the most prominent mechanisms being the displacement of mobile ions[2,3], the formation and rupture of conductive filaments[4,5] and the phase-transitions of an active core[6,7]; a variety of models describing each scenario. In this paper we provide experimental evidence on a switching mechanism that depends upon the expansion/contraction of a TiO$_2$ thin film that serves as the active core for a nanoscale memristor, due to a re-oxidation/de-oxidation (reduction of TiO$_2$) process supported at the top TiO$_2$/Pt interface of the device.

We have observed for a number of pristine samples that the interfaces between the top (TE) and bottom (BE) electrodes and an intervening TiO$_2$ thin film are rather smooth and uniform (Fig. 1a). Nonetheless, the uniformity of the interface between TE and TiO$_2$ is substantially contrived (Fig. 1b), once a device is switched from a low-resistive state (LRS) to a high-resistive state (HRS) and back again to LRS via consecutive voltage sweeps that progressively increase in amplitude ($V_{max}=\pm2.5V$), as illustrated in Fig. 1c. We suggest that the physical mechanism of switching is due to the displacement of Ti^{n+} interstitials towards the TE where these combine with oxygen to form TiO$_2$ that has been extensively studied previously for bulk TiO$_2$[8-10].

During the re-oxidation of the TiO$_2$ layer, the TE is slightly elevated (~1.5-2nm), which could describe the formation of bubbles observed by Yang et al.[11] as well as the topographic changes of bulk TiO$_2$ films after being programmed with C-AFM[12]. The reverse process causes the disassociation of O$_2$ from Ti^{n+} interstitials and the latter are retracted towards the bulk of the TiO$_2$ film. This results in the contraction of the active core, which in turn pulls the Pt TE closer to the BE through the establishment of miniscule metallic filaments, as observed in Fig. 1b. These filaments, in combination with the Ti^{n+} interstitials, gathered in the insulator's bulk, facilitate the establishment of continuous percolation channels between the TE and BE that are responsible for the device's transition from HRS to LRS. This can effectively be modeled by a circuit breaker network[13], comprising a finite number of conductive and capacitive filaments (inset Fig. 1(d)). The establishment of continuous percolation channels diminishes the amount of capacitive filaments, and therefore results into an overall capacitance increase. This functionality is alike to the programming capacity that was observed in perovskite oxide thin films[14] where the effective resistance and capacitance of two terminal devices were modulated simultaneously by appropriate voltage pulsing. The corresponding concurrent capacitance/resistance modulation of our TiO$_2$-based memristor is illustrated in Fig. 1(d), where a bipolar multilevel switching capacity is demonstrated. Finally, we believe that this switching mechanism extends to other systems such as TaO$_x$[15].

1. Chua, L. Memristor - Missing Circuit Element. *IEEE Trans. Circuit Theory* **CT18**, 507–& (1971).
2. Yang, J. J. *et al.* Memristive switching mechanism for metal/oxide/metal nanodevices. *Nature Nanotech* **3**, 429–433 (2008).
3. Lee, M.-J. *et al.* A fast, high-endurance and scalable non-volatile memory device made from asymmetric Ta2O5−x/TaO2−x bilayer structures. *Nature Materials* **10**, 625–630 (2011).
4. Kwon, D.-H. *et al.* Atomic structure of conducting nanofilaments in TiO2 resistive switching memory. *Nature Nanotech* **5**, 148–153 (2010).
5. Kim, K. M., Jeong, D. S. & Hwang, C. S. Nanofilamentary resistive switching in binary oxide system; a review on the present status and outlook. *Nanotechnology* **22**, 254002 (2011).

6. Wuttig, M. & Yamada, N. Phase-change materials for rewriteable data storage. *Nature Materials* **6**, 824–832 (2007).
7. Driscoll, T., Kim, H. T., Chae, B. G., Di Ventra, M. & Basov, D. N. Phase-transition driven memristive system. *Appl. Phys. Lett.* **95**, 043503 (2009).
8. Stone, P., Bennett, R. & Bowker, M. Reactive re-oxidation of reduced TiO2 (110) surfaces demonstrated by high temperature STM movies. *New Journal of Physics* **1**, 8 (1999).
9. Rothschild, A., Komem, Y. & Cosandey, F. Low Temperature Reoxidation Mechanism in Nanocrystalline $TiO_{2-\delta}$ Thin Films. *Journal of The Electrochemical Society* **148**, H85 (2001).
10. Wendt, S. *et al.* The role of interstitial sites in the Ti3d defect state in the band gap of titania. *Science* **320**, 1755–1759 (2008).
11. Joshua Yang, J. *et al.* The mechanism of electroforming of metal oxide memristive switches. *Nanotechnology* **20**, 215201 (2009).
12. Jeong, D. S., Schroeder, H., Breuer, U. & Waser, R. Characteristic electroforming behavior in $Pt/TiO_2/Pt$ resistive switching cells depending on atmosphere. *J. Appl. Phys.* **104**, 123716 (2008).
13. Lee, S. B. *et al.* Interface-modified random circuit breaker network model applicable to both bipolar and unipolar resistance switching. *Appl. Phys. Lett.* **98**, 033502 (2011).
14. Liu, S., Wu, N., Ignatiev, A. & Li, J. Electric-pulse-induced capacitance change effect in perovskite oxide thin films. *J. Appl. Phys.* **100**, 056101 (2006).
15. Yang, J. J. *et al.* High switching endurance in TaO_x memristive devices. *Appl. Phys. Lett.* **97**, 232102 (2010).

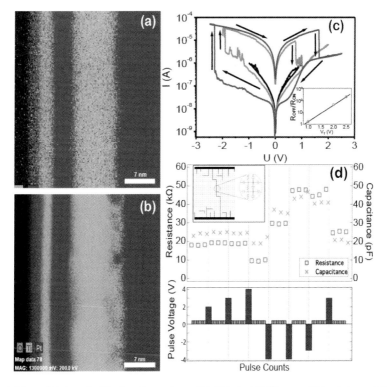

Figure 1: Switching validation of a TiO_2 based memristor. 256x256 pixel EDX maps of (a) pristine and (b) switched device. Both maps were taken at 50μs dwell time, 1.2nA beam current and 8mins acquisition time per map. (c) Current-voltage characteristics by applying 3 progressively increasing bipolar voltage sweeps, with resulting EDX map shown in (b). Inset of (c) demonstrates the increase of R_{OFF}/R_{ON} ratio with applied voltage threshold. Pulsed-induced programming of our device is shown on (d) with both resistive and capacitive states being evaluated with 0.5V 10μs long pulses. Inset of (d) illustrates a conceptual model for elucidating current percolation.

MEM 11

THE MEMORY-CONSERVATION MODEL OF MEMRISTANCE

Ella M. Gale[1]

[1]Unconventional Computing Group, University of the West of England, Bristol, Avon, UK

The memristor, a device which relates charge, q, and magnetic flux, φ, was postulated to exist by Chua in 1971 [1] via the relation: $d\varphi = M(q(t))\, dq$, where $M(q(t))$ is the memristance, which is a charge-dependent. This theory wasn't related to an experimental device (despite several such devices having been fabricated) until 2008 [2] when Strukov et al put forward a phenomenological model of the form: $M(q(t)) = R_{off}(1 - R_{off} R_{on} \beta\, q(t))$ where R_{off} is the resistance of the TiO_2, R_{on} is the resistance of the auto-doped $TiO_{(2-x)}$ and $\beta = \mu_v/D^2$, where β is here called the material parameter. This model expressed the expected hysteretic behaviour of the device but included no magnetic flux. Because of this, and because of the lack of experimental measurements of a magnetic flux which fit the memristor equations, questions have been raised about whether the Strukov memristor is a real Chua memristor at all [3] and even if the Chua memristor might only be a theoretical curiosity with nothing to do with resistive switching memories or ReRAM.

Here it is demonstrated that the relevant charge to put into Chua's equations is not the electronic charge q_{e^-} but the vacancy charge q_v. Using electrodynamical theory, we can then calculate the flux of the vacancy charge flow, J. J is given by:

$$J = (q_v \mu_v L) / (D\,E\,F)$$

Where, μ_v, is the vacancy ion mobility, L is the average electric field and $D\,E\,F$ is the volume of TiO_2 in the memristor. The titanium dioxide layer is D thick, crossed electrodes of width E and F, the boundary between TiO_2 and $TiO_{(2-x)}$ is w and moves over time as a function of the total charge of oxygen vacancies. We can calculate associated magnetic field at as experienced at a point p and time t, $B(p)$, where $p = \{x,y,z\}$, located outside of the memristor:

$$B(p) = (\mu_0/4\pi)\, A\, L\, \mu_v\, q_v\, \{0,\, -x\,z\,P_y,\, x\,y\,P_z\}\,.$$

Where A is the area we calculate B over (ie the side of the memristor) and P_y and P_z are complicated terms which only include the dimensions of the titanium dioxide layer, (D, E and F), in the memristor, these terms are dependant on q_v as the boundary between TiO_2 and $TiO_{(2-x)}$, w, changes with q_v (and time as q_v is time-dependent). Note that the form of the vector arises from the x direction is taken to be the direction of vacancy current flow (ie. w and D are measured on the x axis).

This magnetic field gives rise to the following magnetic flux passing through a surface i-j (again the side of the memristor, note the flux is zero in the plane perpendicular to the vacancy current flow):

$$\varphi = (\mu_0/4\pi)\, i\, j\, L\, \mu_v\, P_k\, q_v\,.$$

From comparison with Chua's definitions, we get the following equation for memristance (note we have explicitly put in the time dependence here):

$$M(q_v(t)) = (\mu_0/4\pi)\, i\, j\, L\, \mu_v\, P_k(q_v(t))\,.$$

which $\mu_v P_k(q_v(t))$ is also called, β, the material parameter. The existence of this equation for memristance, which relates charge and flux, and which is only dependent on one state variable (q_v), in the Strukov memristor shows that it is a true Chua memristor.

The calculated flux for Strukov memristor is 2.44×10^{-29} Wb (approximately 100 000 times smaller than the conducting electron's magnetic flux) and its tiny value explains the lack of experimental measurements.

To understand the measurable *I-V* curves, we need to look at the memristor as a two layer system and include both the effects of the true Chua memristance between the vacancy charge (called the magnetic subsystem as it contains the magnetic flux) and flux as well as the measurable effects of this ionic motion on the electronic current (the electronic subsystem).

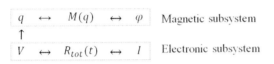

Figure 1: The two layer model of memristance [4]. The magnetic subsystem deals with the Chua memristance that arises as a result of the vacancies, the electronic subsystem deals with the effect of vacancy motion on the conducting electrons and the two systems are related as both the conducting electrons and the oxygen vacancies are responding to the applied voltage.

First we need to calculate the memory function, $M_e(t) = C\, M(q(t))$ where C is an experimentally determined parameter related to the different mobilities for electrons and vacancies and is the Chua memristance as experienced by the conducting electronic current, this covers resistance change in the TiO$_{(2-x)}$ part of the memristor. We then need to include an expression for the time varying resistance of the TiO$_2$, which is called the conservation function (as it keeps space conserved in this model) and it is given by:

$$R_{con}(t) = [\,(D - w(t))\,\rho_{TiO2}\,] / (E\,F).$$

The total time varying resistance of the device, $R_{tot}(t)$, is then given by:

$R_{tot}(t) = M(q_v(t)) + R_{con}(t)$. $R_{tot}(t)$, the memory function and conservation function are all memristances in that their resistance varies with time as a result of a vacancy flow, but only the Chua memristance directly relates charge and magnetic flux.

Further details are given in [4].

[1] L. O. Chua, IEEE Trans. Circuit Theory, **18**, 507-519, (1971).
[2] D. B. Strukov et al, Nature, **453**, 80-83 (2008).
[3] B. Mouttet, ISCAS 2010, http://www.slideshare.net/blaisemouttet/mythical-memristor (downloaded 13/03/12)
[4] E. Gale, The Missing Memristor Flux Found, arXiv:1106.3170v1 [cond-mat.mtrl-sci] (submitted)

ANALOG AND DIGITAL COMPUTING WITH MEMRISTIVE DEVICES

A. Madhavan[1], G. Adam[1], F. Alibart[1], L. Gao[1], and D.B. Strukov[1]

[1] University of California at Santa Barbara, Santa Barbara, CA 93106, U.S.A.

CMOL circuits (Fig. 1b), which combine nanoscale crossbar with extremely dense, passive, crosspoint nanoscale devices and CMOS circuits, are promising candidates for information processing tasks [1]. The key element in such circuitry, the "memristive" device [1, 2], consists of an active thin film layer of some switching material sandwiched between two metal electrodes. Application of relatively large electrical stress (voltage or current) across electrodes changes the resistivity ("memory state") of the thin film material. For many devices the resistance of thin film can be changed continuously so that the memory state is effectively analog. In addition, the memory state of properly engineered devices is nonvolatile and could be read without disturbing it with a relatively small electrical stress. Due to these unique properties of memristive devices, CMOL circuits enable efficient implementation of bottleneck low-level operations in information processing such as digital pattern matching and analog multiply-and-add (MAC) or dot-product computation. To realize pattern matching/MAC circuits, the memristive devices implement high-density configurable digital(binary) or analog weights, while CMOS implements threshold element/summing amplifier which provides signal restoration/gain, and inversion (Figs. 1c, d, e, g).

Figures (1 f, g) show experimental results for configurable 4-input linear threshold gate (Fig. 1d) [3] and digital 2-input/ analog (7-bit) weights MAC circuitry [4] implemented with Ag/a-Si/Pt and Pt/TiO_{2-x}/Pt devices respectively. In both demonstrations the state of memristive devices was tuned by variation tolerant algorithm [4] and discrete chip CMOS D-flip-flop and op-amp were employed. A fully integrated CMOL circuit might enable unprecedented information processing performance. For example, we have preliminary evaluated the case when streaming data in CMOS subsystem is processed in the massively parallel fashion with the help of locally stored data in memristive devices (Figs. 1a, h) [5]. In the simplest case, such processing is simply pattern matching between the CMOS data and the state of memristive devices implemented by configurable diode-logic OR gate connected to a D-flip-flop (Figs. 1c, d). Figure (1 i) shows estimated pattern matching throughputs for 1-cm^2 chip for mapping of fixed length synthetic patterns under an assumption of practical power density. The throughput is at least four orders of magnitude larger as compared to the state-of-the-art implementations even assuming rather conservative values of crossbar half-pitch and CMOS technology [5].

This work is supported by AFOSR and NSF.

Poster: Memristive systems

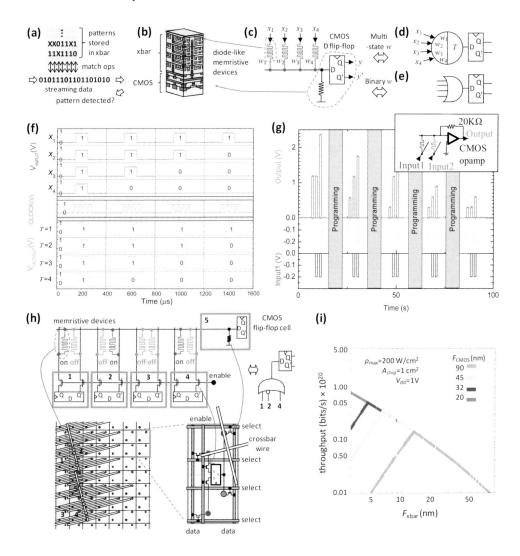

Figure 1: (a) Pattern matching for streaming data with (b) 3D CMOL circuits, (c) a key circuit element consisting of the several nonlinear memristive devices and CMOS flip-flop implementing (d) configurable linear threshold gate and (e) configurable OR gate. (f) Experimental demonstration of symmetric 4-input linear threshold logic gate operation and (g) 2-input analog multiply-and-add circuitry operation. (h) An example of mapping for a specific OR gate (labeled with "5") designed to detect pattern '10X1' onto CMOL fabric and (i) estimates for pattern matching throughput as a function of the crossbar half pitch for several CMOS nodes.

[1] K. K. Likharev, J. Nanoelectronics and Optoelectronics **3**, 203 (2008).
[2] D. B. Strukov and H. Kohlstedt, MRS Bulletin **37**, 108 (2012).
[3] L. Gao, F. Alibart, and D. B. Strukov, submitted (2012).
[4] F. Alibart, L. Gao, B. Hoskins, and D. B. Strukov, Nanotechnology **23**, 075201 (2012).
[5] F. Alibart, T. Sherwood and D. B. Strukov, Proc. AHS'11, 279 (2011).

EFFECT OF VACANCIES ON THE PHASE CHANGE CHARACTERISTICS OF GeSbTe ALLOYS

D. Wamwangi[1], W. Welnic[2], M. Wuttig[2]

[1] School of Physics, University of the Witwatersrand, Private Bag 3, Johannesburg, South Africa;
[2] I. Institute of Physics (IA) and JARA-FIT, RWTH University, 52056 Aachen, Germany

Phase change materials continue to demonstrate their technological importance by the widespread research interest and applications in optical data storage and as active layers in nonvolatile memories. In this work we report on the feasibility of the $Ge_2Sb_1Te_4$, $Ge_3Sb_4Te_8$ and $Ge_2Sb_2Te_4$ alloys and the effect of vacancies on their physical properties [1]. The structural properties, transformation kinetics and the effect of vacancies were studied using temperature dependent van der Pauw measurements, X-ray diffraction, X-Ray Reflectometry as well as a static tester. The electrical contrast between the amorphous and NaCl metastable structure was higher than 10^3 for all three alloys. Alloys with Sb vacancies corresponding to 12.5% produced the highest electrical contrast of 161.8×10^3 compared to 16×10^3 orders for 12.5% Ge vacancies. The activation energy for transport in the amorphous state increased from 0.29 eV to 0.33 eV with decreasing vacancy concentration.

The transition temperatures have been determined for $Ge_2Sb_1Te_4$, $Ge_3Sb_4Te_8$ and $Ge_2Sb_2Te_4$ for a heating rate of 5K/min to 158°C, 169°C and 175°C respectively compared to $Ge_1Sb_2Te_4$, T_c=145°C which has the highest concentration of Ge vacancies in the sub-lattice [2]. This increase in transition temperature is attributed to the increased fraction of the stronger Ge-Te bonds. Using Kissinger analysis, the activation energy against recrystallization has been determined for the $Ge_2Sb_1Te_4$, $Ge_3Sb_4Te_8$ and $Ge_2Sb_2Te_4$ alloys to 2.42 ± 0.15eV, 2.54 ± 0.15 and 2.73 ± 0.13eV, respectively. The energy scales with the fraction of Ge/Sb vacancies as well and the fraction of the Ge-Te bonds in the alloy. X-ray reflectivity measurements have yielded higher density values for the amorphous phase of $Ge_2Sb_1Te_4$ (5.90 ± 0.02g/cm^3) and $Ge_3Sb_4Te_8$ (5.92 ± 0.02g/cm^3) alloys than that of the $Ge_2Sb_2Te_4$ (5.84± 0.02g/cm^3) alloy. These large density values correlate with the Ge-vacancy fraction. The densities in the metastable NaCl state have been determined by XRR to 6.23±0.02g/cm^3, 6.33±0.02g/cm^3, and 6.48 0.02g/cm^3. High density changes corresponding to 9.8% were obtained for the $Ge_2Sb_2Te_4$ alloy which possesses no Ge/Sb structural vacancy. The presence of Ge/Sb vacancies in the sub-lattice of the $Ge_2Sb_1Te_4$ and $Ge_3Sb_4Te_8$ alloys possibly provides sufficient interstitial volume in the metastable stable state thereby ensuring greater atomic rearrangement within a constant volume. The corresponding density changes for the alloys with structural vacancies have been determined to 8.1% and 6.5%, respectively. In addition, such a huge variation in the density change scales with the covalent size of the Sb and Ge atoms and the lower fraction of the Sb-Te bonds for the $Ge_2Sb_1Te_4$ alloy. Despite the large vacancy variations in the sub-lattice, no significant variations have been observed in the lattice constant of the crystalline state as shown by the almost constant values for the three compounds investigated (5.969 ± 002Å, 6.000 ± 0.002Å and 6.003 ± 0.002Å, respectively). The decrease in the lattice constant of the $Ge_2Sb_1Te_4$ alloy is due to the larger covalent size of the Sb atom and the increase in the number of the stronger Ge-Te bonds. Static tester results on the first crystallization of these alloys have yielded the crystallization times for the $Ge_2Sb_1Te_4$, $Ge_3Sb_4Te_8$ and $Ge_2Sb_2Te_4$ alloys to 100 ns, 400 ns and 1 µs respectively. These crystallization times correlate with decreasing vacancy fraction. Our results further indicate that the short crystallization times could possibly be attributed to the strength of the Ge-Te bonds.

Recrystallization is the time limiting step for rewriteable data storage and the corresponding times for the $Ge_3Sb_4Te_8$ and $Ge_2Sb_2Te_4$ alloys have been determined to 20 ns and 300 ns, respectively. The large variation in these values highlights the role of structural vacancies.

[1] M. Wuttig et al., Nat. Mat.**6**, 122 (2007).
[2] C. Rivera-Rodriguez et al., J. Appl. Phys. **96**, 1040 (2004).

ULTRA LOW POWER CONSUMING, THERMALLY STABLE SULPHIDE MATERIALS FOR RESISTIVE AND PHASE CHANGE MEMRISTIVE APPLICATION

<u>Behrad Gholipour</u>[1], Chung-Che Huang[1], Alexandros Anastasopoulos[2,3], Feras Al-Saab[1], Brian E. Hayden[2,3] and Daniel W. Hewak[1]

1. Optoelectronics Research Centre, University of Southampton, Southampton SO17 1BJ, UK
2. School of Chemistry, University of Southampton, Southampton, SO17 1BJ, UK
3. Ilika Technologies Ltd, Kenneth Dibben House, Chilworth, Southampton, SO16 7NS, UK

The use of conventional chalcogenide alloys in rewritable optical disks and the latest generation of electronic memories (phase change and nano-ionic memories) has provided clear commercial and technological advances for the field of data storage, by virtue of the many well-known attributes, in particular scaling, cycling endurance and speed, that these chalcogenide materials offer. While the switching power and current consumption of established germanium antimony telluride based phase change memory cells are a major factor in chip design in real world applications, the thermal stability and high on-state power consumption of these device can be a major obstacle in the path to full commercialization. In this work we describe our research in material discovery and prototype device fabrication and characterization, which through high throughput screening has demonstrated thermally stable, low current consuming chalcogenides for applications in PCRAM and oxygen doped chalcogenides for RRAM which significantly outperform the current contenders.

Our particular interest in the field of electronic data processing and storage is concerned with the discovery of new chalcogenide alloys to outperform the commonly used Ge:Sb:Te (GST) for PCRAM applications. There is a wide range of chalcogenide alloys, range from pure chalcogenides, to pnictogen-chalcogen, tetragen-chalcogen, metal chalcogenides to halogen-chalcogenides [1]. Many of these compounds are covered within high-level patent literature, within which a vast array of potential compounds suitable for phase change memory are proposed, yet with relatively few studied in detail [2]. Indeed, even among phase change memory cells fabricated with the most conventional GST compounds, the reasons why these compositions provide us with useful and attractive physical properties are still veiled [3]. It is therefore our belief that the field and material space is ripe for a thorough analysis of the compositional space to provide both a better understanding of the range of properties the numerous chalcogenide alloys offer and to optimize the compositions to meet the demands of practical solid state memory and information processing devices. Through high throughput synthesis and characterisation techniques, one can quickly reach the optimal composition for a family of alloys for use in different applications and find exciting new physical phenomena in an incredibly efficient manner (Figure 1).

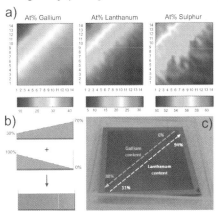

Figure 1: a) EDX results mapping the composition across the sample, b) thickness of individual elemental components across sample for a binary alloy, c) High throughput chip mapping the composition of GLS family of glasses.

High-throughput physical deposition screening techniques provide a unique route to the exploration of thin film media (4). They exploit the deposition of chalcogenide compositional gradients across areas of several square centimetres, and allow rapid analysis of optical, electronic and thermal properties as a function of composition. As such, we show that compositionally optimised phase change media based on the gallium and lanthanum chalcogenides can perform well above the well-known benchmark performance of germanium antimony telluride devices.

The experimental work we have completed shows that these compounds offer set and reset currents over an order of magnitude lower, than an equivalent germanium antimony telluride device, while at the same time offering improved thermal stability and the potential for improved endurance. Our nanoscale devices based on Ga:La:S continues to show the ability to display a measurable threshold up to 400°C higher than equivalent germanium antimony telluride based memory cells Figure 2).

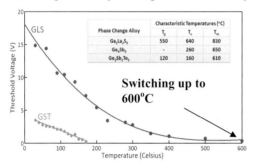

Figure 2: top) The temperature dependence of the threshold voltage of a Ga:La:S and Ge:Sb:Te based memory cells. bottom) The crystallisation temperatures (oC) of GLS compared with

Additionally, through the incorporation of oxygen rich GLS alongside the standard GLS used previously, bipolar resistive switching is observed in the characteristics of such devices. Typical read state current voltage characteristics is shown in Figure 3, with the corresponding resistance memory window of operation, showing a resistance ratio of up to 10^6 Ohms.

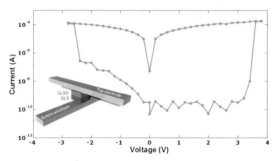

Figure 3: Current voltage characteristics of the device in the on and off states in both polarities.

Summary: Through the combination of high throughput synthesis and nanofabrication techniques we reveal ultra-fast optimisation techniques for data storage, both PCRAM and RRAM applications. We reveal enhanced temperature stability in phase change alloys identified through high throughput screening of families of chalcogenide alloys. Through the incorporation of oxygen into optimised high stability phase change alloys bipolar resistive switching with very large resistance windows is observed. Work in progress is fully quantifying a range of chalcogenides to provide optimized materials for commercial memory based devices.

[1] S. R. Elliott, Chalcogenide Glasses, Vol 9, Materials Science and Technology. VCH, 1991.
[2] D. Semyon et al., Analysis of US Patents in Ovonic Phase Change Memory, Journal of Ovonic Research Vol. 1, No. 3, June 2005, p. 31 – 37.
[3] J. Tominaga et al, Phase Change Meta-material and Device Characteristics, EPCOS'10, Milan Italy (2010)
[4] R.E.Simpson et al, High throughput synthesis of chalcogenide materials for data storage, EPCOS '05, Cambridge, UK (2005).

THERMAL CONDUCTIVITY MEASUREMENTS OF Sb-Te ALLOYS BY HOT STRIP METHOD

Rui Lan[1], Rie Endo[1], Masashi Kuwahara[2], Yoshinao Kobayashi[1], Masahiro Susa[1]

[1]Tokyo Institute of Technology, Tokyo, Japan;
[2]National Institutes of Advanced Industrial Science and Technology, Ibaraki, Japan

1. INTRODUCTION
Phase change random access memory (PCRAM) is a promising nonvolatile data storage technology for next generation memory[1]. Sb-Te binary alloys have been suggested to be suitable candidates for PCM devices because of a dramatic change in electric resistivities associated with transformation between amorphous and crystalline phases, which forms the basis for data storage[2,3]. The phase transformation is controlled by Joule-heating and cooling processes and, thus, accurate data for thermal conductivity of Sb-Te binary alloys are indispensable to optimal designing for PCM devices. The present work aims to determine thermal conductivities of Sb-Te binary alloys as functions of temperature and composition.

2. EXPERIMENTAL
Samples used were Sb-x at% Te (x = 14, 25, 44, 60, 70, and 90). Thermal conductivity was measured by the hot strip method[4]. Cylindrical samples of Sb-Te (20 mm diameter and 40-50 mm length) were prepared from Sb (99.9 mass%) and Te (99.9 mass%) powders except for Sb_2Te_3, which was produced by Kojundo Chemical Laboratory. Thermal conductivity measurements were conducted in argon atmosphere from 298 K up to temperatures just below the respective melting points.

3. RESULTS & DISCUSSION
According to the phase diagram, the alloys fall into three groups: Sb_2Te_3, alloys having x < 60 and alloys having x > 60. The second and last groups are named Sb-rich and Te-rich alloys for convenience, respectively. Fig. 1 shows the temperature dependence of Sb-rich alloys together with the data for Sb obtained by Konno et al.[5]. It can be seen that the thermal conductivities of all Sb-rich alloys keep roughly constant below approximately 600 K and, above 600 K, increase with increasing temperature. The thermal conductivities of the Sb-rich alloys decrease with increasing Te concentration. Fig. 2 shows the temperature dependence of thermal conductivities for the Te-rich alloys together with the data for Sb_2Te_3. It can be seen that the thermal conductivity of Sb_2Te_3 has interesting temperature dependence: it decreases with increasing temperature up to approximately 600 K and then decreases. The thermal conductivities of the Te-rich alloys decrease with temperature increase until their melting points are attained, and this behavior is similar to that of Sb_2Te_3 below 600 K. The temperature dependencies of the Te-rich alloys are quite close to each other although the magnitude of thermal conductivity decreases with the Te concentration increase. It can be seen in Figs. 1 and 2 that the thermal conductivities have quite different temperature dependences below and above 600 K. Below 600 K, free electrons are supposed to dominate the heat conduction and thus the W-F law[6,7] would be applied. The ambipolar diffusion[8] is more effective at higher temperature and contributes to the increase of thermal conductivity.

4. CONCLUSIONS
The thermal conductivities of Sb-x mol%Te alloys (x = 14, 25, 44, 60, 70 and 90) have been measured by the hot strip method. The heat conduction mechanisms have been discussed. It is proposed that free electrons dominate the heat transport below 600 K and ambipolar diffusion contributes to the increase in the thermal conductivity at higher temperatures.

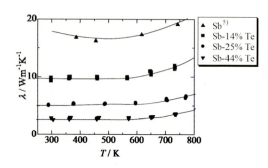 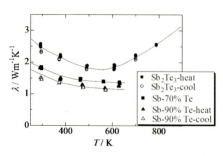

Fig. 1 Thermal conductivities of Sb-rich samples as function of temperature

Fig. 2 Thermal conductivities of Te-rich alloys as function of temperature

1) S. Lai, IEDM Tech. Dig. 2003, p. 10.1.1.
2) H. J. Borg, M. van Schijindel, and J. C. N. Rijpers, Jpn. J. Appl. Phys. **40**, 1592 (2001).
3) N. Oomachi, S. Ashida, and N. Nakamura, Jpn. J. Appl. Phys. **41**, 1695 (2002).
4) R. Lan, R. Endo, M. Kuwahara, Y. Kobayashi, and M. Susa, Jpn. J. Appl. Phys. **49**, 078003 (2010).
5) S. Konno, Sci. Repts. Tohoku Imp. Univ. **8**, 169 (1919).
6) G. Wiedemann and R. Franz, Annalen der Physik und Chemie, **89**, 497 (1853).
7) L. Lorenz, Annalen der Physik und Chemie, **13**, 422 (1881).
8) J. M. Ziman, Electrons and Phonons, (Clarendon Press, Oxford, 1960), pp.427.

juRS – MASSIVELY PARALLEL REAL-SPACE DFT

Paul Baumeister, Daniel Wortmann, Stefan Blügel

Peter Grünberg Institut and Institute for Advanced Simulation,
Forschungszentrum Jülich and JARA, 52425 Jülich, Germany

We present a new Density Functional Theory (DFT) tool developed in Jülich that combines equidistant real-space grids and the Projector Augmented Wave (PAW) method [1]. This DFT code is explicitly designed for the structural relaxation of systems with a large unit cell in real-space consisting of several thousand atoms such as structures of amorphous and disordered systems, impurities or crystal defects, as frequently occurring in oxide materials.

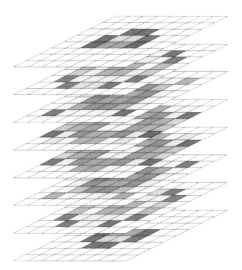

Figure 1: Load balancing of atomic projection operations in a 8 x 8 x 8 real-space domain decomposition for a C_{60} molecule (Buckminster fullerene) visualized by Scalasca [2]. White areas indicate domains that do not intersect with the atomic core regions of the Projector Augmented Wave method [1]. Except for these atomic tasks, the workload per grid point is constant over the uniform real-space grid which enables an efficient parallel treatment.

The real-space representation of wave functions, densities and potentials enable a simple, efficient and strong parallelization with respect to communication and load balancing. The frozen-core PAW method ensures an accurate treatment of the energy contributions and scattering properties of the atom cores at the cost of a pseudopotential approach but with all-electron precision. A domain decomposition of the real-space grid enables us to exploit the local character of the Kohn-Sham equation where we approximate the Laplacian operator with a localized high-order finite-difference stencil. The resulting extremely sparse Hamiltonian favors iterative diagonalization schemes to find the lowest eigenvalues and their eigenvectors.

Besides the distribution of the real-space grid domains (Fig 1), the sampling of k-points and spins is performed in parallel. In addition, a third level of parallelization distributes the evaluation of different energy bands. We will demonstrate the efficiency of the various levels of parallelization on massively parallel supercomputer architectures such as IBM's BlueGene which support cartesian nearest-neighbor communication on the hardware level. Further, we will show the applicability to structural relaxations of large disordered systems, such as the phase-change-alloy Ge:Sb:Te, focusing on the challenges of the accurate and fast determination of the electronic structure with vast numbers of compute cores processing large numbers of atoms.

[1] P.E. Blöchl, Phys. Rev. B **50**, 17953 (1994)
[2] www.scalasca.org

FIRST-PRINCIPLES STUDY OF PHASE-CHANGE MATERIALS DOPED WITH MAGNETIC IMPURITIES

Riccardo Mazzarello[1,2], Yan Li[1,2], Wei Zhang[2], Ider Ronneberger[2]

[1] JARA – Jülich-Aachen Research Alliance, Aachen, Germany;
[2] Institute for Theoretical Solid State Physics, RWTH Aachen, Aachen, Germany

The technological importance of Chalcogenide phase-change materials for storage applications stems from their high crystallization speeds, the stability of the amorphous and crystalline phases at room temperature and the optical and electronic contrast between the two phases [1]. Phase-change materials are currently used to store information in rewritable optical media (CD, DVD, Blu-Ray Disc) and electronic non-volatile Memories. Among them, $Ge_2Sb_2Te_5$ stands out for its superior performance for phase-change memory applications.

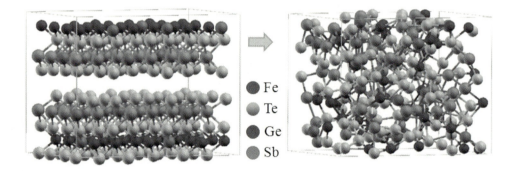

Figure 1: Starting hexagonal model of Fe- doped GST (left) and final amorphous model (right) obtained by melting the crystalline model and subsequent fast quenching from the melt using *ab initio* molecular dynamics.

Recently, the first magnetic phase-change material was synthesized by a pulsed laser deposition method [2]. The authors of this paper showed that $Ge_2Sb_2Te_5$ doped with Fe impurities displays phase-change features; moreover, at low temperatures, both the crystalline and amorphous phases are ferromagnetic but have different saturation magnetization (smaller in the amorphous state), thus exhibiting a pronounced magnetic contrast. These findings open up the perspective of exploiting the phase-change behavior of doped $Ge_2Sb_2Te_5$ for fast magnetic switching in future spintronic devices.

We have carried out an *ab initio* Density Functional Theory study of the structural, electronic and magnetic properties of the crystalline (hexagonal and cubic) phases and amorphous phase of $Ge_2Sb_2Te_5$ doped with Fe impurities to shed light on the microscopic origin of the contrast [3].

Large models of $Ge_2Sb_2Te_5$ containing 199–216 atoms and 7% Fe concentration were employed to describe all the three phases. The model of Fe-doped amorphous $Ge_2Sb_2Te_5$ was generated by fast quenching from the melt using *ab initio* molecular dynamics. The Quantum Espresso [4] and CP2K [5] packages were employed.

We have found that, in the crystalline phases, the most favourable sites for a Fe impurity are the substitutional cation Sb and Ge sites. As regards the amorphous phase, we have shown that the presence of Fe at low concentrations does not affect significantly the structural properties of amorphous $Ge_2Sb_2Te_5$ (presence of four-membered irreducible rings and voids), which have been linked to the fast crystallization, thus suggesting that the crystallization rate should not degrade dramatically upon moderate Fe doping.

We have also shown that our models of crystalline and amorphous Fe-doped $Ge_2Sb_2Te_5$ exhibit a magnetic contrast stemming from a reduction of the magnitude of the magnetic moments of Fe atoms in amorphous $Ge_2Sb_2Te_5$. This behavior can be ultimately related to the differences in the local environment and short-range order between the two phases, which is a peculiar property of phase-change materials.

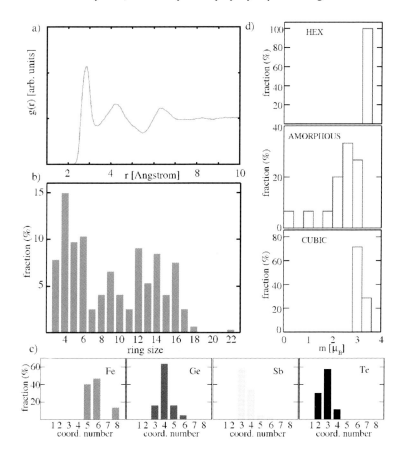

Figure 2: Structural and magnetic properties of the model of Fe-doped amorphous $Ge_2Sb_2Te_5$ generated by quenching from the melt. a) Pair correlation function, $g(r)$. b) Statistics of irreducible n-fold ring configurations: the distribution displays a pronounced maximum at four-membered rings, 75% of which have ABAB alternation (A = Ge,Sb; B = Te). c) Distribution of the coordination numbers of different species: Ge, Sb and Te have an average coordination of 4.1, 3.5 and 2.8 respectively. d) Distribution of the local magnetic moments of Fe atoms for each phase.

We are currently investigating the magnetic properties of $Ge_2Sb_2Te_5$ doped with other $3d$ impurities (namely Cr, Mn, Co and Ni), with the goal of finding the best performing magnetic phase-change materials in terms of phase-change properties and magnitude and stability of the magnetic contrast.

[1] M. Wuttig and N. Yamada, Nature Mater. **6**, 824 (2007).
[2] W.-D. Song, L.-P. Shi, X.-S. Miao, and C.-T. Chong, Adv. Mater. **20**, 2394 (2008); W.-D. Song, L.-P. Shi, and C.-T. Chong, J. Nano Nanotechnol. **11**, 2648 (2011).
[3] Y. Li and R. Mazzarello, Adv. Mater., in press (DOI: 10.1002/adma.201104746).
[4] P. Giannozzi P. *et al.*, J. Phys. Condens. Matter **21**, 395502 (2009).
[5] J. VandeVondele *et al.*, Comput. Phys. Commun. **167**, 103 (2005).

NUCLEAR RESONANCE SCATTERING IN PHASE-CHANGE MATERIALS

<u>Ronnie Simon</u>[1,2], Jens Gallus[1,3], Dimitrios Bessas[1,2], Ilya Sergueev[4], Hans-Christian Wille[5] and *Raphaël* Pierre Hermann[1,2]

[1]Jülich Centre for Neutron Science JCNS and Peter Grünberg Institut PGI, Jara-FIT Forschungszentrum Jülich GmbH, D-52425 Jülich, Germany;
[2]Faculté des Sciences, Université des Liège, B-4000 Liège, Belgium;
[3]RWTH Aachen University, D-52056 Aachen, Germany;
[4]European Synchrotron Radiation Facility, 6 Rue Jules Horowitz, F-38043 Grenoble, France;
[5]Deutsches Elektronen-Synchrotron, D-22607 Hamburg, Germany

Phase-change materials exhibit a significant change of the optical reflectivity and electrical resistivity upon crystallization which renders these materials applicable for optical storage devices and non-volatile electronic memories [1]. In order to fully understand the microscopic mechanisms underlying these unique properties a detailed knowledge of the structure of the amorphous and crystalline phases is necessary. So far, upon crystallization a change of the Ge-coordination from tetrahedral to octahedral has been identified as one of the key features of the phase change [1]. The details and magnitude of this coordination change are, however, still a matter of debate. DFT calculations as well as EXAFS studies yield different Ge-coordinations in the amorphous and crystalline phase [2-4]. Furthermore, Ge was found to be located in a tetrahedral environment in the crystalline phase as well [3].

In order to clarify the Ge-coordination, a quantitative determination of the relative amount of Ge-atoms in tetrahedral and octahedral coordination is necessary. Nuclear forward scattering (NFS), the time domain analogue to Mössbauer spectroscopy, is a suitable technique to accurately determine hyperfine parameters and relate them to different crystal environments, as it probes the interactions of a nucleus with the magnetic field or electric field gradient. NFS utilizing high energy transitions, such as the 68.7 keV nuclear resonance of ^{73}Ge with a short half life of 1.74 ns, has recently become feasible [5]. In order to identify the signature of tetra- and octahedrally coordinated Ge-atoms in Ge-Sb-Te phase-change materials it is necessary to investigate several reference compounds with a distinct Ge-coordination.

Besides the structure of the amorphous and crystalline phases the lattice dynamics have drawn attention [2,7,8]. Nuclear inelastic scattering (NIS) studies in GeSb$_2$Te$_4$ and SnSb$_2$Te$_4$ with ^{121}Sb and ^{125}Te using a high-resolution sapphire backscattering monochromator [6] give access to the element specific density of phonon states (DPS) and reveal a simultaneous hardening of the acoustic phonons and softening of the optical phonons upon crystallization [7,9]. In contrast, NIS spectra of ^{119}Sn, which is isovalent with Ge, reveal no vibrational softening of the optical phonons but a hardening of the acoustic phonons [9]. This vibrational softening upon crystallization is closely connected to the change from covalent to resonant bonding which also causes the high reflectivity contrast between the amorphous and crystalline phase [7]. The element specific DPS in SnSb$_2$Te$_4$ are shown in Figure 1.

Poster: Memristive systems

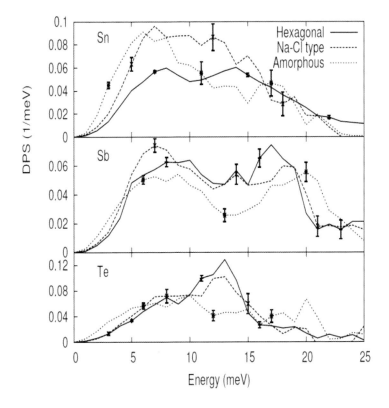

Figure 1: The element specific density of phonon states (DPS) in SnSb$_2$Te$_4$ are shown for the three different phases: amorphous, Na-Cl type and pseudohexagonal [9]. Selected data points with representative measurement uncertainties are displayed.

The authors acknowledge the contribution to this project by the group of Prof. Wuttig, RWTH Aachen University, and support from DFG SFB917 "Nanoswitches". The European Synchrotron Radiation Facility is acknowledged for provision of beamtime at ID18 and ID22N.

[1] M. Wuttig & N. Yamada, Nat. Mater. **6** 824 (2007)
[2] S. Caravati et al., J. Phys. Cond. Mat. **21** 255501 (2009)
[3] X.Q. Liu et al., Phys. Rev. Letters **106** 025501 (2011)
[4] D. Baker et al., J. Mater. Science: Materials in Electronics **18** 399-403 (2007)
[5] I. Sergueev et al., Phys. Rev. Letters **99** 097601 (2007)
[6] I. Sergueev et al., J. Synchrotron Rad. **18**, 802 (2011)
[7] T. Matsunaga et al., Adv. Funct. Mater. **21** 2232-2239 (2011)
[8] R. Mazzarello et al., Physical Review Letters **104**, 1-4(2010)
[9] J. Gallus, Diploma thesis, Lattice Dynamics in the SnSb$_2$Te$_4$ Phase Change Material, RWTH Aachen University (2011)

DEFECT STATES IN AMORPHOUS PHASE-CHANGE MATERIALS

Jennifer Luckas[1], Pascal Rausch[1], Daniel Krebs[2], Peter Zalden[1], Janika Boltz[1], Jean-Yves Raty[3] Martin Salinga[1], Christophe Longeaud[4] and Matthias Wuttig[1]

[1] RWTH Aachen University, I. Physikalisches Institut (IA), 52056 Aachen, Germany
[2] IBM Zürich Research Laboratory, Säumerstraße 4, 8803 Rüschlikon, Switzerland
[3] Physics Department, University of Liege, 4000 Sart-Tilman, Belgium
[4] Laboratoire de Génie Electrique de Paris (CNRS UMR 8507), Plateau de Moulon, 11 rue Joliot Curie, 91190 Gif sur Yvette, France

Phase-change materials show remarkable nonlinear electrical behaviour in their amorphous state. At a critical threshold field of the order of 10 MV/m the amorphous state resistivity suddenly decreases by orders of magnitude. These threshold fields are material dependent [1]. The amorphous state resistivity below the threshold shows thermally activated behaviour and is observed to increase with time. Understanding the physical origins of threshold switching and resistance drift phenomena is crucial to improve non-volatile phase-change memories. Both phenomena are often attributed to localized defect states in the band gap [2-4].

However, little is known about the defect density in amorphous phase-change materials.

This work presents an experimental study of defect states in a-GeTe combining Photothermal Deflection Spectroscopy and Modulated Photo Current Experiments. Based on these experimental findings a defect model for a-GeTe is developed, which is shown to consist of defect bands and band tail states. To get a better understanding of resistance drift phenomena we have studied the evolution of defect state density, optical band gap and resistivity in a-GeTe thin films with time at elevated temperatures. After heating the samples one hour at 140°C the activation energy for electric conduction increases by 30 meV, while the optical band gap increases by 60 meV. This finding demonstrates the impact of the band gap opening on resistance drift. Furthermore, the defect state densities of a-$Ge_{15}Te_{85}$ and a-$Ge_2Sb_2Te_5$ are measured by Modulated Photo Current Experiments to investigate the connection between defect state densities and threshold switching phenomena. For the investigated alloys the measured density of midgap states is observed to decreases with decreasing threshold field known from literature. This is discussed within the frame-work of a generation-recombination model originally proposed by Adler for chalcogenide glasses [5].

[1] D. Krebs, et al., *Appl. Phys. Lett.*, **95**, 082101 (2009)
[2] A. Pirovano et al., *IEEE.*, **51**, 714 (2004)
[3] D. Ielmini, , *Phys. Rev. B*, **78**, 035308 (2008)
[4] D. Ielmini,, et al., *IEEE* , **56**, 1070 (2009)
[5] D. Adler, et al., *J. Appl. Phys*, **51**, 3289 (1980)

ELECTRIC FIELD INDUCED LOCAL SURFACE POTENTIAL MODIFICATION AND TRANSPORT BEHAVIORS OF TiO_2 SINGLE CRYSTALS

Haeri Kim[1], Dong-Wook Kim[1,2], Soo-Hyon Phark[3], and Seungbum Hong[4]

[1] Department of Physics, Ewha Womans University, Seoul 120-750, Korea;
[2] Department of Chemistry and Nano Science, Ewha Womans University, Seoul 120-750, Korea;
[3] Max-Planck-Institut für Mikrostrukturphysik, Weinberg 2, D-06120 Halle, Germany;
[4] Materials Science Division, Argonne National Laboratory, Lemont, IL 60439, USA

Drift and diffusion of oxygen vacancies have been believed to play crucial roles in electric-field induced resistive switching (RS) of metal oxides. We have carried out scanning probe microscopy (SPM) studies to examine local physical properties of the region under a large electric field. We have fabricated micron-size-gapped $Pt/TiO_2/Ti$ planar junctions for simultaneous SPM and transport measurements. The junction resistance decreased after electrical stress regardless of the polarity of the applied field. Surface potential profiles, obtained by Kelvin probe force microscopy (KPFM), revealed that contact resistance and local work function of the TiO_2 surface could be modified by the external bias voltage. These results indicated that surface adsorption/desorption and resulting modification of carrier at the near surface region could occur under an electrical bias. In particular, the tip-induced local charge writing experiments in oxidation and reduction ambient conditions (O_2, Ar, and H_2/Ar) clearly demonstrated that adsorbed oxygen molecules as well as the oxygen vacancies in oxides should be taken into account the field-induced transport behaviors.

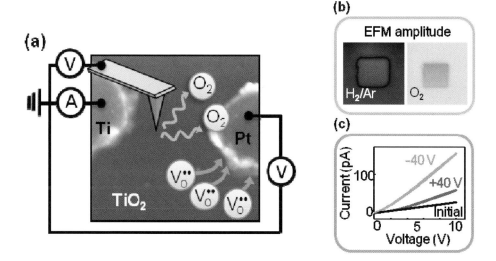

Figure 1: (a) schematic diagram for simultaneous I-V and SPM measurements of $Pt/TiO_2/Ti$ planar junctions including a typical atomic force microscopy topographic image of the sample. Diagrams illustrate drift of oxygen vacancies, adsorption, desorption, and generation of the vacancies during applied electric field to the tip and electrode. (b) EFM amplitude after the tip-induced stress over a square-shaped region at the center region in H_2/Ar and O_2. (c) I-V characteristics before and after electrical stress of application of ± 40 V for 30 min.

[1] H. Kim et al., Appl. Phys. Lett. **100**, 022901 (2012).
[2] H. Kim and D.-W. Kim, Appl. Phys. A. **102**, 949 (2011).
[3] H. Kim et al., J. Phys. D: Appl. Phys. **43**, 505305 (2010).

RESISTIVE SWITCHING IN DIFFERENT FORMING STATES OF Ti/Pr$_{0.48}$Ca$_{0.52}$MnO$_3$ JUNCTIONS

C. Park[1], A. Herpers[1], R. Bruchhaus[1], J. Verbeeck[2], R. Egoavil[2], F. Borgatti[3], G. Panaccione[4], F. Offi[5], and R. Dittmann[1]

[1]PGI-7, FZ Jülich;
[2]EMAT, University of Antwerp, Belgium;
[3]ISMN-CNR, Bologna, Italy;
[4]Laboratorio Nazionale TASC-INFM-CNR, Trieste, Italy;
[5]CNISM and Dipartimento di Fisica, Università Roma Tre, Rome, Italy

Resistance switching (RS) in metal-oxide-metal (MOM) structures has been studied in recent years from the viewpoint of application as a highly scalable future non-volatile memory. [1] For device scaling it has to be taken into account, if a switching current is distributed homogeneously in the device whole area or if it is restricted to one or a few conducting filaments in the device. The interface-type switching operates usually homogeneous RS which is observed at the interface between semiconducting perovskite oxides and the metal electrode with the Schottky contact. Concerning the driving mechanism, recent studies have indicated that the electrochemical migration of oxygen vacancies induces the RS effect. [2]

We investigated the resistive switching behaviors of epitaxial Pr$_{0.48}$Ca$_{0.52}$MnO$_3$ (PCMO) thin films sandwiched between a SrRuO$_3$ (SRO) bottom electrode and a Ti top electrode. Hysteretic current-voltage characteristics, i.e., RS effects, were observed for Ti/PCMO junctions after a first forming procedure, which changes the initial resistance state to the high resistance state (HRS), as shown in fig1. (a) and (b). The initial state and the HRS after the first forming show a clear the area dependence which hints on a homogeneous current distribution in both cases. By performing Hard X-ray Photoelectron Spectroscopy for different resistive states between the pristine state and the first forming state,

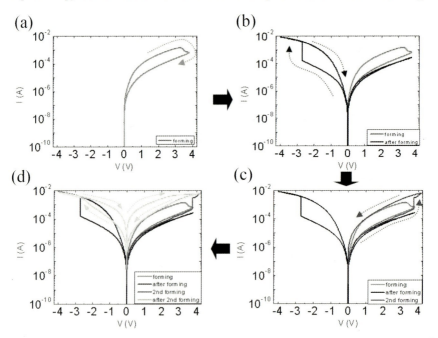

Figure 1: (a) The first forming → (b) the strong asymmetric hysteresis → (c) the second forming →(d) bi-stable resistive switching.

we found a change of the Ti2p peak intensity after the first forming which is associated with the formation of TiO_2 at the interface, as shown in fig. 2 (a, b). Fig. 2 (c, d, e, f) shows that binding energy (BE) of the core levels spectra for all the PCMO elements shifts with respect to the bare PCMO. This is associated to the pinning of the Fermi edge at the Ti/PCMO interface. The formation of TiO_x at the Ti/PCMO interface after the first forming was confirmed by cross-sectional Transmission Electron Microscope investigations. The results indicate that the first forming step is related to a redox process at the Ti/PCMO interface.

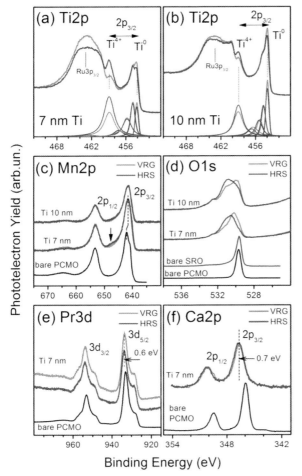

Figure 2: (a,b) Ti2p and (c-f) PCMO-related core-level spectra of the virgin state (VRG, red) and HRS (blue) samples after the first forming and the bare PCMO (black) and SRO (brown). The spectra are normalized to the background intensity. Photon energy was about 5.95 keV.

Moreover, we were able to perform a second forming step which changes the HRS to the low resistance state (LRS), when a higher bias than the bias which is needed for the first forming is applied in the Ti/PCMO junctions, as shown in fig.1 (c). After the second forming step, we observed a stable hysteresis behavior in both bias regions, as shown in fig. 1 (d), but the area dependence disappeared. This implies that conducting filaments might form at the Ti/PCMO interface after the second forming step. We will discuss possible scenarios for the microscopic mechanisms for the observed homogeneous first forming and bipolar RS after the second forming with the conducting filament.

[1] Waser et al., Nature Mater. **5**, 312 (2006)
[2] Sawa et al., Phy. Rev. B. **80**, 235113 (2009).

FIRST PRINCIPLES SIMULATIONS OF OXYGEN DIFFUSION IN RRAM MATERIALS

Sergiu Clima[1], Kiroubanand Sankaran[1,4], Maarten Mees[1,3], Yang Yin Chen[1,3], Ludovic Goux[1], Bogdan Govoreanu[1], Dirk J.Wouters[1,3], Jorge Kittl[1], Malgorzata Jurczak[1], Geoffrey Pourtois[1,2]

[1]imec, B-3001 Leuven, Belgium; [2]PLASMANT, University of Antwerp, B-2610 Antwerpen, Belgium; [3]Katholieke Universiteit Leuven, B-3001 Leuven, Belgium; [4]ETSF and IMCN, Université Catholique de Louvain, B-1348 Louvain-la-Neuve, Belgium

Transition metal oxide valence change Resistor Random Access Memory (RRAM) operation principles are based on oxygen-related defect migration. The switching mechanism is believed to be driven by the Joule heating enhanced drift of O^{2-} ions under the applied electric field through the oxide from/towards the metal electrode.[1,2] The kinetics of the oxygen diffusion is, therefore, a key factor for oxide stoichiometry change, which in turn is responsible for the resistivity of the RRAM cell. Accelerated Ab Initio Molecular Dynamics (AIMD) technique combining the bond-boosted technique[3] is used to compute the diffusion kinetics of O in a series of materials of RRAM interest. In a TiN/Hf/HfO$_2$/TiN stack, it is expected that the sputtering of the Hf capping layer leads to the generation of interfacial amorphous sub-oxides.[4] Beside the electronic barrier function Al$_2$O$_3$ could work as O barrier, therefore pinning the switching layer at a specific location. The knowledge of diffusion data in amorphous phase HfO$_x$ (x=2,1,1/2), Al$_2$O$_3$ and crystalline TiN and Hf metallic electrodes helps us understand where the O comes from and goes into during switching, eventually the impact on the retention of the cell.

In terms of activation energy for the diffusion (E_a) a trend is observed across different materials – hafnium oxides have the lowest activation energies and slightly increasing with the O content/density in the oxide (0.57-0.66 eV), a larger barrier is observed in Al$_2$O$_3$ (1.22 eV) and on the electrodes sides much higher barriers were computed: 2.50 eV in TiN and 4.13 eV in Hf. These activation energies match the experimental window, as measured for HfO$_x$ and Al$_2$O$_3$.[5,6]

The O diffusion prefactors in HfO$_x$ are computed (3.8E-8 – 5.1E-8 m^2/s) to be on the high end of the experimental results (4E-12 – 2E-8 m^2/s). Considering the fact that we have the same computational quality level for all materials, we can conclude that the movement of the O atoms is facilitated by the free volume that is increasingly more available in the sub-stoichiometric Hf oxides but very difficult in the metallic Hf electrode. Somewhat higher diffusion coefficients are computed in TiN and Al$_2$O$_3$ vs Hf, but they are still blocking layers for O, if compared to HfO$_x$. The computed activation energies are in excellent agreement with experimental measurements.[5]

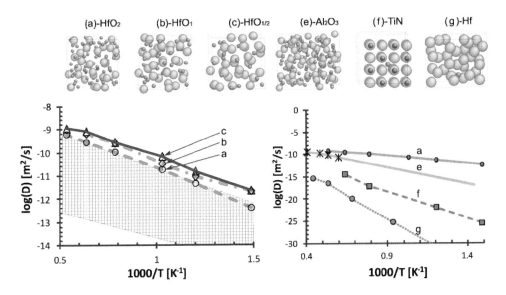

Figure 1. Arrhenius plot of Oxygen self-diffusion in $HfO_{1/2}$, HfO_1, HfO_2, hashed region marks the experimental coefficients for HfO_2 (on the left) and HfO_2, Al_2O_3, TiN and Hf (on the right)

1. Liu, L. F. et al. Engineering oxide resistive switching materials for memristive device application. *Applied Physics a-Materials Science & Processing* **102**, 991-996, doi:10.1007/s00339-011-6331-2 (2011).
2. Waser, R., Dittmann, R., Staikov, G. & Szot, K. Redox-Based Resistive Switching Memories – Nanoionic Mechanisms, Prospects, and Challenges. *Advanced Materials* **21**, 2632-2663, doi:10.1002/adma.200900375 (2009).
3. Miron, R. A. & Fichthorn, K. A. Accelerated molecular dynamics with the bond-boost method. *Journal of Chemical Physics* **119**, 6210-6216, doi:10.1063/1.1603722 (2003).
4. Govoreanu, B. et al. Investigation of forming and its controllability in novel HfO2-based 1T1R 40nm-crossbar RRAM cells. *Ext. Abstr. SSDM Conf., Nagoya, Japan*, pp.1005-1006 (2011).
5. Zafar, S., Jagannathan, H., Edge, L. F. & Gupta, D. Measurement of oxygen diffusion in nanometer scale HfO2 gate dielectric films. *Applied Physics Letters* **98**, 152903, doi:10.1063/1.3579256 (2011).
6. Nabatame, T. et al. Comparative studies on oxygen diffusion coefficients for amorphous and gamma-Al2O3 films using O-18 isotope. *Japanese Journal of Applied Physics Part 1-Regular Papers Short Notes & Review Papers* **42**, 7205-7208, doi:10.1143/jjap.42.7205 (2003).

OBSERVATION OF A CONDUCTIVE REGION IN THE TiN/HfO$_2$ SYSTEM AFTER RESISTANCE SWITCHING

P. Calka, E. Martinez, V. Delaye, D. Lafond, G. Audoit, D. Mariolle, N. Chevalier, H. Grampeix, C. Cagli, V. Jousseaume, C. Guedj

CEA, LETI, MINATEC Campus, 17 rue des Martyrs, 38054 Grenoble Cedex 9, France.

OxRRAMs are promising candidates for future non volatile memories. Data storage principle is based on switching the oxide resistivity reversibly between high and low resistance states by applying a current or a voltage. Usually, a high voltage is required to initiate the conduction in the pristine insulating oxide. However, the resistive switching mechanism is still unclear. Imaging of the conductive path is required to better understand and control this phenomenon. The size of the active region might be small (10-20 nm diameter [3-4]). Furthermore, it is buried beneath the top electrode and randomly generated. Therefore, imaging with Transmission Electron Microscopy (TEM) is challenging. Indeed, the TEM lamella is a cross-section of the M-I-M stack with a total thickness of 100 nm. So, prior to sample preparation, an accurate location of the conductive region must be done.

To tackle the localisation issue, we set up a five steps protocol. We studied a Si/SiO$_2$/TiN(20 nm)/HfO$_2$(10 nm) stack with no permanent top electrode. A nano-probe, namely a conductive AFM tip was used. The advantages are: 1) it is removable, 2) the nano-sized diameter of the contact area. As a result, the localisation of the active region is facilitated.

The five steps in the protocol are as follows:

1) Focused Ion Beam (FIB): marks are made on the HfO$_2$ to facilitate the localisation of the conductive atomic force microscopy measurement (C-AFM).

2) C-AFM: resistance switching is done. (Figure 1a))

3) Scanning Spreading Resistance Microscopy (SSRM): the active region is localised using topography and resistance maps. (Figure 1b))

4) FIB: Cross-section preparation containing the switched area

5) Scanning Transmission Electron Microscopy (STEM): structural and chemical analyses were performed.

Figure 2a) displays the Z-contrast image of the cross-section containing the conductive region. The morphology of the HfO$_2$ layer evolves after resistance switching. A protrusion is formed. Then, chemical analyses are performed using Electron Energy Loss Spectroscopy (EELS). Figure 2b) shows a loss of oxygen atoms in the active region (dark blue region). The oxygen poor region is localised inside the protrusion and has a size of about 20 nm in the lateral direction. On the other hand, as can be inferred from the Ti and N maps, the material of the bottom electrode (TiN) does not migrate into the HfO$_2$ (not shown). Therefore, the resistance switching could be related to the oxygen loss in HfO$_2$. Finally, the oxygen concentration profiles in both poor and rich regions are compared, together with the relative thicknesses of the cross-section. The profiles are plotted in Figure 2c). There is no correlation between the two profiles. Therefore, the oxygen poor region is not related to a hole in the oxide. The resistance switching seems correlated to a local oxygen content modulation in HfO$_2$.

The measurements were performed at the NanoCharacterization Platform of MINATEC.

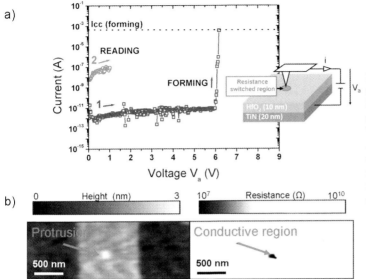

Figure 1: (a) I-V_a characteristics performed by performing C-AFM measurements on the TiN/HfO$_2$ stack. Inset: C-AFM experimental set-up. (b) Topography and resistance maps of the resistance switched region.

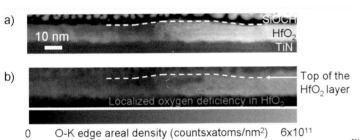

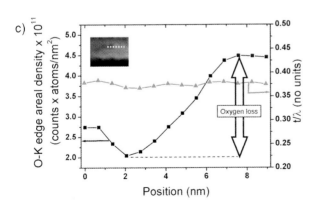

Figure 2: (a) Top: Z-contrast TEM image of the resistance switched region. (b) Corresponding STEM-EELS oxygen map. (c) Relative thickness of the cross-section and oxygen concentration plotted along the dotted line in the inset. Inset: oxygen deficient region in HfO2 corresponding to the red circle in Figure 1.

TRANSIENT CHARACTERISTICS DURING SET OPERATION OF A Ta$_2$O$_5$ SOLID ELECTROLYTE MEMRISTIVE SWITCH

Pragya Shrestha[1,2], Adaku Ochia[1,3], Kin. P. Cheung[1,*], Jason Campbell[1], Helmut Baumgart[2,4] and Gary Harris[3]

[1]Semiconductor and Dimensional Metrology Division, National Institute of Standards and Technology (NIST), Gaithersburg, MD 20899, USA
[2]Department of Electrical and Computer Engineering, Old Dominion University, Norfolk, VA 23529, USA
[3]Department of Electrical Engineering, Howard University, Washington, DC, 20059, USA
[4]The Applied Research Center at Thomas Jefferson National Accelerator Facility, Newport News, VA 23606, USA
*kin.cheung@nist.gov

Recently memristive devices have gained popularity in the field of memory due to its simplicity, small size and low cost [1, 2]. Due to similar reasons these devices also have the potential to work as high performance switches for interconnect architecture [3, 4]. Despite all the ongoing research the actual mechanism behind the switching is yet to be clearly understood.

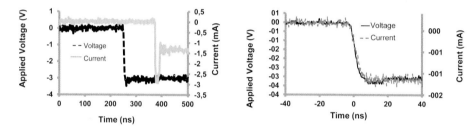

Figure 1: Current through the device a) during SET operation (The measurement details are given in [5]) and b) current through the device when a known pulse of -3 V with rise time of 5 ns is applied to the device.

Transient characteristics during switching operation can be used as a tool to understand the switching mechanism. Figure 1 a) shows the measured current through the device when SET voltage is applied. It can be seen that the current peaks before it stabilizes. This particular behavior is less likely to be a measurement artifact. Figure 1 b) clearly shows that no such measurement artifact exists in the system due to the measurement set up when a known voltage with a rise time as low as 5 ns is applied to the device. Therefore, the spike in the current is likely to be due to the changes that occur during set operation. In this work, we report on our detailed investigation of the switching transients of memristive devices to gain a better understanding of the underlying switching mechanism. The memristive devices tested for this study were fabricated with sputtered Ta$_2$O$_5$ as solid electrolyte sandwiched between the cross point of Cu and Pt metal lines. Measurements are reported for memristive devices of sizes ranging from 25 μm x 25 μm to 20 nm x 20 nm.

[1] J. H. Kriegerand and S. M. Spitzer, "Non-traditional, non-volatile memory based on switching and retention phenomena in polymeric thin films," in *Non-Volatile Memory Technology Symposium, 2004*, 2004, pp. 121-124.
[2] R. Waser, R. Dittmann, G. Staikov, and K. Szot, "Redox-Based Resistive Switching Memories – Nanoionic Mechanisms, Prospects, and Challenges," *Advanced Materials*, vol. 21, pp. 2632-2663, 2009.
[3] A. Gayasen, N. Vijaykrishnan, and M. J. Irwin, "Exploring technology alternatives for nano-scale FPGA interconnects," presented at the Proceedings of the 42nd annual Design Automation Conference, Anaheim, California, USA, 2005.
[4] S. Kaeriyama, T. Sakamoto, H. Sunamura, M. Mizuno, H. Kawaura, T. Hasegawa, K. Terabe, T. Nakayama, and M. Aono, "A nonvolatile programmable solid-electrolyte nanometer switch," *Ieee Journal of Solid-State Circuits*, vol. 40, pp. 168-176, Jan 2005.
[5] P. Shrestha, A. Ochia, K. P. Cheung, J. P. Campbell, H. Baumgart, and G. Harris, "High-Speed Endurance and Switching Measurements for Memristive Switches," *Electrochemical and Solid-State Letters*, vol. 15, pp. H173-H175, 2012.

REMANENT RESISTANCE CHANGES IN METAL- MANGANITE- METAL SANDWICH STRUCTURES

Malte Scherff, Bjoern Meyer, Julius Scholz, Joerg Hoffmann, and Christian Jooss

Institute of Materials Physics, University of Goettingen, Germany

The non-volatile electric pulse induced resistance change (EPIR) seems to be a rather common feature of oxides sandwiched by electrodes. However, possible mechanisms are discussed controversially. In heterostructures based on manganites with a strongly correlated electron system, the study of the interplay between electrochemical and electronic mechanisms seems to be essential. We present a systematic study of dynamic and remanent electrical transport properties of epitaxial hole-doped $Pr_{0.7}Ca_{0.3}MnO_3$ (PCMO) films sandwiched by noble metallic electrodes. The variation of electrode materials, device geometry, electric excitation modes in combination with the study of change of magneto-transport in different resistance states allow us to shed light onto the involved processes.

The thin film heterostructures used in our sample devices are produced by ion beam sputter deposition. First 500 nm thick Pt films are epitaxially deposited on (100) MgO single crystals as the bottom electrode, followed by PCMO grown epitaxially on these templates. For the top electrodes we use different noble metals (Ag, Au, Pt). Two different device geometries are used for examining the influence of symmetry of interface dimensions: In the "pillar device" the top electrode and the PCMO film is structured to a circular diameter of a few μm (Fig. 1 a) in order to get two symmetric interfaces between electrodes and PCMO of same dimensions. An asymmetric setup is realized in the "pad device", where bottom electrode and PCMO film remains extended and only the top electrode is patterned to circular shaped electrodes with diameters between 1 μm and 16 μm.

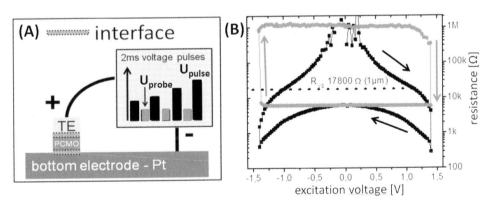

Figure 1: (a) Schematics of pillar-device and measurement approach by variable U_{pulse} and constant U_{probe}. (b) Measured dynamic (black) and remanent (red) resistances as a function of U_{pulse} (voltage cycle +/- 1.3V, pulse length 2 ms) of a pillar device with 1μm diameter and Au top-electrode. U_{probe} for the measurement of the remanent resistance, is fixed at 0.08V. The dotted line represents the initial resistance level of device.

Electrical transport measurements were performed using voltage pulses as excitation and probe pulses. The voltage amplitude U_{pulse} for the excitation pulse is ramped between +/- U_{max}. Pulse duration can be varied between 100 ns and 1 s. During these pulses the excitation current is measured and leads to a "dynamical" resistance. Typical results can be seen in Fig. 1 (b, black curve): A strong dependence of resistance on the applied voltage (decreasing with increasing voltage) and a prominent hysteresis due to the actual switching process at higher voltages. In order to distinguish between dynamic and remanent resistance changes in the samples, we also apply "probe" measurements with a fixed, small voltage U_{probe} after each excitation pulse (red curve in Fig. 1b). The different switching details in devices with symmetric and asymmetric electrode interface dimensions gives evidence for identifying the active, single interface as well as the interplay of two active interfaces in the switching process. This is underpinned by the existence of a switching polarity which depends on the device symmetry and a switching polarity inversion at higher current density (respectively electrical field) regimes [1]. The transition between the two switching regimes agrees well with a theoretical model, where local high resistance areas of the oxide near both interfaces can grow or shrink by thermally assisted electro-diffusion [2].

Figure 2 shows the dependence of the remanent resistance changes on the pulse duration measured at a fixed voltage with alternating polarity. The alternating excitation pulses lead to visible resistance changes at a critical onset pulse length τ_{onset}. Figure 2b shows the switching amplitude for different U_{pulse}: Increasing U_{pulse} lead to a decrease of the τ_{onset}. Our detailed analysis of the results shows the equivalence of voltage amplitude and pulse duration time and gives strong evidence that the size change of RHS domains is induced by thermally assisted diffusion.

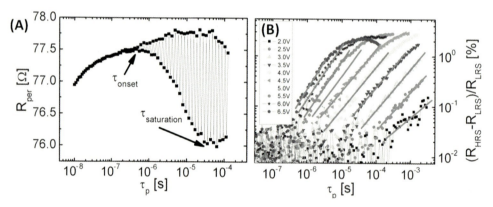

Figure 2: (a) Remanent resistance R_{per} for pulse length variation at fixed, alternating voltage. (b) Switching amplitude as a function of pulse length for various excitation voltages.

In order to study the interplay between resistive switching and the strong electronic and electron-lattice correlations in PCMO, switching cycles at low temperatures under magnetic fields will be presented. The interplay between colossal magneto resistance (CMR), colossal electric resistance (CER) and resistance switching gives fascinating insights into involved changes of cooperative behaviour of polarons (such as charge ordering) and magneto-transport. The induced changes in structural and electronic disorder due to resistive switching are further studied by transmission electron microscopy.

[1] M. Scherff, B.-U. Meyer, J. Hoffmann, and Ch. Jooss, J. Appl. Phys. **110**, 043718 (2011)
[2] M.J. Rozenberg, M. J. Sánchez, R. Weht, C. Acha, F. Gomez-Marlasca, and P. Levy, Phys. Rev. B **81**, 115101 (2010)

THEORETICAL STUDY ON THE CONDUCTIVE PATH IN TANTALUM OXIDE ATOMIC SWITCH

Bo Xiao[1], Tomofumi Tada[1], Tingkun Gu[2], Arihiro Tawara[1], Satoshi Watanabe[1]

[1]Department of Materials Engineering, The University of Tokyo, Tokyo, Japan
[2]School of Electrical Engineering, Shandong University, Jinan, China

Nowadays, "atomic switch" consisting of metal-(oxide or sulfide)-metal heterostructure has attracted attention owing to its potential application in RRAM etc. Its most plausible switching mechanism is the formation/annihilation of metal filaments in the oxide/sulfide between the electrodes. However, microscopic detail of the switching mechanism is still unclear. In the present study, taking the $Cu/Ta_2O_5/Pt$ atomic switch as an example, we have examined possible atomic structures and electronic states of such metallic filaments from first-principles using VASP code. For the Ta_2O_5, we have examined both crystalline and amorphous phases.

In the crystalline case, our simulations on various arrangements of interstitial Cu atoms and oxygen vacancies shows that the alternant Cu-Ta bonding plays a crucial role in the formation of conductive path in $Cu/Ta_2O_5/Pt$ (see Fig. 1(a)) [1]. Moreover, such an alternant Cu-Ta bonding and the resultant conductive path appear only in the ordered Ta_2O_5 crystalline structure.

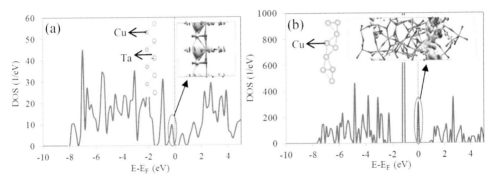

Figure 1: (a) Density of states (DOS) for a Cu doped crystal Ta_2O_5. The inset shows the structure of alternant Cu-Ta bonding, and the charge density corresponding to the defect state around the Fermi level (E_F). (b) DOS for a Cu doped amorphous Ta_2O_5. The inset shows the structure of thinnest doped Cu nanowire in amorphous Ta_2O_5, and the charge density around E_F.

In the amorphous case, we have found that Cu single atomic chains are unstable, and then examined Cu nanowires with various widths. The inset of Fig. 1(b) shows the thinnest Cu nanowire found to be stable within the molecular dynamics simulation of 4 ps at 500 K. This and thicker Cu nanowires can form conductive paths, and Cu-Cu bonding mainly contributes to the conductive, delocalized defect state (see Fig. 1(b)).

This work was partially supported by CREST-JST "Atom transistor", Low Power Electronics Association and Projects, and the grant-in-aid for Innovation Area "Computics", MEXT, Japan.

[1] T. K. Gu, T. Tada and S. Watanabe, ACS Nano **4**, 6477 (2010).

CHARACTERISTIC OF LOW TEMPERATURE FABRICATED NONVOLATILE MEMORY DEVICES OF ZN AND SN NANO THIN FILM EMBEDDED MIS

Tai-Fa Young[1], Ya-Liang Yang[1], Ting-Chang Chang[2,3], Kuang-Ting Hsu[1], and Chao-Yu Chen[1].

[1]Department of Mechanical and Electro-mechanical Engineering, and Research Center for Nanoscienceand Nanotechnology, National Sun Yat-senUniversity, Kaohsiung, Taiwan.
[2]Department of Physics, and
[3]Department of Photonics, National Sun Yat-sen University, Kaohsiung, Taiwan.

Non-volatile memory (NVM) device can keep the data without electrical power, and has applied to many portable electronic products due to the advantage of low power consumption. In current industrial production, high-temperature and long-time process are necessary for the fabrication of NVM, which are heavy loadings on production capacity and cost. Therefore, reducing the temperature of the process is very desired. Recently using the oxidation treatment of supercritical carbon dioxide (SCCD) fluid has showed to efficiently reduce the process temperature.

In this work, a mixture layer of Zn, Sn, and SiO_2 is applied to employ the defects of ZnO and SnO_2 as floating gate for electron storage to fabricate the NVM device. Zn and Sn are applied due to the low temperature melting points. The process of traditional rapid temperature annealing treatment was applied for first step to ensure a successfully fabrication of NVM device.

The co-sputtered Zn-Sn-SiO_2 thin film was deposited on the tunnelling oxide layer, and then the thin film was treated with varied annealing temperature to precipitate ZnO and SnO_2 nanocrystals. After that, the C-V measurement and deep level transition spectroscopy (DLTS) are applied to analyze the change of the electrical and material properties. Using a positive bias, the electrons are injected into the oxide layer, by the threshold voltage the offset is occurred, which is defined as the memory window of the memory effect, and the property of NVM will be applied.

The SCCD fluid technology has been performed to study the memory effect. The capability of electron injection, storages and the defect, in the storage layer were studied by the C-V measurement and DLTS. The experiment confirmed that the Zn-Sn alloy has the memory property after it been treated by the SCCD fluid technology. Zn can promote to the storage capability ability due to the formation of deep level defects of SnO_2 from the DLTS spectra. A new species is found at 0.93 eV with low activation energy and high capability of electron storage. The defect formation mechanism of Zn, ZnO, Zn-O-Si, Sn, and SnO are analyzed by found by the XPS and DLTS. We show the device fabrication using Zn-Sn alloy and SCCD fluid technology has the potential to reduce the process temperature and to improve the memory property of NVM device.

ELECTROCHEMICAL STUDIES ON Al$_2$O$_3$ THIN FILMS FOR RESISTIVE MEMORY APPLICATIONS

A Burkert[1], I. Valov[1,2], G. Staikov[1,2], R. Waser[1,2]

[1]Institute of Electronic Materials in Electrical Engineering (IWE2), Aachen, Germany,
[2]Institute of Electronic Materials (PGI 7), Jülich, Germany

Much research has been done on resistive switching materials as a resistive RAM (RRAM) device. Different effects were found to appear in resistive switching materials: (i) the electrochemical metallization effect (ECM) where an electrochemical driven redox process leads to a formation and rupture of a conducting metallic filament, (ii) the thermochemical mechanism (TCM) where a conducting filament is grown and ruptured by the heat produced by an ionic flux, and (iii) the valence change mechanism (VCM) where the conductivity of a material is changed by a change of the stochiometry of the material composition e.g. oxygen amount [1]. Resistive switching was found in the following systems with thin Al$_2$O$_3$ films: Al|Al$_2$O$_3$|Cu [2], Au|Al$_2$O$_3$|Cu:TCNQ|Cu [3], Pt|Al$_2$O$_3$|Ru [4], Pt|Al$_2$O$_3$|Ti [5]. The formation of a thin Al$_2$O$_3$ film at the Al|Cu:TCNQ-interface of the ECM-system Al|Cu:TCNQ|Cu was found to play an crucial role for the resistive switching in [2].

In this contribution we present electrochemical studies on Al$_2$O$_3$ thin films prepared by anodic oxidation and by sputter deposition. The anodic oxidation of Al was performed on Si|SiO$_2$|TiOx|Pt|Al-substrates in an acetate buffer solution at different voltages in order to produce Al$_2$O$_3$ thin films with different thickness. The sputtered films were deposited on Si|SiO$_2$|TiOx|Pt-substrates. Cu, Au and Pt were deposited to study the influence of the top electrode material. The properties of the anodic Al$_2$O$_3$ films were investigated by impedance spectroscopy and current transient measurements. Figure 1 shows as an example the frequency dependence of the capacitance extracted from impedance measurements of anodic oxide films.

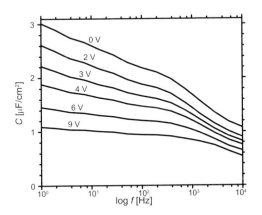

Figure 1: Example of the frequency dependency of the capacitance of anodic oxide foils formed in an aqueous acetate electrolyte with voltages in the range of 0V to 9V. The capacitance shows linear behavior in the range from 1 Hz to 10000 Hz. The change of capacitance is due to the increasing thickness of the Al$_2$O$_3$ film and the slope is due to the frequency dependence of the permittivity.

[1] R. Waser, Microelectronic Engineering **86** 1925 (2009).
[2] T. Kever et al., Appl. Phys.Lett. **91**, 083506 (2007).
[3] A. Hefczyc et al., phys. Stat. sol. (a) **205**, 647 (2008).
[4] K. M. Kim et al., Electrochem. Solid-State Lett., **9**, G343 (2006).
[5] C. Lin et al., J. Electrochem. Soc., **154**, G189 (2007).

RESISTIVE SWITCHING PHENOMENA IN Ag-GeS$_x$ MEMORY CELLS

Jan van den Hurk[1], Ilia Valov[2], Rainer Waser[1,2]

[1]Institut für Werkstoffe der Elektrotechnik II, RWTH Aachen University, Germany
[2]Peter Grünberg Institut 7, Forschungszentrum Jülich, Germany

The ever continuing evolution of materials and technology has been a constant companion of the electronics industry from its early days until now. This is especially true for memory technology where Moore's law acted as a self-fulfilling prophecy pushing memory cells and the surrounding circuitry to today's prevailing nano dimensions. However this development faces its inevitable end as the dimensions reach the ultimate limits of production technology.

One of the most promising discussed alternatives to the current memory technologies (i.e. FLASH and DRAM) are resistively switching memory cells (RRAM), which offer new options in terms of high scalability, speed and power efficiency.[1] Various materials are currently under investigation including germanium based chalcogenides. Ag-GeS$_x$ based cells belong to the class of electrochemical metallization memory cells and already proofed to be a potential candidate for a well working resistive memory material.[2]

It is commonly known that the exact manner of electrical actuation is a critical parameter for the performance of resistively switching memory cells. This is also true for electrochemical metallization memory cells and particularly the case for Ag-GeS$_x$ based cells. In this study we fabricated germanium sulphide based microstructures with silver top electrodes to investigate the switching properties of this type of memory cell.

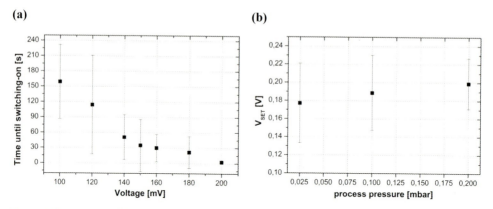

Figure 1: Resistive switching of Pt/70 nm GeS$_x$/Ag microstructures (a) The switching kinetics during constant voltage experiments heavily depends on the selected voltage level. (b) The switching voltage depends on the process pressure during the deposition of the solid electrolyte. By adjusting the process pressures it is possible to generate different S-to-Ge-ratios. The pressures 0.025, 0.100 and 0.200 mbar correspond to S/Ge ratios 1.6, 1.9 and 2.2 respectively.[3]

Besides fundamental aspects of the switching characteristics like for example the possibility of multiple resistive states (multilevel switching) we investigated the dependency between the applied voltage and the resulting switching time. As figure 1 a) indicates we observe shorter switching times for higher voltages. At the same the switching kinetics feature a clear non-linear behaviour what is desirable in resistive memory applications.

Furthermore we had the possibility to change the sulphur to germanium ratio in the solid electrolyte by adjusting the process pressure during deposition. Higher sulphur (i.e. lower germanium) content in the electrolyte gives a higher set voltage as figure 1 b) shows. This measurement directly translates into a decelerated build up of the conducting filament between the two electrodes. As recent results have shown silver in a GeS_x thin film preferably interacts with the germanium part of the solid electrolyte.[4] Our results must be interpreted in the way that a higher germanium content in the solid electrolyte has a stabilizing effect on the filament formation during the switching-on.

[1] R. Waser, R. Dittmann, G. Staikov, and K. Szot, Redox-Based Resistive Switching Memories - Nanoionic Mechanisms, Prospects, and Challenges, *Adv. Mater.* **21**, 2632-2663 (2009)
[2] I. Valov, R. Waser, J. R. Jameson, and M. N. Kozicki, Electrochemical metallization memories-fundamentals, applications, prospects, *Nanotechnology* **22**, 254003/1-22 (2011)
[3] J. van den Hurk, I. Valov, and R. Waser, Characteristics of sputtered germanium sulphide thin-films, *to be published* (2012)
[4] DY Cho, J. van den Hurk, S. Tappertzhofen, I. Valov, and R. Waser, Observation of Ionic Charge Transfer in Solid Electrolytes for Electrochemical Metallization Memory, *to be published* (2012)

STATES AND PROCESSES IN NANO-SCALED CATION BASED RRAM CELLS

Ilia Valov[1,2], Stefan Tappertzhofen[2], Jan van der Hurk[2] and Rainer Waser[1,2]

[1]Research Centre Juelich, Peter Gruenberg Institute (PGI-7), Electronic Materials, 52425 Juelich, Germany;
[2]RWTH Aachen University, Institut für Werkstoffe der Elektrotechnik II, 52074 Aachen, Germany

The electrochemical metallization memory cells (ECM), representing the cation based RRAMs are typical example for nanoscaled electrochemical systems. In a vertical direction the dimensions are in the range of some tens of nanometers and the lateral dimensions are reduced to almost atomic level [1,2,3]. The working principle of the ECM cells is based on a formation and respectively rupture of a nano-sized metallic filament formed and dissolved under applied voltage (of an opposite polarity). The filament short circuits the cell thus, defining the low resistive ON state or if the filament is dissolved the cell is in a high resistive OFF state. Both states are used to read the Boolean 1 and 0, respectively. Despite that the practical functionality of ECM cells has been subject of numerous studies, leading to a development of empirical or semi-empirical models for the operation principles [4] expanding to concepts of multi-bit memories [5,6] and memristive systems [7,8,9], no single report has been published on the thermodynamic properties of the RRAM cells at rest, i.e. the virgin (as deposited) state and the ON and OFF states.

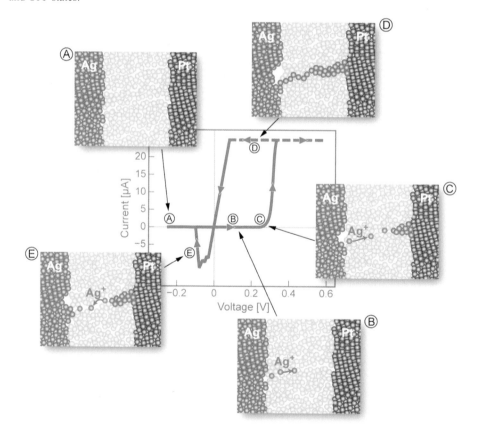

Fig. 1 Schematic presentation of metallic filament formation and dissolution related to the current voltage characteristics of ECM memory cell. Reprinted by ref. 2 ©2011, IOP

The formation and dissolution of the nano-filament seems on first sight to be intuitively understandable but is controversy discussed and no complete microscopic model has been presented even for the simplest systems.

In this contribution we demonstrate that the ON and OFF states in ECM cells are non-equilibrium and are subject of various chemical potential gradients resulting in an effective electromotive force ranging from 20 mV to 350 mV. We studied systematically classes of materials used in ECM cells as cation electrolytes i.e., insulators, mixed conductors, and ion conductors and found that also additional factors e.g. O_2 and H_2O partial pressures determined the properties of these systems.

We also show that the kinetic of the filament formation is mainly influenced by the nucleation, i.e. the formation of the critical nucleus as already demonstrated by us for gap type ECM systems [10]. We discuss the necessity of performing complete analysis for individual systems in order to correctly conclude on the rate determining step. In this respect we found that in highly resistive systems e.g. SiO_2 at higher voltages also the diffusion of metal ions becomes rate limiting.

1. Aono, M. & Hasegawa, T. The Atomic Switch. *Proc. IEEE* **98**, 2228-2236 (2010).
2. Valov, I., Waser, R., Jameson, J. R. & Kozicki, M. N. Electrochemical metallization memories-fundamentals, applications, prospects. *Nanotechnology* **22**, 254003/1-22 (2011).
3. Waser, R., Dittmann, R., Staikov, G. & Szot, K. Redox-Based Resistive Switching Memories - Nanoionic Mechanisms, Prospects, and Challenges. *Adv. Mater.* **21**, 2632-2663 (2009).
4. Jameson, J. R. *et al.* One-dimensional model of the programming kinetics of conductive-bridge memory cells. *Appl. Phys. Lett.* **99**, 063506-063506 (2011).
5. Russo, U., Kamalanathan, D., Ielmini, D., Lacaita, A. L. & Kozicki, M. N. Study of Multilevel Programming in Programmable Metallization Cell (PMC) Memory. *IEEE Trans. Electron Devices* **56**, 1040-1047 (2009).
6. Yu, S. & Wong, H.-S. Compact Modeling of Conducting-Bridge Random-Access Memory (CBRAM). *IEEE Trans. Electron Devices* **58**, 1352-1360 (2011).
7. Takeo Ohno, Tsuyoshi Hasegawa, Tohru Tsuruoka, Kazuya Terabe, Gimzewski, James K. & Masakazu Aono, Short-term plasticity and long-term potentiation mimicked in single inorganic synapses. *Nat. Mater.* **10**, 591-595 (2011).
8. Linn, E., Rosezin, R., Kügeler, C. & Waser, R. Complementary Resistive Switches for Passive Nanocrossbar Memories. *Nat. Mater.* **9**, 403-406 (2010).
9. Strukov, D. B., Snider, G. S., Stewart, D. R. & Williams, R. S. The missing memristor found. *Nature* **453**, 80-83 (2008).
10. Valov, I. *et al.* Atomically controlled electrochemical nucleation at suprionic solid surfaces. *Nature Mater.* accepted (2012)

FIGHTING VARIATIONS IN PT/TIO2-X/PT AND AG/A-SI/PT MEMRISTIVE DEVICES

G. Adam[1], F. Alibart[1], L. Gao[1], B. Hoskins[1], and D.B. Strukov[1]

[1]University of California at Santa Barbara, Santa Barbara, CA 93106, U.S.A.

Resistive ("memristive") switching, the reversible modulation of electronic conductivity in thin films under electrical stress, has been observed in a wide range of material systems and is attributed to diverse physical mechanisms [1]. Research activity in this area has been traditionally fueled by the search for a perfect electronic memory candidate, but recently received additional attention due to a number of other promising applications, such as reconfigurable and neuromorphic computing [2]. One of the major challenges for large scale integration of such devices still remains poor yield and large amount of variations in electrical characteristics. In this work, we report on two solutions to this problem for metal oxide and amorphous silicon memristive devices.

A. Engineering memristive devices with protrusion: Poorly behaved resistive switching is often contributed to the existence of multiple active regions, for example, as a consequence of having several parallel conducting filaments produced by forming process. Because each active region is likely to require unique stress conditions for resistive switching the overall device behavior is poorly controlled. To improve yield and lower variations, the first approach is to make nanoscale metallic protrusion in thin film to localize switching region (Fig. 1a). The initial results are very encouraging showing, e.g., yield improvements from ~60% for the blanket film devices to > 95% for the devices with protrusion (Fig. 1b), as well as significant lowering of variations in switching behavior.

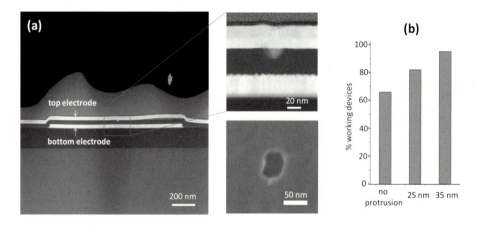

Figure 1: (a) TEM images of 50-nm-thick titanium dioxide devices with e-beam defined protrusion and (b) a comparison of yield for several devices with different protrusion depth.

B. Variation tolerant tuning algorithm: While protrusion technique has resulted in much better devices there is still significant dispersion in *I-V* characteristics. To address these problems, we designed a simple feedback algorithm to tune device conductance at a specific bias point to 1% relative accuracy (which is roughly equivalent to 7-bit precision) within its dynamic range even in the presence of large variations in switching behavior [3]. Our algorithm is based on the fact that large amplitude pulses can be used to reach a desired state faster but also at much cruder precision (Fig. 2a). To take advantage of such switching dynamics the algorithm (Fig. 2c) is based on a sequence of increasing amplitude voltage ramps of appropriate polarity (which depends on the initial and the desired state of the devices). The device state is checked with the read pulse after each write pulse and the voltage ramp is applied until it reaches and overshoots the desired state. At this point the new voltage ramp of opposite polarity begins, which is always started from the same non-disturbing initial voltage with amplitude $|v| < 0.5$ V (Fig. 2a). Because this time the initial state is closer to the desired one the maximum amplitude of the voltage pulse in the new ramp is smaller, which in turn ensures that this new ramp will drive the device to the desired state with even better precision.

This work is supported by AFOSR and NSF.

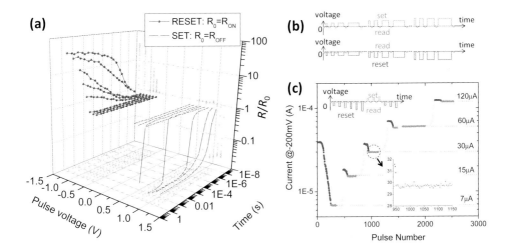

Figure 2. (a) Switching dynamics of microscale TiO_{2-x} devices [3]. Each curve shows the dynamic evolution of the device's state (resistance at -0.2V) as a result of the application of (b) fixed amplitude pulses with exponentially increasing duration (from 200 ns to 1 ms). Before the pulse measurements, the device is set to the same high or low resistance state for set (green) and reset (blue) swtching transitions, correspondingly. (c) Demonstration of the algorithm to tune the resistive state of the device to 7, 15, 30, 60 or 120 µA (at -0.2V) within 1% accuracy of the dynamic range using the algorithm shown in the top inset. The bottom inset is a zoom-in for the particular intermediate state [3].

[1] D. B. Strukov and H. Kohlstedt, MRS Bulletin **37**, 108 (2012).
[2] K. K. Likharev, J. Nanoelectronics and Optoelectronics **3**, 203 (2008).
[3] F. Alibart, L. Gao, B. Hoskins, and D. B. Strukov, Nanotechnology **23**, 075201 (2012).

A STUDY UPON THE SWITCHING CHARACTERISTICS AT RUPTURED CONDUCTING FILAMENTS REGION IN A Pt/TiO$_2$/Pt MEMRISTIVE DEVICE

Kyung Jean Yoon, Seul Ji Song, Gun Hwan Kim, Jun Yeong Seok, Jeong Ho Yoon, and Cheol Seong Hwang

WCU Hybrid Materials Program, Department of Materials Science and Engineering and Inter-university Semiconductor Research Center, Seoul National University, Seoul 151-744, Korea

It has been known through numerous studies that the memristive phenomena in nano-scale devices are originated from the structural as well as electrical defects that reside in materials due to the thermodynamically permissible non-stoichiometry of the materials. Transition metal oxides are such materials that show the high non-stoichiometry in nature and the memristive phenomena in these materials, therefore, are being vigorously reported.

Bipolar resistive switching in TiO$_2$ has been known to be one of the typical charge-controlled memristive phenomena, and studied associated with various kinds of electrode materials, device structures, and electroforming methods. In particular, a Pt/TiO$_2$/Pt device in unipolar resistive switching reset state shows such an intriguing resistive switching behavior involving both the migration of ionic defects and trapping of electronic carriers within the film. The efforts to reveal the precise mechanism for the phenomenon have led to the demonstrations within the context of migration of oxygen ions according to bias polarity as well as electronic carrier trapping at the oxygen vacancies mediated by the space-charge-limited-conduction . Since these different mechanisms share almost the identical I-V characteristics, they are hard to distinguish.

In fact, beside these hardly distinguishable bipolar resistive switching behaviors, other types of switching including bipolar switching with opposite polarity, and even another unipolar resistive switching occur in a single switching sample. The defect configurations at the region of ruptured Magnéli (Ti$_4$O$_7$) conducting filaments, which are derived through the re-oxidation of filaments during the unipolar reset step may substantially contribute to the subsequent resistive switching behavior. In this study, characteristics of different types of switching that arise after unipolar reset step in a Pt/TiO$_2$/Pt device will be examined in accordance with the dominating defect configuration at the filaments-ruptured region. The results are compared with the switching properties of the sample with intentionally fabricated Pt/TiO$_2$/Ti$_4$O$_7$/Pt structure.

[1] Leon Chua, Appl. Phys. A: MATERIALS SCIENCE & PROCESSING. **102**, 4, 765-783 (2011)
[2] R, Waser et al., Adv. Mater. **21**, 25-26, 2632–2663 (2009)
[3] M.H. Lee et al., *Appl. Phys. Lett.* **96**, 152909 (2010)
[4] K.M. Kim et al., Nanotechnology **22**, 254010 (2011)

STUDY ON RESISTIVE SWITCHING OF BINARY OXIDE THIN FILMS USING SEMICONDUCTING $In_2Ga_2ZnO_7$ ELECTRODE

Jun Yeong Seok[1], Gun Hwan Kim[1], Seul Ji Song[1], Jung Ho Yoon[1], Kyung Jin Yoon[1], and Cheol Seong Hwang[1]

[1]WCU Hybrid Materials Program, Department of Materials Science and Engineering and Inter-university Semiconductor Research Center, Seoul National University, Seoul 151-744, Korea

It has been known through numerous studies that the resistive switching (RS) phenomena occur in several binary metal oxides, such as TiO_2, HfO_2, ZrO_2, NiO, using novel metal electrodes. It has been well known that the interface of an electrode/RS oxide plays a crucial role in determining the RS behavior of the memory cell. However, the influence of the semiconductor/RS oxide interface on the memory switching performance has not been reported in detail yet. Combining RS phenomena and accumulation/depletion behaviors of semiconductor could be an interesting research topic considering its diverse functionality. The recent development of amorphous oxide semiconductor, such as In-GaZnO$_x$ (IGZO) which has much higher carrier mobility than that of amorphous Si semiconductor, largely enhances the possible implementation of such combined devices. In this study, therefore, the authors explored the RS behaviors of several metal/IGZO/RS oxide/metal structures in conjunction with the conventional metal/RS oxide/metal structures.

First, the RS behaviors in a Pt/ IGZO/TiO_2/Pt sample were examined for their use in diode-free memory integration. The IGZO layer worked as the semiconductor layer, exhibiting the accumulation and depletion of carriers depending on the bias polarity. A unipolar resistive switching (URS) phenomenon was observed in the TiO_2 sample. The electroforming was possible only under the IGZO depletion condition due to the limited back ground leakage current flow. The repeated set/reset operation was also possible under the depletion condition of the IGZO layer. While the reset was possible, set was impeded by the high back ground current flow of the IGZO layer under the accumulation condition. Another notable merit is the much smaller (by ~ $10^3 - 10^4$ times) reset state current compared to that of a Pt/TiO_2/Pt sample.

As a second example, the bipolar resistive switching (BRS) behavior in a Pt/IGZO/HfO_2/TiN sample was examined and it was compared with the typical BRS shown in a Pt/HfO_2/TiN sample. Detailed switching mechanisms from these two typical RS systems involving semiconductor electrode will be discussed in the presentation.

BIPOLAR RESISTIVE SWITCHING BEHAVIORS OF PLASMA-ENHANCED ATOMIC LAYER DEPOSITED NiO FILMS ON TUNGSTEN SUBSTRATE

Seul Ji Song, Gun Hwan Kim, Jun Yeong Seok, Kyung Jean Yoon, Jung Ho Yoon and Cheol Seong Hwang*

WCU Hybrid Materials Program, Department of Materials Science and Engineering and Inter-university Semiconductor Research Center, Seoul National University, Seoul 151-744, Korea
*cheolsh@snu.ac.kr

Recently, the resistance switching (RS) phenomena in several MIM systems has been studied extensively for applications to the next generation non-volatile memory (NVM) devices. For the specific case of NiO, when the conducting segments make a connected network, the electrical current abruptly jumped to a high level, which is usually controlled to a compliance level to protect the sample from complete breakdown. In contrast, the current was dropped down when the conducting network was disassembled by mostly Joule heating effect.[1, 2] This RS phenomenon was closely correlated with a local non-stoichiometry in the oxide films. Besides, vertical integration is an indispensable factor for fabricating tera-bit scale RS memory, which requires atomic scale control of the film growth as well as highly conformal film morphology on extremely three-dimensional surface.

In this study, therefore, the NiO thin film was deposited using a saturation coverage of bis-methyl cyclopenta dienyl-nickel ($[MeCp]_2Ni$) precursor followed by its reaction with plasma-enhanced O_2 in ALD mode. The structure and electrical properties of the films grown on W substrate were examined in detail, although several other metal film substrates, such as Ru and Pt, were also adopted. This is because the W substrate resulted in the highest growth rate and the most promising electrical properties of NiO films as the RS element.

In the pristine state of W/40nm-thick-NiO/Pt devices, the leakage current was slightly asymmetric with respect to the polarities of voltage, and the leakage properties could be described by the Poole-Frenkel mechanism. Because the bond strength of W-O is very strong and the tungsten oxide is more stable than NiO, many trap sites, which are oxygen vacancies, would be generated at the W/NiO interface.[3] (See right figure below, XPS depth profile) It was recently reported that the conductivity of NiO can be modulated by the amount of oxygen vacancies, because the energy level of oxygen vacancy is not far from the conduction band edge of NiO.[4] Under the electric fields, an oxygen ion seems to be much more mobile compared with a nickel ion, considering their diffusion constants. Therefore, the RS

mechanism in this device could be rigorously explained by oxygen ion drift under electric field, even though NiO is a p-type semiconductor. Left figure below shows the bipolar switching I-V curves when the top Pt electrode was biased while the W bottom electrode was grounded. The performance of W/NiO/Pt stacked RS cells was stable even at the operation temperature of 100°C. The ratio between the bi-stable resistance states was about ~10, but both resistance states can be maintained over 10years as can be understood from extrapolating the retention measurement data. More detailed analysis result on the switching mechanism based on the conduction mechanism analysis of the pristine, set, and reset states will be discussed in conjunction with the structure and chemical properties of the films.

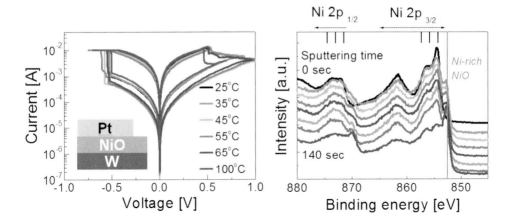

Figure. Typical bipolar RS behaviors of W/40nm-thick NiO/Pt devices at various temperatures and the depth profile of XPS spectra of Ni 2p. At the interface with W, a sudden emergence of Ni-rich NiO signal appears suggesting the chemical interaction between the W and NiO film.

[1] I. K. Yoo et. al., IEEE transactions on nanotechnology, 9, 131, 2010
[2] K. M. Kim et. al., Adv. Funct. Mater., 21, 1587, 2011
[3] CRC Handbook of Chemistry and Physics: A Ready-Reference Book of Chemical and Physical Data
[4] Keisuke Oka et. al., J. Am. Chem. Soc., 134, 2535, 2012

NONVOLATILE RESISTIVE SWITCHING IN Au/BiFeO$_3$ RECTIFYING JUNCTION

Yao Shuai,[1,2] Chuangui Wu,[2] Wanli Zhang,[2] Shengqiang Zhou,[1] Danilo Bürger,[1] Stefan Slesazeck,[3] Thomas Mikolajick,[3] Manfred Helm,[1] and Heidemarie Schmidt[1]

[1] Institute of Ion Beam Physics and Materials Research, Helmholtz-Zentrum Dresden-Rossendorf, P. O. Box 510119, Dresden 01314, Germany;
[2] State Key Laboratory of Electronic Thin Films and Integrated Devices, University of Electronic Science and Technology of China, Chengdu 610054, China;
[3] Namlab gGmbH, Nöthnitzer Strasse 64, 01187 Dresden, Germany

BiFeO$_3$ thin films have been grown on Pt/Ti/SiO$_2$/Si substrates with pulsed laser deposition. RF sputtered Au has been used for the top electrode. An interface-related resistive switching with an ON/OFF ratio of ~300 has been observed in the Au/BiFeO$_3$/Pt structure [1]. The different polarities of the external voltage induce an electron trapping or detrapping process, and consequently change the depletion layer width below the Au Schottky contact, which is revealed by capacitance-voltage measurement. The resistive switching shows a long-term retention and non-destructive read-out character. A dynamic equilibrium process involving the extension of the depletion region can be used to explain the good retention in the Au/BiFeO$_3$/Pt structure. The present work can help to further understand the physical origin of bipolar switching in BiFeO$_3$ and in other thin film oxides with electron trapping centers.

[1] Y. Shuai et al., J. Appl. Phys., 109, 124117 (2011).

DIFFERENT BEHAVIOUR SEEN IN FLEXIBLE TITANIUM DIOXIDE SOL-GEL MEMRISTORS DEPENDENT ON THE CHOICE OF ELECTRODE MATERIAL

Ella Gale[1], David Pearson[2], Stephen Kitson[2], Andrew Adamatzky[1] and Ben de Lacy Costello[1]

[1]Unconventional Computing Group, University of the West of England, Bristol, Avon, UK;
[2]Hewlett-Packard UK, Bristol, Avon, UK

The archetypal memristor was made via atomic deposition of TiO_2 between platinum electrodes [1]. The announcement of flexible solution-processed titanium dioxide memristor was greeted with much excitement as it made it easier to fabricate working memristors [2]. This device consisted of a spun-on sol-gel layer between two aluminium electrodes. Despite several examples in the ReRAM literature of aluminium oxide playing an essential role in resistance switching [3], even when it was just via aluminium electrodes [4,5,6,7], the sol-gel TiO_2 memristor paper did not report a detailed test comparing aluminium electrodes to noble metal electrodes and instead simply stated that the switching was not attributable to the aluminium electrodes as the memristors still switched with noble metal (Au, Pt) contacts.

We have undertaken a study of the effect of changing the electrode metal on TiO_2 sol-gel memristors prepared as in [2] to elucidate possible methods for control of device characteristics. We found that Au-TiO_2-Au memristors did still switch, but in a fundamentally different way. These devices switched in a fuse-like manner, but did not switch back the way that Al-TiO_2-Al devices did (as reported in both [2] and below). The low resistance state, LRS, and first high resistance state, HRS, currents were separated by 5 orders of magnitude and were ohmic, making these devices useful for Write-Once Read Many (times), WORM, memory. There was also a second low resistance separated by an order of magnitude. As all resistance states in these devices were ohmic, thus these devices are best described as resistance switching memory rather than memristors.

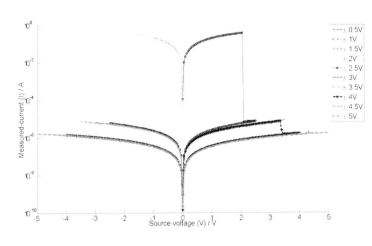

Figure 1: Repeated I-V curves done over an increasing voltage range. Au-TiO2-Au sol-gel resistance switching memory acts like a fuse and could be useful for WORM memory.

We also investigated Al-TiO_2-Al memristors. We found high current triangular memristor profiles that resembled those reported in [2]. The LRS was ohmic and in the mA range and never reached the 10^{-2}A range which the LRS of the Au-TiO_2-Au memristor did (in a comparison of virgin runs of the device between ±3V). These devices had non-linear HRS in the same range as the curved bipolar switching below.

We also found non-linear memristor I-V curves in these devices which resembled both Chua's memristor theory[8] and also ReRAM bipolar resistance switching (they look like a centrosymmetric version of figure 2.). Devices with this mode of operation were higher resistance, (most in the 10^{-6} to 10^{-4} A range over the voltage range tested) and very reproducible. Those with triangular memristor profiles were less reproducible, they had the same behaviour, but the switching voltage was not constant.

Mixed electrode devices, Al-TiO$_2$-Au, (gold as the top electrode) were also fabricated. These showed unidirectional memristance curves, ie. memristor behaviour was only seen when the aluminium electrode was negative with respect to the gold electrode. When the gold electrode was negative the hysteresis was at least 3 orders of magnitude smaller and there was significant deformation of the gold electrode caused by oxygen evolution from the TiO$_2$ gel layer. These devices lacked reproducibility.

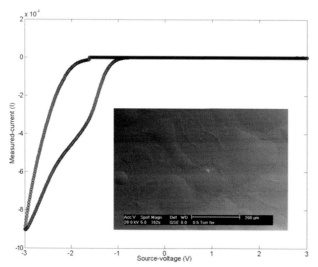

Figure 2: Unidirectional memristance seen in an Al-TiO$_2$-Au sol-gel 'half-memristor'. We see memristance on the negative side of the curve as the aluminium electrode was the source. Hysteresis on the right hand side of the curve has been suppressed. The current scale is in 10^{-4} A, which is towards the top end for the bipolar switching-like memristance seen in these devices. Gold bubbling of the top electrode happened at +1.5V. Insert shows disruption of the gold top electrode via oxygen gas evolution.

Therefore, whilst the spun-coated [2] and HP memristor [1] are both based on TiO$_2$ the mode of operation is subtly different and we have shown the electrode material plays a significant role in the case of sol-gel devices. We expect this work to enable greater control of device characteristics.

[1] D. B. Strukov et al, Nature, **453**, 80-83 (2008).
[2] N. Gergel-Hackett et al, IEEE Electron Device Lett., **30**, 706-708 (2009).
[3] S. Kim et al, Appl. Phys. Lett., **92**, 223508 (2008).
[4] B. J. Choi et al, J. Appl. Phys., **98**, 033715, (2005).
[5] H. Y. Jeong et al, Appl. Phys. Lett., **95**, 162108 (2009).
[6] T. Kever et al, Appl. Phys. Lett., **92**, 083506 (2007).
[7] M. Colle et al, Organic Elec., 7, 305-312 (2006).
[8] L. O. Chua, IEEE Trans. Circuit Theory, **18**, 507-519, (1971).

MEMRISTIVE COGNITIVE COMPUTING

Eero Lehtonen[1], **Jussi Poikonen**[2], **Mika Laiho**[1], **Pentti Kanerva**[3]

[1]Microelectronics Laboratory, University of Turku, Turku, Finland;
[2]Department of Communications and Networking, Aalto University, Espoo, Finland; [3]Redwood Center for Theoretical Neuroscience, University of California, Berkeley, U.S.A

We explore the development of a memristive neuromorphic cognitive computing architecture which is organized into a memory and an arithmetic/logic processor. For the memory we envisage an associative memory and discuss its implementation with memristors in [1]. In software or pure CMOS hardware implementations, associative memory architectures suffer from relatively small storage capacity. On the other hand, memristive technology offers unforeseen density in memory implementations [2], and has been proposed to be used to implement synaptic connections between CMOS neurons within the CMOL neuromorphic architecture [3]. For example, using CMOS/memristor hybrid circuits with feature size F = 50 nm, it is possible to attain a synaptic density of 10^{10} synaptic connections per square centimeter, which is comparable to the density of the human cortex [4]. Consequently, memristive technology seems suitable for realizing associative memories of biologically relevant scale for the first time.

In [1] we proposed memristive implementations of various associative memory architectures, such as the Auto-associative Content-Addressable Memory (ACAM) [5], the Willshaw memory [6], and the Sparse Distributed Memory (SDM) [7]. The first two architectures can be implemented as a CMOS-driven memristor crossbar consisting of binary memristors, while SDM requires an additional crossbar of analog memristors. The CMOL architecture implementing the ACAM and the Willshaw memory is depicted in Figure 1.

In addition to memory, a conventional computer has an arithmetic/logic unit (ALU) for creating new representations from existing ones. Likewise, a cognitive computer needs operations that produce new representations from existing ones, as discussed in the proposal for hyperdimensional computing by Kanerva [8]. The main idea of this computing paradigm is to create associations and other data structures between initially uncorrelated data vectors by applying arithmetic operations. For example, a set of hyperdimensional vectors can be represented by a single vector obtained as a thresholded bitwise sum of the constituent vectors. Other data structures obtained through this arithmetic are, for example, sequences, content-dependent associations, and key-value pairs. For hyperdimensional computing to work, it is crucial that the length of the vectors is large – of the order of thousands of bits – which in turn necessitate the use of large associative memories.

The implementation of hyperdimensional computing requires additional logic at the CMOS neurons, and therefore reduces the neuronal density of the architecture. A possible solution to this problem is to move part of the computation to the memristive crossbar by using memristive implication logic [9] to perform the arithmetic operations.

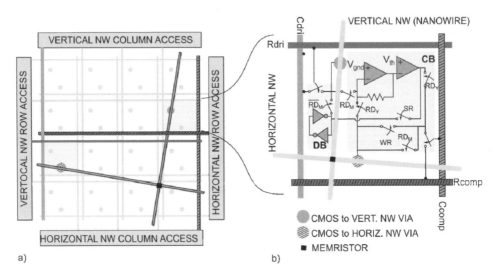

Figure 1: Schematic of a CMOL circuit implementing the ACAM and the Willshaw memories. Left: High-level view of the memory architecture. Two CMOS neurons are highlighted in gray. Their synaptic memristive connection is at the crossing of the horizontal and vertical nanowires. In reality, all CMOS cells are interfaced with a horizontal and a vertical nanowire. Right: CMOS neuron implementing a threshold-type perceptron. The analog OP-AMP components compute the thresholded weighted sum of the inputs, while the switches and the latch are used to configure the neuron into different phases of operation.

We have previously proposed several methods of efficiently implementing Boolean logic using memristors [10, 11, 12], and consider incorporating memristive logic operations into large associative memory structures to be a promising option in large-scale neuromorphic circuit design.

We propose a poster on cognitive computing, as a part of a Nanosession on memristive neuromorphic hardware. In the poster lecture we will describe in detail the memristive associative memory architectures discussed above, and give a tutorial on memristive hyperdimensional computing.

[1] E. Lehtonen et al., "Memristive Associative Memories", submitted to IEEE Trans. on Very Large Scale Integrated Systems, (2012)
[2] D. B. Strukov and R. S. Williams, PNAS **106**, 48, pp. 20155-20158 (2009)
[3] K. Likharev et al., Annals of the New York Academy of Sciences, **1006**, pp. 146-163 (2003)
[4] G. Sinder et al., Computer, **44**, 2, pp. 21-28 (2011)
[5] E. B. Baum et al., Biological Cybernetics **59**, pp. 217-228 (1988)
[6] D. J. Willshaw et al, Nature **222**, pp. 960-962 (1969)
[7] P. Kanerva, Sparse Distributed Memory, MIT Press, (1988)
[8] P. Kanerva, Cognitive Computation **1**, 2, pp. 139-159 (2009)
[9] J. Borghetti et al., Nature **464**, pp. 873-876 (2010)
[10] E. Lehtonen et al., IET Electron. Lett., **46**, pp. 230-231 (2010)
[11] J. Poikonen et al., "On synthesis of Boolean expressions for memristive devices using sequential implication logic", accepted for publication in IEEE Trans. on Computer-Aided Design of Integrated Circuits and Systems (2012)
[12] E. Lehtonen et al., "Implication logic synthesis methods for memristors", accepted for publication in Proc. IEEE International Symposium on Circuits and Systems (2012)

SYNAPTIC PLASTICITY OF ELECTROCHEMICAL CAPACITORS BASED ON TiO_2

Hyungkwang Lim[1,2], Ho-Won Jang[1], Cheol Seong Hwang[2], Doo Seok Jeong[1]

[1]Electronic Materials Research Centre, Korea Institute of Science and Technology, Seoul, Republic of Korea;
[2]WCU Hybrid Materials Program, Department of Materials Science and Engineering, and Inter-university Semiconductor Research Centre, Seoul National University, Seoul, Republic of Korea

In animal brains, the interaction between neighboring neurons via synapses is defined as synaptic weight or strength. The plasticity of synaptic weight, relying on the interaction, plays a key role in learning, which is referred to as the Hebbian learning rule.[1] On the other hand, there have been efforts to find non-biological concepts corresponding to synaptic weight and plasticity in order to achieve artificial synapses based on inorganic materials. The eventual goal of these efforts is to achieve inorganic artificial synapses integrated onto semiconductor devices using semiconductor device fabrication processes. The methodology of investigation on artificial synapses is twofold: (i) building synapse-like functioning circuits using conventional passive and active circuit elements [2], and (ii) utilizing new physical concepts able to mimic synaptic functions, for instance, ferroelectric [3] and phase-change behaviors [4].

A recent sensation of resistive random access memories (RRAMs), utilizing current versus voltage (I-V) hysteresis in mostly transition metal oxides (TMOs), has brought about great interest in the application of these TMOs to artificial synapses.[5] In this case, the resistance of the TMOs and the change of the resistance mimic synaptic weight and synaptic plasticity, respectively.

In this study, we introduce electrochemical-capacitor-type artificial synapses based on TiO_2. The idea is that a redox (faradaic) reaction at the interface between the electrode and the TiO_2 layer of an electrochemical capacitor (EC) is considered to lead to the change of the EC's resistance. Moreover, by controlling the overpotential of the redox reaction analog-type resistance changes can be realized.

The 50 nm thick TiO_2 layer in the ECs was deposited using reactive sputtering at room temperature. And the electrodes were grown using electron-beam evaporation also at room temperature. The electrical characterization of the fabricated ECs was conducted using a Keithley 236 Source Measure Unit and a CHI700 potentiostat. I-V loops of a TiO_2-based EC are plotted in Fig. 1. The loops represent the typical I-V behavior of resistive switching. In Fig. 1, it can be noticed that the positive and the negative voltage are in charge of the resistance's increase (depression) and decrease (potentiation), respectively. However, the resistive switching in this EC obviously differs from filamentary-type resistive switching behavior, most commonly observed in most of resistive switching systems, for the following reasons: (i) the switching in the EC is activated even without an initialization process, often termed "electroforming", and (ii) the resistance of ECs with different pad-sizes nicely scales with the pad-size, implying no filamentary-type switching.

By means of the application of fast rise-time square voltage pulses, changes in the resistance (synaptic weight) of the EC were also successfully measured as shown in Fig. 2. Figure 2(a) represents a gradual decrease in the resistance (potentiation) with regard to ten consecutive voltage pulses whose height was -3 V. A gradual increase in the resistance (depression) was also observed with regard to ten consecutive voltage pulses of which height was 3 V as plotted in Fig. 2(b).

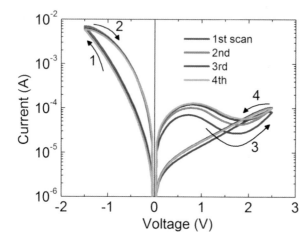

Figure 1: Four consecutive I-V loops of a TiO$_2$-based EC, measured using a Keithley 236 Source Measure Unit. The numbers from 1 to 4 denote the sequence of the hysteresis loops.

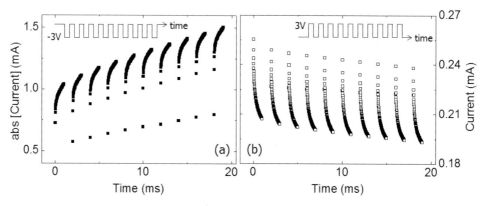

Figure 2: (a) Potentiation of the EC with regard to ten consecutive square pulses (-3 V). The width of each pulse was 1 ms and the time interval between consecutive pulses was also 1 ms. (b) Depression of the same EC with regard to ten consecutive square pulses with the opposite polarity (3 V) and the same width and time interval.

[1] D. O. Hebb, *The organization of behavior*, (Wiley & Sons, New York, 1949).
[2] J. G. Elias and W. T. Rogers, IEEE International Joint Conference on Neural Networks 2490 (1991).
[3] Y. Watanabe, Jpn. Patent Application No. H7-263646 (1995).
[4] M. Suri et al., 2011 IEEE Technical Digest International Electron Devices Meeting 4.4.1 (2011)
[5] G. S. Snider, Nanotechnol. **18**, 365202 (2007).

SIMULATION OF ASYMMETRIC RESISTIVE SWITCHING IN ELECTROCHEMICAL METALLIZATION MEMORY CELLS

Stephan Menzel[1], Ulrich Böttger[1], Rainer Waser[1,2]

[1]Institut für Werkstoffe der Elektrotechnik 2, RWTH Aachen University, 52074 Aachen, Germany
[2]Peter Grünberg Institut (PGI-7), Forschungszentrum Jülich, 52425 Jülich, Germany

Electrochemical metallization memory (ECM) cells are a promising candidate for future non-volatile memory [1]. They combine ultrafast switching, with the possibility of programming different resistance states by adjusting the current compliance [2]. ECM cells typically consist of a Cu or Ag active electrode and ion conducting insulating layer and an inert counter electrode. The resistive switching mechanism is attributed to the electrochemical dissolution and growth of a Cu or Ag filament within the insulating layer. During the SET process a positive voltage is applied to the active electrode. As a result this electrode is oxidized and Cu or Ag cations are driven into the insulating layer. Due to the electric field they migrate towards the inert cathode, where an electrochemical reduction takes place. Subsequently, a metallic filament forms which grows from the inert electrode towards the anode until an electronic contact is achieved. To RESET the cell the voltage polarity is reversed and the filament dissolves.

Based on this switching mechanism we developed a dynamic simulation model for multilevel switching in ECM cells [3]. The origin of multilevel switching is proposed to be direct tunneling between the growing filament and the active electrode. It is shown that the different low resistive states (LRS) are achieved by modulation of the corresponding tunneling gap. The nonlinear switching kinetics is modeled by the electron transfer reaction, which occurs at the electrode/ insulator interfaces. Mathematically, the electron transfer reaction is modeled by the famous Butler-Volmer equation [4]:

$$J_{Me^{z+}} = J_{BV}(\eta) = j_0 \left\{ \exp\left(\frac{(1-\alpha)ez}{kT}\eta\right) - \exp\left(-\frac{\alpha ez}{kT}\eta\right) \right\} \quad (1)$$

Here, j_0 is the exchange current density, α the charge transfer coefficient, z the charge transfer number and η the overpotential. In this study we investigate the influence of the charge transfer coefficient on the resistive switching. Figure 1a shows the simulated *IV*-characteristics for a charge transfer coefficient of $\alpha = 0.1$ and an additional geometrical asymmetry. Apparently, this leads to an asymmetry in the switching currents and voltages. Such an asymmetry is also observed in experiment as illustrated in (b). This asymmetry is further studied by suitable simulations. In addition, the impact on the multilevel switching and the RESET to SET current ratio is studied and compared to experimental data.

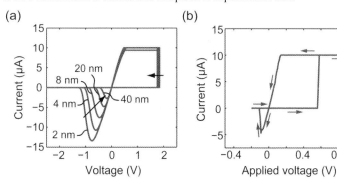

Figure 1: (a) Simulated *IV* characteristics for a charge transfer coefficient of $\alpha = 0.1$ and different radii of the active electrode. The filament radius is 2 nm in all simulations. (b) Experimental IV characteristic.

[1] R. Waser, M. Aono, Nat. Mater., 6 (2007) 833.
[2] I. Valov, R. Waser, J. R. Jameson, M. N. Kozicki, Nanotechnology, 22 (2011) 254003/1.
[3] S. Menzel, U. Böttger, R. Waser, J. Appl. Phys., 111 (2012) 014501.
[4] C. H. Hamann, A. Hamnett, W. Vielstich, Wiley-VCH, Weinheim (2007).

A V_2O_5-BASED RESISTANCE RANDOM ACCESS MEMORY AND IMPROVEMENT OF SWITCHING CHARACTERISTICS BY EMBEDDING A THIN VO_2 INTERFACE LAYER

Xun Cao[1,2], Meng Jiang[1], Feng Zhang[2], Xinjun Liu[3], and Ping Jin[1,2,4]*

[1]Research Center for Industrial Ceramics, Shanghai Institute of Ceramics, Chinese Academy of Sciences, Shanghai 200050, China;
[2]State Key Laboratory of High performance Ceramics and Superfine Microstructures, Shanghai Institute of Ceramics, Chinese Academy of Sciences, Shanghai 200050, China;
[3]School of Materials Science and Engineering, Gwangju Institute of Science and Technology, Gwangju, 500-712, South Korea
[4]National Institute of Advanced Industrial Science&Technology (AIST), Shimoshidami, Moriyama-ku, Nagoya 463-8560, Japan

Resistance random access memory (RRAM) involving a two-terminal resistance device has been recommended as one of the most promising candidates for next-generation nonvolatile memory. A phase-mixed vanadium oxide (VO_x) has been considered to be feasible for such RRAM devices. Although a V_2O_5 single phase which is easy to form and with comparative stable properties comparing to a VO_x mixture, should possibly be a better candidate, however, there is hardly any reports on the use of a V_2O_5 single phase film for such applications.

In this work, resistive switching behaviors of V_2O_5 polycrystalline films of the as-deposited and the annealed have been investigated. Thin films of V_2O_5 were deposited on Pt substrate by dc reactive magnetron sputtering. The films were characterized by X-ray diffraction (XRD) and atomic force microscope (AFM). RRAM devices were formed by depositing an Al top electrode on V_2O_5 film, which present reliable and reproducible switching behaviors. In addition, for an improvement of switching properties of the device, a thin layer of VO_2 was introduced between V_2O_5 and Pt, and it was found that the device with VO_2 exhibits much better resistive hysteresis and switching uniformity. A combined filamentary conduction model was proposed to clarify the role of VO_2 layer on the resistive switching stabilization during the forming process.

METAL–INSULATOR TRANSITION OF ALD VO₂ THIN FILMS FOR PHASE TRANSITION SWITCHING

Kai Zhang[1,3], Madhavi Tangirala[1,3], Pragya Shrestha[1,2,*], Helmut Baumgart[1,3], Salinporn Kittiwatanakul[4], Jiwei Lu[4], Stuart Wolf[4], Venkateswara Pallem[5] and Christian Dussarrat[5]

[1]Department of Electrical and Computer Engineering, Old Dominion University, Norfolk, VA 23529, USA;
[2]Semiconductor and Dimensional Metrology Division, National Institute of Standards and Technology (NIST), Gaithersburg, MD 20899, USA
[3]The Applied Research Center at Thomas Jefferson National Accelerator Facility, Newport News, VA 23606, USA
[4]Department of Materials Science and Engineering, University of Virginia, Charlottesville,Virginia 22904-4745, USA
[5]American Air Liquide, Delaware Research &Technology Center, 200 GBC Drive, Newark, DE 19702, USA
*pragya.shrestha@nist.gov

Among the vanadium oxides, VO_2 has attracted a lot of attention due to its remarkable metal-insulator transition (MIT) or semiconductor-metal transition (SMT) behavior. Recently memristive devices have gained popularity in the field of memory due to its simplicity, small size and low cost [1, 2]. Due to similar reasons these devices also have the potential to work as high performance switches for interconnect architecture. Despite all the ongoing research the actual mechanism behind the switching is yet to be clearly understood. Vanadium dioxide (VO_2), which is a thermochromic material that exhibits a metal-to-semiconductor transition near ~68 °C, has experienced increasing interest due to its technologically useful phase transitions (Zylbersztein A. and Mott N.F. 1975) [2]. This reversible metal-to-semiconductor transition is accompanied by a transformation in crystallographic structure from a monoclinic structure VO_2 (M) below the transition temperature to a tetragonal rutile structure VO_2 (R) above the transition temperature. The interesting effect of this crystallographic transition is the fact that it exhibits a large change in electrical resistivity. This renders the material potential candidate for novel phase transition switches in microelectronics and sensor applications. In addition to its change in resistivity at the phase transition its infrared transmission also changes significantly at the transition temperature rendering this material a good candidate for thermal sensing or optical switching. Most of the film growth of VO_2 has been performed by magnetron sputtering, (MO-CVD), Reactive Bias Target Ion Beam Deposition (RBTIBD) [3], Pulsed Laser Deposition (PLD) and other PVP techniques. Very little is known about ALD of VO_2. In this work, a novel metal-organic ALD precursor, Tetrakis[ethylmethylamino] vanadium {V(NEtMe)₄} [TEMAV], was employed as vanadium precursor source to develop an ALD process for the synthesis of stoichiometric VO_2 films. TEMAV and H_2O were employed as vanadium precursor and oxidizing agent. Generally 20 sccm N_2 was used as a carrier gas for the precursors. The growth temperature was set at 150°C. However, VO_2 thin films obtained by ALD are amorphous, since the growth temperature is lower than the crystallizing temperature. Therefore, post ALD deposition thermal heat treatment has to be used to produce the technologically important polycrystalline stoichiometric VO_2 structure.

Figure 1: AFM analysis of the surface morphology of ALD VO_2 thin film annealed at 450°C yielding an rms roughness of 20 nm.

[1] G. Rampelberg, M. Schaekers, K. Martens, Q. Xie, D. Deduytsche, B. De Schutter, N. Blasco, J. Kittl, and C. Detavernier, Appl. Phys. Lett. 98, (2011), p.162902.
[2] Pritesh Dagur, Anil U. Mane, S.A. Shivashankar, J. Crys. Grow. 275, (2005), p, e1223.
[3] M. A. Mamun, S. Kittiwatanakul, K. Zhang, H. Baumgart, Jiwei Lu, S. Wolf, and A. A. Elmustafa, Supplemental Proceedings: Volume 2: Materials Properties, Characterization, and Modeling (The Minerals, Metals & Materials Society), 2012, p. 753-758

SWITCHING AND LEARNING IN Ni-DOPED GRAPHENE OXIDE THIN FILMS

S. Pinto[1], R. Krishna[2], C. Dias[1], G. Pimentel[1], G. N. P. Oliveira[1], J. M. Teixeira[1], P. Aguiar[3], E. Titus[2], J. Gracio[2], J. Ventura[1], and J. P. Araujo[1]

[1]IFIMUP and IN-Institute of Nanoscience and Nanotechnology, Porto, Portugal;
[2]Nanotechnology Research Division, Aveiro, Portugal
[3]Faculdade de Ciencias da Universidade do Porto, Porto, Portugal

Traditional memory technologies are rapidly approaching miniaturization limits. As an alternative, resistive random access memory (RRAM) relying on the resistive switching (RS) effect occurring in metal-insulator-metal (MIM) cells [1], has attracted scientific and commercial interests as a promising next generation nonvolatile memory [2]. A link between MIM structures and biological synapses was also recently reported, due to the equivalent dynamical reconfiguration in response to inputs [3]. On the other hand, graphene-based materials are also receiving attention due to their large variety of applications. Particularly, graphene oxide (GO), consisting of a single layer of graphene bounded to oxygen in the form of carboxyl, hydroxyl or epoxy groups, is one of the various materials showing RS [4].

Here, we studied two GO-based-devices grown on Si/SiO2 substrates prepared using different methods [polymeric hydrogel protonated with 200 μl HCl (sample 1) or with 0.5 ml N_2H_4 (sample 2)] and probed the influence of the electrodes (W, Cu) on the resistive switching properties. The two samples showed opposite switching polarities, indicating that both the diffusion of metallic ions and oxygen groups can be responsible for RS depending on the fabrication process. We further demonstrate the learning capability of GO-based structures.

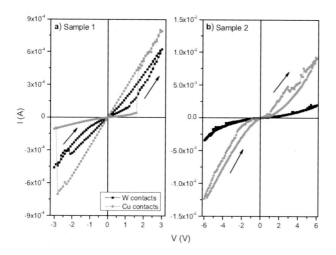

Figure 1: I-V characteristics of (a) sample 1 and (b) sample 2, with W (circles) and Cu (squares) contacts. The arrows indicate the switching polarity.

The I-V characteristics of the two samples are shown in Fig. 1. Substantial hysteresis was found in sample 1 for both W and Cu contacts. With increasing positive voltage, the W-GO-W device gradually switches from the high resistance state (HRS) to a low resistance state (LRS). By sweeping the voltage to negative values, the device recovers the HRS. The use of Cu contacts led to clear sharp resistance variations at the well-defined Set (Reset) voltage of 1.6 V (-2.8 V). Sample 2 showed much smaller hysteresis. Also, the switching polarity of sample 2 is opposite to that of sample 1, a characteristic that is maintained regardless of the deposited contacts on the surface of the samples.

The switching mechanism on graphene oxide based devices is usually attributed to one of two mechanisms. In the first [4], RS originates from the formation and rupture of conductive filaments in an insulating matrix. The switching dependence of sample 1 on the electrode material can then be understood by the formation/rupture of a metallic filament in the GO layer due to the diffusion of metallic ions under a bias voltage. The second mechanism usually mentioned [5] correlates switching with the dependence of the physical properties of GO on bounded oxygen groups. The reversal of the switching polarity of sample 2 shows that its dominant RS mechanism is related with the migration of negative oxygen groups. Both mechanisms however lead to an Ohmic conductance in the low resistance state, indicating the formation of a percolative metallic path connecting the two electrodes.

We also conducted a preliminary study on the learning abilities of graphene-based devices. We observed that consecutive positive (negative) voltage sweeps applied in sample 1 lead to a continuous decrease (increase) of the sample conductance, similarly to unlearning (learning; not shown) [3]. The continuous increase of the current towards a constant value indicates the onset of the long-term-memory (LTM) stage of the device.

In summary, we investigated resistive switching in two Ni-doped graphene devices. We observed that the studied samples have different switching polarities, emphasizing the importance of the fabrication method on device behavior. Resistive switching in graphene oxide structures can then be caused by the diffusion of oxygen or metallic ions from the electrode under a bias voltage. The learning capacity of graphene oxide was also demonstrated, showing that it may be a promising material for future applications in neuromorphic systems.

[1] R. Waser, R. Dittmann, G. Staikov, and K. Szot, Adv. Mater. 21, 2632 (2009).
[2] M.-J. Lee, C. B. Lee, D. Lee, S. R. Lee, M. Chang, J. H. Hur, Y.-B. Kim, C.-J. Kim, D. H. Seo, S. Seo, et al., Nat Mater 10, 625630 (2011).
[3] T. Hasegawa, T. Ohno, K. Terabe, T. Tsuruoka, T. Nakayama, J. K. Gimzewski, and M. Aono, Adv. Mat. 22, 1831 (2010).
[4] F. Zhuge, B. Hu, C. He, X. Zhou, Z. Liu, and R.-W. Li, Carbon 49, 3796 (2011).
[5] O. Ekiz, M. rel, H. Gner, A. Mizrak, and A. Dna, ACS Nano 5, 2475 (2011).

CHARGE CARRIER-MEDIATED FERROMAGNETISM IN $FeSb_{2-x}Sn_xSe_4$

<u>Honore Djieutedjeu</u>[1]; Kulugammana G.S. Ranmohotti[1]; Nathan J. Takas[1]; Julien P. A. Makongo[1]; Xiaoyuan Zhou[2]; Ctirad Uher[2]; N. Haldolaarachchige[3]; D.P. Young[3]; Pierre F. P. Poudeu[1,*]

1) Department of Material Sciences Engineering, University of Michigan, Ann Arbor, MI, 48109, USA
2) Department of Physics, University of Michigan, Ann Arbor, MI, 48109, USA
3) Department of Physics, Louisiana State University, Baton Rouge, LA, 70803-4001, USA
* ppoudeup@umich.edu (PFPP)

The ability to create and manipulate in a cooperative manner within the same crystal lattice, ferromagnetism and semiconductivity, two properties generally difficulty to combine in a convention inorganic compound, is very attractive and might result in new physical phenomena and novel applications such as the spintronic technologies, which utilize both the charge and spin of electrons [1]. Recently, we reported that $FeSb_2Se_4$ exhibits ferromagnetism and semiconductity above 300K and undergoes structural-distortion-driven cooperative magnetic and semiconductor-to-insulator (SI) transitions upon cooling below 130K [2]. We speculated that the observed anisotropic structural distortion is controlled by the stereoactivity of the Sb lone pair oriented along the c-axis. Here, we probe simultaneously through partial Sn substitution at Sb sites, the effect of Sb lone pair on the SI transition temperature as well as how the coupling between magnetic centers within the structure are affected by the change in charge carrier concentration. We find that the SI transition temperature gradually decreases (from 184K to 51K) with increasing Sn concentration (x= 0.01 to 0.25) suggesting decreasing effect of the lone pair due to Sn substitution at Sb sites. For composition with x = 0.15, the SI transition observed at 130K is reversed at ~60K indicating that the stereo-activity of the Sb lone pair maybe vanishing at very low temperatures. Regardless of the temperature, the magnetic susceptibility (ZFC and FC) monotonically increase with Sn concentration reaching peak values at x = 0.2 and decrease with further increase in Sn concentration. The coercivity and the saturation magnetization were also found to be strongly dependent on the Sn concentration. We attempt to correlate the observed alteration of the magnetism and conductivity in $FeSb_{2-x}Sn_xSe_4$ to the variation of charge carrier concentration and mobility upon Sn^{2+} substitution at Sb^{3+} sites.

[1] H. Ohno *et al. Appl. Phys. Lett.* **69**, 363 – 365 (1996).
[2] H. Djieutedjeu et al. *Angew. Chem. Int. Ed.* **49**, 9977 – 9981 (2010).

CHARACTERIZATION OF THE ATOMIC INTERFACE IN HIGH-QUALITY Fe_3O_4/ZnO HETEROSTRUCTURES

O. Kirilmaz[1], M. Paul[1], A. Müller[1], D. Kufer[1], M. Sing[1], R. Claessen[1], S. Brück[1,2,3], C. Praetorius[1], K. Fauth[1], M. Kamp[4], P. Audehm[5], E. Goering[5], J. Verbeeck[6], H. Tian[6], G. Van Tendeloo[6], N.J.C. Ingle[7], M. Przybylski[8], M. Gorgoi[9]

[1]Experimentelle Physik 4, Universität Würzburg, Germany; [2]University of New South Wales, School of Physics, Sydney, Australia; [3]Australian Nuclear Science and Technology Organization, Menai, Australia; [4]Technische Physik, Universität Würzburg, Germany; [5]Max Planck Institute for Intelligent Systems, Stuttgart, Germany; [6]Electron Microscopy for Materials Science, University of Antwerp, Belgium; [7]Advanced Materials and Process Engineering Laboratory, University of British Columbia, Vancouver, Canada; [8]Faculty of Physics and Applied Computer Science, AGH University of Science and Technology, Krakow, Poland; [9]Helmholtz Zentrum Berlin (BESSY II), Berlin, Germany

Magnetite (Fe_3O_4) is ranked among the most promising materials to use as a spin injector into a semiconducting host. We demonstrate epitaxial growth of Fe_3O_4 films on ZnO which presents a further step towards incorporation of magnetic materials into semiconductor technology. Regarding the volume properties of the films [1, 2], X-ray photoelectron spectroscopy evidences that the iron-oxide is phase-pure and nearly stoichiometric magnetite. Diffraction measurements indicate highly oriented epitaxy and almost complete structural relaxation. The magnetic behavior shows a rather slow approach to saturation at high fields in comparison with bulk crystals.

We also have investigated the growth mechanism, the surface structure, and the magnetic interface properties in detail [3]. In scanning transmission electron microscopy (STEM) we observe an atomically sharp interface between Fe_3O_4 and ZnO (Fig. 1). The chemical and the magnetic depth profile of the Fe_3O_4 film near the interface was determined by x-ray resonant magnetic reflectometry (XRMR) and spatially resolved electron energy loss spectroscopy in a scanning transmission electron microscope (STEM-EELS). It was found that only the first Fe layer in Fe_3O_4 on ZnO shows a modification of its valence and thus also of its magnetic properties [4]. While Fe_3O_4 is a mixed valence compound with octahedral $Fe^{2.5+}$ and tetrahedral Fe^{3+} ions, the interface layer contains only Fe^{3+} ions (Fig. 2). Although this 2.5 Angstrom wide region at the interface is very small compared to the spin diffusion length in Fe_3O_4, its presence could nevertheless affect the resistivity matching between the two materials or cause an enhanced spin-scattering at the interface.

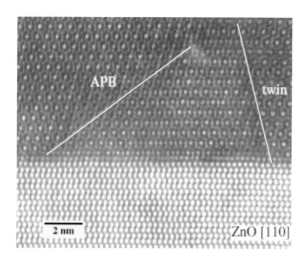

Figure 1: STEM micrograph with marked boundaries [2].

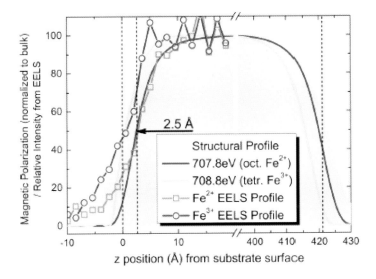

Figure 2: Magnetic and chemical profiles derived from XRMR [4].

[1] A. Müller et al., Thin Solid Films **520**, 368-373 (2011).
[2] M. Paul et al., Appl. Phys. Lett. **98**, 012512 (2011).
[3] M. Paul et al., J. Appl. Phys. **110**, 073519 (2011).
[4] S. Brück et al., Appl. Phys. Lett. **100**, 081603 (2012).

ULTRATHIN Fe OXIDES FILMS: STRUCTURAL, ELECTRONIC AND MAGNETIC PROPERTIES UNDER REDUCED DIMENSIONS

Bernal, Iván[1], Gallego, Silvia[1]

[1]Materials Science Institute of Madrid, Spanish National Research Council, Madrid, Spain

Fe oxides present a rich phase diagram, where different structural orders, chemical compositions and magnetic couplings can be found. At the origin we find a subtle interplay of microscopic interactions, involving electron correlation effects and complex magnetic exchanges. The situation complicates further under low dimensions: both the reduction of film thickness and the presence of interfaces add degrees of freedom which may alter the delicate energy balance governing the structural and magnetic phase transitions.

Here we will focus on FeO. Despite its apparent simplicity, the bulk form is subject to instabilities leading to different metal-insulator, magnetic and structural transitions, whose inter-relation is still the subject of debate. The simplest structure at ambient conditions adopts a non-stoichiometric rock-salt lattice which orders antiferromagnetically along the [111] direction at low temperatures. Ultrathin FeO films with (111) orientation have been grown on several metal substrates, though they transform into other Fe oxide forms (typically magnetite or maghemite) for thicknesses over 2-3 layers. They offer thus a fascinating scenario to study the parameters governing the phase diagram, and provide the opportunity to control it profiting from the additional degrees of freedom under reduced dimensions.

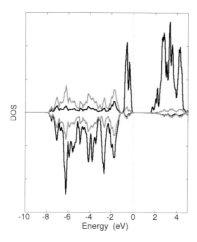
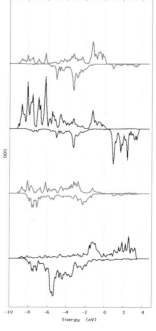

Figure 1: Spin-polarized density of states (DOS) of bulk FeO and an ultrathin film 2 monolayers thick on Ru(0001). Contributions from O (Fe) atoms correspond to the red (black) curves. Positive (negative) DOS refers to minority (majority) spin polarization.

Here we explore the stability and magnetic order of ultrathin FeO films grown on Ru(0001), based on first-principles calculations within the DFT+U (density functional theory including a Hubbard U term) formalism. We determine the influence of different factors in the structural, electronic and magnetic properties: interaction with the substrate, chemical composition, thickness, etc. We address the role of correlation effects and the long-range magnetic order. Finally, we explore the evolution of these films into other Fe oxide forms, i.e. magnetite or maghemite, evaluating the factors that determine the transformation between phases, how to distinghuish between them, and comparing to experimental results.

STRUCTURAL DEPENDENCE OF MAGNETIC PROPERTIES IN TWO DIMENSIONAL NICKEL NANOSTRIPS: MAGNETIC ANOMALY AND MAGNETIC TRANSITION IN NANOSTRIPS

Vikas Kashid[1], Vaishali Shah[2], H. G. Salunke[3], Y. Mokrousov[4] and S. Blügel[4]

[1]Department of Physics, University of Pune, Pune 411 007, INDIA
[2]Interdisciplinary School of Scientific Computing, University of Pune, Pune 411 007, INDIA
[3]Technical Physics Division, Bhabha Atomic Research Centre, Mumbai 400 085, INDIA
[4]Peter Grünberg Institut and Institute for Advanced Simulation, Forschungszentrum, Jülich and JARA, D-52425 Jülich, Germany

We have investigated systematically different geometries of two dimensional (2D) nickel nanostrips using spin density functional theory. The nickel nanostrips are infinite in length and of widths varying from one atomic row to 6 rows of atoms. Our calculations demonstrate that triangular nanostrips are energetically more stable over rectangular nanostrips. The energy difference between triangular and rectangular nanostrips increases as the widths of the strips increase. The energy difference between rhombus structure and rectangular structure decreases as the number of rows increase to 5, however, the rhombus structure becomes more stable as compared to rectangular at 6 rows indicating a cross-over in stability of the structures.

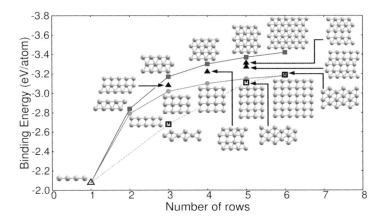

Figure 1: The binding energy as a function of the number of rows of atoms in the Ni nanostrips. The filled squares represent triangular (111) structures, filled circles represent rectangular (100) structures. The open squares connected by dotted lines represent rhombus structures, and all other structures considered are shown by triangles.

The magnetic properties in nickel nanostrips show significant dependence on the geometrical structure. The magnetization in triangular nanostrips is enhanced in comparison with the Ni (111) monolayer and decreases as the number of rows of atoms is increased. However, in rectangular nanostrips, the magnetization initially decreases and then increases in comparison with Ni (100) monolayer. Both Ni (111) monolayer and Ni (100) monolayer show enhanced magnetization of 0.79 μB/atom and 0.83 μB/atom, respectively, in comparison with that of bulk (0.62 μB/atom). The 3 row rectangular structure exhibits a magnetic anomaly with a magnetization of 0.81 μB/atom, which is lower than that of Ni (100) monolayer. We have calculated the magnetocrystalline anisotropy (MAE) in triangular and rectangular nanostrips. All nanostrips have magnetic moments along the infinite direction, except for the 3 row rectangular nanostrip, where they are perpendicular and in the plane of the nanostrip. The linear nanowire shows largest MAE of 4.7 meV/atom due to high asymmetry in the structure. MAE decreases with the increasing number of rows in the nanostrips.

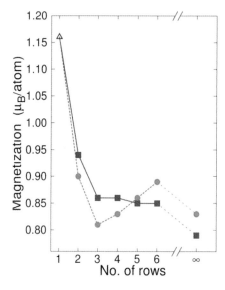

Figure 2: The magnetization in 2D Ni nanostrips with increasing number of rows of atoms. Filled squares represent triangular nanostrips, filled circles represent rectangular nanostrips. Magnetization in monoatomic Nw is shown by empty triangle.

In order to determine the magnetic behavior of Ni nanostrips, we calculated the Stoner's condition [1,2] for (111) and (100) nanostrips. We observed that, triangular nanostrips up to 6 rows of atoms show ferromagnetic solution, however, Ni (111) monolayer shows paramagnetic behavior. This indicates existence of a magnetic transition from ferromagnetic to paramagnetic as the number of rows in triangular nanostrips converge to Ni (111) monolayer. Such a transition is not observed in rectangular nanostrips. All rectangular nanostrips and Ni (100) monolayer show a ferromagnetic ground state. Our result on paramagnetic Ni (111) monolayer is consistent with the experimental result of Bergmann et al., [3].

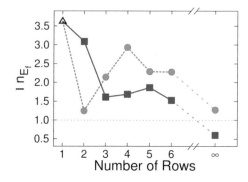

Figure 3: The Stoner criterion for Ni (111) and (100) oriented nanostrips. The product $I\, n_{Ef}$ is plotted against the increasing number of rows. Filled squares show triangular nanostrips and filled circles show rectangular nanostrips. The horizontal line shows the Stoner condition $I\, n_{Ef} = 1$

[1] E. C. Stoner, Proc. R. Soc. A **165**, 372 (1938)
[2] R. Zeller, Computational Nanoscience: Do it yourself! Vol. 31, ISBN 3-00-017350-1 (2006).
[3] G. Bergmann, Phys. Rev. Lett. **41**, 264 (1978).

SRP 5

SPIN BLOCKADE AND MAGNETIC EXCHANGE IN THE LAYERED COBALTATES $La_{1.5}A_{0.5}CoO_4$ (A = Ca, Sr, or Ba)

Dirk Fuchs[1], Michael Merz[1], Levin Dieterle[2], Stefan Uebe[1], Peter Nagel[1], Stefan Schuppler[1], Dagmar Gerthsen[2], and Hilbert von Löhneysen[1,3]

[1]Institut für Festkörperphysik, Karlsruher Institut für Technologie, 76021 Karlsruhe, Germany;
[2]Laboratorium für Elektronenmikroskopie, Karlsruher Institut für Technologie, 76131 Karlsruhe, Germany;
[3]Physikalisches Institut, Karlsruher Institut für Technologie, 76131 Karlsruhe, Germany;

Hole-doping the insulating parent compound of the layered cobaltates, La_2CoO_4, with Sr ions leads to the rapid suppression of the antiferromagnetic ground state (T_N = 275 K). For half-doping, i. e., $La_{1.5}Sr_{0.5}CoO_4$, the simultaneous presence of high-spin (HS, S = 3/2) Co^{2+} ($t_{2g}^5 e_g^2$) and low-spin (LS, S = 0) Co^{3+} ($t_{2g}^6 e_g^0$) leads to the so-called "spin blockade" [1]. Previous measurements have shown that tensile strain is able to suppress a low-temperature spin-state transition towards Co^{3+} LS in $LaCoO_3$, "pinning" the larger room-temperature HS/LS ratio even for low temperature [2]. For this reason it is very likely that the spin blockade in the layered cobaltates can be also suppressed by tensile strain of the CoO_2 layers, i. e., within the ab plane. Possible routes to this may be provided by the epitaxial growth of tensile strained $La_{1.5}Sr_{0.5}CoO_4$ films or by inducing chemical pressure by the substitution of Sr^{2+} by isovalent cations. To this end we prepared polycrystalline bulk samples of $La_{1.5}A_{0.5}CoO_4$ (A = Ca, Sr or Ba) which also served as targets for the thin film preparation. Epitaxially strained (001)-oriented films were grown by pulsed laser deposition on substrates with different lattice matching. The structural and magnetic properties of the samples were characterized by x-ray diffraction, transmission electron microscopy and superconducting quantum interference device magnetometry. Investigations on the spin structure were performed by x-ray absorption spectroscopy on the Co $L_{2,3}$ and O K edge.

[1] C. F. Chang et al., Phys. Rev. Letters **102**, 116401 (2009).
[2] D. Fuchs, et al., Phys. Rev B **75**, 144402 (2007).

EXPERIMENTAL VERIFICATION OF CHIRAL MAGNETIC ORDERS: CHIRAL MAGNETIC SOLITON LATTICE IN CHIRAL HELIMAGNET

Yoshihiko Togawa[1], Tsukasa Koyama[2], Shigeo Mori[2], Yusuke Kousaka[3], Jun Akimitsu[3], Sadafumi Nishihara[4], Katsuya Inoue[4], Alexander Ovchinnikov[5], Jun-ichiro Kishine[6]

[1]Nanoscience and Nanotechnology Research Center (N2RC), Osaka Prefecture University, Sakai, Osaka, Japan; [2]Department of Materials Science, Osaka Prefecture University, Sakai, Osaka, Japan; [3]Department of Physics, Aoyama Gakuin University, Sagamihara, Kanagawa, Japan; [4]Department of Chemistry, Hiroshima University, Higashi-Hiroshima, Hiroshima, Japan; [5]Department of Physics, Ural Federal University, Ekaterinburg, 620083, Russia; [6]Graduate School of Arts and Sciences, The Open University of Japan, Chiba, Japan

The concept of chirality, meaning left- or right-handedness, plays an essential role in symmetry properties of nature at all length scales from elementary particles to biological systems. In materials science, chiral materials are found in molecules or crystals with helical structures, which break mirror and inversion symmetries but combine rotational and translational symmetries. Chiral materials frequently exhibit intriguing functionality because electrons distribute themselves along chiral framework of atomic configurations and their rotational and translational motions couple to give specific physical processes.

In magnetic crystals belonging to chiral space group, orbital motions of localized electrons with spin magnetic moments take helical paths in the chiral framework of atoms and mediate coupling of the neighboring spins of electrons via the relativistic spin-orbit interaction called Dzyaloshinskii-Moriya (DM) interaction. This asymmetric DM exchange competes with ferromagnetic (FM) exchange interaction, which will result in an emergence of chiral magnetic orders and various interesting functions unique to chiral magnets.

In this presentation, we directly show that chiral magnetic orders emerge in chiral magnetic crystals of $Cr_{1/3}NbS_2$ in the presence and absence of small magnetic fields by means of low-temperature Lorenz transmission electron microscopy and small-angle electron diffraction (SAED) technique [1,2]. Based on precise analyses in both real and reciprocal space, we definitely demonstrate that chiral magnetic soliton lattice (CSL) continuously develops from chiral helimagnetic structure (CHM) in rising magnetic fields perpendicular to the helical axis as shown in Fig. 1. CSL is the periodic and nonlinear magnetic order and appears as the ground state in the magnetic field, while CHM is the harmonic and linear magnetic order with 48 nm period in zero magnetic field. Incommensurate CSL undergoes a continuous phase transition to commensurate FM state at the critical field strength. Importantly, upon applying magnetic field, CSL exhibits a continuous growth of the spatial period from 48 nm toward infinity at the incommensurate-to-commensurate (I-C) phase transition (Figs. 1(b)-(d)). Hence, the effective magnetic superlattice potential for the itinerant quantum spins can be tuned by simply changing the field strength. From detailed analysis of the contrast profile of CSL in Lorentz micrographs (i.e., an examination of a sequence of the contrast in Lorentz micrographs in Figs. 1(c) and (d)), the magnetic chiralities of CHM and CSL are identified to be left handed in the present crystals.

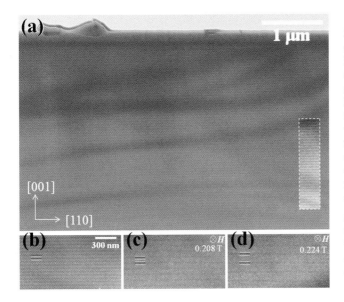

Figure 1: Chiral magnetic orders in a $Cr_{1/3}NbS_2$ crystal. Lorentz images are taken in zero and finite magnetic fields at 110 K. (a) Chiral helimagnetic order (CHM) in zero magnetic field. Defocus value is -512 nm. Continuous transformation from CHM to chiral soliton lattice (CSL) is clearly observed in magnetic fields perpendicular to the helical axis along [001].

(b) CHM at 0 T. (c) CSL at 0.208 T. (d) CSL at 0.224 T. The scale is the same and defocus value is -900 nm in (b) to (c).

Remarkably enough, CHM and CSL are very straight perpendicular to the helical axis and appear very regularly almost all over the specimen as partially shown in Fig. 1. CHM and CSL present no structural dislocation and persist against crystal defects that exist in specimens and scratch defects extrinsically fabricated by an irradiation of focused gallium-ion beams during TEM specimen fabrication. These specific features of high stability and robustness against perturbation indicate that CHM and CSL are macroscopic order of spin magnetic moments. This is because chiral magnetic orders are macroscopically induced by the uniaxial DM exchange interaction that is allowed in $Cr_{1/3}NbS_2$ hexagonal crystals belonging to noncentrosymmetric chiral space group. Namely, CHM and CSL are extremely robust ground states regarded as topological quantum condense of magnetic degrees of freedom.

From theoretical viewpoints, we have already revealed that the CSL would exhibit a variety of interesting functions including spin current induction, nontrivial soliton transport, current-driven collective transport, anomalous magneto-resistance and so on [3]. Present observations will be the first step to explore these functionalities of CHM and CSL for spintronic device applications using chiral magnetic crystals.

[1] Y. Togawa, T. Koyama, K. Takayanagi, S. Mori, Y. Kousaka, J. Akimitsu, S. Nishihara, K. Inoue, A. S. Ovchinnikov, and J. Kishine, Phys. Rev. Lett. **108**, 107202 (2012).
[2] C. Pappas, Physics **5**, 28 (2012).
[3] J. Kishine, I. V. Proskurin, and A. S. Ovchinnikov, Phys. Rev. Lett. **107**, 017205 (2011) and references therein.

FREQUENCY DEPENDENT MAGANOTRANSPORT IN $Sm_{0.6}Sr_{0.4}MnO_3$: USNUAL POSITIVE AND NEGATIVE MAGNETORESISTANCE

Ramanathan MAHENDIRAN[1]

[1]Department of Physics, 2 Science Drive 3, National University of Singapore, Singapore-117542

Although direct current (f = 0 Hz) magnetoresistance of manganites has been extensively studied over a decade, frequency dependence of the magnetoresistance and electrical resistivity in manganites in MHz range received very less attention so far. Recently, we reported that the low field magnetoresistance of the classical double exchange system $La_{0.7}A_{0.3}MnO_3$ (A= Sr, Ba) is dramatically increased in magnitude ($\Delta R/R$ = 40-50% in a field change of ΔH= 0.5- 1 kOe) which was attributed to magnetic field-induced modifications in the radio frequency transverse permeability and associated changes in the magnetic skin depth[1]. The exact origin of the enhanced ac magnetoresistance is yet to be understood. Here, we show a fundamentally different phenomena in the ferromagnetic manganite $Sm_{0.6}Sr_{0.4}MnO_3$. We measured ac impedance of the sample ($Z(T,H,f) = Z'(T,H,f)+iZ''(T,H,f)$) as a function of temperature(T), magnetic field(H), and frequency of the ac current (f = 100 Hz to 5 MHz) in four probe configuration. While the $Z'(T, H= 0 T)$ exhibits a single peak at the Curie temperature (T_C) as like the dc resistivity, it splits into two peaks (α and β) accompanied by a dip at T_C for higher frequencies. With increasing frequency, the two peaks move in opposite direction from T_C in zero H field and also decrease in magnitude. The α peak rapidly shifts upwards with increasing H whereas β peak is insensitive to H. On the other hand, $Z''(T, H= 0 T)$ shows a single minimum around T_C which broadens with increasing frequency. The rapid shift of α-peak with increasing H leads to positive magnetoresistance over a limited temperature range. These results suggest that two competing mechanisms are involved in the dynamical magnetotransport. We suggest that the field-independent high temperature β-peak originates from the dielectric relaxation associated with the polaron hopping in the paramagnetic phase. The field dependent low temperature α-peak is suggested to magnetization dynamics of ferromagnetic clusters just below the Curie temperature. From the isothermal frequency sweep, it is found that the relaxation time associated with the dynamical ac transport shows a divergence like behavior at the Curie temperature. Isothermal field dependence of the ac magnetoresistance shows butterfly wing like features (two positive peaks at a critical field ($\pm H_A$) and a dip at the origin. The field profile of the magnetoresistance shows a systematic change with increasing temperature. We discuss possible origins of the observed behavior.

Poster: Spin-related phenomena

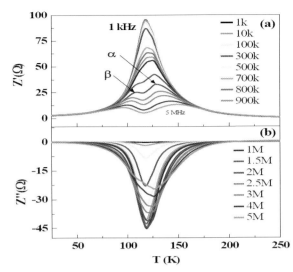

Figure 1: Frequency dependence of the in-phase (Z') and out –of phase (Z'') phase components of the ac impedance as a function of temperature. The behavior Z' at 1 kHz is similar to the dc resistivity. The insulator-metal transition in Z' is accompanied by a dip in Z''. With increasing frequency, the single peak in Z' transforms into two peaks, one above and another below the Curie temperature. The peaks move in opposite direction with increasing frequency and decrease in magnitude. On the other hand, Z'' broadens with increasing frequency and does not show double peaks.

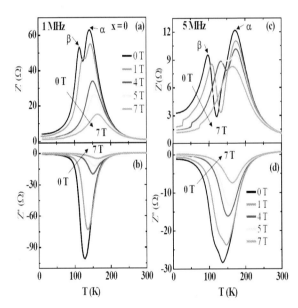

Fig.2: Ac impedance at f = 1 MHz(left column) and at 5 MHz (right column) in superimposed dc magnetic fields. It can be seen that low temperature peak (β) is very sensitive to magnetic fields and shifts rapidly towards high temperature with increasing magnetic field, whereas the position of the high temperature peak (α) is weakly sensitive to magnetic field. It suggest two different mechanisms are responsible for these peak. Over a narrow temperature range, positive magnetoresistance occur. On the other hand, the dip in Z'' shifts progressively to high temperature with increasing magnetic field .

[1] A. Rebelo, V. B. Naik and R. Mahendiran, J. Appl. Phys. **106**, 073905 (2009); A. Rebello and R. Mahendiran, Euro Phys. Letts. 86, 27004 (2009)

FIELD INDUCED SPIN-REORIENTATION AND STRONG SPIN-CHARGE-LATTICE COUPLING IN $EuFe_2As_2$

Y. Xiao[1], Y. Su[2], S. Nandi[1], S. Price[1], Th. Brückel[1,2]

[1]Jülich Centre for Neutron Science JCNS and Peter Grünberg Institute PGI, JARA-FIT, Forschungszentrum Jülich GmbH, 52425 Jülich, Germany;
[2]Jülich Centre for Neutron Science, Forschungszentrum Jülich GmbH, Outstation at FRM II, Lichtenbergstraße 1, 85747 Garching, Germany

Among various iron pnictide parent compounds, $EuFe_2As_2$ stands out due to the presence of both spin density wave of Fe and antiferromagnetic ordering of the localized Eu^{2+} moment [1-4]. A field-induced spin reorientation is observed in $EuFe_2As_2$ with the presence of a magnetic field along both a and c axes. Above critical field, the ground-state antiferromagnetic configuration of Eu^{2+} moments transforms into a ferromagnetic structure with moments along the applied field direction [5]. Interestingly, giant spin lattice coupling has been observed as indicated by the redistribution of the twin population. Since the twin realignment is intimately correlated with the magnetic phase transition, the spin-lattice coupling in $EuFe_2As_2$ can be attributed to the stresses generated by Zeeman energy and magnetic anisotropy energy.

Moreover, in-plane magnetotransport properties of a $EuFe_2As_2$ single crystal are also investigated by angular dependent magnetoresistance measurements. Strong anisotropy in magnetotransport properties is observed below the magnetic ordering temperature and it exhibits intimate correlation with the ordering states of both Eu and Fe spins in $EuFe_2As_2$. Such correlation leads to giant magnetoresistance effect around the magnetic phase boundary. It also suggests that the magnetic order state plays an important role for the determination of transport properties through the strong coupling between itinerant electrons and ordered spins [6].

[1] Y. Kamihara et al., J. Am. Chem. Soc. **130**, 3296 (2008).
[2] David C. Johnston, Advance in Physics **59**, 803 (2010).
[3] R. Marchand et al., J. Solid State Chem. **24**, 351(1978).
[4] Y. Xiao et al., Phys. Rev. B **80**, 174424 (2009).
[5] Y. Xiao et al., Phys. Rev. B **81**, 220406(R) (2010).
[6] Y. Xiao et al., Phys. Rev. B **85**, 094504 (2012).

ANISOTROPY OF SPIN RELAXATION IN hcp OSMIUM AND bcc TUNGSTEN

B. Zimmermann, P. Mavropoulos, S. Heers, N. H. Long, S. Blügel, Y. Mokrousov

Peter Grünberg Institut and Institute for Advanced Simulation, Forschungszentrum Jülich and JARA, 52425 Jülich, Germany

In the field of spintronics, the spin degree of freedom of the electron is used to store, manipulate and transport information [1]. However, a non-equilibrium spin distribution will generally equilibrate on a material-dependent time scale, the spin relaxation time. The dominant spin-relaxation mechanism in non-magnetic metals with space-inversion symmetry is the Elliott-Yafet mechanism [2], where electrons scatter at phonons or impurities and flip their spin with a certain probability. This probability will depend on the choice of the spin-quantization axis, which is in experiments defined by e.g. an external magnetic field. Therefore, the spin-relaxation time exhibits an anisotropy, which was measured e.g. in graphene [3].

The ratio between momentum- and spin-relaxation time can be estimated by means of a single parameter, the Elliott-Yafet or spin mixing parameter b^2, which is a property of the pristine crystal only. Calculations of b^2 have already been performed for various metals and successfully compared to experiments [4]. However, the anisotropy of b^2 was not considered so far, although, as we find, it can be large.

In this contribution, we present first principles density-functional theory (DFT) calculations of b^2 and its anisotropy by means of the Korringa-Kohn-Rostoker (KKR) Green function method. We show that there exists a gigantic anisotropy in hcp osmium (uniaxial crystal structure) and a large anisotropy in bcc tungsten (cubic). We find that so called spin hot-spots [4], points in the band-structure with anomalously high spin-mixing of electronic Bloch states, contribute significantly to the anisotropy (cf. Figure 1). We interpret our findings for the gigantic anisotropy at the hot spots in terms of a simple model.

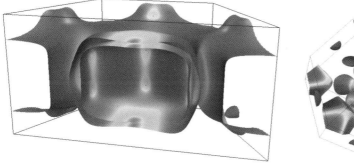
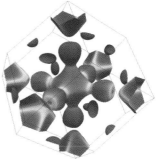

Figure 1: Fermi surface of osmium (left; cut for visualization) and tungsten (right) with the **k**-dependent spin-mixing parameter, shown as color code. The spin hot-spots (red) contribute to a large part to the anisotropy of b^2.

[1] I. Zutic, J. Fabian, and S. Das Sarma, Rev. Mod. Phys. **76**, 323 (2004)
[2] R. J. Elliott, Phys. Rev. **96**, 266 (2004); Y. Yafet, in *Solid State Physics*, Vol. 14 (1963)
[3] N. Tombros *et al.*, Phys. Rev. Lett. **101**, 46601 (2008)
[4] S. Heers, PhD thesis, RWTH Aachen (2011); M. Gradhand *et al.*, Phys. Rev. B **80**, 224413 (2009); D. Steiauf, and M. Fähnle, Phys. Rev. B **79**, 140401 (2009)
[5] J. Fabian, and S. Das Sarma, Phys. Rev. Lett. **83**, 1211 (1999)

SPIN-WAVE DYNAMICS IN TETRAGONAL FeCo ALLOYS

Ersoy Sasioglu, Christoph Friedrich, Stefan Blügel

Peter Grünberg Institut and Institute for Advanced Simulation, Forschungszentrum Jülich and JARA, 52425 Jülich, Germany

Recently, tetragonal FeCo alloys received considerable attention for perpendicular magnetic recording applications due to their large uniaxial magnetic anisotropy energy K_u and large saturation magnetization M_s [1,2], which allows to further increase the recording density in hard disk drives [3]. Besides large K_u and M_s values, another very important issue in device applications is the magnetization reversal processes i.e., magnetic switching time, which impose physical limits on data rates and areal recording densities [4]. Currently the switching speeds have reached the point that dynamical effects are becoming important [4, 5, 6]. This means that the magnetic switching speed is limited by the dissipation of the spin-wave (magnon) energies. Using a recently developed Green-function method [7] in combination with the multiple-scattering T-matrix approach [8] within the full-potential linearized augmented plane-wave (FLAPW) framework [9], we have studied the spin-wave dynamics in tetragonal FeCo alloys considering three different experimental c/a ratios [2], i.e., FeCo grown on Pd, Ir, and Rh with c/a = 1.13, 1.18, and 1.24, respectively. As there are two atoms in the unit cell, the calculated spin-wave dispersions possess two branches: an acoustic branch and an optical one. The former branch persists throughout the Brillouin zone indicating a localized nature of magnetism in FeCo alloys. On the other hand, the optical branch is heavily damped due to the coupling to single-particle Stoner excitations. We find that the tetragonal distortion gives rise to significant magnon softening. The magnon stiffness constant D decreases almost by a factor of two from FeCo/Pd to FeCo/Rh, which is a desired feature to decrease the switching times in devices. The combination of soft magnons, large magnetic anisotropy energy as well as large saturation magnetization suggests the FeCo alloy grown on Rh to be a promising material for the perpendicular magnetic recording applications.

[1] T. Burkert, L. Nordström, O. Eriksson, and O. Heinonen, Phys. Rev. Lett. **93**, 027203 (2004).
[2] F. Yildiz, M. Przybylski, X.-D. Ma, and J. Kirschner, Phys. Rev. B **80**, 064415 (2009).
[3] S. N. Piramanayagam, J. Appl. Phys. **102**, 011301 (2007).
[4] M. Plumer, J. van Ek, and D. Weller (Eds.) The physics of Ultrahigh-Density Magnetic Recording, Springer, Berlin, 2001.
[5] D. A. Garanin and H. Kachkachi, Phys. Rev. B **80**, 014420 (2009).
[6] V. L. Safonov, J. Appl. Phys. **95**, 7145 (2004).
[7] C. Friedrich, S. Blügel, and A. Schindlmayr, Phys. Rev. B **81**, 125102 (2010).
[8] E. Şaşıoğlu, A. Schindlmayr, C. Friedrich, F. Freimuth, and S. Blügel, Phys. Rev. B **81**, 054434 (2010).
[9] http://www.flapw.de

ROLE OF INTERFACES IN MANGANITE PHYSICS

V. Moshnyaga

I. Physikalisches Institut, Georg-August-Universität Göttingen, Friedrich-Hund-Platz 1, 37077 Göttingen, Germany.

In CMR manganites strong electronic correlations may be influenced and/or induced by electron-phonon interaction (Jahn-Teller effect). As a result a unique interplay between spin, charge and lattice degrees of freedom causes a number of interesting and useful phenomena, like colossal (CMR) and tunneling (TMR) magnetoresistance, resistance switching and, finally, multiferroic behavior. The interfacial and polaronic aspect of all these effects, considering competing different electronic/structural phases, will be highlighted and some insights into a common "polaronic" origin of magneto- and electro-resistive properties will be discussed. For a prototypical phase separated CMR system, $(La_{1-y}Pr_y)_{0.7}Ca_{0.3}MnO_3$, very large low-field CMR is due to exchange (AFM) coupling between ferromagnetic nanodomains, actuated by correlated polarons, which are located at the "intrinsic interfaces" between FM domains. Manganite/titanite superlattices (SL), e.g. $[(La_{0.7}Ca_{0.3}MnO_3)_m/(BaTiO_3)_n]_K$ (LCMO/BTO), are interesting as artificial multiferroic system, in which orbital, charge and spin reconstructions at the interfaces play an important role. SL's have been grown by a metalorganic aerosol deposition technique on MgO(100) and $SrTiO_3$ substrates. The structure of the LCMO/BTO interfaces was studied with atomic resolution by TEM and elemental chemical analysis (EELS). We show advantages of interface engineering to control magnetotransport and crystal structure of manganites. In addition a "layer-by-layer" growth of perovskite films is demonstrated.

ISOTHERMAL ELECTRIC CONTROL OF EXCHANGE BIAS NEAR ROOM TEMPERATURE

Christian Binek[1] Xi He[1], Yi Wang[1], N.Wu [1], Aleksander L. Wysocki[1], Takashi Komesu[1], Uday Lanke[2], Anthony N. Caruso[3], Elio Vescovo[4], Kirill D. Belashchenko[1], Peter A. Dowben[1]

[1]Department of Physics & Astronomy and Nebraska Center for Materials and Nanoscience, University of Nebraska, Lincoln, NE, 68588-0111, USA
[2]Canadian Light Source Inc., University of Saskatchewan, Saskatoon, Saskatchewan, Canada S7N 0X4
[3]Department of Physics, 257 Flarsheim Hall, University of Missouri, Kansas City KS 64110, USA
[4]Brookhaven Nat. Laboratory, National Synchrotron Light Source, Upton, NY, 11973, USA

Voltage-controlled spintronics is of particular importance to continue progress in information technology through reduced power consumption, enhanced processing speed, integration density, and functionality in comparison with present day CMOS electronics. Almost all existing and prototypical solid-state spintronic devices rely on tailored interface magnetism, enabling spin-selective transmission or scattering of electrons. Controlling magnetism at thin-film interfaces, preferably by purely electrical means, is a key challenge to better spintronics. Currently, most attempts to electrically control magnetism focus on potentially large magnetoelectric (ME) effects of multiferroics.

I report on the antiferromagnetic (AF) ME Cr_2O_3 (chromia) for voltage-controlled magnetism. Specifically, robust isothermal electric control of exchange bias (EB) is achieved at room temperature in the perpendicular anisotropic EB heterostructure $Cr_2O_3(0001)/CoPd$ [1]. Electrically controllable boundary magnetization (BM) is a key ingredient. First-principles calculations and symmetry considerations show that BM is a generic property at interfaces of ME single domain antiferromagnets. In chromia, an AF single domain can be selected out of two degenerate AF 180 degree domains by magnetoelectrically lifting the degeneracy. Isothermal switching between the single domain states is achieved when an electric field, E, and magnetic field, H, are simultaneously applied such that the magnitude of the product $E H$ overcomes a critical threshold. The emerging BM is coupled to the bulk AF order parameter and follows the latter during switching. Exchange coupling between the BM and the adjacent ferromagnetic thin film gives rise to switching of the EB field. Figure 1 depicts single domain chromia with BM at the rough (0001) surface in contrast to the surface magnetic propertis of conventional antiferromagnets. The order parameter type temperature dependence of the EB is a macroscopic consequence of BM.

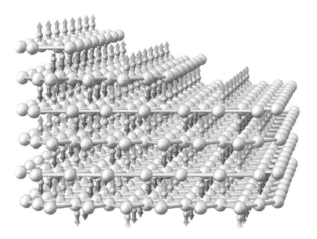

Fig.1: Cartoon of chromia in single domain states. Arrows depict spins of Cr^{3+}-ions. Circles show O^{2-}-ions. Rough surface shows sizable spin polarization.

Measurements of spin-resolved ultraviolet photoemission of a chromia (0001) thin film surface provides additional macroscopically averaged information of the BM. Laterally resolved X-ray PEEM and temperature dependent magnetic force microscopy reveal microscopic information of the chromia surface magnetization [2]. Figure 2 shows the results of the XMCD-PEEM measurements above (Fig. 2a) and below the Néel temperature with (Fig. 2 c) and without (Fig. 2b) field-cooling. PEEM images obtained with left and right circularly polarized light (see Fig. 2d for energy dependence of XMCD signal) were used to generate the images shown in Fig. 2.

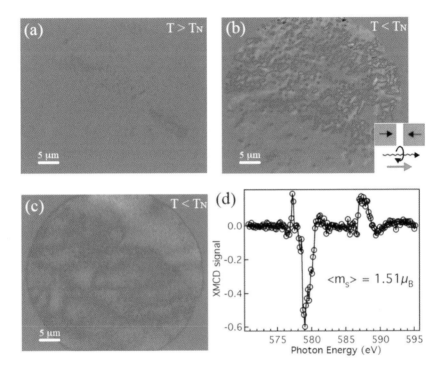

Fig.2: (a-c) Cr_2O_3 (0001) film imaged by XMCD-PEEM at the Cr L-edge. (a) No contrast at 584 K. (b) Multi-domain state after zero-field cooling. Inset shows spin polarization with respect to positively circularly polarized incident light. (c) Nearly single-domain state at 223 K after ME field-cooling. (d) XMCD spectrum recorded from within one domain.

In concert, our data provide a coherent interpretation of our results on isothermal electric control of EB. The latter promise a new route towards purely voltage-controlled spintronics and an exciting way to electrically control magnetism.

Financial support by NSF through MRSEC, CAREER, SRC/NSF Suppl. to MRSEC, and Cottrell Research Corp. is greatly acknowledged.

[1] Xi He, Yi Wang, N. Wu, A. N. Caruso, E. Vescovo., K. D. Belashchenko, P. A. Dowben & Ch. Binek, Nature Mater. 9, 579–585 (2010).
[2] Ning Wu, Xi He, Aleksander L. Wysocki, Uday Lanke, Takashi Komesu, Kirill D. Belashchenko, Christian Binek, and Peter A. Dowben, Phys. Rev. Lett. **106**, 087202 (2011).

STRAIN ENGINEERING OF MULTIFERROIC PHASE TRANSITIONS AND ORDER PARAMETERS IN BiFeO$_3$

Daniel Sando[1], Arsène Agbelele[2], Christophe Daumont[1], Jean Juraszek[2], Maximilien Cazayous[3], Ingrid Infante[4], Wei Ren[5], Sergey Lisenkov[6], Cécile Carretero[1], Laurent Bellaiche[5] and Brahim Dkhil[4], Agnès Barthélémy[1] and <u>Manuel Bibes</u>[1]

[1] Unité Mixte de Physique CNRS/Thales, 1 Av. A. Fresnel, Campus de l'Ecole Polytechnique, 91767 Palaiseau (France) and Université Paris-Sud, 91405 Orsay (France)
[2] Groupe de Physique des Matériaux, Université de Rouen / CNRS, France
[3] Laboratoire MPQ, Université Paris Diderot / CNRS, France
[4] Laboratoire SPMS, Ecole Centrale Paris / CNRS, France
[5] University Of Arkansas, Fayetteville, USA
[6] Universiy Of South Florida, Tampa, USA

The strong coupling of ferroic orders (elastic, electric and magnetic) with the various structural degrees of freedom (notably polar and antiferrodistortive) provides multiferroic BiFeO$_3$ with very rich phase diagrams, as well as with a highly tunable, multifunctional character. Epitaxial strain has recently emerged as a powerful way to tune the various remarkable physical properties of perovskite oxide thin films. Applied to BiFeO$_3$, strain engineering reveals various highly unexpected features as well as novel multifunctional phases with enhanced properties and application potential. For instance, while epitaxial strain controls the Néel temperature in antiferromagnetic films by modifying oxygen octahedral tilt angles, it usually enhances the Curie temperature of ferroelectric films by increasing polar cation shifts. In BiFeO$_3$, we have found that both antiferrodistortive and polar instabilities coexist and respond differently to epitaxial strain. Combining advanced characterization techniques (X-ray and neutron diffraction, Mössbauer spectroscopy and piezoresponse force microscopy) and ab initio calculations, we have established the very rich phase diagram resulting from the influence of epitaxial strain on ferroic phase transitions and order parameters in a strain range spanning over 3% [1]. We will show that these results not only shed light on the interplay between polar and oxygen tilting instabilities but also reveal the possibility to strain-drive the magnetic and ferroelectric transition temperatures close together, offering an original approach to achieve enhanced magnetoelectric responses.

In addition, we will present results on the influence of strain of the ferroelectric and piezoelectric response at room temperature, evidencing an increase of both at large compressive strain [2]. Finally, we will show how Mossbauer and Raman spectroscopy data can reveal strain-driven changes in the spin arrangements and excitations.

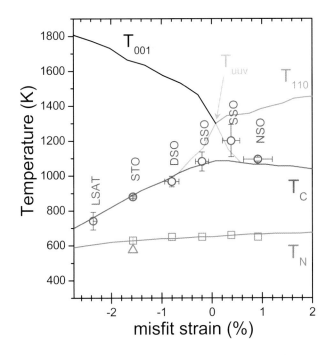

Figure 1. Theoretical results on BiFeO$_3$ film transition temperatures (ferroelectric T$_C$, blue line, and magnetic T$_N$, green line) as a function of misfit strain. The activation temperatures for the antiferrodistortive oxygen tilting along the z direction (tetragonal distortion T$_{001}$, black line), for the antiferrodistortive oxygen tilting within the x-y plane (orthorhombic distortion T$_{110}$, red line), and for the tilting along [uuv] direction (monoclinic distortion T$_{uuv}$, orange line) are also plotted. Experimental T$_C$ (circles) and T$_N$ (squares and triangles) values. Vertical error bars for T$_N$ values correspond to symbol size and for T$_C$ values result from maximum variation of T$_C$ that may be obtained between different fitting processes using the mean field function. Below T$_C$, for any considered strain, the simulations predict that the crystallographic phase is Cc, with a polarization along a [uuv] direction while the oxygen octahedra rotate about a [u$_0$u$_0$v$_0$] axis.

[1] I.Infante et al.; *Phys. Rev. Lett.* 105, 057601 (2010) ; B. Dupé et al. *Phys. Rev B* 81, 144128 (2010)
[2] C. Daumont et al, *J. Phys. Condens. Matter* (in press)

OBSERVATION AND EFFECT OF MAGNETIC DOMAINS IN LATERAL SPIN VALVES

Xianzhong Zhou[1], Julius Mennig[1], Frank Matthes[1], Daniel E. Bürgler[1], Claus M. Schneider[1]

[1]Peter Grünberg Institute, Electronic Properties (PGI-6), Research Center Jülich, D-52425 Jülich, Germany

Co/Cu/Co lateral spin valves (LSV) with Co being the topmost layer are prepared and measured under ultrahigh vacuum (UHV) conditions by combining thin film deposition by thermal evaporation and nanostructuring by means of a dual-beam scanning electron microscopy and focused-ion beam (UHV SEM/FIB) setup. The in-situ measurements comprise separate magnetotransport investigation of the anisotropic magnetoresistance (AMR) for both Co wires, 4-probe non-local measurement of pure spin current signal, and imaging of the magnetic domain structures by scanning electron microscopy with polarization analysis (SEMPA). The spin signal as a function of the in-plane field along the wire axes reflects the magnetic reversal in the Co/Cu contact regions only, whereas the AMR curves are affected by all domains throughout the whole Co wires. The clean fabrication process yields bulk-like resistivities for Co and Cu, ohmic Co/Cu contacts, and –in the case of optimized LSV geometry– a large non-local spin signal of 0.9 mΩ at room temperature. The in-situ SEMPA analysis reveals domain structures in both Co elements indicating magnetic reversals that clearly deviate from the expected single-domain behavior. The presence of the multi-domain state is traced back to substrate roughness inherent to our fabrication process involving a FIB milling step prior to the deposition of the Co wires. The lateral length scale of the FIB-induced roughness is of the order of 250 nm and is related to the polycrystallinity of the films, which gives rise to inhomogeneous, grain-dependent sputter yields. The number of the domains in the Co wires can be reduced by increasing the width of the wires beyond the lateral dimensions of the FIB-induced roughness (Fig. 1).

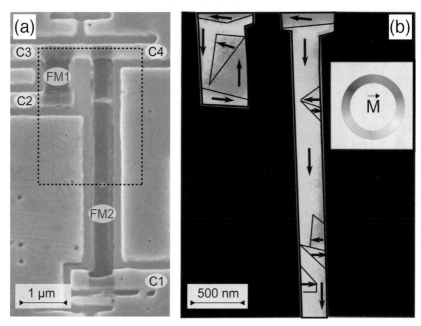

Figure 1:
(a) SEM image of an in-situ fabricated LSV (dark: Co, bright: Cu). FM1 and FM2 denote the Co elements and C1 to C4 the Cu contacts. The Cu channel for the pure current is between C3 and C4.
(b) SEMPA image of the area inside the dashed frame in (a).

For samples with Co wire widths of 350 and 480 nm, respectively, we can correlate the spin signal curve to SEMPA images and to the AMR curves of both Co elements (Fig. 2).

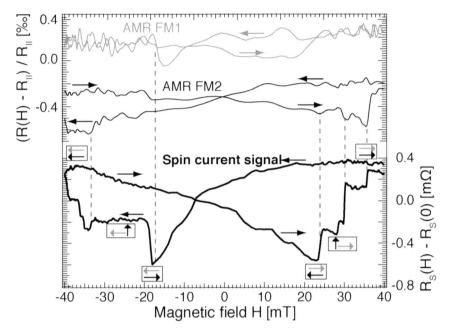

Figure 2: AMR measurements of FM1 (grey) and FM2 (black, vertically offset for clarity) and spin signal (thick black line) of the LSV shown in Fig. 1. Arrows indicate the field sweep direction. Framed pairs of grey and black arrows represent the relative alignment of the magnetizations of FM1 and FM2 at the respective contacts to the Cu channel.

We find that the multi-domain states have a strong impact on the spin signal. Multiple domains in the Co/Cu contact regions impede generation or detection, respectively, of spin accumulation and therefore render measured spin current amplitudes ambiguous [1].

[1] J. Mennig, F. Matthes, D.E. Bürgler, C.M. Schneider, J. Appl. Phys. **111**, 07C504 (2012).

THICKNESS EVOLUTION OF THE STRAIN IN PCMO THIN FILMS

<u>Anja Herpers</u>[1], Chanwoo Park[1], Ricardo Egoavil[2], Jo Verbeeck[2] and Regina Dittmann[1]

[1]Peter Grünberg Institut, Electronic Materials (PGI-7) and Jülich-Aachen Research Alliance, Fundamentals of Future Information Technology (JARA-FIT), Research Center Jülich, Jülich, Germany; [2]EMAT, University of Antwerp, Antwerp, Belgium

As a mixed-valence manganite $Pr_{0.48}Ca_{0.52}MnO_3$ (PCMO) can be used for various applications. It is investigated for fuel cell technology, because it serves as a good ion conductor and can be used as oxygen semipermeable membrane [1]. Being a p-type semiconductor PCMO is also interesting for resistive switching devices [2].

In this study, the epitaxial growth of PCMO by pulsed laser deposition (PLD) has been optimized on $NdGaO_3$ (NGO) and on $SrRuO_3$ (SRO)/ $SrTiO_3$ (STO) for further device application. Under optimized conditions, the rocking curve of the (002) Bragg reflex from the X-ray diffraction (XRD) measurement has a width of $0.045°$ confirming a good crystallinity. On NGO substrates, more than 70 RHEED intensity oscillations have been observed during the growth of PCMO.

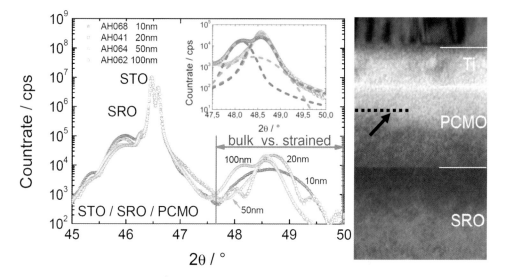

Figure 1: Diffractogram of the thickness series of PCMO on 30nm SRO on STO substrate. The pseudo-cubic (002) PCMO peak first develops with increased thickness, shifts slightly at 50nm towards the balk value and splits up at 100nm suggesting a bilayer structure. In the inset, the PCMO peaks for 50nm and 100nm and each fitted curves (dashed lines) are plotted. The shoulder in the 50nm film peak is due to a second peak.

Figure 2: Cross-sectional TEM image of PCMO stacked between SRO and Ti. Within the PCMO two regions of different strain are visible.

The investigation of a thickness series on SRO/STO by XRD shows that the PCMO peak shifts slightly from 20nm to 50nm towards the bulk value (Figure 1). The peak starts to split up at 50nm and already has a double-peak shape at 100nm pointing on regions with differently relaxed strain (inset). Transmission electron microscopy (TEM) investigations (Figure 2) confirm the existence of a PCMO double layer with different strain. In particular, the thickness of the underlying layer can be determined to 17nm. The reason for this observation might be the kinetically limited growth conditions leading to the formation of misfit dislocations, which do not reach the SRO interface.

From a XRD reciprocal space map (RSM) of a 50nm thick sample grown on NGO substrate (Figure 3) it is visible, that the in-plane lattice constants match perfectly, meaning that the PCMO film is totally strained on NGO. The out-of-plane lattice constant of PCMO is smaller than the bulk value, but larger than it would be expected according to volume conservation, as has already been seen from the thickness series (Figure 1). The reason for the enlargement of the c-axes might be the incorporation of cation vacancies into the lattice, as it has been seen for STO [3].

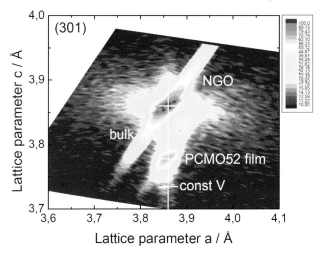

Figure 3: RSM of a 50nm thick PCMO thin film on NGO substrate. As the in-plane lattice parameters are equal, the film is grown strained and the out-of-plane lattice constant is smaller than bulk.

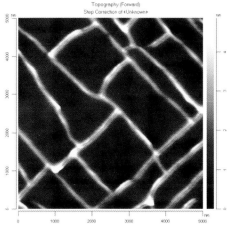

Figure 4: AFM picture of a 50nm thick PCMO film surface showing a perpendicular pattern of self-assembly lines. The lines are due to walls between regions of differently orientated PCMO domains.

In Figure 4, an atomic force microscopy (AFM) scan is depicted showing a few nm thick lines of perpendicular orientation appearing on top of the PCMO surface. The line pattern starts to be formed at a film thickness of about 20nm. A very similar pattern could be found in literature [4], where Fujimoto et al. observe a similar shape on the surface of $Pr_{0.7}Ca_{0.3}MnO_3$ and correlate these lines to walls between differently oriented crystal domains using high-resolution TEM images. These different domains could also be an alternative origin for the peak splitting observed by XRD (Figure 1).

[1] M.A. Peña et al., Chem. Rev. **101**, 1981 (2001)
[2] A. Sawa et al., Appl. Phys. Lett. **85**, 4073 (2004)
[3] D.J. Keeble et al., Phys. Rev. Lett. **105**, 226102 (2010)
[4] M. Fujimoto et al., J. Am. Ceram. Soc. **90**, 2205 (2007)

EFFECT OF MIXED ORTHORHOMBIC/HEXAGONAL STRUCTURE ON MAGNETIC ORDERING IN STRONTIUM DOPED YTTERBIUM MANGANITES

A.I. Kurbakov[1], I.A. Abdel-Latif[2,3], V.A. Trunov[1], H. U. Habermeier4, A. Al-Hajry[3], A.L. Malyshev[1], V.A. Ulyanov[1]

[1]Petersburg Nuclear Physics Institute, Orlova Grove, Gatchina, Leningrad District, 188300, Russia,
[2]Reactor Physics Dept., NRC, Atomic Energy Authority, Abou Zaabal, Cairo, 13759, Egypt,
[3]Physics Dept., Colleage of Science, Najran University, Najran, 1988, Saudia Arabia,
[4]Technology Group, Max Planck Institute of Solid State, Stuttgart,Germany
E-mail; ihab_abdellatif@yahoo.co.uk

Studying crystal and magnetic structures of the complex multi-phase strontium doped ytterbium manganites – multiferroics, $Yb_{1-x}Sr_xMnO_3$ compounds, using high-resolution neutron diffraction at very low temperature are presented for the first time. The partial hole doping allows to pass a range of so-called "geometric" ferroelectrics $YMnO_3$-type to "magnetic" ferroelectrics $TbMnO_3$-type. At the first such compound $Yb_{0.6}Sr_{0.4}MnO_3$, prepared using conventional solid state method, the parameters of the crystal and magnetic structures have been determined and a comparison with un-doped $YbMnO_3$ has been carried out. This compound corresponds to two-phases of crystal structure. Strontium doping partially transforms hexagonal structure to more close-packed orthorhombic perovskite phase. It is shown that the crystal structure of the investigated compound remains invariable in the whole temperature range from helium to room temperature and corresponds a mixture of orthorhombic $Pbnm$ (enclosing Jahn-Teller Mn^{3+} ions) and hexagonal $P6_3cm$ (JT ions free) phases, and the ratio of these phases remains invariable at all temperatures. Such complex crystal structure of $Yb_{0.6}Sr_{0.4}MnO_3$ is explained with a view to the average A-cation size $<r_A> = 1.149$ Å and the local A-cation size mismatch $\sigma^2 = 0.01724$ Å^2 in comparison with the parent compound $YbMnO_3$ ($<r_A> = 1.042$ Å, $\sigma^2 = 0$). The crystal structure of the hexagonal phase is isomorphic undoped hexagonal $YbMnO_3$. Mn ions lie near the center of a trigonal bipyramid. Although the Yb ions lie on threefold axes, apical oxygen atoms are at different distances, which leads to ferroelectric behavior. Detailed structural parameters (lattice constants, the bond Mn-O-Mn angles, Mn-O and Yb-O distances) were obtained as a function of temperature from Rietveld refinements. Based on this result, the distinct differences in temperature behavior between two phases are discussed. Absolutely opposite temperature behavior of unit cell parameters in two crystallographic phases is observed. Unit cell parameter a increases and c parameter remains almost invariable with temperature in the hexagonal phase whereas in the orthorhombic phase: c parameter increases but a and b practically do not change with the temperature.

Possible magnetic structures are estimated from the experimental results and symmetry consideration. We show that both hexagonal and orthorhombic phases show totally different magnetic behaviors. Magnetic ordering of the hexagonal crystallographic phase is revealed at $T \approx 85$ K. This magnetic structure represents a frustrated magnetic structure with a canted-spin ordering of Mn magnetic moments in plane of $\Gamma 2$-type. Mn ions form well-separated triangular layers parallel to (ab) plane, with antiferromagnetic exchange interaction between the spins of the most nearest neighbors, which make the Mn spin subsystem of low-dimensional and frustrated. The phase with orthorhombic crystal structure shows magnetic ordering at $T \approx 130$ K. The magnetic structure presumably C-type is formed. Magnetic ordering of rare earth Yb atom down to $T = 2.6$ K has not been registered. As a result, it is demonstrated the existence of very different physics of manganites on the same compound. The strong crystallographic stability of such a complex system is shown. Also it is not revealed the influence of one C-type spin structure on another strongly frustrated triangular magnetic subsystem. Magnetization measurements showed that there are multiphase transitions $Yb_{0.6}Sr_{0.4}MnO_3$ corresponding to hexagonal and orthorhombic systems as shown in Fig.1.

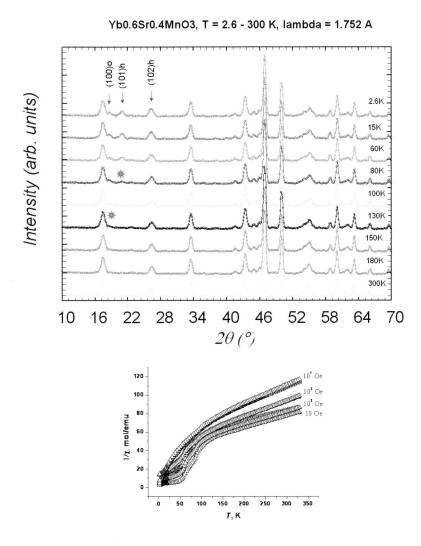

Fig.1 Neutron Diffraction and Magnetic Susceptibility Measurements at very low temperature

QUANTUM TRANSPORT THROUGH TOPOLOGICAL SPIN TEXTURE IN CHIRAL HELIMAGNET

J. Kishine[1], A.S. Ovchinnikov[2], I.V. Proskurin[2]

[1] Graduate School of Arts and Sciences, Open University of Japan [2] Department of Physics, Ural Federal University
E-mail: junkishine@96.alumni.u-tokyo.ac.jp

In a magnetic crystal belonging to chiral space group, competition of the relativistic spin-orbit Dzyaloshinskii-Moriya (DM) interaction and ferromagnetic exchange interaction causes a helical magnetic arrangement with definite vector chirality and a winding period of several tens of nanometers. The most intriguing property of chiral helimagnet is that there appears a nonlinear regular lattice of spin magnetic moments under weak magnetic field applied perpendicular to the helical axis [Fig.]. This state, called chiral soliton lattice (CSL), is an extremely robust topological ground state, where magnetic topological charges condense into regular lattice. The spatial period of the CSL is controlled by magnetic field.

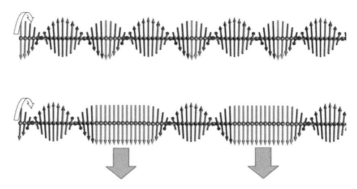

The magnetic moments in chiral helimagnets rotate around in a left-handed helix (top). When an external magnetic field is applied (bottom), a chiral magnetic soliton lattice (CSL) appears, in which the helical shape is stretched out wherever the moments and field are aligned.

When itinerant electrons coexist such as in the case of Cr1/3NbS2[1], the CSL acts on the itinerant electrons as magnetic superlattice potential. On the other hand, the itinerant electrons cause spin torque transfer into the CSL to cause collective translation of the CSL. Consequently, we anticipate that the CSL would exhibit a variety of interesting functions, including spin current induction[2], non-trivial soliton transport[3], new kind of elementary excitations[4], current-driven collective transport[5], anomalous magneto-resistance[6] and so on.

These theoretical proposals are strongly supported by experimental demonstration that the CSL is formed in a single crystal of Cr1/3NbS2[1] and will open up new perspectives in the field of spintronics. In this talk, I will give an overview of our recent progress on this topics.

[1] Y.Togawa, T.Koyama, K.Takayanagi, S.Mori, Y.Kousaka, J.Akimitsu, S.Nishihara, K.Inoue, A.S.Ovchinnikov, and J.Kishine, Phys.Rev.Lett.**108**, 107202 (2012), selected as *APS Spotlighting Exceptional Research* on March 5, 2012.
[2] I.G.Bostrem, J.Kishine and A.S.Ovchinnikov, Phys.Rev.**B78**, 064425(2008)
[3] A.B.Borisov, J.Kishine, I.G.Bostrem, and A.S.Ovchinnikov, Phys.Rev.**B79**, 134436(2009)
[4] J.Kishine and A.S.Ovchinnikov, Phys.Rev.**B79**, 220405(**R**) (2009)
[5] J.Kishine, A.S.Ovchinnikov, and I.V.Proskurin, Phys.Rev.**B82**, 064407 (2010)
[6] J.Kishine, I.V.Proskurin and A.S.Ovchinnikov, Phys.Rev.Lett. **107**, 017205 (2011)

COMPARATIVE INVESTIGATION OF Sb$_2$Te$_3$ NANOSTRUCTURE PROPERTIES WITH RESPECT TO THEIR PRODUCTION METHOD

T. Saltzmann[1], S. Rieß[2], J. Kampmeier[2], G. Mussler[2], T. Stoica[2], B. Kardynal[2], U. Simon[1], H. Hardtdegen[2]

[1]Institut für Anorganische Chemie, RWTH Aachen University, Chemische Institute, Landoltweg 1, D - 52074 Aachen, Germany;
[2]Peter Grünberg Institut 9, Forschungszentrum Jülich & JARA Jülich-Aachen Research Alliance, 52525 Jülich, NRW, Germany

The narrow band gap semiconductor Sb$_2$Te$_3$ belongs to the class of phase change materials, which may be employed for future electrical and optical data storage. They are known to change their electrical conductivity by one or more orders of magnitude as the material changes its structural state. The phase change switching mechanism may be brought on by local heating produced by electrical or optical pulses. However, these materials suffer from an electrical drift of resistance which could be due to stress in the material during its fabrication. A comparative investigation of the material characteristics as a function of their fabrication method – the wet chemical, solvothermal synthesis route and molecular beam epitaxy in this case- may therefore reveal the role of the crystal growth method on properties. To this end suitable characterization techniques were chosen with respect to the dimensionality of the perspective nanostructure

The solvothermal synthesis route produces Sb$_2$Te$_3$ in the form of hexagonal platelets (HP)s. In this method Sb and Te crystallites are first formed by the reduction of Sb$_2$O$_3$ and TeO$_3$ in diethylenglycole in the presence of polyvinylpyrrolidone (PVP) and sodium glyconate. Subsequently they react to the desired alloy which recrystallizes into HPs via an oriented attachment mechanism [2]. The size of the particles varies from 1 two 5 µm in lateral dimension and between a few tens and 250 nm in thickness depending on the reaction conditions.

Sb$_2$Te$_3$ thin films were deposited by means of a solid source molecular beam epitaxy (MBE) on chemically cleaned Si(111) wafers with a typical thickness of 100 nm under ultra high vacuum (UHV) conditions utilizing Knudsen effusion cells. Detailed investigations concerning surface topography, layer thickness, crystalline perfection and band structure were performed utilizing atomic force microscopy (AFM), scanning tunneling microscopy (STM), X-ray reflectivity (XRR), high resolution X-ray diffraction (HRXRD), angle resolved photo emission spectroscopy (ARPES), µRaman and high resolution scanning electron microscopy (HRSEM). The electrical properties of the nanostructures were determined *in situ* in a scanning electron microscope (SEM) on the one hand and by Hall measurements on the other.

It was found that the topology of nanostructures is quite similar as can be seen in Fig.1 even though platelets grow basically strain free and the epitaxial layers exhibit a lattice mismatch of 14 % to the underlying Si substrate. Backscattered electrons (BSE) were recorded in HR-SEM. By using the BSE-mode under perfect conditions dislocations can be imaged [1]. Both nanostructures apparently contain screw type dislocations. A comparison of the electrical properties reveals an anisotropic conductivity behavior in the HPs: they are highly conductive in parallel to the hexagonal surface (*ab*-plane) and highly resistive vertical to this plane (*c*-direction). Figure 2A shows a single HP contacted in c-direction. The corresponding IV characteristic is displayed in the 2B and depicts the dependence of the electrical transport as a function of the compliance current (CC). Figures 2C and D present an HP contacted in the *ab*-plane and the corresponding IV curve. Ohmic behavior is observed. The resistance of the HP was determined to be 26 kΩ. Further results of the comparative study will be presented and discussed.

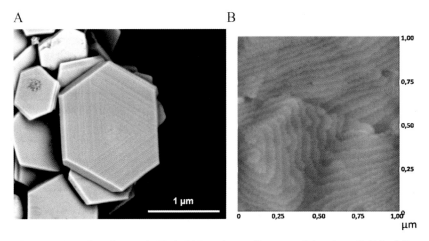

Figure 1: A) HR SEM image of a HP in BSE-mode revealing screw dislocations; B) Epitaxially grown thin films of Sb_2Te_3 measured with AFM illustrate the same structural characteristics as the HPs shown in A)

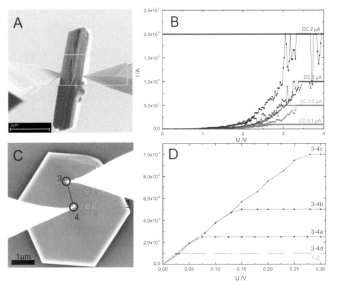

Figure 2: A) HP contacted in c direction. B) Corresponding IV curves measured with increasing compliance currents (CC) from 0.1 to 2 µA resulting in an electrical breakthrough at the highest CC. C) HP contacted in the *ab*-plane. D) Corresponding IV curve, depicting ohmic behavior a resistivity of 26 Kilo ohms.

[1] ZAEFFERER, S.: Orientation microscopy in SEM and TEM: fundamentals, techniques and applications in physical metallurgy; Habilitation, Fakultät für Georessourcen und Materialtechnik, RWTH Aachen, 2010
[2] Zhang, G.; Wang, W.; Lu, X. & Li, X. Cryst. Growth Des, **9**, 145, (2009)
[3] Kim Y., DiVenere A., Womg G., Ketterson J. B., Cho S., Meyer J. R., J. Appl. Phys. **91**, 751, (2002)

TOPOLOGICAL INSULATORS FROM THE VIEW POINT OF CHEMISTRY

C. Felser, L. Müchler, S. Chadov, B. Yan, J. Kübler, HJ Zhang, and SC Zhang

Topological insulators are a hot topic in condensed matter physics. The excitement in the physics community is comparable with the excitement when a new superconductor is discovered. Recently the Quantum Spin Hall effect was theoretically predicted and experimentally realized in quantum wells based on the binary semiconductor HgTe. Many Heusler compounds with C1b structure are ternary semiconductors that are structurally and electronically related to the binary semiconductors. The diversity of Heusler materials opens wide possibilities for tuning the bandgap and setting the desired band inversion by choosing compounds with appropriate hybridization strength (by the lattice parameter) and magnitude of spin-orbit coupling (SOC, by the atomic charge). Based on first-principle calculations we demonstrate that around 50 Heusler compounds show band inversion similar to that of HgTe. The topological state in these zero-gap semiconductors can be created by applying strain or by designing an appropriate quantumwell structure, similar to the case of HgTe. Many of these ternary zero-gap semiconductors (LnAuPb, LnPdBi, LnPtSb and LnPtBi) contain the rare-earth element Ln, which can realize additional properties ranging from superconductivity (for example LaPtBi) to magnetism (for example GdPtBi) and heavy fermion behaviour (for example YbPtBi). These properties can open new research directions in realizing the quantized anomalous Hall effect and topological superconductors. C1b Heusler compounds have been grown as single crystals and as thin films. The control of the defects, the charge carriers and mobilties will be optimized and quantum well structures will be grown. Recently some of the C1b Heusler compounds were predicted to be excellent piezoelectrics. The combination of a piezoelectric Heusler compounds and compounds at the borderline between trivial and topological insulators offers the possibility of a switchable device.

It is also possible to design new topological insulators with strong correlations. In AmN and PuTe a band gap is opened by correlation effects. In a family of semiconductors with the simple NaCl structure band gaps up to 0.4 eV were found. This is not so surprising since the SOC should be large in Actinides.

Heusler compounds are similar to a stuffed diamond, correspondingly, it should be possible to find the "high Z" equivalent of graphene in a graphite-like structure or in other related structure types with 18 valence electrons and with inverted bands. Indeed the ternary compounds, such as LiAuSe and KHgSb with a honeycomb structure of their Au-Se and Hg-Sb layers feature band inversion very similar to HgTe which is a strong precondition for existence of the topological surface states. LiAuSe is a strong TI, whereas KHgSb a weak TI.

We will discuss the necessary and sufficient conditions for new TI materials, based in symmetry and bonding arguments.

PROBING TWO TOPOLOGICAL SURFACE BANDS OF ANTIMONY-TELLURIDE BY SPIN-POLARIZED PHOTOEMISSION SPECTROSCOPY

Christian Pauly[1], Gustav Bihlmayer[2], Marcus Liebmann[1], Martin Grob[1], Alexander Georgi[1], Dinesh Subramaniam[1], Markus Scholz[3], Jaime Sánchez-Barriga[3], Andrei Varykhalov[3], Stefan Blügel[2], Oliver Rader[3], Markus Morgenstern[1]

[1] II. Institute of Physics B and JARA-FIT, RWTH Aachen University, Aachen, Germany;
[2] Peter Grünberg Institute and Institute for Advanced Simulation, Forschungszentrum Jülich and JARA-FIT, Jülich, Germany;
[3] Helmholtz-Zentrum Berlin, Synchrotron BESSY II, Berlin, Germany

Topological insulators are a new phase of quantum matter giving rise to, e.g., a quantum spin Hall phase without external magnetic field [1-3]. Large spin orbit interaction and inversion symmetry lead to nontrivial spin polarized edge or surface states which reside in a bulk energy gap and are protected by time reversal symmetry.

Using high resolution spin- and angle-resolved photoemission spectroscopy, we map the electronic structure and spin texture of the surface states of the topological insulator Sb_2Te_3. Similar to the well explored Bi_2Te_3 and Bi_2Se_3 which possess TI properties with the simplest electronic structure [4], we directly show that the phase change material Sb_2Te_3 [5] exhibits Z_2 topological properties with a single spin-Dirac cone at the Γ-point. Figure 1 shows an ARPES data on a single crystal Sb_2Te_3 revealing the spin polarized nature of the topological surface states. Only the lower part of the Dirac cone is detected due to p-type doping in Sb_2Te_3. The spin is found to be perpendicular to $k_\parallel$ within the surface plane and rotating counterclockwise for the lower part of the Dirac cone in accordance with DFT calculation.

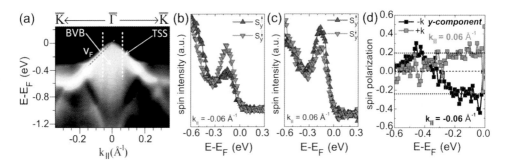

Figure 6: (a) ARPES measurement of the lower Dirac cone around Γ with topological surface states (TSS) and bulk valence band (BVB) marked; $hv = 55$eV. Fermi velocity v_F is deduced from the linear dispersion (dashed line). (b), (c) Spin resolved energy distribution curves (EDCs) for the spin component perpendicular to $k_\parallel$ recorded at $k_\parallel$-values as marked by dashed lines in (a); $hv = 54.5$eV (d) Resulting spin polarization as a function of energy for the two different $k_\parallel$ as indicated.

Beside the topological surface states, an additional strongly spin-orbit split surface state is observed at higher binding energy [Figure 2 (a)]. This band possess Rashba-type characteristics leading to a strong spin splitting (Rashba coefficient is found to be $α = 1.4$ eVÅ). Next to the Rashba properties, the band exhibits further characteristics which are similar to those of the topological surface states. In combination with DFT calculation, we demonstrate that the different spin branches of the band connect the upper and lower band-edges of the gap in which the state is located. Similar to the valence and conduction band which are connected by the topological surface states, these band-edges are also inverted by spin orbit interaction. Furthermore, in Γ-K direction, the band is located within a spin-orbit gap, governing the energy position of the state [Figure 2 (c) and (d)]. So our measurement provides direct evidence for an argument given by Pendry and Gurman [6], which says that there must be at least one surface state inside a SO gap, if the gap is located in the zone. Thus, similar to the topological state, this state is protected by symmetry.

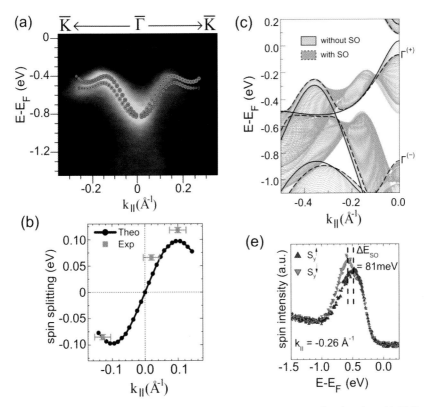

Figure 7: (a) ARPES dispersion of the Rashba type surface state along Γ-K direction; $hv = 22$eV. Band structure from DFT is superposed. (b) Calculated spin splitting of the Rashba state in comparison with measured spin splitting showing good agreement. (c) DFT bulk band structure along Γ-K direction with and without SO interaction; black lines mark the band edge with (dashed) and without (solid) SO interaction. With SO interaction a gap opens around $E = -0.5$eV and $k_\parallel = \pm 0.26$Å^{-1}. (e) Spin resolved EDCs measured at the position of the SO gap revealing the spin splitted Rashba state and confirming the theoretic prediction of Pendry and Gurman [6]; $hv = 22$eV. The resulting spin splitting ΔE_{SO} is marked.

[1] B. A. Bernevig et al., Science **314**, 1757 (2006).
[2] B. A. Bernevig et al., Phys. Rev. Lett. **96**, 106802 (2006).
[3] M. König et al., Science **318**, 766 (2007).
[4] H. Zhang et al., Nature Phys. **5**, 438 (2009).
[5] D. Lencer et al., Nature Mat. **7**, 972 (2008).
[6] J. B. Pendry and S. J.Gurman, Surf. Sci. **49**, 87 (1975).

QUASIPARTICLE CORRECTIONS AND SURFACE STATES OF TOPOLOGICAL INSULATORS Bi_2Se_3, Bi_2Te_3 AND Sb_2Te_3.

Irene Aguilera, Christoph Friedrich, Gustav Bihlmayer, Stefan Blügel

Peter Grünberg Institut and Institute for Advanced Simulation, Forschungszentrum Jülich and JARA, D-52425 Jülich, Germany.

In topological insulators, the spin-orbit interaction leads to band inversion and to nontrivial edge or surface states, which reside in the energy gap of the bulk material and are protected by time-reversal symmetry. Among topological insulators, the family formed by Bi_2Se_3, Bi_2Te_3 and Sb_2Te_3 is one of the most widely studied due to the simplicity of their surface states consisting of a single Dirac cone [1].

Most of the *ab-initio* calculations present in the literature for this family are carried out within density-functional theory (DFT) employing the local-density (LDA) or generalized gradient approximation (GGA), which can describe the ground-state properties accurately but is not appropriate for band gaps and excited-state properties, such as the quasiparticle energies.

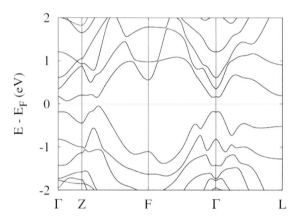

Figure 1: DFT-LDA electronic band structure of Sb_2Te_3 along the high-symmetry lines, calculated using the FLAPW method as implemented in the FLEUR code [2]. Spin-orbit interactions have been included self-consistently in the second-variation method.

To overcome this problem, one can use many-body perturbation theory in the GW approximation to calculate quasiparticle self-energy corrections for the electronic states, which yields results that are directly comparable with photoemission spectroscopy measurements.

Recently, GW calculations on Bi_2Se_3 and Bi_2Te_3 [3,4] have shown that not only a much better agreement of the band gap but also an improvement in the effective masses is found when comparing to experimental results.

We have performed GW calculations for these materials within the all-electron FLAPW formalism, including spin-orbit coupling. While previous GW calculations for these materials were performed in the plane-wave pseudopotential approach, our all-electron calculations avoid the use of pseudopotentials. We have investigated changes in the band structures owing to these two different approaches.

In addition, the GW code SPEX allows us to fully take into account spin-orbit coupling [5], in contrast to previously used approaches where the spin-orbit coupling was treated within LDA or GGA and added *a posteriori* to the quasiparticle spectrum.

While the topological nature of Bi_2Se_3 and Bi_2Te_3 is well established both theoretically and experimentally, work on topological effects in Sb_2Te_3 is scarce. Recently, thin films grown by molecular beam epitaxy have exhibited a Dirac cone around the Fermi level when measured with ARPES [6] and spin-resolved photoemission experiments have mapped the electronic structure and spin texture of the surface states of Sb_2Te_3 [7].

In light of these results, we have extended our GW study to Sb$_2$Te$_3$ and studied in detail the quasiparticle inverted band structure of bulk Sb$_2$Te$_3$ when spin-orbit coupling effects are included.

We have also investigated the surface states and spin chirality of the Dirac cone of Sb$_2$Te$_3$ by simulating films of a thickness of six quintuple layers embedded in vacuum, and compared results with spin- and angle-resolved photoemission spectroscopy measurements.

[1] H. Zhang et al., Nature Phys. **5**, 438 (2009).
[2] http://www.flapw.de
[3] E. Kioupakis et al., Phys. Rev. B, **82**, 245203 (2010).
[4] O. V. Yazyev et al., arXiv: 1108.2088
[5] R. Sakuma et al., Phys. Rev. B, **84**, 085144 (2011).
[6] G. Wang et al., Nano Res. **3**, 874 (2010).
[7] C. Pauly et al., arXiv: 1201.4323v1

ELECTRIC FIELD INDUCED SWITCHING OF MAGNETIZATION IN CoPt$_3$/BaTiO$_3$ HETEROSTRUCTURES

Konstantin Z. Rushchanskii[1], Felipe Garcia-Sanchez[2], Riccardo Hertel[3], Stefan Blügel[1], Marjana Ležaić[1]

[1]Peter Grünberg Institut, Quanten-Theorie der Materialien (PGI-1), Forschungszentrum Jülich GmbH and JARA, 52425 Jülich, Germany;
[2]Peter Grünberg Institut, Elektronische Eigenschaften (PGI-6), Forschungszentrum Jülich GmbH, 52425 Jülich, Germany; [3]Institut de Physique et Chimie des Matériaux de Strasbourg, Université de Strasbourg, CNRS UMR 7504, Strasbourg, France

Multiferroics are materials where two (or even more) ferroic orders coexist simultaneously [1]. They can be single phase, where such a coexistence is an intrinsic property of the material, or of a composite, in which the multiple ferroic order is artificially combined from several phases each of them possessing at least one ferroic order, which are coupled together. Due to the d^0-ness requirement for ferroelectricity in perovskite compounds [2], the coexistence of this property with magnetism is rather rare. Moreover, the superexchange interaction in perovskite-like materials renders them antiferromagnetic. In the so-called improper multiferroics, on the other hand, one finds both magnetic and ferroelectric properties combined. However, here the polarization values are rather small and ordering temperatures low.

A promising approach in multiferroics design is to use composites build from separate materials [1], one with strong ferromagnetic ordering and high Curie temperature, second with ferroelectric properties. An important issue in this design approach is the coupling between two different ferroic orders.

Here we report the theoretical results of our multi-scale studies of the switching properties in CoPt$_3$/BaTiO$_3$ heterostructures exhibiting strong coupling between the ferromagnetic and ferroelectric order. Cobalt-platinum thin-film alloys are mostly studied as materials with large magnetocrystalline anisotropy, making them potential candidates for magnetic recording media for high-density data storage. Here we use ordered CoPt$_3$ alloys with an easy axis located in the film plane. BaTiO$_3$ is, from the other side, a ferroelectric material with high transition temperature to the non-polar phase [3]. We study structural as well as magnetic and electronic properties of the resulting superstructure built from two layers of CoPt$_3$ and one layer of TiO$_2$-terminated BaTiO$_3$. We map out the total-energy surface for different magnetization directions as a function of the direction of the ferroelectric polarization and find that its shape promises a controllable switching not only of the magnetic easy axis but also of the magnetization orientation by an external electric field. Macrospin simulations with input parameters extracted from first-principles calculations predict fast (below 1 ns) as well as controllable switching processes of the magnetization in this system.

[1] M. Fiebig, J. Phys. D: Appl. Phys. **38**, R123 (2005).
[2] N. A. Hill, J. Phys. Chem. B., **104**, 6694 (2000).
[3] M.E. Lines, A.M. Glass, "Principles and applications of ferroelectrics and related materials", Oxford University Press, (2001) 680p.

EVIDENCE OF PARAELECTROMAGNON-LIKE EXCITATIONS IN THz SPECTRA OF HEXAGONAL YMnO$_3$ SINGLE CRYSTAL

Veronica Goian,[1] Stanislav Kamba,[1] Christelle Kadlec,[1] Petr Kužel,[1]
Konstantin Z. Rushchanskii,[2] Marjana Ležaić,[2] Roman V. Pisarev[3]

[1]Institute of Physics of the ASCR Prague, Na Slovance 2, 18221 Praha 8, Czech Republic;
[2]Peter Grünberg Institut, Forschungszentrum Jülich GmbH 52425, Jülich and JARA-FIT, Germany;
[3]Ioffe Physical Technical Institute, St. Petersburg, Russia

Hexagonal YMnO$_3$ is a high temperature ferroelectric (T$_C$ ≈ 1250 K) which becomes antiferromagnetic (AFM) below T$_N$ = 70 K. It exhibits a large spin-phonon coupling which was revealed by temperature dependence of dielectric permittivity[1], thermal conductivity[2] and by inelastic neutron scattering[3]. The last technique proved the existence of three magnon branches below T$_N$.[3,4]

Our time-domain THz spectra reveal a sharp antiferromagnetic resonance below T$_N$ in the spectra of magnetic permeability μ_a. This magnetic excitation softens from 41 to 32 cm^{-1} upon heating and finally disappears above T$_N$. An additional weak and heavily-damped excitation is seen in the spectra of complex dielectric permittivity ε_c within the same frequency region (Fig.1). This excitation contributes to the dielectric spectra in both antiferromagnetic and paramagnetic phase. Its oscillator strength significantly increases upon heating toward room temperature (follows the Bose-Einstein formula), thus providing evidence for piezomagnetic or higher-order coupling to polar phonons. Other heavily-damped dielectric excitations are detected near 100 cm^{-1} in the paramagnetic phase in both ε_c and ε_a spectra, and they exhibit similar temperature behavior. These excitations appearing in the frequency range of magnon dispersion branches well below polar phonons could remind **electromagnons**, however their temperature dependence is quite different.

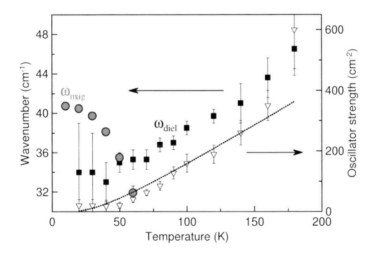

Fig.1. Temperature dependences of parameters of the resonances observed in magnetic and dielectric spectra. Closed circles: frequency of the AFM resonance; Solid squares and open triangles: eigen-frequency and oscillator strength, respectively, of the dielectric excitation. Dotted line is the result of the oscillator strength using the Bose-Einstein formula.

We have used density functional theory for calculating phonon dispersion branches in the whole Brillouin zone. A detailed analysis of these results and of previously published magnon dispersion branches brought us to the conclusion that the **observed absorption bands stem from phonon-phonon and phonon-paramagnon differential absorption processes.** [5] The latter is enabled by interaction of phonons polarized in hexagonal plane with short-range correlated spins in the paramagnetic phase.

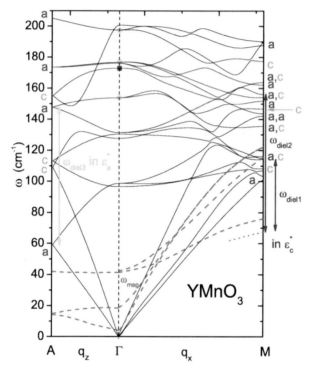

Fig.2. Dispersion branches of phonons (solid black line) and magnons (experimental[4] at 7 K; red dashed lines) The red-dotted line indicates the presumable dispersion of the paramagnon near the M point. The symbols shown at the Brillouin zone edges indicate the polarization of the phonons at the Brillouin zone boundary: a and c stand for phonons polarized within the hexagonal plane and in the perpendicular direction, respectively. In the Γ-point, the E_1 and A_1 phonons observed experimentally[6] are marked by green and blue points, respectively; other modes are silent. Blue arrows with assignment ω_{diel1} and ω_{diel2} indicate phonon-paramagnon excitations observed in the dielectric loss spectra of ε_c. Green arrow marked as ω_{diel3} indicates a broad multiphonon absorption observed in the ε_a loss spectra.

[1] D.G. Tomuta et al., J. Phys : Condens. Matter **13**, 4543 (2001)
[2] P.A. Sharma et al. Phys. Rev. Lett. **93**, 177202 (2004)
[3] S. Petit et al. Phys. Rev. Lett. **99**, 266604 (2007)
[4] T.J. Sato et al. Phys. Rev. B **68**, 014432 (2003)
[5] C. Kadlec, V. Goian, et al. Phys. Rev. 84, 174120 (2011)
[6] M. Zaghrioui et al., Phys. Rev. B, 78, 184305 (2008)

ELECTRICAL BEHAVIOR OF $Bi_{0.95}Nd_{0.05}FeO_3$ THIN FILMS GROWN BY THE SOFT CHEMICAL METHOD.

C. R. Foschini[c], F. Moura[a], M. A. Ramirez[b], J. A. Varela[c], E. Longo[c], A.Z. Simões[a,b]*

[a] Universidade Federal de Itajubá- Unifei - Campus Itabira, Rua São Paulo, 377, Bairro: Amazonas, CEP 35900-37, Itabira, MG, Brazil, Phone: +55 31 3834-6472/6136; Fax: +55 31 3834-6472/6136.
[b] Universidade Estadual Paulista- Unesp - Faculdade de Engenharia de Guaratinguetá, Av. Dr. Ariberto Pereira da Cunha, 333, Bairro Pedregulho, CEP 12516-410– Guaratinguetá-SP, Brazil, Phone +55 12 3123 2765
[c] Laboratório Interdisciplinar em Cerâmica (LIEC), Departamento de Físico-Química, Instituto de Química, UNESP, CEP: 14800-900, Araraquara, SP, Brazil.
*e-mail-alezipo@yahoo.com.

This paper focuses on electrical properties of $Bi_{0.95}Nd_{0.05}FeO_3$ thin film (BNFO05) deposited on $Pt/TiO_2/SiO_2/Si$ (100) substrates by the soft chemical method. A BNFO05 single phase was simultaneously grown at a temperature of 500°C for 2 hours. Room temperature magnetic coercive field indicates that the film is magnetically soft. The remanent polarization (P_r) and the coercitive field (E_c) were 51 µC/cm² and 65.0 kV/cm respectively, and are superior to the values found in the literature. XPS results show that the oxidation state of Fe was purely 3+, which is advantageous for producing a BNFO05 film with a low leakage current. The polarization of the Au/BNFO05 on $Pt/TiO_2/SiO_2/Si$ (100) capacitors with a thickness of 230 nm exhibited no degradation after 1×10^8 switching cycles at a frequency of 1 MHz. Experimental results demonstrated that the soft chemical method is a promising technique to grow films with excellent electrical properties, which can be used in various integrated device applications.

Keywords: A. thin films; B. chemical synthesis; C. electron diffraction; D. ferroelectricity.

THEORETICAL INVESTIGATION OF THE MAGNETOELECTRIC PROPERTIES OF MnPS$_3$

Diana Iuşan[1], Kunihiko Yamauchi[2], Kris Delaney[3], Silvia Picozzi[1]

[1]Consiglio Nazionale delle Ricerche - Superconducting and Innovative Materials and Devices (CNR-SPIN), L'Aquila, Italy; [2]Institute of Scientific and Industrial Research (ISIR-SANKEN), Osaka University, Osaka, Japan; [3]Materials Research Laboratory, University of California, Santa Barbara, USA

Magnetoelectricity - the appearance of a magnetization M under an applied electric field E or of an induced ferroelectric polarization P by an applied magnetic field H - is a cross-coupling effect with potential applications for materials with enriched functionality. Using both a phenomenological approach based on Landau theory and ab initio density functional theory (DFT), we have investigated the magnetoelectric (ME) properties of MnPS$_3$.

The compound has a layered structure with the Mn atoms forming a honeycomb lattice in the (ab) plane. Antiferromagnetic couplings exist among the nearest-neighboring Mn spins (see Figure 1).

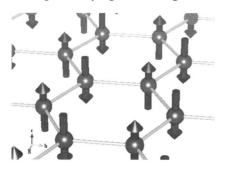

Figure 1: Illustration of the MnPS$_3$ spin structure. Only the Mn atoms are shown.

The symmetry analysis indicates the existence of trilinear magnetoelectric coupling terms of type $P_i M_j L_k$, where the subscripts index different components of the vectors. M and L are the ferromagnetic and antiferromagnetic order parameters, respectively. We have found non-zero xy, yx, yz, and zy elements of the ME tensor by using Landau theory.

Within DFT, the ME tensor was calculated as the induced ferroelectric polarization P upon an applied magnetic field H. In order to simulate the effect of the applied magnetic field, a Zeeman term was introduced in the Kohn-Sham Hamiltonian, as explained in Ref. 1. The new electronic structure was determined self-consistently, and the induced P was determined by using the Berry phase method [2, 3].

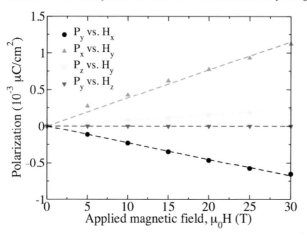

Figure 2: Induced ferroelectric polarization versus applied magnetic field. The x, y, z subscripts index different components of P and H. The slope of the lines determines the ME tensor elements.

In order to separate different contributions to the magnetoelectric tensor we have calculated P(M) in the absence and presence of spin-orbit coupling. Having the spin-orbit coupling turned off, we obtained zero values for all the elements of the ME tensor, indicating that symmetric exchange striction effects play no role in the magnetoelectricity of this compound. Switching on the spin-orbit coupling, a finite linear ME coupling is found. Owing to the relativistic origin of the ME effect, the tensor elements have modest values, of the order of ps/m.

[1] E. Bousquet, N. A. Spaldin, and K. T. Delaney, Phys. Rev. Lett. 106, 107202 (2011).
[2] R. D. King-Smith and D. Vanderbilt, Phys. Rev. B 47, 1651 (1993).
[3] R. Resta, Rev. Mod. Phys. 66, 899 (1994).
[4] E. Ressouche, M. Loire, V. Simonet, R. Ballou, A. Stunault, and A Wildes, Phys. Rev. B 82, 100408 (2010).

HIGH TEMPERATURE MULTIFERROIC COMPOUNDS: FROM BiFeO$_3$ TO Bi$_2$FeCrO$_6$

Jian Yu[*], Linlin Zhang, and Xianbo Hou

Tongji University, 200092, Shanghai, China, [*]E-mail: jyu@tongji.edu.cn

BiFeO$_3$ is well known antiferromagnetic-ferroelectric multiferroics. Since its birth from 1957, the difficulties to obtain high insulating and single-phased samples are like the shadow following BiFeO$_3$ synthesization, for BiFeO$_3$ is not thermal stable with peritectic decomposition of BiFeO$_3 \rightarrow$ Bi$_{25}$FeO$_{39}$ (PM) + Bi$_2$Fe$_4$O$_9$ (AFM) above 675°C.

BiFeO$_3$ belongs to the point group of $3m$, but the presence of spin cycloidal modulation magnetic structure forbids the first-order magnetoelectric coupling effect. In order to destroy this incommensurate cycloidal spin structure for releasing linear magnetoelectric effect with off-diagonal components of $\alpha_{12}=-\alpha_{21}$, the composition modification at A-site or/and B-site are widely adapted, but due to high conductivity and impurities for those synthesized modified-BiFeO$_3$ samples, some confused and debated results are observed experimentally on the magnetoelectric effect.

From Goodenough rule of magnetic superexchange interaction, Fe-O-Fe is antiferromagnetic ordering while Fe-O-Cr is ferromagnetic so that metastable double perovskite-structured Bi$_2$FeCrO$_6$ is a promising candidate for ferromagnetic-ferroelectric multiferroics with both Curie temperature above room temperature. The point group reduces to 3 for Bi$_2$FeCrO$_6$ from $3m$ for BiFeO$_3$ and the point group of 3 is substantially available for linear magnetoelectric effect with both P and $M_\parallel c$ according to Ascher analysis on Heesch-Shubnikov point groups.

Through seven years work, it is found substantial to improve insulating and thermal stability of BiFeO$_3$-related oxides by forming solid solution and choosing suitable heat treatment program. In this talk, the challenges for developing high insulating and single-phased BiFeO$_3$-related multiferroic ceramics are briefly discussed and the experimental results on high insulating Bi$_{1-x}$La$_x$Fe$_{1-y}$Ti$_y$O$_3$ and Bi$_2$FeCrO$_6$-PbTiO$_3$ ceramics presented in details on structural, ferroelectric and magnetic properties. For instance, Bi$_{0.98}$La$_{0.02}$Fe$_{0.99}$Ti$_{0.01}$O$_3$ ceramics can bear more than 5kV/mm DC electrical field at the temperature of 140°C and then take d_{33} = 0.6 pC/N.

HAADF STEM TOMOGRAPHY OF FERRIMAGNETIC $Fe_xCo_{(3-x)}O_4$ NANOSTRUCTURES EMBEDDED IN HIGHLY ORDERED ANTIFERROMAGNETIC Co_3O_4 MESOPOROUS TEMPLATES.

Lluís. Yedra[1], Sònia Estradé[1,2], Eva. Pellicer[3], Moisés Cabo[3], Alberto López-Ortega[4], Marta Estrader[4], Josep Nogués[5], Dolors Baró[3], Zineb Saghi[6], Paul A. Midgley[6] and Francesca Peiró[1]

[1] Laboratory of Electron Nanoscopies (LENS)- MIND/IN2UB, Dept. d'Electrònica, Universitat de Barcelona, c/ Martí Franqués 1, E-08028 Barcelona, Spain
[2] CCiT, Scientific and Technical Centers, Universitat de Barcelona, E-08028, Barcelona, Spain
[3] Departament de Física, Facultat de Ciències, Universitat Autònoma de Barcelona, E-08193 Bellaterra, Spain.
[4] CIN2(CIN-CSIC) and Universitat Autònoma de Barcelona, Catalan Institute of Nanotechnology, Campus de la UAB, E-08193 Bellaterra, Spain.
[5] Institució Catalana de Recerca i Estudis Avançats (ICREA) and Centre d'Investigació en Nanociència i Nanotecnologia (ICN-CSIC), Campus Universitat Autònoma de Barcelona, E-08193 Bellaterra, Spain.
[6] Department of Materials Science and Metallurgy, University of Cambridge, Pembroke Street, Cambridge CB2 3QZ, United Kingdom.

Highly ordered mesoporous materials are gaining interest due to their high surface to volume ratio ant hence their potential applications in a broad range of fields. In this work we present the structural characterization of a series of antiferromagnetic (AFM) Co_3O_4 templates, nanocast replicas of mesoporous KIT-6 silica, filled with ferrimagnetic (FiM) ($Fe_xCo_{(2-x)}O_4$). This novel concept allows increased versatility because of the AFM-FiM exchange interactions as well as the synergic combination of properties of both constituents [1]. High-angle annular dark field (HAADF) tomography in the transmission electron microscope in scanning mode (STEM) [2] was here used in order to assess the ability of different amounts of iron precursor to impregnate the structure of the Co_3O_4 templates (see Figure 1).

3D reconstruction of the tilted series, acquired for the host Co_3O_4 structure and for two different infiltration loadings, provided understanding of the interlaced structure of the mesoporous KIT-6 replicas and allowed to assess their high structural regularity. Table 1 shows that the pore size decreased as the amount of Fe-precursor increased, which suggests the growth of nanotubes within the Co_3O_4 channels, and it was observed that the iron precursor infiltration ability reaches a plateau as the Fe incorporation saturates before completely filling the pores. At higher charges, the iron oxide grew outside of the original framework.

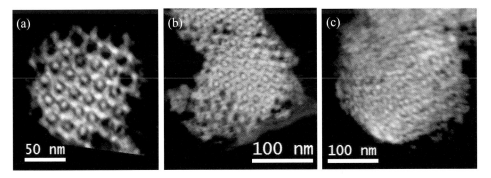

Figure 1: Slices through 3D reconstruction of mesoporous particles with different infiltrations: (a) pure Co_3O_4 KIT-6 replica, (b) template infiltrated with iron charge 4 and (c) template infiltrated with iron charge 6.

Infiltration	Apparent pore wall thickness (nm)
0	5.13 ± 1.07
4	7.77 ± 0.42
6	7.68 ± 0.99

Table 1: Apparent pore wall thickness measured for the three slices in Figure 1.

[1] J. Nogués, J. Sort, V. Langlais, V. Skumryev, S. Suriñach, J.S. Muñoz, M.D. Baró. Exchange bias in nanostructures Phys. Rep. 422 (2005) 65-117.
[2] M. Weyland, P.A. Midgley, J.M. Thomas, Electron Tomography of Nanoparticle Catalysts on Porous Supports: A New Technique Based on Rutherford Scattering, The Journal of Physical Chemistry B. 105 (2001) 7882-7886.

NON-COLLINEAR MAGNETISM IN 3d-5d ZIGZAG CHAINS: TIGHT BINDING MODEL AND AB-INITIO CALCULATIONS

Vikas Kashid[1], Timo Schena[2], Bernd Zimmermann[2], Vaishali Shah[3], H. G. Salunke[4], Y. Mokrousov[2] and S. Blügel[2]

[1]Department of Physics, University of Pune, Pune 411 007, India
[2]Peter Grünberg Institut and Institute for Advanced Simulation, Forschungszentrum Jülich and JARA, D-52425 Jülich, Germany
[3]Interdisciplinary School of Scientific Computing, University of Pune, Pune 411 007, India
[4]Technical Physics Division, Bhabha Atomic Research Centre, Mumbai 400 085, India

We have investigated the anisotropic exchange interactions in one-dimensional free- standing zigzag biatomic 3d(Fe,Co)-5d(Ir,Pt,Au) chains. The calculations were performed using the density functional theory (DFT) within the full-potential linearized augmented plane wave method (FLAPW) as implemented in the Jülich code FLEUR. [1] We calculated isotropic exchange interaction in 3d-5d chains in a chemical unit cell, using generalized Bloch theorem valid without spin-orbit coupling [2]. Our results on isotropic exchange interaction reveal that, 3d-5d chains show ferromagnetic ground state, except for the Fe-Pt and Co-Pt chains. The ground state of Fe-Pt and Co-Pt is the spin spiral with the spiral angle of 10.70° and 25.20°, respectively. The dispersion energy [E($\mathbf{q}$=0)−E($\mathbf{q}$)] increases as the spin-spiral vector $\mathbf{q}$ is increased.

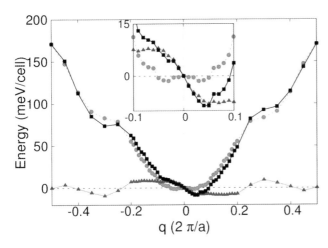

Figure 1: Dzyalonshinskii-Moriya interaction (DMI) contribution to the total energy of the spin-spiral (circles) wihin the 1st order perturbation theory (triangles, along the z-axis). The total corrected by DMI energy is shown with filled squares. The inset shows magnified region close to $\mathbf{q}$ = 0, in the vicinity of the ground state.

The antisymmetric exchange interaction, or the Dzyalonshinskii-Moriya interaction (DMI) was calculated by considering the spin-orbit interaction in the first order perturbation theory [3]. Spin-orbit correction has a non-monotonous behavior and is antisymmetric with respect to the spin-spiral vector $\mathbf{q}$. Our investigations show that, DMI has a significant contribution to the energy of spin spirals up to 10-20 meV/cell. The spin-spiral energy without spin-orbit coupling shows a degenerate ground state, however, the degeneracy is lifted when the DMI correction is included. The resulting system shows a ground state having positive or negative rotational sense of the spin spirals. Among non-collinear ground-states structures, Fe-Ir, Fe-Au and Co-Au are left handed spin spiral, whereas for Co-Pt and Fe-Pt right-handed spirals are stabilized. The DMI along the wire axis and perpendicular to the wire axis direction vanishes due to symmetry in the chain.

In order to understand the results of our *ab initio* calculations for the DMI interaction, we have performed a detailed analysis on a trimer system based on the minimal tight binding model. The model is restricted to d_{xz} and d_{yz} orbitals of the non-magnetic site and d_{xz} of the magnetic site. Our investigations show that, the degree of hybridization between d orbitals of magnetic and non-magnetic sites plays crucial role in determining the DMI in 3d-5d chain. The general features of the DMI observed in 3d-5d chains related with the symmetries and the amplitudes are analogous with that observed in the trimer system.

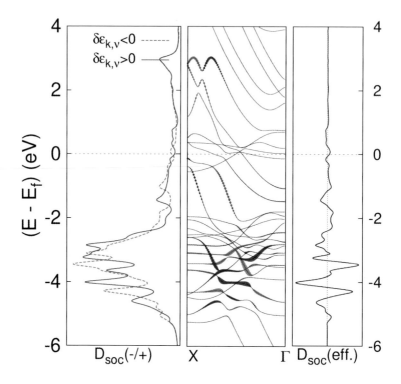

Figure 2: *Left:* The Lorentizian distribution of the spin-orbit coupling (D_{soc}) for Fe-Au, when integrated for all k-kpoints in the first Brillouin zone. The positive difference between the eigenvalues with spin orbit coupling is shown by continuous line and negative difference is shown by dotted line. *Middle:* The difference between the eigenvalues with SOC is shown by filled circles. *Right:* The effective SOC is plotted against the energy spectrum with integration on all **k**-points.

[1] www.flapw.de
[2] L.M. Sandratskii, J. Phys.: Cond. Matter **3**, 8565(1991)
[3] M. Heide *et al.*, Physica B, **404**, 2678 (2009)

DIRECT OBSERVATION OF TRANSIENT NEGATIVE CAPACITANCE IN DOMAIN WALL OF FERROELECTRIC THIN FILMS

Yu Jin Kim[1], Min Hyuk Park[1], Han Joon Kim[1], Doo Seok Jeong[2], Anquan Jiang[3], and Cheol Seong Hwang[1]

[1]WCU Hybrid Materials Program, Department of Material Science & Engineering and Inter-university Semiconductor Research Center, Seoul National University, Republic of Korea,
[2]Electronic Materials Center, Korea Institute of Science and Technology, Republic of Korea,
[3]State Key Laboratory of ASIC & System, Department of Microelectronics, Fudan University, Shanghai, 200433, China

The bi-stable polarization states of ferroelectric (FE) materials received a great deal of attention for its application to electronic devices. However, the switching between the two polarizations states, generally involves formation of domains, which is thermo-dynamically not favored but kinetically favored process. Although many models have been developed to explain domain formation of FE materials, it is still an intriguing research topic for modern electronic devices. One of the most arguable reports in recent years is the involvement of the negative capacitance (NC) effect in the ferroelectrics in FE – dielectrics (DE) stacked system. [1] If NC effect is stably obtained, it may revolutionize the electronic devices because capacitance is not a simple linear function of area anymore. Furthermore, voltage can be amplified without using external circuit. This means that the conventional scaling rule for semiconductor devices must be changed. However, this can be fundamentally difficult because of the very high tendency of domain formation in ferroelectric thin film system; especially the FE-DE system could involve a serious depolarization effect which profoundly favors the poly-domain structure.

In this work, the authors observed NC effect from the domain wall of ferroelectric thin film in a FE-DE stacked system. It is considered that the physical state of domain wall region is prone to show NC because P ~ 0 there but still the crystal structure is the same as FE bulk. In normal FE switching, the domain propagation is generally quite fast, so observing the electric response of the domain wall regions is rather difficult. Therefore, the authors designed several FE-DE systems where the DE controls the compensating charge injection to an appropriate level and enhance the chance to see the responses of domain boundaries to the electric pulses. This must be a thermodynamically reasonable result considering polarization compensated charge contribution to the Landau's phenomenological model. Detailed experimental and model study results will be presented.

[1] S. Salahuddin and S. Datta, *Nano Lett.* **8**, 405 (2008)

STRONTIUM TITANATE ULTRA-THIN FILM CAPACITORS ON SILICON SUBSTRATES FOR APPLICATION IN DYNAMIC RANDOM ACCESS MEMORY (DRAM)

Sebastian Schmelzer[1,2], Ulrich Böttger[1,2], Rainer Waser[1,2]

[1]Institute for Materials in Electrical Engineering and Information Technology (IWE2), RWTH Aachen University of Technology, Aachen, Germany;
[2]JARA-Fundamentals of Future Information Technology, Research Center Jülich, Jülich, Germany

$SrTiO_3$ (STO) is one of the most promising candidates of high permittivity materials in capacitors for future DRAM applications. However, after decades of work on this topic, in this case especially on metal-STO-metal structures with various electrode materials, a typically observed decline of the dielectric constant with decreasing STO film thickness seems to be inevitable. Several existing models show this decline to originate from the formation of interfacial passive layers. The permittivity of these passive layers is by definition significantly below the permittivity of the dielectric itself. Despite these numerous explanations and models, the effect is not completely understood yet [1]. In this work, we present the analysis of non-epitaxial STO ultra-thin film capacitors with $SrRuO_3$(SRO) electrodes showing insignificantly small interfacial passive layers [2]. Additionally, conduction mechanisms of these devices with film thickness down to 4 nm are investigated.

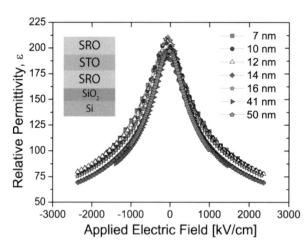

Figure 1: Relative permittivity versus applied electric field for various STO film thickness. The inset shows the schematic of the investigated devices. The Gaussian-type field-dependence is characteristic for perovskite titanates [3]. The widely used SiO_2 capacitance equivalent thickness (CET) therefore is determined to be well below 0.3 nm for both 7 nm and 10 nm STO thickness, meeting the ITRS requirements for future DRAM application by far [4].

Using sputter deposition, the capacitors were grown in situ on oxidized silicon substrates at a temperature of 550 °C and a pressure of 8E-3 mbar. The surface roughness of the layers was investigated with atomic force microscopy and found to be in the range of 0.2 nm (root mean square) or below. By x-ray reflectometry we determined the particular film thickness of each layer and partially confirmed the results by FIB-cut electron microscopy cross sections. Fig.1 shows the permittivity to be in the range of 200 at room temperature for all samples, not depending on STO film thickness. Not displayed temperature dependence of permittivity follows a weak Curie-Weiss behavior.

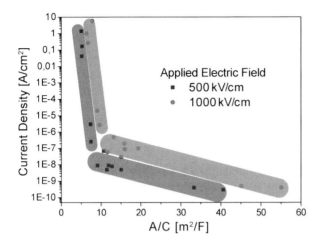

Figure 2: Leakage current density versus reciprocal capacitance density under varying applied electric field, showing a dramatic change in the relation of current density and capacitance. Devices located in the regime with the larger slope all have STO film thickness below 10 nm, the smaller slope regime includes only devices above 10 nm STO film thickness.

A dramatic change in the relation of current density and capacitance is illustrated by the bars in Fig.2. This we believe is originated in the change of the dominant conduction mechanisms from a mixture of several mechanisms like Poole-Frenkel, bulk-limited, and tunneling to mostly thermionic emission. Temperature-dependent IV-measurements on a device with 7 nm STO thickness are shown in Fig.3

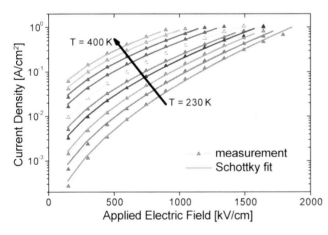

Figure 3: Temperature-dependent current density versus applied electric field for a device with 7 nm STO thickness. Each data set is fitted by the Schottky barrier equation. The characteristic of the field-dependence differs strongly from devices with larger film thickness, which causes the dramatic change in Fig.2.

The Schottky barrier height of the emission strongly depends on the film thickness and is dramatically reduced for the ultra-thin films around 5 nm. Comparing the devices with STO film thickness below 10 nm, the barrier height decreases from 1 eV to below 0.6 eV. Thus, the leakage current requirements by the ITRS cannot be met. Assuming this behavior to be intrinsic for end of roadmap high-k capacitors, this would imply an almost impassable hurdle for future application in DRAM.

[1] L. W. Chang et al., Adv. Mater. 21, 4911 (2009).
[2] S. Schmelzer et al., Appl. Phys. Lett. 97, 132907 (2010).
[3] G. Rupprecht, et al., Phys. Rev. 123, 97 (1961).
[4] ITRS, International Technology Roadmap for Semiconductors, www.itrs.net.

ENHANCEMENT OF FERROELECTRIC POLARIZATION BY INTERFACE ENGINEERING

X. Liu,[1] H. Lu,[1] J. D. Burton,[1] Y. Wang,[1] C. W. Bark,[2] Y. Zhang,[3] D. J. Kim,[1] A. Stamm,[1] P. Lukashev,[1] D. A. Felker,[2] C. M. Folkman,[2] P. Gao,[3] M. S. Rzchowski,[2] X. Q. Pan,[3] C. B. Eom,[2] A. Gruverman,[1] and E. Y. Tsymbal[1]

[1]University of Nebraska, Lincoln, NE, USA.
[2]University of Wisconsin, Madison, USA.
[3]University of Michigan, Ann Arbor, MI, USA.

Utilization of the switchable spontaneous polarization of ferroelectric materials offers a promising avenue for the future of nanoelectronic memories and logic devices, provided nanoscale metal-ferroelectric-metal heterostructures can be engineered to maintain a bi-stable polarization switchable by an applied electric field. The most challenging aspect of this approach is to overcome the deleterious interface effects which tend to render ferroelectric polarization either unstable or unswitchable and which become ever more important as ferroelectric materials are produced thinner and thinner. Here, we demonstrate a synergistic experimental/theoretical approach to enhance ferroelectric polarization in nm-thick ferroelectric films by interface engineering. [1,2]

First-principles density functional calculations and phenomenological modeling demonstrate that a BaO/RuO_2 interface termination sequence in $SrRuO_3/BaTiO_3/SrRuO_3$ epitaxial heterostructures grown on $SrTiO_3$ can lead to a non-switchable polarization state for thin $BaTiO_3$ films due to an unfavorable interface dipole.[1] This dipole at the BaO/RuO_2 interface leads to a pinning of one polarization state and, in thin-film structures, leads to instability of the second state below a certain critical thickness, thereby making the polarization unswitchable. We analyze the contribution of this interface dipole to the energetic stability of these heterostructures. Furthermore, we propose and demonstrate that this unfavorable interface dipole effect can be alleviated by deposition of a thin layer of $SrTiO_3$ at the BaO/RuO_2 terminated interface (Fig. 1a). Our first-principles and phenomenological modeling predict that the associated change of the interface termination sequence to SrO/TiO_2 on both sides of the heterostructure leads to a restoration of bi-stability with a smaller critical thickness, along with an enhancement of the barrier for polarization reversal.

Stimulated by these theoretical predictions we also report on the investigation of polarization dynamics and related behavior in high-quality single-crystalline ultrathin (in the range from 2 to 20 nm) $BaTiO_3$ capacitors by means of piezoresponse force microscopy (PFM), pulsed switching current (PUND) and electrostatic force microscopy (EFM).[2] We show that although polarization is stable in ultrathin $BaTiO_3$ films with no top electrodes, deposition of $SrRuO_3$ top electrodes results in severe polarization relaxation. This effect is a consequence of strong effective depolarizing fields due to unfavorable interface terminations with the deposited electrodes, as opposed to more complete screening in the films by adsorbed charges on the free surface. Based on the first-principle calculations discussed above, we find experimentally that insertion of ultrathin dielectric layers of $SrTiO_3$ between the ferroelectric and the electrode can indeed alleviate stability issues in the case of $SrRuO_3$ electrodes. This approach is confirmed by EFM observations (Fig. 1b), local PFM spectroscopy to determine spatially resolved polarization dynamics and determination of the characteristic relaxation times by PUND.

Poster: Polar dielectrics, optics, and ionics

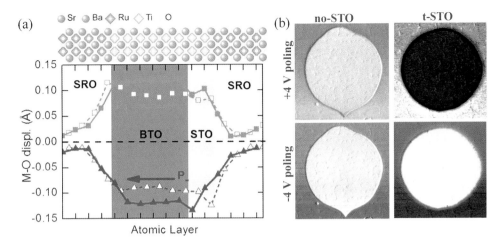

Figure 1: Enhancing ferroelectric polarization by interface engineering. (a) Results of first-principle calculations showing reversible ferroelectric displacements in a 6 u.c. BaTiO$_3$ (BTO) film when 2 u.c. SrTiO$_3$ (STO) interlayer is deposited at the interface. In the absence of STO the polarization of the BTO film has only one stable polarization state. (b) EFM observation of the enhanced polarization retention. Thin STO layer at the interface supports polarization switching evident from the opposite EFM contrast after poling.

Our results demonstrate that interface engineering is a viable approach to enhance ferroelectric properties at the nanoscale.

[1] X. Liu, Y. Wang, P. V. Lukashev, J. D. Burton, and E. Y. Tsymbal, Phys. Rev. B 85, 125407 (2012).
[2] H. Lu, X. Liu, J. D. Burton, C. W. Bark, Y. Wang, Y. Zhang, D. J. Kim, A. Stamm, P. Lukashev, D. A. Felker, C. M. Folkman, P. Gao, M. S. Rzchowski, X. Q. Pan, C. B. Eom, E. Y. Tsymbal, and A. Gruverman, Advanced Materials 24, 1209 (2012).

INFLUENCE OF ADDITIVES WITH LOW MELTING TEMPERATURES ON STRUCTURE, MICROSTRUCTURE, PHASE TRANSITIONS, DIELECTRIC AND PIEZOELECTRIC PROPERTIES OF $BiScO_3 - PbTiO_3$ CERAMICS

E.D. Politova[1], G.M. Kaleva[1], A.V. Mosunov[1], N.V. Sadovskaya[1], A.G. Segalla[2]

[1] Karpov Institute of Physical Chemistry, 105064, Obukha s-str., 3-1/12, b.6
[2] ELPA Company, Panfilovsky pr. 10, Zelenograd, 124460, Moscow, Russia

Piezoelectric materials referred as "smart" ones are widely used in industry [1]. Though most used PZT ceramics have excellent properties, they have rather low T_C values and are considered as ecologically hazardous substances. This stimulates the development of new efficient materials with higher T_C values and with reduced lead content [2]. Enhancement of piezoelectric properties is expected and was discovered in nanostructured materials with reduced domain size [3]. However, dielectric and piezoelectric characteristics usually have non monotonous dependence on average grain size of ceramics. In this work, an approach based on the introduction of the over stoichiometric additives has been used to improve functional properties of perovskite $BiScO_3$-$PbTiO_3$ ceramics. Crystal structure parameters of ceramic solid solutions close to morphotropic boundary $0.36BiScO_3$ - $0.64PbTiO_3$ doped by additives with low melting temperatures in amounts up to 5 w. % were studied (Fig.1).

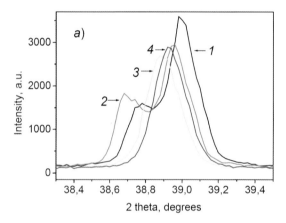

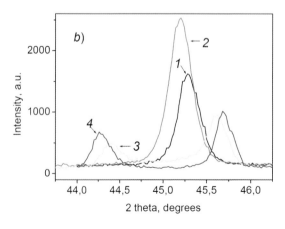

Fig. 1. X-Ray diffraction peaks with $h^2+k^2+l^2=3$ (*a*) and $h^2+k^2+l^2=4$ (*b*) of the samples $0.36BiScO_3$ - $0.64PbTiO_3$ sintered at 1050 K modified by complex additives: 3 w. % Bi_2O_3 + 0.5 w. % Ni_2O_3 (*1*), 3 w. % Bi_2O_3 + 0.5 w. % MnO_2 (*2*), 1 w. % LiF (*3*) and 5 w. % LiF (*4*). The effects of additives on sintering temperature, stoichiometry, structure parameters, microstructure, and functional properties of ceramics were checked. Introduction of these additives influenced the structure of solid solutions, and compositions *1* and *2* had rhombohedral structure, while compositions *3* and *4* had tetragonal structure. Shift of the diffraction peaks positions pointed to small changes in the unit cell parameters of the samples.

Decrease of the phase formation temperature and increase in density of doped ceramics was revealed. Diffraction peaks of samples *1* and *4* sintered at 1223 K were characterized by larger half width values that indicated on small (less than 100 nm) size of coherent length in these ceramics.

The samples were prepared by the solid state reaction method and additionally characterized by SEM, DTA/DSC, SHG, and dielectric spectroscopy methods. Piezoelectric parameters d_{33} and k_t of the preliminary poled samples were measured by the standard methods.

Variation of regimes of temperature treatment allowed us to prepare single phase samples characterized by wide variations in microstructure, with grain size varying from submicron to tens μm (Fig. 2).

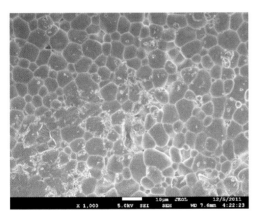

Fig. 2. Microstructure of ceramics $0.36BiScO_3 - 0.64PbTiO_3$ *1* and *2* sintered at T_S = 1223 K, 2 h (*a*) and at T_S =1298 K, 2 h (*б*), respectively.

White bars = 10 μм.

Ceramics *1* have grains with average size ~ 1 μм (T_S = 1223 K) and with average size ~ 10 - 20 μм (T_S = 1323 K). Ceramics *1* prepared at T_S = 1223 K have sharp grain boundaries while those prepared at T_S = 1323 K are characterized by melted boundaries determined by the manifestation of the liquid phase sintering mechanism. Ceramics *2* sintered at T_S = 1298 K have uniform grains with size 5 - 10 μм. Ceramics *3* sintered at T_S = 1298 K have large non uniform grains up to 20 μм. Ceramics *4* prepared at T_S = 1273 K have highly packed grains with size 4 -8 μм. In these ceramics, small crystalline grains with size ~ 100 nm related to the presence of admixture phase were also revealed.

The 1st order sharp ferroelectric phase transitions marked by peaks in dielectric permittivity and dielectric loss versus temperature curves were observed at temperatures near 700 K. In some compositions, effects of dielectric relaxation related to the presence of oxygen vacancies in anion sublattice were observed.

High d_{33} ~ 400 pC/N and k_t ~0.7 values were measured in modified ceramics. The enhancement of piezoelectric properties observed, suppression of the relaxation effect in ceramics *1, 2, 4*, increase in resistance in ceramics *2 - 4* are grounded and discussed in relation to the type and content of additives.

The financial support of the Russian Foundation for Basic Research is acknowledged.

[1] B. Jaffe, W.R. Cook, and H. Jaffe, Piezoelectric ceramics. London & New York: Acad. Press, 1971.
[2] Y. Higuchi et al., Piezoelectric ceramics for high temperature applications // Key Engineering Mater. **421-422**, 375 (2010).
[3] T. Zou et al., Appl. Phys. Lett. **93**, 192913 (2008).

CONTROL OF β- AND γ-PHASE FORMATION IN ELECTROACTIVE P(VDF-HFP) FILMS BY SILVER NANO-PARTICLE DOPING

Dipankar Mandal[1], Karsten Henkel[2], Suken Das[1], Dieter Schmeißer[2]

[1]Department of Physics, Jadavpur University, Kolkata 700032, India; [2]Brandenburg Technical University, Chair Applied Physics and Sensors, 03046 Cottbus, Germany

Poly(vinylidene fluoride) (PVDF) and its copolymers has been widely studied due to their excellent multifunctional properties like ferro-, piezo- and pyro-electric and used in a wide range of applications, such as transducers, touch sensors, nano-generators, non-volatile memories and energy storage devices. PVDF trifluoroethylene copolymer (P(VDF/TrFE)) shows ferroelectric behaviour with high remnant polarization because of the fully electro-active polar β-phase namely the *all*-trans conformation (*TTTT*). In general, it is realized that the polarization of PVDF as well as of P(VDF/TrFE) can be influenced by the chemical modification of the material with chlorofluoroethylene (CFE), chlorotrifluoroethylene (CTFE) or hexafluoropropylene (HFP) groups. This leads for example to ferroelectric relaxor polymers with low remnant polarization [1] suggesting the use as high energy density materials.

Recently, a special attention onto P(VDF-HFP) copolymer has been paid regarding its application in the field of electric energy storage [2, 3]. It was found that mainly the electro-active crystalline phases, such as β- and γ- phases, play the vital role for any electronic and electrical application. For example, the β-phase with edge-on orientations (molecular chains parallel to the substrate) is suitable for non-volatile memory operation because of its easy dipole switching properties. In this orientation, the molecular dipoles (CF_2) are statistically distributed perpendicular to the substrate as the dipoles are perpendicular to the molecular chains. In contrast, a very high electric field is necessary to flip the dipoles in the face-on orientation (molecular chains perpendicular to the substrate). Therefore, the β-phase with edge-on orientations and the semi-electro-active γ- phase might be suitable for energy storage applications because of its stability under high electric fields and late polarization saturation [4]. However, to date, conventional uniaxial stretching technique was applied in order to induce the electro-active β- and γ- phases in P(VDF-HFP).

In this work, we demonstrate the generation of the electro-active β- and γ- phases in P(VDF-HFP) by *in-situ* silver nano-particle (Ag NP) doping and discuss the feasibility of associated applications. The prime advantage of this method is that electro-active films can be achieved in a wide thickness range of ultra-thin (t~80 nm) to thick (t~30 μm) without any further mechanical treatment. The Ag NPs incorporated films were prepared using Dimethylformamide (DMF) as reducing agent of $AgNO_3$ and P(VDF-HFP) as stabilizer. The edge-on orientations are clearly evident in our Ag NPs doped P(VDF-HFP) thin films prepared by spin coating. The ferroelectric response was checked by Dynamic Contact Electrical Force Microscopy [5]. Thick films were produced by solvent casting. For these films we observed that the content of the β- and γ- phases inside the P(VDF-HFP) films can be tuned by the doping concentration (Fig. 1) as well as by the solvent casting temperature. The non-ionic state of silver is evident from the X-ray photo-electron spectroscopy (XPS) spectrum (inset of Fig. 1).

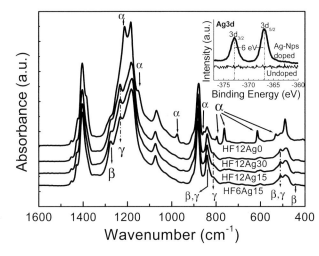

Figure 1: FT-IR spectra of Ag NPs doped P(VDF-HFP) thick films (sample nomenclature: HF*Ag#, where * indicates the w/v concentrations of P(VDF-HFP) and DMF while # gives the AgNO$_3$ concentrations in mM). The top spectra is taken on the undoped P(VDF-HFP) thick film. The inset compares the Ag3d core level XPS spectra (excited with Mg Kα) of an undoped and a doped P(VDF-HP) sample.

Furthermore, *in-situ* temperature dependent FT-IR was performed to determine the melting temperature of the crystalline polymorphs (α-, β-, and γ-phases). These data were cross-checked with Differential Scanning Calorimetric (DSC) results in order to avoid ambiguity and confusion of the melting points. Fig. 2 depicts the *in-situ* FT-IR spectra of a Ag NPs doped P(VDF-HFP) thick film, where the melting behaviour of the β- and γ-phases with respect to the temperature is clearly evident. The inset curves in Fig. 2 indicate that the peak melting temperature of the β-phase (m$_{\beta p}$~125°C) is far below the peak melting temperature of the γ-phase (m$_{\gamma p}$~160°C) and the final melting of the β-phase (m$_{\beta f}$) takes place when the peak melting of the γ-phase occurs. In addition, we have also found that the melting temperature of the α-phase is relatively higher than that of the β-phase but far below the γ-phase melting temperature.

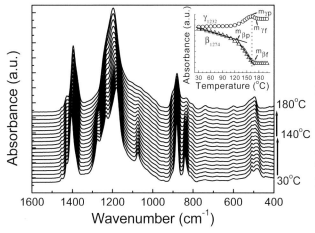

Figure 2: *In-situ* temperature dependent FT-IR spectra of a Ag NPs doped thick P(VDF-HFP) film (thermal ramp. 1°C/min.), where the spectra between 30°C and 140°C are shown in 10°C intervals and above 140°C in 5°C steps. The inset shows the absorbance intensities of the β- (1274 cm^{-1} band) and γ-(1232 cm^{-1}) phases, where the peak and final melting temperatures are evident.

[1] Chu et al., Science **313**, 334 (2006).
[2] Wang et al., IEEE Trans. Dielectr. Electr. Insul. **17**, 1036 (2010).
[3] Chen et al., Macromol. Rapid Commun. **32**, 94 (2011).
[4] Li et al., Appl. Phys. Lett. **96**, 192905 (2010).
[5] Mandal et al., DOI:10.1109/NSTSI.2011.6111801.

SYNTHESIS AND CHARACTERIZATION OF NANOSTRUCTURED MATERIALS FOR REMOVAL OF EXHAUST GASES

K.S. Abdel Halim[1,2], M.H.Khedr[3], A.A.Farghali[3], M.I. Nasr[1], N.K.Soliman[4]

[1] Central Metallurgical Research and Development Institute (CMRDI), Cairo, Egypt
[2] Chemical Eng. Dept., College of Engineering, University of Hail, KSA
[3] Materials Chemistry Dept, Faculty of Science, Beni Suef University, Egypt
[4] Faculty of Oral and Dental Medicine, Nahda University (NUB), Beni-suef, Egypt

The development of nanotechnology is an area of intense research as it providing novel opportunities in synthesizing and designing nanostructured and advanced materials for different applications. Recently, nanostructured materials becoming promising candidates in environmental science, it can be applied to overcome many environmental problems including air and water pollution. However, CO oxidation and hydrocarbon decomposition are considered the most important solutions to get rid of carbon monoxide and hydrocarbon gases emitted from automotive exhaust and consequently control of environmental pollution. In the present work, various metal oxides including CuO/CeO_2 and CuO/Fe_2O_3 with different molar ratio are prepared by co-precipitation techniques. Al_2O_3 was used as support material using wet impregnation technique. The prepared catalysts were characterized physically and chemically through X-ray diffraction, surface area apparatus, scanning electron microscope (SEM) and transmission electron microscope (TEM). The prepared catalysts were tested for the catalytic oxidation of CO to CO_2 and hydrocarbon decomposition. The efficiency of the prepared samples were determined and correlated with operation parameters. Gas analyzer was used for monitoring of the individual gas concentration in the gas mixture during measuring the catalytic activity of the samples. Weight gain technique was applied to determine the decomposition of acetylene over the prepared nanostructured catalysts.

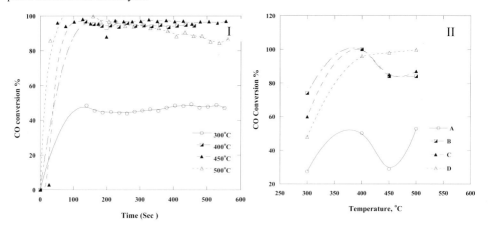

Fig.1.
(I) CO conversion extents as a function of time for CuO -CeO_2-Fe_2O_3-Al_2O_3 catalysts
(II) CO conversion as a function of temperature for;
(A) CuO -Fe_2O_3, (B) CuO -CeO_2-Al_2O_3, (C)) CuO -CeO_2 and (D) CuO -CeO_2-Fe_2O_3-Al_2O_3

It can be reported that the efficiency of CO oxidation and decomposition of hydrocarbon is the highest in case of using nanostructured CuO-CeO_2-Fe_2O_3 catalyst materials supported on Al_2O_3. The results revealed that the rate of oxidation of CO to CO_2 increased by increasing catalytic temperature, increasing catalyst mass and presence of ceria with all the prepared samples.

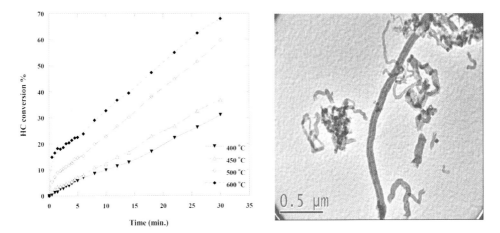

Fig.2. Influence of temperature on the catalytic decomposition of acetylene over CuO-Fe$_2$O$_3$-CeO$_2$-Al$_2$O$_3$ together with TEM image of CNTs produced over the catalyst.

It was found that ceria containing catalysts give the highest efficiency in CO oxidation to CO$_2$ at all the catalytic temperature and lowest efficiency in hydrocarbon decomposition. At relatively lower temperature (300 °C), the maximum CO oxidation occurs over CuO-CeO$_2$-Al$_2$O$_3$, CuO-CeO$_2$ and CuO-CeO$_2$-Fe$_2$O$_3$-Al$_2$O$_3$ while the maximum hydrocarbon decomposition occurs over CuO-CeO$_2$-Fe$_2$O$_3$-Al$_2$O$_3$ and CuO-Fe$_2$O$_3$. With the increasing of temperature to 450 and 500 °C the CO conversion % decrease again and this may be due to the sintering effect, while the increase in hydrocarbon decomposition is maintained.

The experimental data shows that the catalytic oxidation of CO is probably preceded by adsorption mechanism while CNTs occurs by tip growth in case of hydrocarbon decomposition reaction. The reaction temperature (catalytic oxidation and/or catalytic decomposition) is the most important factor controlling the efficiency of catalytic process. The kinetic analysis of catalytic oxidation reaction shows that the value of activation energy is relatively small in all samples containing ceria (3.7-7 kJ/mol). Such lower value of activation energy for catalysts containing ceria might be confirm the high efficiency of the ceria containing catalysts. On the other hand, the Arrhenius plot of decomposition reaction shows lower activation energy value (17.2 kJ mol^{-1} for CuO-CeO$_2$-Fe$_2$O$_3$-Al$_2$O$_3$catalyst), which revealed that the catalyst is very active towards acetylene decomposition. In conclusion, the nanocrystallite CuO-Fe$_2$O$_3$-CeO$_2$-Al$_2$O$_3$ can be recommended as promising catalysts for CO oxidation and hydrocarbon decomposition as well.

This work is supported by the Swedish Research Links Programme (Sweden) and is gratefully acknowledged.

GROWTH AND CHARACTERIZATION OF MAGNETO-OPTICAL CERIUM-DOPED YTTRIUM IRON GARNET FILMS ON SILICON NITRIDE FOR NONRECIPROCAL PHOTONIC DEVICE APPLICATIONS

Mehmet Onbasli[1], T. Goto[1], D. H. Kim[1], L. Bi[2], G. F. Dionne[1], C. A. Ross[1]

[1]Department of Materials Science and Engineering, Massachusetts Institute of Technology, Cambridge, MA, USA [2]Micron Technology, Boise, ID, 83706, USA

Magneto-optical oxides enable the realization of nonreciprocal components such as microwave and optical isolators[1-2], circulators, filters and nonreciprocal optical switches. Previously, our group demonstrated that these oxides can be integrated with resonators to achieve on-chip magneto-optical isolators with 19.5 dB isolation ratio and 18.8 dB insertion loss [1] (Figure 1 inset). Yttrium iron garnet (YIG, $Y_3Fe_5O_{12}$) is often doped with Cerium substitutionally into Yttrium sites, in order to form Cerium-doped YIG (Ce:YIG, $Ce_1Y_2Fe_5O_{12}$) and enhance magneto-optical response[1]. Ce:YIG, YIG and other typical magnetic oxides used in these applications cannot be grown on Si or many other common photonic substrates epitaxially, because of the large mismatch of lattice parameters (Si: 5.431Å, YIG: 12.376 Å, 56.1% mismatch). Integration of these oxides onto amorphous films was challenging due to thermal expansion coefficient and elastic modulus difference between oxide films and amorphous substrates (SiO_2, Si_3N_4 etc.). As a result, integrating these functional oxides onto common substrates involved wafer bonding magnetic oxides onto silicon photonic chips[3] or pulsed laser deposition growth and crystallization by annealing[1,2,4]. In this study, we introduce a method for growing Ce:YIG films onto thick (300 nm) amorphous and transparent silicon nitride films for relieving the thermal and mechanical stress and obtaining strong magneto-optical response and low optical propagation loss in the oxide films. The measured Faraday rotation of the Ce:YIG films exceeds 3500 deg/cm at λ=1525nm, while bulk Faraday rotation of Ce:YIG is ~4500 deg/cm. While the process parameters developed here strictly apply only to Ce:YIG films grown on silicon nitride, the same process flow with different deposition parameters can be applied to the integration of any crystalline film onto any amorphous substrate that can withstand anneals up to 900°C.

First, silicon nitride films were grown onto bare silicon wafers using chemical vapour deposition up to 380 nm thickness. After blanket growth of silicon nitride, 20 to 40 nm thick YIG seed layer films were grown on silicon nitride films using pulsed laser deposition (PLD) method, under at 2 Hz pulse repetition rate, 5 mTorr oxygen pressure and 5×10^{-6} Torr base pressure and at 800°C. YIG films were then rapid thermal annealed for 5 minutes at 30 sccm oxygen flow rate at 850°C. Rapid thermal anneal crystallizes YIG films and allows the subsequent growth of Ce:YIG with polycrystalline phase structure (XRD data not shown). Finally Ce:YIG films have been grown using PLD at pulse rate 10 Hz, P_{oxygen} of 20 mTorr, 5×10^{-6} Torr base pressure and 800°C. YIG seed layer is essential for the structural quality of Ce:YIG films, because highly absorptive and nonmagnetic secondary phases have been observed for Ce:YIG films grown directly on nitride with the above conditions. In order to achieve phase purity and to suppress the formation of nonmagnetic and absorptive phases; a crystallized YIG seed layer is essential. Ce:YIG films on YIG seed layers show in-plane easy axis with saturation magnetization of 135 emu/cc and coercive field of 150 Oe. Optical transmission loss of the entire system is 8.23dB as shown in Figure 1. Faraday rotation of samples at saturation is shown in Figure 2.

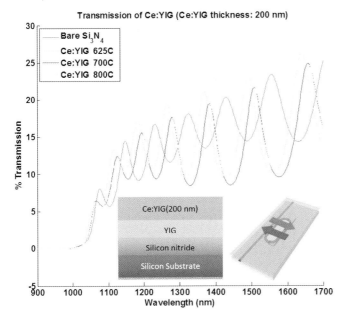

Figure 1: Out-of-plane optical transmission spectra of Ce:YIG samples grown at different temperatures with negligible optical loss. Total optical propagation loss is 8.23 dB, including YIG and nitride layers. Insets: Left is the layers, right is the device topology for the integrated magneto-optical isolator in [1].

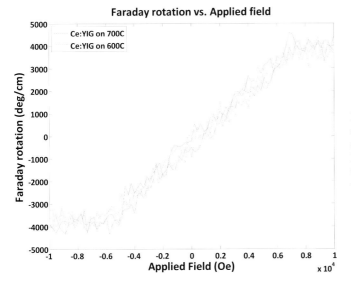

Figure 2: Faraday rotation in out-of-plane configuration (perpendicular to film surface). Blue and red curves indicate the growth temperature of YIG underlayer films.

Support by National Science Foundation and supply of nitride samples from Dr. Steve Spector and Dr. Juan Montoya (MIT Lincoln Laboratory) is gratefully acknowledged.

[1] L. Bi et. al., Nat. Photon. 5, 758 (2011).
[2] L. Bi et. al., Proc. SPIE 7941, 794105 (2011)
[3] S. Ghosh et. al., Opt. Express, 20, 1839 (2012)
[4] S.-Y. Sung et. al., J. Appl. Phys. **109**, 07B738 (2011)

PHOTOELECTROCHEMICAL WATER SPLITTING BY CHEMICAL SOLUTION DEPOSITED NANOSTRUCTURE AND COMPOSITION ENGINEERED HEMATITE FILMS

Theodor Schneller[1], Simon Goodwin[1], Rainer Waser[1,2]

[1]Institut für Werkstoffe der Elektrotechnik II, RWTH Aachen, Aachen, Germany;
[2]Research Centre Jülich, Peter Grünberg Institute, Electronic Materials, Jülich, Germany

The worlds conventional energy sources based on fossil fuels (oil, natural gas, and coal) have a finite lifetime and moreover, probably even more important, their use will intensify the problem of global warming due to the steadily increasing concentration of the greenhouse gas CO_2 in the atmosphere. This insight stimulates the world wide effort towards a renewable (or green) energy economy, predominantly based upon using the energy supplied by our sun either indirectly through wind and water power, or directly by conversion of sun light to electrical power or chemical fuel. Regarding to the latter point, a promising approach is photoelectrochemical water splitting yielding hydrogen which can be stored and used as a clean fuel. Wide bandgap semiconducting metal oxides such as TiO_2, WO_3, and Fe_2O_3 are widely studied materials but each of these materials has its advantages and drawbacks. Due to its abundance, low cost, visible light active bandgap (2.0 - 2.2 eV), nontoxicity and electrochemical stability hematite (α-Fe_2O_3) has received much attention [1]. Nevertheless the photoconversion efficiency of hematite thin films is still low, owing mainly to its short hole diffusion lengths of 2 - 4 nm, low absorption coefficient and very short excited state lifetime. In order to improve the conversion efficiencies doping, nanostructuring and surface treatment of the TCO glass substrate prior the deposition of hematite [2] are promising approaches. Among the multitude of methods used to prepare hematite films, chemical solution deposition (CSD [3]) with its cost efficiency, its possibility of micro-/nanostructuring and its flexibility regarding the variation of composition and dopant addition. In the present work a stable CSD route has been developed [4] which enables the addition of dopants and leads to mesoporous hematite thin films on FTO coated glass (Fig. 1).

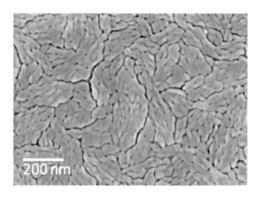

Figure 1: Field emission scanning electron micrograph (FESEM) (top view) of an exemplary sample of hematite coated FTO glass prepared by spin-coating and annealing at 510 °C.

All CSD derived films have been characterized by FESEM, X-ray diffraction and UV/VIS spectroscopy. In order to study the photoelectrochemical activity of the as fabricated thin films a multipurpose solar-simulator combined with an easy to use reaction cell was set up in-house. LED's whose spectra are combined virtually free (Fig. 2) are used as the light source.

Each LEDs current is provided by a high side buck converter. In contrast to pulse width modulation (PWM) based light dimming approaches, the converters current regulation guarantees a flicker free illumination of the sample. The current of each LED is set by a digital analogue converter (DAC) that again is controlled by a microcontroller, which in turn is operated by a PC via the USB-Test and

Measurement Class (USBTMC) protocol. In the current setup (Fig. 3) a spectral range from approximately 300 nm to 1100 nm is achieved. Calculations based on spectra obtained from the LEDs datasheets (Fig. 2) predict the system to meet the requirements for an ASTM Class A spectral math on the ASTM G 173 AM 1.5 spectrum.

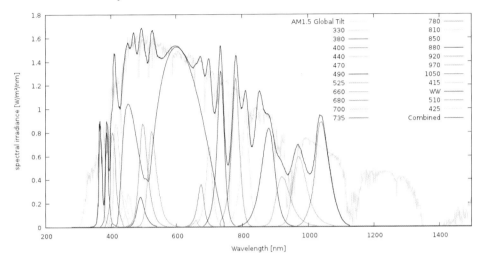

Figure 2: AM1.5 global (grey) and simulated solar spectrum (black line) based on a LED array. The artificial solar spectrum is based on the data sheets of the different LED (maximum wavelengths of each LED are given in the annotation).

Figure 3: Photograph showing the illuminated electrochemical measuring cell, which is filled only ~ half of the volume in order to illustrate the position of the electrolyte in the cell. The sample dimensions are 1*1 inch^2 and the cables in front of the PTFE plate contact the electrodes of the cell. Artificial sunlight is generated by a LED array currently consisting of 20 LEDs (optional up to 24) with different wavelengths, ranging from 300 nm to 1100 nm.

The generated spectrum can be manipulated by computer controlled drivers (background) for each LED. Thus the reaction of the photo anode can be studied depending on individual wavelengths, times, and intensities.

Adjacent to a light guide a PTFE electrochemical measuring cell is placed with the ability to hold and electrically contact 1"*1" samples with an illuminated area of 3 cm² and a working- to counter electrode distance of 14 mm (Fig. 3).

Finally the results of illumination experiments of hematite thin film working electrodes featuring different nanostructures and compositions paired with Pt-coated substrates as counter electrodes will be presented.

[1] K. Sivula, F. Le Formal, M. Grätzel, ChemSusChem **4**, 432 (2011).
[2] F. Le Formal, M. Grätzel, K. Sivula, Adv. Funct. Mater. **20**, 1099 (2010).
[3] R. W. Schwartz, T. Schneller, R. Waser, C. R. Chimie **7**, 433 (2004).
[4] M. Teuber, Bachelor thesis, RWTH Aachen (2011).

A CHEMICAL APPROACH TO THE ESTIMATE OF THE OPTICAL BAND GAP AND BOWING PARAMETER IN MIXED d,d-METAL OXIDES

Francesco Di Quarto, Francesco Di Franco, Monica Santamaria

Electrochemical Materials Science Laboratory - Dipartimento di Ingegneria Civile Ambientale Aeronautica e dei Materiali (DICAM), Università di Palermo, Viale delle Scienze, 90128 Palermo (Italy)

In previous works we have shown that it is possible to correlate the band gap, E_g, of semiconducting and insulating oxides and group III nitrides to the square of the difference of electronegativity between anions and metal [1-2]. More specifically, E_g was found to depend linearly on the square of the difference of electronegativity (Pauling's scale) between anion (χ_{An}) and metal (χ_M), according to the following equation:

$$E_g = A (\chi_{An} - \chi_M)^2 + B \qquad (1)$$

where A and B are determined by the best fitting of the experimental data after adaptation of a regression line to the plot of experimental E_g values vs the parameter $(\chi_{An} - \chi_M)^2$. As detailed described in [1], two different interpolating lines were obtained from the least squares best fitting procedure for of s,p and d-metal oxides.

Owing to the periodic trends of the electronegativity values strictly related to the periodic table of the chemical elements, once A and B are known, the use of eq. (1) allows an easy identification of the expected trend in the optical band gap of oxides. This aspect is one of the most rewarding results of our previous studies on anodic oxides films together with the possibility to tailor, within some extent, the band gap values of mixed oxides.

On the other hand, we have recently shown [3 and refs. therein] that it is possible to extend this correlation to mixed oxides, by substituting χ_M with an average value:

$$\chi_{M,av} = x_i \chi_i + x_j \chi_j \qquad (2)$$

where x_i and $x_j = (1-x_i)$ are the atomic fraction of metals M_i and M_j and $\chi_{i,j}$ their electronegativity. The insertion of eq. (2) in eq. (1) allows to account for a quadratic dependence of E_g on the material composition, which recalls the standard bowing equation, widely used to tailor the band gap, of semiconducting alloys as a function of composition:

$$E_g(x) = E_{g1}(1-x) + E_{g2}x - bx(1-x) \qquad (3)$$

E_{g1} and E_{g2} represent the band gaps of pure compounds (at x = 0 and x = 1, respectively) and b is the bowing parameter, whose value depends on E_{g1} and E_{g2}.

During the years, a large theoretical effort has been devoted by different authors to derive information on the factors influencing the value of the bowing coefficient to be used for different semiconducting alloys. In a series of seminal papers on the influence of the nature of alloying elements and their concentration in determining the value of bowing coefficient of different alloys, Zunger and coworkers have been able, on the basis of DFT and quantum mechanical calculations techniques, to evidence that the value of the bowing coefficient in semiconducting alloys contains several contributions which can be traced out to the chemical nature (difference of electronegativity) of the alloy components or to structural (bond lengths and volume deformation of band structure) properties [4-7]. However such an approach, which is unavoidable in any attempt to put on physically sound basis any theory of electronic properties of materials or to get physical insights on any deviations from observed experimental regularities is rather unpractical in providing general indications on the role that the chemical nature of the components the alloy play in determining the optical band gap value as well as on how this parameter changes with changing the alloy composition. In this work we want to show that for mixed d-d metal oxides (ternary oxides covering a large range of band gap values from 1.9 to 4.6 eV) it is possible to link the bowing parameter of eq. (3) to the electronegativity of the partner metals. In fact after substitution of eq. (2) in eq. (1) and simple algebraic manipulations, we can write for a mixed semiconducting oxide the following relation:

$$E_g(x) = E_{g,i} + 2Ax_j (\chi_i - \chi_j)(\chi_{an} - \chi_i) + Ax_j^2(\chi_j - \chi_i)^2 \quad (4)$$

By direct comparison term to term of eq. (4) with eq. (3) the following expression can be derived for the optical bowing parameter:

$$b = A(\chi_j - \chi_i)^2 \quad (5)$$

In order to test the validity of eq. (4), we have studied the dependence of E_g on the composition for amorphous Ta-Nb and crystalline Ti-Zr and Fe-Ti mixed oxides. As shown in Fig. 1, a quadratic dependence holds with b values following the trend expected according to the electronegativity of the involved metals [8].

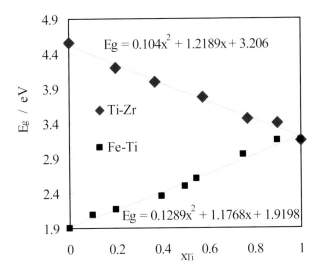

Figure 1
Band gap values as a function of Ti cationic ratio of different oxides, experimentally measured by in situ Photocurrent Spectroscopy. Ti-Zr mixed oxides were grown by anodizing to 5 V sputter-deposited Ti-Zr alloys of several compositions [3]. Fe-Ti mixed oxides were grown by by metalorganic chemical vapour deposition [3]. Estimated electronegativity values for Fe(III), Ti(IV) and Zr(IV) : χ_{Fe}=1.915; χ_{Ti}=1.635; χ_{Zr}= 1.385; χ_O = 3.5 in the Pauling's scale.

[1] F. Di Quarto et al., J. Phys. Chem. B, **101**, 2519 (1997).
[2] F. Di Quarto et al., Electrochem. Solid-State Lett.,**11**, H313 (2008).
[3] F. Di Quarto et al., in "Modern Aspects of Electrochemistry, No. 46: Progress in Corrosion Science and Engineering I", Ed. by Su-Il Pyun and Jomg-Won Lee, Ch. 4, p. 231, New York: Springer (UNITED STATES), 2009.
[4] A. Zunger et al., Phys. Rev. Lett., **51**, 662 (1983).
[5] A. Zunger et al., Phys. Rev. B, **29**, 1882 (1984).
[6] J.E. Bernard et al., Phys. Rev. B, **36**, 3199 (1987).
[7] S-H. Wei et al., J. Appl. Phys., **87**, 1304 (2000).
[8] L. Pauling, The Nature of Chemical Bond, chapter 3, Cornell University Press, Ithaca, NY, 1960.

BAND ALIGNMENT ENGINEERING WITH LIQUID DIELECTRICS

Hongtao Yuan[1,2], Y. Ishida[3], K. Koizumi[3], H. Shimotani[1], K. Kanai[4], K. Akaike[5], Y. Kubozono[5], A. Tsukazaki[1], M. Kawasaki[1,6], S. Shin[3], Y. Iwasa[1,6]

[1]QPEC & Department of Applied Physics, The University of Tokyo, Tokyo, Japan;
[2]Department of Applied Physics, Stanford University, San Francisco, US;
[3]ISSP, The University of Tokyo, Tokyo, Japan;
[4]Tokyo University of Science, Tokyo, Japan;
[5]RLSS, Okayama University, Okayama, Japan;
[6]Correlated Electron Research Group, RIKEN, Wako, Japan.

With a functionalized liquid/solid (L/S) interface, electric-double-layer transistors (EDLTs) recently have attracted intensive attention since the nano-gap EDL capacitors with huge capacitances can accumulate very high-density charges at the interfaces, leading to the amazing progresses not only in practical applications in field-effect transistors[1-2] but also in electrostatic modulation of electron states in oxides, like field induced superconductivity at liquid/semiconductor interfaces[3-5]. However, as the central concept of EDL transistors, basic understanding on the band alignment and band bending of such liquid/semiconductor EDL interface is of great importance but still far from clear.

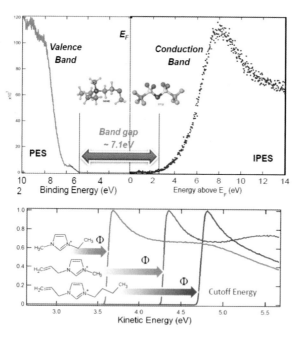

Figure 1, the PES and IPES of an ionic liquid dielectric, providing detail information on band gap, work function, HOMO energy level, LUMO energy level.

In this presentation, by using photoemission spectroscopy (PES) and inversion photoemission spectroscopy (IPES), we figure out the electronic structure of ionic liquids (ILs) and the interfacial band alignment between IL and oxides, which is able to bridge the gap between the electric structure of liquid dielectrics and the modulation of electronic properties of EDL interfaces. Taking the great advantage of the low vapor pressure of ionic liquids, we can perform the photoemission measurement on the electronic structures of a liquid in the same way as those measurements in solids. As shown in the typical PES and IPES spectroscopy of an IL (DEME-TFSI) in Fig. 1, we can confirm the work function Φ, LUMO, HOMO and vacuum energy levels of the IL, which helps us to understand the band alignment of liquid/solid interfaces inside the transistors. The work function Φ can be determinate from 3.57 eV to 4.95 eV with a maximum variation of 1.38 eV, which directly influence the magnitude of the initial band bending before applying external bias. We found that the band gap and HOMO level of the ILs had a direct effect not only on the chemical potential window of the IL but also the maximum attainable carrier densities accumulated on oxide surfaces. As shown in Fig.2, the higher is the HOMO energy level, the larger is the voltage which is available for higher density electron accumulation. With comparing the transport properties of IL/ZnO EDL interfaces to the basic electronic parameters of IL from PES and IPES, the band alignment of ionic-liquid/ZnO interfaces and its influence on the interfacial carrier accumulation were clearly figured out, as shown in Fig.3. Such an clear interfacial band bending diagram can directly tell us which ILs is good for electron/hole accumulation, providing us with a guidance to select the "right" liquids and the "correct" gate metals to tune the band alignment at EDL interface for designing practical devices or for searching new electronic states on oxide surfaces with liquid gating.

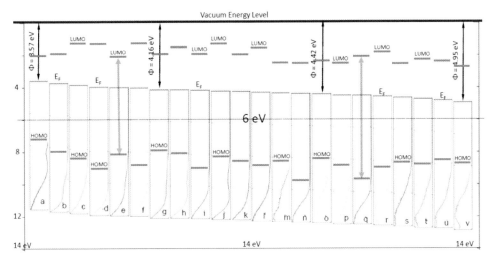

Figure 2, A comprehensive comparison of the electronic band structures of ILs with different molecular structures. Band gap E_g can be varied from 4.9 eV to 8.1 eV, work function Φ from 3.57 eV to 4.95 eV, the lowest HOMO level is 9.8eV, and the highest LUMO level is 1.3 eV.

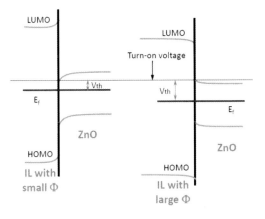

Figure 3, Schematic diagram of the band alignment at liquid/oxide EDL interfaces. The interface band bending can be greatly modulated by simply change the liquid dielectrics.

[1] H. T. Yuan, et al., Adv. Funct. Mater. **19**, 1065 (2009).
[2] H. T. Yuan, et al., J. Am. Chem. Soc. **132**, 6672 (2010).
[3] K. Ueno, et al., Nat. Mater. **7**, 855 (2008).
[4] J. T. Ye, et al., Nat. Mater. **7**, 855 (2008).
[5] K. Ueno, et al., Nat. Nanotech. **6**, 408 (2011).

THE BEHAVIOUR OF OXYGEN VACANCIES IN THE PEROVSKITE OXIDE STRONTIUM TITANATE AND AT ITS EXTENDED DEFECTS

Roger A. De Souza[1], Veronika Metlenko[1], Henning Schraknepper[1], Amr Ramadan[1]

[1]RWTH Aachen University, Institute of Physical Chemistry, Aachen, Germany

There is renewed interest in the behaviour of point defects in bulk $SrTiO_3$ and at its interfaces due to the material's possible application in all-oxide electronics and as a memristive device. The combination of $^{18}O/^{16}O$ exchange and Secondary Ion Mass Spectrometry (SIMS) analysis constitutes a powerful tool for probing the behaviour of oxygen vacancies in oxides. In this contribution, after a brief introduction to the technique and its capabilities and limitations, I demonstrate the application of this method to investigating the behaviour of oxygen vacancies in $SrTiO_3$ and at its extended defects (dislocations, surfaces, hetero-interfaces). Three systems will be examined: (1) single crystal $SrTiO_3$ substrates; (2) low-angle grain boundaries in $SrTiO_3$ comprising periodic arrays of edge dislocations; and (3) heterostructures based on $SrTiO_3$ (e.g. $LaAlO_3|SrTiO_3$, $SrRuO_3|SrTiO_3$). Two aspects regarding the behaviour of oxygen vacancies will be emphasised: their diffusion kinetics and non-uniform distributions near extended defects.

DOPED CERIA: A DFT AND MONTE CARLO STUDY

B. O. H. Grope[1], **S. Grieshammer**[1], **J. Koettgen**[1], **M. Martin**[1]
[1]Institute of Physical Chemistry, RWTH Aachen University, Aachen, Germany

Ionic conductivities of rare-earth oxide doped ceria systems have been investigated for many years now, due to their high oxygen ion conductivity and potential application in solid oxide fuel cells. Theoretical as well as experimental approaches are both vital in understanding the underlying mechanisms in order to provide tailor-made materials specified for individual technical requirements.

The defect association energies for rare-earth oxide (RE_2O_3) doped CeO_2 are calculated by means of density functional theory using a GGA+U functional. Starting from the association energies, the thermodynamic equilibrium distribution of oxygen vacancies and dopant atoms (Y and Sm) within the lattice is simulated using the Monte Carlo algorithm according to Metropolis et al. [1]. From these results the average coordination numbers of the dopants and the radial distribution of defects are determined and compared to literature values. Special focus is here on the alignment of the oxygen vacancies. In order to predict the ionic conductivity in doped CeO_2 the equilibrated lattices are used as starting configurations for simulations employing Kinetic Monte Carlo algorithms [2-4]. The required activation energies for the hopping processes in different environments (see Figure 1) are calculated using the NEB-formalism within density functional theory.

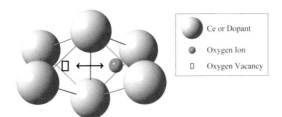

Figure 1: Environment of an oxygen ion and oxygen vacancy involved in one hopping process. The activation energy is changed by variation of the identity of the cations, i.e. cerium or dopant.

The results are compared to a proposed model to split the activation energy into association energies (leading to trapping effects) and migration energies (leading to jump blocking effects) [5]. Furthermore the influence of the vacancy-vacancy interaction on the ionic conductivity is considered and correlation effects between different interactions are investigated. Figure 2 show the results for samaria-doped ceria. The predicted ionic conductivity is compared to experimental values with special regard to the maximum at specific dopant concentrations. In addition scandia-doped ceria will be discussed.

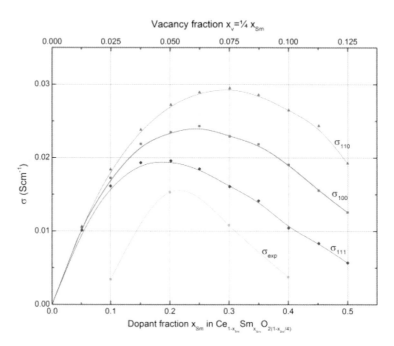

Figure 2: Ionic conductivities of samaria-doped ceria for jump blocking, σ_{100}, combined jump blocking and trapping, σ_{110}, and combined jump blocking, trapping and V–V repulsion, σ_{111}. The experimental results are taken from Ref. [6]; the lines are a guide to the eye. $T = 893$K in a $16 \times 16 \times 16$ lattice.

[1] N. Metropolis et al., J. Chem. Phys. **21**, 1087 (1953).
[2] A. R. Allnatt, A.B. Lidiard, Atomic Transport in Solids, Chap. 10 (Cambridge University Press, Cambridge, 1993).
[3] A. D. Murray et al., Solid State Ionics **18/19**, 196 (1986).
[4] B. O. H. Grope et al., Solid State Ionics, in press (2012).
[5] M. Nakayama, M. Martin, Phys. Chem. Chem. Phys. **11**, 2341 (2009).
[6] D. Y. Wang et al., Solid State Ionics **2**, 95 (1981).

OXYGEN VACANCY FORMATION ON (110) TiO$_2$ SURFACE - A FIRST PRINCIPLES STUDY

Taizo Shibuya[1], Kenji Yasuoka[1], Susanne Mirbt[2] and Biplab Sanyal[2]

[1]Department of Mechanical Engineering, Keio University, Yokohama 223-8522, Japan; [2]Deparment of Physics and Astronomy, Uppsala University, Box 516, 75120 Uppsala, Sweden

The electronic structure and magnetism of rutile TiO$_2$ (110) surface in the presence of oxygen vacancies is a topic of tremendous interest. There are many issues on stable magnetism, defect states, polaron formation etc. In this work, we report a detailed investigation on the electronic structure of this defected systems in terms of a polaron formation by accurate hybrid functional based electronic structure computations.

The rutile (110) surface of TiO$_2$ consists of alternating bridging oxygen (O$_{br}$) rows and 5-coordinated Ti (Ti$_{5c}$) rows. In local approximations of exchange-correlation in DFT, the Ti-3d states lie at conduction band edge and are spatially distributed around Ti sites, occupying 3d states. In contrast, in DFT+U or hybrid functional they create gap states about 1 eV below the conduction band by localizing two electrons at two different Ti sites. In addition, the localized site increases its bond length with nearest oxygen atoms creating a polaron. Experimentally bridging oxygen vacancy is attributed to the gap states about 1 eV below the conduction band, so this localized picture agrees with experimental one. However, the stable sites of these electrons depend on a method or calculation setup. Some calculate two surface polarons [1], some two equivalent polarons on subsurface sites [2,3] and others, a combination with a sub subsurface polaron and a subsurface polaron [4]. Thus it is still an open question. In experimental point of view, while oxygen vacancies are often observed on rutile (110) surface and its maximum coverage is reported to be around 10%, the mechanism behind it is unclear. In this presentation we report that the stable positions of polarons are found at subsurface, independent of the method and reveal the mechanism of maximum coverage of oxygen vacancy.

We performed hybrid functional DFT together with GGA+U calculation with U = 4.2 eV. The p(4x2) 4 trilayer surface cell was used. 10 Å of vacuum was inserted to break the periodic boundary condition. Atoms were relaxed until Hellman-Feynman forces were reduced to 0.05 eV/Å. The bottom layer was fixed at bulk position. In order to investigate the most stable site, we calculated several positions for polaronic sites.

In Fig 1 we show one of the most stable structures. Clearly, electrons localize at Ti below Ti$_{5c}$. Interestingly, this result was independent from method used.

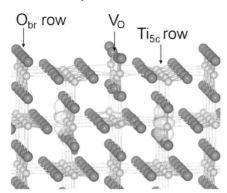

Figure 1: Charge density of the gap states of the most stable structure. The isosurface was plotted at 0.005 e/Å^3. Small circle: Ti, large circle: O.

[1] Cristiana Di Valentin, Gianfranco Pacchioni, and Annabella Selloni, Phys. Rev. Lett. **97**, 166803 (2006).
[2] N. Aaron Deskins, Roger Rousseau, and Michel Dupuis, J. Phys. Chem. C **115**, 7562 (2011).
[3] Piotr M. Kowalski, Matteo Farnesi Canellone, Nisanth N. Nair, Bernd Meyer, and Dominik Marx, Phys. Rev. Lett. **105**, 146405 (2010).
[4] Steeve Chretien and Horia Metiu, J. Phys. Chem. C **115**, 4696 (2011).

FINITE SIZE EFFECT OF PROTON CONDUCTIVITY OF AMORPHOUS OXIDE THIN FILM AND ITS APPLICATION YO HYDROGEN-PERMEABLE MEMBRANE FUEL CELL

Yoshitaka Aoki[1], Manfred Martin[2]

[1]Faculty of Engineering, Hokkaido University, Sapporo, Japan;
[2]Institute of Physical Chemistry, RWTH Aachen University, Aachen, Germany

Recently, several research groups reported on the emerging size enhancement of ionic conduction across very thin glassy layers, where the conductivity exponentially increases with decreasing the thickness to the nanometer range. This finite size effect must be related to the existence of a percolative ionic channel penetrating through the glass matrix. The pathways with relatively high ion-conductivity can be accumulated in the glass network because the heterogeneous nanostructures of a glass provide a distributed potential for ionic diffusion in a nanometer length scale. When the thickness of the glass film is close to the length of the largest pathway inside the film, the channel percolating from edge to edge is involved and thus the overall conductivity across the film must be increased. Hence, it is a motivation to investigate on the mesoscopic situation the ionic conductivity of glassy thin films.

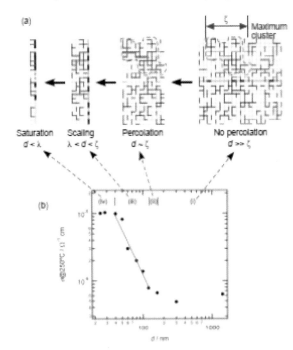

Figure 1: (a) Bond percolation in 20 × 20 square lattice. The calculation is performed with a bond population of 0.35. (b) Log-log plot of σ of $Al_{0.1}Si_{0.9}O_x$ film at 250°C in dry air vs. film thickness d. For $d > \zeta$, the conductive clusters become insulated in a poorly-conducting matrix and the film remains in the low (normal) conductive phase. The conductive pathway penetrating between both electrodes is formed for $d \sim \zeta$, and the conductivity increases for $\zeta < d < \lambda$ since more conductive clusters connect between the both with decreasing d by following $N(l, \zeta) \propto (l/\zeta)^{-\tau}$. When d is close to λ, the increase of σ is terminated. Details of the calculation and the measurement procedures are given in Ref. [1].

Recently, we found that the amorphous aluminosilicate thin film $Al_{0.1}Si_{0.9}O_x$ revealed similar size-enhancement on proton conductivity as reducing the thickness into the sub-100 nm region [1]. The responsible ionic pathways were presumed to be identical to the condensed glassy microdomain which was developed by density fluctuation of the glass network. In addition, it was found that an amorphous zirconium phosphate, $ZrP_{2.6}O_x$, thin film exhibits proton-conductivity transition induced by reducing the thickness to the 100 nm range. The film in the vicinity of the film/electrode interface tends to be more hydrated than other parts of the film, and thus, the proton conductivity of the film in this region is greatly enhanced.

[1] Y. Aoki et al., J. Am. Chem. Soc. **133**, 3471 (2011).

GRAIN BOUNDARIES IN PROTON-CONDUCTING BaZrO$_3$ PEROVSKITES

R. Merkle[1], M. Shirpour[1], B. Rahmati[2], W. Sigle[2], P.A. van Aken[2], J. Maier[1]

[1]Max Planck Institute for Solid State Research, Stuttgart, Germany
[2]Max Planck Institute for Intelligent Systems, Stuttgart, Germany

Y-doped BaZrO$_3$ perovskites exhibit a high bulk proton conductivity [1], but their use e.g. as electrolyte for intermediate-temperature fuel cells is still impeded by the low grain boundary (gb) conductivity (see e.g. [2,3]). Several observations indicate that space charge depletion zones adjacent to the gb core carrying a positive excess charge are responsible for the blocking grain boundaries:

- The gb are free of secondary phases, and the thickness of the low-conducting gb region is larger than the extension of the gb core ($\approx$ 1nm).

- Grain boundaries show a strong decrease of gb resistance as well as gb capacitance under applied DC bias, as expected from the space charge model (Fig. 1a) [4].

- Additional annealing of SPS-sintered samples leads to acceptor dopant accumulation in the grain boundary region (proven by TEM-EDX, Fig. 1b) and results in improved gb conductivity [5]. This can be understood in terms of the space charge model, since dopant segregation into the gb core will directly decrease the core charge, while dopant accumulation in the space charge zone will also decrease the proton depletion (change from Mott-Schottky- to a partially frozen Gouy-Chapman situation).

These findings suggest that a positive gb core charge is typical for acceptor-doped oxides with large bandgap (such as SrTiO$_3$ [6,7], LaGaO$_3$ [8] or ZrO$_2$ [9]).

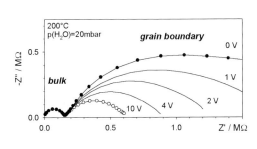 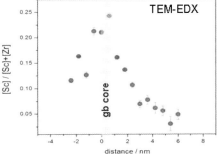

Figure 1a: Impedance spectra of a large-grained (grain size 100µm) Y-doped BaZrO$_3$ sample which allows to apply DC bias values up to 1 V per grain boundary [4].

Figure 1b: TEM-EDX linescan across a GB of annealed SPS-6 at%Sc-doped BaZrO$_3$ showing dopant accumulation at the grain boundary (gb core and adjacent region) [5].

[1] K.D. Kreuer, Solid State Ionics **125**, 285 (1999)
[2] F. Iguchi et al., Solid State Ionics **178**, 691 (2007)
[3] P. Babilo et al., J. Am. Ceram. Soc. **88**, 2362 (2005)
[4] M. Shirpour et al., Phys. Chem. Chem. Phys. **14**, 730 (2012)
[5] M. Shirpour et al., J. Phys. Chem. C **116**, 2453 (2012)
[6] M. Vollmann et al., J. Am. Ceram. Soc. **77**, 235 (1994)
[7] R.A. De Souza et al., J. Am. Ceram. Soc. **86**, 922 (2003)
[8] H.J. Park et al., J. Phys. Chem. C **111**, 14903 (2007)
[9] X. Guo, Solid State Ionics **81**, 235 (1995)

FAST ELECTROMIGRATION OF OXYGEN VACANCIES IN IMPLANTED RUTILE (TiO$_2$)

N.A. Sobolev[1], A.M. Azevedo[1], V.V. Bazarov[2], E.R. Zhiteytsev[2], R.I. Khaibullin[2,3]

[1]Departamento de Física and I3N, Universidade de Aveiro, 3810-193 Aveiro, Portugal
[2]E.K. Zavoisky Physical-Technical Institute of RAS, 420029 Kazan, Russia
[3]Insitute of Physics, Kazan Federal University, 420008 Kazan, Russia

Titanium dioxide (TiO$_2$) had attracted considerable attention due to potential applications in a variety of areas. Recently it was shown that the reduced TiO$_2$ samples reveal ferromagnetic properties up to RT. The observed ferromagnetism of the reduced TiO$_2$ has been associated with spin polarization of electrons trapped by oxygen vacancies, O$_V$ (so-called bounded magnetic polarons, an analogue of F-centers). However, the features of this phenomenon remain unclear up to now.

In this work we report electron paramagnetic resonance (EPR) studies of oxygen vacancies created in TiO$_2$ by 40 keV Ar$^+$ ion implantation to a high fluence of 1.5×10^{17} ion/cm^2. First, it was established that high-energy argon implantation in the colorless plates of (001)-oriented single crystalline rutile (TiO2) results in a blue color induced by high concentration of oxygen vacancies in the implanted region of the sample. Secondly, we developed original methods of oxygen vacancies migration under an applied DC electric field with the aim of changing the concentration of these point defects in the samples under study. Fast radiation-enhanced electromigration of oxygen vacancies became visually observable. Around the negative electrode, the initial blue color of the sample changed into the dark blue-grey one due to an increased density of oxygen vacancies, while the region around the positive electrode lost the coloration. Then the sample was divided in a dark and a light half for subsequent EPR studies. Angular dependences of the EPR spectra were taken in the X- and Q- microwave bands (9.5 GHz and 36.8 GHz, respectively) at different temperatures in the range from 4–300 K. Several types of EPR centers related to oxygen vacancies, trivalent titanium ions and O$_V$-Ti^{3+} defect complexes have been identified. Also we observed a strong redistribution of the EPR line intensities between the dark and light half of the sample. Thus, EPR showed that the concentrations of the oxygen vacancies or their complexes have different values in these sample parts in accordance with our visual observations. The EPR lines of oxygen vacancies vanish when the sample temperature rises above 30 K. We conclude that the EPR lines disappearance is a result of the oxygen vacancy recharging at $T > 30$ K due to the transition of an electron trapped by the vacancy to the conduction band of rutile.

The work has been supported by FCT through project PEst-C/CTM/LA0025/2011, as well as by RFBR through grant 10-02-01130-a.

TAILOR-MADE COMPLEX OXIDE THIN FILMS AS PROTON CONDUCTING ELECTROLYTES FOR LOW TEMPERATURE OPERATING SOLID OXIDE FUEL CELLS

David Griesche[1], Theodor Schneller[1], Rainer Waser[1,2]

[1]Institut für Werkstoffe der Elektrotechnik II, RWTH Aachen, Aachen, Germany
[2]Research Centre Jülich, Peter Grünberg Institute, Electronic Materials, Jülich, Germany

Proton conducting oxides are promising electrolyte materials for fuel cells, because of decreasing the operating temperatures significantly in comparison to classical oxygen-conducting electrolytes. A group of interesting candidates for proton conduction at temperatures below 750 °C are the classic perovskites with the general formula $AB_{1-x}M_xO_{3-\delta}$, in which A represents the divalent cation Ba^{2+} and the tetravalent B-site cation Zr^{4+} is partially replaced by Y^{3+} as trivalent dopant. This yttrium-doped barium zirconate (BZY) exhibits high proton conductivity values at ~ 600 °C as well as chemical stability in different atmospheres. Beside the classic perovskites so called complex perovskites with the general formula $A_3(B_x'B_{2-x}")O_{9-\delta}$, where A is a divalent cation like Sr^{2+} or Ba^{2+} and B' and B" are divalent and pentavalent cations are also established [1]. Derived from the well-known calcium-doped barium niobate system $Ba_3Ca_{1.18}Nb_{1.82}O_{9-\delta}$ (BCN18), Zr^{4+} as tetravalent co-dopant on the B-sites in the corresponding strontium-tantalate was introduced leading to $Sr_3CaZr_{0.5}Ta_{1.5}O_{8.75}$ [2]. This dopant incorporation significantly increased the conductivity values (4.64 x 10^{-4} S cm^{-1}) even at very low temperatures (~300 °C) in comparison to the BCN18 (1.5 x 10^{-4} S cm^{-1} [1]). So far most of the experiments with classic perovskites and all previous experiments with the complex perovskite systems were performed with bulk materials; thus we report on an optimized chemical solution deposition (CSD) [3] route to complex and classic perovskite based electrolyte thin films. CSD-processing combines high flexibility with regard to stoichiometry and choice of substrates with cost efficient working. In the present study tailored CSD-routes for $BaZr_{1-x}Y_xO_{3-\delta}$ (x = 0.1 to 0.3), $Sr_3CaZr_nY_mTa_{2-(n+m)}O_{8.75}$, $Ba_3CaZr_nY_mTa_{2-(n+m)}O_{8.75}$, and $Ba_3SrZr_nY_mTa_{2-(n+m)}O_{8.75}$ (n = 0 to 1, m = 1-n) were developed, where Zr^{4+} is successively replaced by trivalent Y^{3+}. The mixed metal alcoholate/propionate precursor solutions were found to be stable for at least eight months without any detectable precipitation and showed excellent substrate wetting performances on different surfaces. Through careful control of the precursor chemistries, the thermal processing, and the nature of the substrates (oxidized silicon, platinized silicon and lattice matched MgO single crystals [4]) it was possible to prepare crack-free films with polycrystalline fine grained, columnar, and even epitaxial morphologies (examples see Fig. 1). The samples were characterized by X-ray diffraction (XRD), field emission scanning electron microscopy (FE-SEM), time-of-flight secondary ion mass spectroscopy (ToF-SIMS) and also electrochemically with high-temperature electrochemical impedance spectroscopy (HT-EIS).

Figure 8: FE-SEM images of dense
(a) epitaxial $BaZr_{0.9}Y_{0.1}O_{3-\delta}$ films on MgO single crystals and
(b) $Ba_3SrZr_{0.4}Y_{0.4}Ta_{1.2}O_{8.75}$ layers on platinized silicon oxide, exhibiting a polycrystalline morphology. Layers were sintered at 800-900 °C for ten minutes in a diffusion furnace.

The authors would like to acknowledge the BMBF for financial support within the frame of the N-INNER program (grant 03SF0392).

[1] A. S. Nowick, Y. Du, Solid State Ionics **77**, 137 (1995).
[2] D. J. D. Corcoran, J. T. S. Irvine, Solid State Ionics **145**, 307 (2001).
[3] R. W. Schwartz, T. Schneller, R. Waser, C. R. Chimie **7**, 433 (2004).
[4] F. Lenrick, D. Griesche, J.-W. Kim, T. Schneller, L. R. Wallenberg, ECS Trans. **4**, (2012), accepted.

EFFECT OF THE SUBSTITUTION OF OXYGEN BY NITROGEN ON THE CRYSTAL CHEMISTRY OF La-DOPED BaTiO$_3$

Yusuke Otsuka[1,2], Christian Pithan[2], Jürgen Dornseiffer[3], Rainer Waser[2,4]

[1]Materials Development Department, Murata Manufacturing Co. Ltd., Yasu, Shiga, Japan;
[2] Peter Grünberg Institute – Electronic Materials **(PGI-**7), Forschungszentrum Jülich GmbH, Jülich, North-Rhine Westphalia, Germany;
[3]Institut of Energy and Climate Research – Troposphere (IEK-8), Forschungszentrum Jülich GmbH, North-Rhine Westphalia, Jülich, Germany;
[4]Institut for Electronic Materials, RWTH Aachen University, Aachen, North-Rhine Westphalia, Germany

BaTiO$_3$-based electroceramics are widely applied in electronic devices because of their outstanding physical properties, but in particular also because their dielectric, ferroelectric and resistive properties can be tuned in a very large range by compositional or microstructural design. These physical properties are practically often controlled by modifications of the occupancy on the cationic sites of Ba^{2+} and Ti^{4+} within the perovskite lattice of this material system. Only in rather recent years, perovskite-related oxides with a partial substitution of the anionic site O^{2-} by nitrogen N^{3-}, called oxynitrides, show some interesting characteristics making them suitable for applications like non-toxic pigments and photo-catalysts [1]. The question within this context, how the substitution of oxygen by nitrogen affects the crystallography, the electronic, dielectric and ferroelectric characteristics but finally also the crystal chemistry has so far not been thoroughly clarified yet, although the answer is highly relevant not only from the fundamental but also from the technological point of view.

Most widely ammonia (NH$_3$) is used for nitridation of perovskite-related oxides at high temperatures (600~1200 °C) [1]. Under equilibrium conditions in this temperature range ammonia is mostly decomposed at atmospheric pressure. Therefore it is important to realise a possibility to carry out the reaction under non-equilibrium conditions, e.g. by increasing the flow rate of ammonia. Generally during nidridation uncooled ammonia is introduced in a hot reactor that contains the oxide to be converted, resulting in a rather inefficient yield because rather high flow rates have to be adjusted. This disadvantage may be circumvented by cooling the NH$_3$-gas stream before the reaction starts. Eventually this also results in enhanced reaction kinetics and an improved time efficiency.

For this purpose we designed and constructed a new type of ammonolysis reactor, which is presented in Figure 1.

Figure 1: Overview on a new ammonolysis reactor for the nitridation of oxide based perovskites with cooled NH$_3$.

This reactor consists in principle of two zones, both enclosed in a gas tight quartz tube: one for nitridation at elevated temperatures inside the furnace and a second one for cooling the nidridated sample under flowing ammonia outside the furnace. In the hot zone, ammonia is showered over the sample through a quartz tube that is coaxially embedded in another tube, in which compressed cooling air for ammonia is flowing. In the cooling zone, ammonia may be equally showered over the sample while quenching it in order to minimize decomposition.

In figure 2 the dependence of the approximate ammonia temperature, measured directly at the position where the gas streams out of the quartz pipe in the hot zone onto the sample, from the actual sample temperature and for different flow rates of cooling air is represented. Even at an ammonolysis temperature of 950°C at the sample, the ammonia temperature can be kept below 300 °C, where NH_3 is almost completely decomposed (concentration ~ 2 % (table 1)). The present report will present first results on the influence of the nitrification parameters (temperature, NH_3-flow rate, loading time, quenching conditions …) on the phase purity, crystallography, N-content, and diffusivity of N into $BaTiO_3$ at different levels of donor doping.

Temperature (°C)	Ammonia fraction (%)
100	65.7
200	14.7
300	2.1
400	0.4

Table1: The relationship between temperature and ammonia concentration at equilibrium conditions under atmospheric pressure.

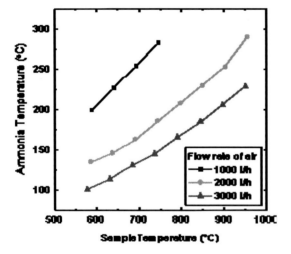

Figure 2: Dependence of the ammonia temperature from the sample temperature in the reactor at different flow rates of cooling air. The flow rate of NH3 was 100 ml/min.

[1] S.G. Ebbinghaus et al., Progress in Solid State Chem. **37**, 173 (2009).

METALLIC ELECTROLYTE COMPOSITES IN THE FRAMEWORK OF THE BRICK-LAYER MODEL

H. Lustfeld[1], C. Pithan[2], M. Reißel[3]

[1]PGI-1 Forschungszentrum Jülich, D52425 Jülich, Germany;
[2]PGI-7 Forschungszentrum Jülich, D52425 Jülich, Germany;
[3]Fachhochschule Aachen, Abteilung Jülich, D52428 Jülich, Germany

It is well known that the already large dielectric constants of some electrolytes like $BaTiO_3$ can be enhanced further by adding metallic (e.g. Ni, Cu or Ag) nanoparticles. The enhancement can be quite large, a factor of more than 1000 is possible. The consequences for the properties will be discussed in the present paper applying a brick-layer model (BLM) for calculating dc-resistivities of thin layers and a modified one (PBLM) that includes percolation for calculating dielectric properties of these materials.

The PBLM results in an at least qualitative description and understanding of the physical phenomena: This model gives an explanation for the steep increase of the dielectric constant below the percolation threshold and why this increase is connected to a dramatic decrease of the breakdown voltage as well as the ability of storing electrical energy. We conclude that metallic electrolyte composites like $BaTiO_3$ are not appropriate for energy storage.

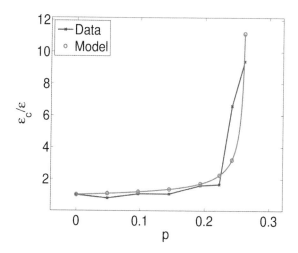

Figure 1: Comparison between measured relative increase of the dielectric constant due to Cu nanoparticles in $BaTiO_3$ and theory. The critical concentration is $p_c \approx 0.28$. Note the huge increase close to p_c. We show that this increase does not lead to an improved storage of electrical energy. On the contrary the ability of storing energy is the inverse of this relative increase and in particular approaches 0 at p_c. Details are given in Ref. [1].

[1] H. Lustfeld, C. Pithan and M Reißel, J. Eur. Ceram. Soc. **32**, 859 (2012).

DETECTION OF FILAMENT FORMATION IN FORMING-FREE RESISTIVELY SWITCHING SrTiO$_3$ DEVICES WITH Ti TOP ELECTRODES

S. Stille[1], Ch. Lenser[2], R. Dittmann[2], A. Köhl[2], I. Krug[3], R. Muenstermann[2], J. Perlich[4], C. M. Schneider[3], U. Klemradt[1], and R. Waser[2,5]

[1]II. Institute of Physics and JARA-FIT, RWTH Aachen University, 52056 Aachen, Germany;
[2]Peter Gruenberg Institute PGI-7 and JARA-FIT, Forschungszentrum Juelich, 52425 Juelich, Germany;
[3]Peter Gruenberg Institute PGI-6 and JARA-FIT, Forschungszentrum Juelich, 52425 Juelich, Germany;
[4]HASYLAB, DESY, 22607 Hamburg, Germany; [5]Institute of Materials in Electrical Engineering and Information Technology II, RWTH Aachen University, 52056 Aachen, Germany

The development of the next memory generation for information technology is based largely on the understanding of new materials. One promising approach consists of the use of resistively switching oxides, which change their electrical resistance when they are exposed to an electric field [1]. In SrTiO$_3$ (STO), stable bipolar resistive switching was first observed around 11 years ago [2]. The electrical conduction in the material takes place along filamentary agglomerations of oxygen vacancies [3, 4].

In this work, we investigated in detail the influence of Ti top electrodes on the oxidation state of epitaxially grown STO thin films and the corresponding resistive switching devices. A sketch of our samples is presented in figure 1a. A capping layer of a few nm Pt was used to avoid oxidation of the Ti layer in the surrounding air.

For 20 nm STO thin films, a Ti layer thickness of above 5 nm strongly reduces the initial resistance of the samples and gives rise to forming-free devices (cf. figure 1b and 1c). Hard X-ray photoemission (HAXPES) experiments on Pt/4 nm Ti/STO heterostructures, carried out at beamline P09 (HASYLAB, Hamburg), reveal the Ti layer to be composed of several oxide phases, which are induced by the reduction of the STO thin film.

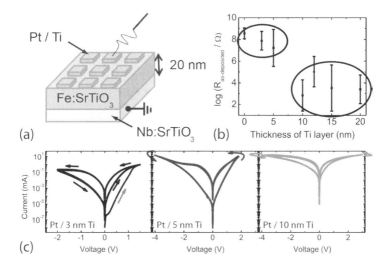

Figure 1:
(a) Sketch of the sample geometry. The STO is doped with Fe or Nb, respectively.
(b) Initial resistance of Pt/Ti/Fe:STO/Nb:STO MIM structures vs. thickness of Ti layer determined at ± 100 mV.
(c) Representative I(V)-curves of samples with different Ti thicknesses. Depending on the Ti thickness and the resulting initial resistivity, samples with Pt/Ti electrodes are forming-free. [6]

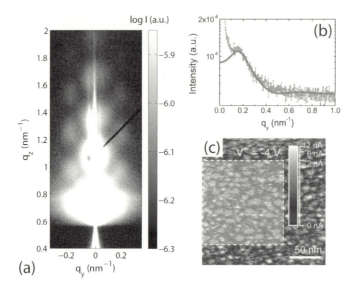

Figure. 2:
(a) GISAXS pattern for sample 5 nm Ti/20 nm Fe:STO/Nb:STO. A laterally formed structure can be seen.
(b) Lateral cut of GISAXS-intensity along $q_z = 0.95$ nm^{-1}. Continuous line: Simulated GISAXS intensity
(c) LC-AFM image of 20 nm Fe:STO showing conductive spots at the surface after a treatment with negative voltage. The highlighted region has been used for statistical analysis. [6]

Grazing incidence small angle X-ray scattering (GISAXS) measurements have been carried out at beamline BW4 (HASYLAB, Hamburg) [5]. The resulting intensity of an as-deposited sample is presented in Fig. 2a. This structure has not been observed in reference samples without Ti. The vertical (q_z) modulation in intensity gives a height of the scatterers of (21.7 ± 2.2) nm. The observed structure is therefore related to the 20 nm thick STO film.

A lateral intensity cut (Fig. 2b) yields a correlation maximum for the observed pattern at $q_y = 0.152$ nm^{-1}. Simulations with the software FitGISAXS [7] (red line in Fig. 2b) have been performed to obtain more quantitative information on the scatterers, which were modeled in a first approximation as cylinders with a different electron density compared to the surrounding STO film. The resulting minimum of the cylindrical form factor led to a mean diameter of the scatterers of D ≈ 15 nm. The correlation peak gives a mean distance between the scatterers of D' = 30 nm. This value agrees very well with conductive AFM measurements of a STO surface after a negative voltage treatment (cf. Fig 2c). A statistical analysis of the selected spots in the scan gives a mean distance of the conductive regions of around 29 nm.

As a result, we could clearly observe that the reduction of the STO thin film occurs in a filamentary way rather than in a homogeneous one. We attribute this behavior to the preferential reduction of STO thin films along highly defective areas.

[1] R. Waser, R. Dittmann, G. Staikov, K. Szot, Adv. Mater. **21**, 2635 (2009).
[2] Y. Watanabe, J. Bednorz, A. Bietsch, C. Gerber, D. Widmer, A. Beck, S. Wind, Appl. Phys. Lett. **78**, 3738 (2001).
[3] T. Menke, R. Dittmann, P. Meuffels, K. Szot. R. Waser, J. Appl. Phys. **106**, 114507/1 (2009).
[4] R. Muenstermann, T. Menke, R. Dittmann, S. Mi, C.-L. Jia, D. Park, J. Mayer, J. Appl. Phys. **108**, 124504 (2010).
[5] S. V. Roth, R. Doehrmann, M. Dommach, M. Kuhlmann, I. Kroeger, R. Gehrke, H. Walter, C. Schroer, B. Lengeler, and P. Mueller-Buschbaum, Rev. Sci. Instrum. **77**, 085106 (2006).
[6] S. Stille, Ch. Lenser, R. Dittmann, A. Koehl, I. Krug, R. Muenstermann, J. Perlich, C. M. Scheider, U. Klemradt, R. Waser, submitted (2012).
[7] D. Babonneau, J. Appl. Crystallogr. **43**, 929 (2010).

FREE-ELECTRON FINAL-STATE CALCULATIONS FOR THE INTERPRETATION OF HARD X-RAY ANGLE-RESOLVED PHOTOEMISSION

L. Plucinski[1], J. Minar[2], J. Braun[2], A. X. Gray[3,4], S. Ueda[5], Y. Yamashita[5], K. Kobayashi[5], H. Ebert[2], C.S. Fadley[3,4], and C.M. Schneider[1,6]

[1]Peter Gruenberg Institut PGI-6, Research Center Juelich, 52425 Juelich, Germany;
[2]Department of Chemistry, Ludwig Maximillian University, 81377 Munich, Germany;
[3]Department of Physics, University of California Davis, Davis, CA 95616, USA;
[4]Materials Sciences Division, Lawrence Berkeley National Laboratory, Berkeley, CA 94720, USA;
[5]NIMS Beamline Station at SPring-8, Nat. Inst. for Mat. Science, Hyogo 679-5148, Japan;
[6]Fakultaet fuer Physik, University of Duisburg-Essen, Duisburg, Germany.

A simple two-step numerical procedure has been developed in order to interpret the electron band dispersions measured by angle-resolved photoemission at energies where free-electron final states are a good approximation. The useful photon energy range is from above approx. 50 eV up until several keV, while for much higher energies the modulations due to band dispersions are smeared out by the photon effects in experimental data.

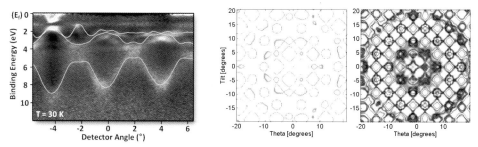

Fig1. *Left*: comparison between the experimental and free-electron band dispersions for tungsten (110) crystal at approx. 6 keV [1]. *Right*: The cut through a Fermi surface and a constant energy surface at $E_B = 2.5$ eV of LaSrMnO as predicted for ARPES at 833.2 eV photon energy, different colors refer to majority and minority electron spin.

In the first step the *k*-point position in the Brillouin zone is calculated taking into account the vector momentums of the emitted electron and incoming photon. A macro written in Matlab allows full 3D geometry by means of matrix rotations and can be automatized for predicting the trajectory in the Brillouin zone or the Fermi or constant energy surfaces.

In the second step the set of *k*-points calculated by macro is plugged into the DFT code which calculates the electron eigenvalues at these points. It can be done by virtually any code, but we found the WIEN2k in particular useful due to its convenient interface, no need to use pseudopotentials, and excellent online support.

The left panel in Fig. 1 presents excellent agreement between the theory and experiment at several keV photon energy [1] where purely bulk states are probed, while the right panel presents the abilities of our procedure in calculating Fermi and constant energy surfaces.

We will present results of the free-electron final state calculations for various materials at various photon energies from VUV up to hard x-rays, and compare them to the experimental results and one-step model photoemission calculations.

[1] A. X Gray, C. Papp, S. Ueda, B. Balke, Y. Yamashita, L. Plucinski, J. Minar, J. Braun, E. R. Ylvisaker, C. M. Schneider, W. E. Pickett, H. Ebert, K. Kobayashi, and C. S. Fadley, Nature Materials 10, 759 (2011),

GROWTH PRESSURE CONTROL OF ELECTRONIC INTERFACE PROPERTIES IN LAO/STO HETEROSTRUCTURES: NEW INSIGHT FROM HIGH-ENERGY SPECTROSCOPY

M. Sing[1], A. Müller[1,2], H. Boschker[2], F. Pfaff[1], G. Berner[1], S. Thiess[3], W. Drube[3], G. Koster[2], G. Rijnders[2], D.H.A Blank[2], M. Sing[1], R. Claessen[1]

[1] Universität Würzburg, Physikalisches Institut, 97074 Würzburg, Germany
[2] Faculty of Science and Technology and MESA+ Institute for Nanotechnology, University of Twente, 7500AE Enschede, The Netherlands
[3] Deutsches Elektronen-Synchrotron DESY, 22607 Hamburg, Germany

Oxide heterostructures display many interesting phenomena, one example being the formation of a two-dimensional electron system at the $LaAlO_3/SrTiO_3$ (LAO/STO) interface beyond a certain critical thickness of the polar LAO [1,2]. One explanation for this behavior is the so-called electronic reconstruction. Within this scenario electrons are transferred to the interface to – at least partially – compensate the potential gradient in LAO and to thus minimize electrostatic energy. Other explanations involve extrinsic effects like oxygen vacancies or La-Sr intermixing.

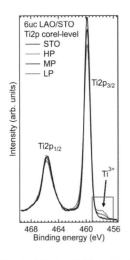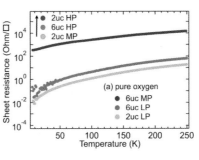

Figure 1:

Left: HAXPES Ti$2p$ core-level spectra exhibiting a small shoulder indicating Ti^{3+} weight.

Right: Transport measurements indicating a metal-insulator transition for samples grown at various oxygen partial pressures.

To shed further light on this issue, we grew samples at various pressure conditions and related the conducting properties of the interface from transport measurements to the charge carrier concentration as derived from hard x-ray photoelectron spectroscopy. Regarding the latter, the amount of interface charge carriers is directly reflected by the Ti$2p$ core-level spectra as shown in Fig. 1. The shoulder at lower binding energy corresponds to emission from Ti ions at the interface with a reduced valency of 3+ due to an additional electron in the $3d$ shell. We find that the oxygen partial pressure determines the amount of extra charge carriers at the interface, while a too high total pressure destroys conductivity. To disentangle the effects of growth kinetics and oxidation, also samples in a mixed argon/oxygen atmosphere have been grown. From this, we conclude that it is actually the changes in the growth kinetics that lead to the insulating behavior at high growth pressures. A subsequent post-oxidation step results for all samples – except for the ones grown at too high pressures – in about the same amount of extra interface charge and the occurrence of a critical overlayer thickness for conductivity, independent of the initial growth conditions.

[1] Ohtomo et al., Nature **427**, 423 (2004).
[2] Thiel et al., Science **313**, 1942 (2006).

ANISOTROPIC MAGNETOELASTIC COUPLING IN FERROPNICTIDES

Haifeng Li[1]

[1] Juelich Centre for Neutron Science JCNS, Outstation at Institut Laue-Langevin (ILL)
BP 156, 38042, Grenoble Cedex 9, France

Single crystal synchrotron x-ray diffraction studies of ferropnictides such as $(Ba_{1-x}K_x)Fe_2As_2$ ($x = 0, 0.1$) and CeFeAsO reveal strong anisotropy in the charge correlation lengths along or perpendicular to the in-plane antiferromagnetic (AFM) wave-vector at low temperatures, indicating an anisotropic two-dimensional magnetoelastic coupling. The high-resolution setup allows to distinctly monitor each of the twin domains by virtue of a finite misfit angle between them that follows the order parameter. In addition, we find that the in-plane correlations, above the orthorhombic (O)-to-tetragonal (T) transition, are shorter than those in each of the domains in the AFM phase, indicating a distribution of the in-plane lattice constants. This strongly suggests that the phase above the structural O-to-T transition is virtually T with strong O-T fluctuations that are probably induced by spin fluctuations.

Note: Nanosessions, Advanced characterization - Spectroscopy, microscopy, scanning probe methods.

DEPTH-RESOLVED ARPES OF BURIED LAYERS AND INTERFACES VIA SOFT X-RAY STANDING-WAVE EXCITATION

<u>Alexander X. Gray</u>[1,2,3], Jan Minár[4], Lukasz Plucinski[5], Mark Huijben[6], Alexander M. Kaiser[1,2], Slavomír Nemšák[1,2], Giuseppina Conti[1,2], Aaron Bostwick[7], Eli Rotenberg[7], See-Hun Yang[8], Jürgen Braun[1], Guus Rijnders[5], Dave H. A. Blank[5], Susanne Stemmer[9], Claus M. Schneider[4], Juergen Braun[4] Hubert Ebert[4], and Charles S. Fadley[1,2]

[1]Department of Physics, University of California Davis, Davis, California, USA;
[2]Materials Sciences Division, Lawrence Berkeley National Laboratory, Berkeley, California, USA;
[3]Stanford Institute for Materials and Energy Science, SLAC National Accelerator Laboratory, Menlo Park, California, USA;
[4]Department of Chemistry and Biochemistry, Ludwig Maximillian University, Munich, Germany;
[5]Peter-Grünberg-Institut PGI-6, Forschungszentrum Jülich GmbH, 52425 Jülich, Germany;
[6]Faculty of Science and Technology, University of Twente, Enschede, The Netherlands;
[7]Advanced Light Source, Lawrence Berkeley National Laboratory, Berkeley, California, USA;
[8]IBM Almaden Research Center, San Jose, California, USA;
[9]Materials Department, University of California Santa Barbara, Santa Barbara, California, USA.

Traditional angle-resolved photoemission spectroscopy (ARPES) in the roughly 50-150 eV regime is a well established and powerful technique, enabling the direct measurement of the valence-band electronic structure of materials. However, in many applications, its high surface sensitivity can make it difficult to discern true bulk or buried layer/interface properties. Here we introduce a new depth-selective photoemission technique by combining higher-energy soft x-ray ARPES with standing-wave (SW) excited photoelectron spectroscopy (SWARPES), wherein the intensity profile of the exciting x-ray radiation is tailored within the sample [1-3]. An x-ray standing-wave field is set up within the sample by growing it on (or as) a synthetic periodic multilayer mirror substrate, which in first-order Bragg reflection acts as a standing-wave generator. The antinodes of the standing wave function as epicenters for photoemission, and can be moved vertically through the buried layers and interfaces by scanning the x-ray incidence angle. We have applied this SWARPES technique to the investigation of the electronic properties of multilayer oxide samples so as to characterize the buried interface within the magnetic tunnel junction system $La_{0.7}Sr_{0.3}MnO_3/SrTiO_3$ (Figures 1 and 2a), an epitaxial $SrTiO_3/LaNiO_3$ superlattice of relevance to low-dimensional heterostructuring and Mott-transition devices (Figure 2b), and an epitaxial $BiFeO_3/La_{0.7}Sr_{0.3}MnO_3$ superlattices of relevance to exchange-bias and multiferroic field-effect devices (Figures 2c-d). The experimental results are compared to the state-of-art one-step photoemission theory including matrix element effects [4,5].

Supported by the U.S. Dept. of Energy under Contract No. DE-AC02-05CH11231 and ARO MURI Grant W911-NF-09-1-0398.

[1] S.-H. Yang *et al.*, J. Phys. Cond. Matt. **14**, L406 (2002).
[2] A. X. Gray *et al.*, Phys. Rev. B **82**, 205116 (2010).
[3] A. M. Kaiser *et al.*, Phys. Rev. Lett. **107**, 116402 (2011).
[4] J. Braun *et al.*, Phys. Rev. B **82**, 024411 (2010).
[5] A.X. Gray *et al.*, to be published.

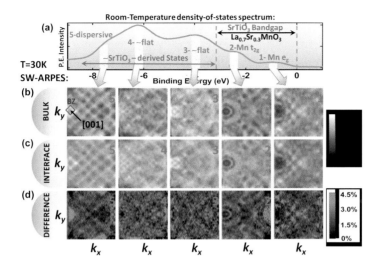

Figure 1. Depth-resolved SW-ARPES measurements for an LSMO/STO superlattice. (a) Angle-integrated density-of-states spectrum, including all the major features of the valence-bands, labeled 1 – 5. (b) Bulk-LSMO sensitive standing-wave measurement geometry. (c) LSMO/STO-interface sensitive measurement geometry. (d) Interface-minus-bulk difference maps, revealing different behavior of the LSMO-derived Mn $3d$ e_g (1) and Mn $3d$ t_{2g} (2) states at the interface between STO and LSMO.

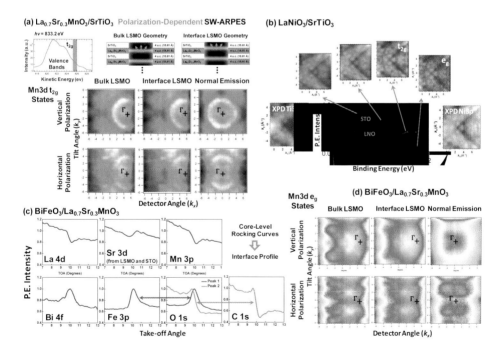

Figure 2. Polarization-dependent SW-ARPES from oxide multilayers. (a) Mn $3d$ t_{2g} states in bulk LSMO and at the interface between LSMO and STO. (b) Interface-sensitive electronic structure for LaNiO$_3$/SrTiO$_3$. (c) Core-level standing-wave rocking curves for the BiFeO$_3$/La$_{0.7}$Sr$_{0.3}$MnO$_3$ superlattice, which reveal the chemical profile of the interface. (d) Bulk and interface-sensitive k-space maps of the Mn $3d$ e_g states at the BiFeO$_3$/La$_{0.7}$Sr$_{0.3}$MnO$_3$ interface, together with normal emission spectra.

STRUCTURING COMPLEX OXIDE HETEROSYSTEMS VIA E-BEAM LITHOGRAPHY

Carsten Woltmann[1], Hans Boschker[1], Rainer Jany[2], Christoph Richter[2], Jochen Mannhart[1]

[1]Max Planck Institute for Solid State Research, 70569 Stuttgart, Germany;
[2]Experimental Physics VI, Augsburg University, 86135 Augsburg, Germany

Electronic correlation in oxides holds a variety of new possibilities for future electronics. Especially the $LaAlO_3$-$SrTiO_3$ system gained much attention over the last years due to the formation of a 2D electron liquid at the heterostructures interface.

Along the path to understanding and exploiting these fascinating materials and bringing them to an applicational stage, the ability to structure them is essential. I will report on the potential and the problems of electron beam lithography for creating oxide electronic nanoscale devices.

LASER HEATING OF OXIDE SUBSTRATES: CHALLENGES AND SOLUTIONS

T. Heeg[1], W. Stein[1]

[1]SURFACE systems+technology GmbH & Co. KG, Hueckelhoven, Germany

The thermal preparation of substrates during advanced oxide deposition processes requires a flexible and precisely controlled heater. In such processes employing high oxygen partial pressures the usual heater materials are very limited and introduce restrictions of the vacuum range or of the temperature. Platinum and silicon carbide are the major heater materials if a conventional radiation or contact heater for temperatures up to 1000-1100 °C is used.

Over the past 20 years applications of laser heating were published in the field of thin film deposition, using CO_2- [1-3], YAG- [4,5], as well as diode [6] lasers. Laser light of high intensity, well focused only to the substrate, is an extremely vacuum friendly tool, especially for the ultra-high vacuum range of applications. The heated zone in the UHV chamber is limited to the substrate only and thus has the smallest possible size, minimizing outgassing and enabling the lowest background pressures and very clean processes.

Using light to transfer energy to the sample requires consideration of the optical properties of the used substrate, because absorption of the laser light is essential for an effective heating process. YAG and diode lasers work in the near infrared region. Many substrate materials commonly used for oxide thin film deposition are partially or fully transparent at near-infrared wavelengths, so direct illumination of the substrate with these lasers is not efficient. Several more or less acceptable workarounds of this basic problem have been developed, but partially foil the advantages of laser heating. In addition, problems with these workarounds are increasing with higher temperatures, so the step to another wavelength is recommended. A CO_2-laser with 10.6 µm wavelength offers a good solution for the absorption problem – but the handling of this light source is not as convenient as that of the frequently used diode laser. In this work diode- and CO_2-laser heater versions are compared and solutions for the existing difficulties are shown. Finally, results of the use of such heaters in oxide film deposition are demonstrated.

[1] P.E. Dyer, A. Issa, P.H. Key, P. Monk, Supercond. Sci. Technol. 3, 472 (1990)
[2] K.H. Wu, C.L. Lee, J.Y. Juang, T.M. Uen, Y.S. Gou, Appl. Phys. Lett. 58, 1089 (1991)
[3] S.J. Barrington, R.W. Eason, Rev. Sci. Instrum. 71, 4223 (2000)
[4] S. Ohashi, M. Lippmaa, N. Nakagawa, H. Nagasawa, H. Koinuma, M. Kawasaki, Rev. Sci. Instrum. 70, 178 (1999)
[5] H. Lippmaa, T. Furumochi, S. Ohashi, M. Kawasaki, H. Koinuma, T. Satoh, T. Ishida, H. Nagasawa, Rev. Sci. Instrum. 72, 1755 (2001)
[6] W. Stein, MRS spring meeting 2006, GG13.6 (Laser Heating, a challenging new Technology for small Substrates in Oxide Deposition Processes)

WET CHEMICAL ETCHING OF STRONTIUM RUTHENIUM OXIDE THIN FILMS BY OXIDATION

Dieter Weber[1], Róza Vőfély[2], Yuehua Chen[3], Ulrich Poppe[1]

[1]Forschungszentrum Jülich, Peter Grünberg Institute, 52425 Jülich, Germany;
[2]Trinity College, University of Cambridge, Cambridge, United Kingdom; [3]Xi'an Jiaotong University, Xi'an, China

$SrRuO_3$ is an oxide with perovskite structure that has a high electronic conductivity. It is therefore used as electrode to investigate the dielectric properties of epitaxial perovskite thin films and build ferroelectric memory cells with superior properties[1]. It can be structured lithographically to form all-oxide contacts and conduction lines for more complex devices. Up to now only sputter etching and aqueous ozone solutions[2] were known to attack this material. Sputter etching is not selective and leaves an amorphous surface layer that has to be removed or recrystallized for subsequent epitaxial deposition, while ozone also oxidizes photo resists and requires a special apparatus. The ozone attacks $SrRuO_3$ by oxidizing Ru(IV) in $SrRuO_3$, corresponding to the insoluble RuO_2, to RuO_4 which is volatile and sufficiently soluble in water.

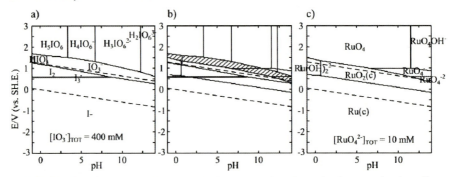

Figure 1: The Pourbaix diagrams of iodine (a) and ruthenium (c) show the predominant species depending on redox potential and pH in water. Overlapping areas where periodate can oxidize RuO_2 are marked in b). The diagrams were calculated with MEDUSA[4].

Based on literature that reports oxidation of RuO_2 to RuO_4 with periodate[3], we developed two chemical etchants that are based on strong oxidizers. An aqueous solution of ceric ammonium nitrate $(NH_4)_2Ce(NO_3)_6$ with perchloric acid as pH stabilizer dissolves $SrRuO_3$ very quickly. The solution is commercially available as an etchant for chromium. Unfortunately, it etches thin films of $SrRuO_3$ too quickly and also degrades oxide substrates because of it's low pH. A sodium periodate solution acts more slowly with an etch rate of a few nm per second and has a nearly neutral pH. It leaves oxide substrates undamaged and allows subsequent epitaxy. The Pourbaix diagrams of iodine and ruthenium (Fig. 1) suggest that etching is possible over the entire pH range.

Oxidation or reduction can potentially also be used to etch other oxides where a metal cation can be brought from a less soluble to a more soluble oxidation state. One known example is etching of $La_{0.33}Sr_{0.67}MnO_3$ (LSMO) with concentrated hydrochloric acid, where Mn(III) and Mn(IV) is reduced to more soluble Mn(II), oxidizing Cl^- to Cl_2 in the process.

[1] C. B. Eom et al., Appl. Phys. Lett. **63**, 2570 (1993).
[2] H. Tomita et al., US Patent No. 6436723 (2002).
[3] W. Griffith, Ruthenium Oxidation Complexes: Their Uses as Homogenous Organic Catalysts, Springer (2010), pp.12-13.
[4] I. Puigdomenech, *Chemical Equilibrium Diagrams*, http://www.kemi.kth.se/medusa/ (2010).

EFFECT OF ALD PROCESSING AND TOP ELECTRODES ON ZrO$_2$ THIN FILMS STRUCTURAL AND RESISTIVE SWITCHING CHARACTERISTICS

<u>Irina Kärkkänen</u>[1*], Mikko Heikkilä[2], Jaakko Niinistö[2], Mikko Ritala[2], Markku Leskelä[2], Susanne Hoffmann-Eifert[1], Rainer Waser[1]

[1] Peter Grünberg Institute and JARA-FIT, Research Center Jülich, 52425 Jülich, Germany
[2] Laboratory of Inorganic Chemistry, University of Helsinki, 00014 Helsinki, Finland

In this work, we studied the influence of ALD processing and different top electrodes on the structural, morphological and electrical properties of ZrO$_2$. The films were deposited from Zr(NEtMe)$_4$ (TEMA-Zr) by atomic layer deposition, using ozone or water as oxygen sources. Deposition temperature was 240°C. The films were characterized with x-ray diffraction (XRD), x-ray reflectometry (XRR), scanning electron microscope (SEM). Differences in structure and electrical properties, like dielectric constants and resistive switching characteristics, were found depending on the type of oxygen source or top electrode, respectively. Ozone grown ZrO$_2$ films showed preferably oriented cubic-tetragonal structure, whereas ones deposited with water were randomly oriented cubic-tetragonal with more pronounced tetragonal phase in thicker films. The difference in microstructure was also confirmed from SEM pictures. Different device structures were built from 20 nm thick ALD ZrO$_2$ films with respect to a variation of an electrode layer: Pt/ZrO$_2$/Pt, Pt/ZrO$_2$/Al/Pt, Pt/ZrO$_2$/Ti/Pt, where the thickness of Ti top electrode was varied from 5 to 90 nm. Resistive switching in the metal-insulator-metal structures showed both types of switching polarity – unipolar (UP) and bipolar (BP) - depending on the top electrode. Samples with Al and Pt top electrodes showed not stable UP switching. By varying Ti top electrode layer we were able to switch from UP to BP behavior. The most stable BP switching was obtained in Pt/ZrO$_2$/Ti/Pt structure with a 40nm Ti top electrode (Fig.1)

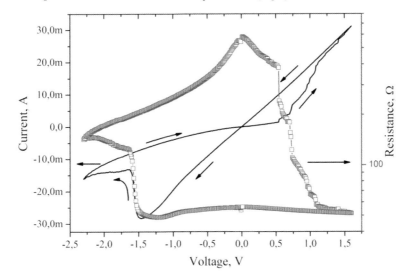

Fig.1. Bipolar resistive switching of Pt/ZrO$_2$/Ti/Pt with 40 nm Ti top electrode. Voltage sweeping was performed up to 100 cycles and showed stable RS. Electroforming was performed at +1,8V. Top electrode pad size is 200 µm.

The research leading to these results has received funding from the European Community's Seventh Framework Program (FP7/2007-2013) under grant agreement number ENHANCE-238409.

RESISTIVE SWITCHING STUDY ON Nb_2O_5 THIN FILMS OF DIFFERENT THICKNESS AND MORPHOLOGY GROWN BY PVD AND ALD

N. Aslam[1], M. Reiners[1], T. Blanquart[2], H. Mähne[3], J. Niinistö[2], M. Leskelä[2], Mikko Ritala[2], T. Mikolajick[3], S. Hoffmann-Eifert[1] and R. Waser[1]

[1]Peter-Grünberg Institute (PGI-7) and JARA-FIT, Forschungszentrum Jülich, Jülich, Germany
[2]Department of Chemistry, University of Helsinki, Helsinki, Finland
[3]NamLab GmbH and Institute of Nanoelectronic Materials, TU Dresden, Dresden, Germany

Metal-Insulator-Metal (MIM) based devices are considered for modern non-volatile resistive switching memories (RRAM). In the field of oxides the focus of materials under investigation has recently been extended towards group (V) transition metal oxides (TMO), especially Ta_2O_5 [1-2]. Beside Ta_2O_5, Nb_2O_5 is wide band gap material which has been successfully integrated into silicon based devices, as for example in dynamic random access memories (DRAM). In 1965 Hiatt [3] reported on resistive switching effects in a MIM-type diode of $Nb/Nb_2O_5/In$. Whereas topical articles discussing the resistive switching (RS) in Nb_2O_5 thin films sandwiched between metal electrodes for applications in RRAM devices are a few [4-5]. In this study we investigated the resistance switching characteristic of Nb_2O_5 thin films integrated into $Pt/Nb_2O_5/Ti/Pt$ micro cross bar structures on Si/SiO_2 substrates. Cross bar structures with lateral dimensions from ($1\mu m \times 1\mu m$) to ($3\mu m \times 3\mu m$) were made by means of photolithography and reactive ion beam etching. The sandwiched Nb_2O_5 thin films with thickness of 8 nm, 15 nm and 35 nm were deposited by atomic layer deposition (ALD) and physical vapor deposition (PVD) processes, respectively. ALD films were grown in an ASM-120 reactor at 275°C from tBuN=Nb(NMeEt)$_3$ and ozone at a rate of about 0.5 Å/cycle. [6]. PVD Nb_2O_5 films were deposited by reactive dc magnetron sputtering from a metallic niobium target. [4] Both types of films were amorphous in the as deposited state. The films changed into single phase polycrystalline structure after post-annealing at 600°C in argon or oxygen.

The resistive switching (RS) behaviour of the Nb_2O_5 films was studied with respect to a dependence on the film growth method, either ALD or PVD, the film thickness, the film structure, i.e. as-deposited amorphous or polycrystalline after post annealing, and with respect to an influence of the atmosphere during annealing, either argon or oxygen. In general, under a defined current compliance the micro cross bar devices show resistance switching after forming. As exemplary results figures 1 and 2 show representative I (V) and R (V) characteristics for polycrystalline Nb_2O_5 films obtained by ALD and PVD, respectively. Figure 1 shows the RS behavior of a $(1\mu m)^2$ cell with a 8 nm thick film of Nb_2O_5 prepared by ALD and crystallized at 600°C under O_2. The cell formed under positive polarity at a voltage of +1.25V. The device shows stable switching cycles at set and reset voltages of +0.67V (SET) and -0.8V (RESET) under a positive current compliance of 450 µA. The respective R_{ON} and R_{OFF} values at a read voltage of +0.1V were about 1.1×10^4 Ω and 1.3×10^5 Ω, giving an R_{OFF}/R_{ON} ratio in the order of 11. Figure 2 shows the RS behavior of a $(1\mu m)^2$ cell with a 15nm thick film of Nb_2O_5 prepared by PVD and crystallized at 600°C in argon. The cell formed at a positive voltage of +2.6V. For a positive current compliance of 1.0 mA the cell shows stable switching. The SET and RESET voltage was about +0 .8V and -0.9V, respectively. At a read voltage of +0.1V the resistance values were $R_{ON} = 2.0 \times 10^3$ Ω and $R_{OFF} = 1.3 \times 10^4$ Ω leading to an R_{OFF}/R_{ON} ratio in the order of 6.5. Further experiments regarding the effects of the changes in the device structure of the $Pt/Nb_2O_5/Ti/Pt$ micro cross bar structures on the RS characteristics are in progress.

Acknowledgment: This work has been supported in parts by the European Community's Seventh Framework Program (FP7/2007-2013) under grant agreement number ENHANCE-238409.

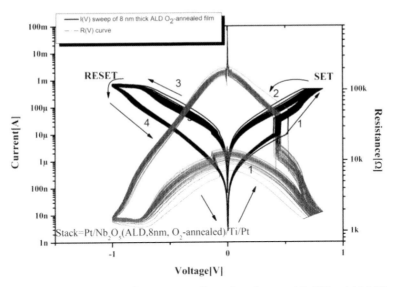

Fig.1. I (V) and R(V) curves for constant cycling voltage between (+1.5 V) and (-1.5 V) at constant positive current compliance of 450µA for an ALD grown 8nm thick polycrystalline Nb_2O_5 thin film sandwiched between Pt bottom and Ti/Pt to electrode.

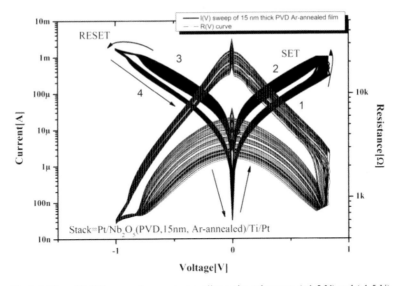

Fig.2. I (V) and R (V) curves for constant cycling voltage between (+1.5 V) and (-1.5 V) at constant positive current compliance of 1.0mA for sputtered 15nm thick Nb_2O_5 thin film crystallized by post annealing in argon atmosphere.

[1] M.-J. Lee et al., Nature Materials 10, (2011), 625-630.
[2] J. J. Yang et al., Appl. Phys. Lett. 97, (2010), 232102.
[3] W. Hiatt et al., Appl. Phys. Lett .6 (1965), 105.
[4] H. Mähne et al., Solid-State Electronics (2012); accepted.
[5] L. Chen et al., Current Applied Physics 11 (2011) 849-85.
[6] T. Blanquart et al., Chem. Mater (2012); accepted.

CALCULATION OF LORENZ TRANSITION ELECTRON MICROSCOPY DIFFRACTION MAP FOR A CHIRAL MAGNETIC SOLITON LATTICE

Yoshihiko Togawa[1], Jun-ichiro Kishine[2], Alexander Ovchinnikov[3], Igor Proskurin[3]

[1]N2RC, Osaka Prefecture University, 1-2 Gakuencho, Sakai, Osaka 599-8570, Japan;
[2]Department of Basic Sciences, Kyushu Institute of Technology, Kitakyushu, 804-8550, Japan;
[3]Department of Physics, Ural Federal University, Ekaterinburg, 620083, Russia

Magnetic crystals belonging to the chiral space group such as MnSi, FeGe, $Cr_{1/3}NbS_2$ and molecular based magnets have attracted significant interest over the last few years due to their promising application in modern technologies. In chiral magnetic crystals structural chirality gives rise to the relativistic spin-orbit Dzyaloshinskii—Moriya (DM) exchange interaction of the form $-\vec{D} \cdot (\vec{S}_i \times \vec{S}_j)$ between the neighboring spin magnetic moments $\vec{S}_i$ and $\vec{S}_j$ of the localized electrons. The DM interaction stabilizes the helical magnetic order.

In recent Lorenz microscopy and small-angle electron diffraction experiments performed on $Cr_{1/3}NbS_2$ crystals there was the first direct observation how the chiral helimagnetic structure in the applied magnetic field continuously evolves into nonlinear magnetic order called chiral soliton lattice (CSL) [1]. The CSL state consists of forced ferromagnetic domains periodically partitioned by 360-degree domain walls. The spatial period of the CSL, which exerts an effective potential for itinerant spins, is tuned by changing the magnetic field strength. Previous theoretical studies revealed that the CSL might support a variety of interesting functions for spintronic applications such as magnetic-phonon-like elementary excitations, current-driven sliding motion [2] and field induced metal-to-insulator transition [3].

To analyze the problem of the observation of the chiral magnetic soliton lattice by transmission electron microscopy, we have performed an analytical calculation of the optical phase shift using the standard Fourier method [4]. The phase shift for the sample in the form of thin slab with the CSL magnetic structure in the sample plane can be calculated from the standard Aharonov—Bohm expression

$$\varphi(x,y) = -\frac{\pi}{\phi_0} \int A_z(x,y,z) dz,$$

where (x,y) is the sample plane, A_z is magnetic vector potential component along the z-axis which is supposed to be the optical axis of the electron microscope, ϕ_0 is the magnetic flux quantum, and integration is over the sample thickness t. The knowledge of the phase shift map allows of restoring the magnetic induction map in the sample plane via the formula $(B_x, B_y) = \frac{\phi_0}{\pi t}\left(-\frac{\partial \varphi}{\partial y}, \frac{\partial \varphi}{\partial x}\right)$.

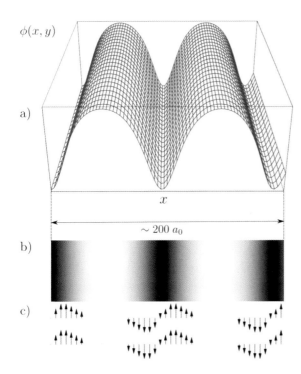

Figure 1: The result for the theoretical simulation of the magnetic phase shift for the case when magnetic field is applied along the optical axis. The magnetic field strength is 95% from its critical value (about 0.230 T in Cr1/3NbS2). a) the surface of constant phase (in arbitrary units); b) the phase shift gradient plot; and c) the magnetic induction map in the sample plane restored from the phase shift.

The result of our calculation (for the case when the magnetic field is applied *along* the optical z-axis and the magnetization modulation is along x-direction) is given by the analytic expression

$$\varphi(x,y) = -\frac{2\pi N_f}{\kappa^2 K}\left[\text{dn}\left(\frac{2Kx}{L_{CSL}}\right) - \frac{\pi}{2K}\right],$$

where "dn" is Jacobi-dn elliptic function, K is the elliptic integral of the first kind with the elliptic modulus κ (controlled by the magnetic field), L_{CSL} is the magnetic soliton lattice period, and N_f denotes the number of magnetic flux quanta trapped in the area tL_{CSL}. The phase shift surface is shown on the Figure 1 a). Figures 1 b), c) show the phase shift gradient and the magnetic induction map respectively. The calculated diffraction pattern has spatial period L_{CSL} which is in good agreement with the diffraction contrasts obtained in [1].

[1] Y. Togawa et al., Phys. Rev. Lett. **108**, 107202 (2012).
[2] J. Kishine et al., Phys. Rev. B **82**, 064407 (2010).
[3] J. Kishine et al., Phys. Rev. Lett. **107**, 017205 (2011).
[4] Zhu (ed.), Modern Techniques for Characterizing Magnetic Materials, Springer (2005).

CHARGE DENSITY WAVE STUDY IN $Dy_5Ir_4Si_{10}$

M. H. Lee[1], C. H. Chen[1,2,3], M.-W. Chu[2], H. D. Yang[4],

1. Department of Physics, National Taiwan University, Taipei, Taiwan
2. Center for Condensed Matter Sciences, National Taiwan University, Taipei, Taiwan
3. Institute of Atomic and Molecular Sciences, Academia Sinica, Taipei, Taiwan
4. Department of Physics, National Sun Yat-sen University, Kaohsiung, Taiwan

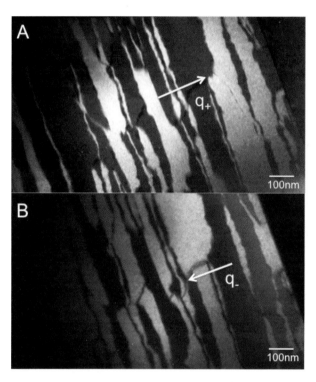

The fig (A) and (B) are dark-field images obtained from $\bar{q}_+ = (0,0,0.25)$ and $\bar{q}_- = (0,0,-0.25)$ superlattice spots in commensurate state, respectively. It is clear that modulation wave vectors $\bar{q}_+$ and $\bar{q}_-$ actually come from different domains. No chemical inhomogeneity was found between these two regions by chemical analysis electron nano-probe. The domain contrast results from breakdown of the crystallographic symmetry along the c-axis, which is also the CDW modulation wave vector.

The tetragonal rare-earth transition-metal silicide system, $R_5T_4Si_{10}$, where R is Dy, Ho, Er, Tm, and Lu, and T = Ir and Rh, with seemingly three-dimensional crystallographic structure, has been shown to exhibit fascinating charge density wave (CDW) phase transitions, a phenomenon found largely in low-dimensional systems. In this study we report the investigations of CDW in $Dy_5Ir_4Si_{10}$ at different temperatures using transmission electron microscopy (TEM) techniques including electron diffraction and dark-field imaging.

Superlattice diffraction spots along c-axis were observed in the electron diffraction pattern when the sample was cooled below the CDW transition temperature ($T_{CDW} \sim 200K$), indicating the presence of incommensurate CDW state with the modulation wave vector of $\bar{q} = (0,0,\pm 0.25 \pm \delta)$. CDW become commensurate with further cooling below 160K. Configurations of CDW dislocations convincingly show that the CDW phase transition is accompanied by a concomitant cell-doubling crystallographic structural phase transition. Furthermore, symmetry breakdown along c-axis observed by convergent beam electron diffraction (CBED) gives rise to two different types of CDW domains. Detailed characteristics of this unusual behavior will also be discussed.

PREPARING INAS NANOWIRES FOR FUNCTIONALIZED STM TIPS

Kilian Flöhr[1,3], H. Yusuf Günel[2,3], Kamil Sladek[2,3], Robert Frielinghaus[2,3], Hilde Hardtdegen[2,3], Marcus Liebmann[1,3], Thomas Schäpers[2,3], and Markus Morgenstern[1,3]

[1] II. Institute of Physics B, RWTH Aachen University, Physikzentrum Melaten, Otto-Blumenthal-Strasse, D-52056 Aachen, Germany,
[2] Peter-Grünberg Institute (PGI), Forschungszentrum Jülich GmbH, D-52425 Jülich, Germany,
[3] JARA Fundamentals of Future Information Technologies

We investigated methods to spacially control InAs nanowires on a substrate using micro-manipulators attached to an optical microscope with the goal of producing InAs tips for scanning tunneling microscopy. The wires, which were grown by selective area metalorganic vapor phase epitaxy (SA-MOVPE) on a GaAs wafer, could be picked up individually using a sharp indium tip exploiting adhesion forces. Later, the wires were placed with submicrometer precision onto a desired position somewhere at the edge of a cleaved GaAs wafer or on other substrates, e.g. a perforated Si_3N_4 membrane [1]. To contact these partly suspended nanowires at the wafer edge, we established a new method to realize ohmic contacts with standard electron beam lithography. We also present STM measurements on Au(111) with our newly developed nanowire STM tip.

[1] K. Flöhr, M. Liebmann, K. Sladek, H.Y. Günel, R. Frielinghaus, F. Haas, C. Meyer, H. Hardtdegen, Th. Schäpers, D. Grützmacher, M. Morgenstern, Rev. Sci. Instrum. **82**, DOI 10.1063/1.3657135 (2011)

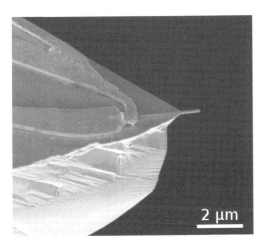
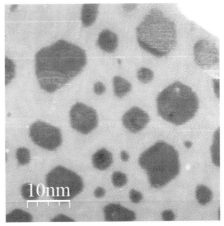

Figure 1: InAs nanowire on GaAs edge, contacted by Ti/Au electrodes

Figure 2: STM image of Au(111) after Ar bombardment at 300 K measured with an InAs nanowire tip

FROM THE MICROSCALE TO THE NANOSCALE: EDX ANALYSIS WITH HIGH SPATIAL RESOLUTION USING SILICON DRIFT DETECTORS (SDD)

Tobias Salge[1], Igor Nemeth[1], Meiken Falke[1]

[1]Bruker Nano GmbH, Berlin, Germany

Continuing technological advances require the elemental analysis of increasingly smaller structures in many fields, including material science, semiconductors, and nanotechnology in general. This confronts the otherwise well proven electron microscope-based energy dispersive X-ray spectroscopy (EDX) with new challenges. One aspect is the limited radiation yield in the low energy range required for the analysis of bulk samples with high spatial resolution in the SEM. Also in TEM/STEM, radiation yield scales with structure size and microscope capabilities. Aberration corrections and high brightness sources enable spectroscopic signals of small structures at voltages below 80 kV needed for reducing beam damage effects. Silicon drift detector (SDD) systems have become state of the art technology in the field of EDX microanalysis outperforming traditional Si(Li) detectors in almost every aspect. One of the main advantages of SDD technology for SEM is the extremely high throughput efficiency combined with good energy resolution (<121 eV Mn-Kα, <38 eV C-K) also in the lower energy range and the high output pulse load capacity of up to 600000 cps. The robustness and liquid nitrogen free cooling, allowing integration into the pole piece, are beneficial for TEM as well. These factors and modern data processing allow a range of advanced analysis options.

We present applications relevant for electronic materials using the FE-SEM and STEM equipped with SDD EDS systems. It will be demonstrated that using low accelerating voltages, the element distribution in bulk structures in the sub-μm scale can be displayed in a short time due to optimized signal processing and solid angle. Fig. 1 demonstrates the fast and easily available composition information using advanced SDD technology on SEM. Standard-based quantification leads to accurate results and can be used to detect sub-μm sized diffusion gradients e.g. in CIGS solar cells (Fig. 2).

In TEM/STEM the element composition of thin electron transparent samples can be analyzed in the nm-range within minutes. Suitable samples and TEM-SDD combinations can deliver atom column and even atomic resolution. Fig. 3 is an example of the analysis of a $BaTiO_3$ sample, demonstrating, that spectra of materials of this type and additives can be deconvolved and quantified. Fig. 4 shows element maps and a quantified line scan extracted from a spectrum image (EDX spectrum for each pixel of the map) of a thermoelectric $CrSi_2$ precipitate. To improve statistics, spectra perpendicular to the line scan were accumulated. This data is part of a study on the tuning and influence of $CrSi_2$ texture on its thermoelectric properties.

In summary, suitable combinations of microscope and SDD technology are valuable tools for fast composition analysis at the front of nanotechnology.

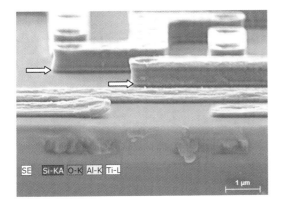

Figure 1: Fracture edge of an etched CMOS chip. Arrows mark 85 nm thick layer of adhesion-promoting agent (Ti/TiN). Measurement conditions: 800x600 pixel, 4 kV, 6 kcps, 23 min.

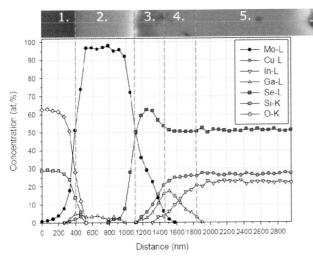

Figure 2: Quantitative line scan of a CIGS [$Cu(In,Ga)Se_2$)] solar cell. A distinct diffusion layer (3.) with enrichment of molybdenum and selenium can be discriminated between the molybdenum layer (2.) and the CIGS absorber layer (4./5.). Measurement conditions: 7 kV, 7 kcps, 46 points (step size 66 nm), ~11 min (15s per point). Reference standards: Mo, Cu, Se, GaP, InP, $KAlSi_3O_8$.

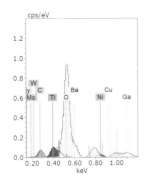
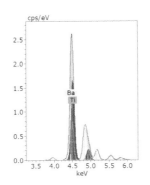

Figure 3: Deconvolution result (colored peaks) of a net intensity EDX spectrum (black line) from a $BaTiO_3$ TEM-standard. Open and versatile software are essential to fully analyze acquired data and deal with system peaks correctly.

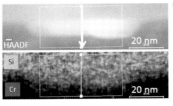
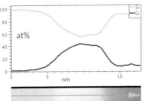

Figure 4: HAADF image and distribution of Cr and Si along a line scan across a $CrSi_2$ precipitate on (001) Si.
To increase statistics, the line scan has been extracted from the spectrum image and then quantified averaging over 9 data points.

NANOPULSED FIELD INDUCED CONTROL OF HYBRID MATERIALS AND ITS FUNCTIONARYTY

Tadachika Nakayama[1], Roman Nowak[2], Koichi Niihara[1]

[1]Nagaoka University of Technology, Nagaoka, Niigata, Japan
[2]Aalto University, Espoo, Uusimaa, Finland

The thermal management (thermal discharge) in compact devices has become an urgent issue as these devices become increasingly more compact. To overcome this problem, research is being conducted in materials that promote heat radiation. Materials conventionally used to dissipate heat include metals such as aluminum, copper, and, to some extent, silver, but they are problematic since it is not possible to insulate them. Consequently, manufacturers need to find inorganic substances that discharge heat to serve as next-generation heat-dissipating materials. Inorganic materials such as aluminum nitride and boron nitride have higher heat conductions than aluminum and copper and they also have the advantage of being electrical insulators". However, these inorganic materials are extremely brittle, which causes reliability problems when they are used in portable devices. Because of this, attention is currently focused on heat-dissipating materials that are composites of organic and inorganic materials (thermal interface materials; TIMs), since they provide a way to overcome the problem of reliability. These inorganic heat-dissipating materials are used as fillers and are dispersed within polymers. They are both pliant and maintain the thermal conducting and electrical insulating properties of the material. We have proposed a new method for realizing orientation that employs a nanosecond pulse power supply as a new way to achieve anisotropic structure control. This method permits characteristic structure control that cannot be realized using other orientation methods.

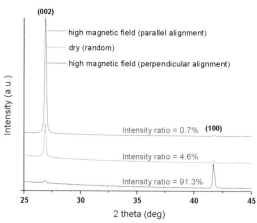

Figure 1. XRD patterns of polysiloxane/BN nanosheet composites produced by applying a high magnetic field (BN content: 5 vol%).

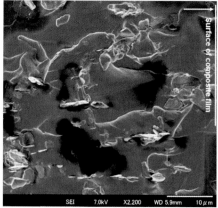

Figure 2. Cross-sectional SEM images of perpendicular alignments of BN nanosheets in polysiloxane/BN nanosheet composite (Film thickness: 255 μm, BN content: 5 vol%).

[1] T. Nakayama et al., *Composites Science and Technology*. **71**, 1046-1052 (2011).
[2] T. Nakayama et al., *Composites Science and Technology*, **72**, 112-118 (2011).
[3] T. Nakayama et al., *Journal of the American Ceramics Society*, **95**, 369–373 (2012)

EUV actinic mask blank defect inspection: results and status of concept realization

Aleksey Maryasov[1], Stefan Herbert[1], Larissa Juschkin[1], Anke Aretz[2], Rainer Lebert[3]

[1]Chair for Technology of Optical Systems, RWTH Aachen University and JARA – Fundamentals of Future Information Technology (FIT), 52074 Aachen;
[2]Central Facility for Electron Microscopy (GFE), RWTH Aachen University and JARA-FIT, 52074 Aachen, Germany;
[3]Bruker Advanced Supercon GmbH, 51069 Cologne-Dellbrueck, Germany

One of the most challenging requirements for the next generation EUV lithography is an extremely low amount of critically sized defects on mask blanks. Fast and reliable inspection of mask blanks is still a challenge. Here we present the current status of the development of our actinic Schwarzschild objective based microscope [1] operating in dark field with EUV discharge produced plasma source. Despite that 193 nm argon fluoride (ArF) immersion lithography with double patterning reaches already feature sizes down to 32 nm [2], there are still numerous advantages to be expected by moving to EUV lithography with a much shorter wavelength of 13.5 nm and possibility of scaling to half-pitch sizes below 22 nm. There are several key challenges on the way to implement this technology into industrial process. Beside the requirements of operation in vacuum, changing all refractive optical components to reflective components, building new stepper and scaling source power to achieve appropriate illumination time, also defect free masks and mask blanks are required.

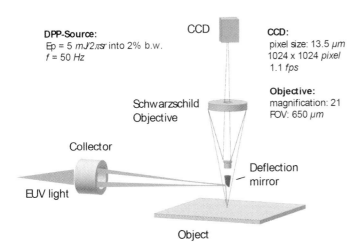

Figure 1: The scheme of the EUV dark field microscope in reflection mode consisting of a xenon discharge produced plasma EUV source, a grazing incidence collector, a deflection mirror, a Schwarzschild objective and a thinned back illuminated CCD camera.

The mask blank and the corresponding reticle must have less than 0.003 printable defects/cm^2 of 25 nm [3], i.e. defects which could affect the quality of structures of the final wafer. This determines the maximal flux inhomogeneity which is allowed to be produced by defects on mask. Correspondent physical dimensions of critical defects on top of the mask blank substrate or inside the multilayer (ML) are still not exactly defined. Formulating critical defect size will determine the required quality level which must be applied to technological process related with mask production. Since the mask scanning for defect inspection is expected to become one of the key routine processes, additional scan speed requirement is applied and currently set up to the level of 7 mm^2/s. This means the total scan of a 142 x 142 mm^2 mask blank has to be done within 1 hour. According to Goldberg et al. [4] most of current projects for mask blank inspection are settled on low-resolution dark-field imaging with large field of view (FOV) and direct EUV detection. Such a system can achieve greater sensitivity and speed for static exposures with a laboratory source. Additionally the inspection has to be done actinically (at the exposure wavelength of the stepper in the lithography process), because the reflecting ML of the mask blank is designed for that wavelength. Inspection with other wavelengths can detect other inhomogeneities, which are not necessarily critical defects or printable under EUV exposure [5].

For microscope performance characterization, several programmed defect structures – artificial pits and bumps were created on top of multilayer mirror surfaces and investigated both with the EUV microscope at RWTH-TOS (see Figure 1) and atomic force microscope (AFM) at GFE. Defect size sensitivity of actinic inspection in dark field mode without resolving the defects is under study. The dependency between defect shape, size and position in relation to the ML surface and its scattering signal as well as results of a defect mapping algorithm are discussed.

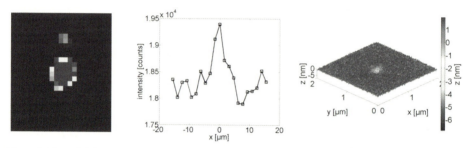

Figure 2: Natural defect: image from EUV microscope (left), profiles from EUV microscope image (middle) and AFM image (right).

The current setup of dark-field EUV microscope at RWTH-TOS (Figure 1) allows for research purposes to scan industrial mask blanks within 25 x 25 mm^2 of its central area. The illumination path was upgraded by installing a grazing incidence collector. With the help of a defect mapping algorithm, detected defects can be easily localized. We have investigated several natural defects down to 45 nm (sphere approximation) on the top of sample ML mirrors (Figure 2) and compared detected signal with theoretical pinhole diffraction model, which are in good agreement. Total performance of future defect inspection tools is defined by a complex optimization of technical parameters of main microscope components. Further research has to be done in order to estimate sensitivity limits with given CCD parameters and flare level.

[1] A. Maryasov, S. Herbert, L. Juschkin, R. Lebert, K. Bergmann, Proc. SPIE **7985**, 79850C (2011)
[2] http://www.asml.com, *Image* ASML's customer magazine (2010)
[3] O. Wood, *EUV Lithography: Approaching Pilot Production*, International Workshop on EUV Lithography, June 21-25, 2010, Maui, Hawaii
[4] K. Goldberg, I. Mochi, J Vac. Sci Technol. B **28**, C6E1-C6E10 (2010)
[5] http://www.imec.be

INVESTIGATIONS OF METALL-FILMS ABSORPTION ON COBALT WITH NVIDIA CUDA TECHNOLOGY

Sergey Seriy[1]

[1]Komsomolsk-on-Amur state technical university, Komsomolsk-on-Amur city, Russia

Thin-film coating hardness and adhesion to the substrate material determined by the value of energy barrier for moving an atom from the cover of the current cell, consisting of the substrate atoms in the neighboring cell [1].

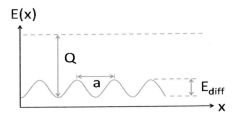

Figure 1: The energy landscape of the surface nanostructures by diffusion of atoms on one dimension (**X** - axis of atoms displacement on substrate surface), **Q** - binding energy of interactions at the boundary between materials on frontier, **E**$_{diff}$ is a diffusion barrier, **a** - distance between the neighboring positions of adsorption.

Migration of coating material is not free, because the presence of energy barriers that exist on the substrate surface with a regular crystal lattice (hexagonal in cobalt). When adsorption occurs at the boundary of the crystals due to the inhomogeneity of the energy periodicity in the arrangement of the elements of the crystal lattice, which prevents these processes is proportional to the energy barrier, which can be calculated using the ab-initio optimization methods (method «QuasiNewton» in this work).

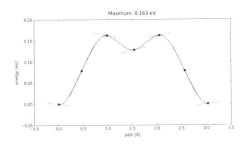

Figure 2: Energy barrier for aluminum on cobalt (Ediff = 0.163 eV). Aluminum atom must be overcome this barrier on the cobalt surface, to get to the neighboring cell, having a distance of 3.1A. Deflection in the central part of barrier caused by the singularity in structure of cobalt surface (orthogonal hexagonal lattice).

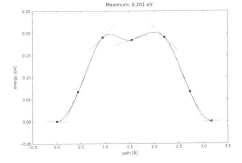

Figure 3: Energy barrier for titanium on cobalt slab (Ediff=0.201 eV). From the value of energy barrier should be the strength of coating positioning on the substrate. In presence of a sufficiently high barrier on the border, the dynamics of any materials is difficult, including external mechanical and thermal influences.

The higher the barrier - the higher the adhesion strength of coating material and substrate material. All calculations performed on GPAW and Abinit software, compiled with CUDA GPU support, to evaluate the speed of calculations.

[1] U. Kabaldin, S. Muraviev, S. Seriy, Information models of assembly systems and nanostructuring of materials under external mechanical stress. KnASTU, Komsomolsk-on-Amur, (2009), - 212 pp

AN OXIDE MBE SYSTEM FOR QUASI IN-SITU NEUTRON REFLECTOMETRY STUDIES

Sabine Pütter[1], Alexandra Steffen[1], Markus Waschk[2], Alexander Weber[2], Stefan Mattauch[1], Thomas Brückel[2]

[1]Jülich Centre for Neutron Science JCNS, Forschungszentrum Jülich GmbH, Outstation at FRM II, Lichtenbergstraße 1, 85747 Garching;
[2]Jülich Centre for Neutron Science JCNS and Peter Grünberg Institut PGI, JARA-FIT, Forschungszentrum Jülich GmbH, 52425 Jülich

Complex transition metal oxides are possible candidates to solve some of todays challenges like energy conversion and information technology applications [1]. Molecular beam epitaxy (MBE) is a low energetic, high purity method and has the potential to produce these materials as high quality epitaxial films with low intrinsic defect concentrations and atomic-layer control.

Therefore, an oxide MBE system was established by the Jülich Centre for Neutron Science on-site of the FRM II Neutron Source in Garching, Germany to allow quasi in-situ neutron scattering studies of the freshly produced samples at the new dedicated MAgnetism Reflectometer with high Incident Angle (MARIA) of the JCNS by subsequent transfer of the sample from the MBE setup to the neutron beam. The MBE setup is open for users and collaborations, i.e. users who do not have access to own MBE facilities are welcome to prepare their samples on-site with our technical support.

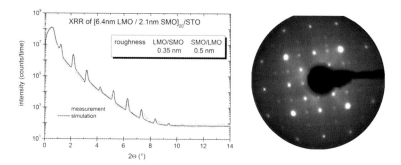

Figure 9: 6.4 nm $LaMnO_3$/2.1 nm $SrMnO_3$ multilayer with 20 bilayer repetitions deposited on $SrTiO_3$. On the left: X-Ray reflectivity. The surface roughness of this sample was determined with AFM and off-specular XRR and is as small as 0.5 nm with an in-plane correlation length of 140 nm. On the right: LEED micrograph at 130 eV of the $LaMnO_3$ top layer. The reconstruction reflects the high quality of the surface.

The MBE is equipped with 6 effusion cells, two electron guns for electron-beam evaporation with 4 crucibles each and an oxygen plasma source. Standard in-situ surface analysis tools like reflective high and low energy electron diffraction (RHEED/LEED), Auger electron spectroscopy (AES) analysis are provided to enable epitaxial growth with high purity.

Our own investigations cover exchange biased and 2DEG systems like $La_{1-x}Sr_xMnO_3/SrTiO_3$, $LaMnO_3/SrTiO_3$, $LaAlO_3/SrTiO_3$ single and multilayers and systems for solid oxide fuel cells (SOFC) like $La_{1-x}Sr_xFeO_3$ and $La_{1-x}Sr_xCoO_3$. First results with this new experimental setup are presented. An example is given in Fig. 1 which reveals the high quality of a $[6.4\ nm\ LaMnO_3/2.1\ nm\ SrMnO_3]_{20}$ multilayer on $SrTiO_3$ substrate with low interface roughness in the X-ray reflectivity and a LEED pattern of the reconstructed surface of the $LaMnO_3$ top layer.

[1] R. Ramesh and N. A. Spaldin, Nature Mater. **6**, 21 (2007).

EXPLORING THE RESOLUTION LIMIT OF THE TALBOT LITHOGRAPHY WITH EUV LIGHT

H. Kim[1], S. Danylyuk[1], L. Juschkin[1], K. Bergmann[2], P. Loosen[1]

[1]Chair for the Technology of Optical Systems, RWTH Aachen University and JARA - Fundamentals of Future Information Technology, 52074 Aachen, Germany;
[2]Fraunhofer Institute for Laser Technology (ILT), 52074 Aachen, Germany

In this article the resolution limit of the Talbot lithography with extreme ultraviolet (EUV) radiation is explored theoretically by employing finite difference time domain (FDTD) electromagnetic field simulation. Lithography has been in a challenge to bring the resolution down to 10 nm level [1]. Self-imaging Talbot lithography is a promising candidate for the high resolution printing [2]. Utilizing EUV radiation with wavelengths around 11 nm increases the achievable resolution due to the much shorter wavelength in comparison to the conventional UV radiation. However as the size of structures on the mask approaches the wavelength of the radiation, diffraction influence needs to be evaluated precisely to estimate the achievable resolution and quality of the patterns. Here we present the results of FDTD simulation of the diffraction on EUV transmission masks in dependence on period (pitch) of the mask, with the aim to determine the resolution that can be realistically achieved with the EUV Talbot lithography.

In the Figure 1 the results of simulation of the contrast and maximal intensity of the first Talbot maxima in dependence on mask period are shown. In this case plane wave illumination with TE-polarized monochromatic light, 1:1 line to space ratio and 100 nm thickness of the nickel absorber are considered. As it can be seen, the contrast stays practically constant down to 40 nm pitch value, where it starts to decrease. The maximal intensity drop accelerates under 40 nm as well, but both contrast and intensity values are more than sufficient for high quality patterning.

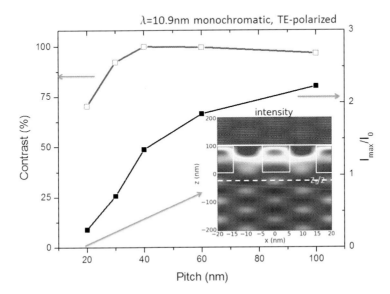

Figure 1: Contrast and intensity curves for several pitches between 20 nm and 100 nm. Inset: 2D picture of simulated light propagation for a mask with 20 nm pitch (the white line is the mask profile). Intensity maxima after the mask appear at distances proportional to half of the Talbot length ($z_T/2$). The intensity and contrast curves are calculated for the first observed maxima at $z_T/2$ position (marked by the dashed line).

As shown in the inset in Figure 1, with monochromatic wave intensity maxima in z_T plane are shifted in x-direction for half of the period of the mask in comparison to the maxima in $z_T/2$ plane. Utilizing the broadband radiation spreads both set of maxima in the propagation direction (z), creating a range of distances where period of the image is half of the mask period. Figure 2 shows the results of the numerical simulations for propagation of broadband radiation through the mask with 20 nm pitch. The TE polarized radiation with 10.9 nm wavelength and 3.2% bandwidth is applied. At the distance (z) around 1.5 µm, intensity maxima of several Talbot images mix (dashed line in the figure) and yield continuous self-imaging [3]. The curve in the figure 2 shows the intensity along x-axis at the dashed line. The resulting image demonstrates the possibility to obtain the 10 nm pitch with this method if the relatively low contrast of the image can be tolerated. Switching to a lower wavelength would allow to improve the contrast of the pattern.

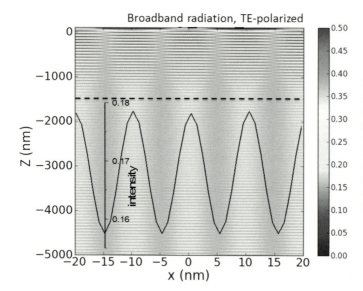

Figure 2: Relative intensity distribution of the light propagation of broadband TE-polarized 10.9 nm radiation with 3.2% bandwidth after 20 nm pitch mask. The black curve shows the intensity distribution along the x-axis at around 1.5 µm in propagation-axis (z).

Resolution limits for the self-imaging with monochromatic wave and continuous self-imaging with broadband EUV radiation have been explored. It is demonstrated that sub-10 nm patterns are theoretically possible but contrast and intensity of the images decrease significantly for patterns with periods below 40 nm.

[1] http://www.itrs.net/Links/2011ITRS/2011Chapters/2011Lithography.pdf
[2] K. Patorski, in *Progress in optics XXVII*, edited by E. Wolf, Elsevier Science Publishers, pp. 2-108 (1989)
[3] Optics Communications **180**, 199–203 (2000)

NON-STOICHIOMETRIC HfO$_{2-x}$ THIN FILMS

Milias Crumbach[1], Manfred Martin[1]

[1]Institute of Physical Chemistry, RWTH Aachen University, Germany

Hafnium oxide, HfO$_2$, is a promising candidate for replacing SiO$_2$ as a high-κ gate dielectric in MOS transistors [1]. Another possible application for oxygen deficient HfO$_{2-x}$ is its use as non-volatile memory. For example, Lee et al. [2] reported of their experiments with Ti-capped hafnium oxide, which exhibited good properties as a fast switching, non-volatile memory material. As they showed, the Ti-capping functions as an oxygen getter to introduce the oxygen deficiency into the hafnium oxide layer. Another way to obtain oxygen deficient HfO$_{2-x}$ thin films is the synthesis by means of Pulsed Laser Deposition (PLD). For this method, a ceramic HfO$_2$ target is ablated with an ultraviolet laser. The evaporated material deposits on a substrate to form the thin film. By executing the procedure under different gas atmospheres, the stoichiometry of the thin film can be influenced. Another advantage of the PLD is the possibility to create amorphous films by depositing at room temperature.

Thin films of hafnium oxide were deposited on amorphous quartz substrates by PLD at room temperature in a) oxygen, b) argon and c) argon/hydrogen atmospheres. As-prepared films deposited under condition a) are colorless, while films deposited under conditions b) and c) are brownish-black. All films show surface roughnesses of less than ±3.0 nm, as determined by Interference Microscopy. Electron Probe Micro Analysis (EPMA) measurements show Hf:O ratios of 1:2.07±0.03 (a), 1:1.83±0.02 (b) and 1:1.75±0.01 (c), respectively. The band gaps of the as-prepared films were determined by UV/VIS spectroscopy yielding values of 5.79±0.23 eV (a), 5.20±0.10 eV (b) and 4.46±0.04 eV (c). X-Ray Diffraction (XRD) shows that the as-prepared films are amorphous; upon heating, they crystallize and form monoclinic hafnium oxide. To further investigate the structural properties and the crystallization process, in-situ Extended X-ray Absorption Fine Structure (EXAFS) measurements were carried out. All as-prepared samples show no signs for long-range order. But as the temperature is increased, the EXAFS Radial Distribution Functions (RDF), shown in Figure 1, exhibit a second peak with increasing intensity, which hints on increasing Hf-Hf order due to the crystallization. To investigate the electrical properties of the films, conductivity measurements using the Van-der-Pauw method were carried out as a function of temperature and in an argon atmosphere, to prevent re-oxidation of the oxygen deficient films. Conductivity measurements were performed during heating and cooling as well. The results of the measurements are depicted in Figure 2. The activation energies determined from an Arrhenius plot yield to 0.63±0.05 eV (heating) and 0.43±0.17 eV (cooling) for a film deposited under conditions a), 0.55±0.01 eV (heating) and 0.78±0.10 eV (cooling) for a film deposited under conditions b) and 0.57±0.03 eV (heating) and 0.67±0.07 eV (cooling) for a film deposited under conditions c). The strong decrease of the conductivities that we found after heating the films prepared in Ar atmosphere (conditions b) may be indicative for re-oxidation.

In summary, we could show that hafnium oxide films prepared by PLD are amorphous and highly non-stoichiometric. The HfO$_{2-x}$ films exhibit semiconducting properties with conductivities as high as $10^{-1.5}$ S·m^{-1} and activation energies between 0.55 and 0.67 eV which may be indicative for excitation of electrons from the Fermi energy to the mobility edge of the amorphous material [3].

[1] G. D. Wilk et al., Journal of Applied Physics **89**, 5243 (2001).
[2] H. Y. Lee et al., Electron Device Letters, IEEE **31**, 44 (2010).
[3] N. F. Mott, Advances in Physics **16**, 49 (1967).

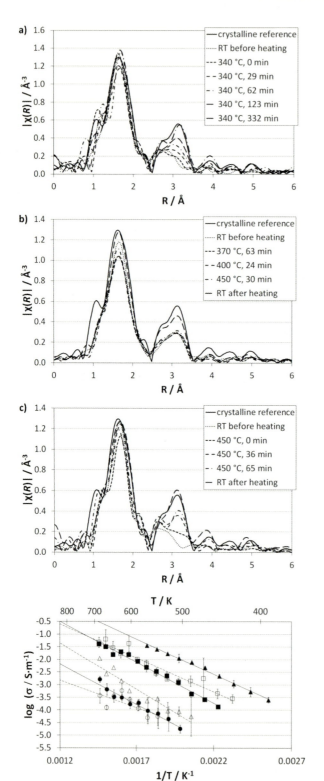

Figure 1: *in-situ* EXAFS RDFs of HfO$_{2-x}$ films deposited in an oxygen (a), an argon (b) and an argon/hydrogen (c) atmosphere at different temperatures and dwell times.

Figure 2: Electrical conductivity of films deposited in an oxygen atmosphere (●), an argon atmosphere (▲) and an argon/hydrogen atmosphere (■). Full symbols depict measurements during heating, while empty symbols depict measurements during cooling.